Pressure Transient Testing

John Lee
Texas A&M University

John B. Rollins
IBM Corporation

John P. Spivey
Phoenix Reservoir Engineering

SPE Textbook Series, Volume 9

Henry L. Doherty Memorial Fund of AIME
Society of Petroleum Engineers
Richardson, TX USA

Dedication

John Lee

To all the Aggie students and former students who have made my teaching career so much fun and so rewarding.

John Rollins

To my family—Becci, Christine, and Cathy—and to my father, J.T. Rollins, a genuine Permian Basin petroleum pioneer.

John Spivey

To my many colleagues at SoftSearch, Dwights Energy Data, S.A. Holditch and Assocs., and Schlumberger Oilfield Technologies who have taught me, challenged me, encouraged me, and inspired me throughout my career.

Disclaimer

This book was prepared by members of the Society of Petroleum Engineers and their well-qualified colleagues from material published in the recognized technical literature and from their own individual experience and expertise. While the material presented is believed to be based on sound technical knowledge, neither the Society of Petroleum Engineers nor any of the authors or editors herein provide a warranty either expressed or implied in its application. Correspondingly, the discussion of materials, methods, or techniques that may be covered by letters patents implies no freedom to use such materials, methods, or techniques without permission through appropriate licensing. Nothing described within this book should be construed to lessen the need to apply sound engineering judgment nor to carefully apply accepted engineering practices in the design, implementation, or application of the techniques described herein.

Printed in the United States of America.

ISBN 978-1-55563-099-7

08 09 10 11 12 13 14 / 10 9 8 7 6 5 4 3 2

Society of Petroleum Engineers
222 Palisades Creek Drive
Richardson, TX 75080-2040 USA

http://store.spe.org
service@spe.org
1.972.952.9393

SPE Textbook Series

The Textbook Series of the Society of Petroleum Engineers was established in 1972 by action of SPE Board of Directors. The Series is intended to ensure availability of high-quality textbooks for the use in undergraduate courses in areas clearly identified as being within the petroleum engineering field. The work is directed by the Society's Books Committee, one of more than 40 Society-wide standing committees. Members of the Books Committee provide technical evaluation of the book. Below is a listing of those who have been most closely involved in the final preparation of this book.

Book Editors

Books Committee (2003)

Introduction

Pressure transient test analysis is a mature technology in petroleum engineering; even so, it continues to evolve. Because of the developments in this technology since the last SPE textbook devoted to transient testing was published, we concluded that students could benefit from a textbook approach to the subject that includes a representative sampling of the more important fundamentals and applications. We deliberately distinguish between a textbook approach, which stresses understanding through numerous examples and exercises dealing with selected fundamentals and applications, and a monograph approach, which attempts to summarize the state-of-the-art in the technology.

Computational methods that transient test analysts use have gone through a revolution since most existing texts on the subject were written. Most calculations are now done with commercial software or by spreadsheets or proprietary software developed by users to meet personal needs and objectives. These advances in software have greatly increased productivity in this technology, but they also have contributed to a "black box" approach to test analysis. In this text, we attempt to explain what's in the box, and we do not include a number of the modern tools that enhance individual engineer productivity. We hope, instead, to provide understanding so that the student can use the commercial software with greater appreciation and so that the student can read monographs and papers on transient testing with greater appreciation for the context of the subject. Accordingly, this text is but an introduction to the vast field of pressure transient test analysis.

Acknowledgments

The contributions of many people were crucial in the preparation of this book. We acknowledge with heartfelt thanks the contributions to the preparation of the subject matter by Tom Blasingame, Jay Rushing, and Jennifer Johnston Blasingame; the contributions to the presentation of the material by Darla-Jean Weatherford; the technical audit by Shah Kabir and Fikri Kuchuk; and the SPE staff, most notably technical editors Valerie Dawe and Jennifer Wegman. To each of you—thanks!

Approximation of Ei — pg 10, Eq. 1.150
dimensionless variables — pg 12, bottom right — if he asks for everything in dimensionless variables look at pg 246 for help
dimensionless pressure — pg 17, table 1.2
Ei TABLE — pg 10
multiple wells — pg 17
productivity index — pg 16, Eq 1.163i
SKIN! — pg 8

CONDITIONS:
pseudo steady + constant rate + cylindrical — pg 14
transient + constant rate + radial flow — pg 10
transient vs. pseudo SS vs steady state (definitions) — pg 10, section 1.5

Contents

Vertical Well Productivity

$$Q_w = \frac{1}{141.2} \frac{kh}{\mu B_o} \frac{(P_e - P_{wf})}{\ln(r_e/r_w)}$$

P_e = Edge Pressure or $\ln\left(\frac{L}{r_1 r_2}\right)$

	Steady State	Pressure Transient
Fundamental Solution (2D)	$\ln\left(\frac{r}{r_w}\right)$	$-Ei\left(-\frac{r^2}{4T}\right)$ or $Ei\left(\frac{r^2}{4T}\right)$, where $T = \frac{kt}{\phi\mu c_t}$
Fundamental Solution (1D to 3D)	$\left(\frac{r}{r_w}\right)$ to $\left(\frac{r_w}{r}\right)$	Erfc to exponential
Skin	$\Delta P = \frac{\mu Q_w}{2\pi kh} \cdot$ skin	same
Anisotropy (xy example)	$k = \sqrt{k_x k_y}$; $r^2 = (k/k_x)x^2 + (k/k_y)y^2$	same
Superposition	in space	in space in time
$Q_w > 0$ producer, $Q_w < 0$ injector		

Skin (dimensionless)

$$S = \left(\frac{k}{k_s} - 1\right) \ln\left(\frac{r_s}{r_w}\right)$$

For Injector/Producers Pair VRR=1

P_{wf_1} = Producer P_{wf_2} = injector

$$P_{wf_1} = P_{in} + \frac{141.2\,\mu Q B_o}{kh} \ln\left(\frac{r_{w1}}{r_{12}}\right)$$

$$P_{wf_2} = P_{in} + \frac{141.2\,\mu Q B_o}{kh} \ln\left(\frac{r_{12}}{r_{w2}}\right)$$

$$\Delta P = \frac{\mu Q_w}{2\pi kh}\left\{S_2 + \ln\left(\frac{L}{r_{w1}}\right) + S_1 + \ln\left(\frac{L}{r_{w2}}\right)\right\} \quad Q_w = \frac{1}{141.2}\frac{kh}{\mu B}\frac{\Delta P}{\ln\left(\frac{L^2}{r_{w1} r_{w2}}\right) + S_1 + S_2}$$

Effective wellbore Radius

$$r_{ws} = r_w e^{-S}$$

Prediction of pressure in reservoir

$$P(r) = P_{wf} + 141.2 \frac{\mu Q_w B_o}{kh} \ln\left(\frac{r}{r_w}\right)$$

Horizontal well productivity

$$\frac{2\pi\sqrt{k_x \cdot k_y} \cdot h_z}{\mu} = \frac{2\pi\sqrt{k_x \cdot k_z}\, h_y}{\mu}$$

h_y = horizontal length

VRR = Voidage Replacement Ratio

$$= \frac{\text{volume of fluid injected}}{\text{volume of fluid produced}}$$

WHEN VRR=1, LIKELY STEADY-STATE CONDITIONS

Diffusivity: Pressure Solution

At wellbore:

$$P_{wf}(t) = P_i - 141.2 \frac{\mu Q_w B}{kh}\left\{\frac{1}{2} Ei\left(\frac{948\,\phi\mu c_t r^2}{kt}\right) + S\right\}$$

Time it takes for drawdown to affect another well

$$t = \frac{948\,\phi c_t \mu r^2}{k}$$

Pressure drop due to skin

$$141.2 \frac{\mu Q_w B_o}{kh}(S)$$

Transient Flow: Fractured Vert Well

$$\Delta P_{wf}(x,t) = 4.06 \frac{\mu Q B}{k x_f h}\sqrt{\frac{kt}{\phi\mu c_t}}$$

Pressure drop vs time at Wellbore

$$P_i - P_{wf}(t) = 70.6 \frac{\mu Q_w B}{kh}\left\{\ln\left(\frac{kt}{1688\,\phi\mu c_t r_w^2}\right) + 2S\right\}$$

Transient Radial Flow

$$\frac{948\,\phi\mu c_t r^2}{kt} = 1$$

Producer Producer well superposition, bottomhole Flowing pressure

$$P_{wf_1} = P_{ref} + 141.2\left(\frac{\mu B_o}{kh}\right) Q_{w_2} \ln\left(\frac{r_{12}}{r_{w2}}\right)$$

$$P_{wf_2} = P_{ref} + 141.2\left(\frac{\mu B_o}{kh}\right) Q_{w_1} \ln\left(\frac{r_{12}}{r_{w2}}\right)$$

$$P_{wf_2} = P_{wf_1} - 141.2 \frac{\mu B_o}{kh}\left\{Q_{w_2} \ln\left(\frac{r_{12}}{r_{w2}}\right) - Q_{w_1} \ln\left(\frac{r_{12}}{r_{w1}}\right)\right\}$$

Chapter 1
Fundamentals of Fluid Flow in Porous Media

1.1 Overview

In this chapter, we develop the equations to describe the flow of slightly compressible liquids and gases and the simultaneous flow of oil, water, and gas in porous media. We then define appropriate dimensionless variables that enable us to simplify the resulting partial-differential equations. We present solutions to those differential equations subject to various inner- and outer-boundary conditions. These solutions are obtained by use of both Laplace transformations and Boltzmann's transformation. We consider radial and linear flow and superposition in space and time.

Appendix A presents a detailed method for finding dimensionless variables. Appendix B details derivations of the different solutions to the diffusivity equation. Appendix C presents the derivations of the diffusivity equation for multiphase flow and for linear flow in detail. Appendix D presents a proof of the validity of the method of images to model boundaries in a reservoir.

This chapter focuses on the mathematical basis for pressure transient test analysis. For those readers with little or no mathematical inclination, we note that it is not necessary to master the material in this chapter to understand the applications in the rest of the book. However, we do think that virtually all readers will derive considerable benefit from browsing through this chapter. The summary in Sec. 1.9 may be especially helpful to browsers.

1.2 Derivation of the Diffusivity Equation

1.2.1 Fundamental Physical Principles. The basic equation to describe the flow of fluid in porous media caused by a potential difference is known as the diffusivity equation. The diffusivity equation is derived from three fundamental physical principles: (1) the principle of conservation of mass, (2) an equation of motion, and (3) an equation of state (EOS).

We derive the diffusivity equation in the radial coordinate system because flow in a simple, homogeneous-acting, cylindrical reservoir takes place radially from the reservoir to the wellbore. We use metric units (implicitly) in derivations; later, we generalize to other systems of units used in the remainder of the text.

Principle of Conservation of Mass. The principle of conservation of mass states that the net rate of creation or destruction of matter is zero. If we consider the control volume, a fixed region in space (illustrated in **Fig. 1.1**), we may write

$$\begin{bmatrix}\text{The mass flow}\\ \text{rate into the}\\ \text{control volume}\\ \text{during time period } \Delta t\end{bmatrix} - \begin{bmatrix}\text{The mass flow}\\ \text{rate out of}\\ \text{control volume}\\ \text{during time period } \Delta t\end{bmatrix}$$

$$+\begin{bmatrix}\text{The mass flow rate}\\ \text{into or out of}\\ \text{control volume}\\ \text{during time period } \Delta t\\ \text{owing to source or sink}\end{bmatrix} = \begin{bmatrix}\text{The rate of mass}\\ \text{accumulation in}\\ \text{the control volume}\\ \text{during time period } \Delta t\end{bmatrix}. \quad\quad (1.1)$$

We now look at each part of the conservation equation, Eq. 1.1, mathematically. The mass flow rate into the system = density × velocity × cross-sectional area of flow.

$$\dot{m}_{in} = -\rho u_r A_{x_1}, \quad\quad (1.2)$$

where the cross-sectional area of flow on the inflow side, A_{x_1}, is given by $A_{x_1} = L_a \times h$ and the minus sign arises because the positive flow direction in the control volume has been chosen in the negative r direction.

For angle θ, the arc length is given by arc length = radius × angle,

$$L_a = (r + \Delta r)\,\theta. \quad\quad (1.3)$$

Therefore,

$$\dot{m}_{in} = -\rho u_r (r + \Delta r)\theta h. \quad\quad (1.4)$$

The mass flow rate out of the system is similarly given by

$$\dot{m}_{out} = -\left[\rho u_r - \Delta(\rho u_r)\right]A_{x_2}, \quad\quad (1.5)$$

where the term $\Delta(\rho u_r)$ is the change in mass flux occurring inside the control volume, and the cross-sectional area of flow on the outflow side, A_{x_2}, is given by

$$A_{x_2} = r\theta h. \quad\quad (1.6)$$

$$\text{Therefore } m_{out} = -\left[\rho u_r - \Delta(\rho u_r)\right]r\theta h. \quad\quad (1.7)$$

We assume that there is neither a source nor a sink in the control volume (i.e., mass is neither being generated nor consumed). Therefore, net mass flow rate owing to source or sink = 0.

$$W_s = 0. \quad\quad (1.8)$$

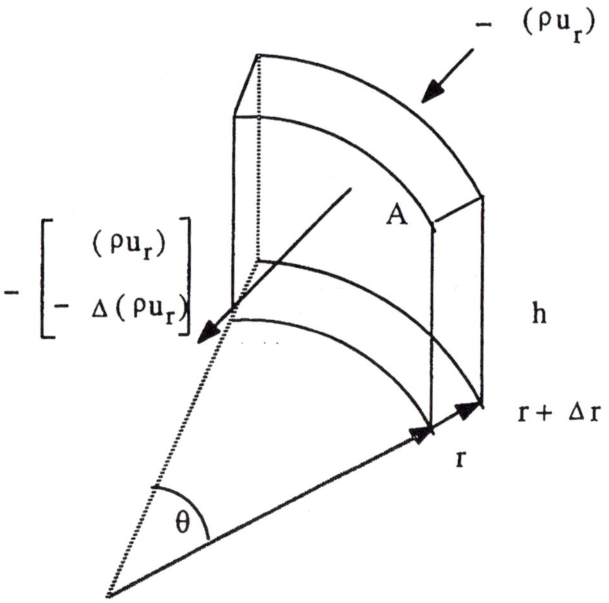

Fig. 1.1—Control volume for deriving the mass-conservation equation.

The mass in the control volume at any time is the product of the pore volume (PV) and the density of the fluid: PV = arc length × width × height × porosity, or

$$V_p = r\theta\Delta rh\phi. \quad (1.9)$$

Therefore $m = r\theta\Delta rh\phi\rho$. (1.10)

The rate of mass accumulation, W_a, in the control volume is given by the change in mass in the control volume from time t to $t+\Delta t$, divided by the change in time, Δt.

$$W_a = \left[\frac{|r\theta\Delta rh\phi\rho|_{(t+\Delta t)} - |r\theta\Delta rh\phi\rho|_t}{\Delta t}\right]. \quad (1.11)$$

We can now express the conservation equation, Eq. 1.1, mathematically by combining Eqs. 1.4, 1.7, 1.8, and 1.11:

$$[-\rho u_r(r+\Delta r)\theta h] - \{-[\rho u_r r\theta h - \Delta(\rho u_r)r\theta h]\}$$

$$= \left[\frac{|r\theta\Delta rh\phi\rho|_{(t+\Delta t)} - |r\theta\Delta rh\phi\rho|_t}{\Delta t}\right]. \quad (1.12)$$

Expanding Eq. 1.12 gives

$$-\rho u_r r\theta h - \rho u_r \Delta r\theta h + \rho u_r r\theta h - \Delta(\rho u_r)r\theta h$$

$$= \frac{1}{\Delta t}\left[|r\theta\Delta rh\phi\rho|_{(t+\Delta t)} - |r\theta\Delta rh\phi\rho|_t\right]. \quad (1.13)$$

Dividing Eq. 1.13 by the bulk volume of the control volume, $hr\theta\Delta r$, we have

$$-\frac{\rho u_r}{r} - \frac{\Delta(\rho u_r)}{\Delta r} = \frac{1}{\Delta t}\left[|\phi\rho|_{(t+\Delta t)} - |\phi\rho|_t\right]. \quad (1.14)$$

Factoring out $1/r\Delta r$ on the left side and multiplying through by -1,

$$\frac{1}{r\Delta r}[\Delta r(\rho u_r) + r\Delta(\rho u_r)] = -\frac{\Delta(\phi\rho)}{\Delta t} \quad (1.15a)$$

$$\text{or } \frac{1}{r}\left\{\rho u_r + r\left[\frac{\Delta(\rho u_r)}{\Delta r}\right]\right\} = -\frac{\Delta(\phi\rho)}{\Delta t}. \quad (1.15b)$$

Taking limits of Eq. 1.15 as $\Delta r, \Delta t \rightarrow 0$, we have

$$\frac{1}{r}\left[(\rho u_r) + \frac{r\partial(\rho u_r)}{\partial r}\right] = -\frac{\partial(\phi\rho)}{\partial t}. \quad (1.16)$$

By the product rule,

$$\frac{\partial}{\partial r}(r\rho u_r) = \rho u_r\frac{\partial r}{\partial r} + r\frac{\partial}{\partial r}(\rho u_r) = \rho u_r + r\frac{\partial}{\partial r}(\rho u_r). \quad (1.17)$$

$$\text{Therefore, } \frac{1}{r}\frac{\partial}{\partial r}(r\rho u_r) = -\frac{\partial}{\partial r}(\phi\rho). \quad (1.18)$$

Eq. 1.18 is known as the continuity equation, a mathematical expression of the principle of conservation of mass in radial coordinates. To this point, the only assumptions we have made are that we have radial flow and that no sources or sinks are in the control volume.

Equation of Motion. An equation of motion, or flux law, relates velocity and pressure or potential gradients within the control volume. Because of the complexity of the flow paths within porous media, we must use empirical relationships for the equation of motion. Liquid flow is generally governed by Darcy's law, which states that the velocity is proportional to the negative of the gradient of the potential. In radial coordinates, with flow in the radial direction only, we write

$$u_r = -\frac{k\rho}{\mu}\frac{\partial\Phi}{\partial r}, \quad (1.19)$$

$$\text{where } \Phi = \int_{p_b}^{p}\frac{dp'}{\rho} + g(Z - Z_0), \quad (1.20)$$

p_b = pressure at a datum, and $Z = Z_0$.

The potential, Φ, consists of two terms: $\int_{p_b}^{p}\frac{dp'}{\rho}$ = flow work and $g(Z-Z_0)$ = potential head.

This form of Darcy's law has two assumptions: (1) flow is in the laminar flow regime (low Reynolds number), and (2) the porous medium is isotropic. For single-phase flow of a slightly compressible liquid in a homogeneous-acting reservoir, these assumptions are generally valid.

We can now combine Eqs. 1.19 and 1.20 to express the velocity in terms of pressure, rather than potential, gradient. From Eq. 1.20,

$$\frac{\partial\Phi}{\partial r} = \frac{\partial}{\partial r}\left[\int_{p_b}^{p}\frac{dp'}{\rho} + g(Z - Z_0)\right]. \quad (1.21)$$

If we assume gravity effects are negligible, $g(Z-Z_0)=0$. Therefore,

$$\frac{\partial\Phi}{\partial r} = \frac{1}{\rho}\frac{\partial p}{\partial r}. \quad (1.22)$$

Substituting Eq. 1.22 into Eq. 1.19 gives

$$u_r = -\frac{k}{\mu}\frac{\partial p}{\partial r}. \quad (1.23)$$

EOS. An EOS relates volume, or density, to the pressure and temperature of the system. We assume isothermal conditions when considering the flow of a slightly compressible liquid in a reservoir because the heat capacity of the fluid is generally negligible compared with the heat capacity of the rock.

The definition of fluid compressibility is

$$c = -\frac{1}{V}\left(\frac{\partial V}{\partial p}\right)_T = \frac{1}{\rho}\left(\frac{\partial \rho}{\partial p}\right)_T. \quad (1.24)$$

Treating $\partial\rho/\partial p$ as a total derivative, $d\rho/dp$, for an isothermal system and rearranging Eq. 1.24 gives

$$c dp = \frac{1}{\rho} d\rho. \quad (1.25)$$

For a fluid of small and constant compressibility, we integrate Eq. 1.25 to obtain

$$c \int_{p_b}^{p} dp' = \int_{\rho_b}^{\rho} \frac{1}{\rho} d\rho, \quad (1.26)$$

where ρ_b = density at base pressure, p_b. Integrating, we obtain

$$c(p - p_b) = \ln\rho - \ln\rho_b \quad (1.27a)$$

$$\text{and } c(p - p_b) = \ln\left(\frac{\rho}{\rho_b}\right). \quad (1.27b)$$

Exponentiating both sides gives

$$\rho = \rho_b \exp[c(p - p_b)]. \quad (1.28)$$

This is the EOS that we use when we assume that the fluid is slightly compressible and the compressibility is constant.

1.2.2 Diffusivity Equation for Radial, Single-Phase Flow of a Liquid With Small, Constant Compressibility. To derive the diffusivity equation, we must combine the continuity equation,

$$\frac{1}{r}\frac{\partial}{\partial r}(r\rho u_r) = -\frac{\partial}{\partial r}(\phi\rho), \quad (1.18)$$

the equation of motion,

$$u_r = -\frac{k}{\mu}\frac{\partial p}{\partial r}, \quad (1.23)$$

and the EOS for the appropriate fluid,

$$\rho = \rho_b \exp[c(p - p_b)]. \quad (1.28)$$

Combining Eqs. 1.18 and Eq. 1.23, we obtain

$$\frac{1}{r}\frac{\partial}{\partial r}\left(r\rho\frac{k}{\mu}\frac{\partial p}{\partial r}\right) = \frac{\partial}{\partial r}(\phi\rho). \quad (1.29)$$

If we assume constant permeability and viscosity, using the product rule gives

$$\frac{1}{r}\frac{\partial}{\partial r}\left(r\rho\frac{\partial p}{\partial r}\right) = \frac{\mu}{k}\left(\phi\frac{\partial\rho}{\partial t} + \rho\frac{\partial\phi}{\partial t}\right). \quad (1.30)$$

We can now expand Eq. 1.30 by use of the chain rule:

$$\frac{\rho}{r}\frac{\partial}{\partial r}\left(r\frac{\partial p}{\partial r}\right) + \frac{1}{r}r\frac{\partial p}{\partial r}\frac{\partial\rho}{\partial r} = \frac{\mu}{k}\left(\phi\frac{\partial\rho}{\partial p}\frac{\partial p}{\partial t} + \rho\frac{\partial\phi}{\partial p}\frac{\partial p}{\partial t}\right)$$

$$\frac{\rho}{r}\frac{\partial}{\partial r}\left(r\frac{\partial p}{\partial r}\right) + \frac{\partial p}{\partial r}\frac{\partial\rho}{\partial p}\frac{\partial p}{\partial r} = \frac{\mu}{k}\phi\rho\left(\frac{1}{\rho}\frac{\partial\rho}{\partial p}\frac{\partial p}{\partial t} + \frac{1}{\phi}\frac{\partial\phi}{\partial p}\frac{\partial p}{\partial t}\right). \quad (1.31)$$

From Eq. 1.28,

$$\frac{\partial\rho}{\partial p} = c\rho_b \exp[c(p - p_b)], \quad (1.32)$$

where compressibility, c, is small and constant. Therefore,

$$\frac{\partial\rho}{\partial p} = c\rho. \quad (1.33)$$

By remembering the definition of compressibility,

$$c = \frac{1}{\rho}\frac{\partial\rho}{\partial p}, \quad (1.34)$$

and defining a formation compressibility,

$$c_f = \frac{1}{\phi}\frac{\partial\phi}{\partial p}, \quad (1.35)$$

we can define a total compressibility, c_t, as

$$c_t = c + c_f, \quad (1.36)$$

and write Eq. 1.31 as

$$\frac{\rho}{r}\frac{\partial}{\partial r}\left(r\frac{\partial p}{\partial r}\right) + c\rho\left(\frac{\partial p}{\partial r}\right)^2 = \phi\frac{\mu c_t}{k}\rho\frac{\partial p}{\partial t}. \quad (1.37)$$

We know that $\rho \neq 0$; therefore, we can divide the equation through by density.

$$\frac{1}{r}\frac{\partial}{\partial r}\left(r\frac{\partial p}{\partial r}\right) + c\left(\frac{\partial p}{\partial r}\right)^2 = \phi\frac{\mu c_t}{k}\frac{\partial p}{\partial t}. \quad (1.38)$$

We now assume for radial flow of a fluid of small, constant compressibility that $c(\partial p/\partial r)^2$ is negligible compared to $\partial/\partial r(r\partial p/\partial r)$ and $\partial p/\partial r$, so the final partial differential equation is

$$\frac{1}{r}\frac{\partial}{\partial r}\left(r\frac{\partial p}{\partial r}\right) = \phi\frac{\mu c_t}{k}\frac{\partial p}{\partial t}. \quad (1.39)$$

Summary of Assumptions for Eq. 1.39.

1. Radial flow.
2. Laminar (or Darcy) flow.
3. Porous medium has constant permeability and compressibility.
4. Negligible gravity effects.
5. Isothermal conditions.
6. Fluid has small, constant compressibility.
7. Compressibility/pressure-gradient-squared product, $c(\partial p/\partial r)^2$, is negligible.

1.2.3 Diffusivity Equation for Radial, Single-Phase Flow of a Gas. The continuity equation and equation of motion for radial single-phase gas flow through porous media are the same as those equations used for slightly compressible liquid flow.

Continuity Equation.

$$\frac{1}{r}\frac{\partial}{\partial r}(r\rho u_r) = -\frac{\partial}{\partial t}(\phi\rho). \quad (1.18)$$

Equation of Motion (Darcy's Law).

$$u_r = -\frac{k}{\mu}\frac{\partial p}{\partial r}. \quad (1.23)$$

EOS. The EOS used for slightly compressible liquids does not, however, model gas behavior. The equation most commonly used to model real-gas pressure/volume/temperature (PVT) behavior is the real-gas law given by

$$\rho = \frac{pM}{zRT}. \quad (1.40)$$

We can now combine the continuity equation, Eq. 1.18, and the equation of motion, Eq. 1.23, to obtain

$$-\frac{1}{r}\frac{\partial}{\partial r}\left(r\frac{\rho k}{\mu}\frac{\partial p}{\partial r}\right) = -\frac{\partial}{\partial t}(\phi\rho) \quad (1.41a)$$

$$\text{or } \frac{1}{r}\frac{\partial}{\partial r}\left(r\frac{\rho k}{\mu}\frac{\partial p}{\partial r}\right) = \frac{\partial}{\partial t}(\phi\rho). \quad (1.41b)$$

Now, substituting the real-gas law, Eq. 1.40, into Eq. 1.41b, we obtain

$$\frac{1}{r}\frac{\partial}{\partial r}\left(r\frac{kpM}{\mu zRT}\frac{\partial p}{\partial t}\right) = \frac{\partial}{\partial t}\left(\phi\frac{pM}{zRT}\right). \quad (1.42)$$

Because R, T, and M are constant and considering the special case with k constant, we find that

$$\frac{1}{r}\frac{\partial}{\partial r}\left(r\frac{p}{\mu z}\frac{\partial p}{\partial r}\right) = \frac{1}{k}\frac{\partial}{\partial t}\left(\phi\frac{p}{z}\right). \quad (1.43)$$

We can expand the right side of Eq. 1.43 using the product rule as follows:

$$\frac{1}{r}\frac{\partial}{\partial r}\left(r\frac{p}{\mu z}\frac{\partial p}{\partial r}\right) = \frac{1}{k}\left[\frac{p}{z}\frac{\partial \phi}{\partial t} + \frac{\phi \partial}{\partial t}\left(\frac{p}{z}\right)\right]. \quad (1.44)$$

We can use the chain rule to obtain another expansion of the right side of Eq. 1.44:

$$\frac{1}{r}\frac{\partial}{\partial r}\left(r\frac{p}{\mu z}\frac{\partial p}{\partial r}\right) = \frac{1}{k}\left[\frac{p}{z}\frac{\partial \phi}{\partial p}\frac{\partial p}{\partial t} + \phi\frac{\partial}{\partial p}\left(\frac{p}{z}\right)\frac{\partial p}{\partial t}\right] \quad (1.45a)$$

or $$\frac{1}{r}\frac{\partial}{\partial r}\left(r\frac{p}{\mu z}\frac{\partial p}{\partial r}\right) = \frac{p\phi}{zk}\frac{\partial p}{\partial t}\left[\frac{1}{\phi}\frac{\partial \phi}{\partial p} + \frac{z}{p}\frac{\partial}{\partial p}\left(\frac{p}{z}\right)\right]. \quad (1.45b)$$

The compressibility of gas is defined similarly to the compressibility of a liquid in terms of the density:

$$c_g = \frac{1}{\rho}\frac{\partial \rho}{\partial p}. \quad (1.46)$$

Substituting density from the real-gas law, Eq. 1.40, into this definition gives

$$c_g = \frac{zRT}{pM}\frac{\partial}{\partial p}\left(\frac{pM}{zRT}\right) = \frac{z}{p}\frac{\partial}{\partial p}\left(\frac{p}{z}\right). \quad (1.47)$$

We define formation compressibility as

$$c_f = \frac{1}{\phi}\frac{\partial \phi}{\partial p}. \quad (1.48)$$

We can now substitute Eqs. 1.47 and 1.48 into Eq. 1.45b, which gives

$$\frac{1}{r}\frac{\partial}{\partial r}\left(r\frac{p}{\mu z}\frac{\partial p}{\partial r}\right) = \frac{p\phi}{zk}\frac{\partial p}{\partial t}\left(c_f + c_g\right). \quad (1.49)$$

If we define total compressibility for this case as

$$c_t = c_g + c_f, \quad (1.50)$$

we have

$$\frac{1}{r}\frac{\partial}{\partial r}\left(r\frac{p}{\mu z}\frac{\partial p}{\partial r}\right) = \frac{p\phi c_t}{zk}\frac{\partial p}{\partial t}. \quad (1.51)$$

Eq. 1.51 is a nonlinear partial-differential equation and cannot be solved directly. We generally consider three limiting assumptions, $p/\mu z$ is constant, μc_t is constant, and the real-gas pseudopressure transformation.

Diffusivity Equation for Gas in Terms of Pressure. If we assume that the term $p/\mu z$ is constant with respect to pressure, and therefore radius, Eq. 1.51 can be written as

$$\frac{1}{r}\frac{p}{\mu z}\frac{\partial}{\partial r}\left(r\frac{\partial p}{\partial r}\right) = \frac{p\phi c_t}{zk}\frac{\partial p}{\partial t} \quad (1.52)$$

or, cancelling terms,

$$\frac{1}{r}\frac{\partial}{\partial r}\left(r\frac{\partial p}{\partial r}\right) = \frac{\phi\mu c_t}{k}\frac{\partial p}{\partial t}. \quad (1.53)$$

Eq. 1.53 is the same as the diffusivity equation for slightly compressible liquids, Eq. 1.39, and can be solved similarly (when μc_t can be considered to be constant). Eq. 1.53 has the following assumptions.

Summary of Assumptions for Eq. 1.53.
1. Radial flow.
2. Laminar (or Darcy) flow.
3. Porous medium has constant permeability and compressibility.
4. Negligible gravity effects.
5. Isothermal conditions.
6. Fluid obeys the real-gas law.
7. The term $p/\mu z$ is constant with respect to pressure.

Diffusivity Equation for Gas in Terms of Pressure Squared. We can write Eq. 1.51 in terms of pressure squared, p^2, by use of the fact that

$$p\frac{\partial p}{\partial r} = \frac{1}{2}\frac{\partial p^2}{\partial r}, \quad (1.54)$$

$$p\frac{\partial p}{\partial t} = \frac{1}{2}\frac{\partial p^2}{\partial t}, \quad (1.55)$$

and $$\frac{1}{r}\frac{\partial}{\partial r}\left(\frac{r}{\mu z}\frac{\partial p^2}{\partial r}\right) = \frac{\phi c_t}{kz}\frac{\partial p^2}{\partial t}. \quad (1.56)$$

If we assume that the term μz is constant with respect to pressure and therefore radius, Eq. 1.56 can be written as

$$\frac{1}{r}\frac{1}{\mu z}\frac{\partial}{\partial r}\left(r\frac{\partial p^2}{\partial r}\right) = \frac{\phi c_t}{kz}\frac{\partial p^2}{\partial t} \quad (1.57)$$

or, multiplying through by the term μz, as

$$\frac{1}{r}\frac{\partial}{\partial r}\left(r\frac{\partial p^2}{\partial r}\right) = \frac{\phi\mu c_t}{k}\frac{\partial p^2}{\partial t}. \quad (1.58)$$

Eq. 1.58 is also similar to the diffusivity equation for slightly compressible liquids, Eq. 1.39, but the dependent variable is pressure squared. Therefore, Eq. 1.58 has solutions similar to those of Eq. 1.39 except these solutions are in terms of pressure-squared. These equations also require that μc_t be constant. Eq. 1.58 has the following assumptions.

Summary of Assumptions for Eq. 1.58.
1. Radial flow.
2. Laminar (or Darcy) flow.
3. Porous medium has constant permeability and compressibility.
4. Negligible gravity effects.
5. Isothermal conditions.
6. Fluid obeys the real-gas law.
7. The term μz is constant with respect to pressure.

Diffusivity Equation for Gas in Terms of Pseudopressure. The assumptions we have discussed so far to obtain the linear diffusivity equation for gas are applicable only under certain conditions. **Figs. 1.2 and 1.3** illustrate the range of applicability of Eqs. 1.53 and 1.58, respectively. Fig. 1.2 shows for gases of different specific gravity when the term $p/\mu z$ is constant with pressure for a constant temperature. The figure shows that we could use Eq. 1.53 for very high pressures. Fig. 1.3 shows for gases of different specific gravity when the term μz is constant with pressure for a constant temperature. This figure shows that we could use Eq. 1.58 for very low pressures.

We prefer to have an accurate solution for all pressure ranges. A more rigorous method of linearizing Eq. 1.51 (at least partially) is by use of the real-gas pseudopressure transformation introduced by Al-Hussainy *et al.*[1] The pseudopressure transformation allows the general gas diffusivity equation, Eq. 1.51, to be solved without the limiting assumptions that certain gas properties are constant with pressure. We define a pseudopressure, p_p, by

$$p_p = 2\int_{p_0}^{p}\frac{p}{\mu z}dp. \quad (1.59)$$

Using Liebnitz's Rule for differentiating an integral,[2]

$$\frac{\partial}{\partial x}\int_{f(x)}^{h(x)} g(u)du = \left\{g[h(x)]\frac{\partial[h(x)]}{\partial x} - g[f(x)]\frac{\partial[f(x)]}{\partial x}\right\}, \quad (1.60)$$

the derivative of pseudopressure is

$$\frac{\partial p_p}{\partial r} = 2\frac{p}{\mu z}\frac{\partial p}{\partial r} \quad (1.61)$$

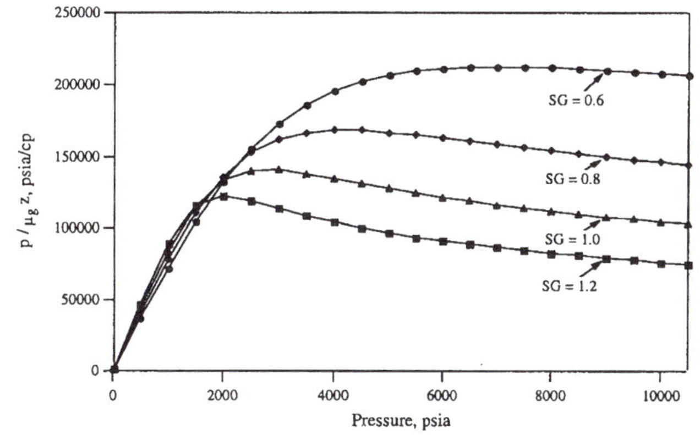

Fig. 1.2—Range of applicability of pressure methods (200°F).

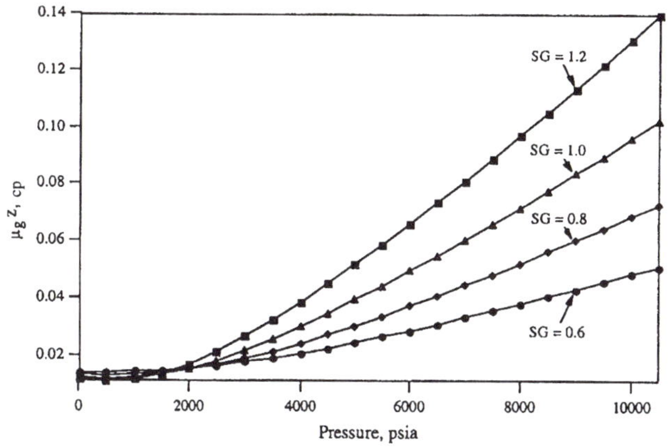

Fig. 1.3—Range of applicability of pressure-squared methods (200°F).

with respect to radius and

$$\frac{\partial p_p}{\partial t} = 2\frac{p}{\mu z}\frac{\partial p}{\partial t} \quad \text{(1.62)}$$

with respect to time.

Rearranging Eqs. 1.61 and 1.62, we can substitute for $\partial p/\partial r$ and $\partial p/\partial t$ into Eq. 1.52 to obtain

$$\frac{1}{r}\frac{\partial}{\partial r}\left[r\frac{p}{\mu z}\left(\frac{\mu z}{2p}\frac{\partial p_p}{\partial r}\right)\right] = \frac{p\phi c_t}{zk}\left(\frac{\mu z}{2p}\frac{\partial p_p}{\partial t}\right) \quad \text{(1.63)}$$

or, simplifying,

$$\frac{1}{r}\frac{\partial}{\partial r}\left(r\frac{\partial p_p}{\partial r}\right) = \frac{\phi\mu c_t}{k}\frac{\partial p_p}{\partial t}. \quad \text{(1.64)}$$

Eq. 1.64 is not completely linear because the μc_t term depends on pressure and therefore on pseudopressure, but we can approximate this quantity as constant[1] and evaluate it at current drainage area pressure, $\bar{p}$.

Eq. 1.64 is also similar to the diffusivity equation for slightly compressible liquids, Eq. 1.39, but in terms of pseudopressure. Therefore, Eq. 1.64 has solutions similar to those of Eq. 1.39, except these solutions are in terms of pseudopressure. Eq. 1.64 has the following assumptions.

Summary of Assumptions for Eq. 1.64.

1. Radial flow.
2. Laminar (or Darcy) flow.
3. Porous medium has constant permeability and compressibility.
4. Negligible gravity effects.
5. Isothermal conditions.
6. Fluid obeys the real-gas law.

1.2.4 Diffusivity Equation for Radial, Multiphase Flow. Martin[3] developed a diffusivity equation for multiphase flow, Eq. 1.65, which looks very similar to the diffusivity equation for single-phase flow, Eq. 1.39.

$$\frac{1}{r}\frac{\partial}{\partial r}\left(r\frac{\partial p}{\partial r}\right) = \left(\frac{\phi c_t}{\lambda_t}\right)\frac{\partial p}{\partial t} \quad \text{(1.65)}$$

$$\text{and } \frac{1}{r}\frac{\partial}{\partial r}\left(r\frac{\partial p}{\partial r}\right) = \phi\frac{\mu c_t}{k}\frac{\partial p}{\partial t}. \quad \text{(1.39)}$$

Appendix C presents the derivation in detail. Note that the only difference between these equations is the use of total mobility, λ_t, and a more general definition of the total compressibility, c_t. We define the total mobility of a three-phase system as the sum of the individual mobilities,

$$\lambda_t = \frac{k_o}{\mu_o} + \frac{k_w}{\mu_w} + \frac{k_g}{\mu_g} = \lambda_o + \lambda_w + \lambda_g. \quad \text{(1.66)}$$

The total compressibility, c_t, for a system with pressure-dependent porosity is defined as

$$c_t = S_o c_o + S_w c_w + S_g c_g + c_f. \quad \text{(1.67)}$$

The similarities between the multiphase flow and single-phase flow equations imply that the solutions to the single-phase diffusivity equation presented later in this chapter also apply to multiphase flow as long as c_t is defined by Eq. 1.67 and we use λ_t instead of k_o/μ_o.

Summary of Assumptions for Eq. 1.65.

1. Radial flow.
2. Laminar (or Darcy) flow.
3. Uniform porous medium.
4. Negligible gravity effects.
5. Isothermal conditions.
6. Effective permeability varies with saturation, but not pressure.
7. Small pressure- and saturation-gradient terms.
8. Negligible capillary pressure.

1.3 Initial and Boundary Conditions

The general diffusivity equation for fluid flow in porous media is a partial-differential equation for pressure with respect to both space (radius) and time.

$$\frac{1}{r}\frac{\partial}{\partial r}\left(r\frac{\partial p}{\partial r}\right) = \left(\frac{\phi\mu c_t}{k}\right)\frac{\partial p}{\partial t}. \quad \text{(1.68)}$$

To solve Eq. 1.68, we must know how the pressure behaves at specific distances and time; that is, we must specify conditions to solve the equation. Conditions specified at different extremes of distance are known as boundary conditions, whereas the condition specified at initial time, $t=0$, is known as the initial condition.

We note that the partial-differential equation is "second order" with respect to space; in other words, we have taken the partial derivative of pressure with respect to radius twice as indicated on the left side of Eq. 1.68. Likewise, the diffusivity equation is "first order" with respect to time. A second-order equation requires two conditions to obtain a solution. Therefore, we must have two boundary conditions. In radial flow, we usually specify a condition on pressure at the wellbore (the inner-boundary condition) and at the edge of the drainage area of the reservoir (the outer-boundary condition). Similarly, a first-order differential equation requires only one condition; therefore, we need only a single condition for time (i.e., the initial condition).

In this section, we will discuss possible initial and boundary conditions for different reservoir models and production schemes.

1.3.1 Initial Condition. We always assume that the reservoir is initially at a uniform, constant pressure throughout the reservoir at a time $t=0$.

$$p(r,0) = p_i. \quad \text{(1.69)}$$

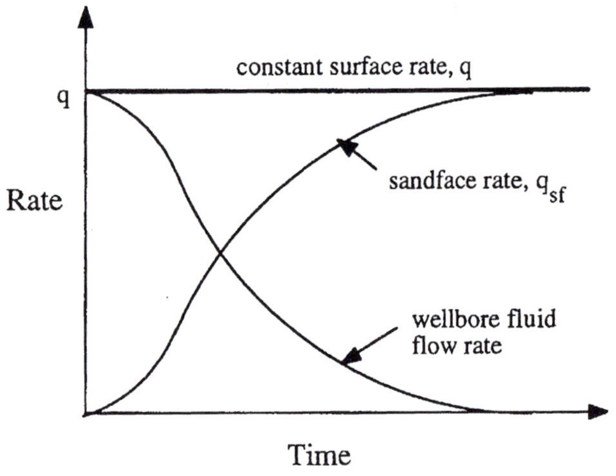

Fig. 1.4—Surface and sandface rates during wellbore storage.

1.3.2 Outer-Boundary Conditions. We consider three cases for the outer boundary of the reservoir. It may be infinite-acting (i.e., it is so large that the outer boundary effects are never felt at points in the reservoir at practical distances from the source or sink). The reservoir may be bounded by a no-flow boundary (i.e., a volumetric reservoir). The reservoir could be bounded by a constant-pressure boundary, such as a reservoir/aquifer system.

Infinite-Acting Reservoir. As the radius becomes very large, approaching infinity, the pressure approaches the initial pressure, p_i, for all times.

$$p(r \to \infty, t) = p_i \quad \text{(1.70)}$$

$$\text{or } \Delta p(r \to \infty, t) = p_i - p(r \to \infty, t) = 0. \quad \text{(1.71)}$$

No-Flow Boundary. For a cylindrical reservoir with a no-flow boundary a distance r_e from the well the flow rate at $r = r_e$ will be $q = 0$ for all times greater than zero.

$$\text{Darcy's law states that } q \propto \frac{\partial p}{\partial r}. \quad \text{(1.72)}$$

$$q = -C\frac{kA}{\mu}\frac{\partial p}{\partial r}, \quad \text{(1.73)}$$

where C = constant $\neq 0$, k = permeability $\neq 0$, A = area (cross-sectional) $\neq 0$, and μ = viscosity $\neq 0$. Therefore,

$$(\partial p/\partial r)_{r_e} = 0. \quad \text{(1.74)}$$

Constant-Pressure Outer Boundary. For a cylindrical reservoir with a constant-pressure boundary at distance r_e from the well, the pressure at the outer boundary will be equal to the initial pressure, p_i, for all times.

$$p(r = r_e, t) = p_i, \quad \text{(1.75)}$$

$$\text{or } \Delta p(r = r_e, t) = 0. \quad \text{(1.76)}$$

1.3.3 Inner-Boundary Conditions. A well may be produced at constant rate or constant pressure and have wellbore storage effects.

Constant-Rate Production. If a well is produced at a constant sandface rate, this rate of flow from the formation into the wellbore of radius r_w may be described by Darcy's law. At $r = r_w$,

$$qB = \left(\frac{kA}{\alpha_1 \mu}\frac{\partial p}{\partial r}\right)_{(r=r_w)}, \quad \text{(1.77)}$$

where α_1 = conversion constant [e.g., in "field units" α_1 = 141.2 (2π), and A = cross-sectional area = 2 $r_w h$ (in square feet)].

If we substitute the definition of area into the inner-boundary condition, we have

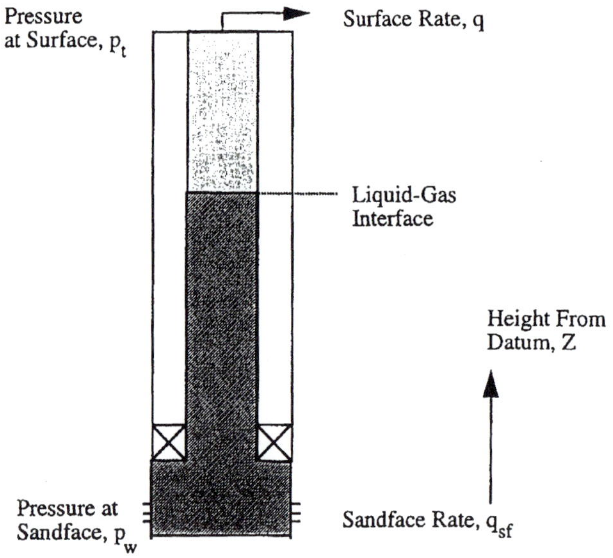

Fig. 1.5—Wellbore diagram for a well with a liquid-gas interface.

$$qB = \left(\frac{2\pi}{\alpha_1}\frac{kh}{\mu}r_w\frac{\partial p}{\partial r}\right)_{(r=r_w)}. \quad \text{(1.78)}$$

Rearranging Eq. 1.78, the constant-rate inner boundary condition becomes

$$\left(r\frac{\partial p}{\partial r}\right)_{(r=r_w)} = \frac{\alpha_1 qB\mu}{2\pi kh}. \quad \text{(1.79)}$$

Constant-Pressure Production. This inner-boundary condition is valid when the reservoir is initially at uniform pressure throughout the reservoir and is produced by simply lowering the wellbore pressure to a constant value, p_{wf}, and producing at a variable sandface rate.

$$p(r_w, t) = p_{wf} = \text{constant}. \quad \text{(1.80)}$$

Wellbore Storage. Wellbore storage may occur if a well is set to produce at constant surface rate after a shut-in period. Initially, fluid will unload from the wellbore with no flow from the formation to the wellbore. As time passes, the sandface rate will equal the surface rate, with the amount of liquid stored being constant; see **Fig. 1.4.**

We call the ability of the wellbore to store or unload fluids per unit change in pressure the wellbore-storage coefficient, C(bbl/psi). The definition of the wellbore-storage coefficient depends on the situation in the wellbore. We consider the following two cases: a liquid/gas interface in the wellbore and a single, compressible fluid in the wellbore.

Liquid/Gas Interface. For a pumping well or a well produced by gas lift, the wellbore will have a column of liquid with a column of gas at the top of the wellbore.

If we let the surface rate, q, be constant, a mass balance for the wellbore shown in **Fig. 1.5** would be

$$\begin{pmatrix}\text{Rate of flow of}\\ \text{mass into wellbore}\end{pmatrix} - \begin{pmatrix}\text{Rate of flow of}\\ \text{mass out of wellbore}\end{pmatrix}$$

$$= \begin{pmatrix}\text{Rate of accumulation}\\ \text{of mass in wellbore}\end{pmatrix}, \text{ which is}$$

$$(q_{sf}B\rho_{sf}) - (qB\rho_{sc}) = \frac{d}{dt}\left(\frac{24\rho_{wb}V_{wb}}{5.615}\right), \quad \text{(1.81)}$$

where time is in hours and the volume of the wellbore, V_{wb}, is expressed in cubic feet.

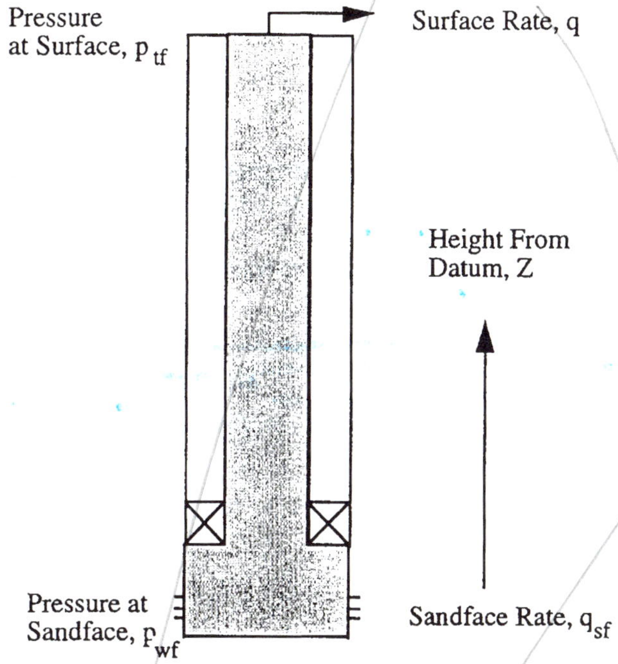

Fig. 1.6—Wellbore diagram for a well producing a single-phase fluid.

$$V_{wb} = A_{wb}\,Z. \quad \text{(1.82)}$$

If we assume a constant wellbore area and constant density ($\rho_{sf} = \rho_{sc} = \rho_{wb}$), we can write the mass balance as

$$\left(q_{sf} - q\right)B = \frac{24}{5.615}A_{wb}\frac{\mathrm{d}Z}{\mathrm{d}t}. \quad \text{(1.83)}$$

The surface pressure, p_t, is related to the bottomhole pressure, p_w, at any time by

$$p_w = p_t + \frac{\rho Z}{144}, \quad \text{(1.84)}$$

where ρ = density of liquid.
Differentiating with respect to time gives

$$\frac{\mathrm{d}}{\mathrm{d}t}(p_w - p_t) = \frac{\rho}{144}\frac{\mathrm{d}Z}{\mathrm{d}t}. \quad \text{(1.85)}$$

Substituting for $\mathrm{d}Z/\mathrm{d}t$ from the mass balance gives

$$\frac{\mathrm{d}}{\mathrm{d}t}(p_w - p_t)\left(\frac{144}{\rho}\right) = \left(\frac{5.615}{24A_{wb}}\right)\left(q_{sf} - q\right)B$$

$$\left(q_{sf} - q\right)B = \frac{(24)(144)}{5.615\rho}A_{wb}\frac{\mathrm{d}}{\mathrm{d}t}(p_w - p_t). \quad \text{(1.86)}$$

We now define a wellbore-storage coefficient, C, as

$$C = \frac{144A_{wb}}{5.615\rho}\frac{\text{bbl}}{\text{psi}}, \quad \text{(1.87)}$$

where A_{wb} is in square feet, the constant 144 in.^{2}/ft^2 converts square feet to square inches, ρ is in lbm/ft^3, and the constant 5.615 ft^3/bbl converts barrels to cubic feet.

If we substitute C into the equation relating sandface and surface rate, we obtain

$$q_{sf} = q + \frac{24C}{B}\frac{\mathrm{d}}{\mathrm{d}t}(p_w - p_t). \quad \text{(1.88)}$$

The sandface rate is given by Darcy's law as

$$q_{sf} = \frac{2\pi kh}{\alpha_1 B\mu}\left(r\frac{\partial p}{\partial r}\right)_{(r=r_w)}, \quad \text{(1.89)}$$

where $\alpha_1 = 141.2(2\pi)$ in field units, thus

$$q = \frac{2\pi kh}{\alpha_1 B\mu}\left(r\frac{\partial p}{\partial r}\right)_{(r=r_w)} - \frac{24C}{B}\frac{\mathrm{d}}{\mathrm{d}t}(p_w - p_t). \quad \text{(1.90)}$$

This is the inner-boundary condition for wellbore storage for a well with a gas-liquid interface. In some cases, we can assume that

$$\frac{\mathrm{d}p_t}{\mathrm{d}t} \approx 0, \quad \text{(1.91)}$$

and the boundary condition becomes

$$q = \frac{2\pi kh}{\alpha_1 B\mu}\left(r\frac{\partial p}{\partial r}\right)_{(r=r_w)} - \frac{24C}{B}\frac{\mathrm{d}p_w}{\mathrm{d}t}. \quad \text{(1.92)}$$

Single Phase in Wellbore. In this case, we consider a well that is producing a single-phase fluid, either liquid or gas, at a constant surface rate, as illustrated in **Fig. 1.6.** The mass balance for this system would be

$$\begin{pmatrix}\text{Rate of mass flow}\\ \text{into wellbore at}\\ \text{sandface}\end{pmatrix} - \begin{pmatrix}\text{Rate of mass flow}\\ \text{out of wellbore at}\\ \text{surface}\end{pmatrix}$$

$$= \begin{pmatrix}\text{Rate of accumulation}\\ \text{of mass in wellbore}\end{pmatrix}$$

$$q_{sf}\,B\rho_{sf} - qB_{sc}\rho_{sc} = \left(24V_{wb}\frac{\mathrm{d}\rho_{wb}}{\mathrm{d}t}\right)$$

$$= \left(24V_{wb}\rho_{wb}c_{wb}\frac{\mathrm{d}p_{wb}}{\mathrm{d}t}\right), \quad \text{(1.93)}$$

because $\dfrac{\mathrm{d}\rho_{wb}}{\mathrm{d}t} = \rho_{wb}c_{wb}\dfrac{\mathrm{d}p_{wb}}{\mathrm{d}t}$. (1.94)

The density/volume factor product is constant and thus the same at both surface and reservoir conditions. Thus, if we define C for the single-phase case as

$$C = V_{wb}c_{wb}, \quad \text{(1.95)}$$

the mass balance becomes

$$q_{sf} = q + \frac{24C}{B}\frac{\mathrm{d}p_w}{\mathrm{d}t}. \quad \text{(1.96)}$$

The wellbore-storage boundary condition is the same, despite the different definition of C.

$$q = \frac{2\pi kh}{\alpha_1 B\mu}\left(r\frac{\partial p}{\partial r}\right)_{(r=r_w)} - \frac{24C}{B}\frac{\mathrm{d}p_w}{\mathrm{d}t}. \quad \text{(1.97)}$$

Skin Factor. To account for the additional pressure drop near the wellbore caused by reduction in permeability owing to adverse drilling and completion conditions, Hawkins[4] developed the idea of a finite skin zone around the wellbore. This skin zone can cause the measured pressure drop to be much greater than the pressure drop calculated from solutions to the diffusivity equation. We assume that the shaded zone in **Fig. 1.7** has a constant permeability, k_s, and extends only a short distance, r_s, from the center of the wellbore into the reservoir. **Fig. 1.8** shows the effect that this altered zone would have on the pressure drop at the wellbore.

Δp_1 represents the pressure drop from a radius r_s to the wellbore radius, r_w, that would normally occur because of flow through the altered zone. Δp_2 represents the pressure drop from a radius r_s to the wellbore radius, r_w, that would have occurred had there been no change in permeability in the altered zone (i.e., if the permeability in this zone remained the average formation permeability, k). The additional pressure drop that results across the skin zone is therefore equal to Δp_s, where

$$\Delta p_s = \Delta p_1 — \Delta p_2. \quad \text{(1.98)}$$

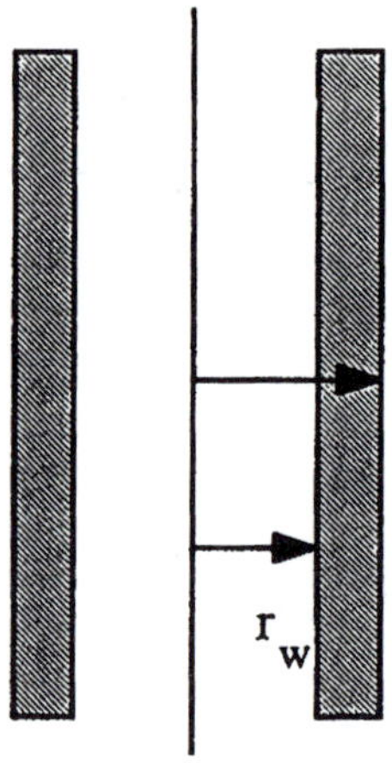

Fig. 1.7—Illustration of the zone of altered permeability around the wellbore.

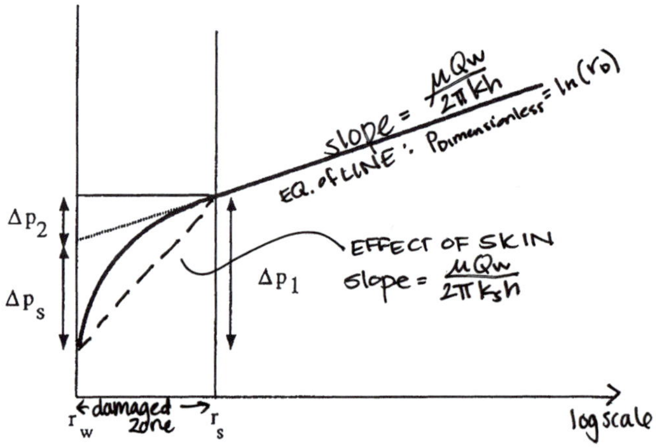

Fig. 1.8—The effect of the skin zone on the wellbore pressure drop.

Because r_s is small, we can assume steady-state flow in the altered zone and write the steady-state radial-flow equations for the pressure drops in this region.

$$\Delta p_1 = \frac{\alpha_1 qB\mu}{2\pi k_s h}\ln\left(\frac{r_s}{r_w}\right) \quad (1.99)$$

and $$\Delta p_2 = \frac{\alpha_1 qB\mu}{2\pi kh}\ln\left(\frac{r_s}{r_w}\right). \quad (1.100)$$

Combining Eqs. 1.98 through 1.100,

$$\Delta p_s = \frac{\alpha_1 qB\mu}{2\pi k_s h}\ln\left(\frac{r_s}{r_w}\right) - \frac{\alpha_1 qB\mu}{2\pi kh}\ln\left(\frac{r_s}{r_w}\right)$$

$$= \frac{\alpha_1 qB\mu}{2\pi h}\ln\left(\frac{r_s}{r_w}\right)\left(\frac{1}{k_s} - \frac{1}{k}\right)$$

$$= \frac{\alpha_1 qB\mu}{2\pi h}\ln\left(\frac{r_s}{r_w}\right)\left[\frac{(k/k_s) - 1}{k}\right]$$

$$= \frac{\alpha_1 qB\mu}{2\pi kh}\left(\frac{k}{k_s} - 1\right)\ln\left(\frac{r_s}{r_w}\right). \quad (1.101)$$

We define a skin factor, s, on the basis of the properties of the altered zone.

$$s = \left(\frac{k}{k_s} - 1\right)\ln\left(\frac{r_s}{r_w}\right). \quad (1.102)$$

From Eqs. 1.101 and 1.102, the definition of skin factor becomes

$$s = \frac{2\pi kh\Delta p_s}{\alpha_1 qB\mu}. \quad (1.103)$$

We note that Eq. 1.102 provides some insight into the physical significance of the sign of the skin factor. If a well is damaged ($k_s < k$), s will be positive; and the greater the contrast between k_s and k and the deeper into the formation the damage extends, the larger the numerical value of s. There is no upper limit for s. Some newly drilled wells will not flow at all before stimulation; for these wells, $k_s \approx 0$ and $s \to \infty$. If a well is stimulated ($k_s > k$), s will be negative; and the deeper the stimulation, the greater the numerical value of s. Rarely does a stimulated well have a skin factor less than -7 or -8, and such skin factors arise only for wells with deeply penetrating, highly conductive hydraulic fractures. We should note finally that, if a well is neither damaged nor stimulated ($k_s = k$), $s = 0$. We caution that Eq. 1.102 is best applied qualitatively; actual wells rarely can be characterized exactly by such a simplified model. We also note that an altered zone near a particular well affects only the pressure near that well; i.e., the pressure in the unaltered formation away from the well is not affected by the presence of the altered zone.

1.4 Dimensionless Groups

We use dimensionless groups to express our equations more simply. Many well-test-analysis techniques use dimensionless variables to depict general trends rather than working with specific parameters (e.g., k and h). To define appropriate dimensionless variables, we find logical groupings of variables that appear naturally in differential equations and initial and boundary conditions.

In this section, we present dimensionless groups used for radial flow of slightly compressible liquids that are being produced at either constant rate, with and without wellbore storage, and constant bottomhole pressure. Appendix A provides a complete explanation of how these dimensionless groups are derived.

1.4.1 Radial Flow—Constant-Rate Production. For this case, we define the following dimensionless variables and use conventional field units.

Dimensionless Pressure.

$$p_D = \frac{kh}{141.2qB\mu}(p_i - p). \quad (1.104)$$

Dimensionless Radius.

$$r_D = \frac{r}{r_w} \quad (1.105)$$

and $$r_{eD} = \frac{r_e}{r_w}. \quad (1.106)$$

Dimensionless Time.

$$t_D = \frac{0.0002637kt}{\phi\mu c_t r_w^2}. \quad (1.107)$$

Dimensionless Wellbore-Storage Coefficient.

$$C_D = \frac{0.8936C}{\phi c_t h r_w^2}. \quad (1.108)$$

Skin Factor.

$$s = \frac{kh\Delta p_s}{141.2qB\mu}. \quad (1.109)$$

The diffusivity equation and various initial and boundary conditions can be rewritten in terms of these dimensionless variables.

Partial-Differential Equation.

$$\frac{1}{r_D}\frac{\partial}{\partial r_D}\left(r_D\frac{\partial p_D}{\partial r_D}\right) = \frac{\partial p_D}{\partial t_D}. \quad (1.110)$$

Initial Condition.

$$p_D(r_D, t_D = 0) = 0. \quad (1.111)$$

Outer Boundary Condition. *Infinite-Acting Reservoir.*

$$p_D(r_D \to \infty, t_D) = 0. \quad (1.112)$$

No-Flow Boundary.

$$\left(\frac{\partial p_D}{\partial r_D}\right)_{r_{eD}} = 0. \quad (1.113)$$

Constant-Pressure Boundary.

$$p_D(r_D = r_{eD}, t_D) = 0. \quad (1.114)$$

Inner-Boundary Condition. *Constant-Rate Production.*

$$\left(\frac{\partial p_D}{\partial r_D}\right)_{(r_D=1)} = -1. \quad (1.115)$$

Constant Rate Production With Wellbore Storage.

$$C_D \frac{dp_{wD}}{dt_D} - \left(r_D \frac{\partial p_D}{\partial r_D}\right)_{(r_D=1)} = 1. \quad (1.116)$$

Skin Factor.

$$p_{wD}(t_D) = p_D(1, t_D) + s. \quad (1.117)$$

1.4.2 Radial Flow—Constant-Pressure Production. This case requires a different definition of dimensionless pressure. Dimensionless time and length are defined the same as for the constant rate case. In addition, we must define dimensionless rate and cumulative production.

Dimensionless Pressure.

$$p_D = \frac{p_i - p}{p_i - p_{wf}}. \quad (1.118)$$

Dimensionless Rate.

$$q_D = \frac{qB\mu}{0.00708kh(p_i - p_{wf})}. \quad (1.119)$$

Dimensionless Cumulative Production.

$$Q_{pD} = \int_0^{t_D} q_D dt' = \frac{B}{1.119\phi c_t h r_w^2 (p_i - p_{wf})} Q_p. \quad (1.120)$$

The diffusivity equation and various initial and boundary conditions can be rewritten in terms of these dimensionless variables.

Partial-Differential Equation.

$$\frac{1}{r_D}\frac{\partial}{\partial r_D}\left(r_D \frac{\partial p_D}{\partial r_D}\right) = \frac{\partial p_D}{\partial t_D}. \quad (1.121)$$

Initial Condition.

$$p_D(r_D, t_D = 0) = 0. \quad (1.122)$$

Outer-Boundary Condition. *Infinite-Acting Reservoir.*

$$p_D(r_D \to \infty, t_D) = 0. \quad (1.123)$$

No-Flow Boundary.

$$\left(\frac{\partial p_D}{\partial r_D}\right)_{r_{eD}} = 0. \quad (1.124)$$

Constant-Pressure Boundary.

$$p_D(r_D = r_{eD}, t_D) = 0. \quad (1.125)$$

Inner-Boundary Condition for Constant-Pressure Production.

$$p_D(r_D = 1, t_D) = 1. \quad (1.126)$$

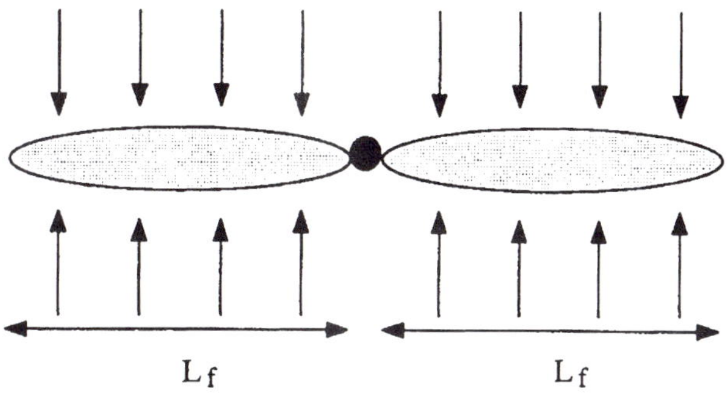

Fig. 1.9—Linear flow to a fractured well system.

1.4.3 Linear Flow—Constant-Rate Production—General Case. For the general linear-flow case, we define the following dimensionless variables on the basis of a cross-sectional area. In Sec. 1.4.4, we will present the specialized case for hydraulically fractured wells. Note that the diffusivity equation that models linear flow may be derived from a shell balance exactly as in the radial-flow diffusivity equation, but with rectangular coordinates. Appendix C presents this derivation.

Dimensionless Pressure.

$$p_D = \frac{k\sqrt{A}}{141.2qB\mu}(p_i - p). \quad (1.127)$$

Dimensionless Length.

$$x_D = \frac{x}{\sqrt{A}}. \quad (1.128)$$

Dimensionless Time.

$$t_{AD} = \frac{0.0002637kt}{\phi \mu c_t A}. \quad (1.129)$$

The diffusivity equation and various initial and boundary conditions can be rewritten in terms of these dimensionless variables.

Partial-Differential Equation.

$$\frac{\partial^2 p_D}{\partial x_D^2} = \frac{\partial p_D}{\partial t_D}. \quad (1.130)$$

Initial Condition.

$$p_D(x_D, t_D = 0) = 0. \quad (1.131)$$

Outer-Boundary Condition. *Infinite-Acting Reservoir.*

$$p_D(x_D \to \infty, t_D) = 0. \quad (1.132)$$

No-Flow Boundary.

$$\left(\frac{\partial p_D}{\partial x_D}\right)_{x_{eD}} = 0. \quad (1.133)$$

Constant-Pressure Boundary.

$$p_D(x_D = x_{eD}, t_D) = 0. \quad (1.134)$$

Inner-Boundary Condition for Constant-Rate Production.

$$\left(\frac{\partial p_D}{\partial x_D}\right)_{x_{D=1}} = -1. \quad (1.135)$$

1.4.4 Linear Flow—Constant-Rate Production—Hydraulically Fractured Wells. Linear flow occurs in hydraulically fractured well systems because the fracture behaves as a "plane source" with the fluid flowing linearly to the fracture. **Fig. 1.9** illustrates this system.

For this case, the cross-sectional area denoted in the general case represents a vertical fracture with two equal-length wings of length L_f and height h. Therefore, $A = 4hL_f$, with flow entering both sides of each wing of the fracture.

The dimensionless pressure for fractured wells is defined the same as in the radial-flow constant-rate cases, but dimensionless length and time are defined on the basis of the fracture half-length, L_f.

Dimensionless Pressure.

$$p_D = \frac{kh}{141.2qB\mu}(p_i - p). \quad (1.136)$$

Dimensionless Length.

$$L_D = \frac{x}{L_f}. \quad (1.137)$$

Dimensionless Time.

$$t_{L_f D} = \frac{0.0002637kt}{\phi\mu c_t L_f^2}. \quad (1.138)$$

The diffusivity equation and various initial and boundary conditions can be rewritten in terms of these dimensionless variables.

Partial-Differential Equation.

$$\frac{\partial^2 p_D}{\partial L_D^2} = \frac{\partial p_D}{\partial t_D}. \quad (1.139)$$

Initial Condition.

$$p_D(L_D, t_D = 0) = 0. \quad (1.140)$$

Outer-Boundary Condition. *Infinite-Acting Reservoir.*

$$p_D(L_D \rightarrow \infty, t_D) = 0. \quad (1.141)$$

No-Flow Boundary.

$$\left(\frac{\partial p_D}{\partial L_D}\right)_{L_{eD}} = 0. \quad (1.142)$$

Constant Pressure Boundary.

$$p_D\left(L_D = L_{eD}, t_D\right) = 0. \quad (1.143)$$

Inner-Boundary Condition for Constant-Rate Production.

$$\left(\frac{\partial p_D}{\partial L_D}\right)_{L_D = 1} = -1. \quad (1.144)$$

1.5 Solutions to the Diffusivity Equation

There are several different solutions to the diffusivity equation, depending on the initial and boundary conditions used to solve the equation. In this chapter, we present the solutions for the following reservoir models.

1. Transient radial flow, constant-rate production from a line-source well, both without skin factor and with skin factor and wellbore storage.
2. Pseudosteady-state radial flow, constant rate production from a cylindrical-source well in a closed reservoir.
3. Steady-state radial flow, constant-rate production from a cylindrical-source well in a reservoir with constant pressure outer boundaries.
4. Transient linear flow, constant rate production from a hydraulically fractured well.

There are numerous possible reservoir models with different boundary conditions, but the solution techniques for all models are similar. Appendix B gives a full explanation of these solution techniques. We also give examples of how to implement these solutions in solving flow problems in reservoirs.

1.5.1 Transient Radial Flow, Constant-Rate Production From a Line-Source Well. In this case, we assume that the well can be represented as a "line source;" in other words, the wellbore is infinitesimally small ($r_w \rightarrow 0$). This well produces at a constant rate with no wellbore storage or skin from an infinitely large reservoir. This does not describe a real situation; all reservoirs have a finite areal extent, and all wells that are drilled have a certain wellbore radius. However, the wellbore radius is small compared with the radius of the reservoir, so a line-source assumption is not unreasonable. Also, at early producing times, the effects of the outer boundaries of the reservoir are not seen and the reservoir acts as if there were no boundaries (i.e., the reservoir is infinite-acting).

The partial-differential equation for this case is given by

$$\frac{1}{r_D}\frac{\partial}{\partial r_D}\left(r_D\frac{\partial p_D}{\partial r_D}\right) = \frac{\partial p_D}{\partial t_D}. \quad (1.110)$$

Initially, pressure in the reservoir is uniform throughout the reservoir, so the initial condition is given by

$$p_D(r_D, t_D = 0) = 0. \quad (1.111)$$

The reservoir is infinite-acting; therefore, as the dimensionless radius tends toward infinity, the pressure at that radius will be the same as initial pressure and the dimensionless pressure function will be zero. The outer boundary condition is therefore written as

$$p_D(r_D \rightarrow \infty, t_D) = 0. \quad (1.112)$$

The reservoir is producing at constant sandface rate with no wellbore storage or skin from a line-source well (i.e., the wellbore radius approaches zero). The inner-boundary condition for this case is

$$\lim_{r_D \rightarrow 0}\left(r_D\frac{\partial p_D}{\partial r_D}\right)_{r_D} = -1. \quad (1.145)$$

Remember, the inner-boundary condition is for a "line-source" well. This is a limiting condition as $r_w \rightarrow 0$ of the constant-rate boundary condition for a finite wellbore.

Using either Laplace transforms or the Boltzmann transformation, as explained in Appendix B, we can derive the line-source solution in dimensionless variables, given here by

$$p_D = -\frac{1}{2}\mathrm{Ei}\left(\frac{-r_D^2}{4t_D}\right), \quad (1.146)$$

where Ei is the exponential integral defined as

$$\mathrm{Ei}(-x) = -\int_x^\infty \frac{e^{-y}}{y}\mathrm{d}y. \quad (1.147)$$

Substituting in the appropriate definitions for dimensionless variables as given in Sec. 1.4, we can write the line-source (or Ei-function) solution in terms of field variables

$$p = p_i + 70.6\frac{qB\mu}{kh}\mathrm{Ei}\left(-\frac{948\phi\mu c_t r^2}{kt}\right). \quad (1.148)$$

The line-source solution is an approximation of the more general cylindrical-source solution, so we must define limits of its applicability. It has been shown to be accurate for the range

$$\frac{(3.975 \times 10^5)\phi\mu c_t r_w^2}{k} < t < \frac{948\phi\mu c_t r_e^2}{k}. \quad (1.149)$$

At times less than the lower limit, the assumption of zero well size limits the accuracy of the equation. At times greater than the upper limit (for a well centered in a circular drainage area), the reservoir boundaries will affect the pressure distribution in the reservoir so that the reservoir is no longer infinite-acting.

When the argument of the Ei function, x, is greater than 0.01, we use **Table 1.1** to estimate the Ei-function value for a given x value. We then use that value in Eq. 1.148 to calculate the pressure.

For values of x less than 0.01, this solution can be further simplified by making an approximation to the exponential integral function, $\mathrm{Ei}(-x)$. This approximation is given by Eq. 1.150.

$$\mathrm{Ei}(-x) \approx \ln(1.781x). \quad (1.150)$$

TABLE 1.1—VALUES OF THE EXPONENTIAL INTEGRAL, $-\mathrm{Ei}(-x)$

$-\mathrm{Ei}(-x)$, $0.000 < x < 0.209$, interval = 0.001

x	0	1	2	3	4	5	6	7	8	9
0.00	+∞	6.332	5.639	5.235	4.948	4.726	4.545	4.392	4.259	4.142
0.01	4.038	3.944	3.858	3.779	3.705	3.637	3.574	3.514	3.458	3.405
0.02	3.355	3.307	3.261	3.218	3.176	3.137	3.098	3.062	3.026	2.992
0.03	2.959	2.927	2.897	2.867	2.838	2.810	2.783	2.756	2.731	2.706
0.04	2.681	2.658	2.634	2.612	2.590	2.568	2.547	2.527	2.507	2.487
0.05	2.468	2.449	2.431	2.413	2.395	2.377	2.360	2.344	2.327	2.311
0.06	2.295	2.279	2.264	2.249	2.235	2.220	2.206	2.192	2.178	2.164
0.07	2.151	2.138	2.125	2.112	2.099	2.087	2.074	2.062	2.050	2.039
0.08	2.027	2.015	2.004	1.993	1.982	1.971	1.960	1.950	1.939	1.929
0.09	1.919	1.909	1.899	1.889	1.879	1.869	1.860	1.850	1.841	1.832
0.10	1.823	1.814	1.805	1.796	1.788	1.779	1.770	1.762	1.754	1.745
0.11	1.737	1.729	1.721	1.713	1.705	1.697	1.689	1.682	1.674	1.667
0.12	1.660	1.652	1.645	1.638	1.631	1.623	1.616	1.609	1.603	1.596
0.13	1.589	1.582	1.576	1.569	1.562	1.556	1.549	1.543	1.537	1.530
0.14	1.524	1.518	1.512	1.506	1.500	1.494	1.488	1.482	1.476	1.470
0.15	1.464	1.459	1.453	1.447	1.442	1.436	1.431	1.425	1.420	1.415
0.16	1.409	1.404	1.399	1.393	1.388	1.383	1.378	1.373	1.368	1.363
0.17	1.358	1.353	1.348	1.343	1.338	1.333	1.329	1.324	1.319	1.314
0.18	1.310	1.305	1.301	1.296	1.291	1.287	1.282	1.278	1.274	1.269
0.19	1.265	1.261	1.256	1.252	1.248	1.243	1.239	1.235	1.231	1.227
0.20	1.223	1.219	1.215	1.210	1.206	1.202	1.198	1.195	1.191	1.187

$-\mathrm{Ei}(-x)$, $0.00 < x < 2.09$, interval = 0.01

x	0	1	2	3	4	5	6	7	8	9
0.0	+∞	4.038	3.335	2.959	2.681	2.468	2.295	2.151	2.027	1.919
0.1	1.823	1.737	1.660	1.589	1.524	1.464	1.409	1.358	1.309	1.265
0.2	1.223	1.183	1.145	1.110	1.076	1.044	1.014	0.985	0.957	0.931
0.3	0.906	0.882	0.858	0.836	0.815	0.794	0.774	0.755	0.737	0.719
0.4	0.702	0.686	0.670	0.655	0.640	0.625	0.611	0.598	0.585	0.572
0.5	0.560	0.548	0.536	0.525	0.514	0.503	0.493	0.483	0.473	0.464
0.6	0.454	0.445	0.437	0.428	0.420	0.412	0.404	0.396	0.388	0.381
0.7	0.374	0.367	0.360	0.353	0.347	0.340	0.334	0.328	0.322	0.316
0.8	0.311	0.305	0.300	0.295	0.289	0.284	0.279	0.274	0.269	0.265
0.9	0.260	0.256	0.251	0.247	0.243	0.239	0.235	0.231	0.227	0.223
1.0	0.219	0.216	0.212	0.209	0.205	0.202	0.198	0.195	0.192	0.189
1.1	0.186	0.183	0.180	0.177	0.174	0.172	0.169	0.166	0.164	0.161
1.2	0.158	0.156	0.153	0.151	0.149	0.146	0.144	0.142	0.140	0.138
1.3	0.135	0.133	0.131	0.129	0.127	0.125	0.124	0.122	0.120	0.118
1.4	0.116	0.114	0.113	0.111	0.109	0.108	0.106	0.105	0.103	0.102
1.5	0.100	0.0985	0.0971	0.0957	0.0943	0.0929	0.0915	0.0902	0.0889	0.0876
1.6	0.0863	0.0851	0.0838	0.0826	0.0814	0.0802	0.0791	0.0780	0.0768	0.0757
1.7	0.0747	0.0736	0.0725	0.0715	0.0705	0.0695	0.0685	0.0675	0.0666	0.0656
1.8	0.0647	0.0638	0.0629	0.0620	0.0612	0.0603	0.0595	0.0586	0.0578	0.0570
1.9	0.0562	0.0554	0.0546	0.0539	0.0531	0.0524	0.0517	0.0510	0.0503	0.0496
2.0	0.0489	0.0482	0.0476	0.0469	0.0463	0.0456	0.0450	0.0444	0.0438	0.0432

$-\mathrm{Ei}(-x)$, $2.0 < x < 10.9$, interval = 0.1

x	0	1	2	3	4	5	6	7	8	9
2	4.89×10^{-2}	4.26×10^{-2}	3.72×10^{-2}	3.25×10^{-2}	2.84×10^{-2}	2.49×10^{-2}	2.19×10^{-2}	1.92×10^{-2}	1.69×10^{-2}	1.48×10^{-2}
3	1.30×10^{-2}	1.15×10^{-2}	1.01×10^{-2}	8.94×10^{-3}	7.89×10^{-3}	6.87×10^{-3}	6.16×10^{-3}	5.45×10^{-3}	4.82×10^{-3}	4.27×10^{-3}
4	3.78×10^{-3}	3.35×10^{-3}	2.97×10^{-3}	2.64×10^{-3}	2.34×10^{-3}	2.07×10^{-3}	1.84×10^{-3}	1.64×10^{-3}	1.45×10^{-3}	1.29×10^{-3}
5	1.15×10^{-3}	1.02×10^{-3}	9.08×10^{-4}	8.09×10^{-4}	7.19×10^{-4}	6.41×10^{-4}	5.71×10^{-4}	5.09×10^{-4}	4.53×10^{-4}	4.04×10^{-4}
6	3.60×10^{-4}	3.21×10^{-4}	2.86×10^{-4}	2.55×10^{-4}	2.28×10^{-4}	2.03×10^{-4}	1.82×10^{-4}	1.62×10^{-4}	1.45×10^{-4}	1.29×10^{-4}
7	1.15×10^{-4}	1.03×10^{-4}	9.22×10^{-5}	8.24×10^{-5}	7.36×10^{-5}	6.58×10^{-5}	5.89×10^{-5}	5.26×10^{-5}	4.71×10^{-5}	4.21×10^{-5}
8	3.77×10^{-5}	3.37×10^{-5}	3.02×10^{-5}	2.70×10^{-5}	2.42×10^{-5}	2.16×10^{-5}	1.94×10^{-5}	1.73×10^{-5}	1.55×10^{-5}	1.39×10^{-5}
9	1.24×10^{-5}	1.11×10^{-5}	9.99×10^{-6}	8.95×10^{-6}	8.02×10^{-6}	7.18×10^{-6}	6.44×10^{-6}	5.77×10^{-6}	5.17×10^{-6}	4.64×10^{-6}
10	4.15×10^{-6}	3.73×10^{-6}	3.34×10^{-6}	3.00×10^{-6}	2.68×10^{-6}	2.41×10^{-6}	2.16×10^{-6}	1.94×10^{-6}	1.74×10^{-6}	1.56×10^{-6}

This approximation simplifies the line-source solution to

$$p = p_i + 70.6\frac{qB\mu}{kh}\left[\ln\left(\frac{1,688\phi\mu c_t r^2}{kt}\right)\right]. \quad \text{(1.151)}$$

Example 1.1—Calculation of Pressures Beyond the Wellbore With the Line-Source Solution. A well and reservoir have the following characteristics. The well is producing only oil at a constant rate of 20 STB/D. The following data describe the well and formation.

$q =$ 20 STB/D

$h =$ 150 ft
$B =$ 1.475 RB/STB
$r_e =$ 3,000 ft
$p_i =$ 3,000 psia
$\phi =$ 0.23
$c_t =$ 1.5×10^{-5} psia^{-1}
$\mu =$ 0.72 cp
$r_w =$ 0.5 ft
$k =$ 0.1 md

Calculate the reservoir pressure at a radius of 1 ft after 3 hours of production; then, calculate the pressure at radii of 10 and 100 ft after 3 hours of production.

Solution.

1. First, we determine whether we have conditions that lie in the range of applicability of the line source solution. From Eq. 1.149, we have the following acceptable range.

$$\frac{(3.975 \times 10^5)\phi\mu c_t r_w^2}{k} < t < \frac{948\phi\mu c_t r_e^2}{k}.$$

We substitute the given well and reservoir conditions into the equation.

$$\frac{3.975 \times 10^5(0.23)(0.72)(1.5 \times 10^{-5})(0.5)^2}{0.1}$$

$$< t < \frac{948(0.23)(0.72)(1.5 \times 10^{-5})(3,000)^2}{0.1},$$

which simplifies to 2.453 hours $< t <$ 211,935 hours. This implies that at 3 hours, the line source solution is a valid solution to the flow equation, and the reservoir is infinite-acting (until a time of 211,935 hours).

2. To calculate the pressure in the reservoir at 3 hours at a radius of 1 ft, we must determine whether the log approximation to the Ei function is valid; therefore we calculate the value of the argument of the Ei function, "x,"

$$x = \frac{948\phi\mu c_t r^2}{kt} = \frac{948(0.23)(0.72)(1.5 \times 10^{-5})(1)^2}{(0.1)(3)}$$

$$= 7.8494 \times 10^{-3}$$

This value is less than 0.01; therefore, we use Eq. 1.151 to estimate the pressure at a radius of 1 ft.

$$p = p_i + 70.6\frac{qB\mu}{kh}\left[\ln\left(\frac{1,688\phi\mu c_t r^2}{kt}\right)\right]. \quad \text{(1.151)}$$

$$p = 3,000 + 70.6\frac{(20)(1.475)(0.72)}{(0.1)(150)}$$

$$\times \left\{\ln\left[\frac{1,688(0.23)(0.72)(1.5 \times 10^{-5})(1)^2}{(0.1)(3)}\right]\right\}$$

$$= 3,000 + 99.97[\ln(0.01398)]$$

$$= 2,573 \text{ psia}.$$

3. To calculate the pressure in the reservoir at 3 hours at a radius of 10 ft, we must determine whether the log approximation to the Ei function is valid. Therefore, we calculate the value of the argument of the Ei function, "x."

$$x = \frac{948\phi\mu c_t r^2}{kt} = \frac{948(0.23)(0.72)(1.5 \times 10^{-5})(10)^2}{(0.1)(3)}$$

$$= 0.7849.$$

This value is greater than 0.01; therefore, we use Eq. 1.148 to estimate the pressure at a radius of 10 ft and we must look up the value of the Ei function from Table 1.1.

From Table 1.1, for $x_2 = 0.7849$, we interpolate between $x_1 = 0.78$ and $x_3 = 0.79$.

$-\text{Ei}(-0.78) = 0.322$ and $-\text{Ei}(-0.79) = 0.316$; therefore, $-\text{Ei}(-0.789) = 0.318$.

We substitute this value into Eq 1.148.

$$p = p_i + 70.6\frac{qB\mu}{kh}\text{Ei}\left(-\frac{948\phi\mu c_t r^2}{kt}\right) \quad \text{(1.148)}$$

$$p = 3,000 - 70.6\frac{(20)(1.475)(0.72)}{(0.1)(150)}(0.318)$$

$$= 3,000 - 99.97(0.318)$$

$$= 2,968 \text{ psia}.$$

4. To calculate the pressure in the reservoir at 3 hours at a radius of 100 ft, we must again determine whether the log approximation to the Ei function is valid. We then calculate the value of the argument of the Ei function, "x."

$$x = \frac{948\phi\mu c_t r^2}{kt} = \frac{948(0.23)(0.72)(1.5 \times 10^{-5})(100)^2}{(0.1)(3)}$$

$$= 78.49.$$

This value is greater than 0.01, so we use Eq. 1.148 to estimate the pressure at a radius of 10 ft, and we must look up the value of the Ei function from Table 1.1.

From Table 1.1, for $x > 10$, the value of $-\text{Ei}(-x)$ approaches zero. If we substitute this value into Eq. 1.148, we have no change in pressure at a radius of 100 ft after 3 hours.

$$p = p_i + 70.6\frac{qB\mu}{kh}\text{Ei}\left(-\frac{948\phi\mu c_t r^2}{kt}\right) \quad \text{(1.148)}$$

$$= 3,000 - 70.6\frac{(20)(1.475)(0.72)}{(0.1)(150)}(0)$$

$$= 3,000 - 0$$

$$= 3,000 \text{ psia}.$$

Skin Factor. To include skin factor in our calculations, we recall that the boundary condition we stated for including skin factor is

$$p_{wD} = p_D(1, t_D) + s.$$

This implies that we can add the skin factor to the dimensionless solution, evaluated at $r_D = 1$, to obtain the pressure at the wellbore. For the line-source well, the solution becomes

$$p_{wD} = -\frac{1}{2}\text{Ei}\left(\frac{-1}{4t_D}\right) + s.$$

We recall the definitions of the dimensionless variables:

$$p_{wD} = \frac{kh(p_i - p_{wf})}{141.2qB\mu}$$

$$t_D = \frac{0.0002637kt}{\phi\mu c_t r_w^2}$$

$$r_D = \frac{r}{r_w} = 1(\text{at } r = r_w).$$

Substituting these definitions into

$$p_{wD} = -\frac{1}{2}\text{Ei}\left(\frac{-1}{4t_D}\right) + s,$$

we obtain an equation to estimate pressures with the line source solution when skin factor is not zero.

$$\frac{kh(p_i - p_{wf})}{141.2qB\mu} = -\frac{1}{2}\text{Ei}\left(-\frac{948\phi\mu c_t r_w^2}{kt}\right) + s$$

Rearranging this equation, we obtain

$$p_i - p_{wf} = -\frac{70.6qB\mu}{kh}\left[\text{Ei}\left(-\frac{948\phi\mu c_t r_w^2}{kt}\right) - 2s\right].$$

When the argument of the Ei function (x) is greater than 0.01, we look up values of $-\text{Ei}(-x)$ in Table 1.1. For values of x less than 0.01, the previous equation can be further simplified by making an approximation to the exponential integral function, $\text{Ei}(-x)$. This approximation is given by

$$\text{Ei}(-x) \approx \ln(1.781x). \quad \text{(1.150)}$$

This approximation simplifies the line-source solution, including skin factor, to

$$p_{wf} = p_i + 70.6\frac{qB\mu}{kh}\left[\ln\left(\frac{1{,}688\phi\mu c_t r_w^2}{kt}\right) - 2s\right].$$

Example 1.2—Using the Line-Source Solution for Damaged or Stimulated Wells. A well and reservoir have the following characteristics. The well is producing only oil at a constant rate of 20 STB/D. The following data describe the well and formation.

q = 20 STB/D
h = 150 ft
B = 1.475 RB/STB
r_e = 3,000 ft
p_i = 3,000 psia
ϕ = 0.23
c_t = 1.5×10^{-5} psia^{-1}
μ = 0.72 cp
r_w = 0.5 ft
k = 0.1 md

Part A. The wellbore pressure was measured to be 1,380 psia after 13 hours of production. Calculate the skin factor.

Part B. The well was then acidized to drop the skin factor to zero. After clean up, the well was put back on production at a constant rate of 20 STB/D and, after 5 hours, a wellbore pressure of 2,380 psia was measured. Was the acidizing treatment successful?

Solution. *Part A.* 1. First, we determine whether we have conditions that lie in the range of applicability of the line-source solution. From Eq. 1.149, we have the following acceptable range.

$$\frac{(3.975 \times 10^5)\phi\mu c_t r_w^2}{k} < t < \frac{948\phi\mu c_t r_e^2}{k}.$$

We substitute the given well and reservoir conditions into the equation

$$\frac{3.975 \times 10^5(0.23)(0.72)(1.5 \times 10^{-5})(0.5)^2}{0.1}$$

$$< t < \frac{948(0.23)(0.72)(1.5 \times 10^{-5})(3{,}000)^2}{0.1},$$

which simplifies to 2.453 hours $< t <$ 211,935 hours. This implies that, at 13 hours, the line-source solution is a valid solution to the flow equation, and the reservoir is infinite-acting (until a time of 211,935 hours).

2. We must determine whether the log approximation to the Ei function is valid; therefore, we calculate the value of the argument of the Ei function, "x," at the wellbore ($r = r_w$).

$$x = \frac{948\phi\mu c_t r_w^2}{kt} = \frac{948(0.23)(0.72)(1.5 \times 10^{-5})(0.5)^2}{(0.1)(13)}$$

$$= 4.529 \times 10^{-4}.$$

This value is less than 0.01; therefore, we use

$$p_{wf} = p_i + 70.6\frac{qB\mu}{kh}\left[\ln\left(\frac{1688\phi\mu c_t r_w^2}{kt}\right) - 2s\right]$$

to estimate the skin factor.

$$1{,}380 = 3{,}000 + 70.6\frac{(20)(1.475)(0.72)}{(0.1)(150)}$$

$$\left[\ln\left(\frac{1688(0.23)(0.72)(1.5 \times 10^{-5})(0.5)^2}{(0.1)(13)}\right) - 2s\right]$$

$$= 3{,}000 + 99.97(-7.123 - 2s)$$

$$= 3{,}000 - 99.97(7.123 + 2s)$$

$$s = \frac{1}{2}\left[\frac{3{,}000 - 1{,}380}{99.97} - 7.123\right]$$

$$= +4.54.$$

Part B. 1. We note that 5 hours falls in the acceptable time range for the line-source solution, as given in the solution to Part A. This means that we can use the line source solution to solve this problem.

2. We must determine whether the log approximation to the Ei function is valid; therefore, we calculate the value of the argument of the Ei function, "x."

$$x = \frac{948\phi\mu c_t r_w^2}{kt} = \frac{948(0.23)(0.72)(1.5 \times 10^{-5})(0.5)^2}{(0.1)(5)}$$

$$= 1.177 \times 10^{-3}.$$

This value is less than 0.01; therefore, we use

$$p_{wf} = p_i + 70.6\frac{qB\mu}{kh}\left[\ln\left(\frac{1688\phi\mu c_t r_w^2}{kt}\right) - 2s\right].$$

$$2{,}380 = 3{,}000 + 70.6\frac{(20)(1.475)(0.72)}{(0.1)(150)}$$

$$\times \left[\ln\left(\frac{1{,}688(0.23)(0.72)(1.5 \times 10^{-5})(0.5)^2}{(0.1)(5)}\right) - 2s\right]$$

$$= 3{,}000 + 99.97(-6.167 - 2s)$$

$$= 3{,}000 - 99.97(6.167 + 2s)$$

$$s = \frac{1}{2}\left[\frac{3{,}000 - 2{,}380}{99.97} - 6.167\right]$$

$$= 0.02 \approx 0.$$

Therefore, the acid treatment was successful.

Consideration of Wellbore Storage and Skin. Wellbore storage is not an additive function like skin factor; therefore, we must solve the flow equation subject to the inner-boundary condition for wellbore storage.

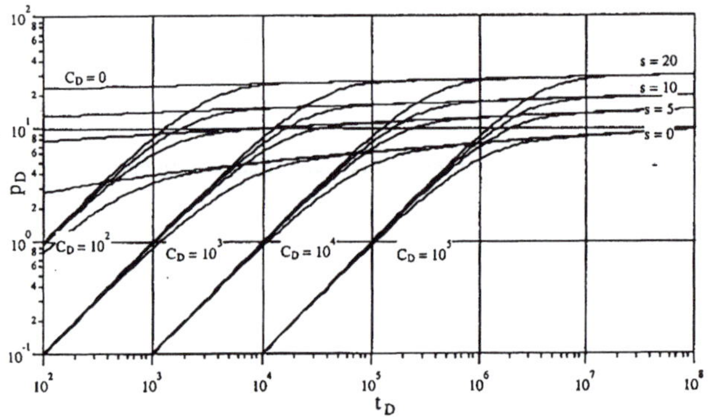

Fig. 1.10—The Ramey type curve (dimensionless-pressure solutions with wellbore storage and skin).

The partial-differential equation is

$$\frac{1}{r_D}\frac{\partial}{\partial r_D}\left(r_D\frac{\partial p_D}{\partial r_D}\right) = \frac{\partial p_D}{\partial t_D}, \quad \ldots\ldots (1.110)$$

with the initial condition

$$p_D(r_D, t_D = 0) = 0. \quad \ldots\ldots (1.111)$$

The outer-boundary condition for an infinite-acting reservoir is

$$p_D(r_D \to \infty, t_D) = 0 \quad \ldots\ldots (1.112)$$

There are two inner boundary conditions. The first is for the constant rate production with wellbore storage, and the second is for skin.

$$C_D\frac{dp_{wD}}{dt_D} - \left(r_D\frac{\partial p_D}{\partial r_D}\right)_{(r_D=1)} = 1 \quad \ldots\ldots (1.116)$$

$$p_{wD} = p_D - s\left(r_D\frac{\partial p_D}{\partial r_D}\right)_{(r_D=1)}.$$

The solution to these equations is given by Agarwal *et al.*,[10]

$$p_{wD}(t_D) = \frac{4}{\pi^2}$$

$$\times \int_0^\infty \left(\left(1 - e^{-x^2 t_D}\right)/x^3\left\{\left[xC_DJ_0(x) - \left(1 - C_Dsx^2\right)J_1(x)\right]^2 + \left[xC_DY_0(x) - \left(1 - C_Dsx^2\right)Y_1(x)\right]^2\right\}\right)dx,$$

where J_0 and J_1 are Bessel functions of the first kind, zero and first order, respectively, and Y_0 and Y_1 are Bessel functions of the second kind, zero and first order, respectively. This solution was derived with Laplace transformations, similar to the line-source solution, and is given in detail in Appendix B.

This solution is difficult to use and led to the development of "type curves," or graphical depictions of analytical solutions. The first such type curve was given by Ref. 10 and is given as **Fig. 1.10.** From this figure, values of p_{wD} (usually written as "p_D" on the vertical axis) and thus p_w can be determined for a well in a formation with given values of t_D, C_D, and s. Because of the similar curve shapes, this type curve does not provide unique analysis; other type curves have been developed for this purpose and will be discussed thoroughly in later chapters. However, there is an important property of this graphical solution that requires special mention at this point. At early times, for each value of C_D, a "unit slope line" (i.e., a line with a 45° slope) is present on the graph. This line appears and remains as long as all production comes from the wellbore and none comes from the formation. Recall the equation that describes the inner-boundary condition for wellbore storage,

$$C_D\frac{dp_{wD}}{dt_D} - \left(r_D\frac{\partial p_D}{\partial r_D}\right)_{(r_D=1)} = 1.$$

The second term in this equation is defined here (from the inner boundary condition for constant-rate production),

$$\left(r\frac{\partial p}{\partial r}\right)_{(r=r_w)} = 141.2\frac{q_{sf}B\mu}{kh}.$$

On the basis of the definitions of dimensionless pressure and radius, this term can be rewritten as follows:

$$\left(r_D\frac{\partial p_D}{\partial r_D}\right)_{(r_D=1)} = -\frac{q_{sf}}{q}.$$

Therefore, we rewrite the inner-boundary condition for wellbore storage as

$$C_D\frac{dp_{wD}}{dt_D} + \frac{q_{sf}}{q} = 1.$$

For $q_{sf}/q = 0$ (i.e., no sandface production—all production from the wellbore), this equation becomes

$$C_D\frac{dp_{wD}}{dt_D} = 1, \text{ or } C_D dp_{wD} = dt_D.$$

Integrating from $t_D = 0$ (where $p_{wD} = 0$) to t_D and p_{wD}, the result is

$$C_Dp_{wD} = t_D.$$

Taking logarithms of both sides of the equation gives

$$\log C_D + \log p_{wD} = \log t_D.$$

Thus, as long as $q_{sf} = 0$, theory leads us to expect that a graph of log p_{wD} vs. log t_D will have a slope of unity; it also leads us to expect that any point (p_{wD}, t_D) on this unit-slope line must satisfy the relation

$$\frac{C_Dp_{wD}}{t_D} = 1.$$

This observation is of major value in well-test analysis.

1.5.2 Pseudosteady-State Flow, Constant-Rate Production From a Cylindrical-Source Well in a Closed Reservoir.

Pseudosteady-state flow occurs when all the boundaries are felt in a closed reservoir system. The conditions we need for pseudosteady state to occur are (1) closed (bounded) reservoir, (2) no-flow boundaries, and (3) constant-rate production at the inner-boundary (wellbore).

We first develop a general solution by use of Laplace transforms, for these conditions (Appendix B gives details). We recall the partial differential equation and the governing boundary conditions.

Partial-Differential Equation.

$$\frac{1}{r_D}\frac{\partial}{\partial r_D}\left(r_D\frac{\partial p_D}{\partial r_D}\right) = \frac{\partial p_D}{\partial t_D}. \quad \ldots\ldots (1.110)$$

Initial Condition.

$$p_D(r_D, t_D = 0) = 0. \quad \ldots\ldots (1.111)$$

Outer-Boundary Condition.

$$\left(\frac{\partial p_D}{\partial r_D}\right)_{r_{eD}} = 0. \quad \ldots\ldots (1.113)$$

Inner-Boundary Condition.

$$\left(\frac{\partial p_D}{\partial r_D}\right)_{(r_D=1)} = -1. \quad \ldots\ldots (1.115)$$

The solution is

$$p_D(r_D, t_D) = \frac{2}{\left(r_{eD}^2 - 1\right)}\left(\frac{r_D^2}{4} + t_D\right) - \frac{r_{eD}^2 \ln r_D}{\left(r_{eD}^2 - 1\right)}$$

$$- \frac{\left(3r_{eD}^4 - 4r_{eD}^4 \ln r_{eD} - 2r_{eD}^2 - 1\right)}{4\left(r_{eD}^2 - 1\right)^2}$$

$$+ \pi \sum_{n=1}^{\infty} \left\{ \frac{e^{-\alpha_n^2 t_D} J_1^2(\alpha_n r_{eD})\left[J_1(\alpha_n) Y_0(\alpha_n r_D) - Y_1(\alpha_n)(J_0)(\alpha_n r_D)\right]}{\alpha_n\left[J_1^2(\alpha_n r_{eD}) - J_1^2(\alpha_n)\right]} \right\}. \quad (1.152)$$

Again, J_0 and J_1 are Bessel functions of the first kind, zero and first order, respectively, and Y_0 and Y_1 are Bessel functions of the second kind, zero and first order, respectively.

Pseudosteady-state flow occurs at large times ($t > 948\phi\mu c_t r_e^2/k$), so we develop a long-time approximation of the general solution to describe pseudosteady-state flow. As $t \rightarrow \infty$, the summation term will drop out because it contains a negative exponential term, and $\lim_{x \to \infty} e^{-x} = 0$.

The solution then becomes

$$p_D(r_D, t_D) = \frac{2}{\left(r_{eD}^2 - 1\right)}\left(\frac{r_D^2}{4} + t_D\right) - \frac{r_{eD}^2 \ln r_D}{\left(r_{eD}^2 - 1\right)}$$

$$- \frac{\left(3r_{eD}^4 - 4r_{eD}^4 \ln r_{eD} - 2r_{eD}^2 - 1\right)}{4\left(r_{eD}^2 - 1\right)^2}. \quad (1.153)$$

At the wellbore, $r_D = r_w/r_w = 1$; therefore,

$$p_{wD}(t_D) = \frac{2}{\left(r_{eD}^2 - 1\right)}\left(\frac{1}{4} + t_D\right) - \frac{r_{eD}^2 \ln 1}{r_{eD}^2 - 1}$$

$$- \frac{\left(3r_{eD}^4 - 4r_{eD}^4 \ln r_{eD} - 2r_{eD}^2 - 1\right)}{4\left(r_{eD}^2 - 1\right)^2}. \quad (1.154)$$

Because $r_{eD} \gg 1$, this equation reduces to

$$p_{wD}(t_D) = \frac{2}{r_{eD}^2}\left(\frac{1}{4} + t_D\right) - \frac{\left(3r_{eD}^4 - 4r_{eD}^4 \ln r_{eD} - 2r_{eD}^2 - 1\right)}{4r_{eD}^4}$$

$$= \frac{2}{r_{eD}^2} + \frac{2t_D}{r_{eD}^2} - \frac{3}{4} + \ln r_{eD} + \frac{1}{2r_{eD}^2} - \frac{1}{4r_{eD}^4}.$$

$$\cong \frac{2t_D}{r_{eD}^2} + \ln r_{eD} - \frac{3}{4}. \quad (1.155)$$

We can define a dimensionless time, t_{AD}, on the basis of on area instead of radius.

$$t_{AD} = \frac{\alpha_2 kt}{\phi\mu c_t A} = \frac{\alpha_2 kt}{\phi\mu c_t \pi r_e^2}. \quad (1.156)$$

In terms of dimensionless time based on wellbore radius, t_{AD} becomes

$$t_{AD} = t_D \frac{r_w^2}{\pi r_e^2} = \frac{t_D}{\pi r_{eD}^2}. \quad (1.157)$$

Therefore,

$$t_D = \pi r_{eD}^2 t_{AD} \text{ or } \frac{t_D}{r_{eD}^2} = \pi t_{AD}. \quad (1.158)$$

Substituting Eq. 1.158 into Eq. 1.155, we obtain the general pseudosteady-state solution,

$$p_{wD}(t_{AD}) = 2\pi t_{AD} + \ln r_{eD} - \frac{3}{4}. \quad (1.159)$$

In this case, we can see that the pressure change with respect to time is independent of time (i.e., the derivative of Eq. 1.159 with respect to t_{AD} is a function of r_{eD} only).

This equation can be rewritten in terms of field variables to solve flow problems.

$$p_{wf} = p_i - 141.2\frac{qB\mu}{kh}\left[0.000527\frac{kt}{\phi\mu c_t r_e^2} + \ln\left(\frac{r_e}{r_w}\right) - \frac{3}{4}\right]. \quad (1.160)$$

Note that we find by differentiating Eq. 1.160,

$$\frac{\partial p_{wf}}{\partial t} = -\frac{0.0744qB}{\phi c_t h r_e^2} \quad (1.161)$$

during this time period. Because the liquid-filled PV of the reservoir, V_p (in cubic feet), is

$$V_p = \pi r_e^2 h\phi, \quad (1.162a)$$

$$\text{then } \frac{\partial p_{wf}}{\partial t} = -\frac{0.234qB}{c_t V_p}. \quad (1.162b)$$

Thus, during this time period, the rate of pressure decline is inversely proportional to the liquid-filled PV, V_p. This result leads to a form of well testing sometimes called reservoir-limits testing, which seeks to determine reservoir size from the rate of pressure decline in a wellbore with time.

Another form of Eq. 1.160 is useful for some applications. It involves replacing original reservoir pressure, p_i, with average pressure, $\bar{p}$, within the drainage volume of the well. The volumetric average pressure within the drainage volume of the well can be found from material balance. The pressure decrease ($p_i - \bar{p}$) results from removal of qB RB/D of fluid for t hours [a total removed of 5.615 qB (t/24) ft^3] is

$$p_i - \bar{p} = \frac{\Delta V}{c_t V} = \frac{5.615qB(t/24)}{c_t(\pi r_e^2 h\phi)} = \frac{0.0744qBt}{\phi c_t h r_e^2}. \quad (1.163a)$$

Substituting in Eq. 1.160,

$$p_{wf} = \bar{p} + \frac{0.0744qBt}{\phi c_t h r_e^2} - \frac{0.0744qBt}{\phi c_t h r_e^2} - 141.2\frac{qB\mu}{kh}\left[\ln\left(\frac{r_e}{r_w}\right) - \frac{3}{4}\right] \quad (1.163b)$$

$$\text{or } \bar{p} - p_{wf} = 141.2\frac{qB\mu}{kh}\left[\ln\left(\frac{r_e}{r_w}\right) - \frac{3}{4}\right]. \quad (1.163c)$$

Eqs. 1.160 and 1.163c become more useful in practice if they include a skin factor to account for the fact that most wells are either damaged or stimulated. For example, in Eq. 1.163c,

$$\bar{p} - p_{wf} = 141.2\frac{qB\mu}{kh}\left[\ln\left(\frac{r_e}{r_w}\right) - \frac{3}{4}\right] + (\Delta p)_s, \quad (1.163d)$$

$$\bar{p} - p_{wf} = 141.2\frac{qB\mu}{kh}\left[\ln\left(\frac{r_e}{r_w}\right) - \frac{3}{4} + s\right], \quad (1.163e)$$

$$\text{and } p_i - p_{wf} = 141.2\frac{qB\mu}{kh}\left[\frac{0.000527kt}{\phi\mu c_t r_e^2} + \ln\left(\frac{r_e}{r_w}\right) - \frac{3}{4} + s\right]. \quad (1.163f)$$

Further, we can define an average permeability, k_J, so that

$$\bar{p} - p_{wf} = 141.2\frac{qB\mu}{k_J h}\left[\ln\left(\frac{r_e}{r_w}\right) - \frac{3}{4}\right]$$

$$= 141.2\frac{qB\mu}{kh}\left[\ln\left(\frac{r_e}{r_w}\right) - \frac{3}{4} + s\right], \quad \text{(1.163g)}$$

from which

$$k_J = k\left[\ln\left(\frac{r_e}{r_w}\right) - \frac{3}{4}\right] \Big/ \left[\ln\left(\frac{r_e}{r_w}\right) - \frac{3}{4} + s\right]. \quad \text{(1.163h)}$$

This average permeability, k_J, proves to be useful in well-test analysis, as we shall see later. Note that, for a damaged well, the average permeability k_J is lower than the true, bulk formation permeability k; in fact, these quantities are equal only when the skin factor, s, is zero. Because we sometimes estimate the permeability of a well from productivity-index (PI) measurements, and because the PI, J (STB/D-psi), of an oil well is defined as

$$J = \frac{q}{\bar{p} - p_{wf}} = \frac{k_J h}{141.2B\mu\left[\ln\left(\frac{r_e}{r_w}\right) - \frac{3}{4}\right]}, \quad \text{(1.163i)}$$

this method does not necessarily provide a good estimate of formation permeability, k. Thus, there is a need for a more complete means of characterizing a producing well than exclusive use of PI information.

Example 1.3—Analysis of a Well From a PI Test. A well produces 100 STB/D oil at a measured flowing bottomhole pressure (BHP) of 1,500 psi. A recent pressure survey showed that average reservoir pressure is 2,000 psi. Logs indicate a net sand thickness of 10 ft. The well drains an area with drainage radius, r_e, of 1,000 ft; the borehole radius is 0.25 ft. Fluid samples indicate that, at current reservoir pressure, oil viscosity is 0.5 cp and formation volume factor is 1.5 RB/STB.

1. Estimate the PI for the tested well.
2. Estimate formation permeability from these data.
3. Core data from the well indicate an effective permeability to oil of 50 md. Does this imply that the well is either damaged or stimulated? What is the apparent skin factor?

Solution.

1. To estimate PI, we use Eq. 1.63i:

$$J = \frac{q}{\bar{p} - p_{wf}} = \frac{100}{(2,000 - 1,500)} = 0.2 \text{ STB/psi-D}.$$

2. We do not have sufficient information to estimate formation permeability; we can calculate average permeability, k_J, only, which is not necessarily a good approximation of formation permeability, k. From Eq. 1.163i,

$$k_J = \frac{141.2JB\mu\left[\ln\left(\frac{r_e}{r_w}\right) - \frac{3}{4}\right]}{h}$$

$$= \frac{(141.2)(0.2)(1.5)(0.5)\left[\ln\left(\frac{1,000}{0.25}\right) - 0.75\right]}{10} = 16 \text{ md}.$$

3. Core data sometimes provide a better estimate of formation permeability than do permeabilities derived from the PI, particularly for a well that is badly damaged. Because cores indicate a permeability of 50 md, we conclude that this well is damaged. Eq. 1.163i provides a method for estimating the skin factor, s.

$$s = \left(\frac{k}{k_J} - 1\right)\left[\ln\left(\frac{r_e}{r_w}\right) - \frac{3}{4}\right] = \left(\frac{50}{16} - 1\right)\left[\ln\left(\frac{1,000}{0.025}\right) - 0.75\right]$$

$$= 16.$$

1.5.3 Steady-State Flow, Constant-Rate Production From a Cylindrical-Source Well in a Reservoir with Constant-Pressure Outer Boundaries. Steady-state flow occurs theoretically at long times in a constant-pressure outer-boundary, constant-rate production case. We present a general solution for the following set of equations. Similar to the pseudosteady-state case, steady-state flow is a long-time approximation to the general solution for these equations.

Partial-Differential Equation.

$$\frac{1}{r_D}\frac{\partial}{\partial r_D}\left(r_D\frac{\partial p_D}{\partial r_D}\right) = \frac{\partial p_D}{\partial t_D}. \quad \text{(1.110)}$$

Initial Condition.

$$p_D(r_D, t_D = 0) = 0. \quad \text{(1.111)}$$

Outer-Boundary Condition.

$$p_D(r_D = r_{eD}, t_D) = 0. \quad \text{(1.114)}$$

Inner-Boundary Condition.

$$\left(\frac{\partial p_D}{\partial r_D}\right)_{(r_D=1)} = -1. \quad \text{(1.115)}$$

The dimensionless-pressure solution, derived with Laplace transformations, is

$$p_D(r_D, t_D) = \left\{\ln r_{eD} - 2\sum_{n=1}^{\infty}\frac{e^{-\beta_n^2 t_D}J_0^2(\beta_n r_{eD})}{\beta_n^2\left[J_1^2(\beta_n) - J_0^2(\beta_n r_{eD})\right]}\right\}. \quad \text{(1.164)}$$

Using the same argument that we used in our previous derivation of the pseudosteady-state flow equation, we derive a long-time approximation of Eq. 1.164. At long times, the summation term drops out, because $\lim_{x\to\infty} e^{-x} = 0$. This leaves

$$p_D(r_D, t_D) = \ln r_{eD}, \quad \text{(1.165)}$$

which is the steady-state solution. We can see here that pressure is independent of time for steady-state flow. In field variables, this equation becomes

$$p_{wf} = p_i - \frac{141.2qB\mu}{kh}\ln\left(\frac{r_e}{r_w}\right). \quad \text{(1.166)}$$

1.5.4 Transient Linear Flow, Constant-Rate Production From a Hydraulically Fractured Well. Theoretically, linear flow occurs in reservoirs with long, highly conductive vertical fractures. This situation is modeled by the following set of equations. We present a general solution for this set of equations and extend the solution to the hydraulic fracture case.

Partial-Differential Equation.

$$\frac{\partial^2 p_D}{\partial x_D^2} = \frac{\partial p_D}{\partial t_D}. \quad \text{(1.130)}$$

Initial Condition.

$$p_D(x_D, t_D = 0) = 0. \quad \text{(1.131)}$$

Outer-Boundary Condition.

$$p_D(x_D \to \infty, t_D) = 0. \quad \text{(1.132)}$$

Inner-Boundary Condition.

$$\left(\frac{\partial p_D}{\partial x_D}\right)_{x_D=1} = -1. \quad \text{(1.135)}$$

TABLE 1.2—DIMENSIONLESS-PRESSURE SOLUTIONS

Reservoir Model	Pressure Response
1. Radial flow Infinite-acting reservoir Constant-rate production Line-source well	$p_D = -\frac{1}{2}Ei\left(-\frac{r_D^2}{4t_D}\right)$
2. Pseudosteady-state flow Cylindrical reservoir No-flow boundaries Constant-rate production Cylindrical-source well	$p_{wD} = \frac{2t_D}{r_{eD}^2} - \frac{3}{4} + \ln r_{eD}$
3. Steady-state radial flow Cylindrical reservoir Constant-pressure boundaries Constant-rate production Cylindrical-source well	$p_{wD} = \ln r_D$
4. Transient linear flow Hydraulically fractured well	$p_{wD} = 2\sqrt{\frac{t_D}{\pi}}$

TABLE 1.3—PRESSURE SOLUTIONS IN FIELD VARIABLES

Reservoir Model	Pressure Response
1. Radial flow Infinite acting reservoir Constant-rate production Line-source well	$p = p_i + 70.6\frac{qB\mu}{kh} Ei\left(-\frac{948\phi\mu c_t r^2}{kt}\right)$
2. Pseudosteady-state flow Cylindrical reservoir No-flow boundaries Constant-rate production Cylindrical-source well	$p_{wf} = p_i - \frac{141.2qB\mu}{kh}\left[\frac{0.000527kt}{\phi\mu c_t r_e^2} + \ln\left(\frac{r_e}{r_w}\right) - \frac{3}{4}\right]$
3. Steady-state radial flow Cylindrical reservoir Constant-pressure boundaries Constant-rate production Cylindrical-source well	$p_{wf} = p_i - \frac{141.2qB\mu}{kh}\ln\left(\frac{r_e}{r_w}\right)$
4. Transient linear flow Hydraulically fractured well	$p_{wf} = p_i - 4.064\frac{qB}{hL_f}\sqrt{\frac{\mu t}{k\phi c_t}}$

We use the Laplace transform technique to develop the following solution to these equations:

$$p_D(x_D, t_D) = 2\sqrt{\frac{t_D}{\pi}}\exp\left(\frac{-x_D^2}{4t_D}\right) - x_D \text{erfc}\left(\frac{x_D}{2\sqrt{t_D}}\right), \quad (1.167)$$

where the complementary error function, erfc(x), is defined as

$$\text{erfc}(x) = \frac{2}{\sqrt{\pi}}\int_x^\infty e^{-\xi^2}d\xi. \quad (1.168)$$

For the special case of $x_D = 0$ (i.e., for a hydraulically fractured well at the fracture face), the solution becomes

$$p_D(0, t_D) = 2\sqrt{\frac{t_D}{\pi}}. \quad (1.169)$$

This can be written in field variables as

$$p_{wf} = p_i - 4.064\frac{qB}{hL_f}\sqrt{\frac{\mu t}{k\phi c_t}}. \quad (1.170)$$

1.5.5 Summary of Dimensionless-Pressure Solutions and Pressure Solutions in Field Variables. **Table 1.2** summarizes important solutions to the diffusivity equation in terms of dimensionless variables. **Table 1.3** presents these same solutions in field variables.

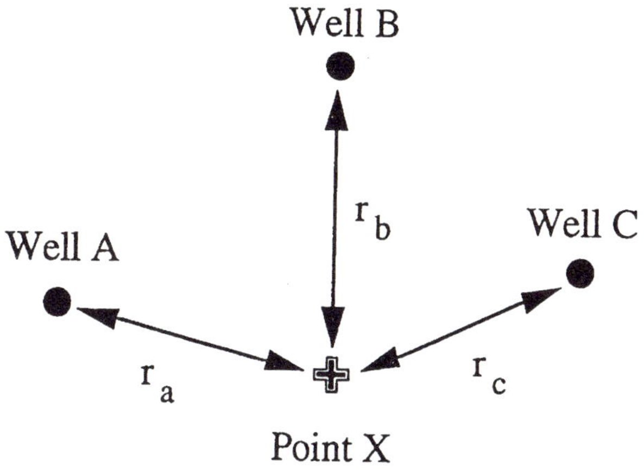

Fig. 1.11—Multiple-well system.

1.6 Superposition in Space

The term superposition simply means a summation of all the individual parts that contribute to the total system. Petroleum engineers use superposition to model complex situations as a sum of several simpler parts. In Sec. 1.5, we solved the diffusivity equation for several "single-well" cases. Superposition allows us to use these solutions to model multiple-well problems.

We can use superposition to develop the method of images to model single or multiple boundaries. Without superposition, we can solve the diffusivity equation only for a completely closed system (i.e., the pseudosteady-state solution). The multiple well problem and the method of images are examples of superposition in space. We can also use superposition in time to solve variable-rate production problems.

1.6.1 Multiple Wells. It is rare to find a reservoir being produced from only a single well. A field usually contains several wells producing from the same reservoir, and each well will have an effect on the pressure at the other wells. In other words, if we have one well producing at a constant rate, the BHP in that well is a function of its own production as well as the production from surrounding wells.

Consider the following system, illustrated by **Fig. 1.11.** A, B, and C are wells that begin to produce at the same time from an infinite-acting reservoir at constant rates, q_A, q_B, and q_C. Point X is any point in the reservoir and is a distance r_a from Well A, r_b from Well B, and r_c from Well C. Superposition states that the pressure drop at Point X is equal to the sum of the pressure drop owing to Well A, the pressure drop owing to Well B, and the pressure drop owing to Well C. In mathematical form, we write

$$(p_i - p)_X = (p_i - p)_A + (p_i - p)_B + (p_i - p)_C. \quad (1.171)$$

If we consider each well singly, we would have the problem of a single well producing at a constant rate in an infinite-acting reservoir. The solution to this single-well problem is given in Sec. 1.5.6 as

$$(p_i - p) = -\frac{70.6qB\mu}{kh}\text{Ei}\left(-\frac{948\phi\mu c_t r^2}{kt}\right). \quad (1.172)$$

So, the pressure drop at Point X would be

$$(p_i - p)_X = \left[-\frac{70.6q_A B\mu}{kh}\text{Ei}\left(-\frac{948\phi\mu c_t r_a^2}{kt}\right)\right] + \left[-\frac{70.6q_B B\mu}{kh}\text{Ei}\left(-\frac{948\phi\mu c_t r_b^2}{kt}\right)\right]$$

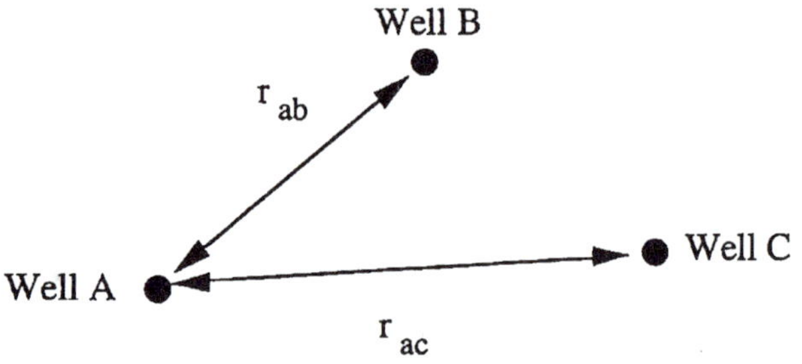

Fig. 1.12—Wellbore pressure in a multiwell system.

$$+\left[-\frac{70.6q_CB\mu}{kh}\mathrm{Ei}\left(-\frac{948\phi\mu c_t r_c^2}{kt}\right)\right], \quad \ldots \quad (1.173)$$

$$\text{or } p_X = p_i + \left[\frac{70.6q_AB\mu}{kh}\mathrm{Ei}\left(-\frac{948\phi\mu c_t r_a^2}{kt}\right)\right]$$

$$+\left[\frac{70.6q_BB\mu}{kh}\mathrm{Ei}\left(-\frac{948\phi\mu c_t r_b^2}{kt}\right)\right]$$

$$+\left[\frac{70.6q_CB\mu}{kh}\mathrm{Ei}\left(-\frac{948\phi\mu c_t r_c^2}{kt}\right)\right]. \quad \ldots\ldots \quad (1.174)$$

Now, consider the pressure drop at Well A, illustrated by **Fig. 1.12.** Superposition states that the pressure drop at Well A is a sum of the pressure drop owing to Well A, the pressure drop owing to Well B, and the pressure drop owing to Well C. In equation form, this is

$$\left(p_i - p_{wf}\right)_A = (p_i - p)_A + (p_i - p)_B$$
$$+ (p_i - p)_C. \quad \ldots \quad (1.175)$$

In this case, the pressure drop owing to Well A would be $(p_i - p_{wf})$, which includes the skin factor, s_A; and because the radius in question is the wellbore radius (which is small compared to the distances from Wells B and C to Well A), the Ei function may be written as its logarithmic approximation. So, the pressure drop owing to Well A becomes

$$(p_i - p)_A = -\frac{70.6qB\mu}{kh}\left[\ln\left(-\frac{1688\phi\mu c_t r_w^2}{kt}\right) - 2s_A\right],$$
$$\ldots \quad (1.176)$$

and the total BHP at Well A is

$$p_{wf,A} = p_i + \frac{70.6qB\mu}{kh}\left[\ln\left(\frac{1,688\phi\mu c_t r_w^2}{kt}\right) - 2s_A\right]$$

$$+\left[\frac{70.6q_BB\mu}{kh}\mathrm{Ei}\left(-\frac{948\phi\mu c_t r_{ab}^2}{kt}\right)\right]$$

$$\left[\frac{70.6q_CB\mu}{kh}Ei\left(-\frac{948\phi\mu c_t r_{ac}^2}{kt}\right)\right]. \quad \ldots \quad (1.177)$$

1.6.2 Method of Images. In this section and in Appendix E, we show how superposition is used to develop the method of images. The method of images states that a fault or a single, no-flow boundary can be represented by an imaginary well, producing at the same rate as the producing well, situated an equal distance on the other side of the fault as the producing well. The fault is thus eliminated, and we are left with a two-well system in an infinite-acting reservoir that can easily be solved with superposition.

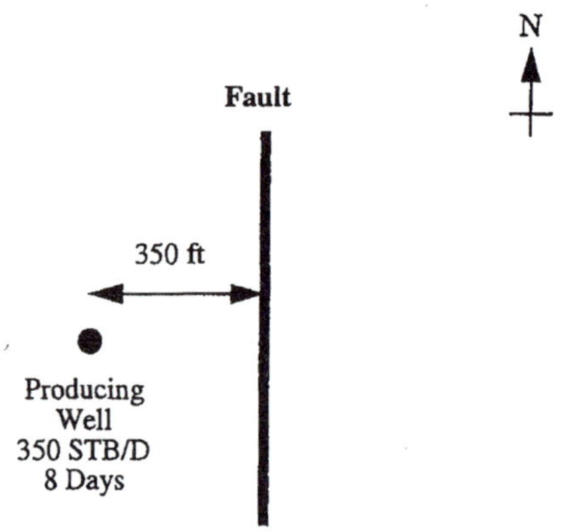

Fig. 1.13—Well mean a fault.

This method also can model a constant-pressure boundary by use of an image well also situated an equal distance on the other side of the boundary as the producing well, but the image well in this case is injecting fluid at the same rate as the producer is producing fluid. We can extend the method of images to multiple boundary configurations by use of superposition.

In Appendix E, we prove the validity of the method of images; i.e., we show that the two-well systems really do model the no-flow and constant pressure boundaries. We will illustrate application of the method of images with an example.

Example 1.4—Superposition in Space—Modeling a Well Near a Fault. Suppose a well is 350 ft due west of a north-south trending fault. From pressure transient tests, the skin factor, s, of this well has been found to be $+5.0$. This well has been flowing at a constant rate of 350 STB/D for 8 days. The following data describe the well and formation.

$r_w = 0.333$ ft
$h = 50$ ft
$B = 1.13$ RB/STB
$r_e = 3{,}000$ ft
$k = 25$ md
$p_i = 3{,}000$ psia
$\phi = 0.16$
$c_t = 2\times10^{-5}$ psia^{-1}
$\mu = 0.5$ cp

Calculate the pressure at the flowing well.

Solution.

1. First, we set up the appropriate image well. **Fig. 1.13** depicts the well and fault configuration.

Note that, to model this fault on the basis of the method of images, we must have the equivalent system of **Fig. 1.14.**

2. We can now consider this as a multiwell problem. From superposition,

$$(p_i - p_{wf})_{Wp} = (p_i - p)_{Wp} + (p_i - p)_{Wi}$$

$$\left(p_i - p_{wf}\right)_{Wp}$$

$$= -70.6\frac{q_{Wp}B\mu}{kh} \times \left[\ln\left(\frac{1688\phi\mu c_t r_w^2}{kt}\right) - 2s\right]$$

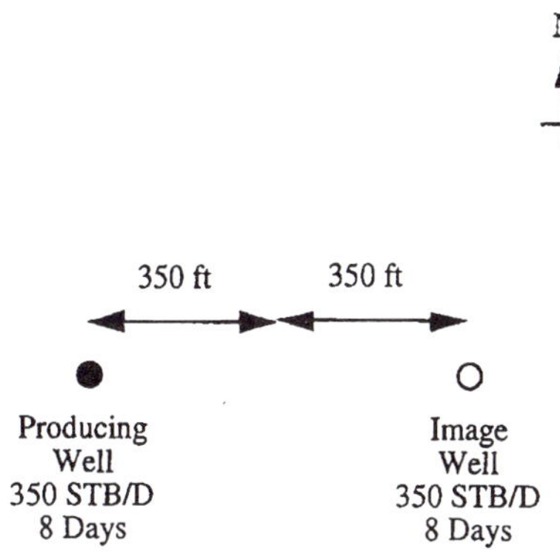

Fig. 1.14—Image to model well mean a fault.

$$- 70.6\frac{q_{Wi}B\mu}{kh}\mathrm{Ei}\left[\frac{-948\phi\mu c_t(2L)^2}{kt}\right]$$

$$= -70.6\frac{(350)(1.13)(0.5)}{(25)(50)}$$

$$\times\left\{\ln\left[\frac{1688(0.16)(0.5)(2\times10^{-5})(0.333)^2}{(25)[(8\ \text{days})\times(24\ \text{hr/D})]}\right]-2(5)\right\}$$

$$-70.6\frac{(350)(1.13)(0.5)}{(25)(50)}$$

$$\times\mathrm{Ei}\left[\frac{-948(0.16)(0.5)(2\times10^{-5})(2\times350)^2}{(25)[(8\ \text{days})\times(24\ \text{hr/D})]}\right].$$

$$(p_i - p_{wf})_{w_p} = -11.17(-16.59-10)$$

$$-11.17\ \mathrm{Ei}(-0.155).$$

We determine that —Ei $[-0.155] = 1.436$ from Table 1.1; therefore,

$$(p_i - p_{wf})_{Wp} = -297 + 16 = 313\ \text{psi}.$$

And, finally,

$$(p_{wf})_{Wp} = 3{,}000 - 313 = 2{,}687\ \text{psia}.$$

1.7 Superposition in Time

As we mentioned previously, superposition is not limited to spatial coordinates; we can also superpose solutions in time. This is especially helpful in variable-rate problems, specifically for modeling buildup tests. In this section, we start with an example to illustrate the procedure. We then generalize and illustrate the following concepts: (1) modeling a variable-rate history with the summation of simpler constant-rate solutions and (2) modeling pressure-buildup tests.

Example 1.5—Use of Superposition in Time. A flowing well is completed in a reservoir that has the following properties.

$p_i =$ 2,500 psia
$h =$ 43 ft
$B =$ 1.32 RB/STB

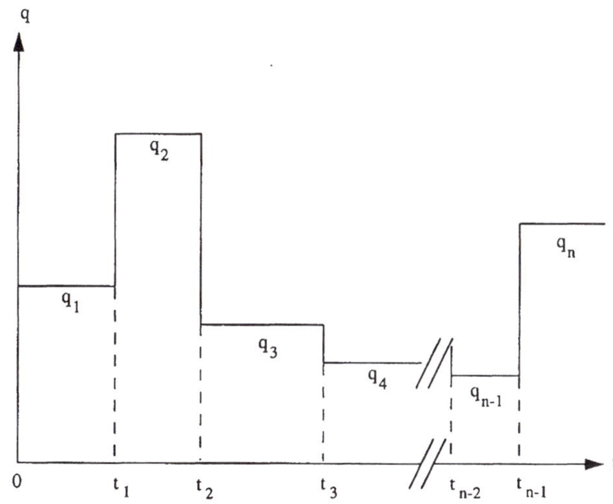

Fig. 1.15—Example rate history.

$c_t =$ 18×10^{-6} psi^{-1}
$\mu =$ 0.44 cp
$\phi =$ 0.16
$k =$ 25 md

What will the pressure drop be in a shut-in well 500 ft from the flowing well when the flowing well has been shut in for 1 day following a flow period of 5 days at 300 STB/D?

Solution. We must superimpose (i.e., add) two drawdowns because of the rate change. The rate term will be "new rate-old" (with $q_o = 0$), and the time term will be the total time for which a rate has been in effect (starting at the time of the rate change and continuing up to total time, t).

$$p_i - p = -\frac{70.6\mu B}{kh}\left\{(q_1 - 0)Ei\left[\frac{-948\phi\mu c_t r^2}{k(t-0)}\right]\right.$$

$$\left.+(q_2 - q_1)Ei\left[\frac{-948\phi\mu c_t r^2}{k(t-t_1)}\right]\right\}.$$

Now,

$$\frac{948\phi\mu c_t r^2}{k} = \left[(948)(0.16)(0.44)(1.8\times10^{-5})(500)^2\right]/$$

$$25 = 12.01.$$

Then,

$$p_i - p = -\frac{(70.6)(0.44)(1.32)}{(25)(43)}$$

$$\left\{(300)Ei\left[\frac{-12.01}{(6)(24)}\right]+(0-300)Ei\left[\frac{-12.01}{(1)(24)}\right]\right\}$$

$$= 11.4[-Ei(-0.0834)+Ei(-0.5)]$$

$$= 11.44(1.989 - 0.560)$$

$$= 16.35\ \text{psi}.$$

1.7.1 Modeling Variable-Rate History. Given the rate history shown in **Fig. 1.15,** wellbore pressure may be modeled by use of the principle of superposition. We know from the concept of superposition in space (Sec. 1.6) that production from two wells will give the following pressure distribution at any point in the reservoir.

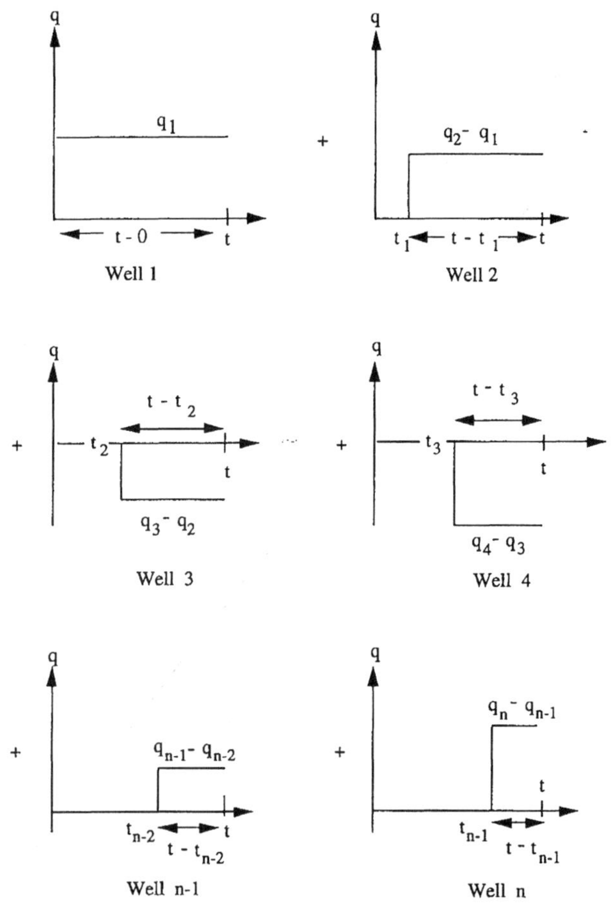

Fig. 1.16—Wells modeling example rate history.

$$p_D(x_D, y_D, t_D) = -\frac{1}{2}\frac{q_1}{q_0}\text{Ei}\left(\frac{-a_{1D}^2}{4t_D}\right) - \frac{1}{2}\frac{q_2}{q_0}\text{Ei}\left(\frac{-a_{2D}^2}{4t_D}\right). \qquad (1.178)$$

Both wells begin to produce at the same time with flow rates of q_1 and q_2, respectively. The term q_0 is a reference flow rate, and $a_{1D}$and a_{2D} are distances between the wells and the point (x_D, y_D).

Suppose Well 1 began production at time t_0 and Well 2 began production at time t_1 (where t_0 and $t_1 \geqq 0$), and we wish to find the pressure distribution at t where $t > t_0$ and $t_1 > t$. If $t < t_1$, Well 2 will not affect the pressure distribution. Therefore,

$$p_D\left(x_D, y_D, t_{D_0} < t_D < t_{D_1}\right) = -\frac{q_1}{2q_0}\text{Ei}\left[\frac{-a_{1D}^2}{4(t - t_0)_D}\right]. \qquad (1.179)$$

Extending this analogy for $t > t_1$, Well 2 will begin to produce and its effect may be added to, or superimposed on, the effect of Well 1.

$$p_D\left(x_D, y_D, t_D > t_{D_1}\right) = -\frac{q_1}{2q_0}\text{Ei}\left[\frac{-a_{1D}^2}{4(t - t_0)_D}\right]$$

$$-\frac{q_2}{2q_0}\text{Ei}\left[\frac{-a_{2D}^2}{4(t - t_1)_D}\right]. \qquad (1.180)$$

Looking at Fig.1.15, we can model the rate schedules as n different "wells" all in the same position. **Fig. 1.16** shows this diagrammatically. Note that, for convenience, we set $t_0 = 0$.

Therefore, Eq. 1.180 becomes

$$p_D\left(x_D, y_D, t_D > t_{D_{n-1}}\right)$$

$$= -\frac{1}{2q_0}\left\{q_1 Ei\left(\frac{-a_{1D}^2}{4t_D}\right) + (q_2 - q_1)\text{Ei}\left[\frac{-a_{2D}^2}{4(t - t_1)_D}\right]\right.$$

$$+ (q_3 - q_2)\text{Ei}\left[\frac{-a_{3D}^2}{4(t - t_2)_D}\right]$$

$$\left. + \ldots + (q_n - q_{n-1})\text{Ei}\left[\frac{-a_{nD}^2}{4(t - t_{n-1})_D}\right]\right\}. \qquad (1.181)$$

Because the "wells" are in the same position, $a_{1D} = a_{2D} \cdot \ldots = a_{nD} = r_D$, we can write

$$p_D(r_D, t_D) = -\frac{1}{2q_0}\left\{q_1\text{Ei}\left(\frac{-r_D^2}{4t_D}\right)\right.$$

$$\left. + (q_2 - q_1)\text{Ei}\left[\frac{-r_D^2}{4(t - t_1)_D}\right] + \ldots\right\}$$

$$\qquad (1.182)$$

If we write the right side of this equation in dimensionless form, where

$$p_D(r_D, t_D) = -\frac{1}{2}\text{Ei}\left(\frac{-r_D^2}{4t_D}\right), \qquad (1.183)$$

$$p_D\left[r_D, (t - t_1)_D\right] = -\frac{1}{2}\text{Ei}\left[\frac{-r_D^2}{4(t - t_1)_D}\right], \qquad (1.184)$$

$$p_D\left[r_D, (t - t_{n-1})_D\right] = -\frac{1}{2}\text{Ei}\left[\frac{-r_D^2}{4(t - t_{n-1})_D}\right], \qquad (1.185)$$

and so on, and define dimensionless rate as

$$q_D = \frac{q}{q_0}, \qquad (1.186)$$

we obtain

$$p_D(r_D, t_D) = q_{1D}p_D(r_D, t_D) + (q_2 - q_1)_D p_D\left[r_D, (t - t_1)_D\right] +$$

$$\ldots + (q_{n-1} - q_{n-2})_D p_D\left[r_D, (t - t_{n-2})_D\right]$$

$$+ (q_n - q_{n-1})_D p_D\left[r_D, (t - t_{n-1})_D\right]. \qquad (1.187)$$

At the wellbore, $r_D = 1$. Also, at the wellbore, skin, s, must be added to each individual dimensionless pressure. We know from the discussion of skin factor in Sec. 1.3 that

$$p_{wD}(t_D) = p_D(1, t_D) + s. \qquad (1.188)$$

Eq. 1.187 becomes

$$p_{wD}(t_D) = q_{1D}[p_D(t_D) + s] + (q_2 - q_1)_D\left[p_D(t - t_1)_D + s\right]$$

$$+ (q_3 - q_2)_D\left[p_D(t - t_2)_D + s\right] + \ldots$$

$$+ (q_{n-1} - q_{n-2})_D\left[p_D(t - t_{n-2})_D + s\right]$$

$$+ (q_n - q_{n-1})_D\left[p_D(t - t_{n-1})_D + s\right], \qquad (1.189)$$

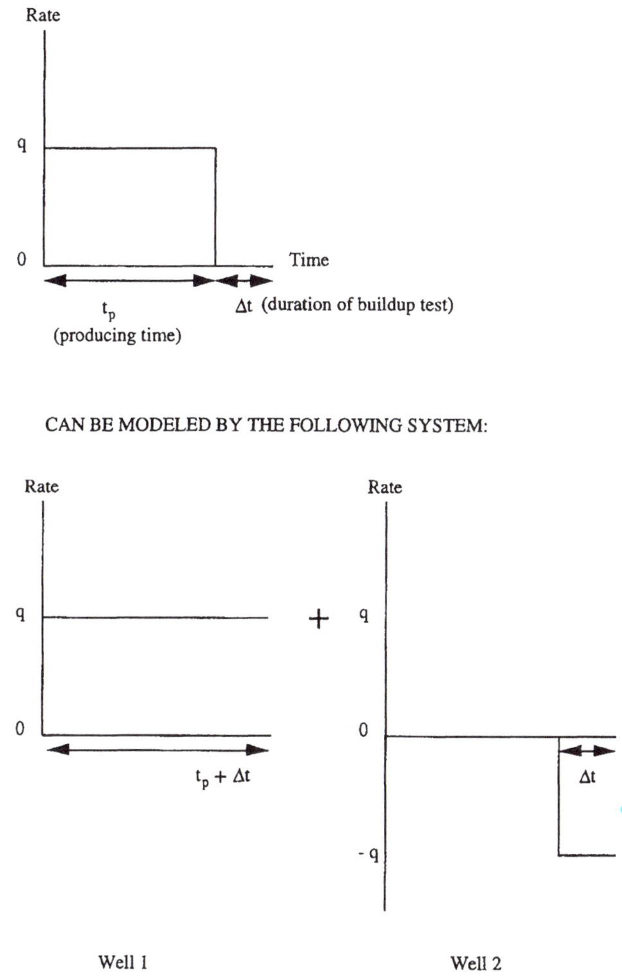

Fig. 1.17—Example rate history for buildup test.

$$\text{where } p_{wD}(t_D) = \frac{\left(p_i - p_{wf}\right)kh}{141.2qB\mu}. \quad \text{(1.190)}$$

Eq. 1.189 may be written as

$$p_{wD}(t_D) = \sum_{i=1}^{n}\left(q_i - q_{i-1}\right)_D\left[p_D\left(t - t_{i-1}\right)_D + s\right]. \quad \text{(1.191)}$$

If we multiply and divide Eq. 1.191 by $\left(t - t_{i-1}\right)_D = \Delta\tau_i$,

$$p_{wD}(t_D) = \sum_{i=1}^{n}\frac{\left(q_i - q_{i-1}\right)_D}{\left(t - t_{i-1}\right)_D}\left[p_D\left(t - t_{i-1}\right)_D + s\right]\left(t - t_{i-1}\right)_D$$

$$= \sum_{i=1}^{n}\frac{\Delta q_{iD}}{\Delta\tau_i}\left[p_D\left(t - t_{i-1}\right)_D + s\right]\Delta\tau_i. \quad \text{(1.192)}$$

Taking limits of Eq. 1.192 as $\Delta\tau_i \rightarrow 0$,

$$p_{wD}(t_D) = \lim_{\Delta\tau_i \rightarrow 0}\sum_{i=1}^{n}\frac{\Delta q_{iD}}{\Delta\tau_i}\left[p_D\left(t - t_{i-1}\right)_D + s\right]\Delta\tau_i. \quad \text{(1.193)}$$

In the limit, the summation becomes an integral, and the "Δ terms" become differentials. Also, we let τ be a dummy variable of integration, corresponding to the t_{i-1} term. Then we obtain

$$p_{wD}(t_D) = \int_0^{t_D}\frac{dq_D(\tau)}{d\tau}\left[p_D(t_D - \tau) + s\right]d\tau. \quad \text{(1.194)}$$

If we let p_{sD} be the symbol for the "constant-rate" solution, where

$$p_{sD} = p_D(t_D) + s \quad \text{(1.195)}$$

and let p_{wD} be the symbol for the "variable-rate" solution, we have the following relationship between the variable and constant-rate dimensionless pressure.

$$p_{wD}(t_D) = \int_0^{t_D}\frac{dq_D(\tau)}{d\tau}\left[p_{sD}(t_D - \tau)\right]d\tau. \quad \text{(1.196)}$$

Eq. 1.196 is known as the convolution integral, and Eq. 1.191, which contains the summation term, is the superposition equation. The uses of convolution are numerous. Suppose we are trying to model variable-rate flow, and we know the solution p_{sD} for constant flow rate (any reservoir type). Given the variation in q, and thus q_D, with time, we may calculate the variable-rate pressure solution from Eq. 1.196 with approximations to evaluate the integral.

1.7.2 Modeling Pressure-Buildup Tests. A pressure-buildup test is conducted by producing a well at constant rate for some time, shutting the well in (usually at the surface), allowing the pressure to build up in the wellbore, and recording the BHP as a function of time. In other words, we can model buildup tests by use of superposition in time as a "two-rate" problem. This is illustrated in **Fig. 1.17** as two wells, one producing at flow rate q for time $(t_p + \Delta t)$ and the other producing at a rate $(0 - q)$ for time $(t_p + \Delta t - t_p)$.

Using Eq. 1.191, we have

$$p_{wD}(t_D) = \sum_{i=1}^{n}\left(q_i - q_{i-1}\right)_D\left[p_{D'}\left(t - t_{i-1}\right)_D + s\right], \quad \text{(1.197)}$$

which becomes

$$p_{wD} = (q - 0)_D\left[p_D\left(t_p + \Delta t - 0\right)_D + s\right]$$

$$+ (0 - q)_D\left[p_D\left(t_p + \Delta t - t_p\right)_D + s\right]$$

$$= q_D\left[p_D\left(t_p + \Delta t\right)_D + s\right] - q_D\left[p_D(\Delta t)_D + s\right]$$

$$= q_D\left[p_D\left(t_p + \Delta t\right)_D - p_D(\Delta t)_D\right]. \quad \text{(1.198)}$$

A buildup test is generally run for a short time, so transient flow is expected. Therefore, we use the transient-radial-flow (line-source) solution for the required constant-rate solutions.

$$p_D(t_D) = -\frac{1}{2}\text{Ei}\left(\frac{-r_D^2}{4t_D}\right). \quad \text{(1.146)}$$

Therefore (at the wellbore where $r_D = 1$),

$$p_D\left(t_p + \Delta t\right)_D = -\frac{1}{2}\text{Ei}\left[\frac{-1}{4\left(t_p + \Delta t\right)_D}\right] \quad \text{(1.199)}$$

$$\text{and } p_D(\Delta t_D) = -\frac{1}{2}\text{Ei}\left[\frac{-1}{4(\Delta t_D)}\right] \quad \text{(1.200)}$$

We substitute Eqs. 1.199 and 1.200 into Eq. 1.198 to obtain

$$p_{wD} = q_D\left\{-\frac{1}{2}\text{Ei}\left[\frac{-1}{4\left(t_p + \Delta t\right)_D}\right] - \left[-\frac{1}{2}\text{Ei}\left(\frac{-1}{4(\Delta t_D)}\right)\right]\right\}$$

$$= \frac{q_D}{2}\left\{Ei\left[\frac{-1}{4(\Delta t_D)}\right] - Ei\left[\frac{-1}{4\left(t_p + \Delta t\right)_D}\right]\right\}. \quad \text{(1.201)}$$

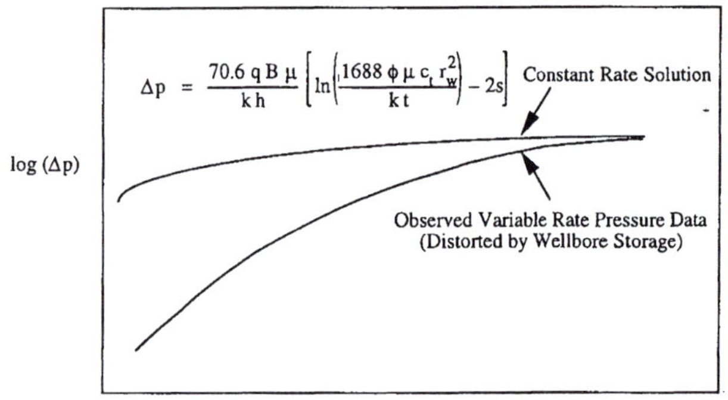

Fig. 1.18—Effect of wellbore storage (variable rate) on pressure profile.

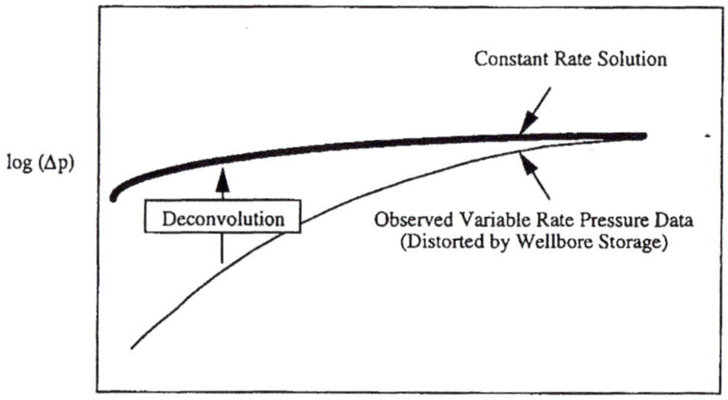

Fig. 1.19—Effect of deconvolution on variable-rate pressure profile.

Recall the log approximation for the Ei function, $\mathrm{Ei}(-x) \approx \ln(1.781x)$. Substituting this into Eq. 1.201 gives

$$p_{wD} = \frac{q_D}{2}\left\{\ln\left[\frac{1.781}{4(\Delta t_D)}\right] - \ln\left[\frac{1.781}{4\left(t_p + \Delta t\right)_D}\right]\right\}. \quad \ldots \quad (1.202)$$

We know that $\ln(a) - \ln(b) = \ln(a/b)$; therefore, Eq. 1.202 becomes

$$p_{wD} = \frac{q_D}{2}\ln\left[\frac{\left(t_p + \Delta t\right)_D}{(\Delta t_D)}\right]. \quad \ldots \quad (1.203)$$

We recall the definitions of the following dimensionless variables,

$$p_{wD} = \frac{kh(p_i - p_{ws})}{141.2qB\mu}, \quad \ldots \quad (1.204)$$

where p_{ws} = shut-in BHP, and

$$q_D = \frac{q}{q_0} = 1 \quad \ldots \quad (1.205)$$

because q_0 = flow rate just before shut-in = q.

$$\Delta t_D = \frac{0.0002637\Delta t}{\phi\mu c_t r_w^2} \quad \ldots \quad (1.206)$$

$$\text{and } \left(t_p + \Delta t\right)_D = \frac{0.0002637k\left(t_p + \Delta t\right)}{\phi\mu c_t r_w^2}. \quad \ldots \quad (1.207)$$

Substitute Eqs. 1.204 through 1.207 into Eq. 1.203 to obtain

$$\frac{kh(p_i - p_{ws})}{141.2qB\mu} = \frac{1}{2}\ln\left\{\left[\frac{0.0002637k\left(t_p + \Delta t\right)}{\phi\mu c_t r_w^2}\right] \times \left(\frac{\phi\mu c_t r_w^2}{0.0002637k\Delta t}\right)\right\}$$

$$p_i - p_{ws} = \frac{70.6qB\mu}{kh}\ln\left[\frac{\left(t_p + \Delta t\right)}{\Delta t}\right]$$

$$p_{ws} = p_i - \frac{70.6qB\mu}{kh}\ln\left[\frac{\left(t_p + \Delta t\right)}{\Delta t}\right]. \quad \ldots \quad (1.208)$$

We know that $\ln(x) = 2.303\log(x)$, therefore Eq. 1.208 becomes

$$p_{ws} = p_i - \frac{70.6(2.303)qB\mu}{kh}\log\left[\frac{\left(t_p + \Delta t\right)}{\Delta t}\right]$$

$$= p_i - \frac{162.6qB\mu}{kh}\log\left[\frac{\left(t_p + \Delta t\right)}{\Delta t}\right]. \quad \ldots \quad (1.209)$$

This equation models a pressure-buildup test and forms the basis for some test-analysis techniques. An important characteristic of this relation is that a plot of p_{ws} (pressure recorded during the buildup test) vs. the logarithm of the function $[(t_p + \Delta t)/\Delta t]$ should be a straight line with a slope inversely proportional to the formation permeability. We discuss pressure-buildup analysis in more detail in Chap. 2.

1.8 Deconvolution

With superposition (discussed in Sec. 1.7), we can calculate the variable-rate solution from a known constant-rate solution. There are times when we know the variable-rate pressure response and wish to calculate a constant-rate pressure profile. This is especially useful when wellbore storage distorts pressure data; wellbore storage causes variable sandface rates, and thus a variable-rate pressure profile. For example, if wellbore storage distorts flow data from an infinite-acting well, the pressures measured at the sandface would not match the infinite-acting solution derived for constant rate. **Fig. 1.18** shows this.

To analyze pressure data distorted by wellbore storage, we would have to use solutions that incorporate wellbore-storage effects. There have been solutions developed with wellbore storage as an inner-boundary condition for many reservoir models; however, we must know what the reservoir model is to use these solutions. If we could calculate a constant-rate pressure profile from the wellbore-storage distorted data, we could eliminate the wellbore-storage effects and have a solution that would indicate the reservoir model, rather than requiring *a priori* knowledge of the model.

Deconvolution is a technique that can be used to remove the effects of wellbore storage from the measured pressure profile. In this section, we discuss some simple methods of deconvolution in rate normalization[12-14] and Laplace transform deconvolution.[15-18] **Fig. 1.19** illustrates the effects of deconvolution.

1.8.1 Rate-Normalization Methods. We can describe discrete changes in rate with superposition in time; however, if the rate is changing smoothly (as a function of time), we can use rate-normalization methods.[12-14] These methods group together the variables dependent on time; for example, the constant-rate, infinite-acting-reservoir solution is given by

$$\Delta p(t) = \frac{70.6qB\mu}{kh}\left[\mathrm{Ei}\left(\frac{-948\phi\mu c_t r_w^2}{kt}\right) - 2s\right]. \quad \ldots \quad (1.210)$$

For variable rates, this equation becomes

$$\Delta p(t) = \frac{70.6q(t)B\mu}{kh}\left[\mathrm{Ei}\left(\frac{-948\phi\mu c_t r_w^2}{kt}\right) - 2s\right]. \quad \ldots \quad (1.211)$$

Grouping together functions of time on the left side of the equation, we have a rate-normalized pressure term that can be used in pressure transient analysis (as described in Chap. 2).

$$\frac{\Delta p(t)}{q(t)} = \frac{70.6B\mu}{kh}\left[\text{Ei}\left(\frac{-948\phi\mu c_t r_w^2}{kt}\right) - 2s\right]. \quad \ldots\ldots (1.212)$$

Ref. 14 gives a more complete discussion of this equation, including its limitations.

1.8.2 Laplace Transform Deconvolution. Kuchuk and Ayestaran[15] presented the idea of deconvolution by use of Laplace transformations to convert the convolution integral into a form that could be solved algebraically for the constant-rate pressure profile. To use this method, they needed to express the sandface flow rate and/or variable-rate pressure profile as approximation functions. On the basis of ideas of van Everdingen[16] and Hurst,[17] the authors developed an exponential series model to fit the flow rate data. Use of this method is limited when the rate profile is not represented accurately by the exponential series model. The use of a numerical-inversion routine is another disadvantage because of inherent instability. Other authors, including Blasingame *et al.*,[18] have introduced more stable Laplace transform methods by use of different approximations to fit the measured data functions. These methods are again limited by the choice of the functions that fit the measured pressure data and sandface rates.

In this section, we present a general development of Laplace transform deconvolution[15] and show how to represent the rate function so that a direct inversion from Laplace space exists.

In Sec. 1.7.1, we developed the convolution integral

$$p_{wD}(t_D) = \int_0^{t_D} \frac{dq_D(\tau)}{d\tau}\left[p_{sD}(t_D - \tau) + s\right]d\tau. \quad \ldots\ldots\ldots (1.196)$$

With variables with dimensions, Eq. 1.196 becomes

$$\Delta p_w(t) = \int_0^t \frac{dq(\tau)}{d\tau}\Delta p_{sf}(t - \tau)d\tau\,, \quad \ldots\ldots\ldots\ldots\ldots\ldots (1.213)$$

where Δp_w = measured pressure drop during test, q_0 = reference rate, $q(\tau)$ = sandface flow rate, and Δp_{sf} = constant-rate pressure behavior of the reservoir at the sandface (i.e., the pressure data that would have been obtained from a constant-rate flow test if wellbore storage had not distorted the test data).

Looking at Eq. 1.213, we have a problem with direct calculation of Δp_{sf} because it is "locked" inside the integral. We can use a technique that we used to develop many of the solutions to the diffusivity equation presented in this chapter—Laplace transforms.

We need to use a theorem that will allow us to take Laplace transforms of Eq. 1.213. This is Duhamel's theorem,[2] which states

$$\mathcal{L}\left[\int_0^t f(\tau)g(t - \tau)d\tau\right] = \mathcal{L}[f(t)]\mathcal{L}[g(t)]. \quad \ldots\ldots\ldots\ldots (1.214)$$

Our equation, however, is in the form

$$\int_0^t f'(\tau)g(t - \tau)d\tau. \quad \ldots\ldots\ldots\ldots\ldots\ldots\ldots\ldots\ldots\ldots\ldots (1.215)$$

We need to use the property of Laplace transforms, which relates the transform of the derivative of a function, $\mathcal{L}[f'(t)]$, to the transform of the function, $\bar{f}(u)$:

$$\mathcal{L}[f'(t)] = u\bar{f}(u) - f(t = 0). \quad \ldots\ldots\ldots\ldots\ldots\ldots\ldots (1.216)$$

Therefore, taking Laplace transforms of Eq. 1.213 by use of Duhamel's theorem gives

$$\Delta\bar{p}_w(u) = \frac{1}{q_0}[u\bar{q}(u) - q(t = 0)]\Delta\bar{p}_{sf}. \quad \ldots\ldots\ldots\ldots (1.217)$$

Because flow rate is zero at initial time, we have

$$\Delta\bar{p}_w(u) = u\bar{q}_D(u)\Delta\bar{p}_{sf}(u)\,, \quad \ldots\ldots\ldots\ldots\ldots\ldots\ldots (1.218)$$

which can be rearranged algebraically as follows to solve for $\Delta\bar{p}_{sf}$:

$$\Delta\bar{p}_{sf} = \frac{\Delta\bar{p}_w}{u\bar{q}_D}. \quad \ldots\ldots\ldots\ldots\ldots\ldots\ldots\ldots\ldots\ldots\ldots\ldots (1.219)$$

We can calculate $\Delta\bar{p}_{sf}$ and then invert to "real space" to obtain Δp_{sf}. This can be done by use of a numerical-inversion routine, such as the Stehfest[7] algorithm, or by choosing functions for Δp_{sf}, Δp_w, and/or q_D so that direct inversion is possible. Many different methods have been presented in the literature; we present a simple direct inversion method in this chapter known as the beta deconvolution[15-17] method.

Beta Deconvolution. van Everdingen[16] and Hurst[17] suggest that the sandface rate for wellbore-storage-dominated data may be approximated by an exponential in the form

$$q_D = 1 - e^{-\beta t}. \quad \ldots\ldots\ldots\ldots\ldots\ldots\ldots\ldots\ldots\ldots\ldots\ldots (1.220)$$

Taking the Laplace transform of Eq. 1.220 gives

$$\mathcal{L}(q_D) = \mathcal{L}(1) - \mathcal{L}\left(e^{-\beta t}\right)$$

$$\bar{q}_D = \frac{1}{u} - \frac{1}{u + \beta} = \frac{\beta}{u(u + \beta)}. \quad \ldots\ldots\ldots\ldots\ldots\ldots (1.221)$$

We can now substitute Eq. 1.221 into Eq. 1.219 and rearrange the equation to obtain a Laplace transform relationship that has a direct inversion.

$$\Delta\bar{p}_s = \frac{1}{u}\frac{u(u + \beta)}{\beta}\Delta\bar{p}_w$$

$$= \frac{u}{\beta}\Delta\bar{p}_w + \Delta\bar{p}_w. \quad \ldots\ldots\ldots\ldots\ldots\ldots\ldots\ldots\ldots (1.222)$$

Inverting this equation gives

$$\Delta p_s = \frac{1}{\beta}\frac{d}{dt}(\Delta p_w) + \Delta p_w. \quad \ldots\ldots\ldots\ldots\ldots\ldots\ldots (1.223)$$

Eq. 1.223 implies that, if we have measured pressure-drop data, Δp_w, that are distorted by wellbore storage and if we can fit the flow-rate vs. time data with an exponential function to find β, we can calculate the constant-rate pressure drop, Δp_{sf} by use of this relationship. Note that we need the derivative of the pressure drop in this equation; this can be calculated numerically by use of finite-difference methods.

1.9 Chapter Summary

In Sec. 1.2, we presented the partial-differential equations describing the flow of fluids in porous media for several systems of interest. These systems include radial flow of a single-phase, slightly compressible liquid (Eq. 1.39), radial flow of a single-phase gas (Eq. 1.51), and multiphase flow (Eq. 1.65). We presented approximate linearizations of the single-phase gas-flow equation in terms of pressure (Eq. 1.53), pressure squared (Eq. 1.58), and pseudopressure (Eq. 1.64). For each of these systems, we arrived at the appropriate partial-differential equation by combining the continuity equation, an equation of motion, and an EOS. Throughout the section, we made the following assumptions.

1. Flow is radial.
2. Darcy's law describes the relationship between flow velocity and pressure gradient.
3. The porous medium is uniform and isotropic.
4. Gravity effects are negligible.
5. Conditions are isothermal.

6. Effective permeability is independent of pressure.

To develop Eq. 1.39 for slightly compressible liquids, we further assumed Points 7 and 8.

7. The fluid has a small, constant compressibility; i.e., the EOS is given by $\rho = \rho_b \exp[c(p - p_b)]$.

8. The compressibility/pressure-gradient squared product, $c(\partial p/\partial r)^2$, is negligible.

For gases, Eq. 1.51, we made Assumptions 1 through 6 and added Assumption 9.

9. The density is given by the real gas law, $\rho = pM/zRT$.

Although Eq. 1.51 is rigorous, it is also nonlinear. To take advantage of the solution methods available for linear PDE's, we would like to linearize Eq. 1.51, even at the expense of making more restrictive assumptions.

10. The term $p/\mu z$ is constant with respect to pressure.

Using Assumption 10, we obtain Eq. 1.53, identical to the expression that describes flow of a slightly compressible liquid (Eq. 1.39). This equation is linear for gases, provided that Assumption 11 follows.

11. The term μc_t is constant with pressure.

If instead we assume the following, we obtain Eq. 1.58.

12. The term μz is constant.

Eq. 1.58 has the same form as Eq. 1.39, with pressure replaced by pressure squared. As with Eq. 1.39, this equation is linear provided that Assumption 11 is true.

By defining the real gas pseudopressure, we obtain Eq. 1.64. This equation depends on Assumptions 1 through 6 and 9, but not 10 or 12. It is linear if Assumption 11 holds.

Finally, for multiphase flow, Assumptions 1 through 6, along with the following additional assumptions, result in Eq. 1.65.

13. Pressure and saturation gradients are small.

14. Capillary pressure is negligible.

In Sec. 1.3, we turned our attention to a description of the more common initial and boundary conditions. A single initial condition, that of uniform pressure at time $t=0$, is used throughout this text. The outer-boundary conditions of interest include the following.

1. Infinite-acting reservoir, where the pressure approaches the initial pressure at distances far from the wellbore for all times.

2. No-flow outer boundary, where the partial derivative of the pressure with respect to r is zero for $r=r_e$.

3. Constant-pressure outer boundary, where the pressure is constant and equal to the initial pressure for all times at some distance r_e from the well.

Inner boundary conditions of interest include (1) constant sandface production rate, (2) constant-pressure production, (3) wellbore storage, and (4) skin factor.

In Sec. 1.4, we developed appropriate dimensionless variables for constant-rate production, radial flow. We stated expressions for dimensionless pressure, dimensionless radius, dimensionless time, dimensionless wellbore-storage coefficient, and skin factor. Using these dimensionless variables, we wrote the diffusivity equation and the associated initial and boundary conditions in dimensionless form.

We also stated dimensionless variables for radial flow, production at constant pressure. We stated expressions for dimensionless pressure, dimensionless rate, and dimensionless cumulative production. As with the constant-rate case, we wrote the diffusivity equation and the associated initial and boundary conditions in dimensionless form by use of these dimensionless variables.

Finally, we stated expressions for dimensionless pressure, dimensionless length, and dimensionless time for constant-rate production from a linear system. We stated these expressions first for the general case, then for a well with a vertical hydraulically induced fracture.

In Sec. 1.5, we presented solutions to the diffusivity equation for several reservoir models, including models (1) with transient radial flow and constant-rate production from a line-source well, (2) without skin factor, (3) with both skin factor and wellbore storage, (4) with pseudosteady-state radial flow and constant-rate production from a cylindrical-source well in a closed reservoir, (5) steady-state radial flow and constant-rate production from a cylindrical-source well in a reservoir with a constant-pressure outer boundary, and (6) transient linear flow and constant-rate production from a hydraulically fractured well. We tied these solutions to field applications through several examples.

In the next three sections, we explored three extremely useful techniques arising from the linearity of the diffusivity equation: superposition in space in Sec. 1.6; superposition in time, or convolution, in Sec. 1.7; and deconvolution in Sec. 1.8.

In Sec. 1.6, we discussed the use of superposition in space. Superposition in space may be used to determine the pressure drop at any given point in a multiwell reservoir as a function of time simply by adding the pressure drops that would result from each well considered independently. In many cases of interest, superposition in space may be used to develop solutions for a single well in a reservoir with one or more no-flow or constant-pressure boundaries by use of the method of images.

Superposition in time, discussed in Sec. 1.7, may be used to calculate the pressure drop for a well where the production rate is a piecewise constant function of time. By approximating an arbitrary rate history as a sequence of constant rates and letting the width of each rate interval go to zero, we obtained an expression for the convolution integral (Eq. 1.196). The convolution integral allows us to calculate the pressure as a function of time for an arbitary rate history, for any reservoir geometry, from the given rate history and the wellbore pressure solution for constant-rate production from the same reservoir geometry. Superposition in time may also be used to model pressure-buildup tests by considering the test as a two-rate problem.

Finally, in Sec. 1.8, we introduced deconvolution, the inverse problem to that of convolution. In convolution, we use a known variable rate and a known pressure response to constant-rate production to calculate the (unknown) pressure response to the known variable rate; in deconvolution, we use a known variable rate and a known pressure response to that variable rate to infer the (unknown) pressure response to constant-rate production. If this deconvolution can be performed without having to make any assumptions as to the reservoir model, the resulting constant-rate pressure response can then be used to infer the reservoir model. It can also be analyzed by use of any pressure transient test analysis methods developed for constant-rate production. This process can be especially useful for analyzing tests distorted by wellbore storage if the sandface flow rates can be measured or calculated.

The simplest deconvolution method we discussed was that of rate normalization, where the pressure change as a function of time is divided by the rate, also as a function of time. This method is applicable whenever the rate is changing slowly and smoothly.

We also introduced a general class of convolution methods referred to Laplace transform deconvolution, along with beta deconvolution, a special case of Laplace transform deconvolution.

1.10 Discussion Questions

For each of the following variables, state the field units, define the variable, and discuss how it might be measured or estimated. What are likely sources of uncertainty in the measurement or estimate of each quantity?

a. porosity, ϕ
b. viscosity, μ
c. formation net pay thickness, h
d. well radius, r_w
e. water saturation, S_w
f. oil saturation, S_o
g. gas saturation, S_g
h. formation compressibility, c_f
i. oil compressibility, c_o
j. gas compressibility, c_g
k. flow rate, q
l. oil formation volume factor, B_o
m. gas formation volume factor, B_g
n. water formation volume factor, B_w
o. time, t
p. pressure, p
q. permeability, k

2. Explain Fig. 1.1. In which direction do we assume fluid flows? What is the flow rate in the θ and z directions?

3. On what assumptions is the continuity equation (Eq. 1.18) based?

4. Why is the differential form of Darcy's law (Eq. 1.19) written in terms of potential Φ instead of pressure? What is the physical significance of Φ?

5. On what assumptions is the equation of state (Eq. 1.28) based?

6. What is the major assumption used in writing the diffusivity equation for gas in terms of pressure (Eq. 1.53)? In what field situations is this assumption reasonable? In what situations is this assumption likely to be violated?

7. What is the major assumption used in writing the diffusivity equation for gas in terms of pressure-squared (Eq. 1.58)? In what field situations is this assumption reasonable? In what situations is this assumption likely to be violated?

8. What is the major assumption used in writing the diffusivity equation for gas in terms of pseudopressure (Eq. 1.64)? In what field situations is this assumption reasonable? In what situations is this assumption likely to be violated?

9. What is the physical significance of the total mobility for a three-phase system (Eq. 1.66)?

10. Explain the significance of each of the terms in the equation for the total compressibility (Eq. 1.67).

11. Which term in the total compressibility equation (Eq. 1.67) is most important for an oil reservoir just above the bubble point pressure of 1500 psi, with a water saturation of 15%? For an oil reservoir just below the bubble point pressure of 1500 psi, with a gas saturation of 5% and a water saturation of 15%? For a low pressure gas reservoir at 150 psi with a water saturation of 25%? For a geopressured gas reservoir at 12000 psi with a water saturation of 20% and a formation compressibility of 30×10^{-6}? Make any assumptions about fluid properties as reasonable as possible.

12. What are the assumptions used in developing the diffusivity equation for multiphase flow (Eq. 1.65)? In what field situations are these assumptions reasonable? In what situations are these assumptions likely to be violated?

13. The diffusivity equation for radial fluid flow in porous media (Eq. 1.68) is first order in time and second order in space. How many boundary conditions must we specify in order to solve this equation?

14. We have assumed permeability is isotropic in this chapter. How would the presence of an anisotropic permeability affect the pressure response for a vertical well producing at constant rate?

15. Under what field conditions would the assumption of constant pressure at the gas-liquid interface (Eq. 1.91) be valid? Under what conditions would it not be valid?

16. Under what field conditions would the wellbore storage coefficient for a single-phase fluid (Eq. 1.95) not be constant?

17. Does the presence of a zone of altered permeability affect the pressure response in the reservoir at some distance from the wellbore for a *fixed* flow rate q? If the pressure response away from the wellbore is not affected by the presence of a zone of altered permeability, why do we ever need to remove skin damage or stimulate a well?

18. Which of the following values appear in the definition of dimensionless pressure for constant rate production (Eq. 1.104)—permeability, porosity, net pay thickness, viscosity, formation volume factor, compressibility, flow rate, wellbore radius? Which appear in the definition of dimensionless time (Eq. 1.107)?

19. The time range of applicability of the line source or Ei function solution is given in Eq. 1.149. What limits the applicability of the line source solution at early times? How will the solution for a finite wellbore differ from the line source solution at times before the line source solution becomes valid? What limits the applicability of the line source solution at late times? How will the solution for a closed circular reservoir differ from the line source solution at times after the line source solution becomes invalid?

20. At very early times, the pressure solution for constant rate production from a finite wellbore is proportional to the square root of time, as is the solution for constant rate production from a linear system. Explain.

21. Is wellbore storage the only phenomenon which can cause a unit slope line on a log-log plot of dimensionless pressure *vs.* dimensionless time? If not, what other conditions could cause such a unit slope line?

22. What fundamental property of the partial differential equation given in Eq. 1.68 allows us to use superposition in space? Superposition in time?

23. Why are there no skin factor terms for Wells B and C in Eq. 1.177, describing the pressure response at Well A due to production Wells A, B, and C?

24. What is "convolution"? What does it allow us to do? How is it applied? Do we need to assume a reservoir model to apply it?

25. What is "deconvolution"? What does it allow us to do? How is it applied? Do we need to assume a reservoir model to apply it?

Exercises

1. Write Darcy's law in field units for the following conditions. State any assumptions for each form of Darcy's law.

a. General differential form, one-dimensional flow. Express flow rate as volume at standard conditions, q. Assume cross-sectional area A.

b. Slightly compressible liquids, linear flow—q in STB/D.

c. Slightly compressible liquids, radial flow—q in STB/D.

d. Real gases, linear flow—q in Mscf/D.

e. Real gases, radial flow—q in Mscf/D.

2. Develop an expression for the general diffusivity equation for fluid flow in isotropic porous media in Cartesian coordinates in the x-y plane similar to that given in Eq. 1.68 for radial flow.

3. Generalize the expression obtained in the previous exercise to allow permeability anisotropy, $k_x \neq k_y$. Show that by using a transformed coordinate system, with $x' \equiv x\sqrt{\frac{\bar{k}}{k_x}}$, and $y' \equiv y\sqrt{\frac{\bar{k}}{k_y}}$, where $\bar{k} \equiv \sqrt{k_x k_y}$, the expression for an anisotropic system takes exactly the same form as that for an isotropic system.

4. Show that the units on the right hand side of the definition of dimensionless pressure for radial flow, constant rate production (Eq. 1.104) cancel, justifying the name dimensionless pressure.

5. Show that the units on the right hand side of the definition of dimensionless time for radial flow, constant rate production (Eq. 1.107) cancel, justifying the name dimensionless time.

6. Using material balance for a slightly compressible liquid, find an expression in field terms for the cumulative production for a closed circular reservoir of radius r_e from an initial pressure p_i to an average reservoir p_{wf}. Using the definition of dimensionless cumulative production (Eq. 1.120), convert your expression to dimensionless form in terms of the dimensionless radius r_D. What is the physical interpretation of your result?

7. Show that the reciprocal of the dimensionless rate (Eq. 1.119) for constant pressure production can be written as $\frac{1}{q_D} = \frac{kh}{141.2qB\mu}\left(p_i - p_{wf}\right)$. What is the difference between this expression and the definition of dimensionless pressure (Eq. 1.104) for constant rate production?

8. Given a reservoir with the following properties:

$q =$ 250 STB/D
$c_t =$ 6.0×10^{-6} psi^{-1}
$\mu =$ 0.75 cp
$r_e =$ 3000 ft
$B =$ 1.25 RB/STB
$r_w =$ 0.25 ft
$k =$ 25 md
$s =$ 5
$h =$ 60 ft
$p_i =$ 3500 psi
$\phi =$ 20%

a. Calculate the time at which the *Ei* function solution first becomes valid.

b. Calculate the time after which the *Ei* function solution is no longer valid.

c. Calculate and plot the pressure as a function of position on semilog graph paper at $t = 100$ hours using the *Ei* function solution (or its logarithmic approximation) for $r = 0.25$ ft, $r = 1$ ft, $r = 2$ ft, $r = 10$ ft, $r = 20$ ft, $r = 50$ ft, $r = 100$ ft, $r = 200$ ft, $r = 500$ ft, $r = 1000$ ft, $r = 2000$ ft, $r = 3000$ ft. *Use the log approximation only if it is valid.*

d. Calculate the additional pressure drop due to the skin factor.

9. Given a reservoir with the following properties:

$q =$ 200 STB/D
$c_t =$ 10.0×10^{-6} psi^{-1}
$\mu =$ 0.75 cp
$r_e =$ 1000 ft
$B =$ 1.25 RB/STB
$r_w =$ 0.25 ft
$k =$ 25 md
$s =$ 3
$h =$ 50 ft
$p_i =$ 3500 psi
$\phi =$ 20%

a) Calculate the time at which the *Ei* function solution first becomes valid.

b) Calculate the time after which the *Ei* function solution is no longer valid.

c) Calculate and plot the pressure as a function of position on semilog graph paper (pressure on linear axis, distance on log axis) at $t = 5$ hours using the *Ei* solution (or its logarithmic approximation) for $r = 0.25$, 1, 2, 10, 20, 50, 100, 150, 200, 250, 300, 400, 500, 750, 1000 ft. *Use the log approximation only if it is valid.*

d) Calculate the additional pressure drop due to the skin factor.

10. Given a reservoir with the following properties:

$q =$ 50 STB/D
$c_t =$ 10.0×10^{-6} psi^{-1}
$\mu =$ 0.15 cp
$r_e =$ 1000 ft
$B =$ 1.25 RB/STB
$r_w =$ 0.25 ft
$k =$ 5 md
$s =$ 4.5
$h =$ 20 ft
$p_i =$ 2500 psi
$\phi =$ 20%

a) Calculate the time at which the *Ei* function solution first becomes valid.

b) Calculate the time after which the *Ei* function solution is no longer valid.

c) Calculate and plot the pressure as a function of position on semilog graph paper (pressure on linear axis, distance on log axis) at t = 5 hours using the *Ei* solution (or its log approximation) for r = 0.25, 1, 2, 5,10, 20, 50, 100, 150, 200, 250, 300, 400, 500, 750, 1000 ft. *Use the log approximation only if it is valid.*

d) Calculate and plot the pressure as a function of time on semilog graph paper (pressure on linear axis, time on log axis) at $r = 10$ ft using the *Ei* solution (or its log approximation) for $t = 0.0001$, 0.0003, 0.001, 0.002,0.005, 0.01, 0.02, 0.05,0.1, 0.3, 1, 3, 10, 30 hours. *Use the log approximation only if it is valid.*

e) Calculate the additional pressure drop due to the skin factor.

11. Given a reservoir with the following properties:

$q =$ 225 STB/D
$c_t =$ 7.5×10^{-6} psi^{-1}
$\mu =$ 0.63 cp
$A =$ 640 acres
$B =$ 1.25 RB/STB
$r_w =$ 0.33 ft
$k =$ 20 md
$s =$ 4.5
$h =$ 63 ft
$p_i =$ 3330 psi
$\phi =$ 18%

Estimate 1) the time at which the Ei solution becomes invalid because of boundary effects; and 2) the time at which the pseudosteady state solution becomes valid for each of the following cases.

a) square reservoir with well ¼ from nearest side;
b) right triangular reservoir with well at centroid;
c) 4×1 rectangle with well in the center.

For each case, assume that the Ei solution is valid until the radius of investigation reaches the nearest boundary for the given shape, and that the pseudosteady state solution becomes valid after the radius of investigation has passed the farthest point from the well for the given shape. In a later chapter, we will consider another, more accurate method for estimating time at which boundary effects become important and the time required to reach pseudosteady state flow.

12. Show that the dimensionless pressure for pseudo-steady state flow (Eq. 1.159) falls on a unit slope line when plotted *vs.* dimensionless time.

13. Calculate the time required to reach pseudosteady state for a well in the center of a circular, 640 acre reservoir under each of the following situations. Express your answers in hours, as well as in either days, months, or years, whichever is most appropriate. For each case, take $h = 75$ feet, $r_w = 0.25$ feet, and $c_f = 3.0 \times 10^{-6}$ psia^{-1}.

a. Undersaturated oil. $p = 4000$ psia, $T = 200$ °F, $\gamma_o = 38$ °API, $\mu_o = 0.46$ cp, $S_o = 1$, $c_o = 1.28 \times 10^{-5}$ psia^{-1}, $\phi = 25\%$, $k = 350$ md.

b. High pressure gas. $p = 12000$ psia, $T = 350$ °F, $\gamma_g = 0.67$, $\mu_g = 0.03636$ cp, $S_g = 1$, $c_g = 3.74 \times 10^{-5}$ psia^{-1}, $\phi = 7\%$, $k = 0.01$ md.

c. Low pressure gas. $p = 600$ psia, $T = 110$ °F, $\gamma_g = 0.72$, $\mu_g = 0.01167$ cp, $S_g = 1$, $c_g = 1.83 \times 10^{-3}$ psia^{-1}, $\phi = 15\%$, $k = 10$ md.

d. Saturated oil. $p = 2500$ psia, $T = 225$ °F, $\gamma_o = 45$ °API, $\mu_o = 0.36$ cp, $S_o = 60\%$, $c_o = 1.8 \times 10^{-4}$ psia^{-1}, $S_g = 25\%$, $c_g = 3.92 \times 10^{-4}$ psia^{-1}, $S_w = 15\%$, $c_w = 4 \times 10^{-6}$ psia^{-1}, $\phi = 21\%$, $k = 100$ md. Calculate c_t using the equation $c_t = c_f + S_o c_o + S_w c_w + S_g c_g$. Assume that oil is the only mobile phase.

14. High Hopes Operating, Inc. has just completed the discovery well in the Yellow Sandstone formation. There are three phases present in the formation, but oil is the only mobile phase. The following formation, well, and fluid properties apply:

depth = 4356 ft
$p_i =$ 950 psia
$\phi =$ 12%
$\gamma_o =$ 32.5 °API
$S_w =$ 18%
$S_g =$ 16%
$S_o =$ 66%
$\mu_o =$ 1.17 cp
$R_{so} =$ 226 scf/STB
$s =$ 0
$h =$ 22 ft
$r_w =$ 9 in.
$k =$ 35.5 md
$\gamma_g =$ 0.893 (air = 1.0)
$c_w =$ 2.6×10^{-5} psi^{-1}
$c_g =$ 1.188×10^{-3} psi^{-1}
$c_o =$ 4.23×10^{-4} psi^{-1}
$c_f =$ 5×10^{-6} psi^{-1}
$B_o =$ 1.146 RB/STB
$q =$ 165 STB/D

Calculate:

A. The total compressibility c_t.
B. The time at which the Ei function solution first becomes valid.
C. The pressure at the wellbore after the well has been producing for 1 hour.
D. The pressure 10 ft from the center of the wellbore after 1 hour.
E. The pressure 100 ft from the center of the wellbore after 1 hour.

15. A well is located in the center of 640 acre circular drainage area. The formation, fluid, and well properties are:

$q =$ 606 STB/D
$c_t =$ 18.0×10^{-6} psi^{-1}
$\mu =$ 0.306 cp
$A =$ 640 acres
$B =$ 1.593 RB/STB
$r_w =$ 0.25 ft

$k =$ 17.5 md
$s =$ 12
$h =$ 103 ft
$p_i =$ 4250 psi
$\phi =$ 12%
$T_f =$ 275°F

Calculate:

A. The additional pressure drop Δp_s due to the skin factor.

B. The pressure at the wellbore p_{wf} when the average reservoir pressure $\bar{p}$ is 4100 psia.

C. The flow rate q which would be obtained for a drawdown $\bar{p} - p_{wf}$ of 500 psia if an acid job were performed, changing the skin factor s from 12 to -2.

16. There are two wells in a formation, Well A and Well B. We are interested at the pressure at a third location in the formation, Point C. Some additional information:

$k =$ 50 md
$B_o =$ 1.345 RB/STB
$\phi =$ 0.18
$\mu_o =$ 0.256 cp
$h =$ 22 ft
$c_w =$ 21×10^{-6} psi^{-1}
$q_A =$ 1240 STB/D
$t_A =$ 4 hours
$r_{AC} =$ 660 ft
$s_A =$ 0
$q_B =$ 1240 STB/D
$t_B =$ 16 hours
$r_{BC} =$ 1320 ft
$s_B =$ 5

a. Calculate the pressure drop Δp_{AC} at Point C due to Well A, a distance 660 feet away from Point C. Well A has been producing at 1240 STB/D for 4 hours.

b. Calculate the pressure drop Δp_{BC} at Point C due to Well B, a distance 1320 feet away from Point C. Well B has been producing at 1240 STB/D for 16 hours.

c. Calculate the pressure p_C at Point C due to the combined influence of Wells A and B, if the initial reservoir pressure is 2776 psi.

17. Three exploration wells have been drilled into a newly discovered oil reservoir. Well B is 1400 ft from Well A, Well C is 2200 ft from Well A, and Well C is 3000 ft from Well B.

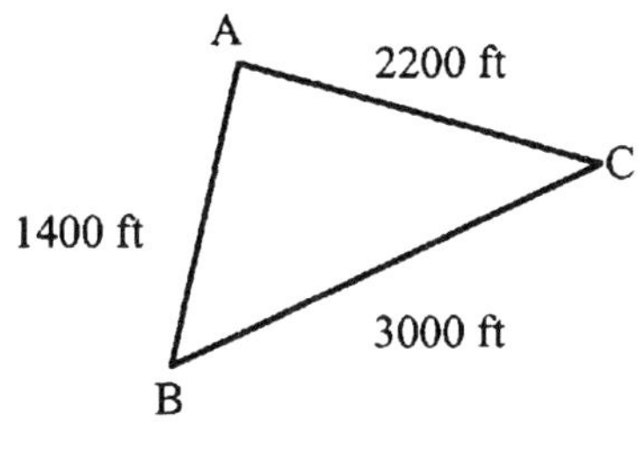

Well A is an observation well (it has a pressure gauge in the hole to measure the pressure response due to production from wells B and C, but is not itself used for production). Well C begins producing at 200 STB/D at time $t = 0$. Well B begins producing at 100 STB/D 48 hours later.

Reservoir data:

$\mu_o =$ 0.27 cp
$c_t =$ 1.8×10^{-5} psi^{-1}
$B_o =$ 1.23 RB/STB
$p_i =$ 2727 psi
$k =$ 42 md
$\phi =$ 22%
$h =$ 7.5 ft

Well data:

	Well A	Well B	Well C
Rate, q	observation well	200 STB/D	200 STB/D
Radius, r_w	0.33 ft	0.25 ft	0.375 ft
Skin factor, s	5.0	7.5	-1.5

a) Calculate the *pressure drop* at Well A due to production from Well B at time $t = 144$ hrs.

b) Calculate the pressure drop at Well A due to production from Well C at time $t = 144$ hrs.

c) Calculate the pressure drop at Well A due to its own skin factor at time $t = 144$ hrs.

d) Calculate the *pressure* at Well A at time t = 144 hrs.

18. Given a well in an infinite-acting reservoir, having the following properties:

$\mu =$ 0.72 cp
$c_t =$ 1.5×10^{-5} psi^{-1}
$B =$ 1.475 RB/STB
$r_w =$ 0.5 ft
$k =$ 10 md
$s =$ 5
$h =$ 150 ft
$p_i =$ 3500 psi
$\phi =$ 23%
$q =$ 1000 STB/D

The well produces for 1000 hours at 1000 STB/D. There is a linear, no-flow boundary 75 feet away from the well. Calculate and plot the pressure at the wellbore as a function of t on semilog graph paper (put time on the logarithmic axis) using superposition in space and the Ei solution (or its logarithmic approximation) for $t = 0.1$, 0.2, 0.5, 1, 2, 5, 10, 20, 50, 100, 200, 500, and 1000 hours.

19. Given a well in an infinite-acting reservoir, having the following properties:

$\mu =$ 0.6 cp
$c_t =$ 1.8×10^{-5} psi^{-1}
$B =$ 1.35 RB/STB
$r_w =$ 0.5 ft
$k =$ 12 md
$s =$ 5
$h =$ 150 ft
$p_i =$ 3750 psi
$\phi =$ 21%
$q =$ 800 STB/D

The well produces for 1000 hours at 800 STB/D. There is a linear, no-flow boundary 100 feet away from the well. Calculate and plot the pressure at the wellbore as a function of t on semilog graph paper (put time on the logarithmic axis) using superposition in space and the Ei solution (or its logarithmic approximation) for $t = 0.1$, 0.2, 0.5, 1, 2, 5, 10, 20, 50, 100, 200, 500, and 1000 hours.

20. Given a well in an infinite-acting reservoir, having the following properties:

$\mu =$ 0.7 cp
$c_t =$ 2×10^{-5} psi^{-1}
$B =$ 1.13 RB/STB
$r_w =$ 0.333 ft
$k =$ 7.5 md
$s =$ 3
$h =$ 25 ft
$p_i =$ 3000 psi
$\phi =$ 20%
$q =$ 100 STB/D

The well produces for 1000 hours at 100 STB/D. There is a linear, no-flow boundary 90 feet away from the well. Calculate and plot the pressure at the wellbore as a function of t on semilog graph paper (put time on the logarithmic axis) using superposition in space and the Ei solution (or its logarithmic approximation) for $t = 0.1$, 0.2, 0.5, 1, 2, 5, 10, 20, 100, 200, 500, and 1000 hours.

21. Given a well in an infinite-acting reservoir, having the following properties:

$\mu =$ 0.72 cp
$c_t =$ 1.5×10^{-5} psi^{-1}
$B =$ 1.475 RB/STB
$r_w =$ 0.5 ft
$k =$ 10 md
$s =$ -3
$h =$ 150 ft
$p_i =$ 3500 psi

$\phi =$ 20%
$q_1 =$ 1000 STB/D
$q_2 =$ 0 STB/D
$t_1 =$ 1000 hours
$t_2 =$ 1010 hours

The well produces for 1000 hours at 1000 STB/D, and is then shut-in. Calculate and plot the pressure at $t = 1010$ hours as a function of r on semilog graph paper (put r on the logarithmic axis) using superposition in time and the Ei solution (or its logarithmic approximation) for $r = 0.5$, 1, 2, 5, 10, 20, 50, 100, 200, 500, 1000, and 2000 feet.

22. Given a well in an infinite-acting reservoir, having the following properties:

$\mu =$ 0.6 cp
$c_t = 1.8 \times 10^{-5}$ psi^{-1}
$B =$ 1.35 RB/STB
$r_w =$ 0.5 ft
$k =$ 12 md
$s =$ 5
$h =$ 150 ft
$p_i =$ 3750 psi
$\phi =$ 21%
$q_1 =$ 800 STB/D
$q_2 =$ 0 STB/D
$t_1 =$ 1000 hours
$t_2 =$ 1010 hours

The well produces for 1000 hours at 800 STB/D, and is then shut-in. Calculate and plot the pressure at $t = 1010$ hours as a function of r on semilog graph paper (put r on the logarithmic axis) using superposition in time and the Ei solution (or its logarithmic approximation) for $r = 0.5$, 1, 2, 5, 10, 20, 50, 100, 200, 500, 1000, and 2000 feet.

23. Given a well in an infinite-acting reservoir, having the following properties:

$\mu =$ 0.7 cp
$c_t = 2 \times 10^{-5}$ psi^{-1}
$B =$ 1.13 RB/STB
$r_w =$ 0.333 ft
$k =$ 7.5 md
$s =$ 3
$h =$ 25 ft
$p_i =$ 3000 psi
$\phi =$ 20%
$q_1 =$ 100 STB/D
$q_2 =$ 0 STB/D
$t_1 =$ 1000 hours
$t_2 =$ 1010 hours

The well produces for 1000 hours at 100 STB/D, and is then shut-in. Calculate and plot the pressure at $t = 1010$ hours as a function of r on semilog graph paper (put r on the logarithmic axis) using superposition in time and the Ei solution (or its logarithmic approximation) for $r = 0.5$, 1, 2, 5, 10, 20, 50, 100, 200, 500, 2000, and 10000 feet.

24. Show that the equation for superposition in time (Eq. 1.189) may be rewritten so that all of the terms involving skin cancel each other except the one involving the last flow rate q_n, giving

$$p_{wD}(t_D) = \sum_{i=1}^{n} (q_i - q_{i-1})_D p_D(t - t_{i-1})_D + q_n s.$$

References

1. Al-Hussainy, R., Ramey, H.J. Jr., and Crawford, P.B.: "The Flow of Real Gases Through Porous Media," *JPT* (May 1966) 637; *Trans.*, AIME, **237.**
2. O'Neil, P.V.: *Advanced Engineering Mathematics*, second edition, Wadsworth Publishing Co., Belmont, California (1987).
3. Martin, J.C.: "The Simultaneous Equations of Flow in Gas Drive Reservoirs and the Theoretical Foundation of Multiphase Pressure Buildup Analyses," *Trans.*, AIME (1959) **216,** 309.
4. Hawkins, M.F. Jr.: "A Note on The Skin Effect," *Trans.*, AIME (1956) **207,** 356.
5. Lee, W.J. *et al.*: *Fundamentals of Fluid Flow Through Porous Media*, Texas A&M U., College Station, Texas (1991), Chap. 1, Appendix.
6. Dake, L.P.: *Fundamentals of Reservoir Engineering*, Elsevier Scientific Publishing Co., New York City, New York (1978).
7. Stehfest, H.: "Numerical Inversion of Laplace Transforms," *Communications ACM* (January 1970) **13,** No. 1, Algorithm 368.
8. Abramowitz, M. and Stegun, I.E.: *Handbook of Mathematical Functions*, Dover Publications Inc., New York City, New York (1972).
9. Hurst, W.: "Unsteady Flow of Fluids," *Physics* (January 1934).
10. Agarwal, R.G., Al-Hussainy, R., and Ramey, H.J. Jr.: "An Investigation of Wellbore Storage and Skin Effects in Unsteady Liquid Flow: I. Analytical Treatment," *SPEJ* (September 1970) 278.
11. Matthews, C.S., Brons, F., and Hazebroek, P.: "A Method for the Determination of Average Pressure in a Bounded Reservoir," *Trans.*, AIME, **201,** 182.
12. Gladfelter, R.E., Tracy, G.W., and Wilsey, L.E.: "Selecting Wells Which Will Respond to Production-Stimulation Treatment," *Drill. & Prod. Prac.* (1955) 117.
13. Kuchuk, F.J.: "Gladfelter Deconvolution," *SPEFE* (September 1990) 285.
14. Winestock, A.G. and Colpitts, G.P.: "Advances in Estimating Gas Well Deliverability," *JCPT* (July–September 1965) 111.
15. Kucuk, F. and Ayestaran, L.: "Analysis of Simultaneously Measured Pressure and Sandface Flow Rate in Transient Well Testing," *JPT* (February 1985) 323.
16. van Everdingen, A.F.: "The Skin Effect and Its Influence on the Productive Capacity of a Well," *JPT* (June 1953) 171; *Trans.*, AIME, **198.**
17. Hurst, W.: "Establishment of the Skin Effect and Its Impediment to Fluid Flow into a Wellbore," *Pet. Eng.* (October 1953) B6.
18. Blasingame, T.A. *et al.*: "The Analysis of Gas Well Test Data Distorted by Wellbore Storage Using an Explicit Deconvolution Method," paper SPE 19099 presented at the 1989 SPE Gas Technology Symposium, Dallas, 7–9 June.
19. Carslaw, H.S. and Jaeger, J.C.: *Conduction of Heat in Solids*, second edition, Oxford U. Press, Oxford, U.K. (1959).

Chapter 2
Introduction to Flow and Buildup-Test Analysis: Slightly Compressible Fluids

2.1 Overview

This chapter presents the underlying theory and practical applications of pressure transient testing of wells completed in reservoirs producing slightly compressible fluids (i.e., liquids). Beginning with the line-source (Ei-function) solution to the diffusivity equation, we develop analysis techniques for flow and buildup tests in homogeneous-acting reservoirs. We then discuss deviations from the ideal pressure behavior predicted by the line-source solution. These deviations are caused by near-wellbore effects and heterogeneities in the drainage area of the tested well. Initially, we develop analysis techniques for single-phase-flow conditions; then, we present a modification for multiphase flow. Application of all analysis techniques presented in this chapter is limited to single-layer formations.

2.2 Analysis of Flow Tests

A flow or pressure-drawdown test is conducted by producing a well at a known rate or rates while measuring changes in bottomhole pressure (BHP) as a function of time. Drawdown tests are designed primarily to quantify the reservoir-flow characteristics, including permeability and skin factor. In addition, when the pressure transient is affected by outer reservoir boundaries, drawdown tests can be used to establish the outer limits of the reservoir and to estimate the hydrocarbon volume in the well's drainage area. These flow tests are called reservoir-limit tests. When economic considerations require a minimum loss of production time, pressure-drawdown tests also can be used to estimate the deliverability of a well and, if conducted and analyzed properly, are viable alternatives to deliverability tests.

The basis of flow-test analysis techniques is the line-source (Ei-function) solution to the diffusivity equation. As Chap. 1 (Eq. 1.151) shows, the relationship between flowing BHP, p_{wf}, and the formation and well characteristics for a well producing at a constant rate is

$$p_{wf} = p_i + \frac{70.6qB\mu}{kh}\left[\ln\left(1,688\phi\mu c_t r_w^2/kt\right) - 2s\right]. \quad \ldots. \quad (2.1)$$

If we change from natural logarithms to base-10 logarithms and simplify, we can rewrite Eq. 2.1 in a more familiar form,[1,2]

$$p_{wf} = p_i - \left(162.6qB\mu/kh\right) \times \left[\log(t) + \log\left(k/\phi\mu c_t r_w^2\right) - 3.23 + 0.869s\right]. \quad \ldots \quad (2.2)$$

The following sections discuss analysis of flow or drawdown well tests. We begin with an analysis technique for a constant-rate flow test. In practice, however, a strictly constant rate is impractical or even impossible to maintain. A more common and probable testing scenario is variable-rate production. Consequently, we also discuss continuously declining rate drawdown tests and multirate tests in infinite-acting reservoirs. Sec. 2.5 addresses flow tests in finite reservoirs.

2.2.1 Constant-Rate Flow Tests. Eq. 2.2 describes the pressure drop at the wellbore as a function of time when a well is produced at a constant rate. Comparing Eq. 2.2 with the equation of a straight line, $y = mx + b$, suggests an analysis technique in which the following terms are analogous[1,2]:

$$y \sim p_{wf}, \quad \ldots \quad (2.3a)$$

$$x \sim \log t, \quad \ldots \quad (2.3b)$$

$$m \sim \left(162.6qB\mu/kh\right), \quad \ldots \quad (2.3c)$$

$$\text{and } b \sim p_i - \frac{162.6qB\mu}{kh}\left[\log\left(\frac{k}{\phi\mu c_t r_w^2}\right) - 3.23 + 0.869s\right]. \quad \ldots \quad (2.3d)$$

These analogies indicate that, if the pressure behavior of a reservoir can be modeled with the line-source solution, a plot of p_{wf} vs. log t will form a straight line from which the slope, m, allows us to estimate formation permeability, k, and skin factor, s. **Fig. 2.1** is an example semilog graph of constant-rate flow-test data. The slope of the line is the difference between two pressures, p_{wf1} and p_{wf2}, one log cycle apart, or $m = p_{wf2} - p_{wf1}$ in psi/cycle. Note that the slope is a negative number; however, in the working equation that follows, we use the absolute value of m for convenience.

For single-phase flow, the effective permeability to the fluid flowing in the drainage area of the well is computed from

$$k = 162.6qB\mu/mh. \quad \ldots \quad (2.4)$$

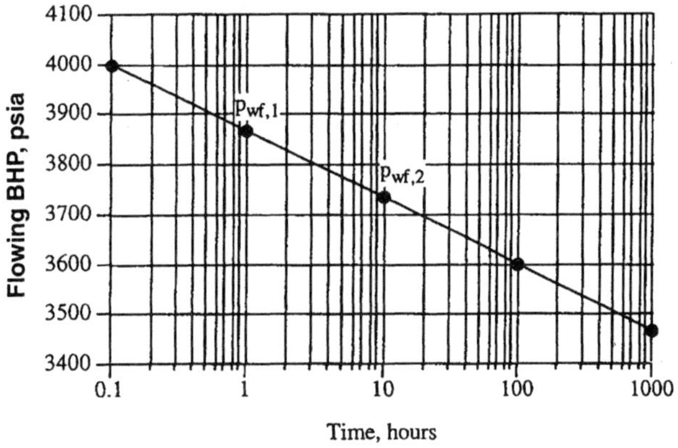

Fig. 2.1—Graphical analysis technique for constant-rate flow test data.

Combining Eqs. 2.2 and 2.4, we can also develop an expression for the skin factor.

$$s = 1.151\left\{\left[(p_i - p_{wf})/m\right] - \left[\log\left(kt/\phi\mu c_t r_w^2\right) + 3.23\right]\right\} \quad (2.5)$$

For convenience, we set the flow time, t, equal to 1 hour and use the symbol p_{1hr} for the flowing BHP, p_{wf}, at this time. Substituting these into Eq. 2.5 yields

$$s = 1.151\left\{\left[(p_i - p_{1hr})/m\right] - \left[\log\left(k/\phi\mu c_t r_w^2\right) + 3.23\right]\right\}. \quad (2.6)$$

Note that the pressure p_{1hr} necessarily lies either on the semilog straight line or its extrapolation.

In summary, we can estimate effective permeability, k, to the fluid flowing in the drainage area of the well and the skin factor, s, from the theoretical straight line on a semilog plot of drawdown test data. This graphical-analysis technique assumes, however, that the pressure behavior can be modeled accurately with the Ei-function solution for constant-rate flow of a single-phase fluid. In subsequent sections, we discuss phenomena that cause the pressure response to deviate from a straight line.

Example 2.1—Determining Permeability and Skin Factor From a Constant-Rate Flow Test. The data summarized here and in **Table 2.1** were recorded during a pressure-drawdown test from an oil well. Estimate the effective permeability to oil and the skin factor by use of the graphical-analysis technique for a constant-rate flow test.

q = 250 STB/D
h = 69 ft
ϕ = 0.039
B = 1.136 RB/STB
p_i = 4,412 psia
$c_t = 17 \times 10^{-6}$ psi^{-1}
r_w = 0.198 ft
μ = 0.8 cp

Solution.

1. Plot flowing BHP, p_{wf}, as a function of time on semilog paper (**Fig. 2.2**). The slope, m, of the resulting straight line is the difference between values of p_{wf} one log cycle apart, or

$$m = \left|\frac{p_{wf2} - p_{wf1}}{\log(t_2) - \log(t_1)}\right| = \left|\frac{3,582 \text{ psia} - 3,652 \text{ psia}}{\log(10) - \log(1)}\right|$$

$$= 70 \text{ psi/cycle},$$

TABLE 2.1—PRESSURE-DRAWDOWN-TEST DATA FOR EXAMPLE 2.1

t (hours)	p_{wf} (psia)	t (hours)	p_{wf} (psia)
0	4,412	14.4	3,573
0.12	3,717	17.3	3,567
1.94	3,633	20.7	3,561
2.79	3,622	24.9	3,555
4.01	3,611	29.8	3,549
4.82	3,605	35.8	3,544
5.78	3,600	43.0	3,537
6.94	3,594	51.5	3,532
8.32	3,588	61.8	3,526
9.99	3,583	74.2	3,521

where, by convention, we use the absolute value of the slope.

2. Effective permeability to oil is estimated by use of Eq. 2.4 and the absolute value of the slope, m, of the line,

$$k = \frac{162.6qB\mu}{mh} = \frac{(162.6)(250)(1.136)(0.8)}{(70)(69)} = 7.6 \text{ md}.$$

3. Noting in Fig. 2.2 that p_{1hr} = 3,652 psi, we calculate the skin factor with Eq. 2.6,

$$s = 1.151\left\{\left[(p_i - p_{1hr})/m\right] - \log\left(k/\phi\mu c_t r_w^2\right) + 3.23\right\}$$

$$= 1.151\left\{\left[(4,412 - 3,652)/70\right] - \log\left[\frac{7.6}{(0.039)(0.8)(17 \times 10^{-6})(0.198)^2}\right] + 3.23\right\}$$

$$= 6.4.$$

As we discuss in Sec. 2.4, a positive value of skin factor indicates a flow restriction (i.e., damage) around the wellbore.

2.2.2 Variable-Rate Testing With Smoothly Changing Rates. The line-source solution given by Eq. 2.2 assumes that rate does not vary with time; however, in many testing situations, a strictly constant producing rate is impractical, or even impossible, to maintain. If rate varies during the flow test, results obtained by use of the technique based on constant flow rate can lead to interpretations that are seriously in error. Winestock and Colpitts[3] show that, even when

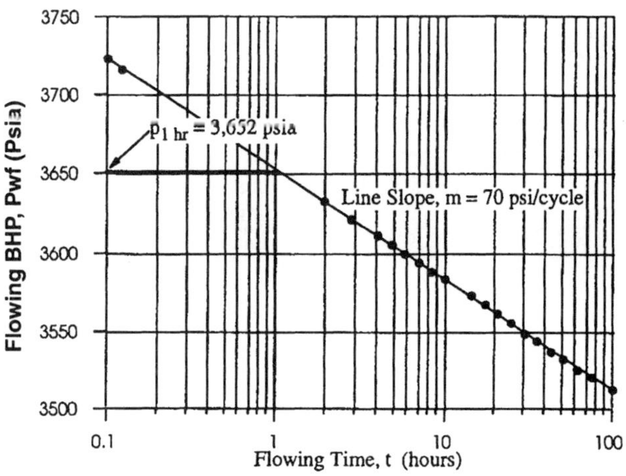

Fig. 2.2—Graphical analysis of time and pressure data from a constant-rate flow test, Example 2.1.

both p_{wf} and q vary with time, the following equation can be used to model variable-rate tests as long as the rate is changing slowly and smoothly.

$$\frac{p_i - p_{wf}}{q} = \frac{162.6\mu B}{kh}\left[\log\left(\frac{kt}{\phi\mu c_t r_w^2}\right) - 3.23 + 0.869s\right]. \qquad (2.7)$$

When compared with the equation of a straight line, $y = mx + b$, the form of Eq. 2.6 suggests that we plot $(p_i - p_{wf})/q$ vs. t on semilog paper and calculate the slope, m', in psi/STB-D-cycle, of the straight line. Gladfelter *et al.*[4] and Ramey[5] earlier suggested the same analysis technique, but for a different application. Effective permeability is calculated by use of the slope of the line from the semilog plot,

$$k = 162.6B\mu/m'h. \qquad (2.8)$$

An equation for the skin factor, s, is obtained by combining Eqs. 2.7 and 2.8,

$$s = 1.151\left[\frac{1}{m'}\left(\frac{p_i - p_{wf}}{q}\right)_{1hr} - \log\left(\frac{k}{\phi\mu c_t r_w^2}\right) + 3.23\right], \qquad (2.9)$$

where $[(p_i - p_{wf})/q]_{1\text{hr}}$ is the value of this quantity on the straight line or its extrapolation at a flowing time of 1 hour.

Note that this analysis technique is applicable only if the producing rate is changing slowly and smoothly. Abrupt rate changes will make the drawdown-test data impossible to interpret accurately with either the constant-rate method discussed earlier or the method Winestock and Colpitts[3] propose. In the next section, we develop an analysis technique for flow tests with discrete rate changes.

Example 2.2—Analysis of a Flow Test With Smoothly Varying Rates. Analyze the drawdown-test data given in **Table 2.2** with those of Winestock and Colpitts'[3] variable-rate method. Known well and formation data are summarized here. Determine the formation permeability and skin factor.

μ = 0.8 cp
h = 69 ft
p_i = 4,412 psia
B = 1.136 RB/STB
ϕ = 0.039
c_t = 17.0×10^{-6} psi^{-1}
r_w = 0.198 ft

TABLE 2.2—VARIABLE-RATE DRAWDOWN-TEST DATA, EXAMPLE 2.2

t (hours)	p_{wf} (psia)	q (STB/D)	t (hours)	p_{wf} (psia)	q (STB/D)
8.32	3,927	147	74.2	3,946	130
9.99	3,928	145	89.1	3,948	129
14.4	3,931	143	107.0	3,950	127
20.7	3,934	140	128.0	3,952	126
29.8	3,937	137	154.0	3,954	125
43.0	3,941	134	185.9	3,956	123
61.8	3,944	132			

TABLE 2.3—PLOTTING FUNCTIONS FOR VARIABLE-RATE DRAWDOWN-TEST ANALYSIS, EXAMPLE 2.2

t (hours)	$(p_i - p_{wf})/q$ (psi/STB-D)	t (hours)	$(p_i - p_{wf})/q$ (psi/STB-D)
8.32	3.30	74.2	3.59
9.99	3.34	89.1	3.60
14.4	3.36	107.0	3.64
20.7	3.41	128.0	3.65
29.8	3.47	154.0	3.66
43.0	3.52	185.9	3.71
61.8	3.55		

Solution.

1. Calculate the pressure plotting functions for each flowing time. **Table 2.3** summarizes these functions, and **Fig. 2.3** shows their plots.

2. Calculate the slope of the line drawn through the data.

$$m' = \frac{\left[\left(p_i - p_{wf2}\right)/q_2\right] - \left[\left(p_i - p_{wf1}\right)/q_1\right]}{\log(t_2) - \log(t_1)}$$

$$= \frac{3.62 \text{ psi/STB/D} - 3.33 \text{ psi/STB/D}}{\log(100) - \log(10)}$$

$$= 0.29 \text{ psi/STB/D-cycle}.$$

3. The permeability is calculated from the slope and with Eq. 2.8,

$$k = \frac{162.6B\mu}{m'h} = \frac{(162.6)(1.136)(0.8)}{(0.29)(69)} = 7.4 \text{ md}.$$

4. The skin factor is estimated by use of Eq. 2.9, where the pressure function evaluated at $t = 1$ hour is 3.0 (Fig. 2.3).

$$s = 1.151\left[\frac{1}{m'}\left(\frac{p_i - p_{wf}}{q}\right)_{1\text{hr}} - \log\left(\frac{k}{\phi\mu c_t r_w^2}\right) + 3.23\right]$$

$$= 1.151\left\{\left(\frac{1}{0.29}\right)(3.0)\right.$$

$$\left. - \log\left[\frac{7.4}{(0.039)(0.8)(17 \times 10^{-6})(0.198)^2}\right] + 3.23\right\}$$

$$= 5.8.$$

Again, a positive skin factor indicates a reduction in permeability near the wellbore.

2.2.3 Flow Tests With Discrete Rate Changes. Consider a well with n rate changes during its production history (**Fig. 2.4**). Our objective is to determine the wellbore pressure of a well producing with this schedule. We use superposition in time (see Sec. 1.7) of the

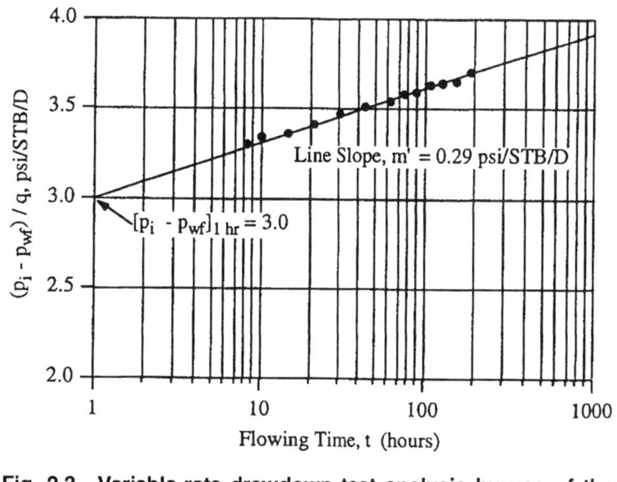

Fig. 2.3—Variable-rate drawdown test analysis by use of the Winestock and Colpitts[3] method, Example 2.2.

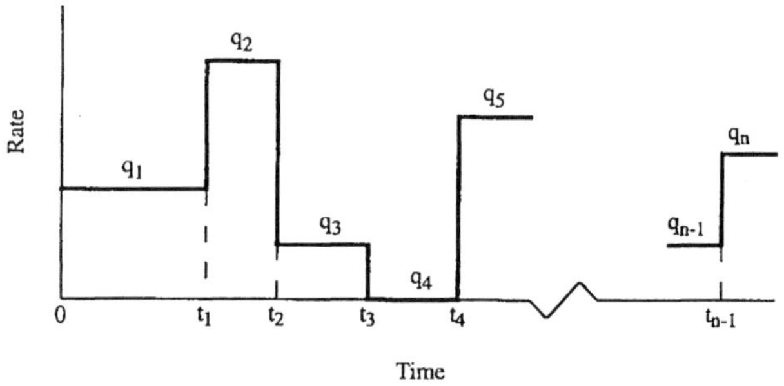

Fig. 2.4—Rate history for a multirate test.

logarithmic approximation to the Ei-function solution. To simplify the algebra, we write the solution given by Eq. 2.2 in the form

$$p_i - p_{wf} = m'q[\log(t) + \bar{s}] \;, \quad \text{(2.10a)}$$

$$\text{where } m' = 162.6B\mu/kh \quad \text{(2.10b)}$$

$$\text{and } \bar{s} = \log\left(k/\phi\mu c_t r_w^2\right) - 3.23 + 0.869s. \quad \text{(2.10c)}$$

With this nomenclature for n rates and for $t > t_{n-1}$, superposition in time yields

$$\begin{aligned} p_i - p_{wf} &= m'q_1[\log(t) + \bar{s}] + m'(q_2 - q_1)[\log(t - t_1) + \bar{s}] \\ &\quad + m'(q_3 - q_2)[\log(t - t_2) + s] + \ldots \\ &\quad \ldots + m'(q_n - q_{n-1})[\log(t - t_{n-1}) + \bar{s}], \end{aligned} \quad \text{(2.11)}$$

which can be written more conveniently as

$$\frac{p_i - p_{wf}}{q_n} = m' \sum_{j=1}^{n} \left[\frac{(q_j - q_{j-1})}{q_n} \log(t - t_{j-1}) \right] + m'\bar{s}, \quad \text{(2.12a)}$$

$$\text{for } q_n \neq 0, \quad \text{(2.12b)}$$

where we define $q_o = 0$ and $t_o = 0$. In terms of more fundamental quantities, Eq. 2.12 can be written as

$$\begin{aligned} \frac{p_i - p_{wf}}{q_n} &= m' \sum_{j=1}^{n} \left[\frac{(q_j - q_{j-1})}{q_n} \log(t - t_{j-1}) \right] \\ &\quad + m'\left[\log(k/\phi\mu c_t r_w^2) - 3.23 + 0.869s\right]. \end{aligned} \quad \text{(2.13)}$$

Eq. 2.13 can be used to model several special cases of practical importance. The form of Eq. 2.13 before dividing by q_n can be used to model pressure-buildup tests that we discuss in Sec. 2.3. We should emphasize, however, that Eq. 2.13 was developed assuming an infinite-acting reservoir for the total time elapsed since the well began producing. Consequently, Eq. 2.13 is not valid when reservoir boundaries affect the pressure response at any time during the producing period.

Two-Rate Flow Tests. For the production history shown in **Fig. 2.5**, a two-rate test can be modeled by[6]

$$\begin{aligned} p_i - p_{wf} &= \frac{162.6q_2B\mu}{kh} \left[\left(\frac{q_1}{q_2}\right)\log(t) + \left(\frac{q_2 - q_1}{q_2}\right)\log(t - t_1) \right. \\ &\quad \left. + \log(k/\phi\mu c_t r_w^2) - 3.23 + 0.869s \right]. \end{aligned} \quad \text{(2.14)}$$

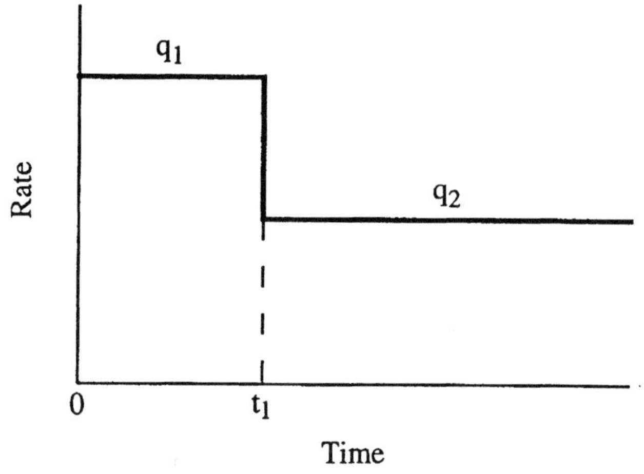

Fig. 2.5—Rate history for a two-rate flow test.

If we rearrange and introduce a special nomenclature, $t_1 = t_{p1}$ and $t - t_{p1} = \Delta t'$, then Eq. 2.14 becomes

$$\begin{aligned} p_{wf} &= p_i - \frac{162.6q_2B\mu}{kh} \left[\log\left(\frac{k}{\phi\mu c_t r_w^2}\right) - 3.23 + 0.869s \right] \\ &\quad - \frac{162.6q_1B\mu}{kh} \left[\log\left(\frac{t_{p1} + \Delta t'}{\Delta t'}\right) + \left(\frac{q_2}{q_1}\right)\log(\Delta t') \right]. \end{aligned} \quad \text{(2.15)}$$

The two-rate test can be used when estimates of permeability, skin factor, or reservoir pressure are needed and economic considerations preclude the well from being shut in (i.e., pressure-buildup tests discussed in Sec. 2.3). Note that, for the two-rate test to be valid, the second rate must be kept strictly constant or the test interpretation may be substantially in error. In addition, Eq. 2.15 is correct only when the reservoir is infinite-acting for the elapsed time $t_{p1} + \Delta t'$.

We recommend the following method of analysis for two-rate flow tests.

1. Plot flowing BHP vs. the time-plotting function; i.e.,

$$p_{wf} \text{ vs.} \left\{ \log\left[(t_{p1} + \Delta t')/\Delta t'\right] + \left(\frac{q_2}{q_1}\right)\log(\Delta t') \right\}, \quad \text{(2.16)}$$

on Cartesian coordinate paper.

2. Determine the slope, m, from the straight line on the plot and use it to calculate permeability, k, from

$$k = 162.6q_1B\mu/mh. \quad \text{(2.17)}$$

3. Calculate the skin factor, s, from

$$s = 1.151\left[\frac{q_1}{(q_1 - q_2)}\left(\frac{p_{1hr} - p_{wf1}}{m}\right) - \log\left(\frac{k}{\phi\mu c_t r_w^2}\right) + 3.23 \right], \quad \text{(2.18)}$$

where p_{1hr} is the flowing pressure at $\Delta t' = 1$ hour on the straight line or its extrapolation, and p_{wf1} is the flowing pressure at the time the rate is changed (i.e., $\Delta t' = 0$). Eq. 2.18 was derived by simultaneous solution of Eqs. 2.15, 2.17, and the drawdown equation for a single rate applied at $t = t_{p1}$, at which time $p_{wf} = p_{wf1}$.

4. The initial reservoir pressure, p_i (or, more generally, p^* in a reservoir with pressure depletion), is obtained by solving for p_i (or p^*) from the drawdown equation written to model conditions at the time of the rate change. (It is implied that s and m are known at this point.)

TABLE 2.4—MULTIRATE FLOW-TEST DATA, EXAMPLE 2.3

t (hours)	p_{wf} (psia)
0	3,000
0.33	999
0.67	857
1.0	778.5
2.0	1,378.5
2.33	2,043
2.67	2,067.5
3.0	2,094

$$p_i = p_{wf1} + m\left[\log\left(kt_{p1}/\phi\mu c_t r_w^2\right) - 3.23 + 0.869s\right]. \quad (2.19)$$

***n*-Rate Flow Tests.** The method presented for analyzing two-rate drawdown tests can be extended to include n different rates. Beginning with Eq. 2.13, an n-rate flow test is modeled by

$$\left(p_i - p_{wf}\right)/q_n = m'\sum_{j=1}^{n}\left\{\left[\left(q_j - q_{j-1}\right)/q_n\right]\log\left(t_n - t_{j-1}\right)\right\}$$

$$+ m'\left[\log\left(k/\phi\mu c_t r_w^2\right) - 3.23 + 0.869s\right]. \quad (2.20)$$

The form of Eq. 2.20 suggests that we prepare a plot on Cartesian coordinate paper of

$$\frac{p_i - p_{wf}}{q_n} \text{ vs. } \sum_{j=1}^{n}\frac{\left(q_j - q_{j-1}\right)}{q_n}\log\left(t_n - t_{j-1}\right), \quad (2.21)$$

where permeability, k, is related to the slope, m', of a straight line on the plot,

$$k = 162.6B\mu/m'h. \quad (2.22)$$

If we let b' equal the value of $(p_i - p_{wf})/q_n$ when the plotting function is zero, then the skin factor is determined from

$$s = 1.151\left[\left(b'/m'\right) - \log\left(k/\phi\mu c_t r_w^2\right) + 3.23\right]. \quad (2.23)$$

Note that the use of Eq. 2.23 and the proposed plotting method require an estimate of p_i from independent measurements. Odeh and Jones[7] discussed this analysis technique and indicated that it can be applied to the analysis of multirate flow tests commonly run on both gas and oil wells, but only when the reservoir is infinite-acting for the total elapsed production period. Consequently, the method is not applicable if either outer reservoir boundaries or interference from adjacent producing wells affect the data during the test. Example 2.3 illustrates analysis of a multirate test with discrete rate changes.

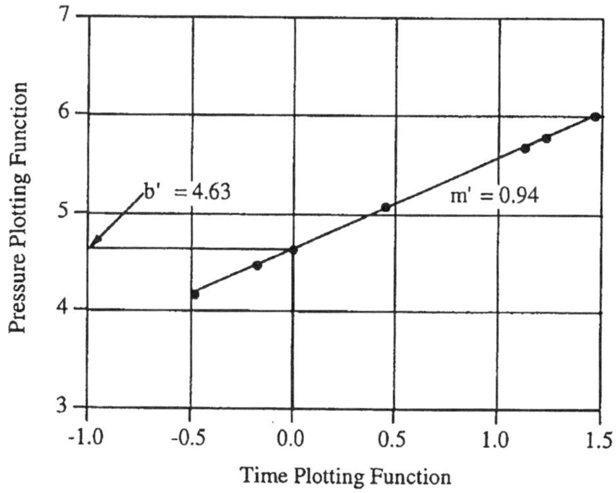

Fig. 2.6—Graphical analysis of a multirate flow test, Example 2.3.

Example 2.3—Analysis of a Multirate Flow Test. A 3-hour drawdown test[7] was conducted in which the rate during the first hour averaged 478.5 STB/D; during the second hour, 319 STB/D; and during the third hour, 159.5 STB/D. **Table 2.4** gives the flowing BHP's measured during the test. Estimate the permeability and skin factor with the multirate-flow-test analysis technique.

μ = 0.6 cp
h = 10 ft
p_i = 3,000 psia
B = 1.2 RB/STB
ϕ = 0.12
c_t = 48×10^{-6} psi^{-1}
r_w = 0.25 ft

Solution.

1. Calculate the plotting functions suggested in the analysis procedure. For a multirate flow test, we calculate and plot

$$\frac{p_i - p_{wf}}{q_n} \text{ vs. } \sum_{j=1}^{n}\frac{\left(q_j - q_{j-1}\right)}{q_n}\log\left(t_n - t_{j-1}\right).$$

Table 2.5 summarizes these functions, and **Fig. 2.6** plots them.

As an example, for $t = 0.33$ hours, the data used in plotting functions are $q_n = q_1 = 478.5$ STB/D and $t_n = 0.333$ hours. This gives

$$\left(p_i - p_{wf}\right)/q_n = (3{,}000 - 999)/478.5 = 4.18$$

TABLE 2.5—PLOTTING FUNCTIONS FOR MULTIRATE-FLOW-TEST ANALYSIS, EXAMPLE 2.3

t (hours)	q_n (STB/D)	p_{wf} (psi)	$(p_i - p_{wf})/q_n$ (psi/STB-D)	$\sum_{j=1}^{n}\frac{\left(q_j - q_{j-1}\right)}{q_n}\log\left(t_n - t_{j-1}\right)$
0	—	3,000	—	—
0.33	478.5	999	4.18	–0.48
0.67	478.5	857	4.48	–0.18
1.0	478.5	778.5	4.64	0.0
2.0	319.0	1,378.5	5.08	0.45
2.33	159.5	2,043	6.00	1.46
2.67	159.5	2,067.5	5.79	1.23
3.0	159.5	2,094	5.68	1.13

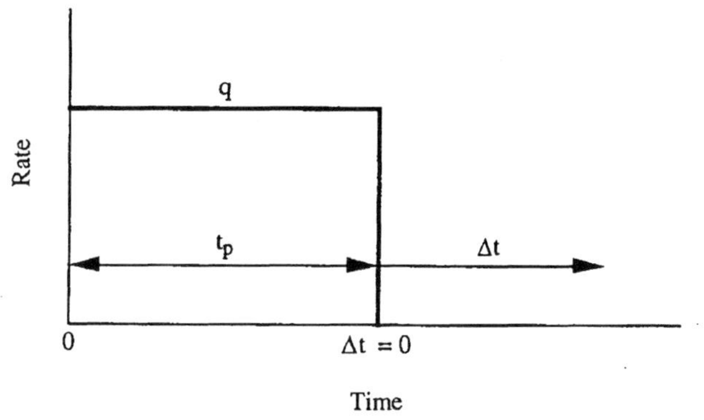

Fig. 2.7—Modeling a pressure-buildup test in terms of variable-rate production.

$$\text{and} \sum_{j=1}^{n} \frac{(q_j - q_{j-1})}{q_n} \log(t_n - t_{j-1}) = \left(\frac{478.5 - 0}{478.5}\right)$$

$$\times \log(0.333 - 0) = -0.48.$$

Similarly, for $t = 3.0$ hours,

$$(p_i - p_{wf})/q_n = (3{,}000 - 2{,}094)/159.5 = 5.68$$

$$\text{and} \sum_{j=1}^{n} \frac{(q_j - q_{j-1})}{q_n} \log(t_n - t_{j-1}) = \frac{1}{159.5}$$

$$\times \begin{bmatrix} 478.5 \log(3.0 - 0) \\ + (319 - 478.5)\log(3.0 - 1.0) \\ + (159.5 - 319)\log(3.0 - 2.0) \end{bmatrix}$$

$$= 1.13.$$

2. The slope of the line drawn through the data points in Fig. 2.6 is

$$m' = (6.0 - 4.2)/[1.46 - (-0.45)] = 0.94.$$

3. The permeability is calculated with the slope of the line.

$$k = \frac{162.6B\mu}{m'h} = \frac{(162.6)(1.2)(0.6)}{(0.94)(10)} = 12.5 \text{ md.}$$

4. The skin factor is estimated with Eq. 2.23, where $b' = 4.63$ (Fig. 2.6).

$$s = 1.151\left[(b'/m') - \log(k/\phi\mu c_t r_w^2) + 3.23\right]$$

$$- 1.151\left\{\frac{4.63}{0.94} - \log\left[\frac{12.5}{(0.12)(0.6)(48.0 \times 10^{-6})(0.25)^2}\right]\right.$$

$$\left. + 3.23\right\} = 0.45.$$

The small positive skin factor suggests very little, if any, reduction in permeability in the formation adjacent to the wellbore.

2.3 Analysis of Pressure-Buildup Tests

Pressure-buildup tests are conducted by first stabilizing a producing well at some fixed rate, placing a BHP measuring device in the well, and shutting in the well. Following shut-in, the BHP builds up as a function of time, and the rate of pressure buildup is used to estimate well and formation properties, such as average drainage area pressure, permeability in the drainage area of the well, and skin factor in the region immediately adjacent to the wellbore. In this section, we discuss analysis techniques for pressure-buildup tests. We begin with analysis techniques for tests preceded by a constant-rate production period, but also discuss the more probable testing scenario—pressure-buildup testing with variable-rate production before shut-in.

2.3.1 Buildup Tests With Constant-Rate Production Before Shut-In. An equation modeling a pressure buildup test can be developed by use of superposition in time. **Fig. 2.7** shows the appropriate rate history. At time $t = 0$, the well begins producing at a constant rate q. Some time t_p later, the well is shut in. This can be modeled with two constant-rate terms, one of rate q beginning at time $t = 0$, the second of rate $-q$ beginning at time t_p. The contribution to the pressure change at time $t_p + \Delta t$ owing to production at rate q beginning at time $t = 0$ is given by

$$\Delta p_1 = \frac{162.6qB\mu}{kh}\left\{\log\left[k\frac{(t_p + \Delta t)}{\phi\mu c_t r_w^2}\right] - 3.23 + 0.869s\right\}, \quad (2.24)$$

while the contribution to the pressure change at time $t_p + \Delta t$ owing to production at rate $-q$ beginning at time $t = t_p$ is given by

$$\Delta p_2 = \frac{162.6(-q)B\mu}{kh}\left\{\log\left[\frac{k\Delta t}{\phi\mu c_t r_w^2}\right] - 3.23 + 0.869s\right\}. \quad (2.25)$$

Thus, the BHP for the rate history shown in Fig. 2.7 is

$$p_i - p_{ws} = \frac{162.6qB\mu}{kh}\left\{\log\left[\frac{k(t_p + \Delta t)}{\phi\mu c_t r_w^2}\right] - 3.23 + 0.869s\right\}$$

$$+ \frac{162.6(-q)B\mu}{kh}\left[\log(k\Delta t/\phi\mu c_t r_w^2) - 3.23 + 0.869s\right], \quad (2.26)$$

where p_{ws} = bottomhole shut-in pressure, t_p = duration of the constant-rate production period before shut-in, and Δt = duration of the shut-in period. If we combine terms and simplify, Eq. 2.26 can be rewritten as

$$p_{ws} = p_i - (162.6qB\mu/kh)\left\{\log\left[(t_p + \Delta t)/\Delta t\right]\right\}. \quad (2.27)$$

Comparing Eq. 2.27 with the equation of a straight line, $y = mx + b$,

$$y \sim p_{ws}, \quad (2.28a)$$

$$b \sim p_i, \quad (2.28b)$$

$$m \sim 162.6qB\mu/kh, \quad (2.28c)$$

$$x \sim \log\left[(t_p + \Delta t)/\Delta t\right]. \quad (2.28d)$$

This suggests that a plot of bottomhole shut-in pressure, p_{ws}, from a buildup test as a function of log $[(t_p + \Delta t)/\Delta t]$, called the Horner[8] time ratio, will exhibit a straight line with slope m, given by

$$m = -(162.6qB\mu)/kh. \quad (2.29).$$

The slope is the difference between two values of pressure, p_{ws1} and p_{ws2}, one log cycle apart. To calculate effective permeability to the fluid flowing in the drainage area of the well, we use the absolute value of the slope, or

$$k = 162.6qB\mu/mh. \quad (2.30)$$

From the semilog graph, the original reservoir pressure, p_i, is estimated by extrapolating the straight line to infinite shut-in time where $(t_p + \Delta t)/\Delta t = 1$ and log $(t_p + \Delta t)/\Delta t = 0$. **Fig. 2.8** illustrates calculation of the slope and original reservoir pressure.

We can also solve for the skin factor, s, from a pressure-buildup test. At the instant a well is shut in, the flowing BHP is

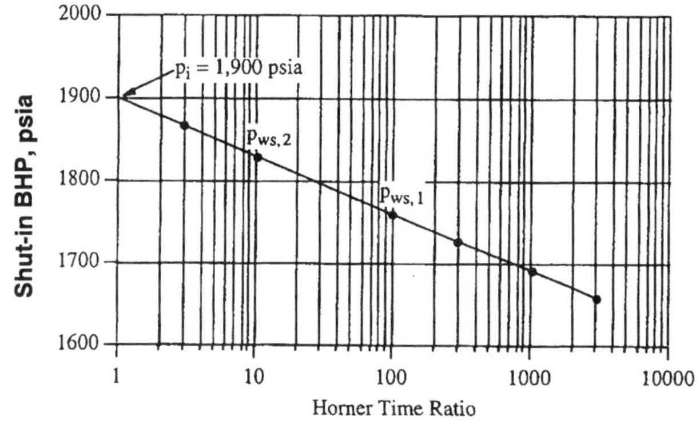

Fig. 2.8—Graphical analysis technique for pressure-buildup-test data.

$$p_{wf} = p_i - \frac{162.6 q_w B\mu}{kh}\left[\log\left(kt_p/\phi\mu c_t r_w^2\right) - 3.23 + 0.869s\right]. \quad (2.31)$$

If we combine Eqs. 2.27 and 2.31, we can derive an expression for the skin factor.

$$s = 1.151\left\{\left[\left(p_{ws} - p_{wf}\right)/m\right] - \log\left(k\Delta t/\phi\mu c_t r_w^2\right) + 3.23 + \log\left[\left(t_p + \Delta t\right)/t_p\right]\right\}, \quad (2.32)$$

where m is the slope of the semilog straight line. Setting shut-in time, Δt, equal to 1 hour, introducing the symbol $p_{1\text{hr}}$ for p_{ws} at $\Delta t = 1$ hour on the semilog straight line, and neglecting the term log $[(t_p + 1)/t_p]$, Eq. 2.32 becomes

$$s = 1.151\left\{\left[\left(p_{1hr} - p_{wf}\right)/m\right] - \log\left(k/\phi\mu c_t r_w^2\right) + 3.23\right\}, \quad (2.33)$$

where p_{wf} = flowing BHP at the instant of shut-in.

In summary, if we plot p_{ws} vs. log $(t_p + \Delta t)/\Delta t$ with information obtained from a pressure-buildup test, we can estimate effective permeability, k; original reservoir pressure, p_i; and the skin factor, s. Example 2.4 illustrates the application of this analysis technique.

Example 2.4—Determining Permeability and Reservoir Pressure From Buildup Tests. A pressure-buildup test (**Table 2.6**) was conducted on a well early in the life of an oil reservoir having the properties summarized next. The well was produced at a constant rate of 500 STB/D for 3 days before being shut in. Determine the effective permeability to oil, the original reservoir pressure, and the skin factor.

μ = 1.0 cp
h = 22 ft
q = 500 STB/D
c_t = 20×10^{-6} psi^{-1}
ϕ = 0.20
t_p = 3 days
r_w = 0.30 ft
B = 1.3 RB/STB
p_{wf} = 1,150 psia

Solution.

1. Construct a semilog plot of shut-in pressure, p_{ws}, as a function of the Horner time ratio, $(t_p + \Delta t)/\Delta t$. **Table 2.7** gives the calculated Horner time ratios, while **Fig. 2.9** shows the semilog plot. The absolute value of the slope of the semilog straight line in Fig. 2.9 is

TABLE 2.6—PRESSURE-BUILDUP-TEST DATA, EXAMPLE 2.4

Shut-In Time, Δt (hours)	BHP, p_{ws} (psia)
0	1,150
2	1,794
4	1,823
8	1,850
16	1,876
24	1,890
48	1,910

TABLE 2.7—HORNER TIME PLOTTING FUNCTIONS, EXAMPLE 2.4

BHP, p_{ws} (psia)	Horner Time Ratio, $(t_p + \Delta t)/\Delta t$
1,150	—
1,794	37.0
1,823	19.0
1,850	10.0
1,876	5.5
1,890	4.0
1,910	2.5

$$m = \left|\frac{p_{ws2} - p_{ws1}}{\log\left(\frac{t_p + \Delta t_2}{\Delta t_2}\right) - \log\left(\frac{t_p + \Delta t_1}{\Delta t_1}\right)}\right|$$

$$= \left|\frac{1,850 - 1,950}{\log(10) - \log(1)}\right|$$

$$= 100 \text{ psi/cycle}.$$

2. Calculate the effective permeability to oil using the slope of the straight line.

$$k = \frac{162.6 qB\mu}{mh} = \frac{(162.6)(500)(1.3)(1.0)}{(100)(22)} = 48 \text{ md}.$$

3. The original reservoir pressure is found by extrapolating the semilog straight line in Fig. 2.9 to infinite shut-in time; i.e., $(t_p + \Delta t)/\Delta t = 1$. The pressure at this point is p_i = 1,950 psia. At $\Delta t = 1$ hour, the Horner time ratio is $(t_p + \Delta t)/\Delta t = (72 + 1)/1 = 73$. From the extrapolated semilog straight line, p_{1hr} = 1,764 psia at this value.

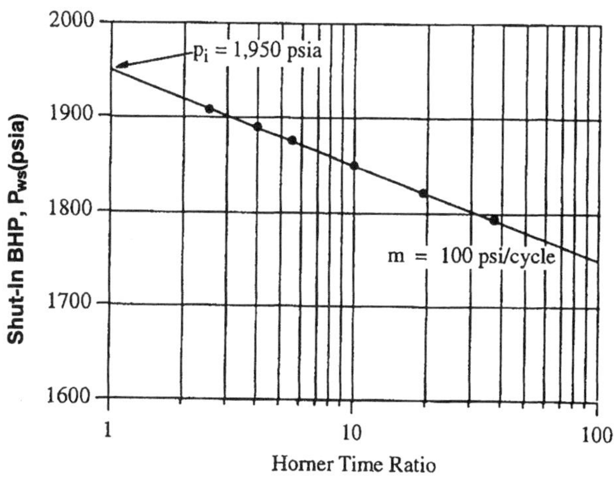

Fig. 2.9—Horner semilog plot, Example 2.4

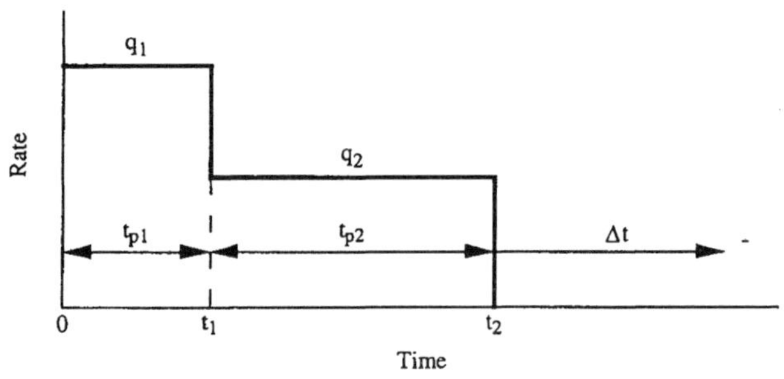

Fig. 2.10—Rate history for buildup test following two different flow rates.

4. The skin factor is

$$s = 1.151\left\{\left[\left(p_{1\text{hr}} - p_{wf}\right)/m\right] - \log\left(k/\phi\mu c_t r_w^2\right) + 3.23\right\}$$

$$= 1.151\left\{[(1,764 - 1,150)/100]\right.$$

$$\left. - \log\left[\frac{48}{(0.20)(1.0)\left(2.0 \times 10^{-6}(0.30)^2\right)}\right] + 3.23\right\}$$

$$= 1.43$$

2.3.2 Buildup Tests Preceded by Two Different Flow Rates. Beginning with the logarithmic approximation to the Ei-function solution derived for multirate flow tests,

$$\frac{p_i - p_{wf}}{q_n} = m'\sum_{j=1}^{n}\left[\frac{\left(q_j - q_{j-1}\right)}{q_n}\log\left(t - t_{j-1}\right)\right]$$

$$+ m'\left[\log\left(k/\phi\mu c_t r_w^2\right) - 3.23 + 0.869s\right], \quad \text{.... (2.13)}$$

we can derive an analysis technique for pressure-buildup tests preceded by two different flow rates. For the special case $q_n = 0$, Eq. 2.13 can be rewritten as

$$p_i - p_{ws} = m'\sum_{j=1}^{n}\left(q_j - q_{j-1}\right)\log\left(t - t_{j-1}\right), \quad \text{........ (2.34a)}$$

where $m' = 162.6\mu B/kh$. .... (2.34b)

In terms of the rate history shown in **Fig. 2.10,** Eq. 2.34a becomes

$$p_i - p_{ws} = m'[q_1\log(t) + (q_2 - q_1)\log(t - t_1) - q_2\log(t - t_2)] \quad \text{.................... (2.35a)}$$

$$\text{or } p_i - p_{ws} = \frac{162.6q_2B\mu}{kh}\left[\frac{q_1}{q_2}\log\left(\frac{t}{t - t_1}\right) + \log\left(\frac{t - t_1}{t - t_2}\right)\right]. \quad \text{.................. (2.35b)}$$

Let $t - t_2 = \Delta t$, $t_1 = t_{p1}$, $t_2 = t_{p1} + t_{p2}$, and $t - t_1 = t_{p2} + \Delta t$. Then,

$$p_i - p_{ws} = \frac{162.6q_2B\mu}{kh}\left\{\left(\frac{q_1}{q_2}\right)\log\left(\frac{t_{p1} + t_{p2} + \Delta t}{t_{p2} + \Delta t}\right)\right.$$

$$\left. + \log\left[\left(t_{p2} + \Delta t\right)/\Delta t\right]\right\}. \quad \text{............... (2.36)}$$

Eq. 2.36 is valid when the producing rate is changed a short time before a buildup test begins, which allows insufficient time for Horner's approximation to be valid. Under these conditions, we frequently can consider all production before time t_1 to have been at rate q_1 for time t_{p1} and production just before the test to have been a rate q_2 for time t_{p2}.

To analyze a pressure-buildup test following two different flow rates, we plot

$$p_{ws} \text{ vs.}\left[\left(\frac{q_1}{q_2}\right)\log\left(\frac{t_{p1} + t_{p2} + \Delta t}{t_{p2} + \Delta t}\right) + \log\left(\frac{t_{p2} + \Delta t}{\Delta t}\right)\right] \quad \text{.................... (2.37)}$$

on Cartesian coordinate paper. The slope, m, of the straight line on this plot is related to effective permeability by

$$k = 162.6q_2B\mu/mh. \quad \text{........................ (2.38)}$$

Extrapolation of the plot to $\Delta t = \infty$ gives $p_{ws} = p_i$ because the plotting function is zero at $\Delta t = \infty$. Note that semilog paper is not to be used; instead, the sum of two logarithms is plotted on an ordinary Cartesian coordinate paper.

To calculate skin factor, s, note that, at the end of the flow period just before shut-in,

$$p_i - p_{wf} = \frac{162.6q_2B\mu}{kh}\left[\left(\frac{q_1}{q_2}\right)\log\left(\frac{t_{p1} + t_{p2}}{t_{p2}}\right) + \log\left(t_{p2}\right)\right.$$

$$\left. + \log\left(k/\phi\mu c_t r_w^2\right) - 3.23 + 0.869s\right]. \quad \text{.... (2.39)}$$

The equation of the straight line on the buildup-test plot is

$$p_i - p_{ws} = \frac{162.6q_2B\mu}{kh}\left\{\left(\frac{q_1}{q_2}\right)\log\left(\frac{t_{p1} + t_{p2} + \Delta t}{t_{p2} + \Delta t}\right)\right.$$

$$\left. + \log\left[\left(t_{p2} + \Delta t\right)/\Delta t\right]\right\}. \quad \text{............... (2.40)}$$

Subtracting Eq. 2.40 from Eq. 2.39 yields

$$p_{ws} - p_{wf} = m\left\{\frac{q_1}{q_2}\log\left[\frac{\left(t_{p1} + t_{p2}\right)\left(t_{p2} + \Delta t\right)}{\left(t_{p1} + t_{p2} + \Delta t\right)\left(t_{p2}\right)}\right]\right.$$

$$\left. + \log\left[\frac{\left(t_{p2}\right)(\Delta t)}{t_{p2} + \Delta t}\right] + \bar{s}\right\}, \quad \text{.............. (2.41a)}$$

where $m = 162.6q_2B\mu/kh$ .... (2.41b)

and $\bar{s} = \log\left(k/\phi\mu c_t r_w^2\right) - 3.23 + 0.869s$. .... (2.41c)

Assuming $t_{p1} + t_{p2} + \Delta t \approx t_{p1} + t_{p2}$ and $t_{p2} + \Delta t \approx t_{p2}$ for small Δt (e.g., $\Delta t = 1$ hour), Eq. 2.41a becomes

$$p_{ws} - p_{wf} = m[\log(\Delta t) + \bar{s}]. \quad \text{.................. (2.42)}$$

If we choose $\Delta t = 1$ hour, $p_{ws} = p_{1\text{hr}}$ (on the straight line or its extrapolation) and

$$\bar{s} = \frac{p_{1\text{hr}} - p_{wf}}{m} = \log\left(\frac{k}{\phi\mu c_t r_w^2}\right) - 3.23 + 0.869s \quad \text{.................... (2.43a)}$$

$$\text{or } s = 1.151\left[\left(\frac{p_{1hr} - p_{wf}}{m}\right) - \log\left(\frac{k}{\phi\mu c_t r_w^2}\right) + 3.23\right] \quad (2.43b)$$

for $t_{p2} \gg 1$.

2.3.3 Buildup Tests Preceded by $(n-1)$ Different Flow Rates. ***Superposition Method.*** We can develop a similar analysis technique by use of superposition in time for $(n-1)$ rates preceding the pressure-buildup test. Beginning with Eq. 2.13 for the general case where $q_n = 0$ and for $(n-1)$ different rates before shut-in, we have

$$p_i - p_{ws} = \frac{162.6q_{n-1}\mu B}{kh}\left[\left(\frac{q_1}{q_{n-1}}\right)\log\left(\frac{t}{t-t_1}\right) + \left(\frac{q_2}{q_{n-1}}\right)\right.$$

$$\left.\times \log\left(\frac{t-t_1}{t-t_2}\right) + \left(\frac{q_{n-2}}{q_{n-1}}\right)\log\left(\frac{t-t_{n-3}}{t-t_{n-2}}\right) + \log\left(\frac{t-t_{n-2}}{t-t_{n-1}}\right)\right]. \quad (2.44)$$

Although we introduce no special nomenclature for this situation, note that $t - t_{n-1} = \Delta t$ (time elapsed since shut-in) and that q_{n-1} is the production rate just before shut-in.

Applications of Eq. 2.44 in which more than three terms are needed are probably rare; sometimes, though, all significant rates are considered to satisfy precise legal contracts (e.g., gas-deliverability contracts). Improvement in accuracy with this approach is questionable; further, the fundamental assumption on which Eq. 2.44 is based (that for $t = t_{p1} + t_{p2} + \ldots + t_{p,n-1} + \Delta t$, the reservoir is infinite-acting) rarely will be valid for large values of t. Nevertheless, we recommend the following analysis procedure when Eq. 2.44 is used to model a buildup test.

1. Calculate the plotting function.

$$X = \left[\left(\frac{q_1}{q_{n-1}}\right)\log\left(\frac{t}{t-t_1}\right) + \ldots + \log\left(\frac{t-t_{n-2}}{t-t_{n-1}}\right)\right]. \quad (2.45)$$

2. Plot bottomhole shut-in pressure, p_{ws}, vs. the plotting function, X, on Cartesian coordinate graph paper.

3. Determine the absolute value of the slope, m, of the straight line on the plot.

$$m = \left|\frac{p_{ws2} - p_{ws1}}{X_2 - X_1}\right| \quad (2.46).$$

4. Calculate effective permeability with the slope from Step 3.

$$k = 162.6q_{n-1}B\mu/mh \quad (2.47)$$

5. Calculate the skin factor, s,

$$s = 1.151\left[\frac{p_{1hr} - p_{wf}}{m} - \log\left(\frac{k}{\phi\mu c_t r_w^2}\right) + 3.23\right]. \quad (2.48)$$

Although not shown, the derivation and assumptions implicit in this equation closely parallel those used for a buildup test preceded by two different flow rates.

6. The initial formation pressure, p_i, is the value of p_{ws} on the straight line extrapolated to the time-plotting function evaluated at zero; i.e., $X = 0$.

Odeh and Selig Method.[9] As an alternative method to superposition, Odeh and Selig suggested that a buildup test following n different rates could be analyzed by a method similar to the Horner method. The shut-in pressure response is given by

$$p_i - p_{ws} = \frac{162.6q^* B\mu}{kh}\log\left(\frac{t_p^* + \Delta t}{\Delta t}\right), \quad (2.49)$$

where the modified production time, t_p^*, and flow rate, q^*, are defined as

$$t_p^* = 2\left[t_n - \frac{\sum_{j=1}^{n} q_j\left(t_j^2 - t_{j-1}^2\right)}{2\sum_{j=1}^{n} q_j\left(t_j - t_{j-1}\right)}\right] \quad (2.50)$$

$$\text{and } q^* = \frac{1}{t_p^*}\sum_{j=1}^{n} q_j\left(t_j - t_{j-1}\right). \quad (2.51)$$

The Odeh-Selig method, approximate but accurate, is applicable only for pressures at values of Δt greater than actual producing time. This condition is likely to occur only in a drillstem test or short production test.

Horner's Approximation. As we discussed in Chap. 1, Horner[8] reported an approximation that can be used in many cases to avoid the use of superposition in modeling the production history of a variable-rate well. He defined a pseudoproducing time, t_{pH}, as

$$t_{pH} = \frac{N_p}{q_{last}}. \quad (2.52)$$

For example, for production from an oil well and for t_{pH} in hours, the cumulative production from the well is N_p (in STB), and the most recent production rate is q_{last} (in STB/D). Eq. 2.52 then becomes

$$t_{pH} = 24N_p/q_{last}. \quad (2.53)$$

Horner proposed to model the effect of the entire rate history by use of Eq. 2.31, where q is replaced with q_{last} and actual producing time, t, is replaced by Horner's pseudoproducing time, t_{pH}.

$$p_{wf} = p_i - \frac{162.6q_{last}B\mu}{kh}\left[\log\left(\frac{kt_{pH}}{\phi\mu c_t r_w^2}\right) - 3.23 + 0.869s\right]. \quad (2.54)$$

An equation modeling a pressure-buildup test can be written by noting that pressure buildup is a special case of variable-rate production. Assuming Horner's[8] approximation adequately models the production history before shut-in, the entire production history can be modeled as production at rate q_{last} for time t_{pH}. If the term Δt denotes time elapsed since shut-in, then superposition in time by use of Eq. 2.2 yields the following equation describing BHP, p_{ws}, after shut-in:

$$p_{ws} = p_i - \frac{162.6q_{last}B\mu}{kh}\left\{\log\left[\frac{k(t_{pH} + \Delta t)}{\phi\mu c_t r_w^2}\right] - 3.23 + 0.869s\right\}$$

$$- \frac{162.6(-q_{last})B\mu}{kh}\left[\log\left(\frac{k\Delta t}{\phi\mu c_t r_w^2}\right) - 3.23 + 0.869s\right]. \quad (2.55)$$

If we combine terms and simplify, Eq. 2.55 can be rewritten as

$$p_{ws} = p_i - \frac{162.6q_{last}B\mu}{kh}\left\{\log\left[\frac{(t_{pH} + \Delta t)}{\Delta t}\right]\right\}. \quad (2.56)$$

Similar to the analysis technique for pressure-buildup tests preceded by a constant-rate production period, we simply plot bottomhole shut-in pressure, p_{ws}, as a function of the log of the Horner[8] time ratio based on the pseudoproducing time, $(t_{pH} + \Delta t)/\Delta t$. To calculate permeability, we use the absolute value of the slope, m, of the semilog straight line,

$$k = 162.6q_{last}B\mu/mh. \quad (2.57)$$

TABLE 2.8—RATE SCHEDULE, EXAMPLE 2.5

Rate (STB/D)	Duration (days)
200	1.5
0	0.5
100	2.0
125	6.0

TABLE 2.9—PRESSURE-BUILDUP-TEST DATA,EXAMPLE 2.5

Shut-in Time, Δt (hours)	Bottomhole Pressure, p_{ws} (psia)
0	1,384
2	1,530
3	1,535
4	1,538
5	1,540
8	1,546
10	1,549
12	1,551
19	1,556
24	1,559
36	1,563

TABLE 2.10—HORNER TIME PLOTTING FUNCTIONS, EXAMPLE 2.5

Shut-In Time, Δt (hours)	BHP, p_{ws} (psia)	Horner Time Ratio
0	1384	—
2	1530	121.0
3	1535	81.0
4	1538	61.0
5	1540	49.0
8	1546	31.0
10	1549	25.0
12	1551	21.0
19	1556	13.6
24	1559	11.0
36	1563	7.67

From the semilog graph, the original reservoir pressure, p_i, is estimated by extrapolating the straight line to infinite shut-in time where $(t_{pH}+\Delta t)/\Delta t=1$ and $\log(t_{pH}+\Delta t)/\Delta t=0$.

The skin factor is estimated from

$$s = 1.151\left[\frac{(p_{1hr}-p_{wf})}{m} - \log\left(\frac{k}{\phi\mu c_t r_w^2}\right) + 3.23\right], \quad (2.58)$$

where m = slope of the semilog straight line, $p_{1\text{hr}}$ equals p_{ws} at $\Delta t=1$ hour on the semilog straight line, and p_{wf} = flowing BHP at the instant of shut-in.

Example 2.5—Determining Permeability and Reservoir Pressure From Buildup Tests. A pressure-buildup test was conducted on a well early in the life of an oil reservoir having the rock and fluid properties summarized next. **Table 2.8** gives rate data, while **Table 2.9** gives pressure/time data. The flowing BHP before shut-in was p_{wf} = 1,384 psia. By use of Horner's approximation, determine the producing time to be used in the Horner plot, the effective permeability to oil, the original reservoir pressure, and skin factor.

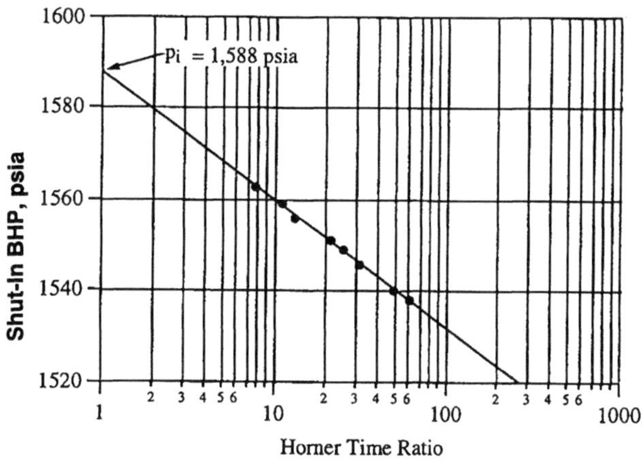

Fig. 2.11—Horner semilog plot, Example 2.5.

μ = 0.8 cp
h = 15 ft
c_t = 15×10^{-6} psi^{-1}
ϕ = 0.25
r_w = 0.333 ft
B = 1.25 RB/STB

Solution.

1. From Eq. 2.37, the Horner pseudoproducing time is

$$t_{pH} = \frac{24N_p}{q_{\text{last}}} = \frac{24[(200)(1.5)+(0)(0.5)+(100)(2)+(125)(6)]}{125}$$

$$= 240 \text{ hours.}$$

Note that the relatively short half-day shut-in period has no effect on the production-time calculation. The longer a shut-in period, however, the less accurate Horner's approximation is.

2. To find permeability, original reservoir pressure, and skin factor, we must make a semilog plot of shut-in pressure, p_{ws}, as a function of the Horner time ratio in terms of the pseudoproducing time, $(t_{pH}+\Delta t)/\Delta t$. **Table 2.10** gives the calculated Horner time ratios, and **Fig. 2.11** shows the semilog plot. The absolute value of the slope of the semilog straight line in Fig. 2.11 is

$$m = \left|\frac{p_{ws2}-p_{ws1}}{\log\left(\frac{t_{pH}+\Delta t_2}{\Delta t_2}\right) - \log\left(\frac{t_{pH}+\Delta t_1}{\Delta t_1}\right)}\right|$$

$$= \left|\frac{1,560 - 1,532}{\log(10)-\log(100)}\right|$$

$$= 28 \text{ psi/cycle.}$$

3. The effective permeability to oil is computed (Eq. 2.57) with the last flow rate (i.e., $q=q_{\text{last}}$ = 125 STB/D) before shutting in the well and with the absolute value of the slope of the straight line.

$$k = \frac{162.6 q_{\text{last}} B\mu}{mh} = \frac{(162.6)(125)(1.25)(0.8)}{(28)(15)} = 48.4 \text{ md.}$$

4. Original reservoir pressure is found by extrapolating the semilog straight line in Fig. 2.11 to infinite shut-in time; i.e., $(t_{pH}+\Delta t)/\Delta t=1$. The pressure at this point is

p_i = 1,588 psia.

5. At $\Delta t=1$ hour, the Horner time ratio is $(t_{pH}+\Delta t)/\Delta t=(240+1)\div 1=241$. From the extrapolated semilog straight line, $p_{1\text{hr}}$ = 1,522 psia at this value, and the skin factor is computed with Eq. 2.58.

TABLE 2.11—VARIABLE-RATE-PRODUCTION HISTORY PRECEDING SHUT-IN PERIOD, EXAMPLE 2.6

Time Interval (hours)	Average Production Rate (STB/D)	Production During Interval (STB)
0 to 3	398.8	49.9
3 to 6	265.8	33.2
6 to 9	132.9	16.6

TABLE 2.12—PRESSURE-BUILDUP-TEST DATA, EXAMPLE 2.6

Δt (hours)	p_{ws} (psi)
2	2,812.5
3	2,838
5	2,872.5
7	2,895
9	2,910
11	2,919
13	2,929.5
15	2,935
17	2,942

TABLE 2.13—PLOTTING FUNCTIONS FOR THE SUPERPOSITION METHOD, EXAMPLE 2.6

Δt (hours)	p_{ws} (psi)	X
2	2,812.5	1.2110
3	2,838	1.0280
5	2,872.5	0.7949
7	2,895	0.6533
9	2,910	0.5563
11	2,919	0.4851
13	2,929.5	0.4305
15	2,935	0.3871
17	2,942	0.3517

$$s = 1.151\left[\frac{\left(p_{1\text{hr}} - p_{wf}\right)}{m} - \log\left(k/\phi\mu c_t r_w^2\right) + 3.23\right]$$

$$= 1.151\left\{\left(\frac{1,522 - 1,384}{28}\right)\right.$$

$$\left. - \log\left[\frac{48.4}{(0.25)(0.8)(15.0\times10^{-6})(0.333)^2}\right] + 3.23\right\}$$

$$= 0.0.$$

A skin factor of zero indicates that the formation near the wellbore is neither damaged nor stimulated.

Example 2.6—Analysis of a Pressure-Buildup Test Following a Multirate Production Period. The following buildup test preceded by a variable-rate production period is adapted from Odeh and Jones.[7] **Table 2.11** summarizes the oil-production history, while **Table 2.12** gives the pressure-buildup-test data. Known reservoir rock and fluid properties follow. Determine effective permeability to oil and initial reservoir pressure with (1) the superposition method, (2) the Odeh-Selig method, and (3) the Horner approximation method.

$\mu = 0.6$ cp

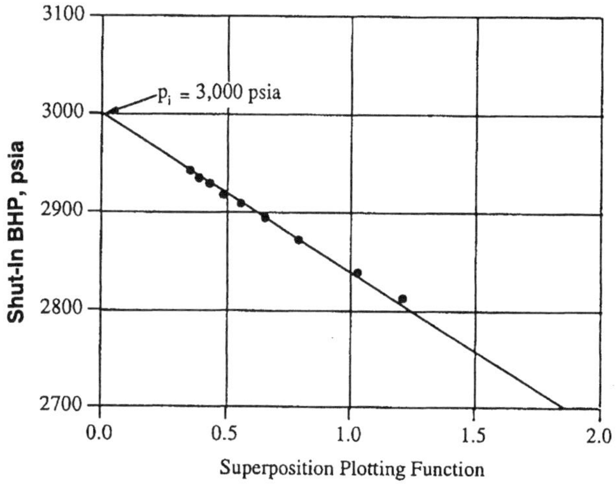

Fig. 2.12—Superposition plot, Example 2.6.

$h = 10$ ft
$B = 1.2$ RB/STB
$\phi = 0.12$
$c_t = 4.8\times10^{-5}$ psi^{-1}
$r_w = 0.25$ ft

Solution. *Superposition Method.*

1. For the superposition method, we first calculate the plotting function, X.

$$X = \frac{q_1}{q_{n-1}}\log\left(\frac{t}{t-t_1}\right) + \frac{q_2}{q_{n-1}}\log\left(\frac{t-t_1}{t-t_2}\right) + \frac{q_3}{q_{n-1}}\log\left(\frac{t-t_2}{t-t_3}\right)$$

$$+ \ldots + \frac{q_{n-2}}{q_{n-1}}\log\left(\frac{t-t_{n-3}}{t-t_{n-2}}\right) + \log\left(\frac{t-t_{n-2}}{t-t_{n-1}}\right)$$

$$= \frac{398.8}{132.9}\log\left(\frac{9+\Delta t}{9+\Delta t-3}\right) + \frac{265.8}{132.9}\log\left(\frac{9+\Delta t-3}{9+\Delta t-6}\right)$$

$$+ \log\left(\frac{9+\Delta t-6}{9+\Delta t-9}\right)$$

$$= 3.0\log\left(\frac{9+\Delta t}{6+\Delta t}\right) + 2.0\log\left(\frac{6+\Delta t}{3+\Delta t}\right) + \log\left(\frac{3+\Delta t}{\Delta t}\right).$$

Table 2.13 gives values of X, while **Fig. 2.12** shows a plot of the data for each value of shut-in time, Δt.

2. The slope of the best-fit line drawn through the data is

$$m = \left|\frac{p_{ws2} - p_{ws1}}{X_2 - X_1}\right| = \left|\frac{2,839 - 2,919.5}{1.0 - 0.5}\right| = 161.$$

3. The permeability is calculated with the absolute value of the slope of the straight line,

$$k = \frac{162.6 q_{n-1}\mu B}{mh} = \frac{(162.6)(132.9)(0.6)(1.2)}{(161)(10)} = 9.7 \text{ md}.$$

4. The initial reservoir pressure is estimated from the extrapolation of the straight line to $X = 0$. We find that

$p_i = 3,000$ psia.

Odeh-Selig Method.

1. For the Odeh-Selig method, we first calculate the production time, t_p^*, defined by Eq. 2.50,

$$t_p^* = 2\left[t_n - \frac{\sum_{j=1}^{n} q_j\left(t_j^2 - t_{j-1}^2\right)}{2\sum_{j=1}^{n} q_j\left(t_j - t_{j-1}\right)}\right].$$

TABLE 2.14—PLOTTING FUNCTIONS FOR THE ODEH-SELIG METHOD, EXAMPLE 2.6

Δt (hours)	p_{ws} (psi)	$(t_p^* + \Delta t)/\Delta t$
2	2,812.5	6.50
3	2,838	4.67
5	2,872.5	3.20
7	2,895	2.57
9	2,910	2.22
11	2,919	2.00
13	2,929.5	1.85
15	2,935	1.73
17	2,942	1.65

TABLE 2.15—PLOTTING FUNCTIONS FOR THE HORNER APPROXIMATION METHOD, EXAMPLE 2.6

Δt (hours)	p_{ws} (psi)	$(t_p + \Delta t)/\Delta t$
2	2,812.5	10.0
3	2,838	7.0
5	2,872.5	4.60
7	2,895	3.57
9	2,910	3.00
11	2,919	2.64
13	2,929.5	2.38
15	2,935	2.20
17	2,942	2.06

The summation in the numerator is

$$\sum_{j=1}^{n} q_j\left(t_j^2 - t_{j-1}^2\right) = 398.8(3^2 - 0^2) + 265.8(6^2 - 3^2)$$

$$+ 132.9(9^2 - 6^2)$$

$$= 16,746.$$

The summation in the denominator is

$$\sum_{j=1}^{n} q_j\left(t_j - t_{j-1}\right) = 398.8(3 - 0) + 265.8(6 - 3)$$

$$+ 132.9(9 - 6)$$

$$= 2,393.$$

The Odeh-Selig producing time is

$$t_p^* = 2\left[9 - \frac{16,746}{2(2,393)}\right] = 11.0 \text{ hours}.$$

2. The Odeh-Selig production rate defined by Eq. 2.51 is

$$q^* = \frac{1}{t_p^*}\sum_{j=1}^{n} q_j\left(t_j - t_{j-1}\right)$$

$$= \frac{1}{11.0}[398.8(3 - 0) + 265.8(6 - 3) + 132.9(9 - 6)]$$

$$= 217.5 \text{ STB/D}.$$

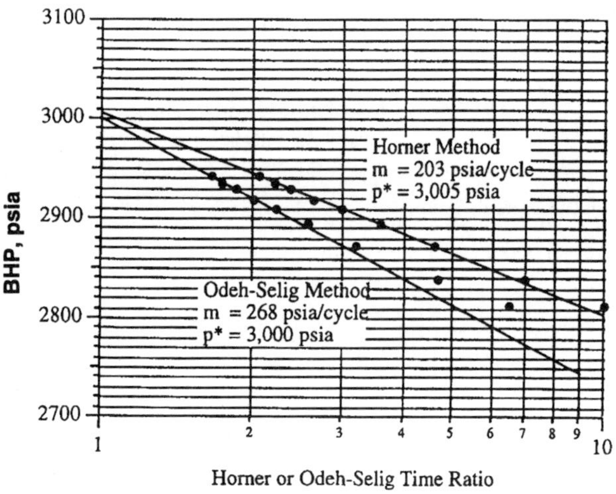

Fig. 2.13—Horner and Odeh-Selig analyses, Example 2.6.

3. Fig. 2.13 shows a semilog plot of p_{ws} as a function of $(t_p^* + \Delta t)/\Delta t$, and **Table 2.14** gives values of the time plotting function.

4. From the plot in **Fig. 2.13**, the slope of the straight line is

$$m = \left|\frac{p_{ws2} - p_{ws1}}{\log\left[\left(t_p^* + \Delta t\right)/\Delta t_2\right] - \log\left[\left(t_p^* + \Delta t\right)/\Delta t_1\right]}\right|$$

$$= \left|\frac{3,000 - 2,732}{\log(1) - \log(10)}\right|$$

$$= 268 \text{ psi/cycle}.$$

The Odeh-Selig method is applicable only when Δt is greater than actual production time (9 hours). Thus, the semilog straight line is drawn through those data points for which $\Delta t > t_p^* = 9$ hours, which includes the last four data points of the test.

5. The permeability is

$$k = \frac{162.6 q^* \mu B}{mh} = \frac{(162.6)(217.5)(0.6)(1.2)}{(268)(10)} = 9.5 \text{ md}.$$

6. Extrapolating the semilog straight line to infinite shut-in time, we estimate that the initial pressure is

$p_i = 3,000$ psia.

Horner's Approximation.

1. For the Horner method, we calculate the pseudoproducing time, t_{pH}, defined by Eq. 2.53,

$$t_{pH} = \frac{24N_p}{q_{\text{last}}} = \frac{24(49.9 + 33.2 + 16.6)}{132.9} = 18 \text{ hours}.$$

2. Fig. 2.13 shows a Horner plot of the buildup data. The Horner time ratio (**Table 2.15**) is calculated with t_p. From Fig. 2.13, the slope of the straight line is

$$m = \left|\frac{p_{ws2} - p_{ws1}}{\log\left[\left(t_{pH} + \Delta t_2\right)/\Delta t_2\right] - \log\left[\left(t_{pH} + \Delta t_1\right)/\Delta t_1\right]}\right|$$

$$= \left|\frac{3,005 - 2,802}{\log(1) - \log(10)}\right|$$

$$= 203 \text{ psi/cycle}.$$

3. The permeability is estimated from the slope of the straight line,

$$k = \frac{162.6 q \mu B}{mh} = \frac{(162.6)(132.9)(1.2)(0.6)}{(203)(10)} = 7.7 \text{ md}.$$

4. Extrapolating the semilog straight line to infinite shut-in time, (i.e., Horner time ratio of one), we find that the initial pressure is

$p_i = 3,005$ psia.

TABLE 2.16—COMPARISON OF RESULTS, EXAMPLE 2.6		
Analysis Method	k (md)	p_i (psia)
Horner	7.7	3,005
Odeh-Selig	9.5	3,000
Superposition	9.7	3,000

Table 2.16 summarizes the results from the three methods. With values of k and p_i with the two variable-rate methods, the Odeh-Selig method and superposition agree. The Horner method, which assumes a single, constant rate, gives a less accurate permeability estimate.

2.4 Complications in Actual Tests

The analysis techniques presented in the previous section were derived assuming a homogenous reservoir model and, therefore, represent ideal conditions. In reality, reservoirs are not homogeneous, and the actual pressure response during a flow or buildup test deviates from the ideal behavior. These deviations are usually caused by conditions in the wellbore and near the drainage radius of the reservoir that are not considered in the simple model described by Eq. 2.2. We use the concept of radius of investigation first introduced in Chap. 1 to help us understand the causes of the nonideal behavior.

2.4.1 Radius-of-Investigation Concept. Consider a graph (**Fig. 2.14**) of pressure as a function of radius for constant-rate flow at various times from the beginning of flow. The pressure in the wellbore continues to decrease as flow time increases. Simultaneously, the size of the area from which fluid is drained increases and the pressure transient moves farther out into the reservoir. The radius of investigation, which is defined as the point in the formation beyond which the pressure drawdown is negligible, is a measure of how far a transient has moved into a formation following any rate change in a well and physically represents the depth to which formation properties are being investigated at any time in a test. The approximate position of the radius of investigation at any time is estimated with Eq. 2.59:

$$r_i = \sqrt{\frac{kt}{948\phi\mu c_t}}. \quad \text{(2.59)}$$

For a buildup test, pressure distributions following shut-in have the profiles illustrated in **Fig. 2.15.** The radius to which the rate of pressure change becomes negligible by a particular shut-in time moves farther into the reservoir with time, and the radius reached by this pressure level is given by

$$r_i = \sqrt{k\Delta t/948\phi\mu c_t}. \quad \text{(2.60)}$$

If the permeability encountered by the radius of investigation near the wellbore at earliest times in a buildup or flow test is different from the permeability encountered later (away from the well), we should not be surprised that the slope of pressure vs. the appropriate time function curve on a semilog graph is different at early and late times. Similarly, because the Ei-function solution assumes an infinite-acting reservoir, we should expect the slope of a buildup- or flow-test plot to change shapes at late times when the radius of investigation reaches the drainage boundaries of a reservoir.

2.4.2 Time Regions on Test Plots. On an actual flow- or buildup-test plot, the straight line predicted by ideal theory rarely occurs over the entire range of test times. Instead, the curve is shaped more like those illustrated in **Figs. 2.16 and 2.17.** To assist in understanding the causes of the nonlinear portions of the curve, we subdivide the flow-test data into three time regions (early, middle, and late) based on the radius-of-investigation concept.

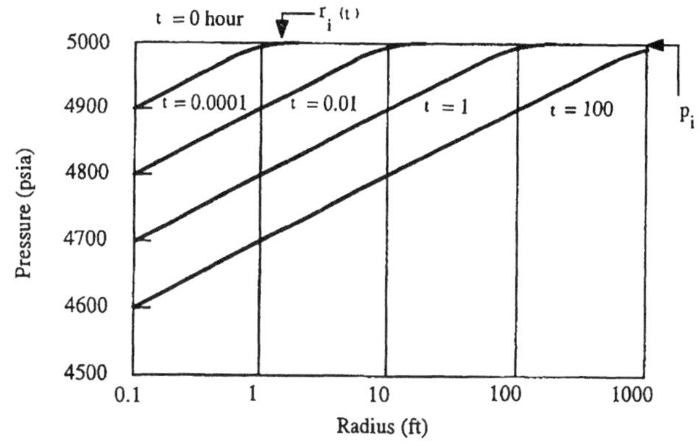

Fig. 2.14—Radius of investigation as a function of flow time during a pressure drawdown test.

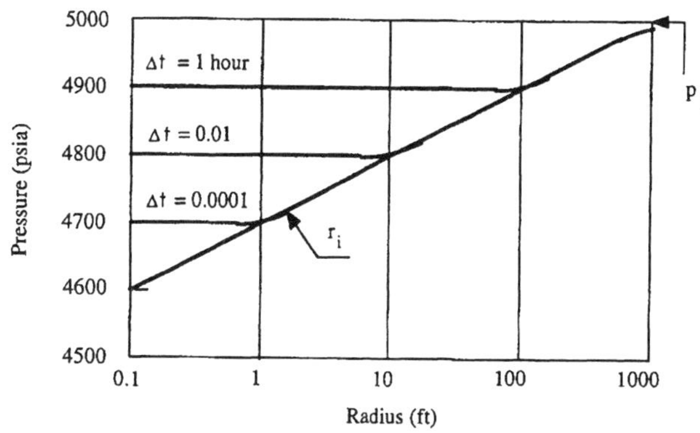

Fig. 2.15—Radius of investigation as a function of shut-in time during a pressure-buildup test.

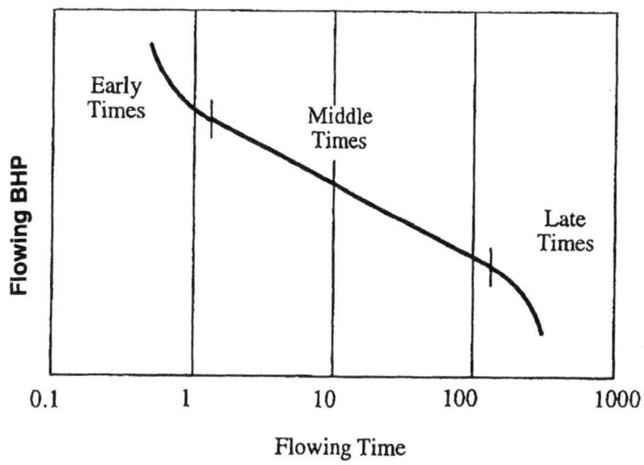

Fig. 2.16—Characteristic curve shapes exhibited during a flow test.

Early Times. The pressure transient is in a damaged or stimulated zone near the wellbore. Wellbore unloading or afterflow of fluid stored in the wellbore also distorts the test data during this period.

Middle Times. The pressure transient has moved into the undamaged formation. A straight line, whose slope is related to the effective permeability of the flowing phase, usually occurs during this period. Often, this flow period is referred to as the middle-time region and the straight line is called the "correct semilog straight line."

Late Times. The pressure transient encounters reservoir boundaries, interference effects from other producing wells, or massive changes in reservoir properties. The flow-test curve deviates from the straight line established during the middle-time region.

2.4.3 Wellbore-Storage Effects. Only in rare cases is the time required for the radius of investigation to move through the altered zone near the wellbore of significant duration. In most cases, the

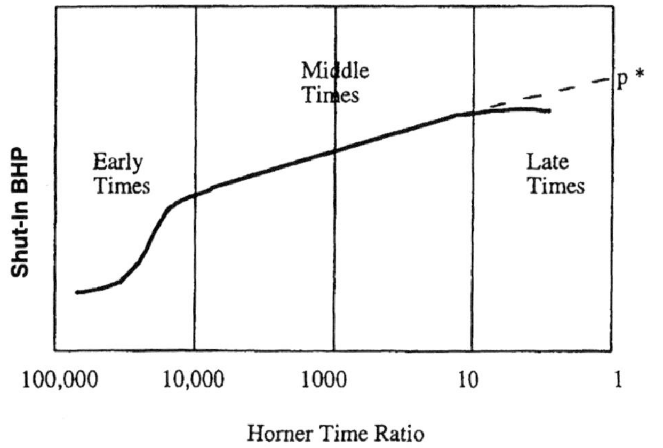

Fig. 2.17—Characteristic curve shapes exhibited during a pressure-buildup test.

duration of the early-time region is determined by the duration of wellbore-storage distortion of test data. In flow tests, the specific case of wellbore storage is called wellbore unloading, which occurs because the initial fluid production measured at the surface originates from fluids stored in the wellbore rather than from the formation. Only after what may be a prolonged time does the bottomhole flow rate approximately equal the surface rate (**Fig. 2.18**). Until then, the assumption of constant bottomhole rate, on which the flow equation and graphing technique are based, is not satisfied.

Wellbore storage also affects the early pressure-buildup reponse. Following shut-in at the surface, fluid continues to flow from the reservoir into the wellbore, compressing the gas and liquid already in the wellbore and also storing more fluid. This continued production, which is also a special case of wellbore storage, is called afterflow (**Fig. 2.19**). Until the rate of afterflow diminishes to less than approximately 1% of the rate before shut-in, the straight line predicted by ideal theory for a Horner graph of buildup-test data does not appear.

Following a mass balance in the wellbore, as illustrated in Sec. 1.3.3, we define a wellbore-storage coefficient, C, in bbl/psi, as

$$C = \Delta V/\Delta p\,, \quad (2.61)$$

where ΔV = change in wellbore fluid volume (in barrels) at wellbore conditions and Δp = change in bottomhole pressure, psi. The form of C depends on the fluid phases in the wellbore. For a well with a liquid/gas interface that is either rising or falling, the wellbore-storage coefficient is calculated from

$$C = 25.64A_{wb}/\rho_{wb}\,, \quad (2.62)$$

where A_{wb} = area of the wellbore and ρ_{wb} = density of the liquid in the wellbore.

If the wellbore contains only a single-phase fluid (either liquid or gas), then

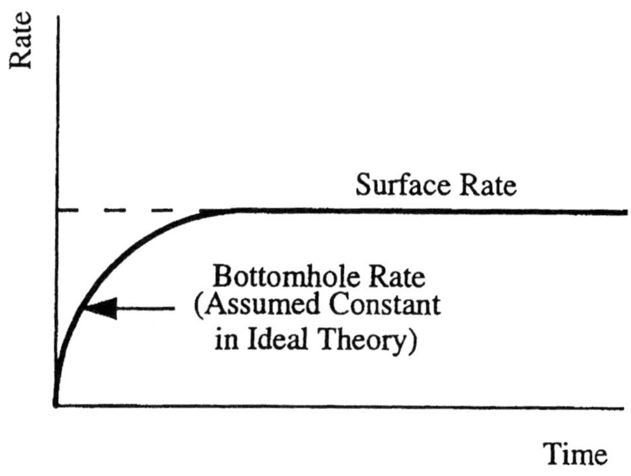

Fig. 2.18—Surface-production-rate schedule during wellbore-storage period.

$$C = V_{wb}\,c_{wb}\,, \quad (2.63)$$

where V_{wb} = volume of the wellbore and c_{wb} = compressibility of the fluid evaluated at average temperature and pressure conditions in the wellbore

2.4.4 Damage and Stimulation Analysis. Many wells either have a zone of reduced permeability near the wellbore resulting from drilling or completion operations or have been stimulated by acidization or hydraulic fracturing. Eq. 2.2 assumes, however, that the formation properties are uniform throughout the drainage area of the well, including the area immediately adjacent to the wellbore. Historically, skin effects have been modeled as an infinitesimally small zone of reduced permeability on the formation face.[10]

Another modeling technique considers the formation to be a two-region reservoir[1] (**Fig. 2.20**) in which the damaged or stimulated zone is considered equivalent to an altered zone of uniform permeability, k_s, extending out to a radius, r_s, while outside of this zone of altered permeability, the formation has permeability, k, unaffected by drilling or completion operations. With this model, the skin factor, s, to quantify either formation damage or stimulation in terms of the properties of the altered zone is

$$s = \left(\frac{k}{k_s} - 1\right)\ln\left(\frac{r_s}{r_w}\right). \quad (2.64)$$

A positive skin factor indicates damage or a permeability reduction, while a negative skin factor indicates an improvement in permeability, resulting from an acidization or hydraulic-fracture treatment. If the formation near the wellbore is neither damaged nor stimulated, the skin factor is zero. If the radial depth of formation damage, r_s, can be determined or assumed, then we can estimate the correspond-

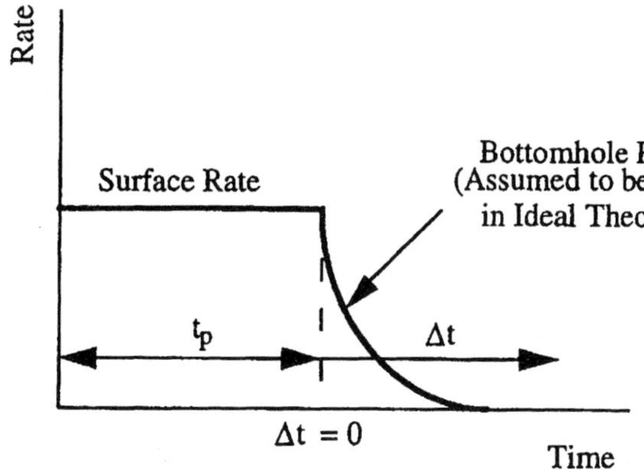

Fig. 2.19—Bottomhole flow rate or afterflow following well shut-in at the surface.

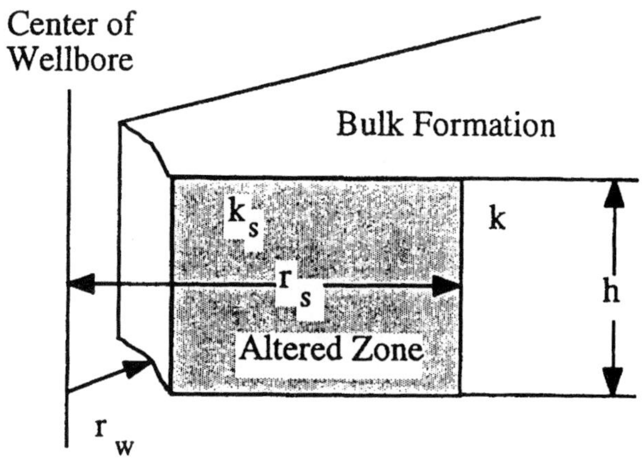

Fig. 2.20—Two-region reservoir model of altered zone near wellbore.

ing value of permeability in the altered zone, k_s, where from an arrangement of Eq. 2.64,

$$k_s = \frac{k}{1 + [s/\ln(r_s/r_w)]}. \quad (2.65)$$

Alternatively, if the permeability-reduction ratio, k_s/k, is available from laboratory measurements, the depth of damage, r_s, can be calculated directly with Eq. 2.64.

Another use of the skin factor is in terms of an effective wellbore radius, r_{wa}, which is defined as[1,11]

$$r_{wa} = r_w e^{-s}. \quad (2.66)$$

Note that, for positive skin factors (i.e., damaged zones), the effective (or apparent) wellbore radius used in flow equations to replace the actual wellbore radius models a well with no damage but with a smaller radius and a larger pressure drop at the well. Conversely, for stimulated wells, the effective wellbore radius models the well as unstimulated but with a very large wellbore. An application of the effective-wellbore-radius concept is based on the observation that, for vertically fractured wells with infinitely conductive fractures having two wings each of equal length L_f, the relationship between r_{wa} and L_f is

$$L_f = 2r_{wa}. \quad (2.67)$$

Thus, if the skin factor for a fractured well can be estimated from a well test and if the fracture is assumed to be infinitely conductive, then the fracture half-length, L_f, can be estimated.

We can also quantify the skin factor in terms of the additional pressure drop associated with the damaged zone, or

$$\Delta p_s = (141.2qB\mu/kh)s. \quad (2.68)$$

For example, if the total drawdown is 1,500 psi and $\Delta p_s = 1{,}000$ psi, then the skin factor has provided some useful information regarding the incentive for well stimulation. Either the pressure drawdown can be reduced by $\approx$ 1,000 psi and the same flow rate maintained, or the same pressure drawdown can be maintained and the rate increased by a factor of approximately three.

Yet another measure of an altered zone near the wellbore is in terms of a flow efficiency, E, which is defined as

$$E = \frac{J_{\text{actual}}}{J_{\text{ideal}}} = \frac{q/(\bar{p} - p_{wf})}{q/(\bar{p} - p_{wf} - \Delta p_s)} = \frac{\bar{p} - p_{wf} - \Delta p_s}{\bar{p} - p_{wf}}, \quad (2.69)$$

where the productivity index (PI), J, with field units of STB/psi, is

$$J = q/(\bar{p} - p_{wf}). \quad (2.70)$$

For a well with neither damage nor stimulation, flow efficiency is unity. For a damaged well, $E < 1$, while for a stimulated well, $E > 1$. For example, if a well has a flow efficiency of 0.1, then the well is damaged and is producing only approximately 10% as much fluid with the drawdown imposed as it would produce if the damage were removed with a stimulation treatment.

Example 2.7—Quantifying Formation Damage and Improvement. Analysis of a pressure-buildup test showed that an oil well with a productivity problem had a skin factor of 6.37 and an effective permeability to oil of 7.65 md. Before shutting in the well, the flowing BHP at the time of shut-in was 3,534 psia, and the well produced oil at a constant rate of 250 STB/D. Estimate the following: (1) altered-zone permeability, k_s, for assumed altered-zone radii of 5 and 10 ft, (2) the effective wellbore radius, r_{wa}, (3) the additional pressure drop near the wellbore owing to formation damage, and (4) the flow efficiency, E.

$\mu =$ 0.8 cp
$h =$ 69 ft
$q =$ 250 STB/D
$B =$ 1.136 RB/STB
$k =$ 7.65 md
$p_{wf} =$ 3,534 psia
$r_w =$ 0.198 ft
$s =$ 6.37
$\bar{p} =$ 4,420 psia

Solution.
1. Calculate the altered-zone permeability with Eq. 2.65.

$$k_s = \frac{k}{1 + [s/\ln(r_s/r_w)]}$$

For $r_s = 5$ ft,

$$k_s = \frac{7.65}{\{1 + [6.37/\ln(5/0.198)]\}} = 2.57 \text{ md}$$

and for $r_s = 10$ ft,

$$k_s = \frac{7.65}{\{1 + [6.37/\ln(10/0.198)]\}} = 2.92 \text{ md.}$$

Note that even though the estimate of the depth of damage, r_s, may be quite uncertain, the altered-zone permeability is relatively insensitive to depth estimates.

2. From Eq. 2.66, the effective wellbore radius is given by

$$r_{wa} = r_w\, e^{-s} = 0.198\, e^{-6.37} = 0.00034 \text{ ft.}$$

The physical interpretation of this result is that the tested well is producing at 250 STB/D with the same drawdown as would a well with a wellbore radius of 0.00034 ft and permeability unaltered up to the sandface.

3. The additional pressure drop near the well caused by the altered-permeability zone is given by Eq. 2.68,

$$\Delta p_s = \frac{141.2qB\mu}{kh}s = \frac{(141.2)(250)(1.136)(0.8)(6.37)}{(7.65)(69)}$$

$$= 387 \text{ psi.}$$

Thus, of the total drawdown of approximately $4{,}420 - 3{,}534 = 886$ psi, $\approx$387 psi is caused by damage. Much of this additional drawdown could be avoided if the skin factor is in fact the result of formation damage (rather some other reason, such as partial penetration of the productive zone) and if the well is to be stimulated.

4. Flow efficiency is estimated from Eq. 2.69,

$$E = \frac{\bar{p} - p_{wf} - \Delta p_s}{\bar{p} - p_{wf}} = \frac{4{,}420 - 3{,}534 - 387}{4{,}420 - 3{,}534} = 0.563.$$

This result indicates that the well is producing $\approx$56% as much fluid with the drawdown imposed as an undamaged well in a completely perforated interval would produce with the same pressure drawdown.

2.4.5 Effect of Incompletely Perforated Interval. The skin factor, s, determined from a buildup or flow test is affected by at least two factors: true formation damage and perforations in only a portion of the total formation thickness. Both these factors cause an additional pressure drop near the well, decreasing the flowing BHP, p_{wf}, for a given flow rate and increasing the value of s calculated from the test. In theory, the effects of perforations and formation damage can be isolated with the equation

$$s = (h_t/h_p)s_d + s_p, \quad (2.71)$$

where s_d = positive skin factor from formation damage, s_p = skin factor from partial penetration of the formation, h_t = total formation thickness (**Fig. 2.21**), and h_p = height of the perforated interval (Fig. 2.21).

Eq. 2.71 is valid only for positive skin, s_d, caused by formation damage. The skin factor caused by partial penetration, s_p, is given by[12]

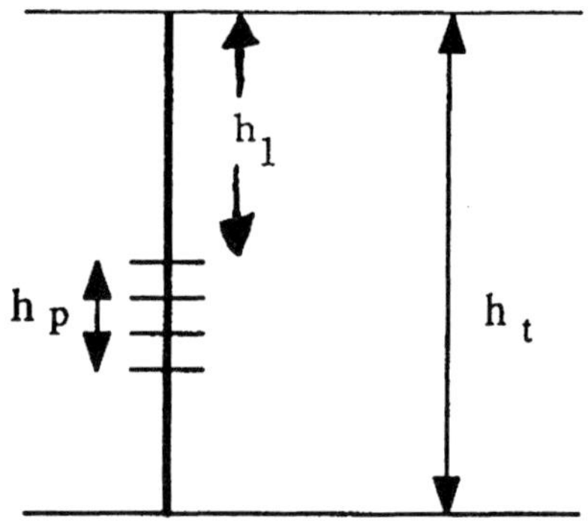

Fig. 2.21—Partial penetration of formation by a well.

$$s_p = \left(\frac{1}{h_{pD}} - 1\right)\ln\frac{\pi}{2r_D} + \frac{1}{h_{pD}}\ln\left[\frac{h_{pD}}{2 + h_{pD}}\left(\frac{A-1}{B-1}\right)^{1/2}\right], \quad (2.72a)$$

where $r_D = (r_w/h_t)(k_v/k_h)^{1/2}$, ... (2.72b)

$$h_{pD} = h_p/h_t, \quad (2.72c)$$

$$h_{1D} = h_1/h_t, \quad (2.72d)$$

$$A = 1/(h_{1D} + h_{pD}/4), \quad (2.72e)$$

and $B = 1/(h_{1D} + 3h_{pD}/4)$. .... (2.72f)

In practice, k_v/k_h is rarely known, so quantitative calculation of s_p may be difficult. The effect of s_p is so important, however, that the best approximation possible with the data available should be made. Values of s_p of 5 to 10 are not uncommon.

Example 2.8—Skin Effect Caused by Partial Penetration. A pressure-buildup test analysis indicated that, for the tested well, the skin factor is $s = 6.37$. Only the upper half of the 69-ft interval was perforated. From core data, vertical permeability is estimated to be $^1/_{10}$ the horizontal permeability. The wellbore radius is 0.198 ft. Estimate the skin effect caused by partial penetration and the skin effect caused by formation damage.

Solution.

1. Calculate r_D, h_{pD}, h_D, A, and B:

$$r_D = (r_w/h_t)(k_v/k_h)^{1/2} = \frac{0.198}{69}(0.1)^{1/2} = 0.000907,$$

$$h_{pD} = h_p/h_t = 0.5,$$

$$h_{1D} = h_1/h_t = 0,$$

$$A = \frac{1}{h_{1D} + h_{pD}/4} = \frac{1}{0 + 0.5/4} = 8.0,$$

$$\text{and } B = \frac{1}{h_{1D} + 3h_{pD}/4} = \frac{1}{0 + (3)(0.5)/4} = 2.67.$$

2. Estimate the skin effect caused by partial penetration with Eq. 2.72a,

$$s_p = \left(\frac{1}{h_{pD}} - 1\right)\ln\frac{\pi}{2r_D} + \frac{1}{h_{pD}}\ln\left[\frac{h_{pD}}{2 + h_{pD}}\left(\frac{A-1}{B-1}\right)^{1/2}\right]$$

$$= \left(\frac{1}{0.5} - 1\right)\ln\frac{\pi}{2(0.000907)}$$

$$+ \frac{1}{0.5}\ln\left[\frac{0.5}{2 + 0.5}\left(\frac{8-1}{2.67-1}\right)^{1/2}\right] = 5.67.$$

3. Finally, estimate the skin effect resulting from formation damage by rearranging Eq. 2.71 to obtain

$$s_d = \frac{h_p}{h_t}(s - s_p) = \frac{0.5(69)}{69}(6.37 - 5.67) = 0.35,$$

which suggests that the well is only slightly damaged. Therefore, most of the skin effect is caused by the incompletely perforated interval ($s_p = 5.67$).

2.4.6 Effect of Deviated Hole. Cinco *et al.*[13] showed that the skin factor, s, determined from buildup or flow-test analysis, includes both the skin factor, s_t, attributable to all causes other than hole deviation, and a pseudoskin factor, s_θ, caused by deviation of the wellbore from the vertical.

$$s = s_t + s_\theta. \quad (2.73)$$

From correlations of numerical solutions to the governing flow equations, Cinco *et al.*[13] found that

$$s_\theta = -\left(\frac{\theta'_w}{41}\right)^{2.06} - \left(\frac{\theta'_w}{56}\right)^{1.865}\log\left(\frac{h_D}{100}\right) \quad (2.74)$$

for $0° \leqq \theta'_w \leqq 75°$, where

$$\theta'_w = \tan^{-1}\left(\sqrt{k_v/k_h}\tan\theta_w\right), \text{ degrees}, \quad (2.75)$$

and where θ_w = angle of deviation from vertical defined (in degrees) and h_D is defined as

$$h_D = (h/r_w)\sqrt{k_h/k_v}. \quad (2.76)$$

Example 2.9—Estimating the Skin Factor for a Deviated Well. A pressure-buildup test indicates that the skin factor for a directionally drilled well is $s = 0.8$, suggesting slight damage. The additional pressure drop owing to the skin factor is $\Delta p_s = 87$ psi. The average angle of inclination is $\theta_w = 45°$, the formation thickness is 100 ft, the wellbore radius is 0.3 ft, and, from core analysis, the ratio of horizontal-to-vertical permeabilities, k_h/k_v, is 5.0. Determine the portion of the skin factor caused by directional drilling.

Solution.

1. From Eqs. 2.75 and 2.76,

$$\theta'_w = \tan^{-1}\left(\sqrt{k_v/k_h}\tan\theta_w\right) = \tan^{-1}\left(\sqrt{\frac{1}{5}}\tan 45°\right) = 24.1°$$

$$h_D = (h/r_w)\sqrt{k_h/k_v} = \frac{100}{0.3}\sqrt{5} = 745.$$

2. From Eq. 2.74,

$$s_\theta = -(\theta'_w/41)^{2.06} - (\theta'_w/56)^{1.865}\log(h_D/100)$$

$$= -(24.1/41)^{2.06} - (24.1/56)^{1.865}\log(745/100)$$

$$= -0.516.$$

3. Rearrangement of Eq. 2.73 shows that the portion of the total skin effect caused by factors other than directional drilling is

$$s_t = s - s_\theta = 0.8 - (-0.516) = 1.32.$$

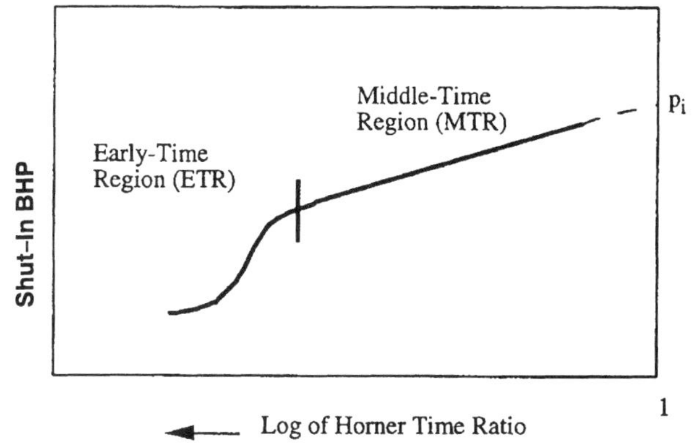

Fig. 2.22—Buildup-test plot for an infinite-acting well in a reservoir with negligible pressure depletion.

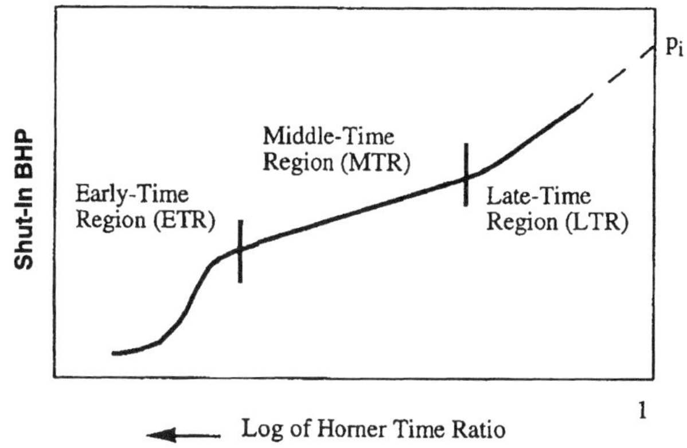

Fig. 2.23—Buildup-test plot for a well near a boundary in a reservoir with negligible pressure depletion.

2.5 Analysis of Late-Time Data in Flow and Buildup Tests

This section focuses on the analysis of late-time well-test data (i.e., data that have been influenced by outer reservoir boundaries). Specifically, we examine methods for estimating drainage-area pressure, distances to reservoir boundaries, and reservoir pore volume (PV) from well tests. We should emphasize that these methods are limited to well tests from homogeneous-acting, single-layer formations producing slightly compressible liquids.

2.5.1 Estimating Drainage-Area Pressure. The average pressure in the drainage area of a well represents the driving force for fluid flow and is useful in material-balance calculations. For a well in a new reservoir with negligible pressure depletion, extrapolation of buildup-test data to infinite shut-in time, $(t_p+\Delta t)/\Delta t=1$, on a Horner semilog plot provides an estimate of original (and current) drainage-area pressure. For a well in a reservoir in which the average pressure has declined from its original value because of fluid production, the pressure extrapolated to infinite shut-in time is called p^*, which is related, but not equal to the current average pressure in the drainage area of the well.

For a well in a reservoir with negligible pressure depletion, we consider two possibilities. First, if the pressure transient data are not influenced by boundaries (either real reservoir boundaries or artificial boundaries created by adjacent producing wells) during the production period before the buildup test, a typical buildup test will have the shape shown in **Fig. 2.22,** which is a Horner semilog plot. The original reservoir pressure, p_i, is obtained by extrapolating the middle-time semilog straight line to $(t_p+\Delta t)/\Delta t=1$.

For a well completed in a reservoir with negligible pressure depletion and having one or more boundaries relatively near the well (and encountered by the radius of investigation during the production period), a buildup test will exhibit the shape shown in **Fig. 2.23,** which is also a Horner semilog plot. The late-time semilog straight line is extrapolated to $(t_p+\Delta t)/\Delta t=1$ to obtain p_i. For a reservoir in which the pressure has been depleted, either the Matthews-Brons-Hazebroek[14] (MBH) method or the modified Muskat method[15] can be used to estimate the average drainage-area pressure.

MBH Method. The MBH method is based on theoretical correlations between the extrapolated pressure, p^*, and current average drainage-area pressure, $\bar{p}$, for various drainage-area configurations. **Fig. 2.24** shows the extrapolated pressure, p^*. We should emphasize that p^* is not the true average drainage-area pressure; however, several correlations have been developed that relate p^* and $\bar{p}$.

Figs. 2.25 and 2.26 show two of the numerous correlation charts available. Fig. 2.25 applies to wells in various locations in square drainage areas. The most common approximation of drainage area is that the well is centered in a square drainage area, with area equal to the acreage assigned to the well. Fig. 2.26 applies to 2:1 rectangular

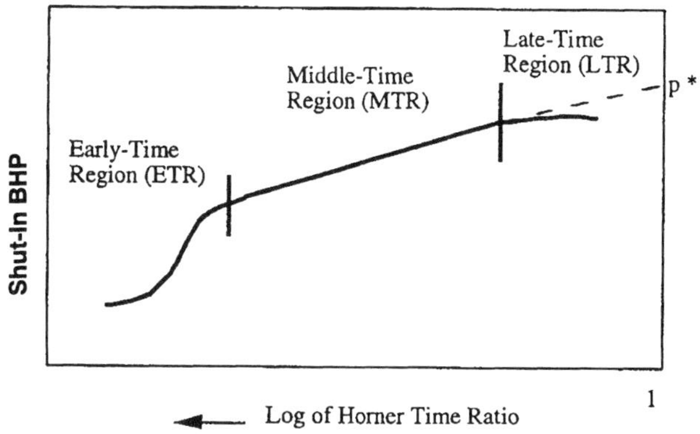

Fig. 2.24—Extrapolated pressure, MBH method.

drainage areas. The symbol $(t_{AD})_{pss}$ indicates the beginning of pseudosteady-state flow, where t_{AD} is defined by Eq. 2.78. Similar charts for other drainage-area shapes are available in several references.[1,2,14] In Figs. 2.25 and 2.26, a dimensionless pressure, $p_{MBH,D}$, is plotted as a function of a dimensionless time, t_{AD}. The dimensionless variables plotted in Figs. 2.24 and 2.25 are defined as

$$p_{\text{MBH,D}} = [kh(p^* - \bar{p})]/70.6qB\mu \quad \text{.................} \quad (2.77)$$

$$\text{and } t_{AD} = 0.0002637kt_p/\phi\mu c_t A, \quad \text{.................} \quad (2.78)$$

where A = drainage area, ft^2, of the tested well.

We recommend the following procedure for applying the MBH method.

1. Extrapolate the middle-time semilog straight line to $(t_p+\Delta t)/\Delta t=1$. The extrapolated pressure is p^*.
2. Estimate the drainage-area shape. If there is insufficient information to estimate the shape, assume a circular drainage area.
3. Select the appropriate MBH chart[1,2,14] for the drainage area.
4. Calculate t_{AD} with Eq. 2.78.
5. From the appropriate MBH chart at the calculated value of t_{AD}, read

$$p_{\text{MBH,D}} = \frac{kh(p^* - \bar{p})}{70.6qB\mu} = \frac{2.303(p^* - \bar{p})}{m}. \quad \text{.........} \quad (2.79)$$

6. Calculate $\bar{p}$ by

$$\bar{p} = p^* - \left(mp_{\text{MBH,D}}/2.303\right). \quad \text{...................} \quad (2.80)$$

The advantages of the MBH method are that it does not require data beyond the middle-time region and that it is applicable to a wide variety of drainage-area shapes. The disadvantages are that the drainage-area size and shape must be known and that reliable estimates of rock and fluid properties, such as c_t and ϕ, must be available. In

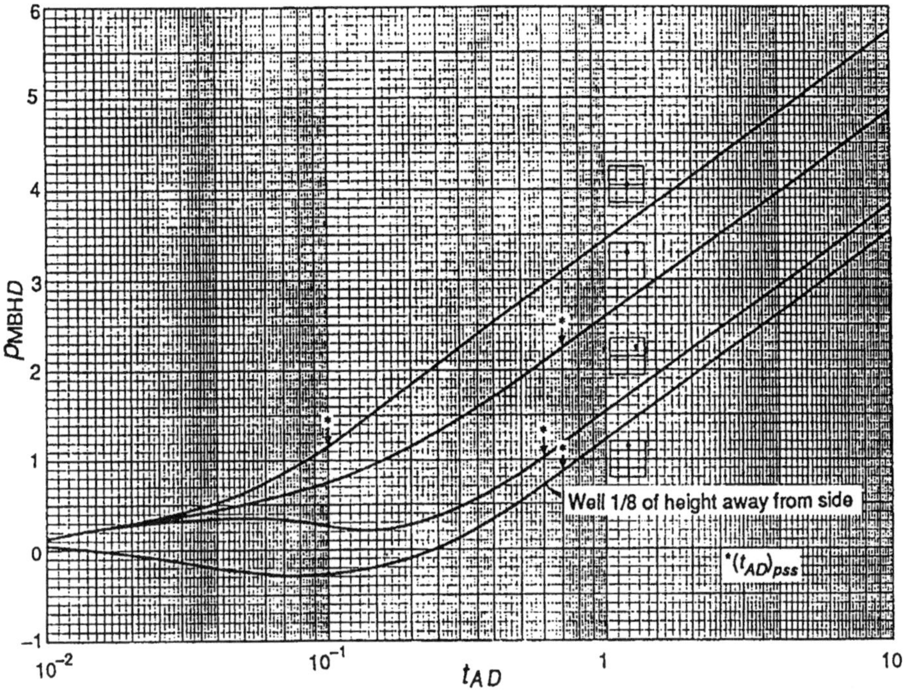

Fig. 2.25—MBH dimensionless pressures for various well locations in a square drainage area (after Earlougher[2]).[14]

addition, the method is limited to well tests in single-layer formations and cannot be applied accurately to multilayer formations.

Modified Muskat Method.[15] The modified Muskat method is based on the theoretical observation that, after boundary effects have been felt for a well centered in its drainage area, the following relationship exists:

$$\log(\bar{p} - p_{ws}) = c_1 + c_2\Delta t\,, \quad\quad (2.81)$$

where c_1 and c_2 are constants. This relationship is valid for Δt approximately in the range

$$\frac{250\phi\mu c_t r_e^2}{k} \leqq \Delta t \leqq \frac{750\phi\mu c_t r_e^2}{k}. \quad\quad (2.82)$$

Thus, late-time, rather than middle-time, data are required for this method. To find $\bar{p}$, we make an initial guess of $\bar{p}$ and plot $\log(\bar{p} - p_{ws})$ vs. Δt. Various values of $\bar{p}$ are chosen until a straight line results (**Fig. 2.27**). The value of $\bar{p}$ that produces a straight line is the correct average reservoir pressure.

Compared with the MBH method, the modified Muskat method has the advantage that no estimates of reservoir properties are required. The method also applies to hydraulically fractured wells and layered reservoirs for which the MBH method is not applicable. The modified Muskat method has the disadvantage that it is limited to

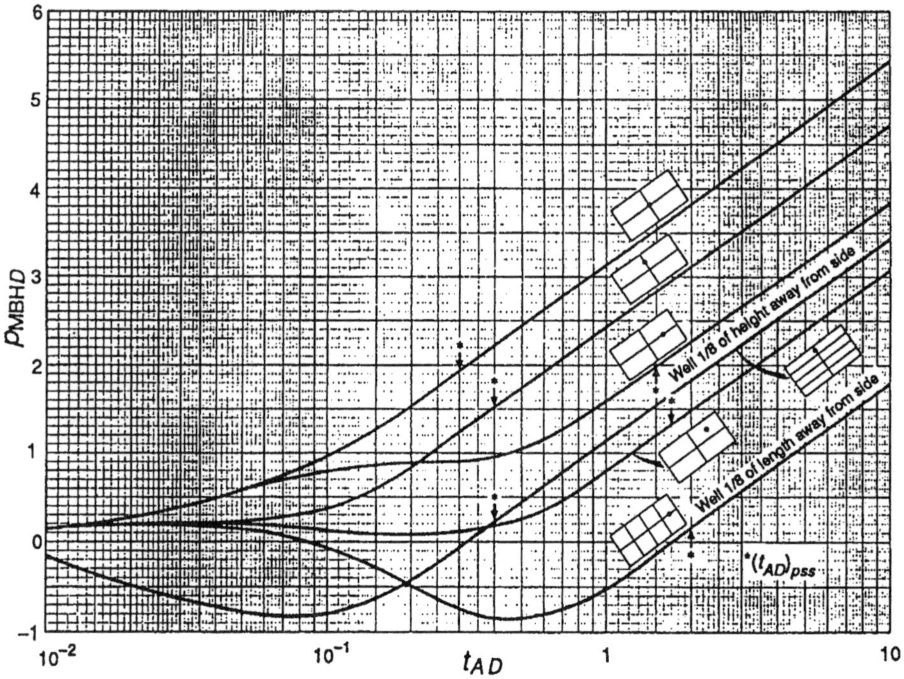

Fig. 2.26—MBH dimensionless pressures for various well locations in a 2:1 rectangular drainage area (after Earlougher[2]).[14]

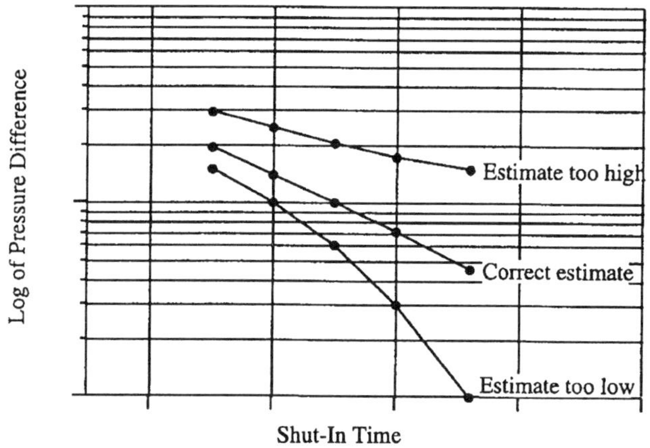

Fig. 2.27—Estimating correct average reservoir pressure with the modified Muskat method.[1]

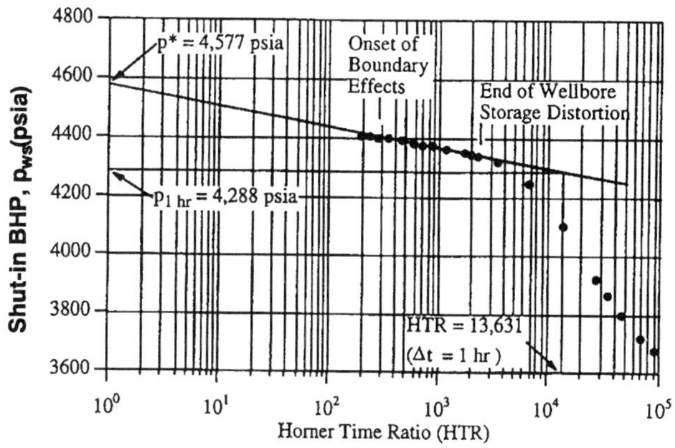

Fig. 2.28—Horner semilog plot, Example 2.10.

wells reasonably centered in their drainage areas. An even more limiting restriction is that shut-in times of $250\phi\mu c_t r_e^2/k$ and greater may be impractically long. We illustrate application of both the MBH and modified Muskat Methods with Example 2.10.

Example 2.10—Estimating Drainage-Area Pressure From a Buildup Test. A buildup test was run on an oil well that is centered in a square drainage area of 160 acres. **Table 2.17** gives time and pressure data. **Fig. 2.28** is a Horner plot of the test data. Data obtained from the Horner plot and other known data are summarized next. Estimate the average pressure in the well's drainage area with the MBH $p*$ method and the modified Muskat method.

$m =$ 70 psi/cycle
$k =$ 7.65 md
$\phi =$ 0.039
$B =$ 1.136 RB/STB
$p_{1hr} =$ 4,288
$q =$ 250 STB/D
$\mu =$ 0.8 cp
$r_e =$ 1,320 ft
$p* =$ 4,577 psia
$t_p =$ 13,630 hours
$c_t = 17 \times 10^{-6}$ psia^{-1}

TABLE 2.17—PRESSURE-BUILDUP-TEST DATA, EXAMPLE 2.10

Time, Δt (hours)	Pressure, p_{ws} (psia)	Horner Time Ratio
0.15	3,680	90,868
0.2	3,723	68,151
0.3	3,800	45,434
0.4	3,866	34,076
0.5	3,920	27,261
1	4,103	13,631
2	4,250	6,816
4	4,320	3,409
6	4,340	2,273
7	4,344	1,948
8	4,350	1,705
12	4,364	1,137
16	4,373	853
20	4,379	683
24	4,384	569
30	4,393	455
40	4,398	342
50	4,402	274
60	4,405	228
72	4,407	190

***Solution.** MBH Method*

1. Extrapolate the middle-time semilog straight line to $(t_p + \Delta t)/\Delta t = 1$. As shown by the Horner plot in Fig. 2.28, $p* = 4{,}577$ psi.

2. For the $p*$ method, we use the MBH chart in Fig. 2.25 for a well centered in a square drainage area. First, we calculate t_{AD}:

$$t_{AD} = 0.0002637kt_p/\phi\mu c_t A$$

$$= \frac{(0.0002637)(7.65)(13,630)}{(0.039)(0.8)(17 \times 10^{-6})(160 \times 43,560)} = 7.44.$$

3. From Fig. 2.25,

$$[2.303(p* - \bar{p})]/m = 5.45.$$

We estimate the average drainage-area pressure, $\bar{p}$, to be

$$\bar{p} = p* - \frac{5.45m}{2.303} = 4,577 - \frac{(5.45)(70)}{2.303} = 4,411 \text{ psi.}$$

Modified Muskat Method.

1. Check whether the modified Muskat method is valid for the test data. In this particular case, because we have estimates of k and r_e, test data outside the applicable time range may be eliminated. Of course, we often do not have these estimates. While lack of data does not limit the applicability of the method, it does require that we assume that the first deviation of the test data from the semilog straight line of the middle-time region indicates the onset of boundary effects.

For the example shown, the lower time limit is

$$\frac{250\phi\mu c_t r_e^2}{k} = \frac{(250)(0.039)(0.8)(17 \times 10^{-6})(1,320)^2}{7.65}$$

$$= 30.2 \text{ hours,}$$

and the upper time limit is

$$\frac{750\phi\mu c_t r_e^2}{k} = \frac{(750)(0.039)(0.8)(17 \times 10^{-6})(1,320)^2}{7.65}$$

$$= 90.6 \text{ hours.}$$

Thus, we can use test data from $\Delta t = 30$ hours until the end of the test at $\Delta t = 72$ hours.

2. Prepare the following data for three trial values of $\bar{p}$ shown in **Table 2.18.** From plots of $(\bar{p} - p_{ws})$ in **Fig. 2.29** for different estimates of $\bar{p}$, we see that $\bar{p} = 4{,}412$ psia is the best estimate of average drainage-area pressure. Note that the method is very sensitive to different values of $\bar{p}$, as evidenced by the distinct curvatures of the plots for only slightly incorrect estimates of $\bar{p}$.

TABLE 2.18—TRIAL VALUES OF $\bar{p}$ FOR MODIFIED MUSKAT METHOD CALCULATIONS, EXAMPLE 2.10

		$\bar{p} - p_{ws}$		
Δt (hours)	p_{ws} (psia)	at $\bar{p}$=4,408 (psi)	at $\bar{p}$=4,412 (psi)	at $\bar{p}$=4,422 (psi)
30	4,393	15	19	29
40	4,398	10	14	24
50	4,402	6	10	20
60	4,405	3	7	17
72	4,407	1	5	15

This example application of the MBH p^* and modified Muskat methods is intended to illustrate only the mechanics of each method. In this particular case, the pressure had built up to within 5 psia of its static value at a shut-in time of 72 hours, in which case there is little value in applying either method. Both methods are of more value when the pressure at the end of a buildup test is far from stabilization.

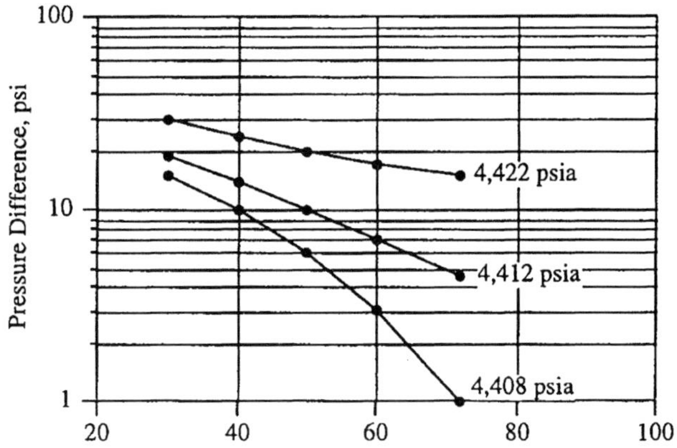

Fig. 2.29—Estimating average reservoir pressure with a modified Muskat method plot, Example 2.10.

2.5.2 Estimating Distance to Boundaries. The pressure history of a well located a distance L from a reservoir boundary can be simulated by adding the drawdown caused by the actual well, acting as if it were in an infinite reservoir, to the drawdown caused by an image well located a distance $2L$ from the actual well and with exactly the same rate history as the actual well. **Fig. 2.30** illustrates this technique, which uses superposition in space. Constant-rate production for a well near a boundary is simulated by

$$p_i - p_{wf} = \frac{162.6qB\mu}{kh}\left[\log\left(kt/\phi\mu c_t r_w^2\right) - 3.23 + 0.869s\right] + \frac{70.6qB\mu}{kh}\left\{-Ei\left[\frac{-948\phi\mu c_t(2L)^2}{kt}\right]\right\}. \quad \text{(2.83)}$$

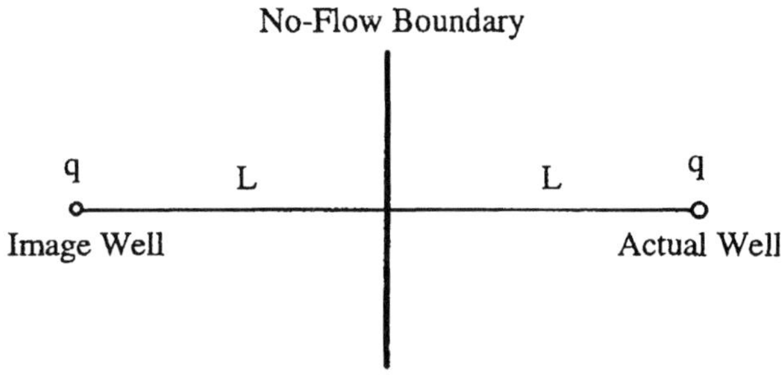

Fig. 2.30—Modeling a no-flow boundary with an image well.

The logarithmic term models the actual well, while the Ei function models the image well. The equation modeling a buildup test is then

$$p_i - p_{ws} = \frac{162.6qB\mu}{kh}\left\{\log\left[\frac{k(t_p + \Delta t)}{\phi\mu c_t r_w^2}\right] - 3.23 + 0.869s\right\} + \frac{162.6(-q)B\mu}{k}\left[\log\left(k\Delta t/\phi\mu c_t r_w^2\right) - 3.23 + 0.869s\right]$$
$$- \frac{70.6qB\mu}{kh}\mathrm{Ei}\left[\frac{-3,792\phi\mu c_t L^2}{k(t_p + \Delta t)}\right] - \frac{70.6(-q)B\mu}{kh}\mathrm{Ei}\left(\frac{-3792\phi\mu c_t L^2}{k\Delta t}\right). \quad \text{(2.84)}$$

Eq. 2.84 can be simplified and written in the form

$$p_{ws} = p_i - (162.6qB\mu/kh) \times \left\{\log\left(\frac{t_p + \Delta t}{\Delta t}\right) - 0.434\mathrm{Ei}\left[\frac{-3792\phi\mu c_t L^2}{k(t_p + \Delta t)}\right] + 0.434\mathrm{Ei}\left(\frac{-3792\phi\mu c_t L^2}{k\Delta t}\right)\right\}. \quad \text{(2.85)}$$

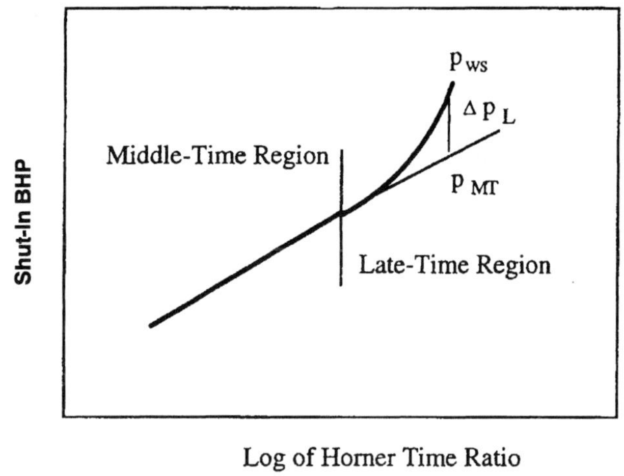

Fig. 2.31—Buildup-pressure behavior affected by a no-flow boundary.

Eq. 2.85 predicts that, initially, the usual middle-time region will develop; however, at later times, the data will deviate from the established semilog straight line of the middle-time region (**Fig. 2.31**). The term p_{MT} is the pressure on the extrapolated middle-time semilog straight line, and Δp_L is the difference between p_{MT} and the measured BHP. To estimate the distance to the fault, we determine Δp_L, which is the difference in the Ei functions,

$$\Delta p_L = 70.6qB\mu/kh \times \left\{-\mathrm{Ei}\left(\frac{-3,792\phi\mu c_t L^2}{k\Delta t}\right) + \mathrm{Ei}\left[\frac{-3792\phi\mu c_t L^2}{k(t_p + \Delta t)}\right]\right.$$

TABLE 2.19—PRESSURE-BUILDUP-TEST DATA, EXAMPLE 2.11			
Δt (hours)	p_{ws} (psia)	Δt (hours)	p_{ws} (psia)
0.0	6,835.6	1.25	6,871.1
0.01	6,856.9	1.64	6,872.2
0.0131	6,857.6	2.14	6,873.4
0.0171	6,858.3	2.80	6,874.7
0.0224	6,859.0	3.66	6,875.9
0.0293	6,859.7	4.79	6,877.2
0.0383	6,860.4	6.27	6,878.5
0.05	6,861.1	8.20	6,879.8
0.0654	6,861.8	10.72	6,881.2
0.0856	6,862.5	14.02	6,882.5
0.1119	6,863.2	18.34	6,883.8
0.1464	6,863.9	23.98	6,885.1
0.1914	6,864.6	31.36	6,886.4
0.2504	6,865.4	41.02	6,887.7
0.3274	6,866.2	53.65	6,888.9
0.4282	6,867.0	70.16	6,890.1
0.5600	6,867.9	91.76	6,891.3
0.7324	6,868.9	120.00	6,892.4
0.9579	6,869.9	—	—

$$- \text{Ei}\left(\frac{-3,792\phi\mu c_t L^2}{kt_p}\right)\Bigg\}. \quad \text{(2.86)}$$

Eq. 2.86 can then be solved for L by trial and error. In the special case when $t_p \gg \Delta t$, Eq. 2.86 reduces to

$$\Delta p_L = \frac{70.6qB\mu}{kh}\left[-\text{Ei}\left(\frac{-3,792\phi\mu c_t L^2}{k\Delta t}\right)\right], \quad \text{(2.87)}$$

which can be solved directly for L. When sufficient shut-in time has elapsed so that the logarithmic approximation applies to the Ei functions, the buildup equation becomes

$$p_{ws} = p_i - (162.6qB\mu/kh)$$

$$\times\left[\log\left(\frac{t_p + \Delta t}{\Delta t}\right)\right] - \frac{162.6qB\mu}{kh}\left[\log\left(\frac{t_p + \Delta t}{\Delta t}\right)\right]$$

$$= p_i - \frac{352.2qB\mu}{kh}\log\left(\frac{t_p + \Delta t}{\Delta t}\right)$$

$$= p_i - 2m\log\left(\frac{t_p + \Delta t}{\Delta t}\right). \quad \text{(2.88)}$$

Eq. 2.88 shows that the slope of a buildup-test plot eventually will double for a well near a barrier, such as a sealing fault. When the slope doubles, the distance to the boundary can be estimated in a simple way. From the Horner graph, we find the shut-in time, Δt_x, corresponding to the point of intersection of the middle and late-time straight lines, which has exactly double the slope of the middle-time line. The distance, L, from the well to the fault is given by the empirical equation[16]

$$L = (0.000148k\Delta t_x/\phi\mu c_t)^{1/2}. \quad \text{(2.89)}$$

This method of estimating the distance to a fault is appealing because of its simplicity. Unfortunately, it is not often applicable because the time required for the slope to double can be long:

$$3,792\phi\mu c_t L^2/k\Delta t < 0.01 \quad \text{(2.90)}$$

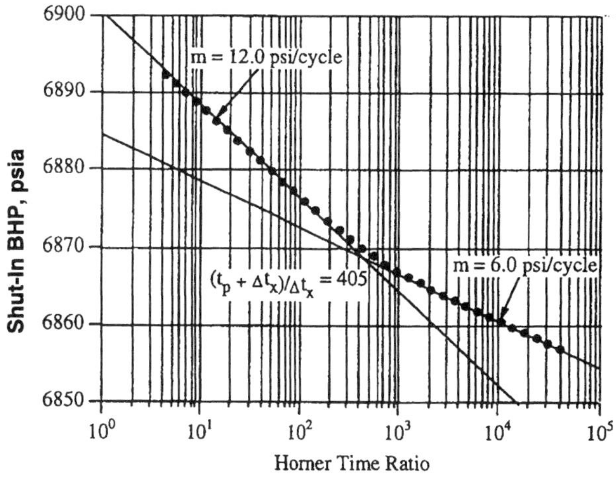

Fig. 2.32—Horner semilog plot, Example 2.11.

$$\text{or } \Delta t > 3.8 \times 10^5\phi\mu c_t L^2/k. \quad \text{(2.91)}$$

For large L or small k, the slope will not double during a typical buildup test. Gray[16] discusses this and other methods for estimating the distance to a no-flow boundary from buildup or drawdown data. We would like to restate that unless the slope of the late-time data is exactly twice the slope of the middle-time data, the double-slope method will not give the correct estimate of the distance to the fault. Furthermore, the reservoir geometry must be such that the nearest no-flow boundary is much closer to the wellbore than the second-nearest boundary. There must be adequate time for the slope to double because of the presence of the nearest boundary before any other boundaries are felt. This restriction often limits the applicability of this method.

Example 2.11—Estimating Distance to a Fault. The following pressure-buildup test was simulated for a well located near a fault. **Table 2.19** summarizes the test data; other known data follow next. Estimate the distance from the well to the fault.

q = 293 STB/D
μ = 0.6 cp
ϕ = 0.22
t_p = 400 hours
B = 1.31 RB/STB
h = 25 ft
c_t = 12.7×10^{-6} psi^{-1}
r_w = 0.5 ft

Solution. Our approach is to plot p_{ws} vs. $(t_p + \Delta t)/\Delta t$, identify the position of the middle-time line, and determine whether the late-time data fall on a line with slope double that of the middle-time line. If so, then the straightforward double-slope method can be used to estimate the distance to the boundary. If not, a more complicated calculation is required.

1. Construct a Horner semilog plot (**Fig. 2.32**) with the plotting functions given in **Table 2.20.**

2. The slope of the best-fit line drawn through the initial data (i.e., middle-time region) is

$$m = \left|\frac{p_{ws2} - p_{ws1}}{\log\left[(t_p + \Delta t_2)/\Delta t_2\right] - \log\left[(t_p + \Delta t_1)/\Delta t_1\right]}\right|$$

$$= \left|\frac{6,854.5 - 6,860.5}{\log(10^5) - \log(10^4)}\right|$$

$$= 6.0 \text{ psi/cycle.}$$

TABLE 2.20—PRESSURE-BUILDUP-TEST DATA, EXAMPLE 2.11

p_{ws} (psia)	$(t_p+\Delta t)/\Delta t$	p_{ws} (psia)	$(t_p+\Delta t)/\Delta t$
6,835.6	—	6,871.1	320.3
6,856.9	40,001	6,872.2	245.1
6,857.6	30,535	6,873.4	187.7
6,858.3	23,393	6,874.7	143.7
6,859.0	17,858	6,875.9	110.1
6,859.7	13,653	6,877.2	84.5
6,860.4	10,445	6,878.5	64.8
6,861.1	8,001	6,879.8	49.8
6,861.8	6,117	6,881.2	38.3
6,862.5	4,674	6,882.5	29.5
6,863.2	3,576	6,883.8	22.8
6,863.9	2,733	6,885.1	17.7
6,864.6	2,091	6,886.4	13.8
6,865.4	1,598	6,887.7	10.8
6,866.2	1,223	6,888.9	8.5
6,867.0	935.1	6,890.1	6.7
6,867.9	715.3	6,891.3	5.4
6,868.9	547.1	6,892.4	4.3
6,869.9	419.6	—	—

3. The permeability is estimated from the slope of the semilog straight line.

$$k = \frac{162.6qB\mu}{mh} = \frac{(162.6)(293)(1.31)(0.6)}{(6.0)(25)} = 249.6 \text{ md}.$$

4. The slope of the best-fit straight line drawn through the later data (i.e., the late-time region) is

$$m = \left| \frac{p_{ws2} - p_{ws1}}{\log\left[(t_p + \Delta t_2)/\Delta t_2\right] - \log\left[(t_p + \Delta t_1)/\Delta t_1\right]} \right|$$

$$= \frac{|6,876.2 - 6,888.2|}{|\log(100) - \log(10)|}$$

$$= 12 \text{ psi/cycle}.$$

Note that this slope is two times greater than that determined in Step 2. From the Horner plot shown in Fig. 2.32, the Horner time ratio at which the middle- and late-time straight lines intersect is $(t_p+\Delta t_x)/\Delta t_x = 405$, from which we calculate $\Delta t_x = 0.99$ hours. Thus, the distance from the well to the fault is estimated to be

$$L = \left(\frac{0.000148k\Delta t_x}{\phi\mu c_t}\right)^{1/2} = \left[\frac{(0.000148)(249.6)(0.99)}{(0.22)(0.6)(12.7 \times 10^{-6})}\right]^{1/2}$$

$$= 147.7 \text{ ft}.$$

The time required for the slope to double (Eq. 2.91) is estimated to be

$$\Delta t > \frac{3.8 \times 10^5 \phi\mu c_t L^2}{k}$$

$$= \frac{(3.8 \times 10^5)(0.22)(0.6)(12.7 \times 10^{-6})(147.7)^2}{249.6}$$

$$= 55.7 \text{ hours},$$

which is much shorter than the duration of the buildup test (i.e., 120 hours). We should note that this pressure-buildup test was simulated without wellbore-storage effects. In practice, however, these effects will be present and may distort or even mask the appearance of the first straight line.

2.5.3 Estimating PV From Flow Tests. For a well centered in a cylindrical drainage area, pseudosteady-state flow of a slightly compressible liquid is modeled by

$$p_i - p_{wf} = \frac{0.0744qBt}{\phi c_t h r_e^2} + \frac{141.2qB\mu}{kh}\left[\ln\left(\frac{r_e}{r_w}\right) - \frac{3}{4} + s\right]. \quad (2.92)$$

Eq. 2.92 predicts that, during pseudosteady-state flow, p_{wf} will plot as a linear function of flowing time, t. The slope, dp_{wf}/dt, of this straight line is

$$\frac{dp_{wf}}{dt} = \frac{-0.0744qB}{\phi h r_e^2 c_t} = \frac{-0.234qB}{c_t V_p}. \quad (2.93)$$

Therefore, PV, V_p (in cubic feet), can be determined from the slope by

$$V_p = \frac{-0.234qB}{c_t\left(dp_{wf}/dt\right)}. \quad (2.94)$$

For wells with drainage-area shapes other than circular (cylindrical), a more general form of the pseudosteady-state-flow equation is

$$p_i - p_{wf} = \frac{0.234qBt}{\phi c_t hA} + \frac{141.2qB\mu}{kh}\left[\frac{1}{2}\ln\left(\frac{10.06A}{C_A r_w^2}\right) - \frac{3}{4} + s\right], \quad (2.95)$$

where A = drainage area of the well, ft^2, and C_A = shape factor,[2] which depends on the drainage-area shape and the location of the well within the drainage area. Similarly, for a generalized reservoir geometry,

$$\frac{dp_{wf}}{dt} = \frac{-0.234qB}{\phi c_t hA} = \frac{-0.234qB}{c_t V_p}. \quad (2.96)$$

Thus, if pseudosteady-state flow is achieved, the reservoir PV can be estimated from the slope of a p_{wf} vs. t graph (Cartesian coordinates) for any drainage-area configuration. The table of shape factors in Appendix D indicates how long a reservoir will be infinite-acting, when the pseudosteady-state-flow equation becomes exact, and when it predicts pressure drawdown within 1%. These times are expressed in the table in terms of dimensionless time, t_{AD}, defined as

$$t_{AD} = 0.0002637kt/\phi\mu c_t A. \quad (2.97)$$

For a well centered in a circular drainage area, the reservoir ceases to become infinite-acting at $t_{AD} = 0.1$. The pseudosteady-state-flow equation becomes exact at $t_{AD} = 0.1$ and is accurate within 1% for $t_{AD} > 0.06$. Thus, for this reservoir shape, there is no transition region, sometimes referred to as the late transient region. By contrast, for a well centered in one quadrant of a square drainage area, the reservoir is infinite-acting only to $t_{AD} = 0.025$, and the pseudosteady-state-flow equation is accurate to within 1% only when $t_{AD} > 0.3$. Thus, a late transient region exists for values of t_{AD} between the limits $0.025 \le t_{AD} \le 0.3$.

Example 2.12—Estimating Reservoir PV From a Constant-Rate Flow Test. A constant-rate pressure-drawdown test (**Table 2.21**) was run on an oil well. During the test, wellbore distortion ended at about 12 hours' flow time. Determine the formation permeability, skin factor, and reservoir PV.

q = 250 STB/D
B = 1.136 bbl/STB
c_t = 17×10^{-6} psi^{-1}
r_w = 0.198 ft
μ = 0.8 cp

TABLE 2.21—CONSTANT-RATE FLOW DATA, EXAMPLE 2.12

t (hours)	p_{wf} (psia)	t (hours)	p_{wf} (psia)
0	4,412	17.3	3,567
0.12	4,272	20.7	3,561
0.20	4,201	24.9	3,555
0.30	4,122	29.8	3,549
0.40	4,057	35.8	3,544
0.50	3,995	43.0	3,537
0.60	3,953	51.5	3,532
0.70	3,912	61.8	3,526
0.80	3,884	74.2	3,521
0.90	3,857	89.1	3,515
1.00	3,822	107	3,509
1.94	3,699	128	3,503
2.79	3,653	154	3,497
4.01	3,636	185	3,490
4.82	3,616	222	3,481
5.78	3,607	266	3,472
6.94	3,600	319	3,460
8.32	3,593	383	3,446
14.40	3,573	460	3,429

$h =$ 69 ft
$\phi =$ 0.039

Solution.

1. Prepare a semilog plot of flowing bottomhole pressure, p_{wf}, as a function of time, t, as **Fig. 2.33** shows. The semilog graph confirms the conclusion reached from type-curve analysis that wellbore storage distortion ended at about 12 hours.

2. Because we have no information about location of boundaries, we assume that boundary effects begin when the drawdown curve in Fig. 2.33 starts to deviate from the established straight line on the semilog graph at a flowing time of 150 hours. The slope of the middle-time line is

$m = 3{,}652 - 3{,}582 = 70$ psi/cycle.

3. The formation permeability is

$$k = \frac{162.6 qB\mu}{mh} = \frac{(162.6)(250)(1.136)(0.8)}{(70)(69)} = 7.65 \text{ md.}$$

4. Now, we check the radius of investigation at the beginning and end of the apparent middle-time straight line to ensure that we are sampling a representative portion of the formation. With Eq. 2.59, the radius of investigation at the beginning ($t = 12$ hours) of the test is

$$r_i = \left(\frac{kt}{948\phi\mu c_t}\right)^{1/2} = \left[\frac{(7.65)(12)}{(948)(0.039)(0.8)(17 \times 10^{-6})}\right]^{1/2}$$

$$= 427 \text{ ft.}$$

Similarly, at the end of the middle-time region ($t = 150$ hours),

$$r_i = \left(\frac{kt}{948\phi\mu c_t}\right)^{1/2} = \left[\frac{(7.65)(150)}{(948)(0.039)(0.8)(17 \times 10^{-6})}\right]^{1/2}$$

$$= 1{,}511 \text{ ft.}$$

A substantial amount of formation has been sampled. Thus, we can be confident that the permeability of 7.65 md is representative of the formation.

5. The skin factor, s, is

$$s = 1.151\left[\frac{(p_i - p_{1hr})}{m} - \log\left(k/\phi\mu c_t r_w^2\right) + 3.23\right]$$

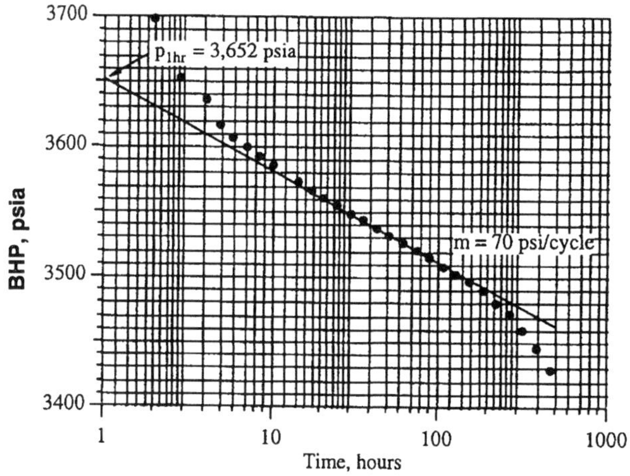

Fig. 2.33—Semilog plot, Example 2.12.

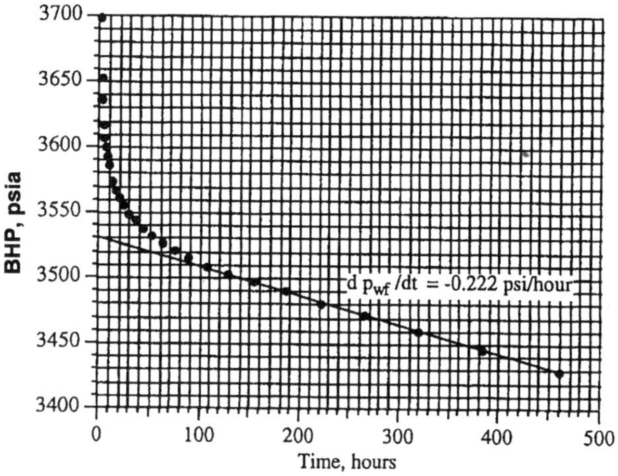

Fig. 2.34—Cartesian plot of test data for estimating reservoir pore volume, Example 2.12.

$$= 1.151\left\{\frac{4{,}412 - 3{,}652}{70} - \log\left[\frac{(7.65)}{(0.039)(0.8)(17 \times 10^{-6})(0.198)^2}\right] + 3.23\right\}$$

$$= 6.4.$$

6. We plot p_{wf} vs. t, as in **Fig. 2.34.** The slope of this curve, dp_{wf}/dt, is constant for $t > 130$ hours and is

$$\frac{dp_{wf}}{dt} = \frac{3{,}531 - 3{,}429}{0 - 460} = -0.222 \text{ psi/hour.}$$

The PV sampled during the test is estimated to be

$$V_p = \frac{-0.234 qB}{c_t\left(dp_{wf}/dt\right)} = \frac{(-0.234)(250)(1.136)}{(17 \times 10^{-6})(-0.222)}$$

$$= 17.61 \times 10^6 \text{ ft}^3$$

or $V_p = 3.14 \times 10^6$ res bbl.

2.6 Analyzing Well Tests With Multiphase Flow

All the previous analysis techniques were derived assuming single-phase flow in the reservoir; however, in well-test analysis, we often encounter situations in which more than one phase is flowing. Therefore, in this section, we discuss a simple yet widely applicable method for analyzing flow and buildup tests complicated by multiphase flow. Perrine[17] developed an approximate method to analyze pressure transient tests in wells that produce oil, gas, and water simultaneously. Martin[18] later provided theoretical justification for Perrine's method, which requires essentially uniform saturations of each phase throughout the drainage area of a well for accurate results. If Perrine's suggestions do not lead to an adequate test analysis, then other methods[19-23] are available. Perrine's method requires the following definitions:

Total rate,

$$(qB)_t = q_o B_o + B_g(q_{gt} - q_o R_s/1{,}000) + q_w B_w. \quad \text{(2.98)}$$

total mobility,

$$\lambda_t = \left(\frac{k}{\mu}\right)_t = \left(\frac{k_o}{\mu_o} + \frac{k_w}{\mu_w} + \frac{k_g}{\mu_g}\right). \quad \text{(2.99)}$$

total compressibility,

$$c_t = c_o S_o + c_g S_g + c_w S_w + c_f, \quad \text{(2.100)}$$

$$\text{where } c_o = \frac{-1}{B_o}\frac{dB_o}{dp} + \frac{B_g}{1000 B_o}\frac{dR_s}{dp}, \quad \text{(2.101)}$$

$$\text{and } c_w = \frac{-1}{B_w}\frac{dB_w}{dp} + \frac{B_g}{1000 B_w}\frac{dR_{sw}}{dp}. \quad \text{(2.102)}$$

The nomenclature defines the other symbols. Oil and water compressibility for a reservoir below the bubblepoint (and, therefore, with R_s and R_{sw} not constant) must be calculated from either experimental B_o, R_s, B_w, and R_{sw} vs. p data or with correlations that include variations of these properties with pressure.

To analyze pressure transient tests from a well producing two or three phases simultaneously, we plot test data just as for a single-phase test. For example, for a pressure-buildup test, we plot

$$p_{ws} \text{ vs. } \log\left[\left(t_p + \Delta t\right)/\Delta t\right].$$

After identifying the end of wellbore-storage distortion with the aid of type curves (used identically as in single-phase test analysis and as discussed in Chap. 3), we identify the middle-time region and determine its slope, m, which is related to the permeability of each phase flowing as

$$k_o = 162.6 q_o B_o \mu_o / mh, \quad \text{(2.103)}$$

$$k_g = \frac{162.6\left(q_{gt} - q_o R_s/1{,}000\right) B_g \mu_g}{mh}, \quad \text{(2.104)}$$

$$\text{and } k_w = 162.6 q_w B_w \mu_w / mh. \quad \text{(2.105)}$$

The total mobility of the system is given by

$$\lambda_t = 162.6 (qB)_t / mh = 162.6/mh$$

$$\times \left[q_o B_o + B_g\left(q_{gt} - q_o R_s/1000\right) + q_w B_w\right]. \quad \text{(2.106)}$$

The total mobility, λ_t, is required for calculations that include the skin factor,

$$s = 1.151\left[\frac{\left(p_{1hr} - p_{wf}\right)}{m} - \log\left(\lambda_t/\phi c_t r_w^2\right) + 3.23\right]. \quad \text{(2.107)}$$

To estimate the current average drainage-area pressure, $\bar{p}$, we proceed just as in a single-phase test, usually the MBH p^* method. For multiphase flow, the dimensionless time and pressure functions are defined, respectively, as

$$t_{AD} = 0.0002637 \lambda_t t_p / \phi c_t A, \quad \text{(2.108)}$$

$$\text{and } p_{\text{MBH,D}} = \frac{2.303(p^* - \bar{p})}{m} = \frac{h(p^* - \bar{p})}{70.6 q B \lambda_t}. \quad \text{(2.109)}$$

Distance to boundaries can be estimated by techniques similar to those used for wells with single-phase flow. For example, when the slope of a semilog graph doubles at late time for a well near a barrier, the distance to the barrier can be estimated with

$$L = \left(0.000148 \lambda_t \Delta t_x / \phi c_t\right)^{1/2}, \quad \text{(2.110)}$$

where Δt_x = shut-in time at which the middle-time line intersects the late-time line, whose slope is double that of the middle-time line. In all these equations used to interpret pressure transient tests, fluid properties should be evaluated at current average drainage-area pressure, $\bar{p}$.

Example 2.13—Analyzing a Pressure-Buildup Test With Multiphase Flow. A pressure-buildup test was run in a well suspected of being in a highly faulted reservoir. In addition to confirming reservoir size, the test also had the objective of confirming a fault quite close to the well. Before shut-in, the well produced oil, gas, and water simultaneously. **Table 2.22** gives the pressure and time data from the buildup test, while other known data are summarized here. Fluid properties were evaluated at p^*, used as an approximation to $\bar{p}$.

Determine the following: (1) the start and end of the middle-time region on a semilog graph, given that type-curve analysis indicates that wellbore-storage distortion ended at about 0.25 hours and that boundary effects began at about 5.3 hours; (2) effective permeabilities k_o, k_w, and k_g; (3) the skin factor, s; (4) the distance, L, to the boundary believed to be near the well; and (5) the current average drainage-area pressure, $\bar{p}$, for this well. Geological evidence suggests that the well is completed one-eighth of the distance from the long edge of a 2 × 1 rectangle.

q_o = 1,100 STB/D
q_w = 4,200 STB/D
q_{gt} = 1,800 Mscf/D
t_p = 20.5 hours
R_s at p^* = 537 scf/STB
R_{si} = 705 scf/STB
μ_o = 0.49 cp
μ_w = 0.231 cp
μ_g = 0.01778 cp
B_o = 1.34 RB/STB
B_w = 1.057 RB/STB
B_g = 1.424 RB/Mscf
c_o = 2.04×10^{-4} psi^{-1}
c_w = 9.79×10^{-6} psi^{-1}

TABLE 2.22—PRESSURE-BUILDUP-TEST DATA, EXAMPLE 2.13

Shut-in Time (hr)	Pressure (psia)	Horner Time Ratio	Shut-in Time (hr)	Pressure (psia)	Horner Time Ratio
0.0	1,658	—	3.0	1,963	7.83
0.1	1,778	206	4.0	1,976	6.13
0.2	1,829	103.5	5.0	1,984	5.10
0.3	1,853	69.3	6.0	1,992	4.42
0.4	1,864	52.3	7.0	1,997	3.93
0.5	1,877	42.0	8.0	2,003	3.56
0.75	1,898	28.3	9	2,008	3.28
1.0	1,912	21.5	10	2,013	3.05
1.5	1,933	14.7	11	2,016	2.86
2.0	1,944	11.3	12	2,021	2.71

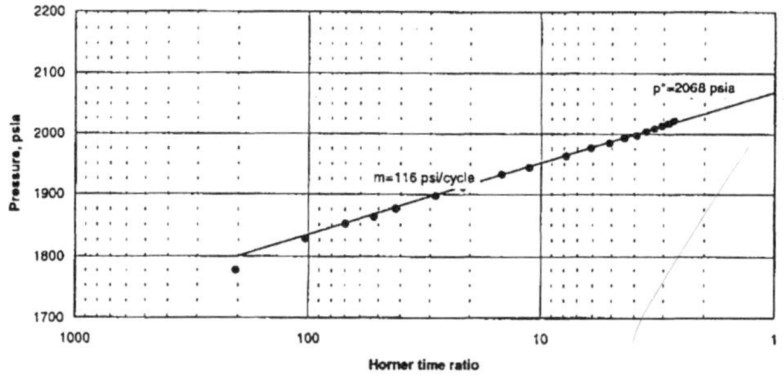

Fig. 2.35—Horner semilog plot, Example 2.13.

$c_g =$ 5.33 × 10^{-4} psi^{-1}
$S_w =$ 0.57
$S_g =$ 0.10
$c_f =$ 3.9 × 10^{-6} psi^{-1}
$\phi =$ 0.165
$h =$ 114 ft
$r_w =$ 0.411 ft
$\gamma_g =$ 0.80 (no N_2, H_2S, or CO_2)
oil gravity = 32°API
$T =$ 250°F
total dissolved solids = 10,000 ppm
$A =$ 23 acres

Solution.

1. Prepare a Horner semilog plot of the data. From **Fig. 2.35,** the middle-time region begins at $(t_p + \Delta t)/\Delta t = 69.3$ or $\Delta t = 0.3$ hours and ends at $(t_p + \Delta t)/\Delta t = 3.93$ or $\Delta t = 7.0$ hours. The slope, m, of the semilog straight line of the middle-time region is

$$m = 116 \text{ psi/cycle}.$$

2. Calculate the total flow rate, $(qB)_t$:

$$(qB)_t = q_o B_o + q_w B_w + \left[q_{gt} - \left(q_o R_s/1,000\right)\right]B_g$$

$$= (1,100)(1.34) + (4,200)(1.057)$$

$$+ \left[1,800 - \frac{(1,100)(527)}{1,000}\right](1.424)$$

$$= 7,635 \text{ RB/D}.$$

3. Calculate the total system mobility and the effective permeabilities to each flowing phase. First, the total system mobility is

$$\lambda_t = \frac{162.6(qB)_t}{mh} = \frac{(162.6)(7,635)}{(116)(114)} = 93.9 \text{ md/cp}.$$

The effective permeabilities to oil, water, and gas are, respectively:

$$k_o = \frac{162.6 q_o B_o \mu_o}{mh} = \frac{(162.6)(1,100)(1.34)(0.49)}{(116)(114)} = 8.9 \text{ md}$$

$$k_w = \frac{162.6 q_w B_w \mu_w}{mh} = \frac{(162.6)(4,200)(1.057)(0.231)}{(116)(114)}$$

$$= 12.6 \text{ md},$$

$$\text{and } k_g = \frac{162.6\left(q_{gt} - \frac{q_o R_s}{1,000}\right)B_g \mu_g}{mh}$$

$$= \frac{(162.6)\left[1,800 - (1,100)(0.537)\right](1.424)(0.01778)}{(116)\ (114)}$$

$$= 0.38 \text{ md}.$$

4. To calculate the skin factor, we need p_{1hr} and c_t. From Fig. 2.35, we find $p_{1hr} = 1,912$ psia at $(t_p + \Delta t)/\Delta t = (20.5 + 1)/1 = 21.5$. Total system compressibility is

$$c_t = c_o S_o + c_g S_g + c_w S_w + c_f$$

$$= \left(2.04 \times 10^{-4}\right)(0.33) + \left(5.33 \times 10^{-4}\right)(0.1)$$

$$+ \left(9.79 \times 10^{-6}\right)(0.57) + \left(3.9 \times 10^{-6}\right)$$

$$= 1.30 \times 10^{-4} \text{psi}^{-1}.$$

The skin factor is computed with the total mobility.

$$s = 1.151\left[\frac{\left(p_{1hr} - p_{wf}\right)}{m} - \log\left(\lambda_t/\phi c_t r_w^2\right) + 3.23\right]$$

$$= 1.151\left\{\frac{(1,912 - 1,658)}{116}\right.$$

$$\left. - \log\left[\frac{93.9}{(0.165)(1.30 \times 10^{-4})(0.411)^2}\right] + 3.23\right\}$$

$$= -2.3.$$

5. Although we see the effects of the boundary on the pressure response, the slope does not double during the test. Thus, we cannot use the methods presented in this chapter to estimate the distance to the boundary. However, we can calculate the radius of investigation at a shut-in time of 7 hours, when boundary effects began to be felt.

$$r_i = \left(\frac{\lambda_t \Delta t}{948\phi c_t}\right)^{1/2} = \left[\frac{(93.9)(7)}{(948)(0.165)(1.3 \times 10^{-4})}\right]^{1/2} = 180 \text{ ft}.$$

We can estimate the distance to the nearest boundary from geological considerations. The well is thought to be one-eighth of the width of a 2 × 1 rectangle from the nearest edge. The length of the short side of a 23-acre, 2 × 1 rectangle is given by

$$2L^2 = 43,560A;$$

$$\text{therefore, } L = \left[\frac{(43,560)(23)}{2}\right]^{1/2} = 88.5 \text{ ft}.$$

We would expect boundary effects to be felt when the radius of investigation reaches the image well; that is, when

$$r_i = 2L = (2)(88.5) = 177 \text{ ft}.$$

Thus, we conclude that the pressure response is consistent with the geological interpretation.

6. To estimate the current average drainage-area pressure, $\bar{p}$, we extrapolate the semilog straight line to obtain $p^* = 2,068$ psia from Fig. 2.35. The MBH dimensionless time group is

$$t_{AD} = \frac{0.0002637\lambda_t t_p}{\phi c_t A} = \frac{(0.0002637)(93.9)(20.5)}{(0.165)(1.3 \times 10^{-4})(1,001,880)}$$

$$= 0.024.$$

From **Fig. 2.36,**

$$p_{MBH,D} = \frac{2.303(p^* - \bar{p})}{m} = -0.55$$

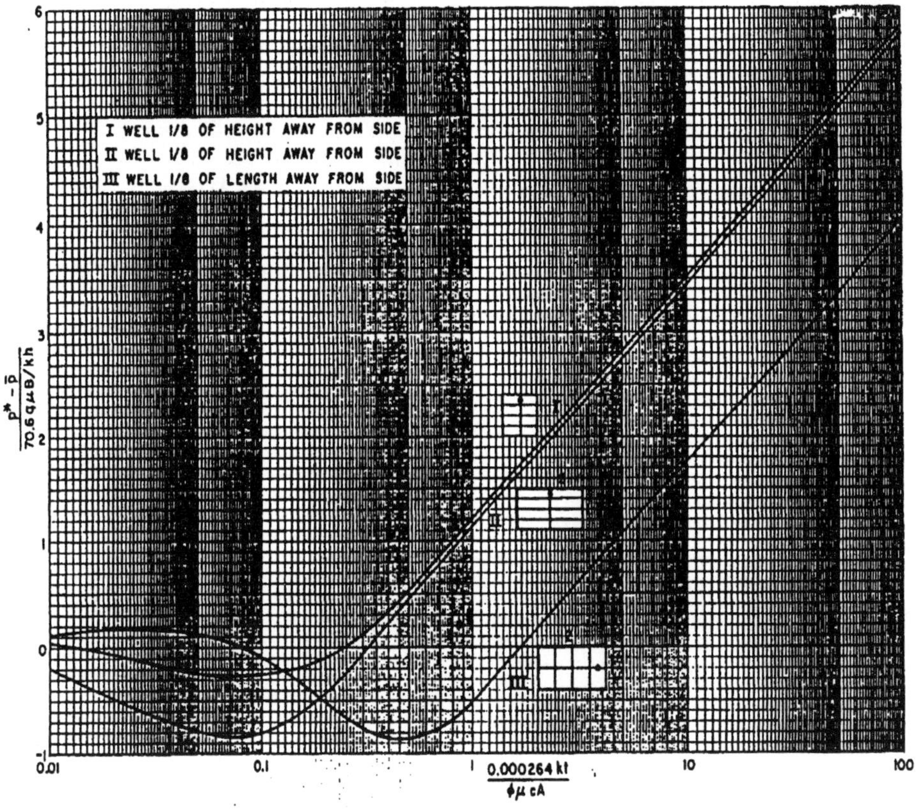

Fig. 2.36—MBH chart for Example 2.13.

$$\text{Then } \bar{p} = p^* - \frac{mp_{\text{MBH,D}}}{2.303} = 2{,}068 - \frac{(116)(-0.55)}{2.303}$$

$$= 2{,}096 \text{ psia.}$$

2.7 Chapter Summary

This chapter introduces the traditional methods for flow- and buildup-test analysis for a well producing a single-phase, slightly compressible liquid from a single-layer reservoir. Sec. 2.1 presents an overview of the chapter.

In Sec. 2.2, we discuss flow tests, also known as drawdown tests, where the pressure response to a known production rate or rates is measured. These tests can be used to estimate permeability, skin factor, reservoir PV, and distance to nearby linear boundaries. Flow tests are often conducted when economic considerations prohibit the use of pressure-buildup tests.

Sec. 2.2.1 presents an analysis method for estimating permeability and skin factor for constant-rate drawdown tests. This method is based on plotting wellbore pressure vs. producing time on a semilog scale and drawing a best-fit straight line through the data.

It is often difficult to maintain a constant flow rate during a flow test. In fact, the inability to maintain a strictly constant flow rate is the major disadvantage of flow tests compared with buildup tests. Accordingly, in the following two subsections, we present methods for analyzing flow tests where the rate is a function of time.

Sec. 2.2.2 introduces the Winestock and Colpitts[3] method of analysis, which is applicable to flow tests where the flow rate is changing slowly and smoothly. In this technique, we plot the pressure change divided by the production rate vs. the production time, again on semilog coordinates. The analysis provides estimates of formation permeability and skin factor.

Sec. 2.2.3 presents an analysis method applicable to a well that has had n discrete rate changes during its production history. This method is based on the assumption that the logarithmic approximation to the Ei function is applicable to each production period; i.e., the reservoir is infinite-acting throughout the entire production history. We considered both the special case of a two-rate test and the more general case of an n-rate test. From the two-rate test, it is possible to solve for the initial pressure, p_i, in addition to permeability and skin factor; for the n-rate test, the initial pressure must be known.

In Sec. 2.3, we turn our attention to pressure-buildup tests. As noted earlier, flow tests are subject to variation in rate, which adversely affects our ability to analyze the tests. By shutting in a well, we can impose a strictly constant surface rate on the well. Thus, pressure-buildup tests traditionally have been used in lieu of flow tests for measuring permeability and skin factor.

Sec. 2.3.1 presents the Horner analysis method, which is applicable to pressure-buildup tests where the well is shut in following a constant-rate flow period. To use this method, we plot the shut-in wellbore pressure vs. the Horner time ratio, defined as $(t_p + \Delta t)/\Delta t$, on semilog coordinates. The Horner method provides estimates of formation permeability, skin factor, and, for infinite-acting reservoirs, original reservoir pressure.

Secs. 2.3.2 and 2.3.3 present analysis techniques for wells where the shut-in period is preceded by two and by $n-1$ different flow rates, respectively. Both these methods are based on the use of superposition, where the logarithmic approximation to the Ei-function solution describes the contribution to the pressure drop from each producing period. Again, both methods provide estimates of formation permeability, skin factor, and initial reservoir pressure.

We also presented two alternatives to the superposition method, for wells in which the number of flow-rate changes makes the superposition analysis inconvenient. The Odeh and Selig method is applicable to tests where the shut-in time is greater than the actual producing time. This method is applied by modifying the production time and flow rate with Eqs. 2.50 and 2.51, then applying Horner's method with these modified values.

Horner's approximation is often used when the flow period immediately preceding the shut-in is comparable in length with the shut-in period. To use this approximation, we compute a pseudoproducing time t_p as

$$t_{pH} = 24N_p/q_{\text{last}}, \qquad (2.111)$$

then use this value in place of the actual producing time in the Horner method.

Actual tests invariably are affected by any of a number of factors that cause the pressure response to deviate from the ideal response assumed in developing the methods considered in Secs. 2.2 and 2.3.

In Sec. 2.4, we discuss a number of these factors, including reservoir heterogeneity, reservoir boundaries, wellbore storage, and wellbore damage and stimulation.

Sec. 2.4.1 introduces the concept of radius of investigation. The radius of investigation is given by

$$r_i = \sqrt{\frac{kt}{948\phi\mu c_t}} \quad \text{(2.112)}$$

for a flow test and by

$$r_i = \sqrt{k\Delta t/948\phi\mu c_t} \quad \text{(2.113)}$$

for a buildup test. The radius of investigation provides a framework for understanding the influence of reservoir heterogeneities on the wellbore-pressure response. Heterogeneities outside the radius of investigation have no effect on the wellbore response. As the duration of the test increases, the radius of investigation also increases, as does the likelihood that significant heterogeneities will be encountered.

In Sec. 2.4.2, we discuss the three time regions (early, middle, and late) into which flow- or buildup-test data are often divided. During the early-time region, the pressure response deviates from the ideal owing to wellbore storage and/or wellbore damage or stimulation. During the middle-time region, the pressure response corresponds to the ideal response on which the analysis methods presented in this chapter are based. Thus, a straight line drawn through the data in this region is often referred to as the "correct semilog straight line." During the late-time region, the pressure response again deviates from the straight-line response, because of the presence of significant reservoir heterogeneities, interference from other wells, or reservoir boundaries.

Sec. 2.4.3 discusses wellbore-storage effects, also known as wellbore unloading for flow tests and afterflow for buildup tests. This phenomenon results from the fact that the sandface flow rate lags the surface flow rate when the wellbore has a finite volume. We characterize the degree of wellbore storage by the wellbore-storage coefficient, C. We present expressions to estimate the wellbore-storage coefficient either for a rising or falling gas/liquid interface (Eq. 2.62) or for a wellbore containing a single-phase fluid (Eq. 2.63).

Sec. 2.4.4 discusses various ways of characterizing the degree of damage or stimulation in the near-wellbore region. The skin factor is often used to characterize an alteration in the near-wellbore region, with positive values representing damage and negative values representing stimulation. The simplest skin-effect model, applicable only to damaged skin, consists of an infinitesimally thin zone of reduced permeability adjacent to the wellbore, causing a step change in the pressure profile across the skin zone. Another model involves a two-zone reservoir, with altered permeability k_s within a radius r_s of the wellbore. This model can represent either damage or stimulation. An alternative model involves the effective wellbore radius, r_{wa}. For damaged wells, r_{wa} is smaller than the actual radius r_w; for stimulated wells, it is larger than r_w. For wells with a high-conductivity hydraulic fracture, the fracture half-length may be related to r_{wa} by $L_f = 2r_{wa}$. The skin factor may also be characterized by the additional pressure drop associated with the damaged zone,

$$\Delta p_s = \left(\frac{141.2qB\mu}{kh}\right)s. \quad \text{(2.114)}$$

A final method of characterizing the skin effect is the flow efficiency, E, where values less than 1 represent damage, and values greater than 1 represent stimulation.

The next two sections discuss the apparent skin factor owing to the presence of either an incompletely perforated interval or partial penetration (Sec. 2.4.5) or a deviated wellbore (Sec. 2.4.6.)

In Sec. 2.5, we turn our attention to the late-time region, discussing methods of estimating drainage-area pressure, distance to boundaries, and reservoir PV.

Sec. 2.5.1 introduces three methods for estimating the drainage-area pressure. If at least one side of the reservoir is still infinite-acting, the Horner straight line may be extrapolated to infinite shut-in time to obtain the initial pressure, p_i. If the reservoir is finite-acting and the drainage-area shape and well placement within the drainage area are known, the MBH method is applicable. The Horner straight line is extrapolated to infinite shut-in time as in the case of an infinite acting reservoir, but the extrapolated pressure p^* must then be corrected by use of the appropriate MBH function to obtain average drainage-area pressure, $\bar{p}$. The modified Muskat method is limited to wells centered in the drainage area. Because it depends on the availability of pressure-buildup data within the transition from the middle-time region to the boundary-dominated region, it is restricted in its applicability.

TABLE 2.23—SOURCES FOR ROCK AND FLUID PROPERTIES

Property	Suggested Sources of Property
Rock Porosity, ϕ (Fraction)	Core analysis or well logs
Net Pay Thickness, h (ft)	Drilling reports or well logs
Formation Compressibility, c_f (psi^{-1}):	Laboratory measurements or correlations (Eq. L-88 or L-89, Appendix L)
Fluid Compressibilities (psi^{-1}):	Laboratory measurements of fluid samples, or use the following correlations from Appendix L:
Oil, c_o	Eq. L-65 or L-66
Water, c_w	Eq. L-82 or L-83
Gas, c_g	Eq. L-44
Fluid Saturations: S_o, S_w, S_g (fraction)	Core analyses or well logs
Fluid Viscosities (cp):	Laboratory measurements of fluid samples, or use the following correlations from Appendix L:
μ_o	Eq. L-68 or L-71
μ_w	Eqs. L-84 and L-87
μ_g	Eq. L-46
Formation Volume Factors:	Laboratory measurements of fluid samples, or calculate as follows:
B_o	Eq. L-63
B_w	Eq. L-73
B_g	Eq. L-8
Total Compressibility, c_t (psi^{-1})	$c_oS_o + c_wS_w + c_gS_g + c_f$

In Sec. 2.5.2, we discuss a method of determining the distance to a linear no-flow boundary when the slope of a buildup-test plot doubles during the test.

Sec. 2.5.3 presents a method for estimating reservoir PV from a Cartesian graph of wellbore pressure vs. time for a well produced at constant rate.

Finally, in Sec. 2.6, we adapt the techniques presented in earlier sections for single-phase flow to the analysis of tests where more than one phase is flowing.

Successful application of the well-test analysis techniques presented in this chapter requires rock and fluid properties. **Table 2.23** summarizes pertinent data requirements and suggests sources for estimating these data.

Exercises

1. Estimate formation permeability and skin factor from the drawdown test data given the following formation and fluid properties: $q = 90$ STB/D; $p_i = 1479$ psia; $h = 17$ ft; $\phi = 25\%$; $r_w = 0.21$ ft; $B = 1.1$ RB/STB; $c_t = 7.65 \times 10^{-6}$ psi^{-1}; and $\mu = 7.86$ cp.

DRAWDOWN TEST DATA FOR EXERCISE 2.1

Time (hours)	Pressure (psi)	Time (hours)	Pressure (psi)
1	283.8	10	140.0
2	239.8	11	134.1
3	214.5	12	128.8
4	196.6	13	123.9
5	182.8	14	119.3
6	171.5	15	115.1
7	162.0	16	111.1
8	153.8	17	107.4
9	146.5	18	103.9

2. Estimate formation permeability and skin factor from the drawdown test data given the following formation and fluid properties: $q = 90$ STB/D; $p_i = 2140$ psia; $h = 5$ ft; $\phi = 21.7\%$; $r_w = 0.49$ ft; $B = 1.091$ RB/STB; $c_t = 7.8 \times 10^{-6}$ psi^{-1}; and $\mu = 2.44$ cp.

DRAWDOWN TEST DATA FOR EXERCISE 2.2

Time (minutes)	Pressure (psi)	Time (minutes)	Pressure (psi)
15	538.8	195	411.6
30	499.2	210	408.3
45	479.1	225	405.2
60	465.4	240	402.4
75	455.0	255	399.7
90	446.6	270	397.1
105	439.6	285	394.7
120	433.5	300	392.4
135	428.1	315	390.3
150	423.4	330	388.2
165	419.1	345	386.2
180	415.2	360	384.3

3. Estimate formation permeability and skin factor from the drawdown test data given the following formation and fluid properties: $q = 777$ STB/D; $p_i = 8294.5$ psia; $h = 98$ ft; $\phi = 18.6\%$; $r_w = 0.24$ ft; $B = 1.342$ RB/STB; $c_t = 11.4 \times 10^{-6}$ psi^{-1}; and $\mu = 0.895$ cp.

TABLE 2.26–DRAWDOWN TEST DATA FOR EXERCISE 2.3

Time (hours)	Pressure (psi)	Time (hours)	Pressure (psi)
1	7811.9	14	7667.6
2	7774.0	16	7660.3
3	7751.8	18	7653.9
4	7736.1	21	7645.4
5	7723.9	24	7638.1
6	7713.9	27	7631.7
8	7698.2	30	7625.9
10	7686.0	33	7620.7
12	7676.0	36	7616.0

4. Estimate formation permeability and skin factor from the drawdown test data given the following formation and fluid properties: $q = 17$ STB/D; $p_i = 788.5$ psia; $h = 37$ ft; $\phi = 16.5$ %; $r_w = 0.24$ ft; $B = 1.046$ RB/STB; $c_t = 6.7 \times 10^{-6}$ psi^{-1}; and $\mu = 13.7$ cp.

DRAWDOWN TEST DATA FOR EXERCISE 2.4

Time (hours)	Pressure (psi)	Time (hours)	Pressure (psi)
1.00	291.1	18.5	248.3
2.10	279.8	21.4	246.3
3.31	273.0	24.5	244.4
4.64	268.1	28.0	242.5
6.11	264.2	31.8	240.7
7.72	260.8	35.9	239.0
9.49	257.9	40.5	237.3
11.4	255.2	45.6	235.6
13.6	252.8	48.0	234.9
15.9	250.5		

5. Estimate formation permeability and skin factor from the drawdown test data given the following formation and fluid properties: $q = 50$ STB/D; $p_i = 1326.6$ psia; $h = 25$ ft; $\phi = 27.6$ %; $r_w = 0.36$ ft; $B = 1.099$ RB/STB; $c_t = 9.4 \times 10^{-6}$ psi^{-1}; and $\mu = 5.28$ cp. Where do you draw the straight line?

DRAWDOWN TEST DATA FOR EXERCISE 2.5

Time (hours)	Pressure (psi)	Time (hours)	Pressure (psi)	Time (hours)	Pressure (psi)
0.100	1109.0	2.60	493.7	15.47	386.1
0.220	937.0	3.22	478.9	18.67	375.8
0.364	805.1	3.96	465.4	22.50	365.6
0.537	707.8	4.85	452.8	27.10	355.5
0.744	638.4	5.92	440.8	32.62	345.5
0.993	589.9	7.20	429.3	39.25	335.5
1.29	555.5	8.74	418.1	47.20	325.6
1.65	530.2	10.59	407.2	48.00	324.7
2.08	510.3	12.81	396.6		

6. Estimate formation permeability and skin factor from the drawdown test data given the following formation and fluid properties: $q = 1200$ STB/D; $p_i = 4474.4$ psia; $h = 26$ ft; $\phi = 21.6$ %; $r_w = 0.22$ ft; $B = 1.52$ RB/STB; $c_t = 16.6 \times 10^{-6}$ psi^{-1}; and $\mu = 0.29$ cp. Where do you draw the straight line?

DRAWDOWN TEST DATA FOR EXERCISE 2.6					
Time (hours)	Pressure (psi)	Time (hours)	Pressure (psi)	Time (hours)	Pressure (psi)
0.010	4449.3	0.592	3447.8	6.82	2340.6
0.022	4419.7	0.720	3310.4	8.19	2319.5
0.036	4384.8	0.874	3171.5	9.84	2299.8
0.054	4343.9	1.06	3035.0	11.8	2281.1
0.074	4296.2	1.28	2905.3	14.2	2263.1
0.099	4240.7	1.55	2786.8	17.0	2245.6
0.129	4176.7	1.87	2682.6	20.5	2228.5
0.165	4103.1	2.25	2595.0	24.6	2211.7
0.208	4019.5	2.71	2524.0	29.5	2195.1
0.260	3925.2	3.26	2468.4	35.4	2178.7
0.322	3820.2	3.92	2425.3	42.5	2162.5
0.396	3704.9	4.72	2391.6	48.0	2151.8
0.485	3580.1	5.67	2364.1		

7. Estimate formation permeability and skin factor from the drawdown test data given the following formation and fluid properties: $p_i = 4521.6$ psia; $h = 13$ ft; $\phi = 3.88$ %; $r_w = 0.25$ ft; $B = 1.30$ RB/STB; $c_t = 8.44 \times 10^{-6}$ psi^{-1}; and $\mu = 1.73$ cp. Note that the pressure is actually rising at the end of the test, in spite of the fact that this is a flow test.

DRAWDOWN TEST DATA FOR EXERCISE 2.7					
Time (hours)	Rate (STB/D)	Pressure (psi)	Time (hours)	Rate (STB/D)	Pressure (psi)
1	249	4329.5	12	238	4310.2
2	248	4322.3	14	236	4310.2
3	247	4318.4	16	234	4310.4
4	246	4315.9	18	232	4310.9
5	245	4314.2	21	229	4311.8
6	244	4312.9	24	226	4313.0
7	243	4312.0	27	223	4314.4
8	242	4311.3	30	220	4316.0
9	241	4310.9	36	214	4319.5
10	240	4310.5	42	208	4323.3
11	239	4310.3	48	202	4327.5

8. Estimate formation permeability and skin factor from the drawdown test data given the following formation and fluid properties: $p_i = 2782.78$ psia; $p_{wf} = 1800$ psi; $h = 19$ ft; $\phi = 7.5$ %; $r_w = 0.37$ ft; $B = 1.166$ RB/STB; $c_t = 12.1 \times 10^{-6}$ psi^{-1}; and $\mu = 2.999$ cp. Note that flowing bottomhole pressure is constant.

DRAWDOWN TEST DATA FOR EXERCISE 2.8					
Time (hours)	Rate (STB/D)	Time (hours)	Rate (STB/D)	Time (hours)	Rate (STB/D)
6	70.0	144	40.2	384	36.0
12	56.9	168	39.5	408	35.8
18	53.2	192	38.9	432	35.6
24	50.9	216	38.4	456	35.4
36	48.0	240	38.0	480	35.2
48	46.1	264	37.6	528	34.9
60	44.8	288	37.2	576	34.5
72	43.8	312	36.9	624	34.3
96	42.2	336	36.6	672	34.0
120	41.1	360	36.3	720	33.7

9. A well is produced at 100 STB/D for 72 hours. The rate is then increased to 200 STB/D for 24 hours. Estimate formation permeability, skin factor, and initial pressure from the two-rate test data given the following formation and fluid properties: $q_1 = 100$ STB/D; $t_{p1} = 72$ hr; $q_2 = 200$ STB/D; $p_{wf1} = 4403.44$ psi; $h = 12$ ft; $\phi = 10.9$ %; $r_w = 0.32$ ft; $B = 1.263$ RB/STB; $c_t = 12.2 \times 10^{-6}$ psi^{-1}; and $\mu = 1.067$ cp.

TWO-RATE TEST DATA FOR EXERCISE 2.9					
Time (hours)	Pressure (psi)	Time (hours)	Pressure (psi)	Time (hours)	Pressure (psi)
0.109	4257.15	0.792	4228.84	6.517	4199.29
0.149	4252.51	0.96	4226.18	8.677	4195.08
0.201	4248.10	1.552	4219.53	10.502	4192.21
0.245	4245.24	2.07	4215.55	15.502	4186.19
0.299	4242.43	2.506	4212.88	20.502	4181.69
0.401	4238.29	3.035	4210.20	24.000	4179.07
0.538	4234.21	4.043	4206.15		
0.653	4231.52	5.384	4202.06		

10. A well is produced at 50 STB/D for 72 hrs. The rate is then decreased to 25 STB/D for 24 hours. Estimate formation permeability, skin factor, and initial pressure from the two-rate test data given the following formation and fluid properties: $q_1 = 50$ STB/D; $t_{p1} = 72$ hr; $q_2 = 25$ STB/D; $p_{wf1} = 1142.24$ psi; $h = 43$ ft; $\phi = 8.2$ %; $r_w = 0.45$ ft; $B = 1.143$ RB/STB; $c_t = 10.5 \times 10^{-6}$ psi^{-1}; and $\mu = 1.278$ cp.

TWO-RATE TEST DATA FOR EXERCISE 2.10					
Time (hours)	Pressure (psi)	Time (hours)	Pressure (psi)	Time (hours)	Pressure (psi)
0.109	1354.12	0.719	1400.01	5.923	1414.25
0.149	1371.25	0.872	1401.46	7.170	1415.34
0.201	1382.65	1.057	1402.87	7.887	1415.87
0.245	1387.69	1.552	1405.61	9.546	1416.90
0.299	1391.29	2.070	1407.58	10.502	1417.39
0.401	1395.01	3.035	1410.12	15.502	1419.25
0.488	1396.87	4.043	1411.94	19.502	1420.20
0.592	1398.50	5.384	1413.68	24.000	1420.96

11. A drawdown test is run on an oil well at a series of three different rates, with each flow period lasting 3 hours. Given the following formation and fluid properties, estimate formation permeability and skin factor from the test data using the multirate analysis method: $q_1 = 15$ STB/D; $q_2 = 30$ STB/D; $q_3 = 45$ STB/D; $p_i = 5883.16$ psi; $h = 13$ ft; $\phi = 17.7$ %; $r_w = 0.42$ ft; $B = 1.231$ RB/STB; $c_t = 9.79 \times 10^{-6}$ psi^{-1}; and $\mu = 3.371$ cp.

TEST DATA FOR EXERCISE 2.11					
Time (hours)	Pressure (psi)	Time (hours)	Pressure (psi)	Time (hours)	Pressure (psi)
0.5	5349.68	3.5	4778.26	6.5	4195.34
1	5335.65	4	4761.74	7	4177.45
1.5	5327.76	4.5	4751.66	7.5	4166.1
2	5322.27	5	4744.21	8	4157.46
2.5	5318.05	5.5	4738.22	8.5	4150.36
3	5314.62	6	4733.19	9	4144.26

12. A drawdown test is run on an oil well at a series of three different rates, with each flow period lasting 4 hours. Given the following formation and fluid properties, estimate formation permeability and skin factor from the test data using the multirate analysis method: $q_1 = 250$ STB/D; $q_2 = 225$ STB/D; $q_3 = 200$ STB/D; $p_i = 2003.80$ psi; $h = 33$ ft; $\phi = 29.5$ %; $r_w = 0.2$ ft; $B = 1.181$ RB/STB; $c_t = 12.0 \times 10^{-6}$ psi^{-1}; and $\mu = 1.661$ cp.

TEST DATA FOR EXERCISE 2.12					
Time (hours)	Pressure (psi)	Time (hours)	Pressure (psi)	Time (hours)	Pressure (psi)
0.5	1810.32	4.5	1779.35	8.5	1789.21
1.0	1794.41	5.0	1778.54	9.0	1789.74
1.5	1785.12	5.5	1777.29	9.5	1789.65
2.0	1778.54	6.0	1775.96	10.0	1789.34
2.5	1773.44	6.5	1774.64	10.5	1788.91
3.0	1769.27	7.0	1773.37	11.0	1788.44
3.5	1765.75	7.5	1772.15	11.5	1787.93
4.0	1762.70	8.0	1770.98	12.0	1787.41

13. Estimate formation permeability and skin factor from the buildup test data given the following formation and fluid properties: $q = 1000$ STB/D; $t_p = 72$ hr; $p_{wf} = 3319.1$ psia; $h = 12$ ft; $\phi = 25.7\%$; $r_w = 0.32$ ft; $B = 1.41$ RB/STB; $c_t = 12.5 \times 10^{-6}$ psi^{-1}; and $\mu = 0.53$ cp.

BUILDUP TEST DATA FOR EXERCISE 2.13					
Time (hours)	Pressure (psi)	Time (hours)	Pressure (psi)	Time (hours)	Pressure (psi)
0.100	5155.7	0.993	5471.2	3.215	5626.1
0.220	5266.4	1.292	5506.3	3.958	5652.9
0.364	5335.5	1.650	5538.8	4.850	5678.7
0.537	5388.3	2.080	5569.3	5.920	5703.7
0.744	5432.4	2.596	5598.3	6.000	5705.4

14. Estimate formation permeability and skin factor from the buildup test data given the following formation and fluid properties: $q = 1000$ STB/D; $t_p = 72$ hr; $p_{wf} = 2096.2$ psia; $h = 55$ ft; $\phi = 28.2\%$; $r_w = 0.37$ ft; $B = 1.155$ RB/STB; $c_t = 10.7 \times 10^{-6}$ psi^{-1}; and $\mu = 2.48$ cp.

BUILDUP TEST DATA FOR EXERCISE 2.14					
Time (hours)	Pressure (psi)	Time (hours)	Pressure (psi)	Time (hours)	Pressure (psi)
0.100	2833.3	0.993	2884.0	3.215	2909.2
0.220	2850.9	1.292	2889.7	3.958	2913.5
0.364	2862.0	1.650	2895.0	4.850	2917.7
0.537	2870.6	2.080	2899.9	5.920	2921.8
0.744	2877.7	2.596	2904.7	6.000	2922.1

15. Estimate formation permeability and skin factor from the buildup test data given the following formation and fluid properties: $q = 7$ STB/D; $t_p = 144$ hr; $p_{wf} = 858.3$ psia; $h = 16$ ft; $\phi = 12.8$ %; $r_w = 0.37$ ft; $B = 1.077$ RB/STB; $c_t = 7.8 \times 10^{-6}$ psi^{-1}; and $\mu = 4.44$ cp.

BUILDUP TEST DATA FOR EXERCISE 2.15					
Time (hours)	Pressure (psi)	Time (hours)	Pressure (psi)	Time (hours)	Pressure (psi)
0.100	1001.55	2.080	1018.79	10.593	1027.51
0.220	1006.20	2.596	1020.00	12.812	1028.48
0.364	1009.08	3.215	1021.16	15.474	1029.43
0.537	1011.27	3.958	1022.29	18.669	1030.36
0.744	1013.10	4.850	1023.38	22.503	1031.27
0.993	1014.71	5.920	1024.45	24.000	1031.58
1.292	1016.16	7.204	1025.49		
1.650	1017.51	8.744	1026.51		

16. Estimate formation permeability and skin factor from the buildup test data given the following formation and fluid properties: $q = 225$ STB/D; $t_p = 144$ hr; $p_{wf} = 6399.66$ psia; $h = 5$ ft; $\phi = 7.9$ %; $r_w = 0.33$ ft; $B = 1.261$ RB/STB; $c_t = 8.45 \times 10^{-6}$ psi^{-1}; and $\mu = 2.43$ cp.

BUILDUP TEST DATA FOR EXERCISE 2.16					
Time (hours)	Pressure (psi)	Time (hours)	Pressure (psi)	Time (hours)	Pressure (psi)
0.05	6552.27	2.42	8021.74	23.6	8503.65
0.11	6707.23	2.96	8078.99	28.4	8533.47
0.182	6864.04	3.60	8130.31	34.1	8562.13
0.268	7020.47	4.37	8177.16	41.0	8589.59
0.372	7173.63	5.30	8220.64	49.2	8615.78
0.496	7320.43	6.41	8261.51	59.1	8640.64
0.646	7457.95	7.74	8300.33	71.0	8664.10
0.825	7583.84	9.33	8337.44	85.2	8686.10
1.040	7696.62	11.25	8373.10	102.3	8706.57
1.298	7795.84	13.55	8407.47	122.8	8725.50
1.608	7882.09	16.31	8440.65	144.0	8740.72
1.979	7956.77	19.62	8472.71		

17. Estimate formation permeability and skin factor from the buildup test data given the following formation and fluid properties: $h = 5$ ft; $\phi = 20.7$ %; $r_w = 0.37$ ft; $B = 1.356$ RB/STB; $c_t = 15.45 \times 10^{-6}$ psi^{-1}; and $\mu = 0.658$ cp.

BUILDUP TEST DATA FOR EXERCISE 2.17					
Time (hours)	Pressure (psi)	Time (hours)	Pressure (psi)	Time (hours)	Pressure (psi)
0.0	2595.04	0.344	3008.87	4.22	3081.57
0.019	2660.54	0.435	3026.09	5.27	3085.13
0.029	2689.87	0.548	3038.36	6.60	3088.65
0.041	2723.10	0.691	3047.01	8.25	3092.13
0.056	2759.94	0.868	3053.34	10.32	3095.58
0.075	2799.64	1.090	3058.35	12.90	3099.01
0.099	2840.96	1.368	3062.68	16.14	3102.41
0.129	2882.19	1.715	3066.69	20.17	3105.79
0.166	2921.28	2.15	3070.53	24.00	3108.40
0.213	2956.28	2.69	3074.28		
0.271	2985.70	3.37	3077.95		

18. Estimate formation permeability and skin factor from the buildup test data given the following formation and fluid properties: $h = 62$ ft; $\phi = 21.5$ %; $r_w = 0.26$ ft; $B = 1.163$ RB/STB; $c_t = 8.38 \times 10^{-6}$ psi^{-1}; and $\mu = 2.19$ cp. Note that the short final flow period makes use of the Horner pseudoproducing time approximation inappropriate.

BUILDUP TEST DATA FOR EXERCISE 2.18

Time (hours)	Rate (STB/D)
144	180
6	90

Time (hours)	Pressure (psi)	Time (hours)	Pressure (psi)	Time (hours)	Pressure (psi)
0	5048.0	0.213	5476.1	2.69	5519.1
0.005	5167.4	0.271	5480.2	3.37	5523.3
0.011	5265.2	0.344	5484.1	4.22	5527.7
0.019	5339.0	0.435	5488.0	5.27	5532.2
0.029	5390.0	0.548	5491.8	6.60	5536.8
0.041	5422.4	0.691	5495.6	8.25	5541.5
0.056	5442.0	0.868	5499.3	10.32	5546.3
0.075	5453.8	1.090	5503.2	12.90	5551.2
0.099	5461.5	1.368	5507.0	16.14	5556.2
0.129	5467.1	1.715	5511.0	20.17	5561.1
0.166	5471.8	2.15	5515.0	24.00	5564.9

19. Estimate formation permeability and skin factor from the buildup test data given the following formation and fluid properties: $h = 11$ ft; $\phi = 19$ %; $r_w = 0.34$ ft; $B = 1.19$ RB/STB; $c_t = 10.5 \times 10^{-6}$ psi^{-1}; and $\mu = 5.91$ cp.

BUILDUP TEST DATA FOR EXERCISE 2.19

Time (hours)	Rate (STB/D)
24	180
72	250
6	0
18	120

Time (hours)	Pressure (psi)	Time (hours)	Pressure (psi)	Time (hours)	Pressure (psi)
0	2526.72	0.516	3258.30	4.36	3338.32
0.020	2645.98	0.665	3274.49	5.47	3345.02
0.045	2765.76	0.851	3286.08	6.86	3351.69
0.076	2881.13	1.084	3295.24	8.59	3358.35
0.115	2986.13	1.375	3303.22	10.76	3365.02
0.164	3075.48	1.739	3310.64	13.47	3371.70
0.225	3146.16	2.19	3317.78	16.86	3378.40
0.301	3198.19	2.76	3324.73	21.1	3385.09
0.397	3234.23	3.47	3331.57	24.0	3388.94

20. Estimate formation permeability and skin factor from the buildup test data given the following formation and fluid properties: $h = 56$ ft; $\phi = 15.6$ %; $r_w = 0.4$ ft; $B = 1.232$ RB/STB; $c_t = 10.1 \times 10^{-6}$ psi^{-1}; and $\mu = 1.35$ cp.

BUILDUP TEST DATA FOR EXERCISE 2.20

Time (hours)	Rate (STB/D)
4	135
4	90
4	45

Time (hours)	Pressure (psi)	Time (hours)	Pressure (psi)	Time (hours)	Pressure (psi)
0	6013.03	0.269	6193.72	3.20	6213.09
0.006	6027.74	0.334	6197.96	3.91	6214.10
0.013	6043.98	0.414	6200.83	4.77	6215.09
0.022	6061.61	0.511	6202.82	5.83	6216.04
0.033	6080.29	0.630	6204.31	7.11	6216.96
0.046	6099.54	0.774	6205.56	8.69	6217.83
0.063	6118.68	0.950	6206.70	10.60	6218.65
0.082	6136.89	1.166	6207.80	12.94	6219.42
0.107	6153.39	1.428	6208.88	15.79	6220.12
0.136	6167.52	1.748	6209.95	19.27	6220.76
0.172	6178.90	2.14	6211.01	23.52	6221.34
0.216	6187.54	2.62	6212.06	24.00	6221.39

21. Given the following formation and fluid properties, calculate the radius of investigation at 1 hour, 10 hours, 100 hours, and 1000 hours: $h = 23$ ft; $\phi = 16.5\%$; $r_w = 0.32$ ft; $B = 1.322$ RB/STB; $c_t = 14.3 \times 10^{-6}$ psi^{-1}; $\mu = 1.53$ cp; $q = 83$ STB/D; and $k = 147$ md.

22. Given the following formation and fluid properties, calculate the radius of investigation at 1 hour, 10 hours, 100 hours, and 1000 hours: $h = 49$ ft; $\phi = 8.9$ %; $r_w = 0.25$ ft; $B = 1.17$ RB/STB; $c_t = 122 \times 10^{-6}$ psi^{-1}; $\mu = 0.35$ cp; $q = 247$ STB/D; and $k = 12$ md.

23. Given the following formation and fluid properties, calculate the time required to reach a radius of investigation of 10 feet, 50 feet, 100 feet, and 500 feet: $h = 63$ ft; $\phi = 19.7$ %; $r_w = 0.23$ ft; $B = 1.26$ RB/STB; $c_t = 12.9 \times 10^{-6}$ psi^{-1}; $\mu = 6.77$ cp; $q = 28$ STB/D; and $k = 252$ md.

24. Given the following formation and fluid properties, calculate the time required to reach a radius of investigation of 10 feet, 50 feet, 100 feet, and 500 feet: $h = 7$ ft; $\phi = 12.2$ %; $r_w = 0.33$ ft; $B = 1.131$ RB/STB; $c_t = 9.8 \times 10^{-6}$ psi^{-1}; $\mu = 12.3$ cp; $q = 7$ STB/D; and $k = 39$ md.

25. Calculate the wellbore storage coefficient for a wellbore filled with water. The well has 2273 feet of 3½-in. tubing (2.992 inches ID), with 327 ft of 6-inch casing (5 inches ID) below the packer. The average compressibility of the water in the wellbore is 3.6×10^{-6} psi^{-1}.

26. Calculate the wellbore storage coefficient for a wellbore filled with gas. The well has 5743 ft of 2⅞-in. tubing, (2.441 inches ID), with 517 ft of 5½-in. casing (4.950 inches ID) below the packer. The average compressibility of the gas in the wellbore is 2.7×10^{-4} psi^{-1}.

27. Calculate the wellbore storage coefficient for a water injection well with a changing liquid level. The well has 3½-in. tubing, (2.992 inches ID) and 6-inch casing (5 inches ID). The gas-liquid interface is within the tubing. The tubing is isolated from the tubing-casing annulus by a packer. The density of the liquid in the wellbore is 1.04 g/cm^3.

28. Calculate the wellbore storage coefficient for a pumping well with a rising liquid level. The well has 2⅞-in. tubing (2.441 inches ID) and 5½-in. casing (4.950 inches ID). Both the tubing and the tubing-casing annulus are in communication with the formation. The density of the liquid in the wellbore is 0.82 g/cm^3.

29. Analysis of a pressure buildup test showed that an oil well with a productivity problem had a skin factor of 8.2 and an effective permeability to oil of 19.1 md. Prior to shutting in the well, the flowing bottomhole pressure at the time of shutin was 2309 psi, and the well produced oil at a constant rate of 105 STB/D. Estimate the following: a) altered zone permeability, k_s, for assumed altered zone radii of 2, 3, 5, 7, and 10 feet; b) the effective wellbore radius, r_{wa}; c) the additional pressure drop near the wellbore due to formation damage; and d) the flow efficiency, E. Use the following data: $m = 3.75$ cp; $B = 1.226$ RB/STB; $r_w = 0.23$ ft; $h = 67$ ft; $k = 19.1$ md; $s = 8.2$; $q = 105$ STB/D; $p_{wf} = 2309$ psi; and $\bar{p} = 3137$ psi.

30. Analysis of a pressure buildup test showed that a recently stimulated oil well had a skin factor of -2.7 and an effective permeability to oil of 8.9 md. Prior to shutting in the well, the flowing bottomhole pressure at the time of shutin was 1441 psi, and the well produced oil at a constant rate of 152 STB/D. Estimate the following: a) the effective wellbore radius, r_{wa}; b) the additional pressure drop near the wellbore due to formation damage; and c) the flow efficiency, E. $\mu = 0.382$ cp; $B = 1.193$ RB/STB; $r_w = 0.33$ ft; $h = 9$ ft; $k = 8.9$ md; $s = -2.7$; $q = 152$ STB/D; $p_{wf} = 1441$ psi; and $\bar{p} = 1992$ psi.

31. A well is completed in the top 6 feet of a 47-foot-thick zone. The wellbore radius is 0.33 feet. The ratio k_v/k_h is 0.08. Calculate the apparent skin factor due to failure to complete the entire interval. What would the apparent skin factor be if the well were completed in the bottom 6 feet of the zone instead? If the perforated interval were centered in the zone?

32. A well is completed in a 192-foot-thick-zone. The perforated interval is 17 feet, with the top of the perforations 21 feet below the top of the zone. The wellbore radius is 0.25 feet. The ratio k_v/k_h is 0.05. Calculate the apparent skin factor due to failure to complete the entire interval.

33. A deviated well is completed in a 92-foot-thick reservoir. The wellbore lies at an angle of 63 degrees from the vertical. The wellbore radius is 0.27 feet. The ratio k_v/k_h is 0.12. Calculate the apparent skin factor due to the deviated wellbore.

34. You are planning a deviated well for a field where you expect to encounter a zone 220 feet thick. The well will be drilled with a 6-inch bit (wellbore radius will be 0.25 feet). The ratio k_v/k_h is not known, so assume a value of 0.1 for planning purposes. Calculate the apparent skin factor due to the deviated wellbore for deviation angles of 0 to 75 degrees in 5 degree increments. Graph your results.

35. A well is centered in a square reservoir. Use Horner analysis to estimate formation permeability and skin factor given the following formation and fluid properties: $q = 27$ STB/D; $t_p = 1440$ hrs; $r_e = 1320$ ft; $h = 51$ ft; $\phi = 29.4$ %; $r_w = 0.29$ ft; $B = 1.136$ RB/STB;

$c_t = 9.68 \times 10^{-6}$ psi^{-1}; and $\mu = 8.7$ cp. Then use the MBH and Modified Muskat methods to estimate average drainage area pressure.

DATA FOR EXERCISE 2.35

Time (hours)	Pressure (psi)	Time (hours)	Pressure (psi)	Time (hours)	Pressure (psi)
0	2162.11	2.42	2317.10	28.4	2335.62
0.050	2202.30	2.96	2318.77	34.1	2336.84
0.110	2234.27	3.60	2320.37	41.0	2338.02
0.182	2258.66	4.37	2321.91	49.2	2339.17
0.268	2276.35	5.30	2323.41	59.1	2340.26
0.372	2288.59	6.41	2324.88	71.0	2341.29
0.496	2296.80	7.74	2326.31	85.2	2342.25
0.646	2302.29	9.33	2327.72	102.3	2343.13
0.825	2306.12	11.25	2329.10	122.8	2343.92
1.040	2309.01	13.55	2330.45	147.4	2344.59
1.298	2311.38	16.31	2331.79	177.0	2345.15
1.608	2313.45	19.62	2333.09	196.0	2345.41
1.979	2315.34	23.60	2334.37		

36. A well is centered in an 80-acre, 2 × 1 rectangular reservoir. Use Horner analysis to estimate formation permeability and skin factor given the following formation and fluid properties: $q = 10$ STB/D; $t_p = 960$ hrs; $A = 80$ acres; $h = 10$ ft; $\phi = 31.9\%$; $r_w = 0.34$ ft; $B = 1.098$ RB/STB; $c_t = 10.9 \times 10^{-6}$ psi^{-1}; and $\mu = 5.11$ cp. Then use the MBH method to estimate average drainage area pressure. Why is the Modified Muskat method not appropriate for this problem?

DATA FOR EXERCISE 2.36

Time (hours)	Pressure (psi)	Time (hours)	Pressure (psi)	Time (hours)	Pressure (psi)
0	1192.45	1.608	1267.15	13.55	1297.24
0.050	1206.92	1.979	1270.37	16.31	1299.66
0.110	1217.90	2.42	1273.43	19.62	1302.04
0.182	1226.92	2.96	1276.36	23.6	1304.39
0.268	1234.52	3.60	1279.19	28.4	1306.71
0.372	1241.02	4.37	1281.93	34.1	1308.99
0.496	1246.66	5.30	1284.61	41.0	1311.23
0.646	1251.62	6.41	1287.22	49.2	1313.42
0.825	1256.04	7.74	1289.79	59.1	1315.56
1.040	1260.05	9.33	1292.31	71.0	1317.65
1.298	1263.73	11.25	1294.80	72.0	1317.81

37. A well is located near a sealing fault in a reservoir with no other nearby boundaries. Use Horner analysis to estimate formation permeability and skin factor, then estimate the distance to the boundary given the following formation and fluid properties: $q = 655$ STB/D; $t_p = 72$ hrs; $h = 11$ ft; $\phi = 17.3$ %; $r_w = 0.23$ ft; $B = 1.301$ RB/STB; $c_t = 10.1 \times 10^{-6}$ psi^{-1}; and $\mu = 0.845$ cp.

DATA FOR EXERCISE 2.37

Time (hours)	Pressure (psi)	Time (hours)	Pressure (psi)	Time (hours)	Pressure (psi)
0	2820.04	0.0318	3256.41	0.737	3291.03
0.00030	3160.59	0.0384	3258.07	0.885	3293.76
0.00066	3213.30	0.0464	3259.73	1.062	3296.54
0.00109	3223.92	0.0560	3261.38	1.274	3299.37
0.00161	3228.68	0.0675	3263.04	1.530	3302.23
0.00223	3232.13	0.0813	3264.73	1.836	3305.12
0.00298	3234.99	0.0979	3266.45	2.20	3308.03
0.00387	3237.50	0.1177	3268.24	2.64	3310.95
0.00495	3239.79	0.1416	3270.09	3.17	3313.87
0.00624	3241.92	0.1702	3272.03	3.81	3316.78
0.00779	3243.93	0.205	3274.07	4.57	3319.69
0.00965	3245.86	0.246	3276.21	5.48	3322.58
0.01187	3247.72	0.295	3278.45	6.58	3325.44
0.01455	3249.53	0.355	3280.79	7.90	3328.27
0.01776	3251.29	0.426	3283.12	9.48	3331.05
0.0216	3253.02	0.511	3285.75	11.37	3333.78
0.0262	3254.73	0.614	3288.35	12.00	3334.57

38. A new well was drilled in a closed reservoir, and a 7-day flow test was run. Estimate formation permeability and skin factor given the following formation and fluid properties: $q = 423$ STB/D; $p_i = 1701.09$ psi; $h = 21$ ft; $\phi = 27.4$ %; $r_w = 0.31$ ft; $B = 1.138$ RB/STB; $c_t = 9.8 \times 10^{-6}$ psi^{-1}; and $\mu = 2.305$ cp. Then estimate the reservoir pore volume.

DATA FOR EXERCISE 2.38

Time (hours)	Pressure (psi)	Time (hours)	Pressure (psi)	Time (hours)	Pressure (psi)
0.0100	1656.73	0.852	1537.15	14.48	1498.50
0.0222	1625.93	1.049	1534.29	17.67	1495.19
0.0371	1604.46	1.290	1531.47	21.6	1491.58
0.0552	1589.60	1.584	1528.70	26.3	1487.58
0.0774	1579.19	1.943	1525.96	32.1	1483.10
0.1044	1571.60	2.38	1523.25	39.2	1478.00
0.1374	1565.75	2.91	1520.56	47.8	1472.08
0.1776	1560.94	3.56	1517.88	58.4	1465.10
0.227	1556.76	4.36	1515.23	71.2	1456.77
0.287	1553.00	5.33	1512.58	86.9	1446.70
0.360	1549.52	6.51	1509.92	106.0	1434.49
0.449	1546.24	7.95	1507.23	129.4	1419.60
0.557	1543.11	9.71	1504.46	157.8	1401.45
0.690	1540.09	11.86	1501.57	168.0	1394.97

39. A new well was drilled in a closed reservoir, and a 7-day flow test was run. Estimate formation permeability and skin factor given the following formation and fluid properties: $q = 75$ STB/D; $p_i = 3394.81$ psi; $h = 11$ ft; $\phi = 9.9$ %; $r_w = 0.48$ ft; $B = 1.303$ RB/STB; $c_t = 14.4 \times 10^{-6}$ psi^{-1}; and $\mu = 1.324$ cp. Then estimate the reservoir pore volume.

DATA FOR EXERCISE 2.39					
Time (hours)	Pressure (psi)	Time (hours)	Pressure (psi)	Time (hours)	Pressure (psi)
0.010	3356.75	0.743	2977.65	13.88	2902.68
0.023	3313.75	0.976	2969.01	18.06	2896.56
0.040	3266.60	1.279	2961.24	23.5	2890.38
0.062	3216.91	1.673	2954.00	30.5	2883.94
0.090	3167.10	2.18	2947.09	39.7	2876.78
0.128	3120.11	2.85	2940.42	51.6	2868.26
0.176	3078.81	3.72	2933.91	67.1	2857.55
0.239	3045.17	4.84	2927.52	87.3	2843.77
0.320	3019.69	6.30	2921.23	113.5	2825.87
0.426	3001.29	8.20	2915.00	147.6	2802.60
0.564	2987.94	10.67	2908.82	168.0	2788.63

40. A well has been producing from a solution gas drive reservoir for 4 years. The well was shut in for a 12-hour buildup test. Estimate total mobility, skin factor, effective permeability to oil, effective permeability to gas, and current average drainage area pressure given the following formation and fluid properties: $q_o = 250$ STB/D; $q_g = 347.6$ Mscf/D; $q_w = 0$ STB/D; $h = 27$ ft; $\phi = 20.2\%$; $r_w = 0.33$ ft; $S_w = 26.7\%$; $S_g = 11.0\%$; $t_p = 3660$ days; $B_o = 1.104$ RB/STB; $c_o = 5.96 \times 10^{-4}$ psi^{-1}; $\mu_o = 1.636$ cp; $B_g = 3.161$ RB/Mscf; $c_g = 1.24 \times 10^{-3}$ psi^{-1}; $\mu_g = 0.013$ cp; $B_w = 1.018$ RB/STB; $c_w = 4.0 \times 10^{-6}$ psi^{-1}; $\mu_w = 0.824$ cp; $R_{so} = 176$ scf/STB; and $A = 306.6$ acres. Assume a circular drainage area shape.

DATA FOR EXERCISE 2.40					
Time (hours)	Pressure (psi)	Time (hours)	Pressure (psi)	Time (hours)	Pressure (psi)
0.0	452.07	0.1348	693.17	1.471	786.41
0.0059	473.71	0.1641	709.23	1.764	789.19
0.0117	497.20	0.1992	723.35	2.12	791.89
0.0176	507.38	0.234	735.30	2.54	794.54
0.0234	531.75	0.281	745.33	3.05	797.14
0.0293	546.21	0.340	753.49	3.66	799.69
0.0352	562.18	0.410	760.09	4.39	802.18
0.0469	579.54	0.492	765.47	5.27	804.64
0.0527	598.08	0.592	769.96	6.32	807.09
0.0645	617.15	0.709	773.84	7.58	809.52
0.0762	636.70	0.850	777.31	9.10	811.94
0.0938	656.28	1.020	780.51	10.92	814.36
0.1113	675.29	1.225	783.53	12.00	815.55

References

1. Matthews, C.S. and Russell, D.G.: *Pressure Buildup and Flow Tests in Wells*, Monograph Series, SPE, Richardson, Texas (1977) **1**, 16, 19, 21, 49, 72–83.
2. Earlougher, R.C., Jr.: *Advances in Well Test Analysis*, Monograph Series, SPE, Richardson, Texas (1977) **5**, 74–89, 191.
3. Winestock, A.G. and Colpitts, G.P.: *Gas Technology*, Reprint Series, SPE, Richardson, Texas (1977) **13**, 122–30.
4. Gladfelter, R.E., Tracy, G.W., and Wilsey, L.E.: "Selecting Wells Which Will Respond to Production-Stimulation Treatment," *Drill. & Prod. Prac.*, API, Dallas (1955) 117–29.
5. Ramey, H.J. Jr.: "Non-Darcy Flow and Wellbore Storage Effects on Pressure Buildup and Drawdown of Gas Wells," *JPT* (February 1965) 223; *Trans.*, AIME, **234.**
6. Russell, D.G.: "Determination of Formation Characteristics From Two-Rate Tests," *JPT* (December 1963) 1,347; *Trans.*, AIME, **228.**
7. Odeh, A.S. and Jones, L.G.: "Pressure Drawdown Analysis, Variable-Rate Case," *JPT* (August 1965) 960; *Trans.*, AIME, **234.**
8. Horner, D.R.: "Pressure Buildup in Wells," *Pressure Analysis Methods*, Reprint Series, SPE, Richardson, Texas (1967) **9,** 25–43.
9. Odeh, A.S. and Selig, F.: "Pressure Buildup Analysis, Variable-Rate Case," *JPT* (July 1963) 790; *Trans.*, AIME (1963) **228.**
10. van Everdingen, A.F. and Hurst, W.: "The Application of the Laplace Transformation to Flow Problems in Reservoirs," *Trans.*, AIME (1949) **186,** 305–24.
11. Prats, M., Hazebroek, P., and Strickler, W.R.: "Effect of Vertical Fractures on Reservoir Behavior—Compressible Fluid Case," *SPEJ* (June 1962) 87; *Trans.*, AIME, **225.**
12. Papatzacos, Paul: "Approximate Partial-Penetration Pseudoskin for Infinite-Conductivity Wells," *SPERE* (May 1987) 227; *Trans.*, AIME (1987) **283.**
13. Cinco, H., Miller, F.G., and Ramey, H.J. Jr.: "Unsteady-State Pressure Distribution Created by a Directionally Drilled Well," *JPT* (November 1975) 1392; *Trans.*, AIME, **259.**
14. Matthews, C.S., Brons, F., and Hazebroek, P.: "A Method for Determination of Average Pressure in a Bounded Reservoir," *Trans.*, AIME (1954) **201,** 182–91.
15. Larson, V.C.: "Understanding the Muskat Method of Analyzing Pressure Buildup Curves," *J. Cdn. Pet. Tech.* (Fall 1963) **2,** 136.
16. Gray, K.E.: "Approximating Well-to-Fault Distance From Pressure Buildup Tests," *JPT* (July 1965) 761.
17. Perrine, R.L.: "Analysis of Pressure Buildup Curves," *Drill. & Prod. Prac.* (1956) 482.
18. Martin, J.C.: "Simplified Equations of Flow in Gasdrive Reservoirs and the Theoretical Foundation of Multiphase Pressure Buildup Analyses," *JPT* (October 1959) 309; *Trans.*, AIME, **216.**
19. Raghavan, R.: "Well Test Analysis: Wells Producing by Solution Gas Drive," *SPEJ* (August 1976) 196; *Trans.*, AIME (1976) **261.**
20. Boe, A., Skjaeveland, S.M., and Whitson, C.H.: "Two-Phase Pressure Test Analysis," *SPEFE* (December 1989) 604.; *Trans.*, AIME (1989) **287.**
21. Al-Khalifah, A-J.A., Aziz, K., and Horne, R.N.: "A New Approach to Multiphase Well Test Analysis," paper SPE 16743 presented at the 1987 SPE Annual Technical Conference and Exhibition, Dallas, 27–30 September.
22. Serra, K.V., Peres, A.M.M., and Reynolds, A.C.: "Well-Test Analysis for Solution-Gas-Drive Reservoirs: Part 2—Buildup Analysis," *SPEFE* (June 1990) 133.
23. Serra, K.V., Peres, A.M.M., and Reynolds, A.C.: "Well-Test Analysis for Solution-Gas-Drive Reservoirs: Part 3—A Unified Treatment of the Pressure-Squared Method," *SPEFE* (June 1990) 141.

Chapter 3
Introduction to Flow and Buildup-Test Analysis: Compressible Fluids

3.1 Overview

Chap. 2 presents the fundamental theory and practical applications of analysis techniques for flow and buildup tests based on the assumption of slightly compressible fluids (i.e., liquids) with relatively constant properties. For flow of a compressible fluid (i.e., gas) in which the properties are strong functions of pressure, however, these analysis techniques usually are not accurate enough to analyze well tests. This chapter extends the concepts presented in Chap. 2 and presents modifications to the slightly compressible liquid solutions that can be used for analysis of gas-well flow and buildup tests. These modifications include pressure, pressure-squared, and pseudopressure transformation variables that account for the variation of gas properties with pressure. In addition, we discuss non-Darcy flow effects that are more pronounced in gas-well testing. The analysis techniques, which we illustrate with several examples, are limited to single-layer formations.

3.2 Pseudopressure and Pseudotime Analysis

The equations developed for slightly compressible fluids (i.e., liquids) can be altered by replacing pressure and time with real-gas pseudopressure and pseudotime variables, respectively.[1-3] These transformations account for variations in gas properties with pressure. Accuracy is improved for both semilog and type-curve analyses of gas-well tests by replacing pressure with the real-gas pseudopressure (or real-gas potential) function, p_p, which is defined as[1]

$$p_p(p) = 2\int_0^p \left[p/\mu_g(p)z(p)\right]\mathrm{d}p. \quad (3.1)$$

For type-curve analysis (discussed in Chap. 4), particularly of wellbore-storage-distorted data from both flow and buildup tests, accuracy also is improved by replacing time with pseudotime, $t_{ap}(p)$, which is defined as[2]

$$t_{ap}(p) = \int_0^t \left[\mathrm{d}t/\mu_g(p)c_t(p)\right]. \quad (3.2)$$

For convenience, although not by necessity, p_p and t_{ap} can be normalized[4] to have units of psia and hours, respectively, like the original variables p and t. Normalization also gives the pseudopressure and pseudotime variables magnitudes comparable with those of the untransformed pressure and time, whereas the unnormalized variables p_p and t_{ap} typically have values on the order of 10^5 to 10^8. Reference values of pressure used for normalization are arbitrary. In this chapter, we define normalized variables as

$$p_n = \frac{1}{2}\left(\frac{\mu_g z}{p}\right)_r p_p = \left(\frac{\mu_g z}{p}\right)_r \int_0^p \frac{p\mathrm{d}p}{\mu_g z} \quad (3.3)$$

and $$t_n = \left(\mu_g c_t\right)_r t_{ap} = \left(\mu_g c_t\right)_r \int_0^t \left(\mathrm{d}t/\mu_g c_t\right), \quad (3.4)$$

where the subscript n refers to the normalized variables and the subscript r refers to the reference values of properties used in the normalization process.

Some engineers prefer properties evaluated at original reservoir pressure, p_i. Because it can be measured directly, others prefer flowing bottomhole pressure (BHP), p_{wf}, at the end of a flow period. In this chapter, we use the current static drainage-area pressure, $\bar{p}$. Although $\bar{p}$ may not be available at the start of an analysis, using the pressure p^* (the pressure on the semilog straight line extrapolated to a Horner time ratio of unity) as an estimate of $\bar{p}$ is quite satisfactory for buildup tests because the choice of a reference pressure is completely arbitrary (i.e., the value of p_r has no effect on results). For a flow test, of course, the pressure at the start of the test is $\bar{p}$. We shall call our normalized variables adjusted pressure, p_a, and adjusted time, t_a, and we define them, respectively, as

$$p_a = \frac{1}{2}\left(\frac{\bar{\mu}_g \bar{z}}{\bar{p}}\right) p_p = \frac{\bar{\mu}_g \bar{z}}{\bar{p}} \int_0^p \frac{p\mathrm{d}p}{\mu_g z} \quad (3.5)$$

and $$t_a = \left(\bar{\mu}\,\bar{c}_t\right) t_{ap} = \left(\bar{\mu}\,\bar{c}_t\right) \int_0^t \frac{\mathrm{d}t}{\mu c_t}, \quad (3.6)$$

where $\bar{p} \approx p^*$ and $\bar{p} = p_i$ for a new well. In terms of adjusted variables, the unsteady-state equation for slightly compressible liquids, Eq. 2.2, becomes

$$p_{a,i} - p_{a,wf} = \frac{162.6 q_g \overline{B}_g \overline{\mu}}{kh} \times \left[\log\left(kt/\phi\overline{\mu}_g\overline{c}_t r_w^2\right) - 3.23 + 0.869s'\right]. \quad \ldots (3.7)$$

For semilog analysis of buildup tests, adjusted pressure and adjusted time should be used, but the adjusted producing time, t_{pa}, used in the Horner time ratio is evaluated at current average drainage-area pressure, and t_{pa} has a very simple form.

$$t_{pa} = \left[\left(\overline{\mu}_g\overline{c}_t\right)/\left(\overline{\mu}_g\overline{c}_t\right)\right] t_p = t_p. \quad \ldots (3.8a)$$

Eq. 3.8a suggests that, for semilog analysis of pressure-buildup tests, the adjusted Horner time ratio is evaluated at the real producing time rather than adjusted producing time, or

$$\left(t_{pa} + \Delta t_a\right)/\Delta t_a = \left(t_p + \Delta t_a\right)/\Delta t_a, \quad \ldots (3.8b)$$

where the adjusted shut-in time, Δt_a, is evaluated from the integral

$$\Delta t_a = \left(\overline{\mu}_g\overline{c}_t\right)\int_0^{\Delta t}\left[d(\Delta t)/\mu_g c_t\right], \quad \ldots (3.9)$$

with μ_g and c_t evaluated at shut-in BHP, p_{ws}, at values of Δt during the test. For semilog analysis of a flow test, adjusted pressure should be used, but adjusted flowing time, t_a, is evaluated at current static drainage-area pressure, $\overline{p}$. Consequently, Eq. 3.8a, which involves no numerical integration, can be used to calculate adjusted flowing time. This is equivalent to using actual flowing time rather than adjusted flowing time, and all our working equations will be written with this result.

The logic behind these rules is that, for semilog analysis (data not distorted by wellbore storage), gas properties should be evaluated at the pressure at the radius of investigation reached at the time under consideration. For flow, the pressure at the radius of investigation is $\overline{p}$. For buildup tests, the pressure at the radius of investigation is p_{ws}, the current shut-in pressure in the wellbore. For type-curve analysis, which deals particularly with data distorted by wellbore storage, we defer further discussion about the pressure at which gas properties are evaluated in the pseudotime transformation until Chap. 4.

3.3 Pressure and Pressure-Squared Analysis

The use of adjusted time and adjusted pressure in formulating equations for analysis of transient tests in gas wells is not always necessary. For cases where use of transformed variables is not needed, two approximations are possible. The first of these approximations is based on the observation that, for some (although certainly not all) gases at high pressure (greater than 3,000 psia, for example), an adequate approximation is $p/\mu_g z = \text{constant} = \overline{p}/\overline{\mu}_g\overline{z}$ (Chap. 1, Sec. 1.2.3 discusses the conditions under which these approximations are valid). When this approximation is valid, Eq. 3.5 becomes

$$p_a = \frac{\overline{\mu}_g\overline{z}}{\overline{p}}\int_0^p \frac{p dp}{\mu_g z} = p. \quad \ldots (3.10)$$

Thus, in the unsteady-state-flow equation given by Eq. 3.7, adjusted pressure, p_a, can be replaced by ordinary pressure, p, and the unsteady-state-flow equation becomes

$$p_i - p_{wf} = \left(162.6 q_g\overline{B}_g\overline{\mu}_g/kh\right)\left[\log\left(kt/\phi\overline{\mu}_g\overline{c}_t r_w^2\right) - 3.23 + 0.869s'\right], \quad \ldots (3.11)$$

where the average gas formation volume factor (FVF), $\overline{B}_g$, viscosity, $\overline{\mu}_g$, and total compressibility, $\overline{c}_t$, are evaluated at the average pressure in the drainage area of the well.

The second approximation is based on the observation that, for some gases at low pressures (less than 2,000 psia, for example), as an adequate approximation, $\mu_g z = \text{constant} = \overline{\mu}_g\,\overline{z}$ (see Sec. 1.2.3). When this approximation is valid, Eq. 3.5 becomes

$$p_a = \frac{\overline{\mu}_g\overline{z}}{\overline{p}}\int_0^p \frac{p dp}{\mu_g z} = \frac{p^2}{2\overline{p}} \quad \ldots (3.12)$$

and the unsteady-state-flow equation can be written as

$$p_i^2 - p_{wf}^2 = \frac{57{,}910 q_g p_{sc} T\overline{z}\overline{\mu}_g}{khT_{sc}}\left[\log\left(kt/\phi\overline{\mu}_g\overline{c}_t r_w^2\right) - 3.23 + 0.869s'\right]. \quad \ldots (3.13)$$

For $p_{sc} = 14.7$ psia and $T_{sc} = 520°\text{R}$, Eq. 3.12 becomes

$$p_i^2 - p_{wf}^2 = \frac{1{,}637 q_g T\overline{z}\overline{\mu}_g}{kh}\left[\log\left(kt/\phi\overline{\mu}_g\overline{c}_t r_w^2\right) - 3.23 + 0.869s'\right], \quad \ldots (3.14)$$

where s' = apparent skin that includes non-Darcy-flow effects, which we discuss in more detail in a later section.

These modified unsteady-state-flow equations for gas wells serve as the basis for buildup- and flow-test analysis techniques for gas wells. **Table 3.1** provides summaries of interpretation and analysis equations for the variables frequently used in gas-well-test analysis. For comparison, we also include variables for well-test analysis of slightly compressible liquids.

Aziz *et al.*[5] showed that neither the p nor p^2 approximation should be assumed to be adequate in any given case. Plots of $(p/\mu_g z)$ and $\mu_g z$ as functions of pressure should always be prepared to determine the pressure range, if any, in which $(p/\mu_g z)$ or $\mu_g z$ is constant in a particular case. If any question exists concerning which transformed variables to use, we recommend adjusted pressures and times.

3.4 Non-Darcy Flow

The transient-pressure response of a gas well may be affected by high-velocity or non-Darcy flow near the wellbore.[6,7] The commonly used unsteady-state-flow equation for pressure drawdown analysis with constant-rate production of a gas is based on the solution for slightly compressible liquid flow with pressure replaced by pseudopressure, or

$$p_p\left(p_{wf}\right) = p_p(p_i) - \frac{1{,}637 q_g T}{kh}\left[\log t_p + \log\left(kt/\phi\overline{\mu}_g\overline{c}_t r_w^2\right) - 3.23 + 0.869s'\right]. \quad \ldots (3.15)$$

In terms of normalized pseudopressure (adjusted pressure), p_a, the unsteady-state-flow equation becomes

$$p_{a,wf} = p_{a,i} - \frac{162.6 q_g\overline{B}_g\overline{\mu}_g}{kh}\left[\log t_p + \log\left(kt/\phi\overline{\mu}_g\overline{c}_t r_w^2\right) - 3.23 + 0.869s'\right], \quad \ldots (3.16)$$

where $s' = s + Dq_g$ is an effective skin factor that includes true formation damage (or stimulation) and the effects of non-Darcy flow. In dimensionless form, Eq. 3.15 at $r_D = 1$ can be written

$$p_{pD}(t_D) = 0.5\left[\ln(t_D) + 0.80907\right] + s + Dq_g, \quad \ldots (3.17)$$

where $p_{pD}(t_D)$ = dimensionless pseudopressure (or normalized pseudopressure) at the wellbore and dimensionless time, t_D, is defined by

$$t_D = 0.0002637kt/\phi\overline{\mu}_g\overline{c}_t r_w^2, \quad \ldots (3.18)$$

where μ_g and c_t are evaluated at $\overline{p}$. Chaps. 1 and 4 discuss the development of dimensionless variables.

The usual assumption in conventional calculations is that the non-Darcy-flow effect can be represented as a rate-dependent pseudoskin

TABLE 3.1—SUMMARY OF WORKING EQUATIONS, RADIAL OR PSEUDORADIAL FLOW

	Case: Oil	Case: Gas, Using Adjusted Variables
Flow Test		
Semilog graph variables	p_{wf} vs. t	$p_{a,wf}$ vs. t
Permeability from slope, m, of semilog straight line	$k = \frac{162.6 q_o B_o \mu_o}{mh}$	$k = \frac{162.6 q_g \bar{B}_g \bar{\mu}}{mh}$
Skin-factor calculation	$s = 1.151\left[\frac{p_i - p_{1hr}}{m} - \log\left(\frac{k}{\phi \mu_o c_t r_w^2}\right) + 3.23\right]$	$s' = 1.151\left[\frac{p_{a,i} - p_{a,1hr}}{m} - \log\left(\frac{k}{\phi \bar{\mu} \bar{c}_t r_w^2}\right) + 3.23\right]$
Buildup Test		
Semilog graph variables	p_{ws} vs. $(t_p + \Delta t)/\Delta t$	$p_{a,ws}$ vs. $(t_p + \Delta t_a)/\Delta t_a$
Permeability from slope, m, of semilog straight line	$k = \frac{162.6 q_o B_o \mu_o}{mh}$	$k = \frac{162.6 q_g \bar{B}_g \bar{\mu}}{mh}$
Skin-factor calculation	$s = 1.151\left[\frac{p_{1hr} - p_{wf}}{m} - \log\left(\frac{k}{\phi \mu_o c_t r_w^2}\right) + 3.23\right]$	$s' = 1.151\left[\frac{p_{a,1hr} - p_{a,wf}}{m} - \log\left(\frac{k}{\phi \bar{\mu} \bar{c}_t r_w^2}\right) + 3.23\right]$
Definition of $p_{MBH,D}$	$\frac{kh(p^* - \bar{p})}{70.6 q_o B_o \mu_o}$	$\frac{kh(p_a^* - \bar{p}_a)}{70.6 q_g \bar{B}_g \bar{\mu}}$

	Case: Gas, Using Pressure and Time	Case: Gas, Using Pressure Squared and Time
Flow Test		
Semilog graph variables	p_{wf} vs. t	p_{wf}^2 vs. t
Permeability from slope, m, of semilog straight line	$k = \frac{162.6 q_g \bar{B}_g \bar{\mu}}{mh}$	$k = \frac{1637 q_g T \bar{z} \bar{\mu}}{mh}$
Skin factor calculation	$s' = 1.151\left[\frac{\bar{p} - p_{a,1hr}}{m} - \log\left(\frac{k}{\phi \bar{\mu} \bar{c}_t r_w^2}\right) + 3.23\right]$	$s' = 1.151\left[\frac{\bar{p}^2 - p_{1hr}^2}{m} - \log\left(\frac{k}{\phi \bar{\mu} \bar{c}_t r_w^2}\right) + 3.23\right]$
Buildup Test		
Semilog graph variables	p_{ws} vs. $(t_p + \Delta t)/\Delta t$	p_{ws}^2 vs. $(t_p + \Delta t)/\Delta t$
Permeability from slope, m, of semilog straight line	$k = \frac{162.6 q_g \bar{B}_g \bar{\mu}}{mh}$	$k = \frac{1637 q_g T \bar{z} \bar{\mu}}{mh}$
Skin factor calculation	$s' = 1.151\left[\frac{p_{1hr} - p_{wf}}{m} - \log\left(\frac{k}{\phi \bar{\mu} \bar{c}_t r_w^2}\right) + 3.23\right]$	$s' = 1.151\left[\frac{\bar{p}_{1hr}^2 - p_{wf}^2}{m} - \log\left(\frac{k}{\phi \bar{\mu} \bar{c}_t r_w^2}\right) + 3.23\right]$
Definition of $p_{MBH,D}$	$\frac{kh(p^* - \bar{p})}{70.6 q_g \bar{B}_g \bar{\mu}}$	$\frac{kh\left(p^{*2} - \bar{p}^2\right)}{711 q_g T \bar{z} \bar{\mu}}$

	Case: Gas, Using Pseudopressure and Time
Flow Test	
Semilog graph variables	p_p vs. t
Permeability from slope, m, of semilog straight line	$k = \frac{1637 q_g T}{mh}$
Skin factor calculation	$s' = 1.151\left[\frac{\bar{p}_p - p_{p,1hr}}{m} - \log\left(\frac{k}{\phi \bar{\mu} \bar{c}_t r_w^2}\right) + 3.23\right]$
Buildup Test	
Semilog graph variables	p_p vs. $(t_p + \Delta t)/\Delta t$
Permeability from slope, m, of semilog straight line	$k = \frac{1637 q_g T}{mh}$
Skin factor calculation	$s' = 1.151\left[\frac{p_{p,1hr} - p_{p,wf}}{m} - \log\left(\frac{k}{\phi \bar{\mu}_g \bar{c}_t r_w^2}\right) + 3.23\right]$
Definition of $p_{MBH,D}$	$\frac{kh\left(p_p^* - \bar{p}_p\right)}{711 q_g T}$

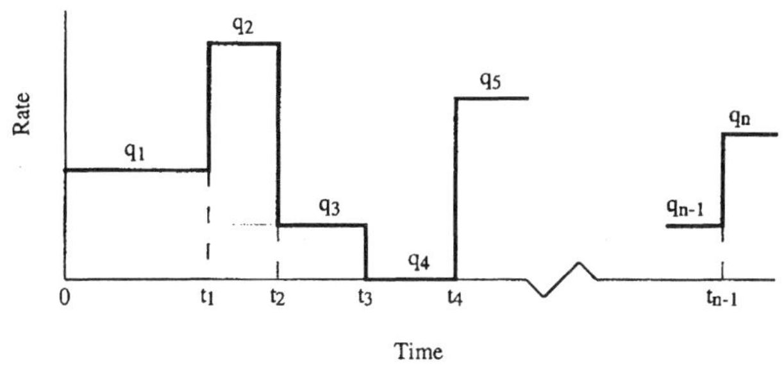

Fig. 3.1—Variable-rate production history.

defined as Dq_g, where D is a constant known as the non-Darcy-flow coefficient with units of D/Mscf, and q_g is the flow rate with units of Mscf/D. The true skin factor, s, that reflects formation damage or stimulation near the wellbore cannot be determined from a single constant-rate drawdown or buildup test. Rather, the apparent skin factor, $s' = s + Dq_g$, is obtained. If s and D are to be determined separately, two flow tests may be run at different rates so that two equations (given by Eq. 3.15) can be solved simultaneously[8] for the two unknowns, s and D. If only one test is available, D may be estimated by

$$D = \frac{2.715 \times 10^{-15} \beta k_g M p_{sc}}{h r_w T_{sc} \mu_{g,wf}}, \quad \text{(3.19)}$$

where k_g = effective gas permeability, md; p_{sc} = pressure at standard conditions, psia; T_{sc} = temperature at standard conditions, °R; M = gas molecular weight, lbm/lbm-mol; $\mu_{g,wf}$ = pressure-dependent gas viscosity evaluated at flowing BHP, cp; and β = turbulence parameter. The turbulence parameter, β, is approximately inversely proportional to effective gas permeability and may be determined experimentally or from a correlation,[9] such as

$$\beta = 1.88 \times 10^{10} k^{-1.47} \phi^{-0.53}. \quad \text{(3.20)}$$

3.5 Analysis of Gas-Well Flow Tests

In most gasfield operations, the wells are produced at conditions approximating constant wellhead pressure or variable bottomhole rates, rather than at constant bottomhole rates. In addition, many gas-well tests, especially deliverability tests, are conducted under variable-rate conditions. In this section, we begin with a brief discussion of constant-rate gas-flow tests, but we concentrate on analysis techniques for variable gas rates, including gas-well tests with discrete rate changes and tests in which the rates are smoothly changing. We also address non-Darcy-flow effects in flow tests. Finally, we discuss the effects of reservoir boundaries on gas-well testing.

3.5.1 Constant-Rate Gas-Flow Tests. Eq. 3.7, developed in terms of adjusted pressures, describes the pressure drop at the wellbore as a function of time when a gas well is produced at a constant rate. Like the solution for production of slightly compressible liquids, comparing Eq. 3.7 with the equation of a straight line suggests that, if the pressure behavior of a reservoir can be modeled with Eq. 3.7, a plot of $p_{a,wf}$ vs. log t should result in a straight line from which the slope, m, allows us to estimate effective gas permeability, k, and apparent skin factor, s'. This method of analysis assumes the non-Darcy-flow coefficient, D, is constant. When non-Darcy-flow losses are large, this assumption is not valid, because Eq. 3.19 shows D is inversely proportional to gas viscosity evaluated at flowing BHP. Sec. 3.5.4 addresses the more complicated situation in which D cannot be assumed constant.

For single-phase gas flow, the effective gas permeability in the drainage area of the well is computed from

$$k = 162.6 q_g \overline{B}_g \overline{\mu}_g / mh, \quad \text{(3.21)}$$

where the absolute value of the slope of the semilog straight line is used. Rearranging Eq. 3.7, we also can develop an expression for the apparent skin factor, s'.

$$s' = 1.151\left\{\left[\left(p_{a,i} - p_{a,wf}\right)/m\right] - \left[\log\left(kt/\phi\overline{\mu}\overline{c}_t r_w^2\right) + 3.23\right]\right\}. \quad \text{(3.22)}$$

Again, for convenience, we set the flow time, t, equal to 1 hour, and use the symbol $p_{a,\,1\text{hr}}$ for the flowing BHP, $p_{a,wf}$, at this time. Note that the pressure $p_{a,\,1\text{hr}}$ necessarily lies on the semilog straight line or its extrapolation. Substituting these into Eq. 3.22 yields

$$s' = 1.151\left\{\left[\left(p_{a,i} - p_{a,1\text{hr}}\right)/m\right] - \log(kt/\phi\overline{\mu}\overline{c}_t r_w^2) + 3.23\right\}. \quad \text{(3.23)}$$

In summary, from the straight line predicted by theory for a plot of adjusted pressure vs. time on semilogarithmic paper, we can estimate effective gas permeability, k, and the apparent skin factor, s'. Similar calculations are possible when either pressure or pressure-squared variables are used. Table 3.1 summarizes the appropriate plotting functions and equations.

3.5.2 Gas-Flow Tests With Discrete Rate Changes. Similar to the variable-rate solution for slightly compressible fluids presented in Chap. 2, we model a variable-rate gas-flow test with superposition in time.[10,11] First, we consider the pressure drawdown caused by a single production rate in terms of adjusted pressures in an infinite-acting gas reservoir. If we assume, for now, negligible non-Darcy-flow effects, Eq. 3.7 can be rewritten as

$$p_{a,i} - p_{a,wf} = \left(162.6 q_g \overline{B}_g \overline{\mu}_g / kh\right)\left[\log(t) + \log\left(k/\phi\overline{\mu}_g \overline{c}_t r_w^2\right) - 3.23 + 0.869s\right] \quad \text{(3.24)}$$

or $p_{a,i} - p_{a,wf} = m' q_g(\log t + \overline{s})$, (3.25)

where $m' = 162.6\overline{B}_g \overline{\mu}_g / kh$ (3.26)

and $\overline{s} = \log\left(k/\phi\overline{\mu}_g \overline{c}_t r_w^2\right) - 3.23 + 0.869s.$ (3.27)

For the variable-rate production history illustrated in **Fig. 3.1,** the pressure drawdown resulting from n discrete rate changes and for time $t > t_{n-1}$ is[12]

$$p_{a,i} - p_{a,wf} = m' q_1(\log t + \overline{s}) + m'(q_2 - q_1)\left[\log(t - t_1) + \overline{s}\right] + m'(q_3 - q_2)\left[\log(t - t_2) + \overline{s}\right] + \ldots + m'(q_n - q_{n-1})\left[\log(t - t_{n-1}) + \overline{s}\right]. \quad \text{(3.28)}$$

Eq. 3.28 can be rewritten as

$$\left(p_{a,i} - p_{a,wf}\right)/q_n = m' \sum_{j=1}^{n}\left[\left(q_j - q_{j-1}\right)/q_n\right]\log\left(t - t_{j-1}\right) + m'\overline{s} \quad \text{(3.29a)}$$

for $q_n \neq 0$.

The form of Eq. 3.29 suggests that we plot

$$\left(p_{a,i} - p_{a,wf}\right)/q_n \text{ vs. } \frac{1}{q_n}\sum_{j=1}^{n}\Delta q_j \log\left(t_n - t_{j-1}\right) \quad \text{(3.29b)}$$

on Cartesian coordinate paper where $p_{a,i}$ = adjusted initial reservoir pressure, psia; $p_{a,wf}$ = adjusted flowing BHP at time t_n, psia; q_n = last of n different flow rates, STB/D; $\Delta q_j = q_j - q_{j-1}$, $(q_0 = 0)$; t_n = total (cumulative) flowing time for n constant-rate flow periods, hours; and Δt_j = time at which rate was changed, hours. A straight line with slope, m', inversely proportional to k should result from this plot. Specifically,

TABLE 3.2—FLOW-AFTER-FLOW TEST DATA, EXAMPLE 3.1

Cumulative Flow Time, t (hours)	Bottomhole Flowing Pressure, p_{wf} (psia)	Adjusted Flowing Pressure, $p_{a,wf}$ (psia)	Gas Flow Rate, q (Mscf/D)
0	6,180	4,709.4	0
6	566	66.4	2,711
8	832	142.2	2,607
9.5	1,130	258.9	2,504
12.5	1,646	535.2	2,309

$$k = 162.6 q_g \overline{B}_g \overline{\mu}_g / m'h, \quad \text{(3.30)}$$

where m' is defined by Eq. 3.26.

The skin factor also can be determined from this plot. If the intercept of the plot (the value of $\Delta p/q$ at which the time-plotting function is zero) is denoted by b, then the skin factor, s, is given by

$$s = 1.151\left[\left(b/m'\right) - \log\left(k/\phi\overline{\mu}_g \overline{c}_t t r_w^2\right) + 3.23\right]. \quad \text{(3.31)}$$

The plotting method assumes that the reservoir is infinite-acting at all times up to t_n. Once boundary effects are felt by the pressure transient, the method is invalid. This method of analysis can be particularly valuable for gas-well testing. The method can be used with four-point deliverability or backpressure tests, which are simply variable-rate tests with discrete changes in flow rate.

To include non-Darcy flow effects in a gas-well test, we plot

$$\left[\left(p_{a,i} - p_{a,wf} - D'q_n^2\right)/q_n\right] \text{ vs. } \frac{1}{q_n}\sum_{j=1}^{n} \Delta q_j \log\left(t_n - t_{j-1}\right),$$

where $D' = 141.2\overline{B}_g \overline{\mu}_g D/kh$ is a constant, and D = non-Darcy flow coefficient. With the non-Darcy-flow term, $D'q_n^2$, we are forced either to assume $D' = 0$ or to find, by trial and error, the value of D' that results in the best straight-line plot. The slope, m', of the straight line provides an estimate of permeability and skin factor by use of Eqs. 3.30 and 3.31, respectively.

In principle, this method is potentially useful because backpressure data are available for virtually all gas wells. Thus, we have a useful method to estimate permeability and skin factor when buildup or drawdown tests are not available. However, the method has at least two limitations. First, a well may not be cleaned up at the time of testing. Second, the method assumes wellbore-storage distortion is negligible; however, wellbore storage almost certainly distorts some of the test data and may distort all or most of the data in short-duration flow periods.

Example 3.1—Determining Permeability and Skin Factor From a Multirate Test. A gas well was tested with a conventional flow-after-flow backpressure test. **Table 3.2** summarizes the measured pressures and rates. Assuming non-Darcy-flow effects ($D'q_g^2$) are negligible, estimate effective gas permeability and skin factor with the following well, rock, and gas properties.

γ_g = 0.7
T = 200°F
ϕ = 0.176
r_w = 0.255 ft
$\overline{c}_t$ = 73.3×10^{-6} psia^{-1}
$\overline{\mu}_g$ = 0.0286 cp
h = 19 ft
$\overline{B}_g$ = 0.5878 RB/Mscf

Solution.

1. Our objective is to prepare a plot of

$$\left[\left(p_{a,i} - p_{a,wf} - D'q_n^2\right)/q_n\right] \text{vs.} \frac{1}{q_n}\sum_{j=1}^{n} \Delta q_j \log\left(t_n - t_{j-1}\right).$$

Thus, the first step is to calculate the pressure and time-rate plotting functions for each of the four flow periods. For example, at $t=6$ hours, $n=1$, $t_1=6$ hours, $q_1=2{,}711$ Mscf/D, and $D'=0$:

$$\frac{p_{a,i} - p_{a,wf}}{q_n} = \frac{4{,}709.4 - 66.4}{2{,}711} = 1.713 \text{ psi/Mscf-D}.$$

$$\frac{1}{q_n}\sum_{j=1}^{n} \Delta q_j \log\left(t_n - t_{j-1}\right) = \frac{1}{q_1}\left[\Delta q_1 \log\left(t_1 - t_0\right)\right]$$

$$= \frac{1}{2{,}711}[(2{,}711 - 0)\log(6 - 0)]$$

$$= 0.7782.$$

At $t=9.5$ hours, $n=3$, $t_3=9.5$ hours, $q_3=2{,}504$ Mscf-D, and $D'=0$:

$$\frac{p_{a,i} - p_{a,wf}}{q_n} = \frac{4{,}709.4 - 258.9}{2{,}504} = 1.777 \text{ psi/Mscf-D}.$$

$$\frac{1}{q_n}\sum_{j=1}^{n} \Delta q_j \log\left(t_n - t_{j-1}\right)$$

$$= \left(1/q_3\right)\left\{\left[\Delta q_1 \log\left(t_3 - t_0\right)\right] + \left[\Delta q_2 \log\left(t_3 - t_1\right)\right]\right.$$

$$\left. + \left[\Delta q_3 \log\left(t_3 - t_2\right)\right]\right\}$$

$$= (1/2{,}504) - \{[(2{,}711 - 0)\log(9.5 - 0)]$$

$$+ [(2{,}607 - 2{,}711)\log(9.5 - 6)]$$

$$+ [(2{,}504 - 2{,}607)\log(9.5 - 8)]\}$$

$$= 1.0287.$$

Fig. 3.2 plots calculated values of the pressure and time/rate plotting functions (**Table 3.3**).

Using least-squares regression analysis, calculate the best-fit straight line through the data. From this line, we determine the slope, m', and intercept, b, to be

$m' = 0.236$ and $b = 1.532$.

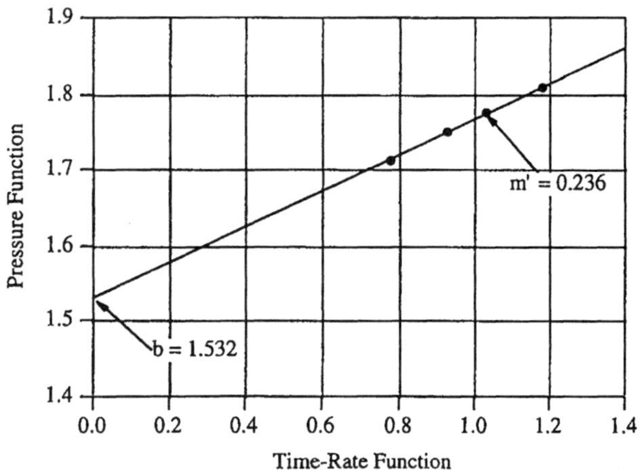

Fig. 3.2—Cartesian plot of multirate test data, Example 3.1.

TABLE 3.3—PRESSURE AND TIME FUNCTIONS, EXAMPLE 3.1

Pressure Function (psi/Mscf-D)	Time-Rate Function
—	—
1.713	0.7782
1.752	0.9271
1.777	1.0287
1.808	1.1819

2. The effective gas permeability is calculated with Eq. 3.30.

$$k = \frac{162.6\overline{B}_g\overline{\mu}_g}{m'h} = \frac{(162.6)(0.5878)(0.0286)}{(0.236)(19)} = 0.61 \text{ md.}$$

3. Next, calculate the skin factor with Eq. 3.31.

$$s = 1.151\left[(b/m') - \log\left(k/\phi\overline{\mu}_g\overline{c}_t r_w^2\right) + 3.23\right]$$

$$= 1.151\Big((1.532/0.236) - \log\Big\{0.61/\big[(0.176)$$

$$\times (0.0286)\left(7.33 \times 10^{-5}\right)(0.255)^2\big]\Big\} + 3.23\Big)$$

$$= 2.6.$$

A positive skin factor indicates damage resulting from a reduction in permeability in the formation adjacent to the wellbore.

3.5.3 Variable-Rate Gas-Flow Tests With Smoothly Changing Rates. In many testing situations, a strictly constant producing rate is impractical or even impossible to maintain. A more probable mode of operation is production at a constant surface pressure; if tubing friction effects are negligible, the BHP also is constant. At early times, however, both BHP and rate may be changing rapidly. Data obtained under these nonideal test conditions can be analyzed accurately with a simple modification of the transient-flow equation for constant-rate production.

Winestock and Colpitts[13] showed that, even when both p_{wf} and q vary with time, the following equation (written to model slightly compressible liquid flow) may be used as long as the rate is changing slowly and smoothly rather than abruptly:

$$\frac{p_i - p_{wf}}{q} = \frac{162.6B\mu}{kh}\left[\log\left(kt/\phi\mu c_t r_w^2\right) - 3.23 + 0.869s\right]. \quad \text{.....................} \quad (3.32)$$

To analyze transient data, we prepare a semilog graph of $(p_i - p_{wf})/q$ as a function of time, t. For analyzing variable-rate tests in gas wells where non-Darcy effects are important, we can rewrite Eq. 3.32 in terms of adjusted pressures, or

$$\left(p_{a,i} - p_{a,wf} - D'q_g^2\right)/q_g$$

$$= \frac{162.6\mu\overline{B}_g}{kh}\left[\log\left(kt/\phi\overline{\mu}_g\overline{c}_t r_w^2\right) - 3.23 + 0.869s\right], \quad \text{.....................} \quad (3.33)$$

where $D' = 141.2\overline{B}_g\overline{\mu}_g D/kh.$

Either the non-Darcy-flow effects (the $D'q_g^2$ term) must be neglected or the value of D' that leads to the best straight line in the middle-time region must be found with an iterative process.

Once we have identified the semilog straight line indicative of the middle-time region, the formation permeability is estimated from the slope, m', of this line or from

TABLE 3.4—TIME AND FLOW DATA FOR EXAMPLE 3.2

Flowing Time, t (hours)	Gas Flow Rate, q (Mscf/D)	Flowing Time, t (hours)	Gas Flow Rate, q (Mscf/D)
2.4	432.0	43.2	353.5
4.8	405.9	48.0	351.7
7.2	392.9	60.0	348.1
9.6	385.4	72.0	345.1
12.0	380.3	84.0	342.8
14.4	376.3	96.0	340.8
16.8	372.9	108.0	339.6
19.2	370.0	120.0	337.4
21.6	367.5	132.0	335.9
24.0	365.2	144.0	334.5
28.8	361.4	168.0	332.1
33.6	358.3	192.0	330.4
38.4	355.7	216.0	328.1

$$k = 162.6\overline{B}_g\overline{\mu}_g/m'h. \quad \text{.........................} \quad (3.34)$$

The skin factor is

$$s' = 1.151\left\{\frac{1}{m'}\left[\left(p_{a,i} - p_{a,wf}\right)/q\right]_{1\text{hr}} - \log\left(k/\phi\overline{\mu}_g\overline{c}_t r_w^2\right) + 3.23\right\} \quad \text{.....................} \quad (3.35)$$

where $[(p_{a,i} - p_{a,wf})/q]_{1\text{hr}}$ must lie on the semilog straight line or its extrapolation to 1 hour. Similar to constant-rate production data, the variable rate data may be affected by wellbore storage at early times and reservoir boundaries at late times, which distorts the pressure response and possibly masks the correct semilog straight line indicative of the middle-time region.

Example 3.2—Analysis of Variable-Rate Gas-Flow Tests. A gas well was produced at a constant BHP of 1,000 psia. We summarize known data next and in **Table 3.4.** Gas and formation properties are evaluated at the initial reservoir pressure, p_i. Assuming non-Darcy-flow effects ($D'q_g^2$) are negligible, estimate effective gas permeability and skin factor with the semilog analysis technique for variable-rate gas-flow tests.

r_w = 0.365 ft
h = 23 ft
ϕ = 0.14
γ_g = 0.7
T = 180°F
$\overline{B}_g$ = 0.668 RB/Mscf
p_i = 4,650 psia
$p_{a,i}$ = 3,251.5 psia
μ_g = 0.0247 cp
p_{wf} = 1,000 psia
$p_{a,wf}$ = 213.0 psia
c_t = 11.53×10^{-5} psia^{-1}

Solution.

1. Prepare a plot of

$$\left[\left(p_{a,i} - p_{a,wf} - D'q_g^2\right)/q_g\right] \text{ vs. } \log t.$$

For this problem, we assume that non-Darcy-flow effects are negligible (i.e., $D' = 0$). With the adjusted initial and BHP's given above, the plotting functions are calculated as illustrated with the following example. For example, at $t = 4.8$ hours, the plotting function is

$$\left(p_{a,i} - p_{a,wf}\right)/q_g = (3,251.5 - 213.0)/405.9 = 7.49.$$

Similarly, **Table 3.5** summarizes plotting functions for each measured rate.

TABLE 3.5—PLOTTING FUNCTIONS, EXAMPLE 3.2

t (hours)	q (Mscf/D)	$\frac{p_{a,i}-p_{a,wf}}{q_g}$ (psi/Mscf-D)	t (hours)	q (Mscf/D)	$\frac{p_{a,i}-p_{a,wf}}{q_g}$ (psi/Mscf-D)
2.4	432.0	7.03	43.2	353.5	8.60
4.8	405.9	7.49	48.0	351.7	8.64
7.2	392.9	7.73	60.0	348.1	8.73
9.6	385.4	7.88	72.0	345.1	8.80
12.0	380.3	7.99	84.0	342.8	8.86
14.4	376.3	8.07	96.0	340.8	8.92
19.2	370.0	8.21	120.0	337.4	9.01
21.6	367.5	8.27	132.0	335.9	9.05
24.0	365.2	8.32	144.0	334.5	9.08
28.8	361.4	8.41	168.0	332.1	9.15
33.6	358.3	8.48	192.0	330.4	9.20
38.4	355.7	8.54	216.0	328.1	9.26

2. **Fig. 3.3** also plots the data from Table 3.5. From type-curve analysis (discussed in Chap. 4), we have identified the beginning of the middle-time region at approximately 30 hours. Therefore, the slope of the line drawn through data points after this time is

$m' = 0.947$ psi/cycle.

3. From Eq. 3.34, the effective gas permeability is

$$k = \frac{162.6\overline{B}_g\overline{\mu}_g}{m'h} = \frac{(162.6)(0.668)(0.0247)}{(0.947)(23)} = 0.12 \text{ md.}$$

4. Next, we calculate the skin factor. From Fig. 3.3, we see the value of the pressure function at the extrapolation of the straight line to $t = 1$ hour is

$$\left[\left(p_{a,i} - p_{a,wf}\right)/q_g\right]_{1\text{hr}} = 7.04 \text{ psi/Mscf-D.}$$

From Eq. 3.35, the skin factor is

$$s' = 1.151\left\{\frac{1}{m'}\left[\left(p_a - p_{a,wf}\right)/q\right]_{1\text{hr}} - \log\left(k/\phi\overline{\mu}_g\overline{c}_t r_w^2\right) + 3.23\right\}$$

$$= 1.151\Big((7.04)\ (1/0.947) - \log\Big\{0.12/\big[(0.14)(0.0247)$$

$$\times\ (11.53 \times 10^{-5})(0.365)^2\big]\Big\} + 3.23\Big)$$

$$= 5.0.$$

Because we assume that non-Darcy effects are negligible, we also assume that the skin factor represents true formation damage.

3.5.4 Gas-Flow Tests With Variable Non-Darcy-Flow Coefficient.

In Sec. 3.5.1, we discussed analysis procedures for a constant-rate flow test in which the non-Darcy-flow coefficient, D, could be assumed constant. In this section, we show how to analyze a constant-rate-flow test taking into account that D can vary throughout the test.

Unsteady-state, constant-rate gas flow can be modeled in terms of dimensionless pseudopressure as

$$p_{pD}(t_D) = 0.5\ [\ln(t_D) + 0.80907] + s + Dq_g\ , \quad \text{(3.36)}$$

where the non-Darcy-flow coefficient, D, which is frequently treated as a constant, is in fact variable because it depends on gas viscosity near the well. Because the gas viscosity near the wellbore depends on the flowing BHP, which continuously decreases during a flow test, the gas viscosity also changes throughout the test.

Fig. 3.4 illustrates the effects of non-Darcy flow and skin damage on the pressure response. The dimensionless pseudopressure responses A, B, and C are simulated by use of Eq. 3.36 for the conditions shown. The lowermost solid line is the liquid-flow response with a slope of 1.151. Response B, with a slope of 1.163, shows only the effect of non-Darcy flow. Response C is a translation of response A with a positive skin factor, assuming that the non-Darcy pseudo-skin effect is an additive term to the pressure response. Response D, with a slope of 1.183, is the actual response with non-Darcy flow and a positive skin factor, indicating that the non-Darcy term is not additive and thereby confirming that D is not constant.

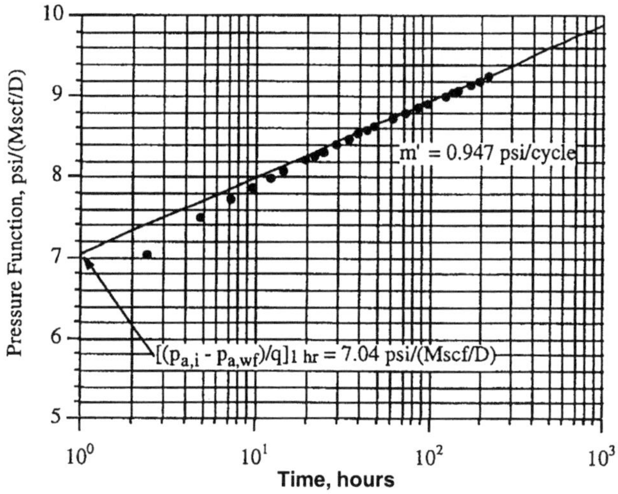

Fig. 3.3—Semilog plot of variable-rate data, Example 3.2.

Fligelman *et al.*[6] present a method to estimate formation permeability, k, and true skin factor, s, from pressure drawdown data for a gas well. Their equations, written in terms of pseudopressures, are modified here for use with adjusted pressures. They define the dimensionless group

$$Bq_D = \frac{2.715 \times 10^{-15}\beta kMp_{sc}q_g}{T_{sc}hr_w\overline{\mu}_g}, \quad \text{(3.37)}$$

where $\overline{\mu}_g$ = gas viscosity evaluated at $\overline{p}$. For a constant skin factor, the dimensionless intercept, b_D, of the semilog straight line in a plot of dimensionless pseudopressure as a function of the log of dimensionless time is defined as

$$b_D = \frac{p_{a,i} - m\log\left(0.0002637k/\phi\overline{\mu}_g\overline{c}_t r_w^2\right) - p_{a,1\text{hr}}}{q_D p_{a,i}}, \quad \text{(3.38)}$$

where m and $p_{a,\,1\text{hr}}$ = the slope and intercept, respectively, of the semilog straight line in a plot of p_a vs. $\log t$. The dimensionless form of the slope m is given by

$$m_D = m/q_D p_{a,i}\ . \quad \text{(3.39)}$$

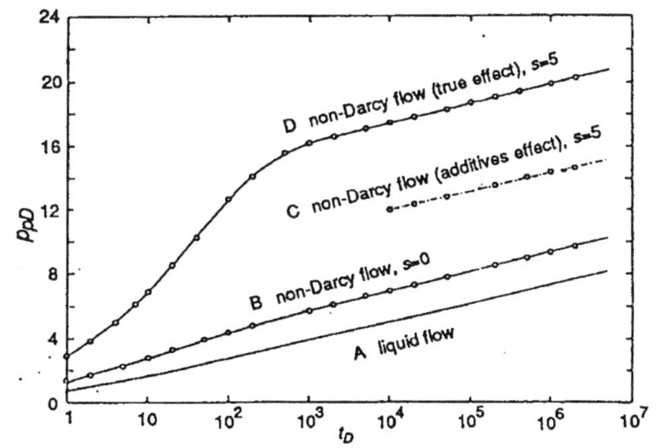

Fig. 3.4—Effects of non-Darcy flow and skin factor on transient-pressure response. After Fligelman *et al.*[6]

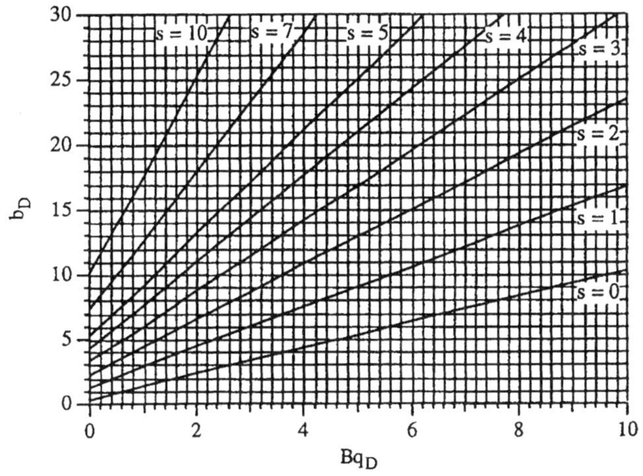

Fig. 3.5—Correlation of b_D and Bq_D. After Fligelman *et al.*[6]

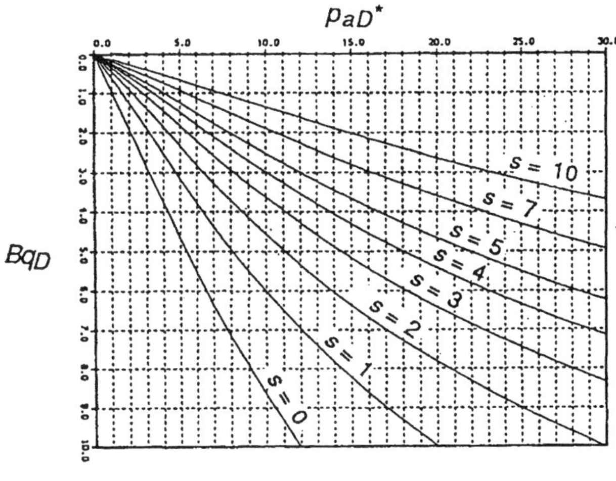

Fig. 3.6—Correlation of p^*_{aD} and Bq_D. After Fligelman *et al.*[6]

The dimensionless flow rate, q_D, is given by

$$q_D = 0.138 q_g p_{sc} T / kh T_{sc} p_{a,i} \text{ . . . . } (3.40)$$

The intercept b_D is related to Bq_D by the correlation

$$b_D = c_1(Bq_D) + c_2 \text{ , ... } (3.41)$$

where selected values of the constants, c_1 and c_2, are summarized next. **Fig. 3.5** shows the correlation of b_D and Bq_D for a range of skin factors.

Skin Factor, s	c_1	c_2
0	1.0013	0.3205
5	3.96273	5.2411
10	7.4733	10.1993

A second correlation relates the pseudopressure response with non-Darcy flow to the group Bq_D. The dimensionless-wellbore-pressure response as a function of dimensionless time is given by

$$p_{aD} = (m_D - 1.151) \log t_D^* + b_D - 0.4045 - s \text{ , ... } (3.42)$$

where m_D is the slope of the straight line in a plot of p_{pD} vs. log t_D. The time t_D^* is defined as

$$t_D^* = 0.0002637 kt / \overline{\mu}_g \overline{c}_t r_w^2 . \text{ . . . . } (3.43)$$

The pressure response with turbulent flow, p^*_{aD}, is related to Bq_D by the correlation

$$p^*_{aD} = c_3(Bq_D) + c_4(Bq_D)^2 + c_5(Bq_D)^3 \text{ , ... } (3.44)$$

where selected values of the constants, c_3, c_4, and c_5, are summarized next. **Fig. 3.6** shows the correlation of p^*_{aD} and Bq_D for a range of skin factors.

Skin Factor, s	c_3	c_4	c_5
0	0.002	–0.002	0.002
5	4.168	–0.202	0.049
10	7.624	–0.366	0.134

Although Fligelman's method allows for correction for non-Darcy-flow effects, the method has certain limitations. First, only non-Darcy-flow effects are considered. The effects of the non-Darcy-flow coefficient, D, on a phase change (i.e., condensation) as gas approaches the wellbore, if such occurs, are not taken into account. Second, because the correlations are based on positive skin factors, the procedure applies only when $s > 0$. Third, the correlations were developed for a damaged-region/wellbore-radius ratio greater than 10. For a higher ratio and a given value of s, the slope of the semilog straight line is unaffected but the non-Darcy effect is smaller. The greatest limitation is that the accuracy of the method depends on the applicability of the empirical correlation for β used in the study, which is subject to considerable uncertainty.[9]

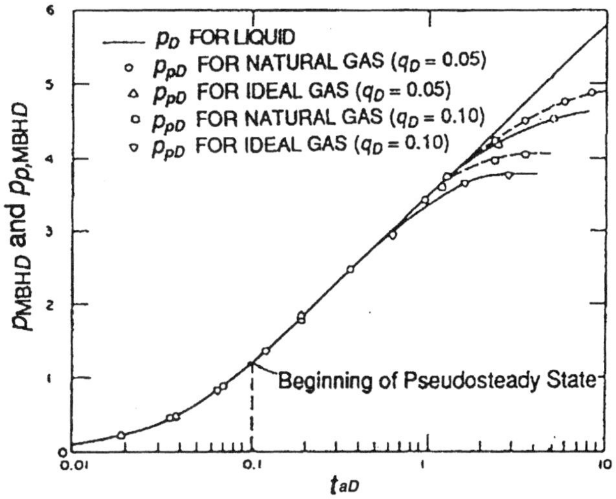

Fig. 3.7—Comparison of dimensionless pressure responses for liquid and gas solutions. After Al-Hussainy.[14]

3.5.5 Gas-Flow Tests in Bounded Reservoirs. When a pressure transient encounters reservoir boundaries during a gas-well test, the liquid solution no longer describes the pressure behavior.[1,14,15] With boundary effects, variations in μ_g and c_t affect the pressure response. In the semilog plot of **Fig. 3.7,** dimensionless-pseudopressure solutions are compared with the liquid solution. Dimensionless time, t_{aD}, is defined by Eq. 3.45:

$$t_{aD} = 0.0002637 k t_p / \phi \overline{\mu}_g \overline{c}_t A \text{ , ... } (3.45)$$

where fluid properties are evaluated at $\overline{p}$. About 1 log cycle of dimensionless time after the pseudosteady-state flow begins, the gas solutions begin to deviate from the liquid solution. The cause of the deviation is the variation in the μc_t product for gas; for a liquid, this product is constant. As a result, the pore volume cannot be found as described in Chap. 2 for slightly compressible fluids (i.e., liquids). However, Fraim and Wattenbarger[16] developed an iterative procedure that accounts for variations in gas properties during boundary-dominated flow in a gas well producing at constant BHP.

3.6 Analysis of Gas-Well Buildup Tests

This section discusses analysis techniques for pressure-buildup tests in wells completed in gas reservoirs. We begin with buildup tests with constant-rate production before shut-in and show how the

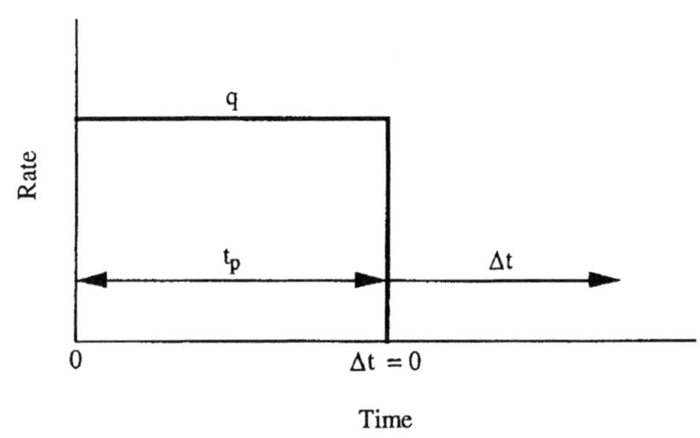

Fig. 3.8—Modeling a pressure-buildup test with rate superposition in time.

analysis techniques for slightly compressible liquids can be extended to gas wells when transformed variables are used. We then discuss the more probable production scenarios—discrete rate changes before shut-in or constant-pressure production before shut-in. Finally, we illustrate how average drainage-area pressure is determined from a gas-well buildup test.

3.6.1 Buildup Tests With Constant-Rate Production Before Shut-In. Similar to the method presented in Chap. 2 for analyzing pressure-buildup tests with slightly compressible liquids, we can also use superposition in time to develop an analysis technique for gas-well buildup tests. Again, assuming that Horner's approximation is valid, the entire production history before shut-in can be modeled as production for a period t_p at the last production rate, q_{last}.

Referring to **Fig. 3.8,** the term Δt denotes the time elapsed since shut-in. In terms of adjusted pressures and times, the variation of the adjusted shut-in BHP with time is

$$p_{a,ws} = p_{a,i} - \left(162.6q\bar{B}_g\bar{\mu}_g/kh\right)\log\left[\left(t_p + \Delta t_a\right)/\Delta t_a\right], \quad (3.46)$$

where t_p = adjusted producing time and Δt_a = adjusted shut-in time defined by Eqs. 3.8 and 3.9, respectively, and $q = q_{last}$. In addition, note that B_g and μ_g are evaluated at the average drainage-area pressure.

The form of Eq. 3.46 suggests that permeability can be estimated from the slope of the semilog line on a plot of $p_{a,ws}$ vs. $\log[(t_p + \Delta t_a)/\Delta t_a]$, or

$$k = 162.6q\bar{B}_g\bar{\mu}_g/mh. \quad (3.47)$$

We also can develop an equation for calculating the skin factor.

$$s' = 1.151\left\{\left[\left(p_{a,1hr} - p_{a,wf}\right)/m\right] - \log\left(k/\phi\bar{\mu}_g\bar{c}_t r_w^2\right) + 3.23\right\}, \quad (3.48)$$

where $p_{a,1hr}$ = adjusted pressure at Δt_a equal to 1 hour and $p_{a,wf}$ = adjusted flowing BHP before shut-in. Similar equations can be developed with pressure and pressure-squared variables (Table 3.1).

Example 3.3—Analysis of a Gas-Well Pressure-Buildup Test. A pressure-buildup test was run on a gas well in a newly discovered reservoir. We summarize known and calculated data here and in **Table 3.6.** Because the pressure has not been depleted in the reservoir, the semilog straight line should extrapolate to original reservoir pressure at a Horner time ratio of unity; i.e., $p^* = p_i$. Prepare Horner plots of (1) p_{ws} vs. $(t_p + \Delta t)/\Delta t$; (2) p_{ws}^2 vs. $(t_p + \Delta t)/\Delta t$; and (3) $p_{a,ws}$ vs. adjusted Horner time ratio, $(t_p + \Delta t_a)/\Delta t_a$. Determine effective permeability to gas and skin factor from each plot.

h = 28 ft
r_w = 0.3 ft
ϕ = 0.18
t_p = 2,000 hours
γ_g = 0.7
$\bar{T}$ = 640°R (180°F)
$\bar{\mu}_g$ = 0.01885 cp
q_g = 5,256 Mscf/D
$\bar{B}_g$ = 0.962 RB/Mscf
p_i = 2,906 psia
$p_{a,i}$ = 1,718.59 psia
$\bar{z}$ = 0.8678
$\bar{c}_t$ = 2.238×10^{-4} psia^{-1}

Solution. **Figs. 3.9 through 3.11** give Horner plots in terms of pressure, pressure-squared, and adjusted pressures, respectively. Note that semilog straight lines passing through the later data and through p_i (equal to p^* in this case) appear in each case.

Analysis With Pressure Variables **(Fig. 3.9):**

1. Calculate the slope of the line drawn through the middle-time region:

$m = 2{,}375 - 2{,}329.9 = 45.1$ psi/cycle.

2. The effective permeability to gas is

$$k = \frac{162.6q_g\bar{B}_g\bar{\mu}_g}{mh} = \frac{(162.6)(5,256)(0.962)(0.01885)}{(45.1)(28)}.$$

$$= 12.27 \text{ md.}$$

3. Calculate the skin factor. The pressure p_{1hr} at $(t_p + \Delta t)/\Delta t = (2000 + 1)/1 = 2{,}001$ is 2,226 psia. Therefore,

$$s' = 1.151\left\{\left[\left(p_{1hr} - p_{wf}\right)/m\right] - \log\left(k/\phi\bar{\mu}_g\bar{c}_t r_w^2\right) + 3.23\right\}$$

$$= 1.151\Big(\left[(2,226 - 1,800)/45.1\right] - \log\Big\{12.27/\big[(0.18) \times (0.01885)(0.0002238)(0.3)^2\big]\Big\} + 3.23\Big)$$

$$= 5.09.$$

Analysis With Pressure-Squared Variables **(Fig. 3.10).**

1. Calculate the slope of the line drawn through the middle-time region:

$m = 5.63 \times 10^6 - 5.425 \times 10^6 = 0.205 \times 10^6$ psia²/cycle.

2. The effective permeability to gas is

$$k = \frac{1637q_g\bar{T}\bar{z}\bar{\mu}_g}{mh} = \frac{(1637)(5,256)(640)(0.868)(0.01885)}{(2.05 \times 10^5)(28)}$$

$$= 15.7 \text{ md.}$$

3. Calculate the skin factor. The term p_{1hr}^2 at $(t_p + \Delta t)/\Delta t = 2{,}001$ equals 4.95×10^6 psia², so

$$s' = \left\{\left[\left(p_{1hr}^2 - p_{wf}^2\right)/m\right] - \log\left(k/\phi\bar{\mu}_g\bar{c}_t r_w^2\right) + 3.23\right\}$$

$$= 1.151\left(\frac{4.95 \times 10^6 - 3.24 \times 10^6}{2.05 \times 10^5} - \log\left\{15.7/\left[(0.18)(0.01885)(0.0002238)(0.3)^2\right]\right\} + 3.23\right)$$

$$= 3.69.$$

Analysis With Adjusted Pressure Variables **(Fig. 3.11).**

1. Calculate the slope of the line drawn through the middle-time region:

$m = 39.7$ psi/cycle.

TABLE 3.6—GAS WELL PRESSURE BUILDUP TEST DATA, EXAMPLE 3.3					
Δt (hours)	p_{ws} (psia)	p_{ws}^2 (psia$^2 \times 10^{-6}$)	Adjusted Pressure (psia)	Horner Time Ratio	Adjusted Horner Time Ratio
0	1800	3.24	726.47	—	—
0.001	1,842.4	3.39	758.94	2,000,001	3,031,200
0.002	1,878.3	3.53	786.86	1,000,001	1,502,000
0.003	1,909.2	3.65	811.21	666,668	993,280
0.004	1,935.8	3.75	832.42	500,001	739,530
0.005	1,959	3.84	851.09	400,001	587,710
0.006	1,979.3	3.92	867.54	333,334	486,800
0.007	1,997.1	3.99	882.11	285,715	414,950
0.008	2,012.7	4.05	894.91	250,001	361,260
0.009	2,026.6	4.11	906.41	222,223	319,660
0.01	2,038.8	4.16	916.5	200,001	286,500
0.02	2,108.3	4.44	974.88	100,001	139,470
0.03	2,135.2	4.56	997.78	66,668	91,578
0.04	2,148.6	4.62	1,009.3	50,001	68,034
0.05	2,156.7	4.65	1,016.2	40,001	54,074
0.06	2,162.4	4.68	1,021.1	33,334	44,848
0.07	2,166.7	4.69	1,024.8	28,572	38,300
0.08	2,170.2	4.71	1,027.9	25,001	33,414
0.09	2,173.2	4.72	1,030.5	22,223	29,630
0.1	2,175.8	4.73	1,032.7	20,001	26,612
0.2	2,191.7	4.80	1,046.5	10,001	13,162
0.3	2,200.4	4.84	1,054.1	6,667.7	8,729.2
0.4	2,206.4	4.87	1,059.3	5,001.0	6,525.5
0.5	2,211	4.89	1,063.3	4,001.0	5,208.2
0.6	2,214.8	4.91	1,066.6	3,334.3	4,332.2
0.7	2,217.9	4.92	1,069.3	2,858.1	3,707.8
0.8	2,220.6	4.93	1,071.7	2,501.0	3,240.3
0.9	2,223	4.94	1,073.8	2,223.2	2,877.2
1	2,225.2	4.95	1,075.7	2,001.0	2,587.1
2	2,239.1	5.01	1,087.9	1,001.0	1,286.4
3	2,247.1	5.05	1,095.0	667.67	855.08
4	2,252.8	5.08	1,100.0	501.00	640.07
5	2,257.2	5.09	1,103.9	401.00	511.34
6	2,260.8	5.11	1,107.1	334.33	425.67
7	2,263.8	5.12	1,109.7	286.71	364.56
8	2,266.4	5.14	1,112	251.00	318.78
9	2,268.7	5.15	1,114.1	223.22	283.22
10	2,270.8	5.16	1,115.9	201.00	254.79
20	2,284.3	5.22	1,127.9	101.00	127.24
30	2,292.1	5.25	1,134.9	67.667	84.903
40	2,297.6	5.28	1,139.8	51.000	63.791
50	2,301.9	5.30	1,143.6	41.000	51.149
60	2,305.3	5.31	1,146.7	34.333	42.734
70	2,308.2	5.33	1,149.2	29.571	36.731
80	2,310.7	5.34	1,151.5	26.000	32.234

2. The effective permeability to gas is

$$k = \frac{162.6 q_g \overline{B}_g \overline{\mu}}{mh} = \frac{(162.6)(5,256)(0.962)(0.01885)}{(39.7)(28)}$$

$$= 13.9 \text{ md.}$$

3. Calculate the skin factor. At an adjusted Horner time ratio of $(t_p + \Delta t_a)/\Delta t_a = (2{,}000 + 1)/1 = 2{,}001$, $p_{a,\,1\text{hr}} = 1{,}079.9$ psia.

$$s' = 1.151\left\{\left[\left(p_{a,1\text{hr}} - p_{a,wf}\right)/m\right] - \log\left(k/\phi\overline{\mu}_g\overline{c}_t r_w^2\right) + 3.23\right\}$$

$$= 1.151\left(\frac{(1,079.96 - 726.47)}{39.7}\right.$$

$$\left. - \log\left\{13.9/\left[(0.18)(0.01885)(0.0002238)(0.3)^2\right]\right\} + 3.23\right)$$

$$= 4.4.$$

In this example, little difference exists in the results of the pressure, pressure-squared, or adjusted-pressure analysis on a Horner semilog graph. We should emphasize, however, that when wellbore-storage distorted data are analyzed on a type curve, the differences can be significant. We discuss type-curve analysis in Chap. 4.

3.6.2 Buildup Tests With Discrete Changes in Rate Before Shut-In. When Horner's approximation is not an accurate representation of the rate history before shut-in, we can use superposition in time to develop an alternative analysis technique for a gas-well buildup

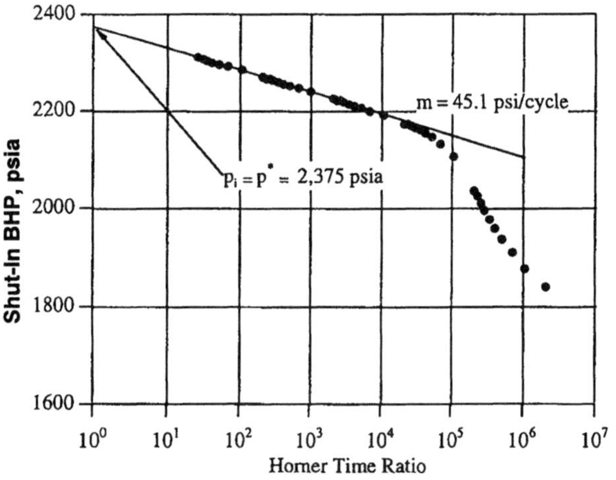

Fig. 3.9—Horner plot with pressure, Example 3.3.

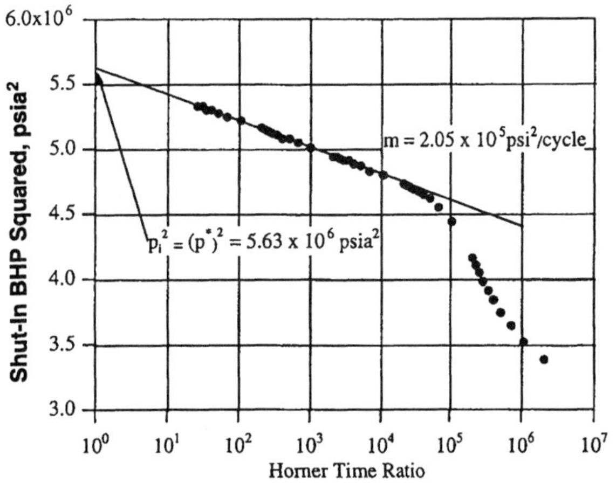

Fig. 3.10—Horner plot with pressure-squared, Example 3.3.

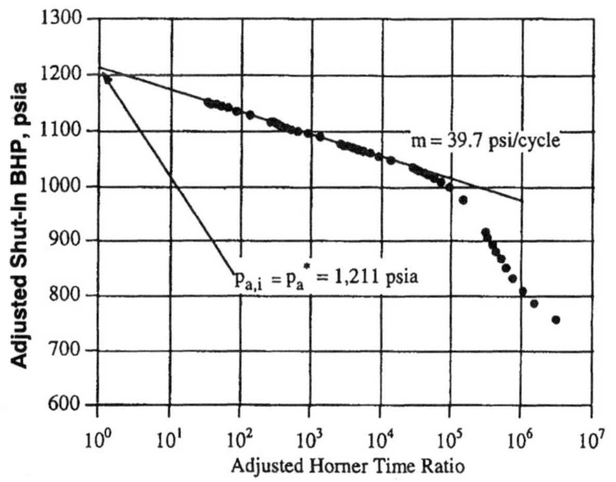

Fig. 3.11—Horner plot with adjusted pressure and adjusted Horner time ratio, Example 3.3.

test following n distinct rates. Similar to the development presented in Chap. 2, we can write

$$p_{a,i} - p_{a,ws} = \frac{162.6 q_n \overline{B}\overline{\mu}_g}{kh} \sum_{j=1}^{n+1} \left(\frac{q_j}{q_n}\right) \log\left(\frac{t - t_{j-1}}{t - t_j}\right)$$

$$= m' q_n \left\{ \left[\frac{q_1}{q_n} \log\left(\frac{t}{t - t_1}\right) + \ldots + \frac{q_{n-1}}{q_n} \log\left(\frac{t - t_{n-2}}{t - t_{n-1}}\right)\right] + \left[\log\left(\frac{t - t_{n-1}}{t - t_n}\right)\right]\right\}, \quad (3.49)$$

where $q_{n+1} = 0$ by definition. Note that $t - t_n = \Delta t$, shut-in time. The following analysis procedure, based on Eq. 3.49, is similar to that presented in Chap. 2. Although presented in terms of adjusted variables, the same procedure is applicable with either pressure or pressure-squared variables.

1. Calculate the plotting function, X.

$$X = \frac{q_1}{q_n} \log\left(\frac{t}{t - t_1}\right) + \ldots + \log\left(\frac{t - t_{n-1}}{t - t_n}\right). \quad (3.50)$$

2. Plot $p_{a,ws}$ as a function of X on Cartesian coordinate graph paper.
3. Determine the slope, m, of the plot and calculate effective gas permeability.

$$k = \left(-162.6 q_n \overline{B}_g \overline{\mu}_g\right) / mh. \quad (3.51)$$

4. Calculate the skin factor, s, from the equation

$$s = 1.151 \left\{ \left[\left(p_{a,1\text{hr}} - p_{a,wf}\right)/m\right] - \log\left(k/\phi \overline{\mu}_g \overline{c}_t r_w^2\right) + 3.23 \right\}. \quad (3.52)$$

5. Extrapolate the middle-time line to $X = 0$ and read the value of $p_{a,ws}$. This extrapolated pressure is $p_a{}^*$, which is equal to $p_{a,i}$ for an infinite-acting reservoir.

Three important limitations apply. First, the improvement in accuracy by use of more than three terms in the summation in Eq. 3.48 is negligible. Second, the plotting function, X, is based on the fundamental assumption that, for a producing time $t = t_{p1} + t_{p2} + \ldots + t_{pn} + \Delta t$, the reservoir is infinite-acting. This assumption will rarely be true for large values of t. Third, the plotting function is based on the assumption that, at the end of each flow period, the data are not distorted by wellbore storage. This assumption will not be satisfied in some cases with short-duration flow periods.[10]

Odeh and Selig[17] suggested that a buildup test following n different rates could be analyzed by a method similar to the Horner method. The shut-in pressure response is given by

$$p_{a,i} - p_{a,ws} = \left(162.6 q^* B \mu_g / kh\right) \log\left[\left(t_p^* + \Delta t\right)/\Delta t\right], \quad (3.53)$$

where the modified production time, t_p^*, and flow rate, q^*, are defined, respectively, as

$$t_p^* = 2 \left[t_n - \frac{\sum_{j=1}^{n} q_j \left(t_j^2 - t_{j-1}^2\right)}{2 \sum_{j=1}^{n} q_j \left(t_j - t_{j-1}\right)} \right] \quad (3.54)$$

and $$q^* = \frac{1}{t_p^*} \sum_{j=1}^{n} q_j \left(t_j - t_{j-1}\right). \quad (3.55)$$

The Odeh-Selig method, approximate but accurate, is applicable only for pressures at values of Δt greater than the producing time, t_p. Unfortunately, this condition is likely to occur only in a drillstem test or short production test. Chap. 2, Example 2.6 illustrates application of the superposition and Odeh-Selig methods.

3.6.3 Buildup Tests With Constant-Pressure Production Before Shut-In. Conventional buildup-test-analysis techniques have been developed primarily for wells producing at a constant rate before shut-in. However, some common situations involve production at constant BHP rather than at constant rate. Examples include declining-rate production during reservoir depletion, fluid flow into a constant-pressure separator, or open wells flowing at atmospheric pressure. With slight modification, conventional Horner and type-curve analysis can be used to analyze buildup test data following production at constant BHP.[1,20] The Horner method gives the correct semilog straight line from which formation permeability, skin factor, and average drainage-area pressure can be found.

For Horner analysis of a buildup test following constant-pressure production, the actual producing time, t_{actual}, rather than t_p defined by Eq. 3.56, should be used in the Horner plotting function. This is in contrast to conventional Horner analysis, in which we use a producing time defined by

$$t_p = \frac{\text{Cumulative Production}}{\text{Most Recent Production Rate}}, \qquad (3.56)$$

where the most recent production rate is the last established flow rate before shut-in.

To illustrate the difference in plotting functions, consider an example: suppose that a gas well has been produced at constant BHP for 1 year (8,760 hours). At the time of shut-in for a buildup test, cumulative production is 100,000 Mscf. The last established flow rate is 100 Mscf/D. The conventional Horner producing time is

$$t_p = \frac{\text{Cumulative Production}}{\text{Most Recent Production Rate}}$$

$$= \frac{(24)(100,000)}{100} = 24,000 \text{ hours.}$$

At a shut-in time of $\Delta t = 10$ hours, the Horner plotting function is

$$\frac{t_p + \Delta t}{\Delta t} = \frac{24,000 + 10}{10} = 2,401.$$

With $t_{actual} = 8,760$ hours rather than $t_p = 24,000$ hours, the Horner plotting function is

$$\frac{t_{actual} + \Delta t}{\Delta t} = \frac{8,760 + 10}{10} = 877.$$

A plot of shut-in pressure, $p_{a,ws}$, vs. the adjusted Horner time ratio based on t_{actual} gives the correct semilog straight line with slope m. Effective permeability to gas is

$$k = 162.6 q_{g,last} \bar{B}_g \bar{\mu}_g / mh, \qquad (3.57)$$

where $q_{g,last}$ = last established producing rate, not the average rate over the producing period. The skin factor is

$$s = 1.151\left\{\left[\left(p_{a,1hr} - p_{a,wf}\right)/m\right] - \log\left(k/\phi\bar{\mu}_g\bar{c}_t r_w^2\right) + 3.23\right\}. \qquad (3.58)$$

Extrapolation of the semilog straight line to infinite shut-in time (adjusted Horner time equal to 1) gives $p_{a,i}$ for an infinite-acting reservoir or p_a^* for a bounded reservoir with partial pressure depletion (boundary effects evident before shut-in). For the latter case, the Matthews-Brons-Hazebroek (MBH) method[19] provides an estimate of average drainage-pressure in a reservoir.

For buildup tests following constant-pressure production, the effects of wellbore storage, skin damage, and non-Darcy (high-velocity) gas flow usually are short-lived and do not affect the slope of the semilog straight line. However, in gas reservoirs with $k > 0.1$ md, non-Darcy-flow effects may cause substantial error in estimates of the apparent skin factor, s', from match-point data in type-curve analysis. Under these conditions, semilog analysis must be used for reliable formation evaluation. Furthermore, when non-Darcy flow effects are significant, a buildup test gives more reliable results than does a drawdown test.[6]

3.6.4 Determining Average Drainage-Area Pressure for Gas Wells. The average drainage-area pressure, $\bar{p}$, for a gas well can be found with the MBH p^* method discussed earlier for the liquid case. To find the correct $\bar{p}$, pressure-dependent fluid properties must be evaluated at that pressure, $\bar{p}$. This suggests an iterative procedure. For gas, we use adjusted pressure, p_a. The procedure, which is outlined here, involves making an initial estimate of $\bar{p}_a$ (the adjusted pressure evaluated at $\bar{p}$), evaluating fluid properties at the pressure associated with $\bar{p}_a$, calculating the MBH pressure function, $p_{a,MBHD}$, on the basis of those properties, and obtaining another estimate of from $p_{a,MBHD}$. This same procedure is applicable when either pressure or pressure-squared variables are used.

1. Assume a value of $\bar{p}$; calculate the adjusted pressure and time, and make a Horner plot of the buildup data using values of adjusted pressure, $p_{a,ws}$, and adjusted Horner time ratio, $(t_p + \Delta t_a)/\Delta t_a$.
2. Extrapolate $p_{a,ws}$ to $(t_p + \Delta t_a)/\Delta t_a = 1$. The extrapolated adjusted pressure, p_a^*, may be an improved estimate of $\bar{p}_a$; however, if p_a^* is clearly larger than the apparent level value of $\bar{p}_a$, continue to use the initial estimate of $\bar{p}_a$ in subsequent steps.
3. Estimate the drainage-area shape. If there is insufficient information to estimate the shape, assume a circular drainage area.
4. Select the appropriate MBH chart[10,11,19] for the drainage area.
5. Calculate t_{aD} with Eq. 3.45 with μ_g and c_t evaluated at the pressure corresponding to the most recent estimate of $\bar{p}_a$.
6. From the MBH chart at the calculated value of t_{aD}, read the value of the MBH pressure function, which equals

$$p_{a,\text{MBHD}} = \frac{kh\left(p_a^* - \bar{p}_a\right)}{70.6 q \bar{B}_g \bar{\mu}_g} = \frac{2.303\left(p_a^* - \bar{p}_a\right)}{m}. \qquad (3.59)$$

7. Calculate the next estimate of $\bar{p}_a$ with

$$\bar{p}_a = p_a^* - \frac{m p_{a,\text{MBH},D}}{2.303}. \qquad (3.60)$$

8. If the estimate of $\bar{p}_a$ from Eq. 3.61 is different from the previous estimate, then evaluate the gas properties at the pressure associated with the new estimate of $\bar{p}_a$. Repeat Steps 5 through 7, recalculating t_{aD} at the new values of the fluid properties, until the estimate of $\bar{p}_a$ from Eq. 3.60 converges. The correct $\bar{p}$ is the pressure corresponding to the final value of $\bar{p}_a$.

Kazemi[15] and Reynolds *et al.*[21] present other methods to estimate $\bar{p}$ for gas reservoirs. Kazemi's method requires an iterative procedure, while the method of Reynolds *et al.* avoids iteration and requires a Horner plot in terms of real time and pseudopressure rather than adjusted time and adjusted pressure.

3.7 Chapter Summary

In this chapter, we introduced the reader to semilog analysis of gas-well pressure transient tests.

In Sec. 3.2, we reviewed the use of pseudopressure and pseudotime introduced in Chap. 1. We introduced the concept of normalized pressure as an alternative to the use of pseudopressure. Normalized pressure is simply pseudopressure multiplied by a factor $\mu_g(p_r)z(p_r)/2p_r$, for some suitable reference pressure p_r, to give units of psi rather than psi^2/cp. We defined the adjusted pressure by taking p_r to be the current average drainage area pressure, $\bar{p}$. Similarly, we defined the adjusted time as the pseudotime multiplied by the factor $\mu_g(\bar{p})c_t(\bar{p})$. Finally, we defined the Horner time ratio for semilog analysis of gas-well pressure-buildup tests as $\left(t_p + \Delta t_a\right)/\Delta t_a$ by use of the actual producing time rather than an adjusted producing time.

Analysis in terms of pseudopressure or adjusted pressure is the most general formulation for gas wells. However, the practicing engineer needs to be acquainted with analysis methods in terms of pressure and pressure-squared simply because of the widespread use of these methods in industry and in the literature. In particular, the limitations of these two methods to high and low pressures, respectively, should be kept in mind. In Sec. 3.3 we wrote the semilog analysis equations in terms of pressure and pressure-squared. In Table 3.1, we summarized the semilog analysis equations for both flow and buildup tests—for oil wells in terms of pressure and time; and for gas wells in terms of adjusted pressure and adjusted time, pressure and time, pressure-squared and time, and pseudopressure and time.

Gas wells are frequently subject to flow where the pressure drop deviates from Darcy's law. In Sec. 3.4, we discussed the effect of non-Darcy flow on the pressure response and modified the semilog analysis equations to take into account non-Darcy flow.

In Sec. 3.5, we discussed the analysis of flow tests in gas wells. We discussed the simplest case, that of pressure response caused by constant-rate production, in Sec. 3.5.1. We presented equations for obtaining both apparent skin factor and effective permeability from the pressure drop resulting from constant-rate production. The expression for apparent skin factor, Eq. 3.23, now includes a contribution from non-Darcy flow, in addition to the true skin factor from damage.

In Sec. 3.5.2, we modified the method developed in Chap. 2 for analyzing oil well tests with discrete rate changes to apply to gas wells. As was the case with oil wells, this method assumes that the reservoir is infinite-acting throughout the test and that wellbore storage is negligible. Further, we must assume that the non-Darcy-flow coefficient, D, is either negligible or determine D by trial and error.

In Sec. 3.5.3 we discussed gas-well flow tests where the flow rate changes slowly and smoothly. As in the case of discrete rate changes, we must assume no non-Darcy flow or estimate the degree of non-Darcy flow by trial and error.

In Sec. 3.5.4, we discussed a method introduced by Fligelman *et al.* for estimating formation permeability and true skin factor, rather that the apparent skin factor. Fligelman's method is limited to single-phase gas flow to wells that have positive skin factors and that have a fairly large damaged zone. Finally, Fligelman's method depends on the validity of the empirical correlation he used for β.

In Sec. 3.5.5, we discussed the differences between the liquid solution and the gas solution for bounded reservoirs. Although the solution to the gas-flow equation in terms of pseudopressure and time agrees with the liquid solution as long as the reservoir is infinite-acting, this does not hold true once boundary-dominated flow is established because of the variation of the $\mu_g c_t$ product with average drainage-area pressure.

In Sec. 3.6, we discussed the analysis of pressure-buildup tests in gas wells. We introduce buildup-test analysis for shut-in following a period of constant-rate production, in Sec. 3.6.1. We presented the analysis in terms of adjusted pressure and adjusted Horner time ratio, $\left(t_p + \Delta t_a\right)/\Delta t_a$. We then analyzed an example buildup test with pressure, pressure-squared, and adjusted-pressure analysis and compared the results. For this particular example, all three analysis methods give comparable results.

In Sec. 3.6.2, we discussed analysis of pressure-buildup tests where a series of discrete changes in rate occur during the production period before shut-in. In Sec. 3.6.3, we discuss analysis of buildup tests following a period of constant-pressure production. In this case, we find that the actual producing time, t_{actual}, should be used instead of the Horner producing time t_p (defined in Eq. 3.56) in calculating the Horner time ratio. It is also necessary to use the last producing rate in the analysis rather than the average rate over the producing period.

In Sec. 3.6.4 we modified the MBH method for estimating average drainage-area pressure to allow it to be used for analyzing gas-well tests.

Exercises

1. Given the following formation and fluid properties, estimate formation permeability and skin factor from the drawdown test data. $p_i = 278.3$ psia; $h = 10$ ft; $\gamma_g = 0.89$ (air = 1.0); $r_w = 0.21$ ft; $T_f = 60$°F; $\phi = 29.6\%$; $c_w = 3.6 \times 10^{-6}$ psi^{-1}; $q = 200$ Mscf/D; $S_w = 10.6\%$; $c_f = 4 \times 10^{-6}$ psi^{-1}.

FLOW TEST DATA FOR EXERCISE 3.1

Time (hours)	Pressure (psi)	Time (hours)	Pressure (psi)	Time (hours)	Pressure (psi)
0.25	113.3	8.75	77.8	17.75	69.4
0.75	102.5	9.75	76.6	18.75	68.7
1.75	94.5	10.75	75.5	19.75	68.0
2.75	90.1	11.75	74.4	20.75	67.4
3.75	87.0	12.75	73.4	21.75	66.7
4.75	84.5	13.75	72.5	22.75	66.1
5.75	82.5	14.75	71.7	23.75	65.6
6.75	80.7	15.75	70.9	24	65.4
7.75	79.2	16.75	70.1		

2. Given the following formation and fluid properties, estimate formation permeability and skin factor from the drawdown test data. $p_i = 2618$ psia; $h = 11$ ft; $\gamma_g = 0.68$ (air = 1.0); $r_w = 0.35$ ft; $T_f = 147$°F; $\phi = 21.3\%$; $c_w = 3.6 \times 10^{-6}$ psi^{-1}; $q = 2100$ Mscf/D; $S_w = 26.8\%$; $c_f = 4 \times 10^{-6}$ psi^{-1}.

FLOW TEST DATA FOR EXERCISE 3.2

Time (hours)	Pressure (psi)	Time (hours)	Pressure (psi)
2	1386.56	14	1255.42
4	1337.06	16	1246.79
6	1310.24	18	1239.17
8	1291.54	20	1232.33
10	1277.12	22	1226.14
12	1265.36	24	1220.47

3. Given the following formation and fluid properties, estimate formation permeability and skin factor from the multirate drawdown test data. $p_i = 968.9$ psia; $h = 16$ ft; $\gamma_g = 0.63$ (air = 1.0); $T_f = 76$°F; $\phi = 16.7\%$; $c_w = 3.6 \times 10^{-6}$ psi^{-1}; $r_w = 0.27$ ft; $S_w = 14.3\%$; $c_f = 4 \times 10^{-6}$ psi^{-1}.

FLOW TEST DATA FOR EXERCISE 3.3

Time (hours)	Pressure (psi)	Rate (Mscf/D)	Time (hours)	Pressure (psi)	Rate (Mscf/D)
0.27	862.4	550	6.25	686.2	1100
0.49	857.0	550	7.20	680.8	1100
1.11	849.7	550	8.00	677.3	1100
2.25	843.5	550	8.27	516.3	1650
3.20	840.4	550	8.49	506.1	1650
4.00	838.5	550	9.11	490.2	1650
4.27	713.6	1100	10.25	474.5	1650
4.49	706.6	1100	11.20	465.3	1650
5.11	696.1	1100	12.00	459.0	1650

4. Given the following formation and fluid properties, estimate formation permeability and skin factor from the multirate drawdown test data. $p_i = 3140.48$ psia; $h = 74$ ft; $\gamma_g = 0.78$ (air = 1.0); $T_f = 122$°F; $\phi = 26.0\%$; $c_w = 3.6 \times 10^{-6}$ psi^{-1}; $r_w = 0.45$ ft; $S_w = 10.1\%$; $c_f = 4 \times 10^{-6}$ psi^{-1}.

FLOW TEST DATA FOR EXERCISE 3.4

Time (hours)	Pressure (psi)	Rate (Mscf/D)	Time (hours)	Pressure (psi)	Rate (Mscf/D)
0.24	2740.18	3150	6.25	2301.31	5975
0.55	2726.50	3150	7.20	2295.41	5975
1.11	2716.89	3150	8.00	2291.52	5975
2.25	2707.80	3150	8.24	1960.43	8500
3.20	2703.35	3150	8.55	1942.70	8500
4.00	2700.56	3150	9.11	1931.55	8500
4.24	2338.24	5975	10.25	1919.54	8500
4.55	2322.89	5975	11.20	1912.78	8500
5.11	2312.32	5975	12.00	1908.21	8500

5. Given the following formation and fluid properties, estimate formation permeability and skin factor from the buildup test data. $p_{wf}=1378.84$ psia; $h=10$ ft; $\gamma_g=0.741$ (air $=1.0$); $r_w=0.46$ ft; $T_f=108.9$°F; $\phi=10.6\%$; $c_w=3.6\times10^{-6}$ psi^{-1}; $q=8000$ Mscf/D; $S_w=13.5\%$; $c_f=4\times10^{-6}$ psi^{-1}; $t_p=72$ hr.

BUILDUP TEST DATA FOR EXERCISE 3.5

Time (hours)	Pressure (psi)	Time (hours)	Pressure (psi)	Time (hours)	Pressure (psi)
0.0020	1404.55	0.0851	2056.86	0.859	2417.79
0.0045	1435.71	0.1084	2146.08	1.076	2423.72
0.0076	1473.23	0.1375	2224.89	1.347	2429.36
0.0115	1518.03	0.1739	2288.38	1.686	2434.79
0.0164	1570.97	0.219	2334.55	2.11	2440.06
0.0225	1632.74	0.276	2364.99	2.64	2445.20
0.0301	1703.63	0.347	2383.87	3.30	2450.22
0.0397	1783.36	0.436	2395.84	4.13	2455.13
0.0516	1870.73	0.547	2404.40	5.16	2459.94
0.0665	1963.25	0.686	2411.45	6.00	2463.11

6. Given the following formation and fluid properties, estimate formation permeability and skin factor from the buildup test data. $p_{wf}=1219.1$ psia; $h=27$ ft; $\gamma_g=0.725$ (air $=1.0$); $r_w=0.41$ ft; $T_f=134.7$°F; $\phi=26.7\%$; $c_w=3.6\times10^{-6}$ psi^{-1}; $q=12500$ Mscf/D; $S_w=11\%$; $c_f=4\times10^{-6}$ psi^{-1}; $t_p=144$ hr.

BUILDUP TEST DATA FOR EXERCISE 3.6

Time (hours)	Pressure (psi)	Time (hours)	Pressure (psi)	Time (hours)	Pressure (psi)
0.0010	1259.13	0.0542	2099.97	0.674	2630.90
0.0023	1305.43	0.0688	2177.37	0.843	2663.17
0.0038	1358.89	0.0869	2246.31	1.055	2694.76
0.0058	1420.06	0.1097	2306.98	1.319	2725.79
0.0082	1489.17	0.1381	2360.40	1.650	2756.34
0.0113	1566.02	0.1736	2407.95	2.06	2786.47
0.0151	1649.82	0.218	2451.03	2.58	2816.24
0.0198	1739.08	0.274	2490.78	3.23	2845.66
0.0258	1831.62	0.343	2528.11	4.03	2874.76
0.0333	1924.63	0.430	2563.63	5.04	2903.53
0.0426	2015.04	0.538	2597.79	6.00	2925.64

7. Given the following formation and fluid properties, estimate formation permeability, skin factor, and average reservoir pressure from the buildup test following a multirate flow period. $p_{wf}=459.0$ psia; $h=16$ ft; $\gamma_g=0.63$ (air $=1.0$); $T_f=76$°F; $\phi=16.7\%$; $c_w=3.6\times10^{-6}$ psi^{-1}; $r_w=0.27$ ft; $S_w=14.3\%$; $c_f=4\times10^{-6}$ psi^{-1}.

BUILDUP TEST DATA FOR EXERCISE 3.7

Time (hours)	Rate Mscf/D
4	550
4	1100
4	1650

Time (hours)	Pressure (psi)	Time (hours)	Pressure (psi)	Time (hours)	Pressure (psi)
0.0036	567.71	0.0917	853.27	1.247	922.18
0.0060	617.67	0.1166	861.23	1.579	927.12
0.0093	670.48	0.1480	868.65	1.999	931.79
0.0123	704.79	0.1879	875.71	2.53	936.09
0.0161	736.43	0.238	882.46	3.20	940.10
0.0209	764.26	0.302	888.76	4.06	943.81
0.0269	787.77	0.383	894.80	5.13	947.22
0.0346	806.59	0.485	900.64	6.50	950.31
0.0443	821.79	0.614	906.29	8.22	953.10
0.0565	834.26	0.778	911.76	10.41	955.58
0.0720	844.45	0.985	917.06	12.00	956.97

8. Given the following formation and fluid properties, estimate formation permeability, skin factor, and average reservoir pressure from the buildup test following a multirate flow period. $p_{wf}=1908.21$ psia; $h=74$ ft; $\gamma_g=0.78$ (air $=1.0$); $T_f=122$°F; $\phi=26.0\%$; $c_w=3.6\times10^{-6}$ psi^{-1}; $r_w=0.45$ ft; $S_w=10.1\%$; $c_f=4\times10^{-6}$ psi^{-1}.

BUILDUP TEST DATA FOR EXERCISE 3.8

Time (hours)	Rate Mscf/D
4	3150
4	5975
4	8500

Time (hours)	Pressure (psi)	Time (hours)	Pressure (psi)	Time (hours)	Pressure (psi)
0.0015	1942.99	0.0917	2871.62	1.247	3073.35
0.0030	1974.92	0.1166	2928.53	1.579	3080.20
0.0051	2018.72	0.1480	2967.18	1.999	3086.69
0.0081	2079.34	0.1879	2993.34	2.53	3092.81
0.0123	2159.88	0.238	3010.30	3.20	3098.54
0.0183	2264.51	0.302	3022.69	4.06	3103.85
0.0269	2395.00	0.383	3032.93	5.13	3108.75
0.0346	2493.96	0.485	3042.07	6.50	3113.21
0.0443	2598.64	0.614	3050.55	8.22	3117.22
0.0565	2701.32	0.778	3058.54	10.41	3120.80
0.0720	2793.80	0.985	3066.13	12.00	3122.80

References

1. Al-Hussainy, R., Ramey, H.J. Jr., and Crawford, P.B.: "The Flow of Real Gases Through Porous Media," *JPT* (May 1966) 624; *Trans.*, AIME, **237.**
2. Agarwal, R.G.: "Real Gas Pseudotime—A New Function for Pressure Buildup Analysis of MFH Gas Wells," paper SPE 8279 presented at the 1979 SPE Annual Technical Conference and Exhibition, Las Vegas, Nevada, 23–26 September.
3. Lee, W.J. and Holditch, S.A.: "Application of Pseudotime to Buildup-Test Analysis of Low-Permeability Gas Wells With Long-Duration Wellbore-Storage Distortion," *JPT* (December 1982) 2877.
4. Meunier, D.F., Kabir, C.S., and Wittmann, M.J.: "Gas Well Test Analysis: Use of Normalized Pressure and Time Functions," *SPEFE* (December 1987) 629.
5. Aziz, K. *et al.*: "Use of Pressure, Pressure-Squared or Pseudo-Pressure in the Analysis of Transient Pressure Drawdown Data from Gas Wells," *J. Cdn. Pet. Tech.* (April–June 1976) 58.
6. Fligelman, H. *et al.*: "Pressure-Drawdown Test Analysis of a Gas Well-Application of New Correlations," *SPEFE* (Sept. 1989) 406.

7. Ramey Jr., H.J.: "Non-Darcy Flow and Wellbore Storage Effects in Pressure Buildup and Drawdown of Gas Wells," *JPT* (February 1965) 223; *Trans.*, AIME (1965) **234.**
8. Dake, L.P.: *Fundamentals of Reservoir Engineering*, Elsevier Scientific Publishing, Amsterdam (1978) 249, 264–286.
9. Jones, S.C.: "Using the Inertial Coefficient, β, To Characterize Heterogeneity in Reservoir Rock," paper SPE 16949 presented at the 1987 SPE Annual Technical Conference and Exhibition, Dallas, September 27–30.
10. Earlougher Jr., R.C.: *Advances in Well Test Analysis*, Monograph Series, SPE, Richardson, Texas (1977) **5,** 74–89, 191.
11. Matthews, C.S. and Russell, D.G.: *Pressure Buildup and Flow Tests in Wells*, Monograph Series, SPE, Richardson, Texas (1977) **1,** 16, 19, 21, 49, 72–83.
12. Odeh, A.S. and Jones, L.G.: "Pressure Drawdown Analysis, Variable-Rate Case," *Trans.*, AIME (1965) **234,** 960.
13. Winestock, A.G. and Colpitts, G.P.: "Advances in Estimating Gas Well Deliverability," *Gas Technology*, Reprint Series, SPE, Richardson, Texas (1977) **13,** 122–130.
14. Al-Hussainy, R.: "Transient Flow of Ideal and Real Gases Through Porous Media," PhD dissertation, Texas A&M U., College Station, Texas (1967).
15. Kazemi, H.: "Determining Average Reservoir Pressure From Pressure Buildup Tests," *SPEJ* (February 1974) 55; *Trans.*, AIME (1974) **257.**
16. Fraim, M.L. and Wattenbarger, R.A.: "Gas Reservoir Decline-Curve Analysis Using Type Curves With Real Gas Pseudopressure and Normalized Time," *SPEJ* (December 1987) 671.
17. Odeh, A.S. and Selig, F.: "Pressure Build-Up Analysis, Variable-Rate Case," *JPT* (July 1963) 790; *Trans.*, AIME (1963) **228.**
18. Martin, J.C.: "Simplified Equations of Flow in Gasdrive Reservoirs and the Theoretical Foundation of Multiphase Pressure-Buildup Analyses," *JPT* (October 1959) 309; *Trans.*, AIME, **216.**
19. Matthews, C.S., Brons, F., and Hazebroek, P.: "A Method for Determination of Average Pressure in a Bounded Reservoir," *Trans.*, AIME (1954) **201,** 182.
20. Behrumen-C., S., Samaniego-V., F., and Cinco-Ley, H.: "Transient Pressure Analysis and Performance of Gas Wells Producing Under Constant Pressure Conditions," paper SPE 19098 presented at the 1989 SPE Gas Technology Symposium, Dallas, 7–9 June.
21. Reynolds, A.C., Bratvold, R.B., and Ding, W.: "Semilog Analysis of Gas Well Drawdown and Buildup Data," *SPEFE* (Dec. 1987) 657.

Chapter 4
Well-Test Analysis by Use of Type Curves

4.1 Overview

Type curves—plots of theoretical solutions to flow equations—are very useful in well-test analysis, especially when used in concert with semilog-analysis techniques. Type curves can help estimate reservoir properties, identify the appropriate reservoir model, and identify various flow patterns during a test. In this chapter, we explain what type curves are, explore some of their more useful properties, and illustrate how they can be used to improve pressure transient test analysis. We begin with type curves[1–6] developed for slightly compressible fluids, then present modifications for applications to gas-well-test analysis. We also discuss the development and application of pressure-derivative type curves. Systematic analysis procedures illustrate the use of type curves, and we present worksheets for analysis of test data and reservoir-model identification.

4.2 Development of Type Curves

Because type curves are plots of theoretical solutions to flow equations, we can generate them for virtually any kind of reservoir model for which a general solution describing the flow behavior is available. To apply a type curve correctly, one must understand the assumptions underlying the solution. Furthermore, those assumptions must model accurately the well or reservoir conditions being analyzed. As a matter of convenience, type curves are usually presented in terms of dimensionless, rather than real, variables. The definition of the dimensionless variables varies according to the reservoir model. For example, consider the line-source or Ei-function solution for slightly compressible liquids developed in Chap. 1,

$$p_i - p = -\frac{70.6qB\mu}{kh}\mathrm{Ei}\left(\frac{-948\phi\mu c_t r^2}{kt}\right). \quad \text{(4.1)}$$

Eq. 4.1 implies that pressure, p, at a point in the reservoir depends on the numerical value of many other variables. The solution can be made more compact if the equation is rearranged and some dimensionless variables are defined. Rearrangement of Eq. 4.1 as

$$\frac{kh(p_i - p)}{141.2qB\mu} = -\frac{1}{2}\mathrm{Ei}\left[\frac{-(r/r_w)^2}{4(0.0002637kt/\phi\mu c_t r_w^2)}\right] \quad \text{(4.2)}$$

suggests the following dimensionless pressure, radius, and time variables, respectively:

$$p_D = \frac{kh(p_i - p)}{141.2qB\mu}, \quad \text{(4.3)}$$

$$r_D = \frac{r}{r_w}, \quad \text{(4.4)}$$

and $$t_D = \frac{0.0002637kt}{\phi\mu c_t r_w^2}. \quad \text{(4.5)}$$

Combining Eqs. 4.2 through 4.5 yields the Ei-function solution in dimensionless form,

$$p_D = -\frac{1}{2}\mathrm{Ei}\left(\frac{-r_D^2}{4t_D}\right). \quad \text{(4.6)}$$

At the well where $r_D = 1$, the solution simplifies to

$$p_D = p_{wD} = -\frac{1}{2}\mathrm{Ei}\left(\frac{-1}{4t_D}\right), \quad \text{(4.7)}$$

where the dimensionless pressure evaluated at the wellbore is

$$p_{wD} = \frac{kh(p_i - p_{wf})}{141.2qB\mu}. \quad \text{(4.8)}$$

Eq. 4.7 implies that we can develop a type curve from a plot of p_{wD} as a function of the single variable t_D. Generating a single graph in terms of p_{wD} is much simpler than attempting to plot bottomhole flowing pressure, p_{wf}, as a function of time, t, for all reasonable values of the variables that appear in the dimensional form of the line-source solution (Eq. 4.1). Thus, with this type curve, we can analyze any pressure transient test conducted under conditions that satisfy the assumptions made in deriving the Ei-function solution. For more complex reservoirs than that modeled by the line-source solution, the solutions to the flow equations can be expressed in a general functional form as

$$p_D = \mathrm{f}(t_D, r_D, s, \ldots). \quad \text{(4.9)}$$

The function represented by Eq. 4.9 may be very complex depending on the wellbore and reservoir conditions being modeled. Again, before using any type curve, the engineer must fully understand the inherent assumptions of the model.

4.3 Application of Type Curves—Homogeneous Reservoir Model, Slightly Compressible Liquid Solution

In this section, we introduce four common type curves and illustrate their application with systematic-analysis procedures and examples. All the type curves were developed assuming homogeneous-acting

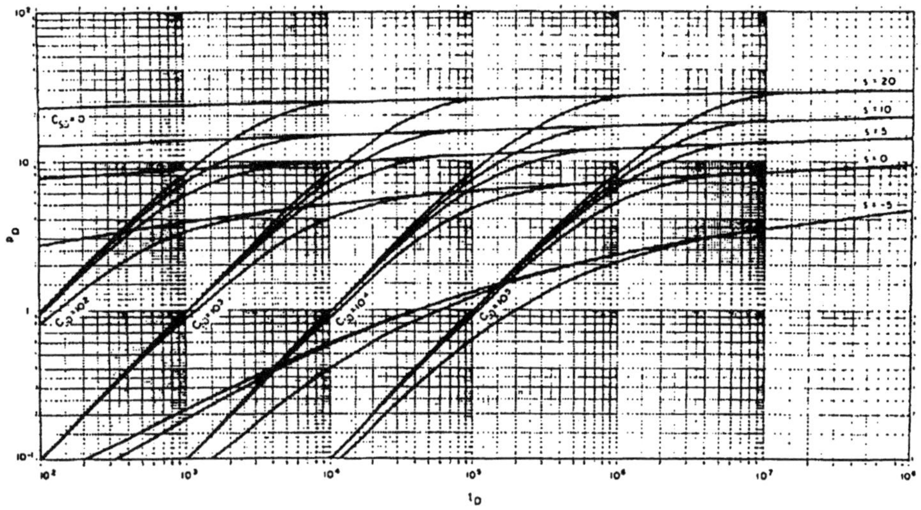

Fig. 4.1—Type curves for constant production rate, infinite-acting reservoir (Ramey[6]).

formations producing slightly compressible fluids (i.e., liquids); however, the type curves can be used for gas-well-test analysis with appropriate plotting functions discussed in later sections.

4.3.1 Ramey's Type Curves. Agarwal *et al.* and Ramey[5,6] generated type curves for the situation of a constant-rate pressure-drawdown test in a reservoir with the following characteristics: slightly compressible, single-phase liquid flowing; sufficient homogeneity so that the radial-diffusivity equation adequately models flow in the reservoir; uniform pressure in the drainage area of the well before production; infinite-acting reservoir (no boundary effects during the flow period of interest for test-analysis purposes); constant withdrawal rate at the surface; and wellbore storage and concentrated wellbore damage or stimulation characterized by a skin factor, s. This list of assumptions is tedious but important. When one or more of these assumptions is not valid in a specific case, there is no assurance that use of the type curves can lead to a valid test interpretation. (Some of these limitations can be removed, as we note later in this chapter. Of major importance is that the curves can be used for buildup tests and for gas-well tests.) **Fig. 4.1** shows the result of Ramey's work.

Some important properties of these curves follow.

1. Examination of the analytical solution on which the type curves are based shows that, at earliest times when wellbore unloading is responsible for 100% of the flow in a drawdown test (or afterflow rate equals rate before shut-in in a buildup test), Δp is a linear function of Δt (Δp is pressure change and Δt is time elapsed since test start). Thus, the log Δp-log Δt curve is also linear with a slope of unity (a 45° line), and the wellbore-storage coefficient C, can be determined from any point (Δt,Δp) on this line (**Fig. 4.2**) from the relation

$$C = \frac{qB}{24}\left(\frac{\Delta t}{\Delta p}\right)_{USL} \quad \text{.......} \quad (4.10)$$

Note that, in a well with a liquid/gas interface in the wellbore,

$$C = \frac{25.65 A_{wb}}{\rho_{wb}} \text{ bbl/psi}, \quad \text{.......} \quad (4.11)$$

and for a wellbore filled with a single-phase liquid or with gas,

$$C = c_{wb} V_{wb} \text{ bbl/psi} \quad \text{.......} \quad (4.12)$$

and $C_D = 0.8936\ C/\phi c_t h r_w^2$. .... (4.13)

Successful application of Ramey's type curves for quantitative analysis depends significantly on our ability to establish the correct value of C_D to be used for curve-matching type curves for a given value of s. Curves for different values of C_D have very similar shapes, which makes it difficult to find the best fit without prior knowledge of C_D. Direct calculation of C, and thus C_D, from known values of A_{wb} and ρ_{wb} or c_{wb} and V_{wb} does not characterize test conditions as well as the value of C determined from actual test performance as reflected in the unit-slope lines.[3]

2. Wellbore storage has ceased distorting the pressure transient test data when the type curve for the value of C_D characterizing the test becomes identical to the type curve for $C_D = 0$ (**Fig. 4.3**). This usually occurs about one and a half to two cycles from the end of the unit-slope line. Thus, these type curves can be used to determine how much data (if any) can be analyzed by conventional methods, such as the Horner plot for buildup tests.

3. The type curves, which were developed for drawdown tests, also can be used for buildup-test analysis under certain circumstances if an equivalent shut-in time, $\Delta t_e = \Delta t/(1 + \Delta t/t_p)$, is used as the time variable. The equivalent time is valid under either of two conditions.

First, equivalent time is valid for shut-in times where the pressure responses from both the flow period and the shut-in period lie within the middle-time region. Specifically, the pressure responses from both the flow period and the shut-in period must be described by the logarithmic approximation to the Ei function; that is, the assumption of homogeneous, infinite-acting radial flow must be applicable.

If $\Delta t \geqq 0.1 t_p$ and the pressure response from either the flow period or the shut-in period do not lie within the middle-time region, the buildup-test data will not lie on the drawdown type curve, even

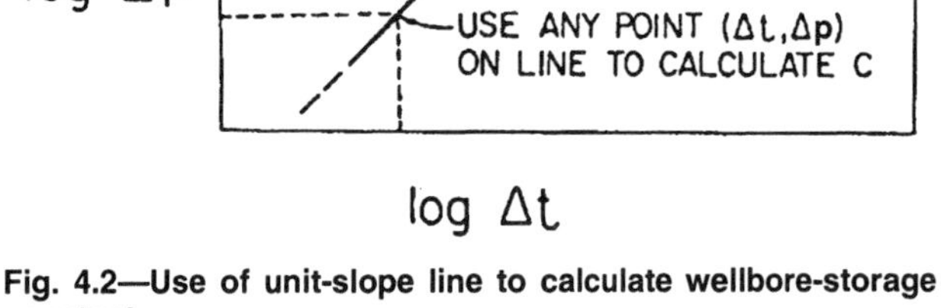

Fig. 4.2—Use of unit-slope line to calculate wellbore-storage constant.

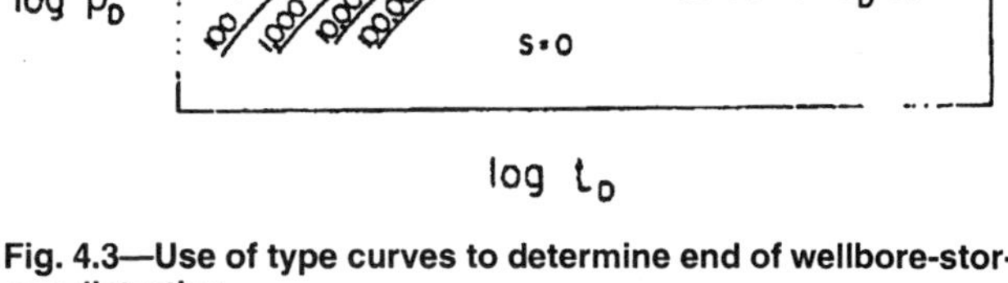

Fig. 4.3—Use of type curves to determine end of wellbore-storage distortion.

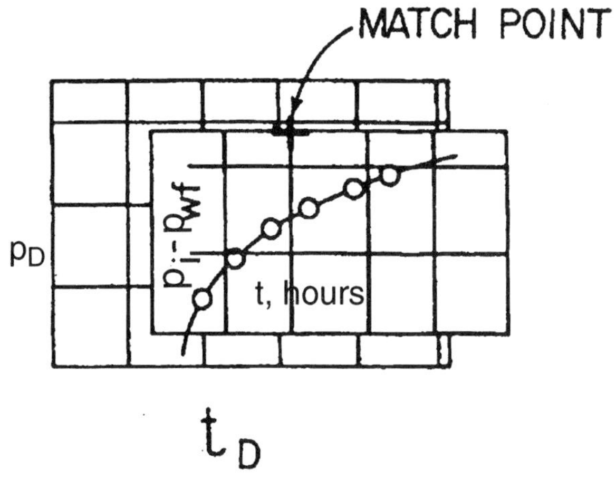

Fig. 4.4—Horizontal and vertical shifting to find position of fit and match points.

when equivalent time is used. A number of factors can cause this behavior, including wellbore storage, boundary effects, nonradial-flow patterns in hydraulically fractured wells (Chap. 6), and naturally fractured reservoirs (Chap. 7). If any of these conditions are suspected, equivalent time should be used with caution.

Equivalent time is also valid for short shut-in times; that is, when $\Delta t \ll t_p$. In this case, $\Delta t_e \approx \Delta t$. When $\Delta t \ll t_p$, the pressure response from the flow period is approximately constant during the shut-in period: $p_{wf}(t_p + \Delta t) \approx p_{wf}(t_p)$. For such times, either time or equivalent time may be used to analyze buildup tests by use of drawdown type curves.

For large shut-in times, $\lim_{\Delta t \to \infty} \Delta t_e = t_p$. Thus, for shut-in times much larger than the producing time, the use of equivalent time compresses the time scale. Tests with a very short production period followed by an extended shut-in period cannot be analyzed with the equivalent-time function. Horner analysis is also impractical for these tests because of the compression of the time scale. Methods known as impulse tests (Chap. 9) are available to analyze such tests.

4. A log-log plot of p_D vs. t_D differs from a log-log plot of $(p_i - p_{wf})$ vs. t (for a drawdown test) only by a shift in the origin of the coordinate system; i.e., $\log t_D$ differs from $\log t$ by a constant and $\log p_D$ differs from $\log (p_i - p_{wf})$ by another constant. Eqs. 4.14 and 4.15 show this.

$$t_D = \frac{0.0002637kt}{\phi\mu c_t r_w^2} \quad \text{(4.14)}$$

$$\text{and } p_D = \frac{kh(p_i - p_{wf})}{141.2qB\mu}; \quad \text{(4.15)}$$

$$\text{thus, } \log t_D = \log t + \log\frac{0.000264k}{\phi\mu c_t r_w^2} \quad \text{(4.16)}$$

$$\text{and } \log p_D = \log(p_i - p_{wf}) + \log\frac{kh}{141.2qB\mu} \quad \text{(4.17)}$$

The significance of this result is that the plot of an actual drawdown test ($\log t$ vs. $\log \Delta p$) should have a shape identical to that of a plot of $\log t_D$ vs. $\log p_D$, but we have to displace both the horizontal and vertical axes (i.e., shift the origin of the plot) to find the position of best fit (**Fig. 4.4**).

Once a fit is found by vertical and horizontal shifting, we choose a match point to determine the relationship between actual and dimensionless time and between actual pressure drawdown and dimensionless pressure for the test being analyzed. Any point on the graph paper will suffice as a match point (i.e., the result is independent of the choice of match point). For the match point chosen, we determine the corresponding values of (t, t_D) and $[(p_i - p_{wf}), p_D]$. Then, from definition of p_D and t_D,

$$k = 141.2\frac{qB\mu}{h}\left(\frac{p_D}{p_i - p_{wf}}\right)_{MP} \quad \text{(4.18)}$$

$$\text{and } \phi c_t = \frac{0.000264k}{\mu r_w^2}\left(\frac{t}{t_D}\right)_{MP}. \quad \text{(4.19)}$$

Use of Ramey's Type Curves. The theory of Ramey's type curves leads to the following procedure for using the curves for test analysis. The procedure is given for a slightly compressible liquid; **Tables 4.1 and 4.2** show the changes necessary when a gas-well test is analyzed.

1. Plot $(p_i - p_{wf})$ vs. t (drawdown test) or $(p_{ws} - p_{wf})$ vs. $\Delta t_e = \Delta t/(1 + \Delta t/t_p)$ (buildup test) on log-log paper the same size as Ramey's type curve. Caution: Unless a type curve is used that is not distorted in the reproduction process, it will not have the same dimensions as commercially available graph paper and finding a fit may be misleading or impossible. The best solution is to use an undistorted type curve; an acceptable alternative is to plot test data on tracing paper using the grid on the distorted type curve as a plotting aid.

2. If the test has a uniform-slope region (45° line at earliest times), choose any point $[t, (p_i - p_{wf})]$ or $[\Delta t, (p_{ws} - p_{wf})]$ on the unit-slope line and calculate the wellbore-storage coefficient C:

$$C = \frac{qB}{24}\left(\frac{t}{p_i - p_{wf}}\right)_{USL}. \quad \text{(4.20)}$$

Then calculate the dimensionless wellbore-storage coefficient:

$$C_D = \frac{0.894C}{\phi c_t h r_w^2}. \quad \text{(4.21)}$$

Note that estimates of ϕ and c_t are required at this point; the implications of this are discussed later.

If a unit-slope line is not present, C and C_D must be calculated from wellbore properties, and inaccuracies may result if these properties do not describe actual test behavior.[14]

3. Using type curves with C_D as calculated in Step 2, find the curve that most nearly fits all the plotted data. This curve will be characterized by some skin factor, s; record its value. Interpolation between curves should improve the precision of the analysis but may prove difficult. Even for fixed C_D from the unit-slope curve, the analyst may experience difficulty in determining that one value of s provides a better fit than another, particularly if all data are distorted by wellbore storage or if the "scatter" or "noise" that characterizes much actual field data is present. If C_D is not known with certainty, the possible ambiguity in finding the best fit is even more pronounced.

4. With the actual test-data plot placed in the position of best fit, record corresponding values of $(p_i - p_{wf}, p_D)$ and (t, t_D) from any convenient match point.

5. Calculate k and ϕc_t with Eqs. 4.18 and 4.19. Eq. 4.19 does not establish ϕc_t on the basis of test performance unless C_D can be established without assuming values for ϕc_t; it simply reproduces the values assumed in Step 2.

In summary, the procedure outlined in Steps 1 through 5 provides estimates of k, s, and C.

Example 4.1—Drawdown Test Analysis by Use of Ramey's Type Curves. ***Problem.*** Determine k, s, and C from the data below and in **Table 4.3**, which were obtained in a pressure drawdown test on an oil well.

q = 500 STB/D
ϕ = 0.2
μ = 0.8 cp
c_t = 10×10^{-6} psi^{-1}
r_w = 0.3 ft
h = 56 ft
B_o = 1.2 RB/STB
p_i = 3,000 psia

Solution. We must first prepare the data for plotting (**Table 4.4**). **Fig. 4.5** shows the plot of the data.

TABLE 4.1—EQUIVALENT VARIABLES IN LIQUID AND GAS FLOW

Liquid	Gas, Using Pseudopressure and Pseudotime	Gas, Using Adjusted Pressure and Adjusted Time
Pressure, p	Pseudopressure, $p_p(p) = 2\int_{p_0}^{p} \frac{pdp}{\mu z}$	Adjusted pressure, $p_a(p) = \frac{\overline{\mu z}}{2\bar{p}} p_p(p)$
Dimensionless pressure, $p_D = \frac{kh\Delta p}{141.2qB\mu}$	Dimensionless pseudopressure, $p_{pD} = \frac{khT_{sc}\Delta p}{50,300qp_{sc}T}$	Dimensionless adjusted pressure, $p_{aD} = \frac{kh\Delta p_a}{141.2q\overline{B}_g\overline{\mu}_g}$
Time, t	Pseudotime, $t_{ap}(p) = \int_{t_0}^{t} \frac{dt}{\mu c_t}$	Adjusted Time, $t_a(p) = (\overline{\mu_g C_t}) t_{ap}(p)$
Radius, r	Radius, r	Radius, r
Dimensionless radius, $r_D = r/r_w$	Dimensionless radius, $r_D = r/r_w$	Dimensionless radius, $r_D = r/r_w$
Skin factor, s	Apparent skin factor, $s' = s + Dlql$	Apparent skin factor, $s' = s + Dlql$
Wellbore-storage coefficient, $C = 25.65\,A_{wb}/\rho_{wb}$	Modified wellbore-storage coefficient, $C_e = C\,(T\bar{c}_t)/(T_{wb}\,c_{wb})$ $= V_{wb}T\bar{c}_t/T_{wb}$	Modified wellbore-storage coefficient, $C_e = C\,(T\bar{c}_t)/(T_{wb}\,c_{wb})$ $= V_{wb}T\bar{c}_t/T_{wb}$
Dimensionless wellbore-storage coefficient, $C_D = 0.8936C/\phi c_t h\, r_w^2$	Modified dimensionless wellbore-storage coefficient, $C_{eD} = 0.8936V_{wb}T/\phi h r_w^2\, T_{wb}$	Modified dimensionless wellbore-storage coefficient, $C_{eD} = 0.8936V_{wb}T/\phi h r_w^2\, T_{wb}$

TABLE 4.2—SUMMARY OF ANALYSIS TECHNIQUES FOR TYPE-CURVE PLOTS

Case	Interpretation of Unit Slope Line	Definition of C_D	Interpretation of Pressure Match Point
	Slightly Compressible Fluids (Oil)		
Δp vs. Δt	$C_D = \frac{0.0372q_oB_o}{\phi hc_tr_w^2}\left(\frac{\Delta t}{\Delta p}\right)_{USL}$	$C_D = \frac{22.92A_{wb}}{\rho_{wb}\phi c_t h r_w^2}$	$k = \frac{141.2q_oB_o\mu_o}{h}\left(\frac{p_D}{\Delta p}\right)_{MP}$
	Compressible Fluids (Gas)		
Δp vs. Δt	$C_D = \frac{0.0372q_g\overline{B}_g}{\phi h\bar{c}_tr_w^2}\left(\frac{\Delta t}{\Delta p}\right)_{USL}$	$C_D = \frac{0.8936V_{wb}c_{wb}}{\bar{c}_t r_w^2}$	$k = \frac{141.2q_g\bar{\mu}_g\overline{B}_g}{h}\left(\frac{p_D}{\Delta p}\right)_{MP}$
Δp_a vs. Δt_a	$C_D = \frac{0.0372q_g\overline{B}_g}{\phi h\bar{c}_tr_w^2}\left(\frac{\Delta t}{\Delta p}\right)_{USL}$	$C_D = \frac{0.8936V_{wb}T}{\phi h\, r_w^2 T_{wb}}$	$k = \frac{141.2q_g\bar{\mu}_g\overline{B}_g}{h}\left(\frac{p_D}{\Delta p}\right)_{MP}$
$\Delta(p^2)$ vs. Δt	$C_D = \frac{0.375q_gT\bar{z}}{\phi h\bar{c}_tr_w^2}\left[\frac{\Delta t}{\Delta(p^2)}\right]_{USL}$	$C_D = \frac{0.8936V_{wb}c_{wb}}{\phi\bar{c}_t h r_w^2}$	$k = \frac{1422q_gT\bar{z}\bar{\mu}_g}{h}\left[\frac{p_D}{(\Delta p^2)}\right]_{MP}$
Δp_p vs. Δt	$C_D = \frac{0.375q_gT}{\phi h\bar{c}_tr_w^2}\left(\frac{\Delta t}{\Delta p_p}\right)_{USL}$	$C_D = \frac{0.8936V_{wb}c_{wb}}{\phi\bar{c}_t h r_w^2}$	$k = \frac{1422q_gT}{h}\left[\frac{p_D}{(\Delta p_p)}\right]_{MP}$

1. For drawdown test analysis, the plotting functions are defined as $\Delta p = \bar{p} - p_{wf}$ (or equivalent) and $\Delta t = t$ = flow time (or equivalent).
2. For buildup test analysis, the plotting functions are defined as $\Delta p = p_{ws} - p_{wf}$ ($\Delta t = 0$) (or equivalent) and $\Delta t = \Delta t_e = \Delta t/(1 + \Delta t/t_p)$ (or equivalent).

From the unit-slope line (on which the data lie for $t \leqq 0.0218$ hour),

$$C = \frac{qB}{24}\left(\frac{t}{p_i - p_{wf}}\right)_{USL}$$

$$= \frac{(500)(1.2)}{(24)} \times \frac{(0.0218)}{(47)}$$

$$= 0.0116 \text{ RB/psi}.$$

TABLE 4.2—SUMMARY OF ANALYSIS TECHNIQUES FOR TYPE-CURVE PLOTS (continued)

Case	Definition of C_D	Interpretation of Time for Hydraulically Fractured Wells
	Slightly Compressible Fluids (Oil)	
Δp vs. Δt	$C_D = \dfrac{0.0002637k}{\phi\mu_o c_t h r_w^2}\left(\dfrac{\Delta t}{t_D/C_D}\right)_{MP}$	$L_f = \left[\left(\dfrac{0.0002637k}{\phi\mu_o c_t}\right)\left(\dfrac{\Delta t}{t_{L_fD}}\right)_{MP}\right]^{1/2}$
	Compressible Fluids (Gas)	
Δp vs. Δt	$C_D = \dfrac{0.0002637k}{\phi\overline{\mu_g c_t}\, r_w^2}\left(\dfrac{\Delta t}{t_D/C_D}\right)_{MP}$	$L_f = \left[\left(\dfrac{0.0002637k}{\phi\overline{\mu_g c_t}}\right)\left(\dfrac{\Delta t}{t_{L_fD}}\right)_{MP}\right]^{1/2}$
Δp_a vs. Δt_a	$C_D = \dfrac{0.0002637k}{\phi\overline{\mu_g c_t}\, r_w^2}\left(\dfrac{\Delta t}{t_D/C_D}\right)_{MP}$	$L_f = \left[\left(\dfrac{0.0002637k}{\phi\overline{\mu_g c_t}}\right)\left(\dfrac{\Delta t}{t_{L_fD}}\right)_{MP}\right]^{1/2}$
$\Delta(p^2)$ vs. Δt	$C_D = \dfrac{0.0002637k}{\phi\overline{\mu_g c_t}\, r_w^2}\left(\dfrac{\Delta t}{t_D/C_D}\right)_{MP}$	$L_f = \left[\left(\dfrac{0.0002637k}{\phi\overline{\mu_g c_t}}\right)\left(\dfrac{\Delta t}{t_{L_fD}}\right)_{MP}\right]^{1/2}$
Δp_p vs. Δt	$C_D = \dfrac{0.0002637k}{\phi\overline{\mu_g c_t}\, r_w^2}\left(\dfrac{\Delta t}{t_D/C_D}\right)_{MP}$	$L_f = \left[\left(\dfrac{0.0002637k}{\phi\overline{\mu_g c_t}}\right)\left(\dfrac{\Delta t}{t_{L_fD}}\right)_{MP}\right]^{1/2}$

1. For drawdown test analysis, the plotting functions are defined as $\Delta p = \bar{p} - p_{wf}$ (or equivalent) and $\Delta t = t$ = flow time (or equivalent).
2. For buildup test analysis, the plotting functions are defined as $\Delta p = p_{ws} - p_{wf}$ ($\Delta t = 0$) (or equivalent) and $\Delta t = \Delta t_e = \Delta t/(1 + \Delta t/t_p)$ (or equivalent).

TABLE 4.3—CONSTANT-RATE DRAWDOWN-TEST DATA

t (hours)	p_{wf} (psi)	t (hours)	p_{wf} (psi)	t (hours)	p_{wf} (psi)
0.0109	2,976	0.164	2,693	2.18	1,768
0.0164	2,964	0.218	2,611	2.73	1,734
0.0218	2,953	0.273	2,536	3.28	1,712
0.0273	2,942	0.328	2,469	3.82	1,696
0.0328	2,930	0.382	2,408	4.37	1,684
0.0382	2,919	0.437	2,352	4.91	1,674
0.0437	2,908	0.491	2,302	5.46	1,665
0.0491	2,897	0.546	2,256	6.55	1,651
0.0546	2,886	1.09	1,952	8.74	1,630
0.109	2,785	1.64	1,828	10.9	1,614
				16.4	1,587

TABLE 4.4—DRAWDOWN DATA TABULATED FOR PLOTTING

t (hours)	$p_i - p_{wf}$ (psi)	t (hours)	$p_i - p_{wf}$ (psi)	t (hours)	$p_i - p_{wf}$ (psi)
0.0109	24	0.164	307	2.18	1,232
0.0164	36	0.218	389	2.73	1,266
0.0218	47	0.273	464	3.28	1,288
0.0273	58	0.328	531	3.82	1,304
0.0328	70	0.382	592	4.37	1,316
0.0382	81	0.437	648	4.91	1,326
0.0437	92	0.491	698	5.46	1,335
0.0491	103	0.546	744	6.55	1,349
0.0546	114	1.09	1,048	8.74	1,370
0.109	215	1.64	1,172	10.9	1,386
				16.4	1,413

Then,

$$C_D = \frac{0.894C}{\phi c_t h r_w^2}$$

$$= \frac{(0.894)(0.0116)}{(0.2)(1\times10^{-5})(56)(0.3)^2}$$

$$= 1.03\times10^3$$

$$\cong 1\times10^3.$$

For $C_D = 10^3$, the best-fitting type curve is for $s = 5$. A time match point is $t = 1$ hour when $t_D = 1.93\times10^4$. A pressure match point is $(p_i - p_{wf})$ 100 psi, when $p_D = 0.85$.

From the match, we also note that wellbore-storage distortion ends at $t = 5.0$ hours (i.e., the type curve for $C_D = 10^3$ becomes identical to the type curve for $C_D = 0$).

From the pressure match point,

$$k = 141.2\frac{qB\mu}{h}\left(\frac{p_D}{p_i - p_{wf}}\right)_{MP}$$

$$= \frac{(141.2)(500)(1.2)(0.8)}{(56)}\left(\frac{0.85}{100}\right)$$

$$= 10.3 \text{ md}.$$

From the time match point,

$$\phi c_t = \frac{0.000264k}{\mu r_w^2}\left(\frac{t}{t_D}\right)_{MP}$$

$$= \frac{(0.000264)(10.3)}{(0.8)(0.3)^2}\left(\frac{1}{1.93\times10^4}\right)$$

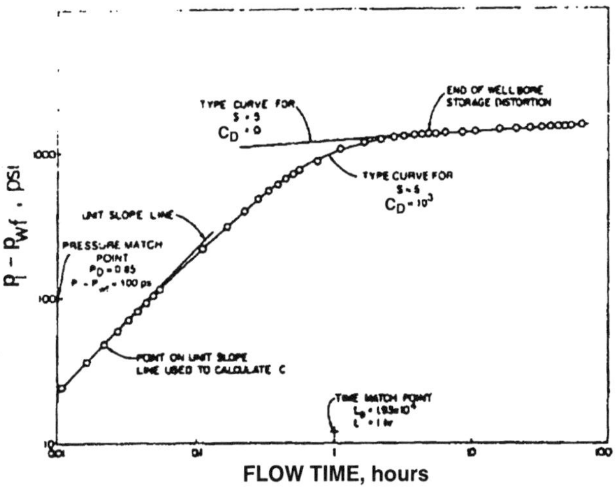

Fig. 4.5—Drawdown-test analysis with Ramey's type curve.

$$= 1.96 \times 10^{-6} \text{ psi}^{-1}.$$

Compare these with the values used to determine C_D from C:

$$\phi c_t = (0.2)(1 \times 10^{-5})$$

$$= 2 \times 10^{-6}.$$

This is the same value we used to calculate C_D. As noted previously, the time match point does not provide an independent estimate of ϕc_t.

4.3.2 McKinley Type Curve. McKinley[1] developed type curves with the primary objective of characterizing damage or stimulation in a drawdown or buildup test in which wellbore storage distorts most of or all the data, thus making this characterization possible with relatively short-term tests. In constructing his type curves, McKinley observed that the ratio of pressure change, Δp, to the flow rate causing the change, qB, is a function of several dimensionless quantities:

$$\frac{\Delta p}{qB} = f\left(\frac{kh\Delta t}{\mu C}, \frac{k\Delta t}{\phi\mu c_t r_w^2}, \frac{r_e}{r_w}, \frac{\Delta t}{t_p}\right). \quad (4.22)$$

Type curves with so many parameters would be difficult, if not impossible, to use. Accordingly, McKinley simplified the problem with the following assumptions.

1. The first is that the well has produced long enough (essentially to stabilization) so that the last group, $\Delta t/t_p$, is not important. In other words, he assumed that the producing time, t_p, is much greater than the test duration, Δt. Consequently, the type curves may not give accurate results for pressure-buildup tests with short producing periods before shut-in.

2. He ignored boundary effects, thus eliminating the variable r_e/r_w in the logic used to construct the type curves (see, however, Assumption 6).

3. His analysis of simulated buildup and drawdown curves showed that, during the wellbore-storage-dominated portion of a test, the parameter $kh\Delta t/\mu C$ was much more important in determining $\Delta p/qB$ than was $k\Delta t/\phi\mu c_t r_w^2$. Accordingly, he let $k/\phi\mu c_t r_w^2 = 10 \times 10^6$ md-psi/cp-ft^2 (an average value) for all his type curves. Even when $k/\phi\mu c_t r_w^2$ varies from this average value by one or two orders of magnitude, the shape of the type curves is not affected significantly. The reason for this approximation was McKinley's judgment that the loss of accuracy is more than compensated for by the

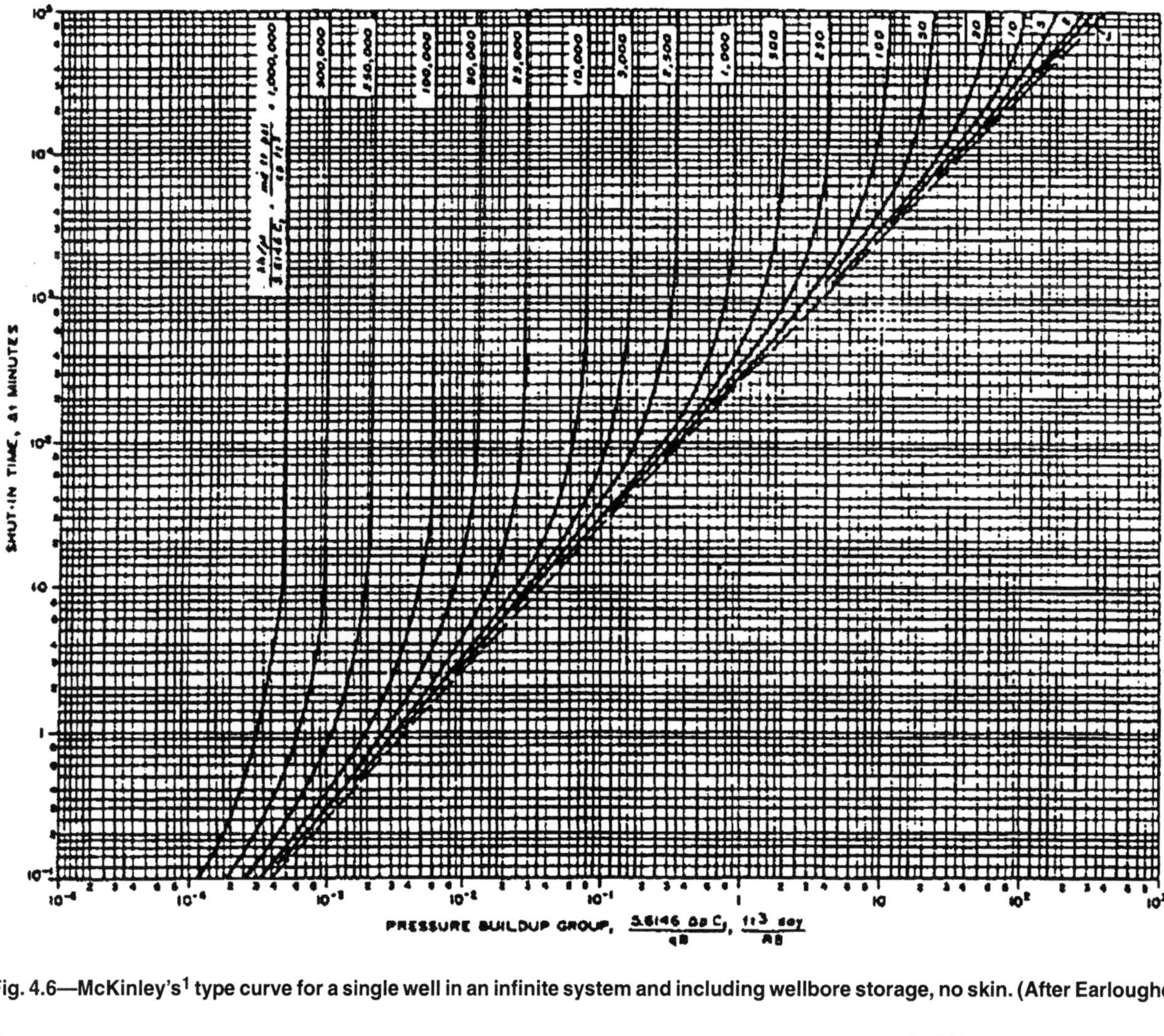

Fig. 4.6—McKinley's[1] type curve for a single well in an infinite system and including wellbore storage, no skin. (After Earlougher[3]).

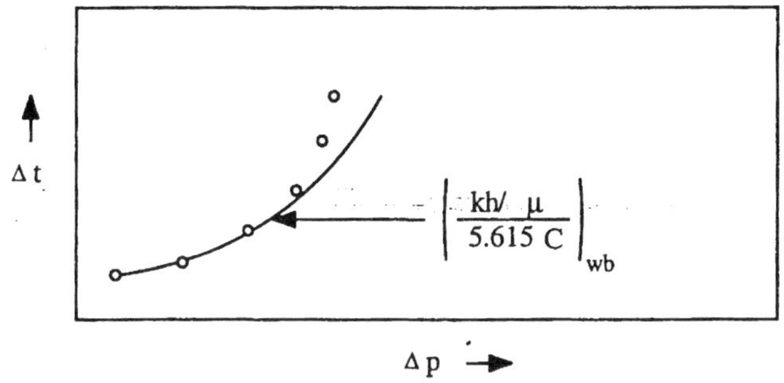

Fig. 4.7—Early-time data fit on McKinley's[1] type curve.

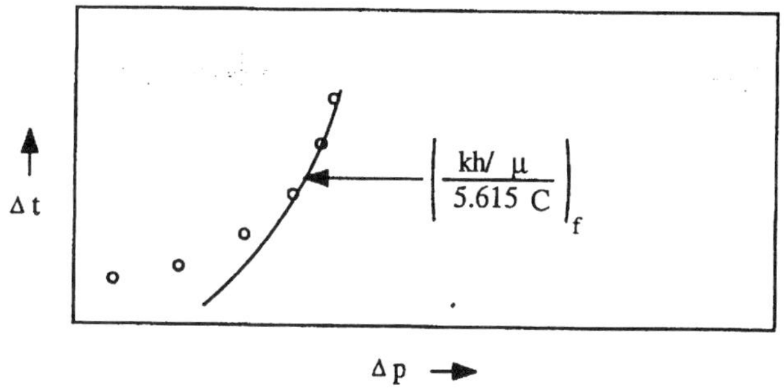

Fig. 4.8—Later-time data fit on McKinley's[1] type curve.

gain in sensitivity in the type curves (i.e., the shape of each curve is distinctly different at earliest times).

4. To account for the remaining parameters that do have a significant influence on test results, McKinley plotted his type curves as Δt (ordinate) vs. $5.615C\Delta p/qB$ (abscissa) with the single correlating parameter $kh/5.615C\mu$. **Fig. 4.6** shows a small-scale version of McKinley's curves.

5. The skin factor, s, does not appear as a parameter in his curves. Instead, they assess damage or stimulation by noting that the earliest wellbore-storage-distorted data are dominated by the effective near-well transmissibility, $(kh/\mu)_{wb}$. Thus, $(kh/\mu)_{wb}$ can be calculated from a type-curve match of the earliest test data. After wellbore-storage distortion has diminished, pressure/time behavior is governed by the transmissibility in the formation, $(kh/\mu)_f$, which can be estimated from a type-curve match of later data.

6. McKinley approximated boundary effects by plotting the simulator-generated type curves for approximately one-fifth log cycle beyond the end of wellbore-storage distortion and then making the curves vertical. This step roughly simulates drainage conditions of 40-acre spacing. Note that this gives the curves early-, middle-, and late-time regions, but the curves were designed to be used primarily to analyze early-time data. Hence, when the curves are applied to drawdown tests, they must be applied to data not affected by boundaries because they do not simulate boundary effects in drawdown tests properly.

McKinley prepared three different type curves: one for 0.01 to 10 minutes, one for 1 to 1,000 minutes, and one for 10^3 to 10^6 minutes. The curve for 1 to 1,000 minutes is by far the most useful and, accordingly, is the only one shown here. Earlougher[3] provides the other curves. The following steps outline a recommended procedure for using McKinley's type curves.

1. Plot Δt (minutes) as the ordinate vs. $\Delta p = (\bar{p} - p_{wf})$ for a drawdown test or $(p_{ws} - p_{wf})$ for a buildup test as the abscissa either on tracing paper or on 3 × 5 cycle log-log paper the same size as McKinley's type curve. The time range on the axis should correspond exactly to one of the type curves; i.e., it should span 0.01 to 10, 1 to 1,000, or 10^3 to 10^6 minutes.

2. Match the time axis of the test-data plot with that of the appropriate McKinley type curve. Move the data horizontally only until the earliest data match one of the type curves (**Fig. 4.7**).

3. Record the correlating parameter value $[(kh/\mu)/5.615C]_{wb}$ for the matched type curve.

4. Choose a match point, any Δp from the test data plot and the corresponding value of $5.615\Delta pC/qB$ from the type curve, or $(\Delta p, 5.615\Delta pC/qB)_{MP}$.

5. Determine the wellbore-storage coefficient, C, from the match point:

$$C = \left(\frac{5.615\Delta pC/qB}{\Delta p}\right)_{MP}\left(\frac{qB}{5.615}\right). \qquad (4.23)$$

6. Calculate near-well transmissibility, $(kh/\mu)_{wb}$, from the parameter value recorded in Step 3 and the wellbore-storage coefficient determined in Step 5:

$$(kh/\mu)_{wb} = 5.615C\left(\frac{kh/\mu}{5.615C}\right)_{wb}. \qquad (4.24)$$

7. If the data trend away from the type curve matching the earliest data, then shift the data plot horizontally to find another type curve that better fits the later data. A shift to a higher value of $(kh/\mu)/5.615C$ indicates damage, while a shift to a lower value indicates stimulation (**Fig. 4.8**).

8. Calculate formation transmissibility:

$$(kh/\mu)_f = 5.615C\left(\frac{kh/\mu}{5.615C}\right)_7 (C)_5, \qquad (4.25)$$

where 7 and 5 = Steps 7 and 5, respectively. Note that we do not find a new pressure match point to determine C again. The storage coefficient, C, is found ultimately in Step 5. In fact, if only data reflecting formation transmissivity (after wellbore-storage distortion has disappeared) are analyzed, error will result with the McKinley method. However, in such a case, other methods, such as semilog analysis, are accurate. The match point must be found with early wellbore-storage-distorted data.

9. Flow efficiency also can be estimated with McKinley's type curves.[2] A working definition of flow efficiency, E, is

$$E \approx \frac{p^* - p_{wf} - \Delta p_s}{p^* - p_{wf}} = \frac{\Delta p^* - \Delta p_s}{\Delta p^*}. \qquad (4.26)$$

The quantities Δp^* and Δp_s are estimated from the McKinley type curves. In Eq. 4.26, Δp^* = the vertical asymptote approached by Δp in the McKinley plot (**Fig. 4.9**), while Δp_s is calculated from Δp_d, the pressure change at the intersection of the early-fit type curve and the late-fit type curve. Δp_s is calculated from

$$\Delta p_s = \left(1 - \frac{k_{wb}}{k_f}\right)\Delta p_d. \qquad (4.27)$$

Example 4.2—Drawdown Test Analysis With McKinley's Type Curve. ***Problem.*** A drawdown test was run on an oil well. **Table 4.5** summarizes known data. Estimate near-well permeability, forma-

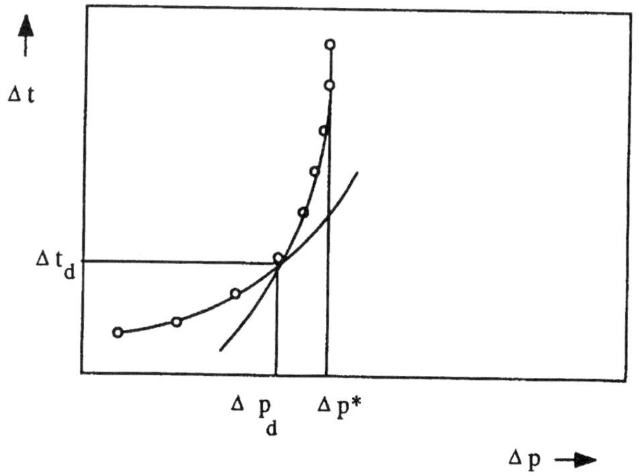

Fig. 4.9—Flow efficiency calculation from McKinley's[2] type curve.

TABLE 4.5—PRESSURE-DRAWDOWN-TEST DATA, EXAMPLE 4.2

t (minutes)	p_{wf} (psi)	t (minutes)	p_{wf} (psi)
1.31	2,953	32.8	2,256
1.64	2,942	65.4	1,952
1.97	2,930	98.4	1,828
2.29	2,919	131	1,768
2.62	2,908	164	1,734
2.95	2,897	197	1,712
3.28	2,886	229	1,696
6.54	2,785	262	1,684
9.84	2,693	295	1,674
13.1	2,611	328	1,665
16.4	2,536	393	1,651
19.7	2,469	524	1,630
22.9	2,408	654	1,614
26.2	2,352	984	1,587

TABLE 4.6—PRESSURE-DRAWDOWN-TEST PLOTTING FUNCTIONS, EXAMPLE 4.2

t (minutes)	p_i-p_{wf} (psi)	t (minutes)	p_i-p_{wf} (psi)
1.31	47	32.8	744
1.64	58	65.4	1,048
1.97	70	98.4	1,172
2.29	81	131	1,232
2.62	92	164	1,266
2.95	103	197	1,288
3.28	114	229	1,304
6.54	215	262	1,316
9.84	307	295	1,326
13.1	389	328	1,335
16.4	464	393	1,349
19.7	531	524	1,370
22.9	592	654	1,386
26.2	648	984	1,413

tion permeability, and flow efficiency with the McKinley type curves for the following conditions.

$q =$ 500 STB/D
$c_t = 10^{-5}$ psi^{-1}
$\phi =$ 0.2
$h =$ 56 ft
$B_o =$ 1.2 RB/STB
$p_i =$ 3,000 psia
$r_w =$ 0.3 ft
$\mu =$ 0.8 cp

Solution. **Example 4.2—Drawdown Test Analysis by Use of McKinley's Type Curve.**

1. First, plot flowing time, t vs. pressure change, $\Delta p = (\bar{p} - p_{wf}) = p_i - p_{wf}$ with the data summarized in **Table 4.6.** Because all the data lie in the 1- to 1,000-minute range, we choose the type curve corresponding to this time range.

2. Match the data with the type curve by shifting the data plot horizontally (**Fig. 4.10**).

3. From the match with the best-fitting McKinley[1] curve for the early data, the type-curve parameter is

$$\left(\frac{kh/\mu}{5.615C}\right)_{wb} = 5,000.$$

4. A match point for the early-time fit is

$\Delta p = 107$ psi and $5.615\Delta pC/qB = 0.010$.

5. From the match point, the wellbore-storage coefficient is

$$C = \left(\frac{5.615\Delta pC/qB}{\Delta p}\right)_{MP}\left(\frac{qB}{5.615}\right) = \left(\frac{0.010}{107}\right)\left(\frac{500 \times 1.2}{5.615}\right)$$

$$= 0.01 \text{ RB/psi}.$$

6. The near-well transmissibility is

$$\left(\frac{kh}{\mu}\right)_{wb}\left(\frac{kh/\mu}{5.615C}\right)_{wb}(5.615C) = (5,000)(5.615)(0.01)$$

$$= 281 \text{ md-ft/cp}.$$

The apparent near-well permeability to oil, k_{wb}, is

$$k_{wb} = \left(\frac{kh}{\mu}\right)_{wb}\frac{\mu}{h} = \frac{(281)(0.8)}{(56)} = 4.01 \text{ md}.$$

7. At Δt_d of approximately 100 minutes, the data depart from the type curve matched with the early data. Accordingly, we shift the data plot horizontally to match the later data with a type curve. The best fit of the later data is for

$$\left(\frac{kh/\mu}{5.615C}\right)_{wb} = 10,000.$$

The shift to a higher value indicates that the formation near the wellbore is damaged (i.e., the near-wellbore formation permeability is lower than that in the formation away from the wellbore).

8. The formation transmissibility is

$$\left(\frac{kh}{\mu}\right)_f = \frac{[(kh/\mu) \times 5.615C]_f}{[(kh/\mu) \times 5.615C]_{wb}}\left(\frac{kh}{\mu}\right)_{wb} = \frac{(10,000)}{(5,000)}(281)$$

$$= 562 \text{ md-ft/cp}.$$

The effective permeability to oil is

$$k = \left(\frac{kh}{\mu}\right)_f\left(\frac{\mu}{h}\right) = (562)\left(\frac{0.8}{56}\right) = 8.03 \text{ md}.$$

9. At the point of intersection of the early- and late-time matched type curves, $\Delta p_d = 1{,}180$ psi. The data become asymptotic at approximately $\Delta p^* = 1{,}500$ psi. The parameter Δp_s is

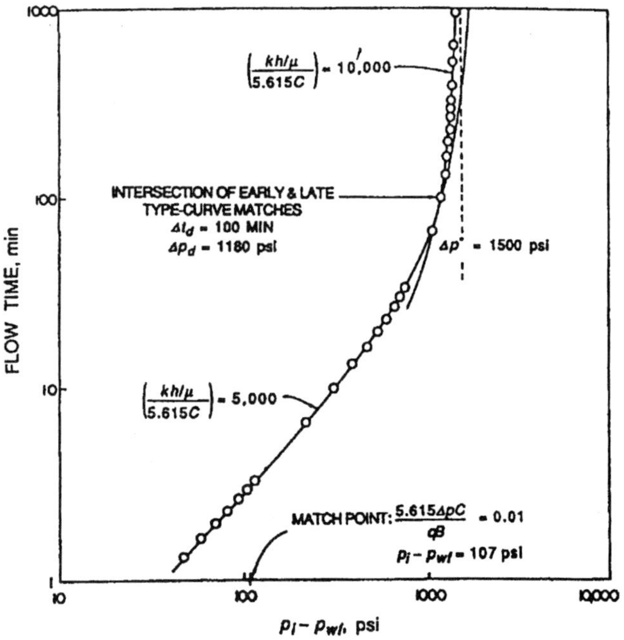

Fig. 4.10—Drawdown test analysis with McKinley's[1,2] type curve.

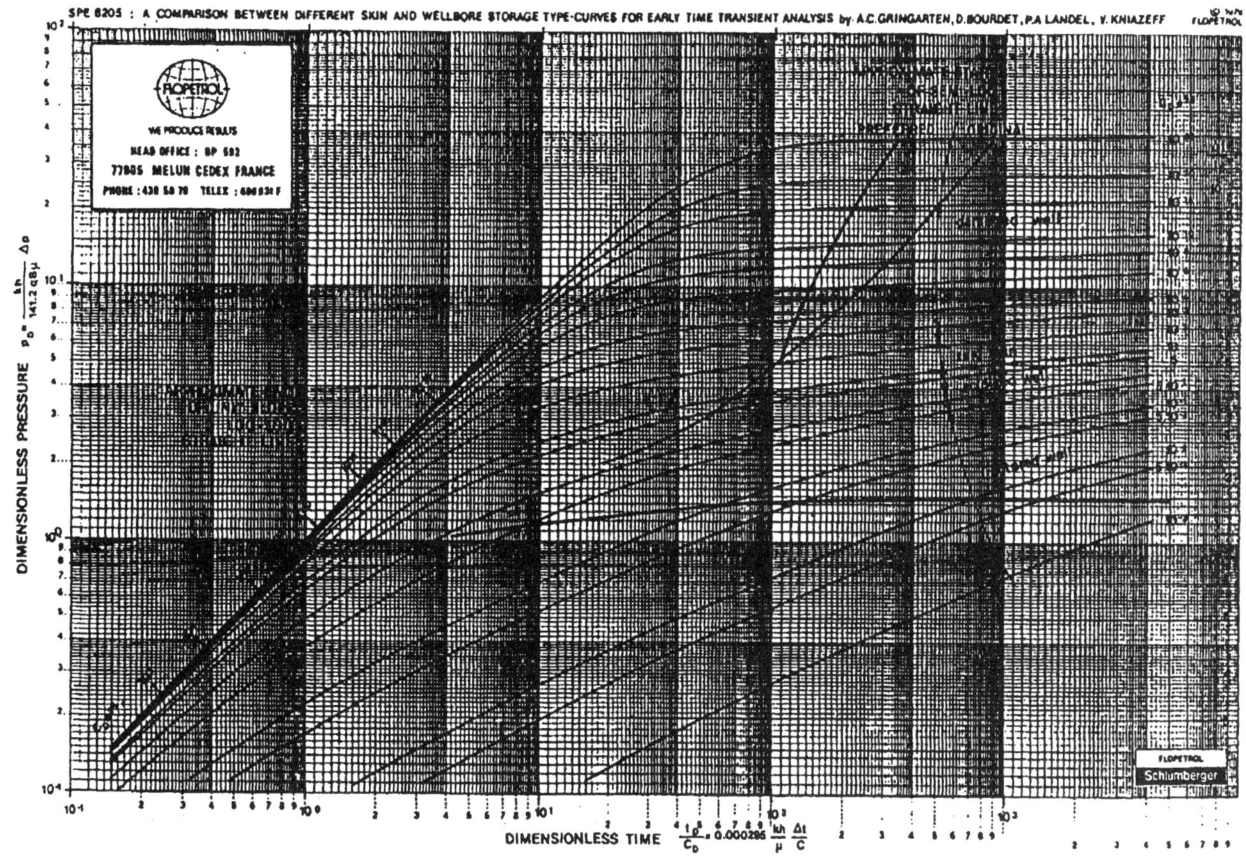

Fig. 4.11—Gringarten *et al.*[4] type curve for a well with wellbore-storage and skin.

$$\Delta p_s = \left(1 - \frac{k_{wb}}{k}\right)\Delta p_d = \left(1 - \frac{4.01}{8.03}\right)(1,180) = 590 \text{ psi}.$$

The flow efficiency is

$$E \approx \frac{\Delta p^* - \Delta p_s}{\Delta p^*} = \frac{1,500 - 590}{1,500} = 0.607 \text{ or } 60.7\%.$$

4.3.2 Gringarten *et al.* Type Curve. The type curve developed by Gringarten *et al.*[4] is based on solutions to the diffusivity equation modeling the flow of a slightly compressible liquid in a homogeneous-acting formation. The initial condition is uniform pressure throughout the drainage area of the well. The outer-boundary condition specifies an infinite-acting or unbounded reservoir, while the inner boundary condition is constant-rate flow with wellbore storage and skin effect. These initial and boundary conditions are the same conditions assumed for the solutions plotted on the Ramey[5,6] type curve. Gringarten *et al.*[4] replotted Ramey's solutions to facilitate application of the type curve. In **Fig. 4.11,** p_D is plotted vs. the time plotting function, t_D/C_D, and as a function of the correlating parameter C_De^{2s}. We discuss some important properties of type curves in general and of the Gringarten type curve in particular in the following section.

1. As long as wellbore unloading accounts for all the flow during a drawdown test or while afterflow accounts for 100% of the flow rate before shut-in in a buildup test, a line having a slope equal to one will occur at early times on a log-log plot. In dimensionless variables, the unit-slope line has the property

$$\frac{t_D}{C_Dp_D} = 1, \quad \text{(4.28)}$$

where C_D = dimensionless wellbore-storage coefficient defined by

$$C_D = \frac{0.8936C}{\phi c_t h r_w^2}, \quad \text{(4.29)}$$

and C = wellbore-storage coefficient (bbl/psi). For a rising or falling liquid level in the wellbore, C is calculated from

$$C = \frac{25.65A_{wb}}{\rho_{wb}}. \quad \text{(4.30)}$$

For a wellbore filled with a single-phase liquid or gas,

$$C = V_{wb}\, c_{wb}. \quad \text{(4.31)}$$

The variables in Eqs. 4.30 and 4.31 are defined as A_{wb} = wellbore area, ft^2; ρ_{wb} = density of the liquid in the wellbore, lbm/ft^3; V_{wb} = wellbore volume, bbl; and c_{wb} = compressibility of the fluid in the wellbore, psi^{-1}.

The wellbore-storage coefficient, C, can be determined from any point $(\Delta t, \Delta p)_{USL}$ on the unit-slope line. Substituting the definitions of p_D, t_D, and C_D (Eqs. 4.3, 4.5, and 4.29, respectively) for a slightly compressible liquid into Eq. 4.28, we find that, for a point on the unit-slope line,

$$C = \frac{qB}{24}\left(\frac{\Delta t}{\Delta p}\right)_{USL}. \quad \text{(4.32)}$$

The dimensionless wellbore-storage coefficient, C_D, is

$$C_D = \frac{0.03723qB}{\phi c_t h r_w^2}\left(\frac{\Delta t}{\Delta p}\right)_{USL}. \quad \text{(4.33)}$$

2. The "preferred" curve on the Gringarten type curve indicates the end of wellbore-storage distortion of the pressure transient data. The "original" curve is a much more conservative indication of the end of wellbore storage. Accordingly, when a plot of test data crosses the preferred curve in a type-curve match, a straight line begins at approximately the same time on a semilog graph of the test data. This straight line commonly is referred to as the "semilog straight line" or the "middle-time region." The more pronounced the effect of wellbore storage on the pressure response during a test, the longer the delay in the onset of the middle-time region.

A nearly horizontal line that crosses the type curves for fractional values of C_De^{2s} lies in the "fractured-well" region on the Gringarten curve. As before, this line indicates the beginning of a straight line on a semilog graph of test data for the fractured-well solutions plotted on the Gringarten type curves. However, wellbore storage does not delay the onset of the middle-time region. Rather, the nearly horizontal line marks the transition from a flow regime characterized by linear flow from the reservoir into a vertical, high-conductivity hydraulic fracture (Chap. 6) to essentially radial flow (called pseudoradial flow). Neither the transition from linear to pseudoradial flow is complete nor will a semilog straight line appear until the the test-data plot crosses this line.

3. The Gringarten type curves are based on solutions to equations modeling constant-rate flow. For a drawdown test, a plot is made of $\Delta p = p_i - p_{wf}$ as a function of flowing time, t. The type curves also can be used to analyze buildup tests if the producing time before shutting in the well is much greater than the duration of the buildup test; i.e., $t_p \gg \Delta t$. However, for situations in which the producing time before shutting in the well is less than $^1/_{10}$ the maximum shut-in time to be analyzed (i.e., $\Delta t_{max} > 0.1t_p$), the drawdown type curves cannot be used to analyze pressure-buildup data accurately.

To account for the effects of producing time on pressure-buildup tests, Agarwal[7] proposed an equivalent time defined by

$$\Delta t_e = \left[\frac{\Delta t}{1 + \left(\frac{\Delta t}{t_p}\right)}\right] . \quad \ldots \quad (4.34)$$

The meaning of equivalent time is that a given pressure change, Δp, that occurred at shut-in time Δt during a buildup test would have occurred at an equivalent time, Δt_e, during a constant-rate flow test. For analyzing buildup tests, we recommend plotting $\Delta p = p_{ws} - p_{wf}$ as a function of equivalent time, Δt_e. The definition of equivalent time is rigorous only for radial flow in homogeneous, infinite-acting reservoirs with test data undistorted by wellbore storage. It may be used to analyze radial-flow data distorted by wellbore storage, but test data affected by boundaries are usually best plotted vs. Δt. For linear flow, which occurs at early test times in many hydraulically fractured wells (Chap. 6), another expression for equivalent time is more appropriate and is presented later.

4. For a drawdown test from a well producing a slightly compressible liquid, a log-log plot of p_D vs. t_D/C_D differs from a log-log plot of $(p_i - p_{wf})$ vs. t only by displacement of both coordinates by constants. Combining Eqs. 4.5 and 4.29, the dimensionless time group is

$$\frac{t_D}{C_D} = \left(\frac{0.0002637kt}{\phi\mu c_t r_w^2}\right)\left(\frac{\phi\mu c_t r_w^2}{0.8936C}\right) = \frac{0.0002951kht}{\mu C} . \quad \ldots \quad (4.35)$$

Taking the logarithms of each side of Eq. 4.35 yields

$$\log\left(\frac{t_D}{C_D}\right) = \log t + \log\left(\frac{0.0002951kh}{\mu C}\right). \quad \ldots \quad (4.36)$$

Similarly, taking the logarithms of each side of the dimensionless pressure defined by Eq. 4.3 yields

$$\log p_D = \log\left(p_i - p_{wf}\right) + \log\left(\frac{kh}{141.2qB\mu}\right). \quad \ldots \quad (4.37)$$

Thus, a plot of drawdown-test data (log Δp vs. log t) should have a shape identical to that of a plot of log p_D vs. log t_D/C_D but with the horizontal and vertical axes displaced.

To use the Gringarten type curve, we compare a plot of test data with the type curve. The test data are plotted either on tracing paper or on log-log graph paper with the same size log cycles as the type curve. Next, the data plot is overlaid on the type curves and matched with the curve having the same shape as the test data plot. When the match is found, the correlating parameter, C_De^{2s}, describing the matched curve is recorded and corresponding values of the pressure and time match points, $(p_D, \Delta p)$ and $(t_D/C_D, t)$, respectively, are selected. Any convenient dimensionless values and the corresponding real values can be used when choosing the match point. Finally, the formation properties can be estimated by use of the match points and the definitions of the dimensionless variables. We recommend the following procedure for analyzing buildup or flow tests with the Gringarten type curve.

1. Plot pressure change, $\Delta p = (\bar{p} - p_{wf})$, vs. flowing time, t, for a drawdown test or pressure change, $\Delta p = p_{ws} - p_{wf}$ $(\Delta t = 0)$, vs. equivalent shut-in time, Δt_e, for a buildup test. Make the plot either on tracing paper or on log-log graph paper with the same size log cycles as the Gringarten type curve.

2. If a unit-slope line is present on the test-data plot at early times, calculate the dimensionless wellbore-storage coefficient, C_D, with a data point (t or Δt_e,Δp)$_{USL}$ from the unit-slope line.

$$C_D = \frac{0.03723qB}{\phi hc_t r_w^2}\left(\frac{t \text{ or } \Delta t_e}{\Delta p}\right)_{USL} . \quad \ldots \quad (4.33)$$

If desired, calculate the dimensional storage coefficient, C, with the same point from the unit-slope line,

$$C_D = \frac{qB}{24}\left(\frac{t \text{ or } \Delta t_e}{\Delta p}\right)_{USL} . \quad \ldots \quad (4.32)$$

Note that, even if no unit-slope line is present at early time, we can still determine C_D from the time match point (Step 6).

3. Overlay the test data on the type curves and find the type curve that most nearly fits all the plotted test data. Record the value of the type-curve correlating parameter, C_De^{2s}, corresponding to the matched type curve. This match will probably not be unique; i.e., other curves may match the data equally well.

4. With the test-data plot still fitted to the type curve, select convenient pressure and time match points. Record the values of $(\Delta p, p_D)$ and $(t, t_D/C_D)$ for a drawdown test or $(\Delta t_e, t_D/C_D)$ for a buildup test at the match point.

The pressure-match-point values $(\Delta p, p_D)$ are corresponding values of the pressure variables on both the test data plot and on the type curve, while the time-match-point values $(t, t_D/C_D)$ or $(\Delta t_e, t_D/C_D)$ are corresponding values of the time variables on the test-data plot and on the type curve.

5. Using the definition of dimensionless pressure given by Eq. 4.3, calculate permeability from the pressure match point.

$$k = \frac{141.2qB\mu}{h}\left(\frac{p_D}{\Delta p}\right)_{MP}, \quad \ldots \quad (4.38)$$

where $\Delta p = (p_i - p_{wf})$ for a drawdown test or $\Delta p = (p_{ws} - p_{wf})$ for a pressure buildup test.

6. Calculate the dimensionless wellbore-storage coefficient, C_D, from the time match point.

$$C_D = \frac{0.0002637k}{\phi hc_t r_w^2}\left(\frac{t \text{ or } \Delta t_e}{t_D/C_D}\right)_{MP} . \quad \ldots \quad (4.39)$$

This value should be comparable with that calculated in Step 2 by use of a data point on the unit-slope line. Inconsistencies between the two values indicate possible errors in the analysis.

7. Calculate the skin factor, s, with the type-curve correlating parameter, C_De^{2s}, from Step 3 and the dimensionless wellbore-storage coefficient, C_D, determined from the time match point in Step 6.

$$s = 0.5\ln\left(\frac{C_De^{2s}}{C_D}\right). \quad \ldots \quad (4.40)$$

The pressure-buildup test in Example 4.3 illustrates the type-curve-matching procedure with the Gringarten type curve.

Example 4.3—Analyzing a Pressure-Buildup Test With the Gringarten Type Curve. ***Problem.*** The data summarized in **Table 4.7** are from a buildup test on an oil well. Because the reservoir pressure is still above the original bubblepoint pressure, only oil is flowing in the reservoir. Estimate the effective permeability to oil, the skin factor, and the wellbore-storage coefficient with the Gringarten

TABLE 4.7—PRESSURE-BUILDUP-TEST DATA FOR EXAMPLE 4.3			
Δt (hours)	p_{ws} (psi)	Δt (hours)	p_{ws} (psi)
0	250	1.71	2,043.1
0.0001	254.09	2.73	2,072.1
0.0002	258.16	4.36	2,100.7
0.0005	270.30	6.50	2,124.9
0.0008	282.33	10.5	2,153.7
0.0010	290.29	15.1	2,175.4
0.0030	367.39	20.0	2,192.0
0.0050	440.40	25.0	2,205.2
0.0080	542.99	30.0	2,215.9
0.0100	607.11	35.0	2,225.0
0.0160	780.68	40.0	2,232.8
0.0255	1,005.1	45.0	2,239.6
0.0406	1,263.2	50.0	2,245.7
0.0649	1,515.6	55.0	2,251.2
0.104	1,714.0	60.0	2,256.2
0.165	1,837.0	65.0	2,260.8
0.264	1,907.4	70.0	2,265.0
0.421	1,950.0	72.0	2,266.6
0.672	1,983.2	80.0	2,272.6
1.07	2,013.5		

TABLE 4.8—PLOTTING FUNCTIONS FOR GRINGARTEN TYPE-CURVE ANALYSIS, EXAMPLE 4.3			
Δt_e (hours)	Δp (psi)	Δt_e (hours)	Δp (psi)
—	—	1.7079	1,793.1
0.0001	4.0900	2.7247	1,822.1
0.0002	8.1600	4.3465	1,850.7
0.0005	20.300	6.4700	1,874.9
0.0008	32.330	10.422	1,903.7
0.0010	40.290	14.939	1,925.4
0.0030	117.39	19.718	1,942.0
0.0050	190.40	24.561	1,955.2
0.0080	292.99	29.371	1,965.9
0.0100	357.11	34.146	1,975.0
0.0160	530.68	38.889	1,982.8
0.0255	755.10	43.599	1,989.6
0.0406	1,013.2	48.276	1,995.7
0.0649	1,265.6	52.921	2,001.2
0.1040	1,464.0	57.534	2,006.2
0.1650	1,587.0	62.116	2,010.8
0.2640	1,657.4	66.667	2,015.0
0.4209	1,700.0	68.478	2,016.6
0.6717	1,733.2	75.676	2,022.6
1.0692	1,763.5		

type curve. In addition, analyze the data with the Horner semilog plotting method and compare the results.

$h =$ 78 ft
$\mu =$ 1 cp
$q =$ 600 STB/D
$p_{wf} =$ 250 psia
$\phi =$ 0.20
$B =$ 1.1 RB/STB
$\bar{p} =$ 2,447 psia
$r_w =$ 0.365 ft
$c_t = 1.61 \times 10^{-5}$ psi^{-1}
$t_p =$ 1,400 hours

Solution. *Gringarten Type-Curve Analysis.*

1. First, we plot the pressure/time data. Because our test is a pressure-buildup test, we plot pressure change, $\Delta p = p_{ws} - p_{wf}$, as a function of the Agarwal[7] equivalent time, Δt_e, on a log-log graph (**Fig. 4.12**). **Table 4.8** summarizes the type-curve plotting functions.

2. Because the earliest data fall on a unit-slope (45°) line, the wellbore-storage coefficient can be estimated with a point on the line. One point on the unit-slope line is $\Delta p = 40.29$ psi, $\Delta t_e = 0.001$ hour. With this point, the dimensionless wellbore-storage coefficient is

$$C_D = \frac{0.03723qB}{\phi h c_t r_w^2}\left(\frac{\Delta t_e}{\Delta p}\right)_{USL}$$

$$= \frac{(0.03723)(600)(1.1)}{(0.2)(78)(1.61 \times 10^{-5})(0.365)^2}\left(\frac{0.001}{40.29}\right)$$

$$= 18.2.$$

The dimensional wellbore-storage coefficient is

$$C = \frac{qB}{24}\left(\frac{\Delta t_e}{\Delta p}\right)_{USL}$$

$$= \frac{(600)(1.1)}{(24)}\left(\frac{0.001}{40.29}\right)$$

$$= 6.84 \times 10^{-4} \text{ RB/psi.}$$

3. Next, we match the log-log plot of the pressure buildup data with the Gringarten type curve (**Fig. 4.13**) by overlaying the test data on the type curve. A good fit is found for the type curve defined by the correlating parameter $C_D e^{2s} = 10^{10}$. With the "preferred" curve on the type-curve plot, the approximate start of the semilog

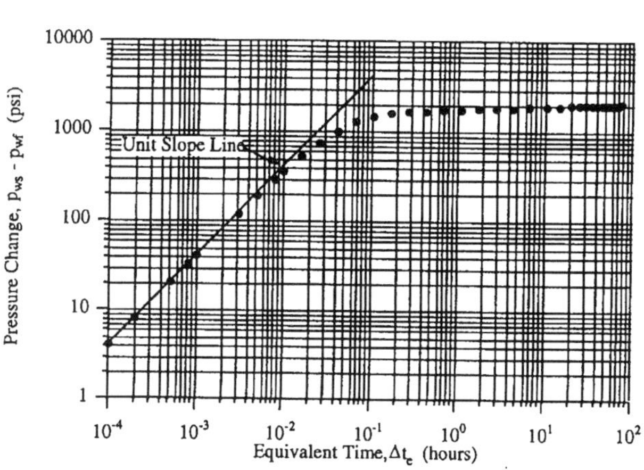

Fig. 4.12—Log-log plot of pressure buildup test data, example 4.3.

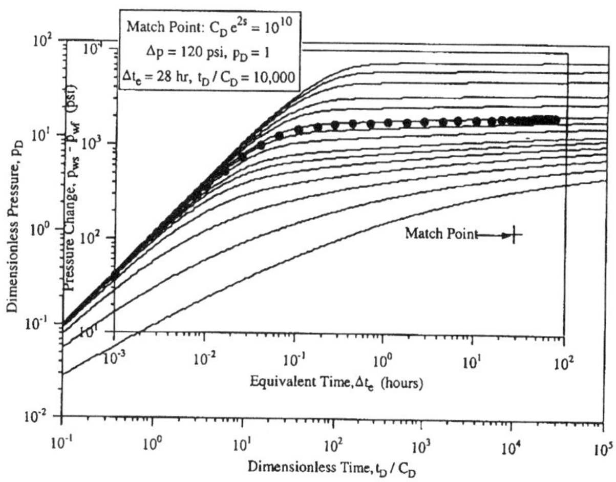

Fig. 4.13—Type curve match with Gringarten type curve, Example 4.3.

p_{ws} (psia)	Horner Time Ratio	p_{ws} (psia)	Horner Time Ratio
250	—	2,043.1	819.71
254.09	14,000,000	2,072.1	513.82
258.16	7,000,000	2,100.7	322.10
270.30	2,800,000	2,124.9	216.38
282.33	1,750,000	2,153.7	134.33
290.29	1,400,000	2,175.4	93.715
367.39	466,670	2,192.0	71.000
440.40	280,000	2,205.2	57.000
542.99	175,000	2,215.9	47.667
607.11	140,000	2,225.0	41.000
780.68	87,501	2,232.8	36.000
1,005.1	54,903	2,239.6	32.111
1,263.2	34,484	2,245.7	29.000
1,515.6	21,573	2,251.2	26.455
1,714.0	13,463	2,256.2	24.333
1,837.0	8,485.8	2,260.8	22.538
1,907.4	5,304.0	2,265.0	21.000
1,950.0	3,326.4	2,266.6	20.444
1,983.2	2,084.3	2,272.6	18.500
2,013.5	1,309.4		

TABLE 4.9—PLOTTING FUNCTIONS FOR HORNER SEMILOG ANALYSIS, EXAMPLE 4.3

straight line of the middle-time region is $\Delta t_e = 0.53$ hour. Because the test data do not deviate from the type curve before the end of the test, we conclude that the reservoir was infinite-acting throughout the test.

4. From the match with the type curve, we select a pressure match point of $\Delta p = 120$ psi, $p_D = 1$ and a time match point of $\Delta t_e = 28$ hour, $t_D/C_D = 10{,}000$. Note that the match points are completely arbitrary. Any convenient values can be selected.

5. From the pressure match point, the effective permeability to oil is

$$k = \frac{141.2qB\mu}{h}\left(\frac{p_D}{\Delta p}\right)_{MP}$$

$$= \frac{(141.2)(600)(1.1)(1.0)}{(78)}\left(\frac{1}{120}\right)$$

$$= 9.96 \text{ md.}$$

6. The dimensionless wellbore-storage coefficient is

$$C_D = \frac{0.0002637k}{\phi h c_t r_w^2}\left(\frac{\Delta t_e}{t_D/C_D}\right)_{MP}$$

$$= \frac{(0.0002637)(9.96)}{(0.2)(1.0)(1.61 \times 10^{-5})(0.365)^2}\left(\frac{28}{10{,}000}\right)$$

$$= 17.1,$$

which is consistent with $C_D = 18.2$ calculated from the unit-slope line in Step 2 and suggests that our analysis is correct.

7. The skin factor, s, is estimated with the matching parameter, C_De^{2s}, from the type-curve match and the dimensionless wellbore-storage coefficient, C_D, from Step 6, giving

$$s = 0.5\ln\left(\frac{C_De^{2s}}{C_D}\right) = 0.5\ln\left(\frac{10^{10}}{17.1}\right) = 10.1.$$

Horner Semilog Analysis.

1. **Table 4.9** summarizes plotting functions for the Horner[8] semilog analysis, and **Fig. 4.14** plots the data. The results from the Gringarten type-curve analysis suggest the middle-time region begins at $\Delta t_e = 0.53$ hour or a Horner time of 2,642. From Fig. 4.14, the slope of the best-fit straight line drawn through the last 20 data points is

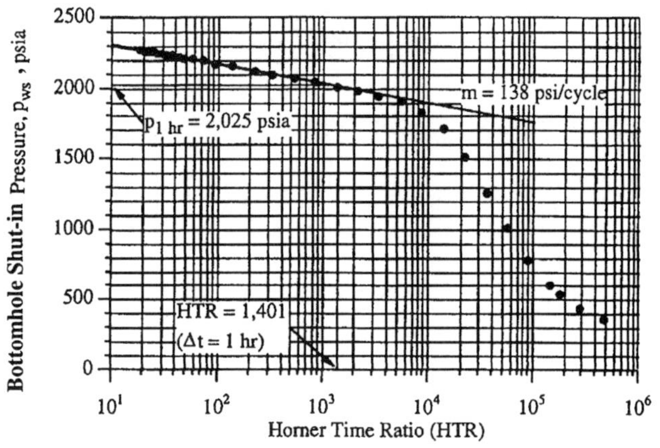

Fig. 4.14—Horner[8] semilog analysis of pressure-buildup–test data, Example 4.3.

$$m = \left|\frac{p_{ws2} - p_{ws1}}{\log[(t_p + \Delta t_2)/\Delta t_2] - \log[(t_p + \Delta t_1)/\Delta t_1]}\right|$$

$$= \left|\frac{2,192.0 - 2,272.6}{\log(71) - \log(18.5)}\right| = 138 \text{ psi/cycle.}$$

2. Next, calculate the effective permeability to oil with Eq. 2.30. By convention, we use the absolute value of the slope for calculating the effective permeability.

$$k = \frac{162.6qB\mu}{mh} = \frac{(162.6)(600)(1.1)(1.0)}{(138)(78)} = 9.97 \text{ md,}$$

which agrees with the value estimated from the Gringarten type-curve analysis, suggesting that our initial type-curve analysis is correct.

3. The skin factor is calculated from Eq. 2.33. From Fig. 4.14, the value of p_{ws} at $\Delta t = 1$ hour (i.e., a Horner time ratio = 1,401) is $p_{1hr} = 2{,}025$ psia.

$$s = 1.151\left[\frac{(p_{1hr} - p_{wf})}{m} - \log\left(\frac{k}{\phi\mu c_t r_w^2}\right) + 3.23\right]$$

$$= 1.151\left\{\frac{(2,025 - 250)}{138}\right.$$

$$\left. - \log\left[\frac{10}{(0.20)(1.0)(16.1 \times 10^{-6})(0.365)^2}\right] + 3.23\right\}$$

$$= 10.0,$$

which also agrees with the result from the type-curve analysis.

One of the inherent problems with type-curve analysis is the difficulty in finding a unique match of the data. Because of the similar shapes of the type curves for a wide range of the correlating parameter, the field data often can be matched with more than one type curve; consequently, more than one solution is possible. To eliminate this problem, we can iterate between type-curve and semilog analyses until we obtain consistent results. This iterative technique is a viable method, however, only if semilog analysis is possible; i.e., wellbore storage or reservoir boundary effects may distort the middle-time region, thus making semilog analysis impossible.

Although the agreement between the results obtained with the Gringarten type-curve and Horner semilog plotting techniques gives us confidence in our analyses, the pressure-derivative plotting technique provides an alternative to pressure/time plots. The pressure derivative has found great utility in well-test analysis and usually is used simultaneously with pressure/time plots to reduce the ambiguity of

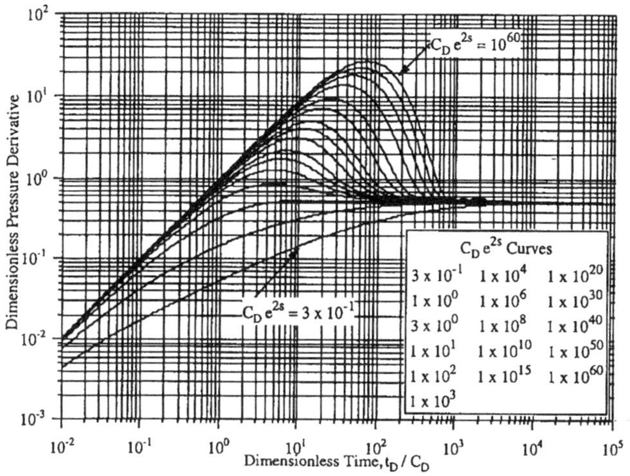

Fig. 4.15—Pressure-derivative type curve (after Bourdet *et al.*[10]).

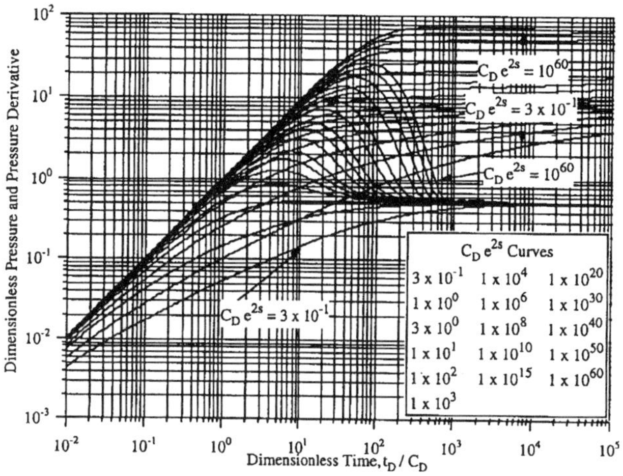

Fig. 4.16—Pressure-change and –derivative type curves (after Bourdet *et al.*[10]).

type-curve analysis. In the next section, we introduce pressure-derivative type curves and illustrate their use with an example.

4.3.3 Bourdet *et al.* Derivative Type Curve. Tiab and Kumar[9] first introduced the use of pressure derivatives for well-test analysis to the petroleum industry by using pressure derivatives to identify sealing faults uniquely from interference tests. Similarly, Bourdet *et al.*[10] developed a type curve, which includes a pressure-derivative function, based on the analytical solution derived by Agarwal *et al.*[5] and plotted on the Gringarten type curves. The dimensionless pressure-derivative function $p_D'(t_D/C_D)$ is plotted as a function of t_D/C_D for various values of the correlating parameter $C_D e^{2s}$. For these type curves, the derivative is defined by

$$p_D' = \frac{dp_D}{d(t_D/C_D)}. \quad (4.41)$$

The derivative type curve has the useful properties summarized next. Ref. 11 presents more details about the utility of derivative curves.

1. For test data on the unit-slope line, $p_D = t_D/C_D$; thus,

$$\frac{dp_D}{d(t_D/C_D)} = p_D' = 1 \text{ and } p_D'(t_D/C_D) = t_D/C_D. \quad (4.42)$$

Then, $\log p_D'(t_D/C_D) = \log t_D/C_D$, and the slope of a plot of p_D' (t_D/C_D) vs. t_D/C_D on a log-log graph is unity. Consequently, at early times a type-curve plot of p_D vs. t_D/C_D should coincide with the plot of $p_D'(t_D/C_D)$ vs. t_D/C_D if the early data are distorted by wellbore storage and are characterized by a unit-slope line.

2. For test data on the semilog straight line, dimensionless pressure can be modeled with the logarithmic approximation to the line-source solution,

$$p_D = 0.5(\ln t_D + 0.80907 + 2s). \quad (4.43)$$

Adding and subtracting $\ln C_D$ inside the brackets in Eq. 4.43 yields

$$p_D = 0.5\left[\ln t_D - \ln C_D + 0.80907 + \ln C_D + \ln\left(e^{2s}\right)\right] \quad (4.44a)$$

or $$p_D = 0.5\left[\ln(t_D/C_D) + 0.80907 + \ln\left(C_D e^{2s}\right)\right]. \quad (4.44b)$$

Thus, $$\frac{dp_D}{d(t_D/C_D)} = p_D' = \frac{0.5}{(t_D/C_D)} \quad (4.45)$$

and $$p_D'(t_D/C_D) = 0.5, \quad (4.46)$$

which indicates that the dimensionless pressure derivatives from the semilog straight line or middle-time region will form a horizontal line at $p_D'(t_D/C_D) = 0.5$ on the derivative type curve.

3. **Fig. 4.15** illustrates derivative curves with unit-slope lines at early times, a horizontal line at late times, and more complex shapes at intermediate times. As we see in later sections, reservoirs exhibit distinct shapes at these intermediate times, making the derivative curve a powerful tool for identifying the correct reservoir model for well-test analysis.

4. The p_D and $p_D'(t_D/C_D)$ type curves can and should be included together on a single plot, permitting simultaneous type-curve analysis with both pressure and pressure-derivative curves (**Fig. 4.16**) and reducing the ambiguity (nonuniqueness) of the Gringarten type curves.

Procedure for Test Analysis With the Derivative Type Curve. The following procedure is recommended for using the Bourdet *et al.*[10] derivative type curve in analyzing test data.

1. Calculate pressure-derivative functions of the well-test data. For drawdown tests, calculate

$$-\frac{dp_{wf}}{d(\ln t)} = -t\left[\frac{dp_{wf}}{d(t)}\right] = t\Delta p'. \quad (4.47)$$

For a buildup test, calculate the pressure derivative in terms of the equivalent time,

$$\frac{dp_{ws}}{d(\ln \Delta t_e)} = \Delta t_e\left[\frac{dp_{ws}}{d(\Delta t_e)}\right] = \Delta t_e \Delta p'. \quad (4.48)$$

2. Plot $t\Delta p'$ (or $\Delta t_e \Delta p'$) and Δp as functions of t (or Δt for tests with boundary effects) either on tracing paper or on log-log graph paper with the same size log cycles as the Bourdet type-curve graph.

3. If possible, force a match of the data to the type curve in the vertical direction by aligning the flat region of the test data with the $p_D'(t_D/C_D) = 0.5$ line on the type curve.

4. If possible, force a match in the horizontal direction by aligning the unit-slope regions of the test-data derivative plot and the derivative type curve. If a unit-slope line is present on the test-data plot, calculate the dimensionless wellbore-storage coefficient, C_D, using a data point $(t \text{ or } \Delta t_e, \Delta p)_{USL}$ from the unit-slope line.

$$C_D = \frac{0.03723 qB}{\phi c_t h r_w^2}\left(\frac{\Delta t}{\Delta p}\right)_{USL}. \quad (4.33)$$

If desired, calculate the dimensional storage coefficient, C, with the same point from the unit-slope line,

$$C = \frac{qB}{24}\left(\frac{\Delta t}{\Delta p}\right)_{USL}. \quad (4.32)$$

5. Determine $C_D e^{2s}$ from the matching parameter of the derivative type-curve match. This same matching parameter also charac-

TABLE 4.10—PRESSURE-BUILDUP-TEST DATA, EXAMPLE 4.4

Δt (hours)	p_{ws} (psi)	Δt (hours)	p_{ws} (psi)
0	250	1.71	2,043.1
0.0001	254.09	2.73	2,072.1
0.0002	258.16	4.36	2,100.7
0.0005	270.30	6.50	2,124.9
0.0008	282.33	10.5	2,153.7
0.0010	290.29	15.1	2,175.4
0.0030	367.39	20.0	2,192.0
0.0050	440.40	25.0	2,205.2
0.0080	542.99	30.0	2,215.9
0.0100	607.11	35.0	2,225.0
0.0160	780.68	40.0	2,232.8
0.0255	1,005.1	45.0	2,239.6
0.0406	1,263.2	50.0	2,245.7
0.0649	1,515.6	55.0	2,251.2
0.104	1,714.0	60.0	2,256.2
0.165	1,837.0	65.0	2,260.8
0.264	1,907.4	70.0	2,265.0
0.421	1,950.0	72.0	2,266.6
0.672	1,983.2	80.0	2,272.6
1.07	2,013.5		

terizes the fit on the pressure-change type curve (p_D vs. t_D/C_D). Select pressure (Δp,p_D) and time (Δt or Δt_e,t_D/C_D) match points.

6. Calculate permeability, k, from the pressure match point.

$$k = \frac{141.2qB\mu}{h}\left(\frac{p_D}{\Delta p}\right)_{MP} . \quad (4.38)$$

7. Calculate the dimensionless wellbore-storage coefficient, C_D, from the time match point and compare with the value computed from the unit-slope line if a unit-slope line is present.

$$C_D = \frac{0.0002637k}{\phi h c_t r_w^2}\left(\frac{t \text{ or } \Delta t_e}{t_D/C_D}\right)_{MP} . \quad (4.39)$$

8. Calculate the skin factor, s, with C_D from Step 7 and C_De^{2s} from Step 5.

$$s = 0.5\ln\left(\frac{C_De^{2s}}{C_D}\right). \quad (4.40)$$

Determining derivatives requires such tedious arithmetic that a computer program is essential. Several programs to perform this calculation, as well as other computations and test analysis, are marketed for use on microcomputers. The algorithm[11] outlined in Appendix F can also be implemented with a spreadsheet.

Example 4.4—Analyzing a Buildup Test With the Derivative Type Curve. ***Problem.*** Use p_D and p_D' type curves (Gringarten and Bourdet) to determine permeability and skin factor for the pressure-buildup test described in Example 4.3. **Table 4.10** repeats the pressure-buildup test data.

h = 78 ft
μ = 1 cp
q = 600 STB/D
p_{wf} = 250 psia
ϕ = 0.20
B = 1.1 RB/STB
$\bar{p}$ = 2,447 psia
r_w = 0.365 ft
c_t = 1.61×10^{-5} psi^{-1}
t_p = 1,400 hours

TABLE 4.11—PRESSURE AND DERIVATIVE PLOTTING FUNCTIONS, EXAMPLE 4.4

Δt_e (hours)	Δp (psi)	$\Delta t_e \Delta p'$	Δt_e (hours)	Δp (psi)	$\Delta t_e \Delta p'$
0.0001	4.0900	5.8718	1.7079	1,793.1	62.642
0.0002	8.1600	9.0490	2.7247	1,822.1	61.664
0.0005	20.300	21.410	4.3465	1,850.7	61.021
0.0008	32.330	35.762	6.4700	1,874.9	60.641
0.0010	40.290	44.832	10.422	1,903.7	60.329
0.0030	117.39	119.84	14.939	1,925.4	60.127
0.0050	190.40	182.17	19.718	1,942.0	60.014
0.0080	292.99	268.64	24.561	1,955.2	60.031
0.0100	357.11	317.26	29.371	1,965.9	59.989
0.0160	530.68	425.64	34.146	1,975.0	59.941
0.0255	755.10	518.27	38.889	1,982.8	59.858
0.0406	1,013.2	546.56	43.599	1,989.6	59.940
0.0649	1,265.6	479.59	48.276	1,995.7	59.787
0.1040	1,464.0	342.82	52.921	2,001.2	59.774
0.1650	1,587.0	208.69	57.534	2,006.2	59.799
0.2640	1,657.4	120.45	62.116	2,010.8	59.808
0.4209	1,700.0	81.174	66.667	2,015.0	59.914
0.6717	1,733.2	68.094	68.478	2,016.6	59.980
1.0692	1,763.5	64.192	75.676	2,022.6	59.834

Solution.

1. Calculate pressure-derivative functions of the well-test data. For a buildup test, calculate

$$\frac{dp_{ws}}{d(\ln \Delta t_e)} = \Delta t_e\left[\frac{dp_{ws}}{d(\Delta t_e)}\right] = \Delta t_e \Delta p'.$$

Table 4.11 summarizes both pressure and pressure-derivative log-log plotting functions.

2. Plot $\Delta t_e \Delta p'$ and Δp as functions of Δt_e either on tracing paper or on log-log graph paper with the same size log cycles as the type-curve graph (**Fig. 4.17**). Note that both the pressure and pressure-derivative test data form a unit-slope line at early times, indicating wellbore-storage effects. In addition, note that the later-time derivative data flatten, suggesting the presence of the middle-time region.

3. To find the best fit of the data to the type curve, we align the horizontal portion of the later-time derivative data for $\Delta t_e > 0.672$ hour with $(t_D/C_D)p_D' = 0.5$ on the derivative type curve (**Fig. 4.18**).

4. We then align the data points exhibiting a unit slope, $\Delta t_e < 0.005$ hour, with the unit-slope line of the pressure-change and -derivative type curve (Fig. 4.18).

5. Both pressure and pressure-derivative data fit the type curve for the correlating parameter $C_De^{2s} = 10^{10}$. With the data in the matched position, we select the following pressure and time match points: ressure match point of $\Delta p = 120$ psi and $p_D = 1$ and a time match point of $\Delta t_e = 28$ hours and $t_D/C_D = 10{,}000$. Again, the selection of both the pressure and time match points is completely arbitrary.

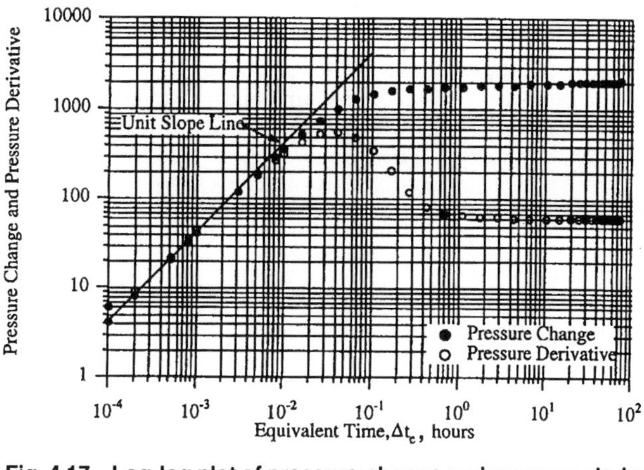

Fig. 4.17—Log-log plot of pressure-change and pressure-derivative data, Example 4.4.

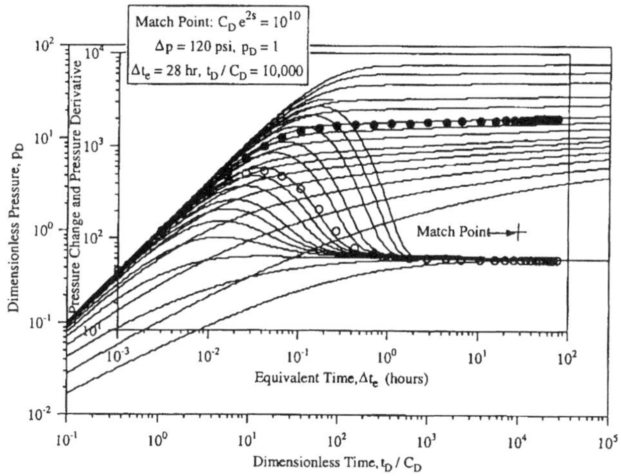

Fig. 4.18—Type-curve match by use of the Bourdet *et al.*[10] pressure-derivative model, Example 4.4.

6. The effective permeability to oil is computed with the pressure match point:

$$k = \frac{141.2qB\mu}{h}\left(\frac{p_D}{\Delta p}\right)_{MP}$$

$$= \frac{(141.2)(600)(1.1)(1.0)}{(78)}\left(\frac{1}{120}\right)$$

$$= 9.96 \text{ md.}$$

7. The dimensionless wellbore-storage coefficient is computed with the time match point:

$$C_D = \frac{0.0002637k}{\phi h c_t r_w^2}\left(\frac{\Delta t_e}{t_D/C_D}\right)_{MP}$$

$$= \frac{(0.0002637)(10.0)}{(0.2)(1.0)(1.61 \times 10^{-5})(0.365)^2}\left(\frac{28}{100}\right)$$

$$= 17.21.$$

8. Calculate the skin factor, s, with C_D from Step 7 and C_De^{2s} from Step 5.

$$s = 0.5\ln\left(\frac{C_De^{2s}}{C_D}\right) = 0.5\ln\left(\frac{10^{10}}{17.2}\right) = 10.1.$$

4.4 Application of Type Curves—Homogeneous Reservoir Model, Compressible Fluids

Spivey and Lee[12] showed that, for both buildup- and flow-test analysis in gas wells, the use of pseudotime and pseudopressure (or adjusted time and adjusted pressure) is essential for analyzing early-time data distorted by wellbore storage. Without this transformation of variables, conventional type curves cannot model the early pressure behavior. For a gas well with no liquid production, the wellbore-storage coefficient, C, is

$$C = V_{wb}c_{wb}, \quad \text{(4.31)}$$

where gas compressibility, c_{wb}, is a strong function of pressure. Thus, in a test where the pressure changes by one to two orders of magnitude, the storage coefficient also changes drastically. Lee and Holditch[13] showed that, with the change in variables to pseudopressure and pseudotime, the modified dimensionless wellbore-storage coefficient becomes independent of pressure and can be treated as a constant. Therefore, type curves (such as the Gringarten *et al.*[4] and Bourdet *et al.*'s[10]) derived on the assumption of a constant wellbore-storage coefficient can be used for gas-well-test analysis.

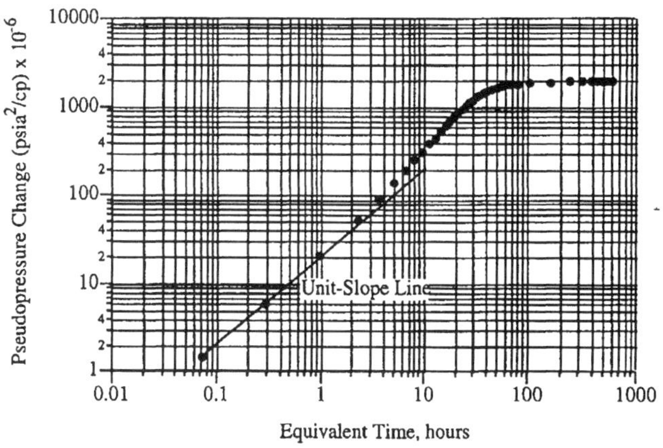

Fig. 4.19—Effect of pseudopressure and equivalent time variables on curve shape.

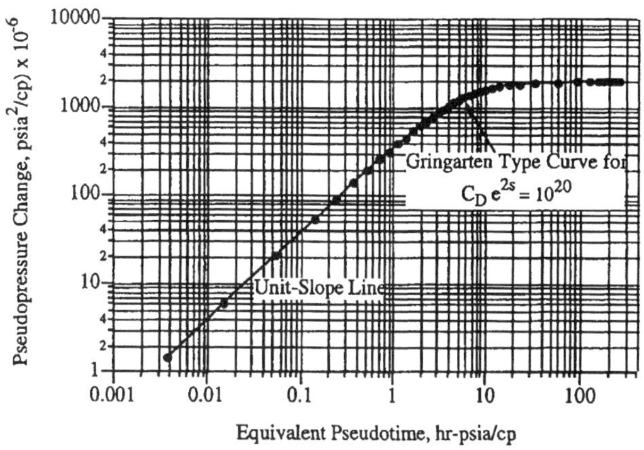

Fig. 4.20—Effect of pseudopressure and equivalent pseudotime variables on curve shape.

Table 4.1 summarizes analogies between variables used in type-curve analysis with the pseudopressure and pseudotime transformations for slightly compressible liquids and gases. When adjusted time and adjusted pressure are used for gas-well-test analysis, the same modified wellbore-storage coefficients arise. The definitions of dimensionless pressure and time variables have the same form as for slightly compressible liquids, except that the gas properties are evaluated at average drainage-area pressure, $\bar{p}$.

Figs. 4.19 and 4.20 illustrate the effect of replacing time with pseudotime on a logarithmic plot of a gas-well buildup test. Fig. 4.19 is a plot of pseudopressure change as a function of equivalent time for a simulated gas-well test. The early-time data fall on a line with a slope greater than unity and followed by a sharp flattening at later times, indicating a changing wellbore-storage coefficient during the test. Because the Gringarten type curves were developed assuming a constant wellbore-storage coefficient, no single curve fits all the test data.

When the same data are plotted as a function of adjusted equivalent time, the earliest data fall on a unit-slope line, indicating a constant wellbore-storage coefficient. In addition, all test data are fit by a single type curve (Fig. 4.20). Table 4.2 summarizes the interpretation of the unit-slope line and the pressure and time-match points for type-curve analysis of gas-well tests that use the common plotting functions, including adjusted pressure and adjusted time. For comparison, we also include interpretation equations for type-curve analysis of oil-well tests.

The pressure-derivative approach presented for well-test analysis of slightly compressible liquids is also applicable to gas wells. The following type-analysis procedure has been adapted from slightly compressible liquid analysis techniques by use of pressure and pressure-derivative type curves. Simultaneous application of pressure and pressure derivatives, either with liquid or gas-well-test analysis, helps to reduce the ambiguity associated with type curves. Although we present the procedure in terms of adjusted pressure and time variables, the procedure also can be used with other gas-well-test plotting functions.

Procedure for Gas-Well-Test Analysis With Type Curves.

1. From test data, calculate the following derivatives. For a drawdown test,

$$-\frac{dp_{a,wf}}{d(\ln t_a)} = -t_a\left[\frac{dp_{a,wf}}{d(t_a)}\right] = t_a \Delta p_a'. \quad (4.49)$$

For a buildup test,

$$\frac{dp_{a,ws}}{d(\ln \Delta t_{ae})} = \Delta t_{ae}\left[\frac{dp_{a,ws}}{d(\Delta t_{ae})}\right] = \Delta t_{ae} \Delta p_a'. \quad (4.50)$$

2. For a drawdown test, plot $t_a \Delta p_a'$ and $\Delta p_a = (p_{a,i} - p_{a,wf})$ vs. t_a. For a buildup test, plot $\Delta t_{ae} \Delta p_a'$ and $\Delta p_a = (p_{a,ws} - p_{a,wf}(\Delta t = 0))$ vs. Δt_{ae} where $\Delta t_{ae} = \Delta t_a/(1 + \Delta t_a/t_p))$. Make the plots either on tracing paper or on log-log graph paper the same size as the type-curve graph.
3. If possible, force a match of the pressure and pressure-derivative data to the type curve in the vertical direction. Attempt to align the flat region of the derivative test data with the $p_D'(t_D/C_D) = 0.5$ line on the type curve. The flat region is the middle-time region. Note that, if this flat region is not present, semilog analysis is not possible.
4. If possible, force a match in the horizontal direction by aligning the early-time regions of the pressure-change and -derivative test data with the unit-slope regions of the type curves. If a unit-slope line is present, calculate the dimensionless wellbore-storage coefficient, C_D, using a point (t_a or Δt_{ae}, Δp_a)$_{USL}$ from the unit-slope line of the pressure-change data.

$$C_D = \frac{0.0372 q_g \overline{B}_g}{\phi h \overline{c}_t r_w^2}\left(\frac{t_a \text{ or } \Delta t_{ae}}{\Delta p_a}\right)_{USL}, \quad (4.51)$$

where $\overline{B}_g$ and $\overline{c}_t$ are evaluated at the average reservoir pressure.

5. Record the value of the correlating parameter C_De^{2s} corresponding to the matched type curve. The value of C_De^{2s} should be the same from the match of both the pressure and pressure-derivative type curves.
6. With the test-data plot fitted to the type curve, record a pressure match point $(\Delta p_a, p_D)_{MP}$ and a time match point $(t_a, t_D/C_D)_{MP}$ for a drawdown test or $(\Delta t_{ae}, t_D/C_D)_{MP}$ for a buildup test.
7. Calculate permeability from the pressure match point.

$$k = \frac{141.2 q_g \overline{B}_g \overline{\mu}_g}{h}\left(\frac{p_D}{\Delta p_a}\right)_{MP}, \quad (4.52)$$

where $\overline{\mu}_g$ is evaluated at the average reservoir pressure.

8. Calculate C_D from the time match point. Compare this value of C_D with that calculated from the unit-slope line.

$$C_D = \frac{0.0002637k}{\phi \overline{\mu}_g \overline{c}_t r_w^2}\left(\frac{t_a \text{ or } \Delta t_{ae}}{t_D/C_D}\right)_{MP}. \quad (4.53)$$

9. Calculate the skin factor, s, with the matching parameter, C_De^{2s}, and C_D from Step 8.

$$s = 0.5 \ln\left(\frac{C_De^{2s}}{C_D}\right). \quad (4.54)$$

Example 4.5—Gas-Well Buildup-Test Analysis With Pressure and Pressure-Derivative Type Curves. ***Problem.*** With the gas-well buildup-test data presented in **Table 4.12,** estimate the formation permeability and skin factor with the Bourdet type curves. Compare the results with those obtained with the Horner semilog analysis method. Use adjusted pressure and time variables as the plotting functions.

TABLE 4.12—PRESSURE BUILDUP TEST DATA, EXAMPLE 4.5

Δt (hours)	p_{ws} (psia)	Δt (hours)	p_{ws} (psia)
0	6,287.1	3.86	7,586.0
0.0100	6,296.6	5.74	7,738.7
0.0149	6,301.1	8.53	7,864.9
0.0221	6,307.8	12.7	7,971.4
0.0329	6,317.7	18.9	8,065.6
0.0489	6,332.1	28.1	8,153.2
0.0728	6,353.1	41.8	8,234.4
0.108	6,383.5	62.1	8,313.4
0.161	6,427.1	92.4	8,389.8
0.240	6,488.6	137	8,463.7
0.356	6,573.6	204	8,534.9
0.530	6,687.9	304	8,602.9
0.788	6,834.7	452	8,666.6
1.17	7,011.8	672	8,725.3
1.74	7,208.3	1,000	8,777.6
2.59	7,405.9		

h = 21 ft
t_p = 2,000 hours
μ_g = 0.03274 cp
p_i = 9,000 psia
r_w = 0.365 ft
q_g = 100 Mscf/D
c_t = 45.4×10^{-6} psia^{-1}
$p_{a,i}$ = 7,366.1 psia
ϕ = 0.10
T = 670°R (210°F)
γ_g = 0.6
z = 1.310
$\overline{B}_g$ = 0.491 RB/Mscf

Solution.

1. The first step is to calculate the pressure-derivative functions, which are tabulated in **Table 4.13** in terms of adjusted pressure change and adjusted equivalent time.
2. Next, we plot adjusted pressure change (Δp_a) and adjusted pressure derivative ($\Delta t_{ae} \Delta p_a'$) vs. adjusted equivalent time (Δt_{ae}) on log-log paper (**Fig. 4.21**). Note that both the pressure-change and -derivative data at early times form a unit-slope line, indicating wellbore-storage effects. In addition, the pressure-derivative data at the end of the test flatten, suggesting the presence of a middle-time region.
3. Using the Bourdet *et al.* type curves, match both Δp_a and $\Delta t_{ae} \Delta p_a'$ in the vertical direction (**Fig. 4.22**). Several adjusted pressure-derivative data points at the end of the test begin to flatten and appear to form a middle-time region. In addition, we can match several of the last few pressure-derivative points on the type curve for $p_D'(t_D/C_D) = 0.5$, suggesting the presence of the middle-time region.
4. Because the early-time data, both pressure change and pressure derivative, form a unit-slope line indicative of wellbore-storage effects, we can obtain a match in the horizontal direction. In addition, we can calculate C_D using a point $(\Delta t_{ae}, \Delta p_a)_{USL}$ from the unit-slope line of the pressure-change data. We choose the point ($\Delta t_{ae} = 0.01528$ hour $\Delta p_a = 21.010$ psi) and calculate

$$C_D = \frac{0.0372 q_g \overline{B}_g}{\phi h \overline{c}_t r_w^2}\left(\frac{\Delta t_{ae}}{\Delta p_a}\right)_{USL}$$

$$= \frac{(0.0372)(100)(0.491)}{(0.10)(21)(45.4 \times 10^{-6})(0.365)^2}\left(\frac{0.01528}{21.010}\right)$$

$$= 105.$$

TABLE 4.13—PRESSURE AND PRESSURE-DERIVATIVE PLOTTING FUNCTIONS, EXAMPLE 4.5

Δp_a (psia)	Δt_e (hours)	Δt_{ae} (hours)	$\Delta t_{ae}\Delta p_a'$
—	—	—	—
9.6426	0.0100	0.0069	11.442
14.210	0.0149	0.0103	14.350
21.010	0.0221	0.0153	21.194
31.059	0.0329	0.0228	30.993
45.675	0.0489	0.0339	45.022
66.990	0.0728	0.0505	65.568
97.850	0.1080	0.0751	93.656
142.12	0.1610	0.1125	132.03
204.56	0.2400	0.1685	184.69
290.87	0.3559	0.2519	249.29
406.93	0.5299	0.3789	323.68
556.02	0.7877	0.5711	399.43
736.00	1.1693	0.8631	456.53
935.76	1.7385	1.3114	476.88
1,136.6	2.5867	1.9993	454.02
1,319.6	3.8526	3.0545	400.08
1,474.6	5.7236	4.6507	337.57
1,602.5	8.4938	7.0594	282.93
1,710.5	12.620	10.704	245.20
1,805.8	18.723	16.167	224.20
1,894.3	27.711	24.310	209.96
1,976.5	40.944	36.439	201.32
2,056.3	60.230	54.310	197.48
2,133.4	88.320	80.629	194.43
2,207.9	128.22	118.42	191.57
2,279.6	185.12	172.93	188.26
2,348.0	263.89	249.23	186.43
2,412.0	368.68	351.85	228.46
2,498.1	502.99	484.97	173.32
2,523.5	666.67	648.68	182.19

5. A good match of both adjusted pressure-change and -derivative data is obtained with the type curves characterized by the correlating parameter $C_D e^{2s} = 10^2$.

6. From the match shown in Fig. 4.17, we obtain a pressure match point of $\Delta p_a = 3{,}800$ psi and $p_D = 10$ and a time match point of $\Delta t_{ae} = 0.27$ hour and $t_D/C_D = 0.1$.

7. From the pressure match point, the permeability is

$$k = \frac{141.2 q_g \bar{B}_g \bar{\mu}}{h}\left(\frac{p_D}{\Delta p_a}\right)_{MP}$$

$$= \frac{(141.2)(100)(0.491)(0.03274)}{21}\left(\frac{10}{3,800}\right)$$

$$= 0.03 \text{ md},$$

which agrees with the Horner analysis (not shown) and confirms our initial selection of the middle-time region.

8. Calculate the dimensionless wellbore-storage coefficient, C_D, from the time match point.

$$C_D = \frac{0.0002637k}{\phi \bar{\mu} \bar{c}_t r_w^2}\left(\frac{\Delta t_{ae}}{t_D/C_D}\right)_{MP}$$

$$= \frac{(0.0002637)(0.03)}{(0.10)(0.03274)(45.4 \times 10^{-6})(0.365)^2}\left(\frac{0.027}{0.1}\right)$$

$$= 108,$$

which agrees with the value calculated from the unit-slope line.

9. Estimate the skin factor with the value of C_D from Step 8 and the correlating parameter, $C_D e^{2s} = 100$.

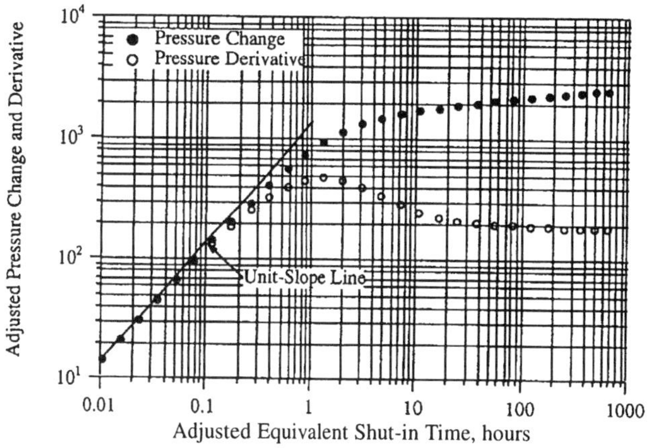

Fig. 4.21—Log-log plot of adjusted-pressure-change and -pressure-derivative data, Example 4.5.

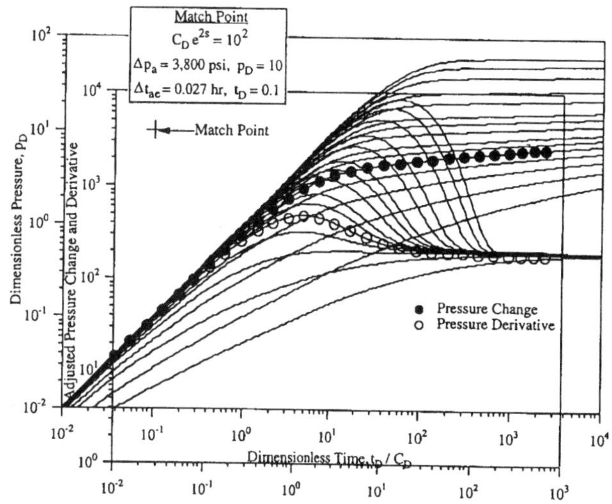

Fig. 4.22—Type curve match of a gas-well pressure-buildup test by use of the Bourdet *et al.*[10] type curve, Example 4.5.

$$s = 0.5\ln\left(\frac{C_D e^{2s}}{C_D}\right) = 0.5\ln\left(\frac{100}{108}\right) = -0.04.$$

4.5 Correcting Initial Pressure in a Well Test

The pressure at the beginning of a test (i.e., p_i for a drawdown test or p_{wf} ($\Delta t = 0$) for a buildup test) is sometimes read or recorded incorrectly from the pressure-measuring device. An incorrect initial pressure distorts the shape of the pressure-change-data plot for type-curve matching. In semilog analysis, an incorrect initial pressure results in an incorrect estimate of skin factor. When the derivative plot is smooth, an incorrect initial pressure is clearly evident. Because the derivative is not affected by an incorrect estimate of initial pressure, the derivative- and pressure-change-data plots will have inconsistent shapes on a log-log graph if the initial pressure is wrong. Derivative data, however, frequently scatter at early times. Consequently, we need an alternative to the pressure derivative to determine not only whether the initial pressure is correct but also correct the initial pressure to its correct value. A related problem is that zero time in a test is not always known accurately. Thus, we need a method to correct initial time to its actual value.

A suitable method to correct initial pressure or time is to prepare a Cartesian-coordinate plot of pressure as a function of elapsed test time. Data falling on a unit-slope line on a log-log plot will also fall on a straight line on a Cartesian-coordinate plot. Therefore, if any early test data fall on a unit-slope straight line, we can plot these data on Cartesian coordinates and extrapolate the straight line back to $\Delta t = 0$.

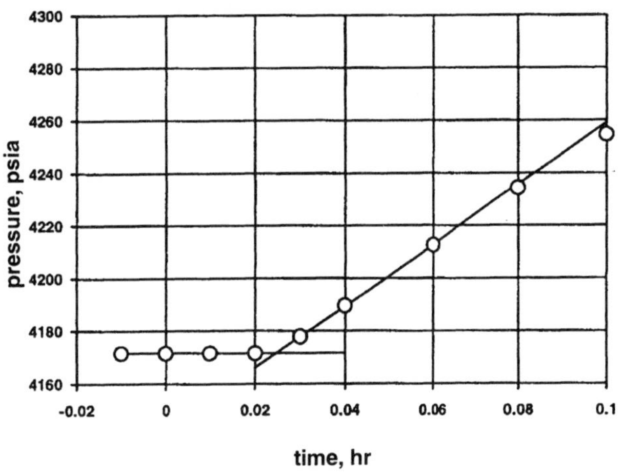

Fig. 4.23—Example 4.6, correcting initial pressure and time.

Pressures recorded during the flow period normally will be changing slowly, and we can extrapolate these data forward to $\Delta t=0$. The intersection of these two straight lines gives the time and pressure at $\Delta t=0$. **Fig. 4.23** illustrates this process.

Example 4.6—Correcting Initial Time and Pressure. ***Problem.*** Given the times and pressures in **Table 4.14,** determine the time at which $\Delta t=0$ and the correct flowing bottomhole pressure at shut-in.

Solution.

1. Inspection of the data shows that that well was not shut in until a few moments after the recorded shut-in time. To correct the initial time and pressure, we plot the early data on a convenient Cartesian-coordinate scale. Then, we draw straight lines through both the flow data and the initial straight-line portion of the buildup data, as Fig. 4.23 shows.
2. Next, we determine the location of the intersection of the two straight lines to be $\Delta t=0.025$ hour, $p_{wf}=4{,}171.58$ psia.
3. Finally, we correct the recorded shut-in time by subtracting the recorded shut-in time at $\Delta t=0.025$ hour, as **Table 4.15** shows.

TABLE 4.14—PRESSURE-BUILDUP-TEST DATA, EXAMPLE 4.6

t hours	p_{ws} psia	t hours	p_{ws} psia	t hours	p_{ws} psia
−0.01	4,171.58	0.2	4,341.86	6	4,777.77
0.0	4,171.58	0.3	4,410.04	8	4,790.91
0.01	4,171.58	0.4	4,464.11	10	4,800.73
0.02	4,171.58	0.6	4,542.64	15	4,817.95
0.03	4,177.74	0.8	4,595.12	20	4,829.81
0.04	4,189.73	1	4,631.46	30	4,846.15
0.06	4,212.60	1.5	4,684.62	40	4,857.51
0.08	4,234.17	2	4,712.41	60	4,873.22
0.1	4,254.56	3	4,741.42	72	4,880.17
0.15	4,301.01	4	4,757.71		

4.6 Reservoir Identification With Type Curves

In the previous sections, we presented analysis techniques for well tests. For each technique, we made the implicit assumption that we know *a priori* the correct reservoir model to use for the analysis. In many cases, however, the engineer may not have sufficient information from which to choose a reservoir model, especially in newly discovered fields. Fortunately, wells exhibit characteristic pressure responses that vary depending on the near-wellbore conditions and heterogeneities in the drainage area of the well. In fact, the basis of type-curve-analysis techniques is recognition of a curve shape representative of some reservoir model.

The underlying principle of type-curve analysis is that, if the test data exhibit the same shape as a type curve in all time regions (i.e., early, middle, and late times), the reservoir is of the same type as that characterized by the type curves. Unfortunately, this principle is not infallible. Different reservoir types sometimes exhibit essentially the same shape on a type-curve plot. In addition, both semilog and log-log plots of pressure/time data often are insensitive to pressure changes characteristic of a specific reservoir model. The pressure-derivative type curve is the most definitive of the type curves for identifying reservoir type. It can identify subtle but characteristic changes in slope that may be masked or that are not apparent on a pressure/time type curve. However, both the derivative and the pressure/time type curves are better than a semilog graph, such as a Horner graph, for identifying reservoir type. A type curve spans all time regions, while usually only the semilog straight line (middle-time region) is examined on the Horner plot. Furthermore, Horner analysis generally assumes a homogeneous reservoir, while a type curve reflects a particular reservoir type.

The best approach for identifying the correct reservoir model incorporates three major plotting techniques—the ordinary type curve, the derivative type curve, and a specialized graph for a test. On a specialized graph, properties can be deduced when a straight line develops during a certain time region. These graphs include the Horner plot for a homogeneous reservoir, a square-root-of-time plot for a well with a high-conductivity fracture, and a fourth-root-of-time plot for a well with a low-conductivity fracture (Chap. 6). When the reservoir type is identified correctly, all three plots will confirm or at least be consistent with the hypothesized reservoir type.

TABLE 4.15—PRESSURE-BUILDUP-TEST DATA, EXAMPLE 4.6

Recorded t (hours)	Corrected t (hours)	p_{wf} and p_{ws} (psia)	Recorded t (hours)	Corrected t (hours)	p_{wf} and p_{ws} (psia)
−0.01	—	4,171.58	1	0.975	4,631.46
0	—	4,171.58	1.5	1.475	4,684.62
0.01	—	4,171.58	2	1.975	4,712.41
0.02	—	4,171.58	3	2.975	4,741.42
0.03	0.005	4,177.74	4	3.975	4,757.71
0.04	0.015	4,189.73	6	5.975	4,777.77
0.06	0.035	4,212.6	8	7.975	4,790.91
0.08	0.055	4,234.17	10	9.975	4,800.73
0.1	0.075	4,254.56	15	14.975	4,817.95
0.15	0.125	4,301.01	20	19.975	4,829.81
0.2	0.175	4,341.86	30	29.975	4,846.15
0.3	0.275	4,410.04	40	39.975	4,857.51
0.4	0.375	4,464.11	60	59.975	4,873.22
0.6	0.575	4,542.64	72	71.975	4,880.17
0.8	0.775	4,595.12			

Now, we consider the specific application of the derivative type curve to establish near-well conditions and reservoir type. The following characteristics of the derivative curve are useful for determining the appropriate reservoir model for well-test analysis.

1. A maximum in the curve at early times indicates wellbore storage and skin. The greater the maximum, the more severely damaged the well. Conversely, the absence of a maximum suggests a stimulated well (i.e., acidized or hydraulically fractured).

2. A minimum in the curve at intermediate times indicates a deviation from homogeneous reservoir behavior (i.e., a reservoir heterogeneity). Examples include dual-porosity (naturally fractured, Chap. 7) or layered reservoirs.

3. Stabilization or flattening at later times indicates radial flow and corresponds to the semilog straight line or middle-time region on a Horner graph. Once we have identified this region on a derivative plot, we can use the Horner semilog analysis technique to estimate permeability and skin factor.

4. An upward or downward trend of the data at the end of the test indicates the presence of a reservoir boundary. An upward trend is characteristic of one or more boundaries having been encountered with the reservoir still open in at least one direction. An example of this situation is a single well centered in a rectangular-shaped reservoir. Similarly, a downward trend is an indicator of reservoir closure (i.e., that all boundaries, either no-flow or constant-pressure, have been felt). No-flow boundaries include sealing faults or interference effects from adjacent wells, while water influx from a large, active aquifer can be modeled as a constant-pressure boundary.

Reservoir Identification Worksheets (Appendix G) summarize these concepts and facilitate the identification process.

4.7 Systematic Analysis Procedures for Flow and Buildup Tests

In this section, we recommend a procedure for systematically analyzing pressure-drawdown and -buildup well tests. Although illustrated with homogeneous-acting reservoirs, these same analysis procedures, with slight modifications, are applicable to well tests with data distorted by wellbore-phase redistribution (Chap. 5), well tests from hydraulically fractured wells (Chap. 6), tests from wells completed in naturally fractured reservoirs (Chap. 7), and injectivity tests (Chap. 10).

One of the inherent problems with type-curve analysis is the difficulty in finding a unique match of the data. Because of the similar shapes of the type curves, the data often can be matched with more than one curve; consequently, more than one solution is possible. Note that simultaneously matching both the pressure-change and -derivative curves eliminates much of the ambiguity associated with conventional type-curve analysis that uses only pressure-change data. In addition, if a semilog straight line indicative of the radial flow or middle-time region is present and can be identified correctly, semilog analysis can provide the correct estimate of formation permeability. These observations suggest that we should not rely upon either type curves or semilog techniques as the only analysis method but should use the qualities of each method to complement and verify the results from the other.

On the basis of these observations, we have developed the following analysis procedure. Although presented in terms of the variables required for analyzing a test in a well producing a slightly compressible liquid, the method is applicable for gas-well-test analysis with the appropriate variables for compressible fluids, such as adjusted pressures and times.

Data Plots. The first step in well-test analysis is to prepare specific plots of the data. The following suggested plots should always be made; however, additional specialized plots may be required depending on the reservoir model. These specialized plots will be discussed in subsequent chapters.

1. Prepare log-log plots of pressure change and pressure derivative vs. elapsed time during the test. For drawdown tests, plot $\Delta p = (\bar{p} - p_{wf})$ and $t\,(dp_{wf}/dt)$ vs. t on log-log graph paper. For pressure-buildup tests, plot $\Delta p = p_{ws} - p_{wf}$ and $\Delta t_e\,(dp_{ws}/d\Delta t_e)$ vs. Δt_e (or Δt if boundary effects are present).

2. Prepare special plots of the data. Typically, we prepare a semilog plot of the data. For drawdown tests, plot p_{wf} vs. log t, while for pressure buildup tests, plot p_{ws} vs. the Horner time ratio, $(t_p + \Delta t)/\Delta t$. Depending on the reservoir model, additional special plots may be required. For example, when analyzing tests from hydraulically fractured wells, we would plot pressure vs. either square or fourth root of time functions.

Qualitative Type-Curve Analysis. The objectives of the preliminary or qualitative type-curve analysis are (1) to identify the appropriate reservoir model, (2) to obtain an initial match of the data by use of the type curves developed for the selected reservoir model, and (3) to identify any characteristic flow regimes that can be analyzed with special analysis techniques (e.g., the middle-time region can be analyzed with a semilog plotting technique).

1. The engineer often has reservoir data from which the reservoir model can be identified; however, when the type of reservoir model is unknown, use the reservoir identification procedure (Sec. 4.6) and the data plots to select a reservoir model.

2. Once the appropriate reservoir model has been identified, we select the correct type curves and proceed with a preliminary type-curve analysis.

First compare the log-log graph of test data to the pressure-change and pressure-derivative type curves to identify specific flow regimes. Specifically, we hope to identify the early-, intermediate-, and late-time regions on the plot. Early-time data, which have been distorted by wellbore storage, will exhibit a unit-slope line on the log-log plot. For pressure-buildup data distorted by wellbore phase redistribution (Chap. 5), we can sometimes see a gas hump develop following the unit-slope line. The shape of the intermediate-data will vary, depending on the reservoir model. For homogeneous-acting reservoirs, we hope to identify the radial-flow or middle-time region from which we can calculate permeability. For hydraulically fractured reservoirs, we may see several different flow regimes, such as bilinear and formation linear flow. Naturally fractured reservoirs also exhibit unique shapes at intermediate times. Late-time data, indicative of outer-boundary effects, will deviate from type curves developed for infinite-acting reservoirs. As we see in subsequent chapters, some type curves are generated for finite reservoirs.

Once we have identified tentative flow regimes, we attempt an initial match of the test data with the type curves. From this match, record pressure and time match points and the type-curve correlating parameter. This correlating parameter will vary, depending on the type-curve model. For example, the correlating parameter for the Gringarten and Bourdet type curves is $C_D e^{2s}$.

Semilog or Specialized Analysis. The purpose of the semilog or specialized analysis is to estimate formation properties with the specific characteristics of the flow regimes identified in the qualitative type-curve analysis. For example, a semilog analysis of the radial-flow or middle-time region for a homogeneous-acting reservoir provides estimates of permeability and skin factor. Similar analysis is possible for the pseudoradial-flow period in hydraulically fractured wells and for homogeneous-acting periods in naturally fractured reservoirs. In addition, we can estimate average pressures in the drainage area of the well from a pressure-buildup test.

Specialized analysis techniques provide additional estimates of well and reservoir properties. For example, hydraulic-fracture half-length can be estimated from a bilinear flow analysis in a finite-conductivity, vertically fractured well. Specialized analysis techniques are discussed in subsequent chapters.

Quantitative Type-Curve Analysis. The purpose of the quantitative type-curve analysis is either to confirm the results obtained from the semilog or specialized analysis or to estimate formation properties when no specialized analysis is possible.

1. If an estimate of permeability is available from the specialized analysis, we can confirm the analysis by precalculating a pressure match point. If this type-curve match with the precalculated pressure match point agrees with the qualitative type-curve analysis and confirms the middle-time region used in the semilog analysis, we have obtained a certain confidence in our analysis. If agreement cannot be achieved among the various analyses, we repeat the procedure and iterate until agreement is obtained.

2. If formation property estimates are not available from specialized analysis, we still attempt a type-curve analysis. Formation and wellbore properties are estimated with the pressure and time match points, and the definitions of the dimensionless parameters are used to develop the type curves. We remember, however, that unique solutions are difficult to obtain from type-curve analysis alone.

4.8 Well-Test-Analysis Worksheets

To facilitate application of the recommended well-test-analysis procedure, we developed worksheets. Appendix H presents examples of worksheets for analysis of homogeneous-acting reservoirs producing either oil or gas. The oil-well-test-analysis worksheet may be used for either buildup- or drawdown-test analysis in an oil well. Calculated results include permeability and skin factor. Although the worksheet does not have sections for estimating current reservoir pressure or distance to boundaries, such calculations can be appended easily. In later chapters, we present similar worksheets for other types of reservoirs.

The first section of the form summarizes well information and rock and fluid properties needed in the analysis. Pressure-gauge readings taken at the depth where the pressure bomb is hung in the well are corrected to the desired reservoir datum. For a single well, test pressures are usually corrected to the midpoint of the perforations by use of the wellbore fluid gradient. For a pressure-buildup test, the static wellbore gradient, measured as the pressure bomb is being pulled out of the wellbore at the conclusion of the buildup test, should be used. When tests from more than one well are being analyzed, test pressures are first corrected to the midpoint of the perforations in each well (using each well's respective wellbore gradient), then corrected to a common reservoir datum with the reservoir fluid gradient. If a fluid contact is crossed when correcting a particular well's test pressures to the reservoir datum, then both fluid gradients must be taken into account.

The section at the bottom of the first page leads us through an important preliminary phase of the analysis. We must determine when, if ever, the semilog straight line begins. This determination is made by preliminary (i.e., qualitative rather than quantitative) comparison of a log-log plot of pressure-change and -derivative data to a type curve. The onset of boundary effects, if encountered during the test, is also determined qualitatively.

When a semilog straight line is present, we perform a conventional semilog analysis, proceeding through the steps on the second page of the worksheet. The semilog analysis gives estimates of permeability and skin factor.

We then proceed with quantitative type-curve analysis, following the top section on the third page of the form. When semilog analysis is possible, the semilog analysis can be confirmed by type-curve analysis. Type-curve analysis may be somewhat ambiguous in that several curves may fit the test data equally well. We want to determine whether, among several possible data fits, there is one match that confirms the results of the semilog analysis, thereby increasing our confidence in the results and possibly inducing us to fine tune the semilog analysis slightly to improve the consistency of results.

To confirm the semilog analysis when a semilog straight line exists, we precalculate a pressure match point on the type curve, using the known permeability from the semilog analysis. We choose a convenient p_D from the type curve and calculate the corresponding Δp. Having fixed the match in the vertical direction, we then find the best fit by sliding the data plot only horizontally over the type curve, ensuring that the calculated Δp always overlays the corresponding p_D. This forced match should, of course, leave the flat portion of the test data plot aligned with $(t_D/C_D)\,p_D'=0.5$ on the derivative type curve. If the data are not so aligned, then the semilog analysis may be in error.

When semilog analysis is not possible because the middle-time region is not present in the test data, the pressure match point cannot be precalculated. In this case, we must find the best type-curve match possible and estimate permeability from the pressure match point. The derivative curve can be of considerable value in finding the best type-curve match as long as good-quality derivative data are available. The derivative curve is especially helpful when the derivative data clearly lie on a unit-slope line at early times and on a horizontal line in the middle-time region.

The final calculation on the third page of the worksheet is a production-time check. If the production time preceding a buildup test, $t_{p,\mathrm{actual}}$, does not exceed the duration of wellbore storage, $t_{p,\min}$, then the test is invalid and the analysis results are incorrect. The test must be rerun with a production period sufficiently long so that the test data are not all distorted by wellbore storage.

The last page of the worksheet summarizes the radii of investigation achieved during the production period before shut-in and at various times during the buildup test. A key consideration is that reservoir heterogeneities and boundaries apparent during a shut-in test are highly questionable if $r_i > r_{i,prod}$. If the drawdown before shut-in did not reach beyond a certain distance in the reservoir, $r_{i,prod}$, then the buildup test will not sample reservoir properties beyond that distance.

4.9 Chapter Summary

In this chapter, we have introduced the reader to the fundamentals of type-curve analysis. We began the chapter with a brief overview in Sec. 4.1. Then, in Sec. 4.2, we discussed the reasons for presenting solutions to the diffusivity equation in terms of dimensionless variables. We noted that it is much simpler to present a dimensionless solution in terms of a rather limited number of dimensionless parameters than to present solutions for all possible different combinations of physical variables.

In Sec. 4.3, we introduced and discussed the most commonly used type curves for radial flow in an infinite reservoir: the Ramey type curve, the McKinley type curve, the Gringarten type curve, and the Bourdet derivative type curve. All these type curves share the same basic assumptions—radial flow of a slightly compressible liquid produced at constant rate from a line-source well—with wellbore storage and skin, in an infinite-acting, homogeneous, reservoir.

In Sec. 4.3.1, we discussed the Ramey type curve. The Ramey type curve presents the dimensionless pressure, p_D, as a function of dimensionless time, t_D, dimensionless wellbore-storage coefficient, C_D, and skin factor, s. The biggest drawback to the use of the Ramey type curve is the difficulty in finding the best match because of the similarity of the shapes of the curves and the need to determine both time and pressure match points and two correlating parameters simultaneously. As a result, the Ramey type curve is seldom used for manual type-curve matching.

In Sec. 4.3.2, we introduced the McKinley type curve. This type curve is closely related to the Ramey type curve but is plotted in a nonstandard way. By fixing the value of one of the parameters of the solution at a typical value, McKinley made it much easier to find the best match of the data by requiring identification of only the pressure match point and a single correlating parameter.

In Sec. 4.3.3, we presented the Gringarten type curve. Although this type curve is essentially a replotting of the Ramey type curve, it combines the two correlating parameters of the Ramey type curve, C_D and s, into a single correlating parameter, C_De^{2s}. The dimensionless time t_D in the Ramey type curve is replaced by the time group t_D/C_D. With the Gringarten type curve, all wellbore-storage-dominated data fall on a single unit-slope line. As the transition to the middle-time region occurs, the data follow one of several curves characterized by the parameter C_De^{2s}. With this type curve, the engineer can be much more confident that the best match has been identified with the Ramey type curve.

In Sec. 4.3.4, we introduced the Bourdet derivative type curve. In contrast to the preceding three type curves, this type curve is a graph of the derivative of the dimensionless pressure with respect to the natural logarithm of dimensionless time, rather than the dimensionless pressure itself. By taking the logarithmic derivative, the pressure and pressure-derivative type curves can be plotted on a common scale. Thus, field-pressure and pressure-derivative data can be matched to the pressure and pressure-derivative type curves at the same time. The Bourdet derivative type curve has two interesting features: first, field-pressure-derivative data from the wellbore-storage-dominated period fall on a unit-slope line that coincides with the unit-slope line followed by the pressure data; second, data from the middle time region fall on a horizontal line at $(t_D/C_D)\,p_D'=\frac{1}{2}$.

The Bourdet derivative type curve is much more sensitive to the reservoir model than the Gringarten type curve and is often used in reservoirmodel identification. If all the available pressure and pressure-derivative data can be matched with a single set of pressure and pressure-derivative type curves, the well-test analyst is reassured that a reasonable interpretion model has been selected and that the analysis is correct.

In Sec. 4.4, we discussed the application of type curves developed for slightly compressible liquids to the analysis of gas-well tests. We presented compared analysis methods for gases in terms of pseudopressure/pseudotime or adjusted pressure/adjusted time with the pressure/time analysis for slightly compressible liquids. We pointed out that a pressure-buildup test in terms of pseudopressure/time can deviate substantially from the corresponding liquid type curve because of the changing wellbore-storage coefficient, while the same data analyzed in terms of pseudopressure/pseudotime (or adjusted pressure/adjusted time) follow the liquid type curve almost exactly. Finally, we illustrated the use of the gas-well-analysis methods by analyzing a pressure-buildup test in terms of adjusted pressure/adjusted time by use of the Gringarten/Bourdet type curves.

In Sec. 4.5, we noted that the recorded shut-in time is often inaccurate and must be corrected. We presented a simple graphical method for estimating both the time of shut-in and the flowing wellbore pressure at that time.

In Sec. 4.6, we discussed the use of pressure and pressure-derivative type curves in identifying the reservoir model. We discussed specific features of the derivative type curve that signal certain characteristics of the reservoir model, such as damage, heterogenous behavior, or the presence of boundaries.

In Sec. 4.7, we presented systematic analysis procedures for analyzing both flow and buildup tests. The first step is to make the necessary data plots. For virtually all tests, two plots should be made: a log-log plot of both pressure and pressure derivative for type-curve matching and a semilog plot of pressure vs. time for semilog analysis. The next step is a preliminary type-curve match. The primary objectives of this match are to identify the reservoir model and to identify any characteristic flow regimes that can be analyzed with straight-line methods. The third step is to apply straight-line analysis to the data identified in the preliminary type-curve match. The analysis is complete when a consistent interpretation with all applicable straight-line-analysis methods and the corresponding type-curve match is obtained.

In Sec. 4.8, we discussed the use of worksheets designed to facilitate application of the procedures presented in Sec. 4.7.

Exercises

1. Identify the middle time region and find the permeability, skin factor, and WBS coefficient for the drawdown test data in Exercise 5, Chapter 2, using type curve analysis.

2. Identify the middle time region and find the permeability, skin factor, and WBS coefficient for the drawdown test data in Exercise 6, Chapter 2, using type curve analysis.

3. Identify the middle time region and find the permeability, skin factor, and WBS coefficient for the drawdown test data in Exercise 38, Chapter 2, using type curve analysis.

4. Identify the middle time region and find the permeability, skin factor, and WBS coefficient for the drawdown test data in Exercise 39, Chapter 2, using type curve analysis.

5. Identify the middle time region and find the permeability and skin factor for the buildup test data in Exercise 16, Chapter 2, using type curve analysis.

6. Identify the middle time region and find the permeability and skin factor for the buildup test data in Exercise 17, Chapter 2, using type curve analysis. Can you use the Horner pseudoproducing time approximation? Why or why not?

7. (Difficult) Identify the middle time region and find the permeability and skin factor for the buildup test data in Exercise 18, Chapter 2, using type curve analysis. Can you use the Horner pseudoproducing time approximation? Why or why not?

8. (Difficult) Identify the middle time region and find the permeability and skin factor for the buildup test data in Exercise 19, Chapter 2, using type curve analysis. Can you use the Horner pseudoproducing time approximation? Why or why not?

9. (Difficult) Identify the middle time region and find the permeability and skin factor for the buildup test data in Exercise 20, Chapter 2, using type curve analysis. Can you use the Horner pseudoproducing time approximation? Why or why not?

10. Identify the middle time region and find the permeability and skin factor for the buildup test data in Exercise 5, Chapter 3, using type curve analysis.

11. Identify the middle time region and find the permeability and skin factor for the buildup test data in Exercise 6, Chapter 3, using type curve analysis.

12. Identify the reservoir model for the data in Exercise 16, Chapter 2, using the reservoir identification worksheets.

13. Identify the reservoir model for the data in Exercise 35, Chapter 2, using the reservoir identification worksheets. Find the permeability, skin factor, and WBS coefficient using type curve analysis.

14. Identify the reservoir model for the data in Exercise 36, Chapter 2, using the reservoir identification worksheets. Can you correctly identify the model? Why or why not? Find the permeability, skin factor, and WBS coefficient using type curve analysis.

15. Identify the reservoir model for the data in Exercise 37, Chapter 2, using the reservoir identification worksheets. Find the permeability, skin factor, and WBS coefficient using type curve analysis.

16. Find the permeability and skin factor for the buildup test in Exercise 40, Chapter 2, using type curve analysis.

References

1. McKinley, R.M.: "Wellbore Transmissivity From Afterflow-Dominated Pressure-Buildup Data," *JPT* (July 1971) 863; *Trans.*, AIME, **251.**
2. McKinley, R.M.: "Estimating Flow Efficiency From Afterflow-Distorted Pressure-Buildup Data," *JPT* (June 1974) 696.
3. Earlougher, R.C. Jr.: *Advances in Well Test Analysis*, Monograph Series, SPE, Richardson, Texas (1977) **5,** 74–89.
4. Gringarten, A.C. *et al.*: "A Comparison Between Different Skin and Wellbore Storage Type-Curves for Early-Time Transient Analysis," paper SPE 8205 presented at the 1979 SPE Annual Technical Conference and Exhibition, Las Vegas, Nevada, 23–26 September.
5. Agarwal, R.G., Al-Hussainy, R., and Ramey, H.J. Jr.: "An Investigation of Wellbore Storage and Skin Effect in Unsteady Liquid Flow—I. Analytical Treatment," *SPEJ* (September 1970) 279; *Trans.*, AIME **249.**
6. Ramey, H.J. Jr.: "Short-Time Well Test Data Interpretation in the Presence of Skin Effect and Wellbore Storage," *JPT* (January 1970) 97; *Trans.*, AIME **249.**
7. Agarwal, R.G.: "A New Method To Account for Producing Time Effects When Drawdown Type Curves are Used To Analyze Pressure-Buildup and Other Test Data," paper SPE 9289 presented at the 1980 SPE Annual Technical Conference and Exhibition, Dallas, 21–24 September.
8. Horner, D.R.: "Pressure Buildup in Wells," *Pressure Analysis Methods*, Reprint Series, SPE, Richardson, Texas (1067) **9,** 25–43.
9. Tiab, D. and Kumar, A.: "Detection and Location of Two Parallel Sealing Faults Around a Well," *JPT* (October1980) 1701.
10. Bourdet, D. *et al.*: "A New Set of Type Curves Simplifies Well Test Analysis," *World Oil* (May 1983) 95.
11. Bourdet, D., Ayoub, J.A., and Pirard, Y.M.: "Use of Pressure Derivative in Well-Test Interpretation," *SPEFE* (June 1989) 293.
12. Spivey, J.P. and Lee, W.J.: "A Comparison of the Use of Pseudotime and Normalized Time for Gas-Well Buildup Analysis in Various Geometries," paper SPE 15580 presented at the 1986 SPE Annual Technical Conference and Exhibition, New Orleans, 5–8 October.
13. Lee, W.J. and Holditch, S.A.: "Application of Pseudotime to Buildup-Test Analysis of Low-Permeability Gas Wells With Long-Duration Wellbore-Storage Distortion," *JPT* (December 1982) 2877.
14. Ramey, H.J. Jr.: "Practical Use of Modern Well Test Analysis," paper SPE 5878 presented at the 1976 Annual Technical Conference and Exhibition, New Orleans, 3–6 October .

Chapter 5
Analysis of Pressure-Buildup Tests Distorted by Phase Redistribution

5.1 Overview

In phase redistribution, the gas and liquid phases in the production string of a well that has been shut in at the surface segregate. The effects of phase redistribution during a buildup test cause the wellbore-pressure response to deviate from its expected behavior, thus complicating the analysis of the test data. We describe the wellbore-phase-redistribution phenomenon and present a mathematical model of phase redistribution. We present an approach to analyze pressure-buildup data graphically with phase redistribution and illustrate this approach with several examples. And finally, we identify situations where graphical analysis fails and automatic history matching of buildup data is necessary.

Phase redistribution is one example of a more general phenomenon of changing wellbore storage. Other examples include increasing wellbore storage resulting from changing gas compressibility in a gas well with a large drawdown before shut-in, discussed in Chap. 3, and a step increase or decrease in wellbore storage, discussed in Chap. 11 of Ref. 1.

5.2 Description of Phase Redistribution

The phenomenon of wellbore phase redistribution may occur during shut-in of a well with multiphase flow of gas and liquid.[2] Gravity forces cause the free-gas phase to rise through the liquid column and the liquid phase to flow downward. The gas bubbles near the bottom of the wellbore are at a pressure comparable with formation pressure. As the bubbles rise to the surface following shut-in, they cannot expand if the wellbore is in poor communication with the formation. Thus, a high-pressure gas column develops at the top of the wellbore and a column of liquid develops below the gas. Under these conditions, the pressure at the sandface is the summation of the pressure in the free-gas phase and the hydrostatic head of liquid below it.

In extreme cases, the bottomhole pressure (BHP) temporarily exceeds formation pressure and liquid in the wellbore flows back into the formation until equilibrium is reached and BHP declines to formation pressure. This process results in the characteristic hump in the BHP response during a buildup test (see **Figs. 5.1 and 5.2**). At early times during the test, the pressure increase will be abnormally large, complicating the determination of permeability, skin factor, and wellbore-storage coefficient.[2-4] The wellbore diagram in **Fig. 5.3** depicts the phase-distribution phenomenon.

A gas hump is most likely to occur during a buildup test in a well completed in a moderately permeable (10 to 100 md in oil or 0.1 to 1.0 md in gas) formation and having a large positive skin factor.[2] In low-permeability formations, wellbore pressure builds up slowly, so the hump usually does not develop. In such wells, the formation pressure is always higher than the pressure increase caused by bubble rise in the wellbore. Conversely, when permeability is high and skin damage is slight, fluid can flow back into the formation fast enough to prevent a significant increase in BHP resulting from rising gas bubbles.

Wells without packers tend to exhibit smaller gas humps from phase redistribution than wells with packers.[2] The BHP increase caused by rising gas bubbles is inversely related to the volume of gas initially in the wellbore. Because wells without packers generally have much more gas in the wellbore initially than do wells with packers, the pressure increase from bubble rise builds up more slowly and to a lesser degree than in wells with packers. Using a bottomhole shut-in device instead of shutting in the well at the surface can minimize severe phase-redistribution effects in buildup tests.

5.3 Phase-Redistribution Model

Wellbore phase redistribution is closely associated with wellbore storage.[3] The wellbore must store fluids of finite compressibility for phase redistribution to cause a wellbore-pressure increase. Although wellbore storage is a limiting case of phase redistribution, phase redistribution coupled with wellbore storage is a more complex condition than wellbore storage alone.

Fair[3] developed a mathematical model for phase redistribution. Because phase redistribution is associated with wellbore-storage effects, he developed the phase redistribution model using the classic wellbore-storage model that van Everdingen and Hurst[5] propose. The effects of wellbore storage can be described in dimensionless form in terms of the sandface flow rate after shut-in (afterflow), q_{sf}, the constant flow rate at the surface just before shut-in, q, the wellbore-storage coefficient, C_D, and the rate of wellbore-pressure change as

$$\frac{q_{sf}}{q} = 1 - C_D \frac{dp_{wD}}{dt_D}, \quad \text{(5.1)}$$

where dimensionless pressure, time, and wellbore-storage coefficient, respectively, are defined, respectively:

$$p_{wD} = \frac{kh(p_{ws} - p_{wf})}{141.2 qB\mu}, \quad \text{(5.2)}$$

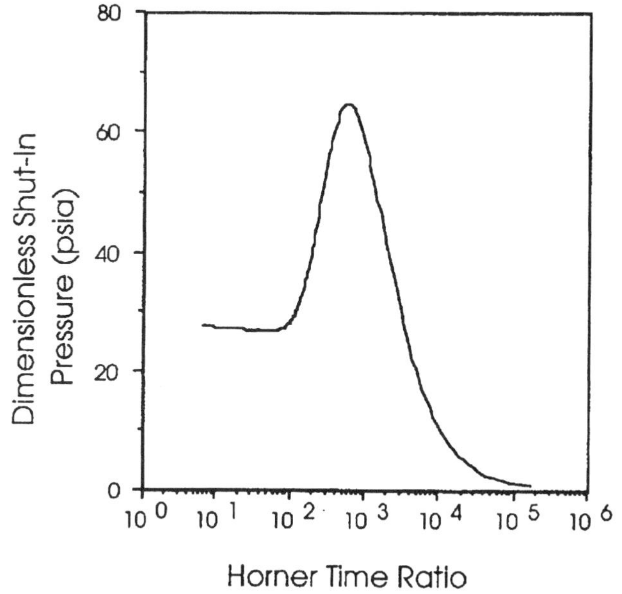

Fig. 5.1—Horner plot of Type 1 pressure response with wellbore storage and phase redistribution.

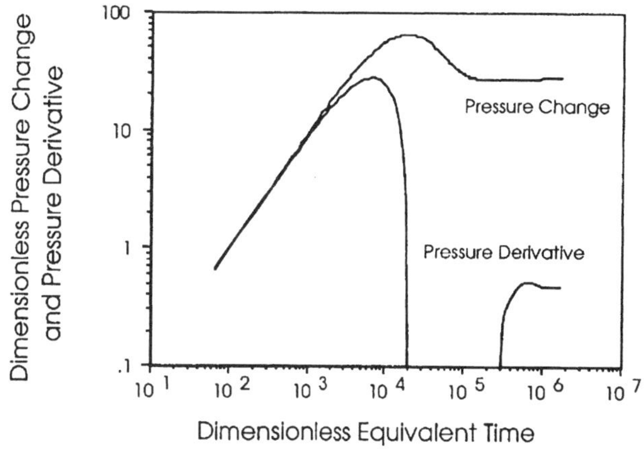

Fig. 5.2—Log-log plot of Type 1 pressure response with wellbore storage and phase redistribution.

$$t_D = \frac{0.0002637kt}{\phi\mu c_t r_w^2}, \quad \ldots\ldots (5.3)$$

and $$C_D = \frac{0.8936C}{\phi c_t h r_w^2}. \quad \ldots\ldots (5.4)$$

To account for the contribution of phase redistribution on the measured pressure change in the wellbore, Fair[3] included an additional term in Eq. 5.1,

$$\frac{q_{sf}}{q} = 1 - C_D\left(\frac{dp_{wD}}{dt_D} - \frac{dp_{\phi D}}{dt_D}\right), \quad \ldots\ldots (5.5a)$$

where $p_{\phi D}$ = the pressure component attributed to phase redistribution given in dimensionless form by

$$p_{\phi D} = \frac{khp_\phi}{141.2qB\mu}. \quad \ldots\ldots (5.5b)$$

The pressure drop, p_ϕ, is the change in pressure caused by phase redistribution and has the following properties. First, p_ϕ must be equal to zero at shut-in ($t=0$); i.e.,

$$\lim_{t\to 0} p_\phi = 0. \quad \ldots\ldots (5.5c)$$

Second, p_ϕ must increase monotonically to its maximum value, C_ϕ (a constant); i.e.,

$$\lim_{t\to \infty} p_\phi = C_\phi. \quad \ldots\ldots (5.5d)$$

During phase redistribution in the wellbore, gas bubbles or slugs rise through the liquid column. When the first bubble or slug reaches the surface after shut-in, the pressure in the wellbore must increase by some amount. This pressure increase causes a volume decrease and a density increase of all other gas bubbles or slugs. These effects cause a decrease in the rise velocity of the remaining gas, so the rate of pressure rise must decrease. The same situation holds for all subsequent bubbles or slugs reaching the surface after shut-in. Furthermore, because the initial sizes of gas bubbles and slugs in a wellbore may widely differ, their rise velocities will likewise vary widely. Thus, when the well is shut in, the phase-redistribution pressure rises quickly from its initial value of zero and slowly approaches its maximum value, C_ϕ. A function satisfying the physical constraints of the situation is

$$p_\phi = C_\phi\left(1 - e^{-t/\alpha}\right), \quad \ldots\ldots (5.5e)$$

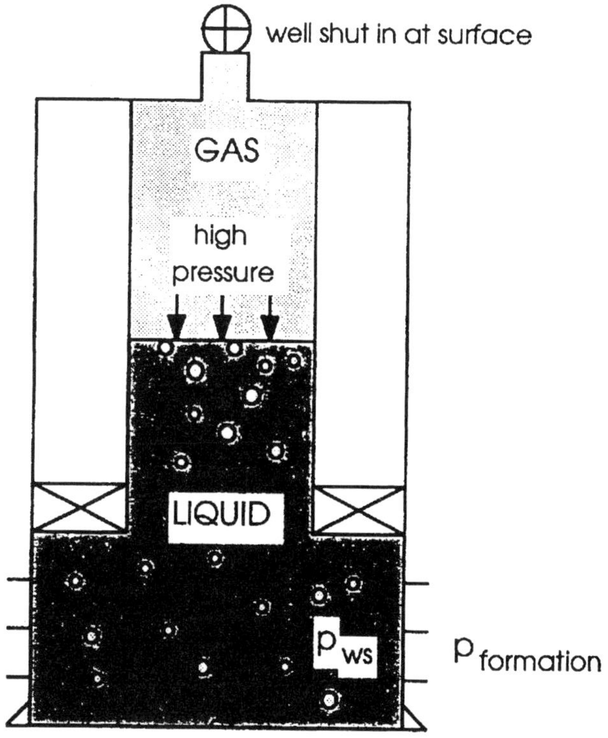

Fig. 5.3—Schematic of wellbore showing segregation of liquid and gas phases during pressure-buildup test.

where C_ϕ = maximum phase-redistribution pressure (a constant) and α = time at which 63% of the total phase-redistribution pressure change has occurred.

The total volume of gas in the wellbore remains constant because of the assumed incompressibility of the liquid in the wellbore. Thus, C_ϕ can be estimated from the following equation, which is valid if (1) the gas/oil ratio in the wellbore is constant, (2) temperature effects are neglected, (3) the liquid is incompressible and the gas is weightless and ideal-acting, and (4) wellbore pressure increases linearly with depth:

$$C_\phi = \frac{p_{gef} - p_{whf}}{\ln\left(p_{gef}/p_{whf}\right)}, \quad \ldots\ldots (5.5f)$$

where p_{gef} = flowing pressure at the point of gas entry into the wellbore and p_{whf} = flowing wellhead pressure.

Fair[3] defined a pseudo-wellbore-storage coefficient, C_{eD}, as

$$C_{eD} = C_D\left(1 - \frac{dp_{\phi D}}{dt_D} \Big/ \frac{dp_{wD}}{dt_D}\right), \quad \ldots\ldots (5.6)$$

which permits us to write Eq. 5.5 as

$$\frac{q_{sf}}{q} = 1 - C_{eD}\frac{dp_{wD}}{dt_D}. \qquad (5.7)$$

Eq. 5.7 has the same form as Eq. 5.1, implying that the wellbore storage case is a limiting form of phase redistribution. For the condition $dp_{\phi D}/dt_D \geqq 0$, phase redistribution causes an apparent reduction in the pseudo-wellbore-storage coefficient, C_{eD}. In this case, the true wellbore-storage coefficient, C_D, is the upper limit of C_{eD}. Furthermore, for the condition $\left(dp_{\phi D}/dt_D\right) \geqq \left(dp_{wD}/dt_D\right)$, the pseudo-wellbore-storage coefficient becomes negative, indicating flow from the wellbore back into the formation. This situation results in the gas hump typically associated with phase redistribution in buildup tests.

We define the dimensionless phase-redistribution pressure, $p_{\phi D}$, as an exponential function similar to Eq. 5.5e, given by

$$p_{\phi D} = C_{\phi D}\left(1 - e^{-t_D/\alpha_D}\right), \qquad (5.8)$$

where $C_{\phi D}$ = some maximum dimensionless phase-redistribution pressure change defined by

$$C_{\phi D} = \frac{khC_{\phi}}{141.2qB\mu}. \qquad (5.9)$$

The term α_D is the dimensionless time at which 63.2% of this maximum pressure change occurs and is defined by

$$\alpha_D = \frac{0.0002637k\alpha}{\phi\mu c_t r_w^2}. \qquad (5.10)$$

We should note that Fair's[3] exponential phase-redistribution model is intuitive and is consistent with only limited laboratory data; therefore, it may not model all field conditions accurately.

If we define an apparent wellbore-storage coefficient, C_{aD}, as

$$\frac{1}{C_{aD}} = \frac{1}{C_D} + \frac{C_{\phi D}}{\alpha_D}, \qquad (5.11)$$

the early-time solution given by Fair is

$$p_{wD} = \frac{t_D}{C_{aD}}. \qquad (5.12)$$

Eq. 5.12 has the same form as the solution for the early-time pressure response, characterized by a unit-slope line on a logarithmic plot of test data, in a line-source well with wellbore storage and skin. This is given by Agarwal *et al.*[6] as

$$p_{wD} = \frac{t_D}{C_D}. \qquad (5.13)$$

At early times, if phase-redistribution effects are negligible ($C_{\phi D} = 0$), the pressure behavior represents true wellbore storage ($C_{aD} = C_D$). If phase redistribution affects the pressure response ($C_{\phi D} > 0$), the apparent wellbore-storage coefficient is less than the true wellbore-storage coefficient ($C_{aD} < C_D$). This effect appears as a deviation of the early-time data from the theoretical unit-slope line. The dimensional apparent wellbore-storage coefficient, C_a, is found with any point ($\Delta t_e, \Delta p$) on the unit-slope line or its extrapolation,

$$C_a = \frac{qB}{24}\left(\frac{\Delta t_e}{\Delta p}\right)_{USL}, \qquad (5.14)$$

and the dimensionless apparent wellbore-storage coefficient, C_{aD}, is found from

$$C_{aD} = \frac{0.0372qB}{\phi hc_t r_w^2}\left(\frac{\Delta t_e}{\Delta p}\right)_{USL}. \qquad (5.14a)$$

The pressure solution in the radial-flow or middle-time region in which the true semilog straight line occurs is[7]

$$p_{wD} = 1.151(\log t_D + 0.351) + s. \qquad (5.15)$$

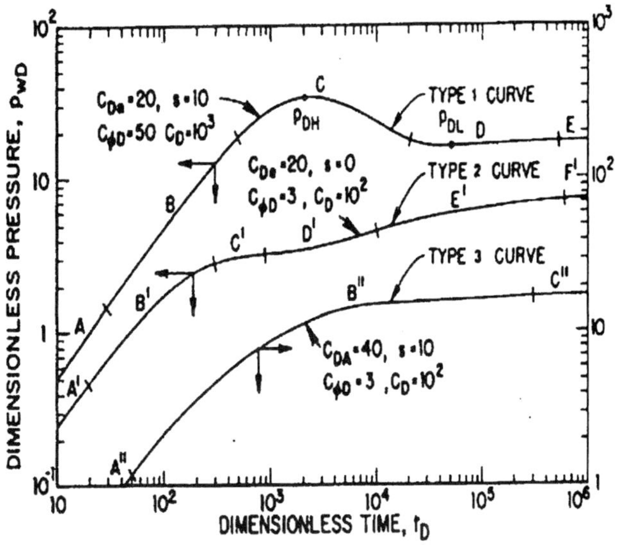

Fig. 5.4—Types of pressure responses resulting from phase redistribution (after Thompson *et al.*[7]).

When phase redistribution affects the measured pressure response during a buildup test, a log-log plot of wellbore pressure as a function of shut-in time displays one of three typical shapes[7] (**Fig. 5.4**). The Type 1 curve represents the greatest degree of phase redistribution as wellbore pressure increases above formation pressure and forces fluid from the wellbore back into the formation. Region A is the unit-slope line of Eq. 5.12, followed by a transitional period, Region B, where the curve falls off from the unit-slope line. Region C is the gas hump typically associated with phase redistribution as free gas rises in the wellbore. The maximum dimensionless pressure change, p_{DH}, occurs in this region. Region D has a local minimum dimensionless pressure change, p_{DL}, and is a transitional period between the gas hump and the semilog straight line in Region E.

The Type 2 curve represents a smaller degree of phase redistribution than the Type 1 curve. Region A′ is the unit-slope line (Eq. 5.12) followed by a transitional period denoted by Region B′. The curve flattens in Region C′, then begins to rise in the transitional Regions D′ and E′. The curve is concave upward throughout Region D′ and concave downward throughout Region E′; therefore, an inflection point separates the two regions. Region F′ is the semilog-straight-line region.

The Type 3 curve displays the smallest effects of phase redistribution on the pressure response. The curve has a shape similar to those of the wellbore storage and skin solutions[6] with no phase-redistribution effects. The unit-slope line in Region A″ is followed by a transitional period, Region B″, and the semilog-straight-line flow regime in Region C″. While a Type 3 curve shows no obvious effects of phase redistribution, it cannot be fit accurately with a conventional wellbore-storage and skin type curve.

Conditions on the parameters s, $C_{\phi D}$, C_{aD}, and C_D determine which form the pressure response takes during a buildup test.[7] The Type 1 curve appears when the conditions $s \leqq C_{\phi D}/3$, $C_{\phi D} \geqq 10$, and $C_{aD} \leqq C_D/5$ are satisfied. The pressure response will be of Type 2 if either set of the following conditions holds: $C_{\phi D} \leqq s \leqq 5C_{\phi D}$ and $C_{aD} \leqq C_D/5$ or $C_{\phi D} \leqq 5$, $C_D \geqq 10^4$ and $C_{aD} \leqq C_D/5$. If the conditions for a Type 1 or Type 2 response are not met, the pressure response will be of Type 3.

Figs. 5.1 and 5.2 and 5.5 through 5.8 show examples of Types 1, 2, and 3 pressure responses during buildup tests. The semilog plots show shut-in pressure as a function of the Horner time ratio. The log-log plots show dimensionless pressure change, given by Eq. 5.2, and dimensionless pressure derivative as functions of dimensionless time, given by Eq. 5.3. As wellbore-storage and phase-redistribution effects decay over time, the pressure derivative flattens at a value of 0.5, indicating the development of the true semilog straight line. Thompson *et al.*[7] developed a set of rules to predict the start of the semilog straight line, t_{sl}, on a Horner plot of buildup data.

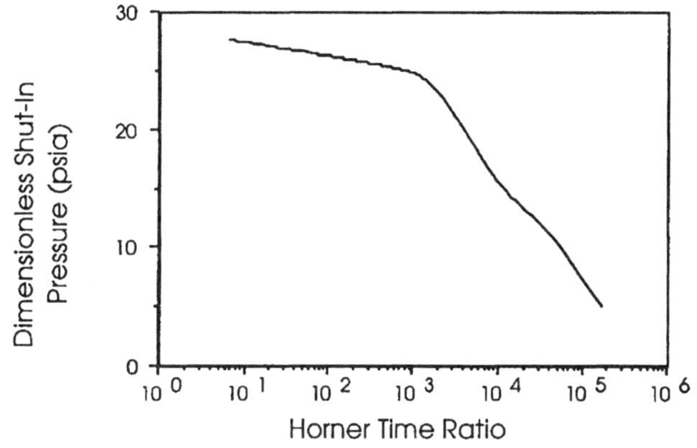

Fig. 5.5—Horner plot of Type 2 pressure response with wellbore storage and phase redistribution.

Derived from simulated solutions for various combinations of parameters, these rules may require *a priori* estimates of permeability and skin factor, which themselves depend on the semilog straight line. Therefore, the major use of these rules is as guidelines to confirm results from other methods of analysis. We illustrate their use in subsequent examples.

For a Type 1 pressure response, the shut-in time at the start of the semilog straight line is estimated by[7]

$$t_{sl} \approx 50\left(1 - \frac{s}{C_{\phi D}}\right)t_L , \qquad (5.16)$$

where t_{sl} is in hours and t_L = shut-in time (in hours) when the local minimum pressure, p_{DL}, occurs in Region D. In many cases, $C_{\phi D} \approx p_{DH}$ when the pressure response is of Type 1. Shut-in time, t_{sl}, can then be approximated with field data

$$t_{sl} \approx 50\left(1 - \frac{s}{p_{DH}}\right)t_L$$

$$\approx \left[1 - \frac{141.2qB\mu s}{kh\left(p_{ws,H} - p_{wf}\right)}\right]t_L , \qquad (5.17)$$

where p_{DH} = dimensionless pressure (defined by Eq. 5.2) at the top of the gas hump with the maximum pressure change at the hump, $(p_{ws,H} - p_{wf})$, replacing $(p_{ws} - p_{wf})$. The approximation given by Eq. 5.17 is most accurate when $C_D \geqq 10^4$. Note that Eq. 5.17 may overestimate the time when the middle-time region begins.[7]

For a Type 2 pressure response, the shut-in time at the start of the semilog straight line is estimated by

$$t_{sl} = 75\ t_{inf} , \qquad (5.18)$$

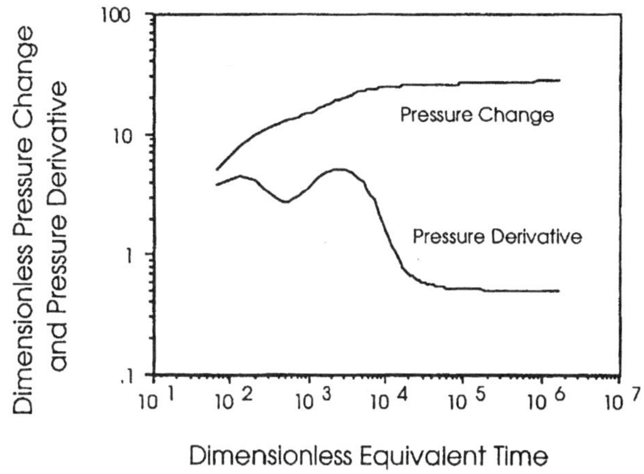

Fig. 5.6—Log-log plot of Type 2 pressure response with wellbore storage and phase redistribution.

where t_{inf} = time (in hours) when the inflection point separating Regions D′ and E′ occurs. Eq. 5.3 gives the dimensionless form of t_{inf} with t_{inf} replacing t.

For a Type 3 pressure response, the shut-in time at the start of the semilog straight line is estimated by

$$t_{sl} = 10^6\ t_e \qquad (5.19)$$

for the condition $s > 5C_{\phi D}$, $C_D \leqq 10^3$, and $C_D \geqq 2C_{aD}$, where t_e = time at which the unit-slope line ends on a log-log plot of Δp vs. Δt_e.

The shut-in time, t_{sl}, is estimated by

$$t_{sl} = 10^3\ t_e \qquad (5.20)$$

for the condition $s \leqq 3C_{\phi D}$, $C_{\phi D} \geqq 1$, and $C_D \leqq 1.5C_{aD}$. Eqs. 5.19 and 5.20 can be regarded as six- and three-log-cycle rules, respectively, similar to the 1.5-log-cycle rule for the wellbore-storage case.[9]

The Thompson *et al.*[7] rules for estimating the beginning of the middle-time region can result in substantial inaccuracies, particularly for a Type 3 pressure response. For this reason, these rules should be used only as very rough guidelines to validate results from the semilog and type-curve analyses.

5.4 Analysis Procedure

The basis of pressure-buildup test analysis is the correct identification of the semilog straight line whose slope is related to effective permeability. Phase-redistribution effects, however, can delay or even completely suppress the onset of the radial-flow or middle-time period during a buildup test. As a result, neither conventional

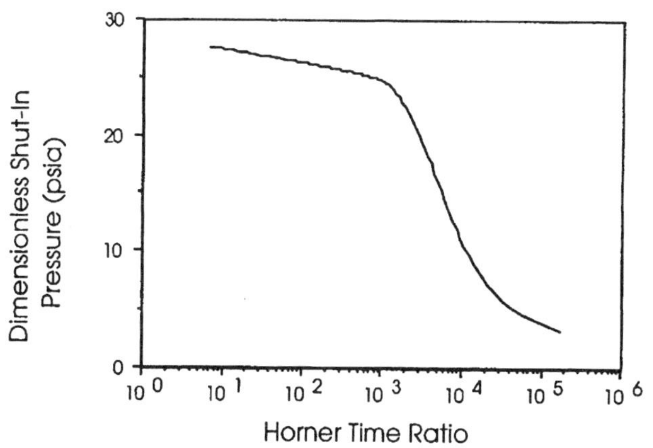

Fig. 5.7—Horner plot of Type 3 pressure response with wellbore storage and phase redistribution.

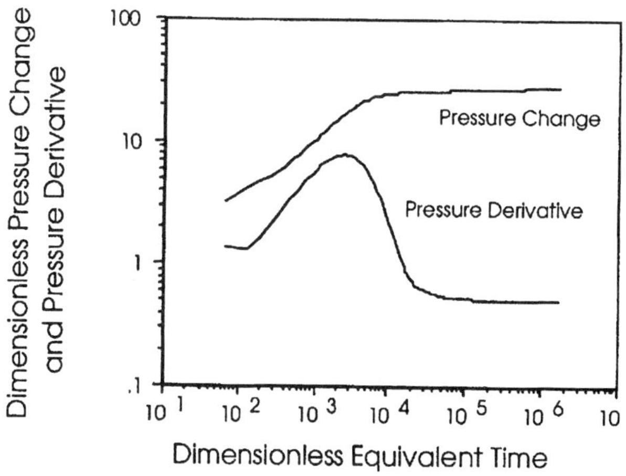

Fig. 5.8—Log-log plot of Type 3 pressure response with wellbore storage and phase redistribution.

Horner semilog nor type-curve analyses may be possible when phase redistribution distorts buildup data. Specifically, data plots do not fit conventional type curves until phase-redistribution effects cease or are negligible, and apparent or false straight lines with slopes unrelated to formation permeability may appear before the true semilog straight line on a Horner plot. If phase-redistribution effects are sufficiently pronounced, the pressure transient may encounter boundaries before the semilog straight line appears on a Horner plot.

Fair[3] developed type curves on the basis of his mathematical model for analyzing pressure-buildup data distorted by wellbore phase redistribution. His type curves, however, are functions of several parameters, and unambiguous type-curve matches are difficult to obtain unless most of the parameters are known. Unfortunately, obtaining estimates of many of the parameters is an objective of the well test.

As we discussed in Chap. 4, the shapes of log-log and semilog plots of buildup test data are useful in identifying characteristic flow regimes. The pressure-derivative curve is particularly helpful in identifying the presence of phase-redistribution effects and in determining the presence and correct position of the semilog straight line. If the semilog straight line is present, the middle-time region begins when the derivative plot becomes constant at a dimensionless value of 0.5. Any apparent straight line on the Horner plot before this time does not represent the correct semilog straight line. The middle-time region most likely will not occur with a Type 1 or 2 pressure response. The Type 1 pressure response is identified readily by its characteristic gas hump. The Type 2 and especially Type 3 responses, however, are not so easily identified from initial examination of the data plots, so the presence of phase redistribution during the buildup test may not be apparent.

A qualitative type-curve analysis by use of the Bourdet[8] type curves can help verify the presence of phase redistribution, particularly when the pressure response is Type 2 or 3, and can confirm the presence and correct position of the semilog straight line. The pressure-change and -derivative data plots are matched with the Bourdet type curves by aligning the horizontal portion of the test data derivative plot with the corresponding horizontal portion of the derivative type curve at a dimensionless value of 0.5 and sliding the data plots horizontally to obtain the best fit of the pressure-change curve to the type curves. When phase redistribution has affected the pressure response, a log-log plot of the early-time data will not exhibit the unit-slope line characteristic of a constant wellbore-storage coefficient and, thus, cannot be matched on conventional type curves. The match must be made on the basis of middle-time data after phase-redistribution effects have ceased distorting the pressure response.

When the correct semilog straight line indicative of the middle-time region exists, formation permeability and skin factor can be found with conventional Horner semilog analysis. Wellbore-storage and phase-redistribution effects have ceased by the time the semilog straight line develops, so a Horner semilog analysis can be used to find the formation permeability and skin factor from the slope, m, of the semilog straight line.

Although the correct permeability and skin factor can be found when the correct semilog straight line exists, the wellbore-storage coefficient determined from the unit-slope portion of the pressure-change curve depends on whether phase redistribution affects the early-time data. If phase-redistribution effects are not present, the coefficient obtained from the unit-slope line is the true wellbore-storage coefficient, C. If phase-redistribution effects are present, this estimate represents the apparent wellbore-storage coefficient, C_a. In the latter case, C must be found independently. When phase redistribution affects the pressure response, the true wellbore-storage coefficient must be estimated from wellbore characteristics. During phase redistribution, the liquid level in the wellbore is falling. The wellbore-storage coefficient for a changing liquid level and with constant tubing pressure is given by

$$C = 25.65\frac{A_{wb}}{\rho_{wb}}, \qquad (5.21)$$

where A_{wb} is the cross-sectional area in square feet of the wellbore and ρ_{wb} is the density in pounds mass per cubic foot of the liquid in the wellbore. Because tubing pressure is not constant in wells undergoing phase redistribution, C calculated from Eq. 5.21 and C_D calculated from Eq. 5.4 are, at best, approximations.

The apparent wellbore-storage coefficient, C_a, is obtained from the unit-slope line or its extrapolation if phase redistribution causes the early-time data to deviate from the line. For any point $(\Delta t_e,\Delta p)$ on the unit-slope line, the coefficient is given by

$$C_a = \frac{qB}{24}\left(\frac{\Delta t_e}{\Delta p}\right)_{USL} \qquad (5.22)$$

and

$$C_{aD} = \frac{0.0372qB}{\phi\mu c_t r_w^2}\left(\frac{\Delta t_e}{\Delta p}\right)_{USL}. \qquad (5.23)$$

C_{aD} also can be obtained with the time match point in type-curve analysis. The smaller the ratio of C_{aD} to C_D, the more pronounced are the phase-redistribution effects.

The maximum pressure change owing to phase redistribution, C_ϕ, can be estimated from field data.[2,3] Under the assumptions of a constant liquid/gas ratio in the wellbore, negligible temperature effects, incompressible liquid and ideal gas of negligible weight in the wellbore, and a linear increase in wellbore pressure with depth, C_ϕ is estimated by

$$C_\phi = \frac{p_{gef} - p_{whf}}{\ln\left(p_{gef}/p_{whf}\right)}, \qquad (5.5f)$$

where p_{gef} = flowing pressure at the point of gas entry into the wellbore, psia, and p_{whf} = flowing wellhead pressure just before shut-in, psia. Although C_ϕ is not required for reservoir characterization, the estimate from Eq. 5.5f can be used as a good initial guess of C_ϕ if automatic history matching[4,9] with Fair's[3] model is used as another means of obtaining k, s, and C.

The Bourdet type curves can be used in a quantitative type-curve analysis to confirm the semilog results and the estimate of C. The pressure-change data are force-matched with the type curves, ensuring that the pressure-derivative data of the semilog straight line region are aligned with the horizontal portion of the derivative type curve at a dimensionless value of 0.5. The forced match is done with a convenient value of Δp and the corresponding dimensionless pressure calculated with k from the semilog analysis. Because of phase-redistribution effects, we do not expect early-time data to match any type curve. Only middle-time data (after phase redistribution and wellbore-storage effects have disappeared and before the onset of boundary effects) can be used to match the pressure-change and pressure-derivative curves to the type curves.

In moving the data plot horizontally to obtain the best match, the early pressure-change data are fitted to the early portion of the type curves. If the early data fall on a unit-slope line, those data points should be forced to match the unit-slope portion of the type curves. Even if no unit-slope data points are present, the early data should still be aligned with the type curves. Observing this constraint in matching the data will usually give the correct C_{aD} from the time match point.

Because phase-redistribution effects may distort the pressure response for much or all of a buildup test, possibly resulting in little or no middle-time data, a unique match often is difficult to obtain. Under these conditions, automatic history matching[4,9] of the buildup data is a viable alternative analysis technique. Automatic history matching generally gives the most accurate estimates of parameters even when semilog analysis can be performed and, when available, should be used in analyzing pressure-buildup data.

As a final confirmation of the results, the guidelines established by Thompson *et al.*[7] can be applied to estimate the beginning of the semilog straight line, t_{sl}. The estimate of t_{sl} should be consistent with the beginning of the semilog straight line from the semilog and type-curve analysis. Any inconsistencies in the results indicate that the semilog analysis and type-curve matching should be repeated to refine the estimates.

We recommend the following step-by-step procedure to analyze pressure-buildup tests when the pressure response is distorted by wellbore-phase redistribution. Although presented in terms of pressure and time variables for slightly compressible fluids, the same procedure is applicable to gas-well buildup tests with adjusted variables.

1. Construct the following plots of the pressure-buildup test data: Prepare a semilog plot of shut-in pressure, p_{ws}, as a function of the Horner time ratio, $(t_p + \Delta t)/\Delta t$, and a log-log plot of pressure change, $\Delta p = p_{ws} - p_{wf}$, and pressure derivative, $\Delta t_e \Delta p'$, as functions of Agarwal's[10] equivalent time, $\Delta t_e = \Delta t/(1 + \Delta t/t_p)$.

2. Perform a preliminary or qualitative type-curve analysis. The purpose of this analysis is to determine the existence of phase redistribution, the type of pressure response (Type 1, 2, or 3),[7] and boundary effects. In addition, the preliminary type-curve analysis can determine whether the semilog straight line (indicative of the middle-time region) is present and, if possible, identify its position. The horizontal portion of the pressure-derivative curve is aligned with the horizontal portion of the Bourdet[8] type curves for radial flow in a homogeneous reservoir. The type-curve match is obtained by sliding the pressure-change- and pressure-derivative-data curves horizontally. Early-time pressure-change data lying on a unit-slope line are forced to match the unit-slope portion of the type curves. Even if no unit-slope data are present, the early data are still aligned with the early portion of the type curves. The middle-time region corresponds to the data points lying on the horizontal portion of the derivative type curve.

3. If the middle-time region is present, perform a Horner semilog analysis to estimate formation permeability and skin factor. Formation permeability and skin factor can be estimated from the slope of this line (Chap. 2).

4. Perform a quantitative type-curve analysis to confirm the semilog analysis. If a middle-time region is present, then use the permeability determined from a Horner semilog analysis to precalculate a pressure match point $(\Delta p, p_D)$. Choose a convenient but arbitrary value of p_D, and calculate Δp:

$$(\Delta p)_{MP} = \frac{141.2 qB\mu}{kh}(p_D)_{MP}. \quad (5.24)$$

The type-curve match is found by aligning Δp and p_D from the pressure match point and sliding the pressure-change- and pressure-derivative-data curves horizontally. As in the qualitative type-curve match, early pressure-change data are forced to align with the early portion of the type curves. In particular, data points on the unit-slope line are forced to fit the unit-slope line of the type curves. Once the match is found, select a time match point $(\Delta t_e, t_D/C_D)_{MP}$ and calculate the dimensionless apparent wellbore-storage coefficient, C_{aD}.

$$C_{aD} = \frac{0.0002637k}{\phi r_w^2 c_t \mu_o}\left(\frac{\Delta t_e}{t_D/C_D}\right)_{MP}. \quad (5.25)$$

C_{aD} also can be calculated from a point on the unit-slope line $(\Delta p, \Delta t_e)_{USL}$ by

$$C_{aD} = \frac{0.0372 qB}{\phi \mu c_t r_w^2}\left(\frac{\Delta t_e}{\Delta p}\right)_{USL}. \quad (5.23)$$

Calculate the skin factor with C_{aD} and the type-curve correlating parameter, $C_D e^{2s}$:

$$s = 0.5 \ln\left(\frac{C_D e^{2s}}{C_{aD}}\right). \quad (5.26)$$

This skin factor should agree with the value obtained from the Horner semilog analysis. Determine C_D from wellbore characteristics and

$$C_D = \frac{0.8936C}{\phi c_t h r_w^2}, \quad (5.4)$$

where the wellbore-storage coefficient, C, is estimated with

$$C = 25.65\frac{A_{wb}}{\rho_{wb}}. \quad (5.21)$$

The smaller the ratio of C_{aD} to C_D, the more pronounced is the effect of phase redistribution on the pressure-buildup response.

5. Confirm the results with the Thompson *et al.* guidelines (Eqs. 5.16 to 5.20) for estimating the beginning of the semilog straight line, t_{sl}. The estimate of t_{sl} should be consistent with the beginning of the semilog straight line from the semilog and type-curve analyses; otherwise, the semilog analysis and type-curve matching should be repeated to refine the estimates.

The following examples, which were generated with a numerical simulator, illustrate the correct procedure for buildup-test analysis when the pressure response is distorted by wellbore phase redistribution. Results agree closely with simulation results for the same conditions, confirming the procedure.

TABLE 5.1—PRESSURE-BUILDUP TEST DATA, EXAMPLE 5.1

Shut-In Time, Δt (hours)	Pressure, p_{ws} (psia)	Shut-In Time, Δt (hours)	Pressure, p_{ws} (psia)
0.000417	1,009.06	3.4788	1,067.52
0.00417	1,078.31	4.0478	1,067.99
0.02081	1,216.79	4.6865	1,068.45
0.03741	1,232.73	5.5705	1,069.00
0.06638	1,195.64	6.3189	1,069.40
0.1076	1,142.90	6.9607	1,069.71
0.1608	1,104.46	7.6449	1,070.02
0.2258	1,082.59	8.4844	1,070.36
0.3660	1,067.75	9.0764	1,070.58
0.5232	1,064.54	9.6780	1,070.79
0.6770	1,064.02	12.00	1,071.49
0.8835	1,064.14	25.00	1,073.94
1.1021	1,064.48	40.00	1,075.51
1.3143	1,064.84	80.00	1,077.84
1.6707	1,065.42	120.0	1,079.20
2.1495	1,066.10	144.0	1,079.81
2.8494	1,066.92	168.0	1,080.33

Example 5.1—Buildup Test With Wellbore Phase Redistribution, Type 1 Pressure Response. A pressure-buildup test was run on an oil well completed with 2.875-in. tubing inside 7.625-in. casing. **Table 5.1** gives pressure and time data. Other known data are summarized next. Estimate formation permeability, skin factor, and the wellbore-storage coefficient.

h = 100 ft
r_e = 3,000 ft
μ_o = 0.8 cp
p_i = 2,000 psia
ϕ = 0.10
q = 250 STB/D
r_w = 0.365 ft
B_o = 1.2 RB/STB
p_{wf} = 1,000 psia
ρ_{wb} = 48 lbm/ft^3
t_p = 14,400 hours
r_t = 0.1017 ft
c_t = 15.0×10^{-6} psia^{-1}

Solution. The following illustrates the application of the recommended analysis procedure.

Construct Data Plots. **Fig. 5.9** shows a Horner semilog plot, while **Fig. 5.10** plots pressure-change and pressure-derivative data. **Table 5.2** gives plotting functions for these figures.

Qualitative Type-Curve Analysis. A gas hump is evident in each of the plots, suggesting that phase redistribution affected the pressure response during the test. In particular, the shapes of the pressure-change

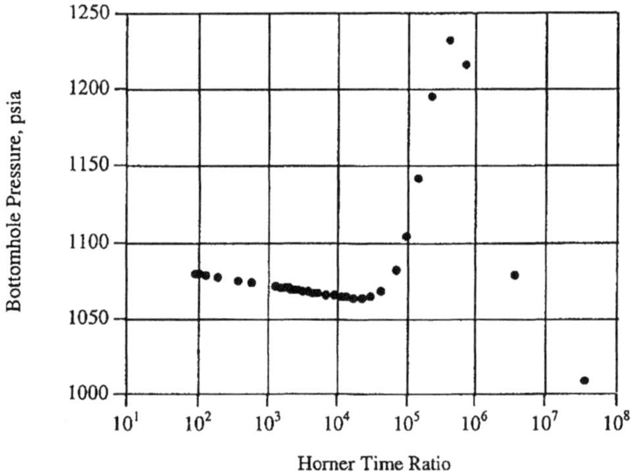

Fig. 5.9—Horner plot of buildup test data, Example 5.1.

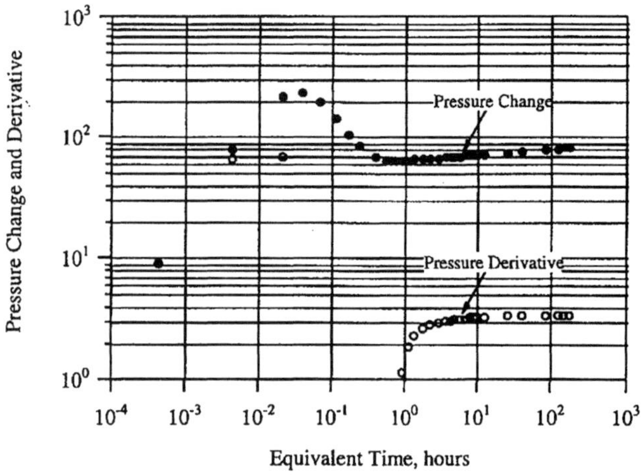

Fig. 5.10—Log-log plot of pressure-change and -derivative data, Example 5.1.

and pressure-derivative curves in Fig. 5.10 are characteristic of a Type 1 pressure response. In addition, the shape of the pressure-derivative curve at the end of the test suggests that the middle-time region is present; however, we should confirm its presence with a qualitative type-curve analysis before attempting a semilog analysis.

1. Qualitative type-curve analysis (**Fig. 5.11**) is performed by matching the pressure-change and -derivative data with the Bourdet[8] type curves for radial flow in a homogeneous reservoir. The match is obtained by aligning the horizontal portion of the pressure-derivative curve with the corresponding horizontal portion of the type curves at a dimensionless pressure of 0.5. The data are moved horizontally to find the best fit with the type curves.

2. The first two early-time data points appear to lie on a unit-slope line and are aligned with the unit-slope portion of the type curves. The best fit is obtained with the type curve defined by the correlating parameter $C_D e^{2s} = 10^4$. The last 12 data points on the derivative curve lie on the horizontal portion of the derivative type curve, suggesting that the middle-time region (semilog straight line) begins at a shut-in time of approximately 6.96 hours (see Table 5.1).

Semilog Analysis.

1. **Fig. 5.12** shows the Horner plot. The semilog straight line is drawn through the last 12 data points, giving a slope of $m = 7.70$ psi/cycle.

2. The effective permeability to oil is

$$k = \frac{162.6 q_o B_o \mu_o}{mh} = \frac{(162.6)(250)(1.2)(0.8)}{(7.70)(100)} = 50.7 \text{ md.}$$

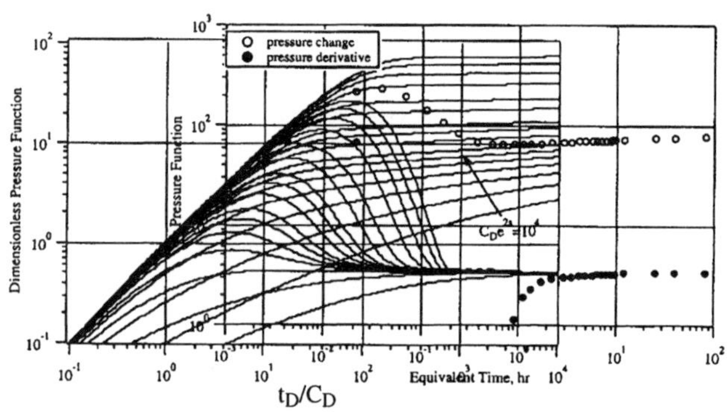

Fig. 5.11—Qualitative type-curve match, Example 5.1.

3. The skin factor is estimated to be

$$s = 1.151\left[\left(\frac{p_{1\text{hr}} - p_{wf}}{m}\right) - \log\left(\frac{k}{\phi\mu_o c_t r_w^2}\right) + 3.23\right]$$

$$= 1.151\left\{\left(\frac{1,063.22 - 1,000}{7.70}\right)\right.$$

TABLE 5.2—SEMILOG AND TYPE-CURVE PLOTTING FUNCTIONS, EXAMPLE 5.1

Δt_e (hours)	Δp, (psi)	$\Delta t_e\,\Delta p'$ (psi)	Horner Time Ratio	Δt_e (hours)	Δp, (psi)	$\Delta t_e\,\Delta p'$ (psi)	Horner Time Ratio
0.000417	9.06		34,569,265	3.4780	67.52	3.05508	4,140
0.00417	78.31	65.9123	3,456,926	4.0466	67.99	3.09311	3,559
0.02081	216.79	67.1860	692,141	4.6850	68.45	3.12853	3,074
0.03741	232.73	−19.8932	384,940	5.5683	69.00	3.16909	2,586
0.06638	195.64	−100.450	216,940	6.3162	69.40	3.19734	2,280
0.1076	142.90	−108.130	133,864	6.9574	69.71	3.21710	2,070
0.1608	104.46	−79.6891	89,556	7.6409	70.02	3.23492	1,885
0.2258	82.59	−48.9904	63,777	8.4794	70.36	3.25340	1,698
0.3660	67.75	−15.7708	39,340	9.0707	70.58	3.26381	1,588
0.5232	64.54	−4.13181	27,525	9.6715	70.79	3.27339	1,489
0.6770	64.02	−0.55966	21,270	11.99	71.49	3.30079	1,201
0.8834	64.14	1.13528	16,300	24.96	73.94	3.35258	577
1.1020	64.48	1.89826	13,067	39.89	75.51	3.36759	361
1.3141	64.84	2.30500	10,958	79.56	77.84	3.37773	181
1.6705	65.42	2.66656	8,620	119.0	79.20	3.38155	121
2.1492	66.10	2.87676	6,700	142.6	79.81	3.38266	101

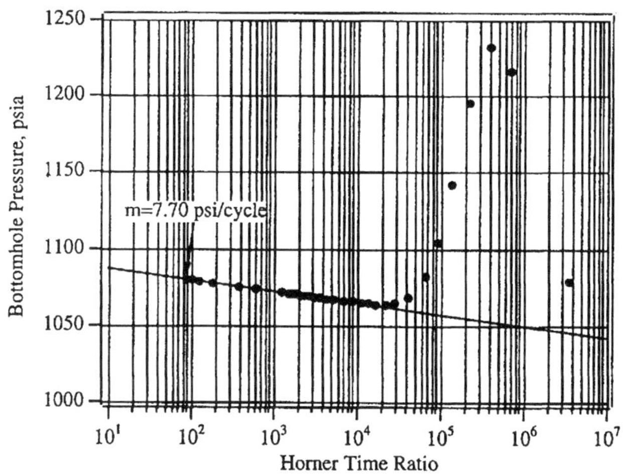

Fig. 5.12—Horner plot, Example 5.1.

$$- \log\left[\frac{50.7}{(0.10)(0.8)(15 \times 10^{-6})(0.365)^2}\right] + 3.23\Bigg\}$$

$$= 3.38.$$

Quantitative Type-Curve Analysis. Quantitative type-curve analysis, illustrated in **Fig. 5.13,** confirms the results from the semilog analysis.

1. Precalculate a pressure match point. With $k = 50.7$ md from the semilog analysis and $p_D = 100$,

$$(\Delta p)_{MP} = \frac{141.2qB\mu}{kh}(p_D)_{MP}$$

$$= \frac{(141.2)(250)(1.2)(0.8)}{(50.7)(100)}(100)$$

$$= 668 \text{ psi}.$$

2. The type-curve match is found by aligning Δp and p_D from the pressure match point and sliding the pressure-change and pressure-derivative curves horizontally. Similar to the qualitative type-curve match, the early pressure-change data are forced to match the early portion of the type curves. Interpolating, the best fit is obtained with the $C_De^{2s} = 2 \times 10^4$ curve. Because of phase-redistribution effects on the pressure response, the early test data do not match the type curve. The match is made with the middle-time data, after phase-redistribution effects no longer distort the pressure response.

3. Selecting a time match point of ($\Delta t_e = 2.62$ hours, $t_D/C_D = 10{,}000$), the correct value of the apparent wellbore-storage coefficient, C_{aD}, is estimated to be

$$C_{aD} = \frac{0.0002637k}{\phi r_w^2 c_t \mu_o}\left(\frac{\Delta t_e}{t_D/C_D}\right)_{MP}$$

$$= \frac{(0.0002637)(50.7)}{(0.10)(0.365)^2(15 \times 10^{-6})(0.8)}\left(\frac{2.62}{10{,}000}\right)$$

$$= 21.9.$$

A similar value is obtained from the unit-slope line. Assuming that the first data point ($\Delta t_e = 0.000417$ hour, $\Delta p = 9.06$ psia) lies on the unit-slope line, C_{aD} is

$$C_{aD} = \frac{0.0372qB}{\phi h c_t r_w^2}\left(\frac{\Delta t_e}{\Delta p}\right)_{USL}$$

$$= \frac{0.0372(250)(1.2)}{(0.1)(15 \times 10^{-6})(100)(0.365^2)}\left(\frac{0.000417}{9.06}\right) = 25.7,$$

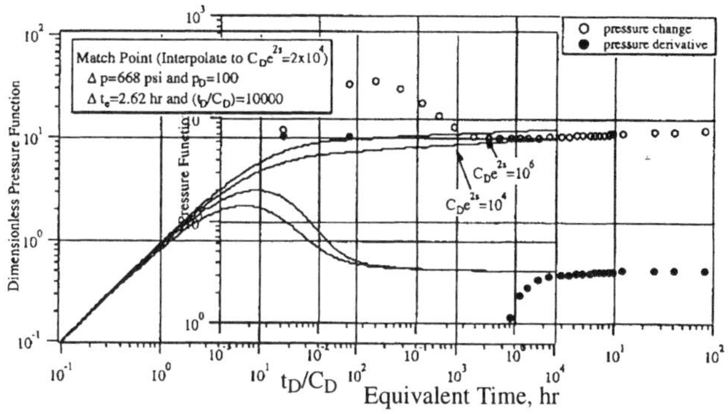

Fig. 5.13—Quantitative type-curve match, Example 5.1.

which qualitatively confirms the value of 21.9 from the type-curve match.

4. With $C_{aD} = 21.9$ and $C_De^{2s} = 2 \times 10^4$ from the type-curve match, the skin factor is

$$s = 0.5\ln\left(\frac{C_De^{2s}}{C_{aD}}\right) = 0.5\ln\left(\frac{2 \times 10^4}{21.9}\right) = 3.41,$$

which is in close agreement with $s = 3.38$ from the semilog analysis.

5. The true dimensionless wellbore-storage coefficient, C_D, is determined from wellbore characteristics. For a falling liquid level in the wellbore, the true wellbore-storage coefficient is

$$C = 25.65\frac{A_{wb}}{\rho_{wb}} = 25.65\frac{\pi(0.1017^2)}{48} = 0.01736 \text{ bbl/psi}.$$

From Eq. 5.4, the dimensionless wellbore-storage coefficient is

$$C_D = \frac{0.8936C}{\phi c_t h r_w^2} = \frac{(0.8936)(0.01736)}{(0.10)(15 \times 10^{-6})(100)(0.365)^2} = 776.$$

The significant effect of phase redistribution on the pressure buildup during the test is indicated by C_D being much greater than C_{aD}.

Confirmation of Results by Use of Thompson et al.*'s Rules.*

1. As a final check of the results, the rules given by Thompson *et al.*[7] can be used to confirm the beginning of the semilog straight line. From Fig. 5.10 and Table 5.2, the maximum pressure change at the top of the gas hump is

$$\Delta p_H = p_{wsH} - p_{wf} = 1232.73 - 1000 = 232.73 \text{ psi}.$$

2. With Eq. 5.2, Δp_H in dimensionless terms is

$$p_{DH} = \frac{kh\Delta p_H}{141.2qB\mu} = \frac{(50.7)(100)(232.73)}{(141.2)(250)(1.2)(0.8)} = 34.82.$$

3. Again from Fig. 5.10 and Table 5.2, the shut-in time at the local minimum pressure change following the gas hump is $t_L = 0.677$ hour. The shut-in time at the beginning of the semilog straight line is given by Eq. 5.17,

$$t_{sl} \approx 50\left(1 - \frac{s}{p_{DH}}\right)t_L = 50\left(1 - \frac{3.38}{34.82}\right)(0.677) = 30.6 \text{ hours}.$$

Recognizing that Eq. 5.17 tends to overestimate the beginning of the middle-time region, this estimate is somewhat consistent with the time of 6.96 hours from the qualitative type-curve match and somewhat confirms our identification of the correct semilog straight line. Note that the straight line that began at 6.96 hours continues to have the same slope at times greater than 30.6 hours.

Example 5.2—Buildup Test With Wellbore Phase Redistribution, Type 2 Pressure Response. A pressure-buildup test was run on an oil well completed with 2.875-in. tubing inside 7.625-in. casing. **Table 5.3** gives pressure and time data. Other known data are summarized next. Estimate formation permeability, skin factor, and the wellbore-storage coefficient.

TABLE 5.3—PRESSURE-BUILDUP TEST DATA, EXAMPLE 5.2

Shut-In Time, Δt (hours)	Pressure, p_{ws} (psia)	Shut-In Time, Δt (hours)	Pressure, p_{ws} (psia)
0.000417	602.582	3.4788	705.226
0.00417	621.702	4.0478	705.680
0.02081	657.839	4.6865	706.115
0.03741	667.218	5.5705	706.624
0.06638	673.459	6.3189	706.994
0.1076	679.144	6.9607	707.276
0.1608	684.581	7.6449	707.549
0.2258	689.219	8.4844	707.851
0.3660	694.872	9.0764	708.046
0.5232	697.856	9.6780	708.232
0.6770	699.409	12.00	708.851
0.8835	700.647	25.00	710.949
1.1021	701.508	40.00	712.283
1.3143	702.134	80.00	714.240
1.6707	702.938	120.0	715.381
2.1495	703.746	144.0	715.892
2.8494	704.621	168.0	716.323

h = 100 ft
r_e = 3,000 ft
μ_o = 0.8 cp
p_i = 2,000 psia
ϕ = 0.20
q = 250 STB/D
r_w = 0.365 ft
B_o = 1.2 RB/STB
ρ_{wb} = 600 psia
r_{wb} = 48 lbm/ft^3
t_p = 14,400 hours
r_t = 0.1017 ft
c_t = 15.0×10^{-6} psia^{-1}

Solution. Construct Data Plots. **Fig. 5.14** shows a Horner semilog plot, while **Fig. 5.15** plots pressure-change and -derivative data. **Table 5.4** gives plotting functions for these figures.

Qualitative Type-Curve Analysis. The pressure-change and pressure-derivative curves of the log-log plot in Fig. 5.15 display the characteristic shapes of a Type 2 pressure response with phase redistribution. In addition, the shape of the pressure-derivative curve at the end of the test suggests that the middle-time region is present; however, we should confirm its presence before attempting a semilog analysis.

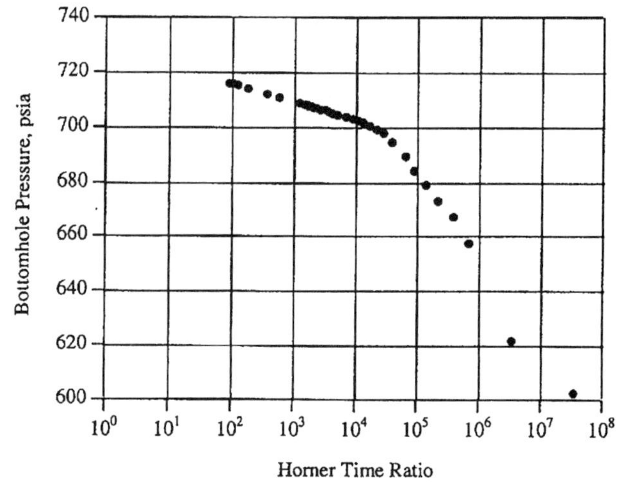

Fig. 5.14—Horner plot of buildup test data, Example 5.2.

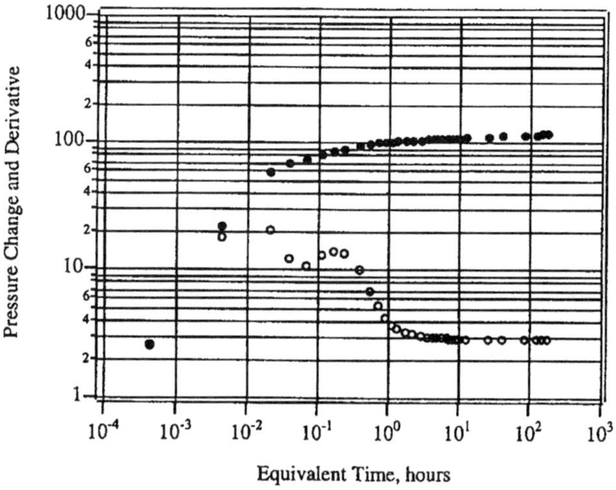

Fig. 5.15—Log-log plot of pressure change and pressure derivative, Example 5.2.

1. Qualitative type-curve analysis is performed by matching the pressure-change and -derivative data with the Bourdet type curves

TABLE 5.4—SEMILOG AND LOG-LOG PLOTTING FUNCTIONS, EXAMPLE 5.2

Δt_e (hours)	Δp (psi)	$\Delta t_e\,\Delta p'$ (psi)	Horner Time Ratio	Δt_e (hours)	Δp (psi)	$\Delta t_e\,\Delta p'$ (psi)	Horner Time Ratio
0.000417	2.582		34,569,265	3.4780	105.226	3.00122	4,140
0.00417	21.702	17.66610	3,456,926	4.0466	105.680	2.97386	3,559
0.02081	57.839	20.08040	692,141	4.6850	106.115	2.95362	3,074
0.03741	67.218	12.27360	384,940	5.5683	106.624	2.93452	2,586
0.06638	73.459	10.73690	216,940	6.3162	106.994	2.92278	2,280
0.1076	79.144	12.92660	133,864	6.9574	107.276	2.91442	2,070
0.1608	84.581	13.87880	89,556	7.6409	107.549	2.90708	1,885
0.2258	89.219	13.16310	63,777	8.4794	107.851	2.89973	1,698
0.3660	94.872	9.86433	39,340	9.0707	108.046	2.89483	1,588
0.5232	97.856	6.90051	27,525	9.6715	108.232	2.89106	1,489
0.6770	99.409	5.23901	21,270	11.99	108.851	2.87808	1,201
0.8834	100.647	4.16778	16,300	24.96	110.949	2.84904	577
1.1020	101.508	3.68630	13,067	39.89	112.283	2.83932	361
1.3141	102.134	3.46534	10,958	79.56	114.240	2.83127	181
1.6705	102.938	3.28087	8,620	119.0	115.381	2.82964	121
2.1492	103.746	3.15134	6,700	142.6	115.892	2.82771	101
2.8489	104.621	3.05005	5,055	166.1	116.323		87

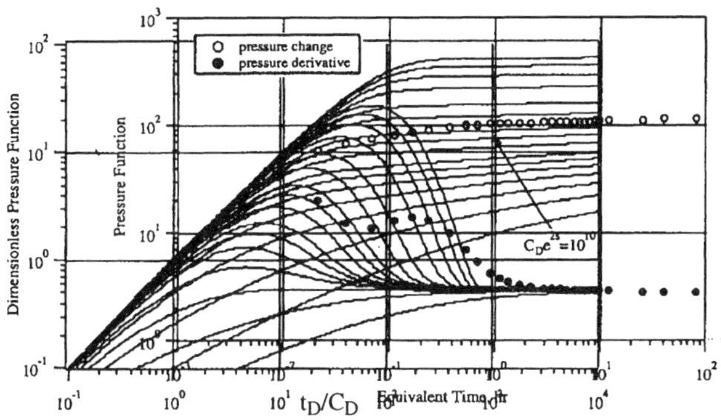

Fig. 5.16—Qualitative type-curve match, Example 5.2.

for radial flow in a homogeneous reservoir. The match is obtained by aligning the horizontal portion of the pressure-derivative curve with the corresponding horizontal portion of the type curves at a dimensionless pressure of 0.5. The data are moved horizontally to find the best fit of the pressure-change curve to the type curves, as **Fig. 5.16** shows. In finding the match, early pressure-change data are forced to match the early portion of the type curves. The first two data points appear to lie on a unit-slope line and are aligned with the unit-slope portion of the type curves. The best fit is obtained with the $C_De^{2s}=10^{10}$ curve.

2. The last 14 data points on the derivative curve lie on the horizontal portion of the derivative type curve, suggesting that the semilog straight line of the middle-time region begins at a shut-in time of approximately 5.57 hours.

Horner Semilog Analysis.

1. The Horner plot is given in **Fig. 5.17.** The semilog straight line is drawn through the last 14 data points of the test, giving $m=6.58$ psi/cycle.

2. With this slope, the effective permeability to oil is

$$k=\frac{162.6q_oB_o\mu_o}{mh}=\frac{(162.6)(250)(1.2)(0.8)}{(6.58)(100)}=59.3\text{ md}.$$

3. The skin factor is estimated to be

$$s=1.151\left[\left(\frac{p_{1\text{hr}}-p_{wf}}{m}\right)-\log\left(\frac{k}{\phi\mu_oc_tr_w^2}\right)+3.23\right]$$

$$=1.151\left\{\left(\frac{701.74-600}{6.58}\right)\right.$$

$$\left.-\log\left[\frac{59.3}{(0.20)(0.8)(15\times10^{-6})(0.365)^2}\right]+3.23\right\}$$

$$=12.$$

Quantitative Type-Curve Analysis. Quantitative type-curve analysis (**Fig. 5.18**) confirms the results from the semilog analysis.

1. With $k=59.3$ md from the semilog analysis and an arbitrary value of $p_D=10$, precalculate a pressure match point.

$$(\Delta p)_{MP}=\frac{141.2qB\mu}{kh}(p_D)_{MP}$$

$$=\frac{(141.2)(250)(1.2)(0.8)}{(59.3)(100)}(10)$$

$$=57\text{ psi}.$$

2. The type-curve match is found by aligning Δp and p_D from the pressure match point and sliding the pressure-change and pressure-derivative curves horizontally. Similar to the qualitative type-curve

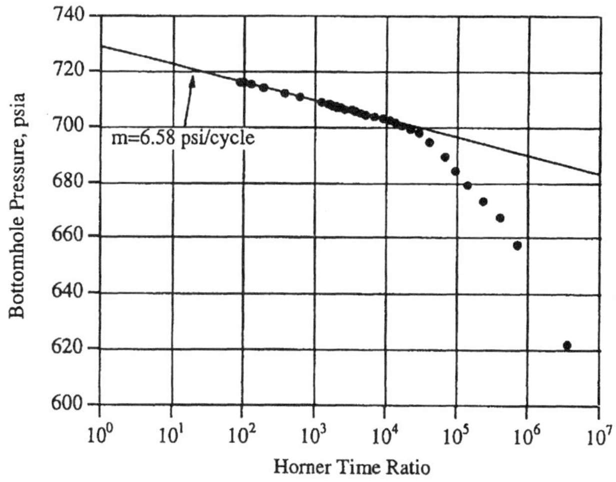

Fig. 5.17—Horner plot, Example 5.2.

match, the early pressure-change data are forced to match the early portion of the type curves. Interpolating, the best fit is obtained with the $C_De^{2s}=10^{12}$ curve. Because of phase-redistribution effects on the pressure response, the early test data do not match the type curve. The match is made with the middle-time data, after phase-redistribution effects no longer distort the pressure response.

3. Selecting a time match point of ($\Delta t_e=9.26$ hours, $t_D/C_D=10{,}000)_{MP}$ from Fig. 5.18, the dimensionless apparent wellbore-storage coefficient, C_{aD}, is estimated to be

$$C_{aD}=\frac{0.0002637k}{\phi r_w^2c_t\mu_o}\left(\frac{\Delta t_e}{t_D/C_D}\right)_{MP}$$

$$=\frac{(0.0002637)(59.3)}{(0.20)(0.365)^2(15\times10^{-6})(0.8)}\left(\frac{9.26}{10{,}000}\right)$$

$$=45.3.$$

A similar value is obtained from the unit-slope line. With the first data point ($\Delta t_e=0.000417$ hour, $\Delta p=2.582$ psia) that appears to lie on the unit-slope line, C_{aD} is

$$C_{aD}=\frac{0.0372qB}{\phi hc_tr_w^2}\left(\frac{\Delta t_e}{\Delta p}\right)_{USL}$$

$$=\frac{0.0372(250)(1.2)}{(0.2)(15\times10^{-6})(100)(0.365^2)}\left(\frac{0.000417}{2.582}\right)=45.1,$$

which confirms the value of 45.3 from the type-curve match.

4. Using $C_{aD}=45.3$ and $C_De^{2s}=10^{12}$ from the type-curve match, the skin factor is

$$s=0.5\ln\left(\frac{C_De^{2s}}{C_D}\right)=0.5\ln\left(\frac{1\times10^{12}}{45.3}\right)=11.9,$$

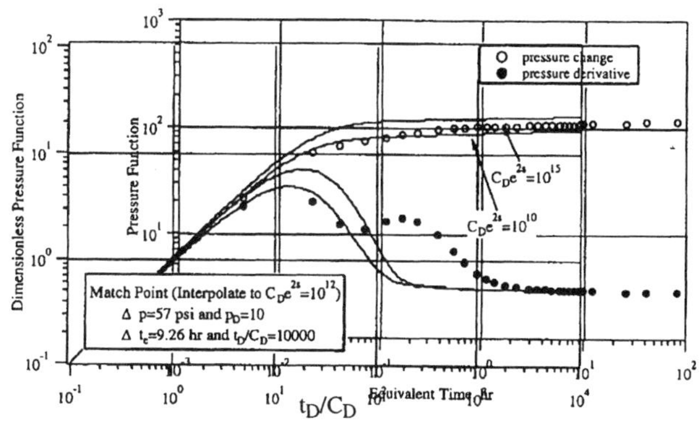

Fig. 5.18—Quantitative type-curve match, Example 5.2.

TABLE 5.5—PRESSURE-BUILDUP TEST DATA, EXAMPLE 5.3

Shut-in Time, Δt (hours)	Pressure, p_{ws} (psia)	Shut-in Time, Δt (hours)	Pressure, p_{ws} (psia)
0.000417	804.671	3.4788	865.414
0.00417	812.057	4.0478	865.798
0.02081	820.227	4.6865	866.167
0.03741	826.823	5.5705	866.599
0.06638	835.546	6.3189	866.913
0.1076	843.772	6.9607	867.153
0.1608	850.075	7.6449	867.385
0.2258	854.290	8.4844	867.643
0.3660	858.196	9.0764	867.809
0.5232	859.922	9.6780	867.967
0.6770	860.860	12.00	868.496
0.8835	861.703	25.00	870.289
1.1021	862.353	40.00	871.430
1.3143	862.849	80.00	873.107
1.6707	863.505	120.0	874.084
2.1495	864.173	144.0	874.522
2.8494	864.905	168.0	874.891

which is in close agreement with $s=12$ from the semilog analysis. This agreement seems to confirm our selection of the middle-time region.

5. The true dimensionless wellbore-storage coefficient, C_D, is determined from wellbore characteristics. For a falling liquid level in the wellbore, the true wellbore-storage coefficient, C, is

$$C = 25.65\frac{A_{wb}}{\rho_{wb}} = 25.65\frac{\pi(0.1017^2)}{48} = 0.01736 \text{ bbl/psi.}$$

From Eq. 5.4, the dimensionless wellbore-storage coefficient is

$$C_D = \frac{0.8936C}{\phi c_t h r_w^2} = \frac{(0.8936)(0.01736)}{(0.20)(15\times10^{-6})(100)(0.365)^2} = 388.$$

The pronounced effect of phase redistribution on the pressure response during the test is indicated by C_D much greater than C_{aD}.

Confirmation of Results by Use of Thompson et al.*'s Rules.* As a final check of the results, the rules given by Thompson *et al.*[7] confirm the beginning of the semilog straight line. From Fig. 5.15 and Table 5.4, the shut-in time at the inflection point separating Regions D′ and E′ (Fig. 5.4) is 0.1608 hour. The inflection point on the pressure-change curve corresponds to the local maximum on the second hump of the derivative curve. The shut-in time at the beginning of the semilog straight line is given by Eq. 5.18,

$$t_{sl} = 75\ t_{inf} = 75\ (0.1608) = 12 \text{ hours.}$$

This estimate is reasonably consistent with the time of 5.57 hours from the qualitative type-curve match and roughly confirms our selection of the middle-time region and our analysis. Note that the straight line that began at 5.57 hours continues to have the same slope for times greater than 12 hours.

Example 5.3—Buildup Test With Wellbore Phase Redistribution, Type 3 Pressure Response. A pressure buildup test was run on an oil well completed with 2.875-in. tubing inside 7.625-in. casing. **Table 5.5** gives pressure-buildup test data. Other known data are summarized next. Estimate formation permeability, skin factor, and the wellbore-storage coefficient.

$h =$ 100 ft
$r_e =$ 3,000 ft
$\mu_o =$ 0.8 cp
$p_i =$ 1,500 psia
$\phi =$ 0.30
$q =$ 250 STB/D
$r_w =$ 0.365 ft

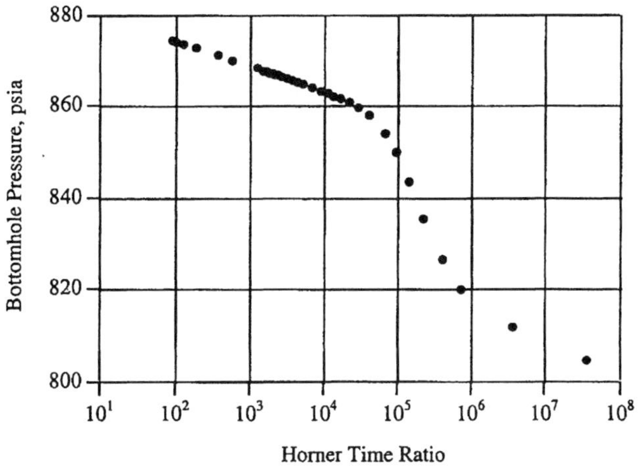

Fig. 5.19—Horner plot of buildup test data, Example 5.3.

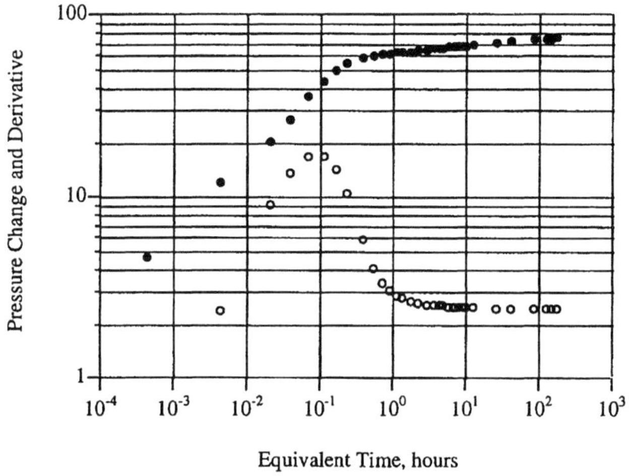

Fig. 5.20—Log-log plot of pressure change and pressure derivative, Example 5.3.

$B_o =$ 1.2 RB/STB
$p_{wf} =$ 800 psia
$\rho_{wb} =$ 48 lbm/ft^3
$t_p =$ 14,400 hours
$r_t =$ 0.1017 ft
$c_t =$ 15.0×10^{-6} psia^{-1}

Solution. *Construct Data Plots.* **Fig. 5.19** shows a Horner semilog plot, while **Fig. 5.20** plots pressure-change and pressure-derivative data. **Table 5.6** gives plotting functions for both plots.

Qualitative Type-Curve Analysis. Although phase-redistribution effects are not obvious from the plots, the data do not match the Bourdet type curves over the complete time range, so we suspect phase redistribution. The shape of the derivative curve during the latter half of the test suggests that the middle-time region was reached during the test; however, we should confirm its existence with further analysis. No boundary effects appear, indicating that the reservoir was infinite-acting throughout the test duration.

1. Qualitative type-curve analysis is performed by matching the pressure-change and -derivative data with the Bourdet type curves for radial flow in a homogeneous reservoir. Again, the match is obtained by aligning the horizontal portion of the pressure-derivative curve with the corresponding horizontal portion of the type curves at a dimensionless pressure of 0.5. The data are moved horizontally to find the best fit of the pressure-change curve to the type curves, as **Fig. 5.21** shows.

2. In finding the match, early pressure-change data are forced to match the early portion of the type curves. Assuming that the first data point on the pressure-change plot lies on a unit-slope line and aligning that point with the unit-slope portion of the type curves, the best fit is obtained with the $C_De^{2s}=10^8$ curve.

TABLE 5.6—SEMILOG AND LOG-LOG PLOTTING FUNCTIONS, EXAMPLE 5.3

Δt_e (hours)	Δp (psi)	$\Delta t_e\,\Delta p'$ (psi)	Horner Time Ratio	Δt_e (hours)	Δp (psi)	$\Delta t_e\,\Delta p'$ (psi)	Horner Time Ratio
0.000417	4.671		34,569,265	3.4780	65.414	2.53983	4,140
0.00417	12.057	2.39135	3,456,926	4.0466	65.798	2.52546	3,559
0.02081	20.227	9.15455	692,141	4.6850	66.167	2.51231	3,074
0.03741	26.823	13.37680	384,940	5.5683	66.599	2.49857	2,586
0.06638	35.546	16.69880	216,940	6.3162	66.913	2.48984	2,280
0.1076	43.772	16.80930	133,864	6.9574	67.153	2.48337	2,070
0.1608	50.075	14.15930	89,556	7.6409	67.385	2.47814	1,885
0.2258	54.290	10.57720	63,777	8.4794	67.643	2.47181	1,698
0.3660	58.196	5.89135	39,340	9.0707	67.809	2.46873	1,588
0.5232	59.922	4.00006	27,525	9.6715	67.967	2.46530	1,489
0.6770	60.860	3.36444	21,270	11.99	68.496	2.45650	1,201
0.8834	61.703	3.03704	16,300	24.96	70.289	2.43695	577
1.1020	62.353	2.87322	13,067	39.89	71.430	2.43091	361
1.3141	62.849	2.77731	10,958	79.56	73.107	2.42618	181
1.6705	63.505	2.67956	8,620	119.0	74.084	2.42376	121
2.1492	64.173	2.61208	6,700	142.6	74.522	2.42378	101
2.8489	64.905	2.56417	5,055	166.1	74.891	-	87

3. The last 17 data points on the derivative curve lie on the horizontal portion of the derivative type curve, indicating that the middle-time region begins at a shut-in time of approximately 3.48 hours (see Table 5.6).

Horner Semilog Analysis.

1. **Fig. 5.22** shows the Horner plot. The semilog straight line is drawn through the last 17 data points of the test, giving $m = 5.65$ psi/cycle.

2. Using this slope, the effective permeability to oil is

$$k = \frac{162.6 q_o B_o \mu_o}{mh} = \frac{(162.6)(250)(1.2)(0.8)}{(5.65)(100)} = 69.1 \text{ md.}$$

3. The skin factor is

$$s = 1.151\left[\left(\frac{p_{1\text{hr}} - p_{wf}}{m}\right) - \log\left(\frac{k}{\phi \mu_o c_t r_w^2}\right) + 3.23\right]$$

$$= 1.151\left\{\left(\frac{862.38 - 800}{5.65}\right)\right.$$

$$\left. - \log\left[\frac{69.1}{(0.30)(0.8)(15 \times 10^{-6})(0.365)^2}\right] + 3.23\right\}$$

$$= 7.04.$$

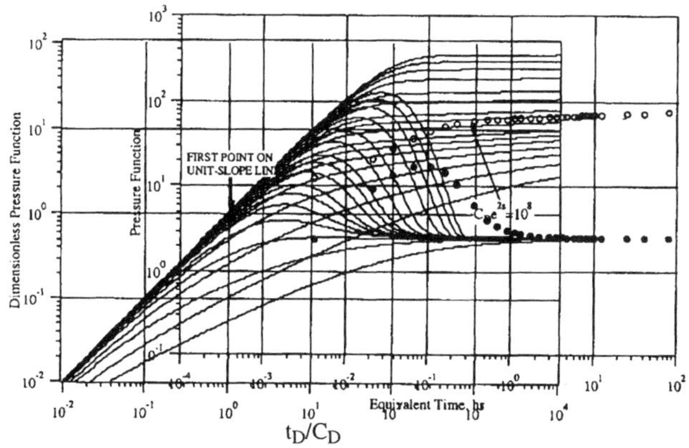

Fig. 5.21—Qualitative type-curve match, Example 5.3.

Quantitative Type-Curve Analysis. Quantitative type-curve analysis, illustrated in **Fig. 5.23,** confirms the results from the semilog analysis.

1. Precalculate a pressure match point using $k = 69.1$ md from the semilog analysis and $p_D = 10$.

$$(\Delta p)_{MP} = \frac{141.2 q B \mu}{kh} (p_D)_{MP}$$

$$= \frac{(141.2)(250)(1.2)(0.8)}{(69.1)(100)}(10)$$

$$= 49 \text{ psi.}$$

2. The type-curve match is found by aligning Δp and p_D from the pressure match point and sliding the pressure-change and pressure-derivative curves horizontally. Similar to the qualitative type-curve match, the early pressure-change data are forced to match the early portion of the type curves. Interpolating, the best fit is obtained with the $C_D e^{2s} = 2 \times 10^7$ curve. Because of phase-redistribution effects on the pressure response, the early test data do not match the type curve. The match is made with the middle-time data, after phase-redistribution effects have ceased distorting the pressure response.

3. Selecting a time match point of ($\Delta t_e = 3.92$ hours, $t_D/C_D = 10{,}000$), the correct value of the apparent wellbore-storage coefficient, C_{aD}, is estimated to be

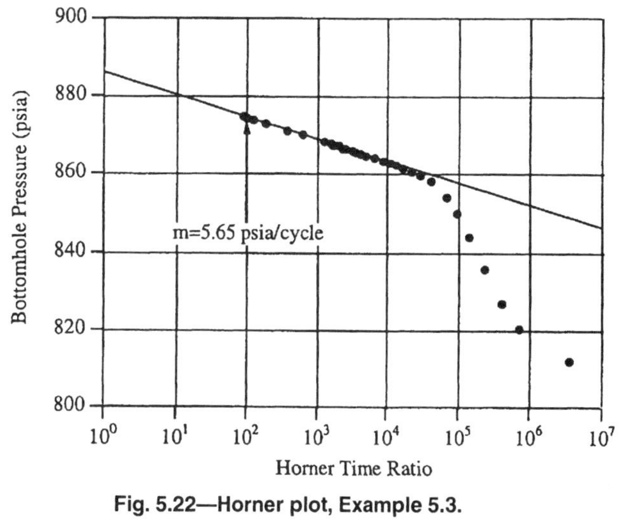

Fig. 5.22—Horner plot, Example 5.3.

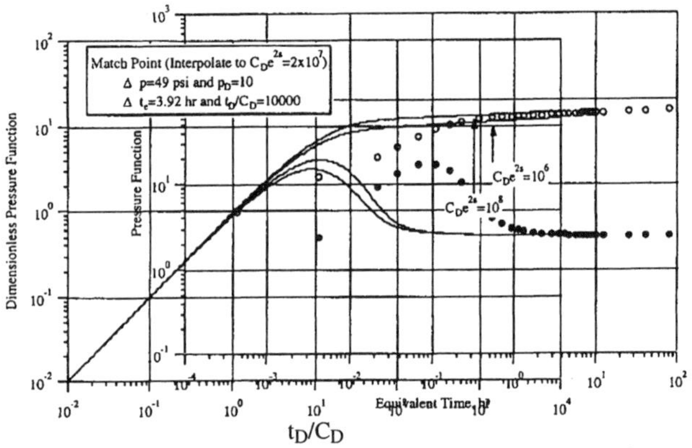

Fig. 5.23—Quantitative type-curve match, Example 5.3.

$$C_{aD} = \frac{0.0002637k}{\phi r_w^2 c_t \mu_o}\left(\frac{\Delta t_e}{t_D/C_D}\right)_{MP}$$

$$= \frac{(0.0002637)(69.1)}{(0.30)(0.365)^2(15 \times 10^{-6})(0.8)}\left(\frac{3.92}{10,000}\right)$$

$$= 14.9.$$

A similar value is obtained from the unit-slope line. Assuming that the first data point ($\Delta t_e = 0.000417$ hour, $\Delta p = 4.671$ psi) lies on the unit-slope line,

$$C_{aD} = \frac{0.0372qB}{\phi hc_t r_w^2}\left(\frac{\Delta t_e}{\Delta p}\right)_{USL}$$

$$= \frac{(0.0372)(250)(1.2)}{(0.3)(15 \times 10^{-6})(100)(0.365^2)}\left(\frac{0.000417}{4.671}\right) = 16.6,$$

which agrees with 14.9 from the type-curve match.

4. Using $C_{aD} = 14.9$ and $C_De^{2s} = 2 \times 10^7$ from the type-curve match, the skin factor is

$$s = 0.5\ln\left(\frac{C_De^{2s}}{C_D}\right) = 0.5\ln\left(\frac{2 \times 10^7}{14.9}\right) = 7.05,$$

which agrees with $s = 7.04$ from the semilog analysis.

5. The true wellbore-storage coefficient, C, is determined from wellbore characteristics. For a falling liquid level in the wellbore,

$$C = 25.65\frac{A_{wb}}{\rho_{wb}} = 25.65\frac{\pi(0.1017^2)}{48} = 0.01736 \text{ bbl/psi}.$$

From Eq. 5.4, the dimensionless wellbore-storage coefficient is

$$C_D = \frac{0.8936C}{\phi c_t h r_w^2} = \frac{(0.8936)(0.01736)}{(0.30)(15 \times 10^{-6})(100)(0.365)^2} = 259.$$

The difference between C_D and C_{aD} (i.e., $C_D > C_{aD}$) confirms that phase redistribution has affected the buildup pressure response; however, the small difference suggests a Type 3 pressure response.

Confirmation of Results With Thompson et al.*'s Rules.* As a final check of the results, the rules given by Thompson *et al.*[7] confirm the beginning of the semilog straight line. Noting that $C_D \geqq 2C_{aD}$ and $C_D \leqq 10^3$, the six-cycle log rule of Eq. 5.19 applies. In Fig. 5.23, the unit-slope portion of the pressure change curve appears to end with the first data point at $\Delta t_e = 0.00417$ hours. From Eq. 5.19, the shut-in time at the beginning of the semilog straight line is

$$t_{sl} = 10^6 \Delta t_e = 10^6(0.00417) = 417 \text{ hours}.$$

Thus, the beginning of the semilog straight line, t_{sl}, is estimated to be 417 hours compared with 3.48 hours from the qualitative type-curve analysis. This discrepancy illustrates the inaccuracy that is possible with the Thompson *et al.* rules for estimating the beginning of the middle-time region for a Type 3 pressure response. Consequently, these rules should be used only as very rough guidelines to validate results from the semilog and type-curve analyses.

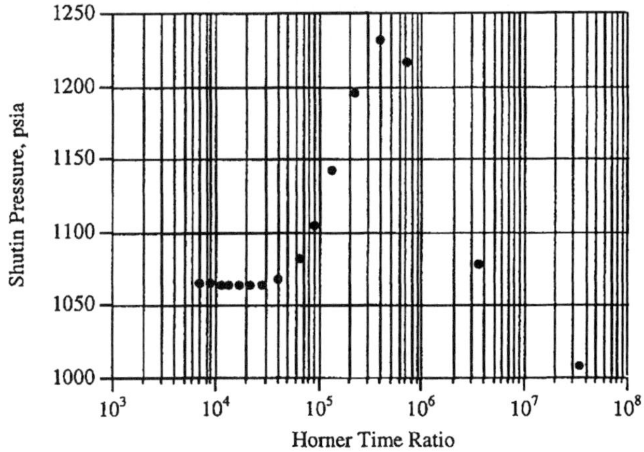

Fig. 5.24—Horner Plot of buildup test data, Example 5.4.

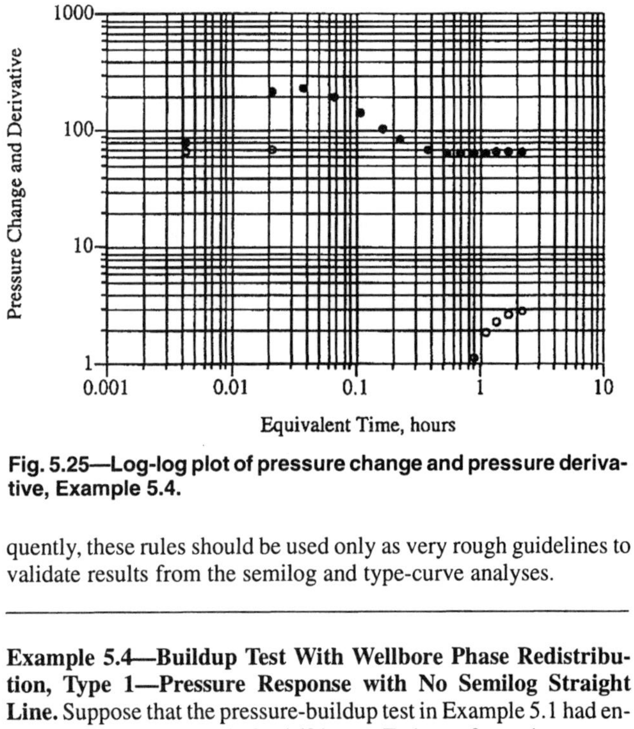

Fig. 5.25—Log-log plot of pressure change and pressure derivative, Example 5.4.

Example 5.4—Buildup Test With Wellbore Phase Redistribution, Type 1—Pressure Response with No Semilog Straight Line. Suppose that the pressure-buildup test in Example 5.1 had ended at 2.15 hours instead of at 168 hours. Estimate formation permeability, skin factor, and the wellbore-storage coefficient from these more limited data.

Solution. *Construct Data Plots.* **Fig. 5.24** shows a Horner semilog plot, while **Fig. 5.25** plots pressure change and pressure-derivative data. **Table 5.7** summarizes the plotting functions, and Table 5.1 gives pressure-buildup data.

Qualitative Type-Curve Analysis. The gas hump characteristic of a Type 1 pressure response is evident in the plots and is confirmed by the typical shape of the pressure-derivative curve in Fig. 5.25. In this case, however, the derivative curve does not become horizontal, suggesting that the middle-time region was not reached during the test. Hence, the straight line formed by the last several data points in Fig. 5.24 is not the true semilog straight line of the middle-time region and cannot be used to estimate permeability and skin factor.

Because the derivative curve does not become constant, the pressure-change and -derivative curves cannot be aligned uniquely with the Bourdet type curves. As shown in **Figs. 5.26 through 5.28**, the late-time data fit many different type curves equally well, giving a wide range of values for permeability, skin, and apparent wellbore-storage coefficient. Therefore, the only way to obtain accurate estimates of reservoir properties is to use automatic history matching.[4,9]

TABLE 5.7—SEMILOG AND TYPE CURVE PLOTTING FUNCTIONS, EXAMPLE 5.4

Δt_e (hours)	Δp, (psi)	$\Delta t_e\, \Delta p'$ (psi)	Horner Time Ratio
0.000417	9.06		34,569,265
0.00417	78.31	65.9123	3,456,926
0.02081	216.79	67.1860	692,141
0.03741	232.73	−19.8932	384,940
0.06638	195.64	−100.450	216,940
0.1076	142.90	−108.130	133,864
0.1608	104.46	−79.6891	89,556
0.2258	82.59	−48.9904	63,777
0.3660	67.75	−15.7708	39,340
0.5232	64.54	−4.13181	27,525
0.6770	64.02	−0.55966	21,270
0.8834	64.14	1.13528	16,300
1.1020	64.48	1.89826	13,067
1.3141	64.84	2.30500	10,958
1.6705	65.42	2.66656	8,620
2.1492	66.10	2.87676	6,700

5.5 Chapter Summary

In this chapter, we discussed phase redistribution or phase segregation, a common example of the more general phenomenon of changing wellbore storage. Phase redistribution results in an apparent decrease in the wellbore-storage coefficient, while other conditions may either increase or decrease wellbore storage.

In Sec. 5.2, we described the physical mechanism that causes phase redistribution. Here, a flowing well with a wellbore initially filled with a uniform mixture of liquid and gas experiences segregation of the liquid and gas following shut-in. This segregation results in a pressure change that distorts the reservoir response to the shut-in. Phase redistribution is most likely to occur in formations of moderate permeability (10 to 100 md for oil, 0.1 to 1 md for gas) and a large positive skin factor.

In Sec. 5.3, we presented the mathematical model of phase redistribution introduced by Fair. Although this model is intuitive, it does describe the general shape of the pressure response caused by phase redistribution. Specifically, this model predicts the existence of three different shapes for the pressure response, which are referred to as Types 1, 2, and 3. Type 1 phase redistribution is the most obvious. Here, the pressure response exhibits a "hump," where the phase redistribution is severe enough that a fluid flows back into the formation during shut-in. Eventually, the flow reversal ceases and the pressure response approaches that of a well with no phase redistribution. The flow reversal causes the pressure derivative to become negative. The Type 2 pressure response is more subtle. Here, no flow reversal occurs, hence the derivative is always positive. The pressure flattens before approaching the undistorted pressure response, while the pressure derivative has a local minimum. This type of response is unlikely to be mistaken for homogenous reservoir behavior, but it may be confused with dual-porosity behavior, discussed in Chap. 7. The Type 3 pressure response is the most subtle. There is no apparent hump in the pressure or local minimum in the pressure derivative. It may be difficult, however, to find a single type curve that fits all the pressure and pressure-derivative data. For each of the typical responses for phase redistribution, we presented equations to estimate the shut-in time necessary to reach the semilog straight line.

In Sec. 5.4, we presented a technique for analyzing pressure-buildup tests that have been distorted by phase redistribution that uses only standard wellbore storage and skin type curves and Horner semilog analysis. If the test is of sufficient duration, the middle-time region will appear and the test can be analyzed with the methods discussed in Chaps. 2 through 4, ignoring the portion of the data affected by phase redistribution. We illustrated this technique with four examples.

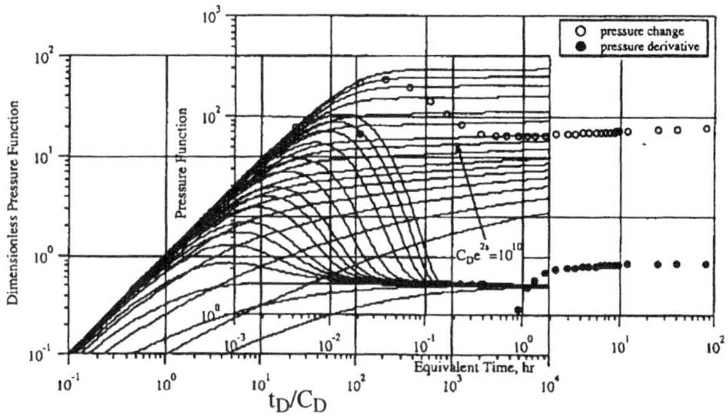

Fig. 5.26—Qualitative type-curve match, $C_De^{2s}=10^{10}$, $k=87$ md, $s=10$, and $C_{aD}=38$, Example 5.4.

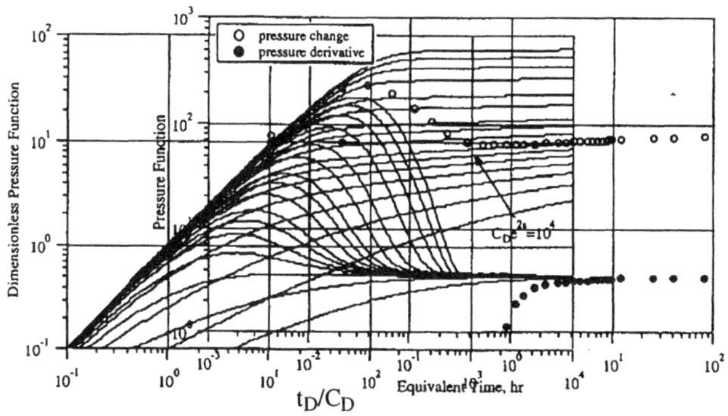

Fig. 5.27—Qualitative type-curve match, $C_De^{2s}=10^4$, $k=50$ md, $s=3$, and $C_{aD}=41$, Example 5.4.

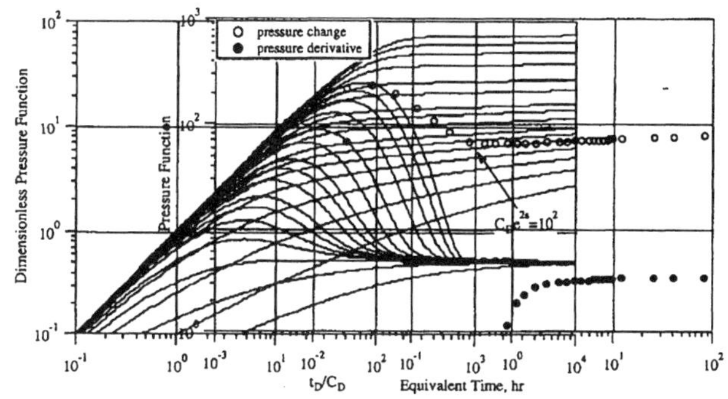

Fig. 5.28—Qualitative type-curve match, $C_De^{2s}=10^2$, $k=37$ md, $s=0$, and $C_{aD}=39$, Example 5.4.

Exercises

1. Given the following formation and fluid properties, estimate formation permeability, skin factor, WBS coefficient, and initial reservoir pressure from the buildup test data. What type of phase redistribution does this test exhibit? $q=175$ STB/D; $h=6$ ft; $B=1.157$ RB/STB; $t_p=600$ hr; $\phi=6.0\%$; $c_t=13.0\times10^{-6}$ psi^{-1}; $p_{wf}=1067.79$ psia; $r_w=0.26$ ft; $\mu=0.82$ cp.

PRESSURE-TIME DATA FOR EXERCISE 5.1					
Time (hours)	Pressure (psi)	Time (hours)	Pressure (psi)	Time (hours)	Pressure (psi)
0.000	1067.79	0.485	2147.92	5.67	1714.69
0.010	1112.41	0.592	2208.69	6.82	1714.68
0.022	1164.08	0.720	2241.24	8.19	1716.82
0.036	1223.51	0.874	2241.27	9.84	1719.86
0.054	1291.32	1.059	2208.00	11.82	1723.33
0.074	1367.89	1.281	2145.37	14.19	1727.06
0.099	1453.30	1.547	2062.22	17.04	1730.97
0.129	1547.07	1.867	1970.97	20.5	1734.97
0.165	1648.06	2.25	1884.79	24.6	1739.02
0.208	1754.20	2.71	1814.15	29.5	1743.08
0.260	1862.29	3.26	1764.19	35.4	1747.13
0.322	1967.89	3.92	1734.23	42.5	1751.17
0.396	2065.32	4.72	1719.65	48.0	1753.87

2. Given the following formation and fluid properties, estimate formation permeability, skin factor, WBS coefficient, and initial reservoir pressure from the buildup test data. What type of phase redistribution does this test exhibit? $q = 75$ STB/D; $h = 14$ ft; $B = 1.126$ RB/STB; $t_p = 720$ hr; $\phi = 24.6\%$; $c_t = 11.3 \times 10^{-6}$ psi^{-1}; $p_{wf} = 1662.62$ psia; $r_w = 0.46$ ft; $\mu = 1.00$ cp.

PRESSURE-TIME DATA FOR EXERCISE 5.2					
Time (hours)	Pressure (psi)	Time (hours)	Pressure (psi)	Time (hours)	Pressure (psi)
0.000	1662.62	1.739	1698.64	41.279	1731.42
0.020	1666.13	2.194	1701.41	51.619	1732.06
0.045	1669.83	2.762	1704.49	64.544	1732.67
0.076	1673.58	3.473	1707.83	80.699	1733.24
0.115	1677.21	4.361	1711.32	100.894	1733.79
0.164	1680.54	5.471	1714.82	126.138	1734.31
0.225	1683.41	6.859	1718.19	157.692	1734.81
0.301	1685.77	8.594	1721.25	197.135	1735.28
0.397	1687.67	10.762	1723.88	246.439	1735.73
0.516	1689.25	13.473	1726.01	308.069	1736.15
0.665	1690.74	16.861	1727.67	385.106	1736.53
0.851	1692.31	21.096	1728.92	481.402	1736.89
1.084	1694.11	26.390	1729.90	601.773	1737.22
1.375	1696.21	33.007	1730.71	720.000	1737.46

3. Given the following formation and fluid properties, estimate formation permeability, skin factor, WBS coefficient, and initial reservoir pressure from the buildup test data. What type of phase redistribution does this test exhibit? $q = 2300$ STB/D; $h = 87$ ft; $B = 1.33$ RB/STB; $t_p = 1440$ hr; $\phi = 13.1\%$; $c_t = 12.0 \times 10^{-6}$ psi^{-1}; $p_{wf} = 2773.20$ psia; $r_w = 0.44$ ft; $\mu = 0.66$ cp.

PRESSURE-TIME DATA FOR EXERCISE 5.3					
Time (hours)	Pressure (psi)	Time (hours)	Pressure (psi)	Time (hours)	Pressure (psi)
0.000	2773.20	0.846	5813.10	16.009	4054.35
0.015	2874.73	1.115	6078.31	20.827	4057.10
0.035	3002.96	1.464	6181.88	27.090	4064.32
0.060	3163.64	1.919	6081.66	35.232	4073.76
0.093	3362.78	2.509	5779.27	45.817	4084.39
0.136	3606.08	3.277	5337.08	59.577	4095.64
0.191	3897.66	4.275	4865.91	77.465	4107.20
0.264	4238.23	5.573	4477.21	100.719	4118.91
0.358	4622.25	7.260	4228.40	130.950	4130.66
0.480	5034.42	9.452	4106.98	170.250	4142.31
0.639	5446.07	12.303	4063.56	196.000	4148.48

4. Given the following formation and fluid properties, estimate formation permeability, skin factor, WBS coefficient, and initial reservoir pressure from the buildup test data. What type of phase redistribution does this test exhibit? $q = 1800$ STB/D; $h = 24$ ft; $B = 1.213$ RB/STB; $t_p = 720$ hr; $\phi = 25.2\%$; $c_t = 14.7 \times 10^{-6}$ psi^{-1}; $p_{wf} = 1518.53$ psia; $r_w = 0.37$ ft; $\mu = 0.70$ cp.

PRESSURE-TIME DATA FOR EXERCISE 5.4					
Time (hours)	Pressure (psi)	Time (hours)	Pressure (psi)	Time (hours)	Pressure (psi)
0.000	1518.53	0.696	1701.76	16.51	1983.72
0.008	1528.76	0.877	1708.18	20.6	2012.94
0.018	1540.69	1.105	1715.63	25.8	2038.38
0.031	1554.35	1.389	1724.62	32.3	2059.22
0.046	1569.66	1.744	1735.54	40.4	2075.38
0.066	1586.38	2.19	1748.71	50.5	2087.46
0.090	1604.05	2.74	1764.46	63.1	2096.52
0.121	1622.00	3.44	1783.05	78.9	2103.61
0.159	1639.38	4.31	1804.71	98.6	2109.54
0.206	1655.28	5.39	1829.49	123.2	2114.78
0.266	1668.96	6.74	1857.24	154.0	2119.55
0.341	1680.04	8.44	1887.51	192.6	2123.95
0.434	1688.68	10.56	1919.51	240.7	2128.02
0.550	1695.59	13.20	1952.06	288.0	2131.06

5. Given the following formation and fluid properties, estimate formation permeability, skin factor, WBS coefficient, and initial reservoir pressure from the buildup test data. What type of phase redistribution does this test exhibit? $q = 112$ STB/D; $h = 5$ ft; $B = 1.182$ RB/STB; $t_p = 1440$ hr; $\phi = 29.2\%$; $c_t = 17.2 \times 10^{-6}$ psi^{-1}; $p_{wf} = 1189.66$ psia; $r_w = 0.49$ ft; $\mu = 0.69$ cp.

PRESSURE-TIME DATA FOR EXERCISE 5.5					
Time (hours)	Pressure (psi)	Time (hours)	Pressure (psi)	Time (hours)	Pressure (psi)
0.000	1189.66	2.06	1464.77	39.6	1702.95
0.030	1198.43	2.61	1487.24	49.5	1714.21
0.068	1209.05	3.29	1506.53	61.9	1722.84
0.114	1221.79	4.14	1523.40	77.4	1729.46
0.173	1236.95	5.21	1539.18	96.8	1734.78
0.246	1254.75	6.54	1555.22	121.0	1739.32
0.338	1275.34	8.21	1572.48	151.3	1743.41
0.452	1298.71	10.29	1591.27	189.2	1747.21
0.595	1324.65	12.89	1611.35	236.5	1750.78
0.774	1352.63	16.14	1632.07	295.7	1754.15
0.998	1381.82	20.2	1652.52	369.7	1757.32
1.277	1411.08	25.3	1671.71	462.1	1760.31
1.626	1439.13	31.6	1688.73	480.0	1760.80

6. Given the following formation and fluid properties, estimate formation permeability, skin factor, WBS coefficient, and initial reservoir pressure from the buildup test data. What type of phase redistribution does this test exhibit? $q = 75$ STB/D; $h = 8$ ft; $B = 1.076$ RB/STB; $t_p = 720$ hr; $\phi = 19.6\%$; $c_t = 7.64 \times 10^{-6}$ psi^{-1}; $p_{wf} = 664.69$ psia; $r_w = 0.34$ ft; $\mu = 1.50$ cp.

PRESSURE-TIME DATA FOR EXERCISE 5.6					
Time (hours)	Pressure (psi)	Time (hours)	Pressure (psi)	Time (hours)	Pressure (psi)
0.000	664.69	1.626	1160.57	25.3	1505.88
0.030	677.22	2.06	1245.01	31.6	1476.96
0.068	692.68	2.61	1330.44	39.6	1450.45
0.114	711.66	3.29	1412.09	49.5	1427.95
0.173	734.87	4.14	1484.41	61.9	1410.65
0.246	763.09	5.21	1542.03	77.4	1399.03
0.338	797.18	6.54	1580.96	96.8	1392.66
0.452	837.99	8.21	1599.80	121.0	1390.35
0.595	886.34	10.29	1600.23	151.3	1390.59
0.774	942.83	12.89	1586.38	189.2	1392.07
0.998	1007.73	16.14	1563.29	236.5	1393.95
1.277	1080.71	20.2	1535.40	288.0	1395.63

7. Given the following formation and fluid properties, estimate formation permeability, skin factor, WBS coefficient, and initial reservoir pressure from the buildup test data. What type of phase redistribution does this test exhibit? $q = 675$ STB/D; $h = 82$ ft; $B = 1.361$ RB/STB; $t_p = 72$ hr; $\phi = 13.8\%$; $c_t = 14.0 \times 10^{-6}$ psi^{-1}; $p_{wf} = 3581.24$ psia; $r_w = 0.3$ ft; $\mu = 0.47$ cp.

PRESSURE-TIME DATA FOR EXERCISE 5.7					
Time (hours)	Pressure (psi)	Time (hours)	Pressure (psi)	Time (hours)	Pressure (psi)
0.000	3581.24	0.1355	3666.15	2.11	3735.54
0.0025	3592.97	0.1719	3676.49	2.64	3736.89
0.0056	3601.31	0.217	3686.90	3.30	3738.17
0.0095	3607.13	0.274	3696.90	4.13	3739.40
0.0144	3611.60	0.345	3705.99	5.16	3740.58
0.0205	3615.75	0.434	3713.76	6.45	3741.73
0.0281	3620.30	0.545	3719.97	8.07	3742.84
0.0377	3625.59	0.684	3724.66	10.09	3743.91
0.0496	3631.80	0.857	3728.07	12.61	3744.94
0.0645	3638.99	1.074	3730.56	15.77	3745.94
0.0831	3647.18	1.345	3732.48	19.71	3746.89
0.1064	3656.29	1.684	3734.09	24.00	3747.68

8. Given the following formation and fluid properties, estimate formation permeability, skin factor, WBS coefficient, and initial reservoir pressure from the buildup test data. What type of phase redistribution does this test exhibit? $q = 900$ STB/D; $h = 12$ ft; $B = 1.325$ RB/STB; $t_p = 600$ hr; $\phi = 11.8\%$; $c_t = 14.7 \times 10^{-6}$ psi^{-1}; $p_{wf} = 2447.24$ psia; $r_w = 0.24$ ft; $\mu = 0.49$ cp.

PRESSURE-TIME DATA FOR EXERCISE 5.8					
Time (hours)	Pressure (psi)	Time (hours)	Pressure (psi)	Time (hours)	Pressure (psi)
0.000	2447.24	0.1279	3079.52	5.16	3247.12
0.0010	2463.61	0.1673	3098.25	6.71	3254.99
0.0023	2484.26	0.219	3108.73	8.73	3261.58
0.0040	2510.12	0.285	3115.70	11.35	3267.44
0.0062	2542.11	0.372	3122.47	14.76	3272.90
0.0090	2581.13	0.484	3130.55	19.18	3278.11
0.0128	2627.81	0.630	3140.43	24.9	3283.14
0.0176	2682.30	0.820	3152.21	32.4	3288.02
0.0239	2743.89	1.067	3165.76	42.2	3292.78
0.0320	2810.66	1.389	3180.67	54.8	3297.40
0.0426	2879.28	1.806	3196.23	71.2	3301.88
0.0564	2945.09	2.35	3211.48	72.0	3302.06
0.0743	3002.87	3.05	3225.43		
0.0976	3048.28	3.97	3237.38		

References

1. Earlougher, R.C. Jr.: *Advances in Well Test Analysis*, Monograph Series, SPE, Richardson, Texas (1977) **5.**
2. Stegemeier, G.L. and Matthews, C.S.: "A Study of Anomalous Pressure-Buildup Behavior," *Trans.*, AIME (1958) **213**, 44.
3. Fair, W.B., Jr.: "Pressure-Buildup Analysis With Wellbore Phase Redistribution," *SPEJ* (April 1981) 259.
4. Rushing, J.A. and Lee, W.J.: "Use of an Automatic History-Matching Technique To Analyze Pressure-Buildup Data Affected by Wellbore Phase Segregation: Case Histories," paper SPE 18837 presented at the 1989 SPE Production Operations Symposium, Oklahoma City, Oklahoma, 13–14 March.
5. Van Everdingen, A.F. and Hurst, W.: "The Application of the Laplace Transformation to Flow Problems in Reservoirs," *Trans.*, AIME (1949) **186,** 305.
6. Agarwal, R.G., Al-Hussainy, R., and Ramey, H.J. Jr.: "An Investigation of Wellbore Storage and Skin Effect in Unsteady Liquid Flow: I. Analytical Treatment," *SPEJ* (September 1970) 278; *Trans.*, AIME (1970) **249.**
7. Thompson, L.G., Jones, J.R., and Reynolds, A.C.: "Analysis of Pressure-Buildup Data Influenced by Wellbore Phase Redistribution," *SPEFE* (October 1986) 435.
8. Bourdet, D. *et al.*: "A New Set of Type Curves Simplifies Well Test Analysis," *World Oil* (May 1983) 95.
9. Olarewaju, J.S. and Lee, W.J.: "Effects of Phase Segregation on Buildup Test Data From Gas Wells," paper SPE 19100 presented at the 1989 SPE Gas Technology Symposium, Dallas, 7–9 June.
10. Agarwal, R.G.: "A New Method To Account for Producing-Time Effects When Drawdown Type Curves Are Used To Analyze Pressure Buildup and Other Test Data," paper SPE 9289 presented at the 1980 SPE Annual Technical Conference and Exhibition, Dallas, 21–24 September.

Chapter 6
Well-Test Interpretation in Hydraulically Fractured Wells

6.1 Overview

Many wells, particularly gas wells in low-permeability formations, require hydraulic fracturing to be commercially viable. Interpreting pressure transient data in hydraulically fractured wells is important in evaluating the success of fracture treatments and for predicting future performance of fractured wells. In this chapter, we describe graphical techniques, including semilog, log-log, and Cartesian coordinate, for analyzing post-fracture pressure transient tests. Because identification of specific flow patterns often can aid in well-test analysis, we first identify several flow patterns that are characteristic of hydraulically fractured wells.

6.2 Flow Patterns in Hydraulically Fractured Wells

Five distinct flow patterns (**Fig. 6.1**) occur in the fracture and formation around a hydraulically fractured well.[1-3] Successive flow patterns, often separated by transition periods, include fracture linear, bilinear, formation linear, elliptical, and pseudoradial flow. Fracture linear flow (Fig. 6.1a) is very short-lived and may be masked by wellbore-storage effects. During this flow period, most of the fluid entering the wellbore comes from fluid expansion in the fracture, and the flow pattern is essentially linear. Because of its extremely short duration, the fracture-linear-flow period often is of no practical use in well-test analysis. The duration of the fracture-linear-flow period is estimated by[1,3]

$$t_{L_fD} \approx \frac{0.1C_{rD}^2}{\eta_{fD}^2}, \quad \text{(6.1)}$$

where t_{L_fD} = dimensionless time in terms of fracture half-length,

$$t_{L_fD} \approx \frac{0.0002637kt}{\phi \mu c_t L_f^2}. \quad \text{(6.2)}$$

The dimensionless fracture conductivity, C_{rD}, is

$$C_{rD} = \frac{wk_f}{\pi k L_f}, \quad \text{(6.3)}$$

and η_{fD} = dimensionless hydraulic diffusivity defined by

$$\eta_{fD} = \frac{k_f \phi c_t}{k \phi_f c_{ft}}. \quad \text{(6.4)}$$

Bilinear flow (Fig. 6.1b) evolves only in finite-conductivity fractures as fluid in the surrounding formation flows linearly into the fracture and before fracture-tip effects begin to influence well behavior. Fractures are considered to be finite conductivity when $C_{rD} < 100$. Most of the fluid entering the wellbore during this flow period comes from the formation. During the bilinear-flow period, bottomhole pressure (BHP), p_{wf}, is a linear function of $t^{1/4}$ on Cartesian coordinate paper. A log-log plot of $(p_i - p_{wf})$ as a function of time exhibits a slope of one fourth. The pressure derivative also has a slope of one fourth during this same time period. The bilinear-flow period may last for a significant time, as **Fig. 6.2** shows.

The duration of bilinear flow depends on dimensionless fracture conductivity and is given by Eqs. 6.5a through 6.5c for a range of dimensionless times and fracture conductivities[1]:

$$t_{L_fD} \approx \frac{0.01}{C_{rD}^2} \text{ for } C_{rD} > 3, \quad \text{(6.5a)}$$

$$t_{L_fD} \approx 0.0205(C_{rD} - 1.5)^{-1.53} \text{ for } 1.6 \leqq C_{rD} \leqq 3, \quad \text{(6.5b)}$$

$$\text{and } t_{L_fD} \approx \left(\frac{4.55}{\sqrt{C_{rD}}} - 2.5\right)^{-4} \text{ for } C_{rD} < 1.6. \quad \text{(6.5c)}$$

Formation-linear flow (Fig. 6.1c) occurs only in high-conductivity ($C_{rD} \geqq 100$) fractures. This period continues to a dimensionless time of $t_{L_fD} \approx 0.016$. The transition from fracture-linear flow to formation linear flow is complete by a time of $t_{L_fD} = 10^{-4}$. On Cartesian coordinate paper, p_{wf} is a linear function of $t^{1/2}$, and a log-log plot of both $(p_i - p_{wf})$ and the pressure derivative as a function of time exhibits a slope of one half.

Elliptical flow (Fig. 6.1d) is a transitional flow period and occurs between a linear- or near-linear-flow pattern at early times and a radial- or near-radial-flow pattern at late times.

Pseudoradial flow (Fig. 6.1e) occurs with fractures of all conductivities. After a sufficiently long flow period, the fracture appears to the reservoir as an expanded wellbore (consistent with the effective wellbore radius concept that Prats *et al.*[6] suggest). At this time, the drainage pattern can be considered a circle for practical purposes. The larger the fracture conductivity, the later the development of an essentially radial drainage pattern. If the fracture length is large relative to the drainage area, then boundary effects distort or entirely

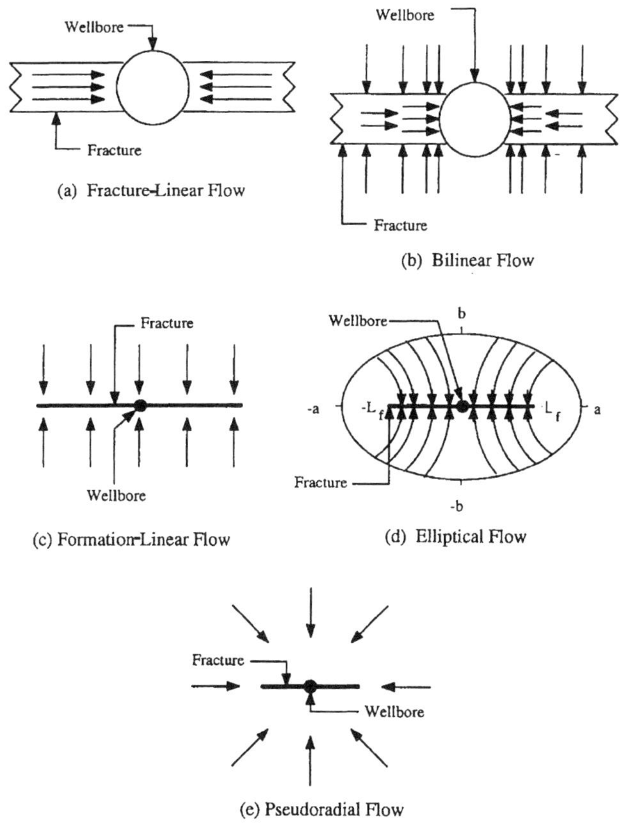

Fig. 6.1—Flow periods in a vertically fractured well (after Cinco-Ley and Samaniego-V.[1]).

mask the pseudoradial flow regime. Pseudoradial flow begins at $t_{L_fD} \approx 3$ for high-conductivity fractures ($C_{rD} \geqq 100$) and at slightly smaller values of t_{L_fD} for lower values of C_{rD}. **Fig. 6.3** shows the beginning of pseudoradial flow, which is indicated by the start of a straight line on a graph of p_{wf} vs. log t, for several values of C_{rD}. The pressure derivative flattens when pseudoradial flow begins. This characteristic is often used to identify this flow period.

These flow patterns also appear in pressure-buildup tests and occur at about the same dimensionless times as in flow tests. The physical interpretation is that the pressure has built up to an essentially uniform value throughout a particular region at a given time during a buildup test. For example, at a given time during bilinear or formation-linear flow, pressure has built up to a uniform level throughout an approximately rectangular region around the fracture. At a later time during elliptical flow, pressure has built up to a uniform level throughout an approximately elliptical region centered at the wellbore. At a given time during pseudoradial flow, pressure has built up to a uniform level throughout an approximately circular region centered at the wellbore. The area of the region and the pressure level within that area increase with increasing shut-in time. Example 6.1 illustrates how to estimate the duration of flow periods for hydraulically fractured wells.

Example 6.1—Estimating Duration of Flow Periods in a Hydraulically Fractured Well. For each of the following cases, estimate the end of the linear-flow period and the time at which pseudoradial-flow period begins. Assume that pseudoradial flow begins when $t_{L_fD} = 3$ (see **Table 6.1**).

$\phi = 0.15$
$C_{rD} = 100$
$m = 0.03$ cp
$c_t = 1 \times 10^{-4}$ psi^{-1}

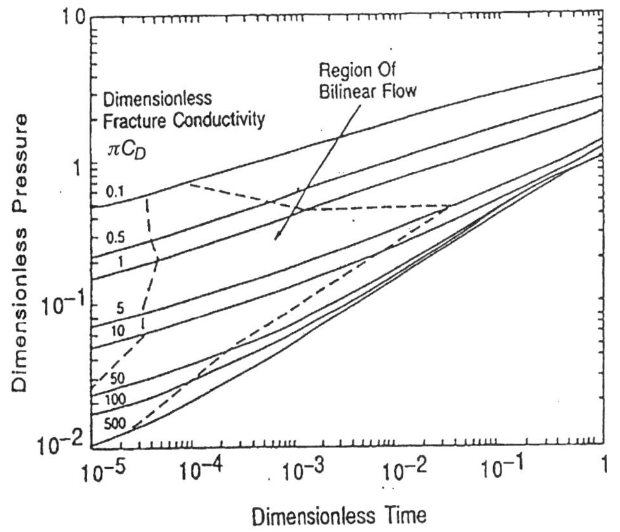

Fig. 6.2—Bilinear-flow region[4] for a finite-conductivity vertical fracture[5] (after Economides[2]).

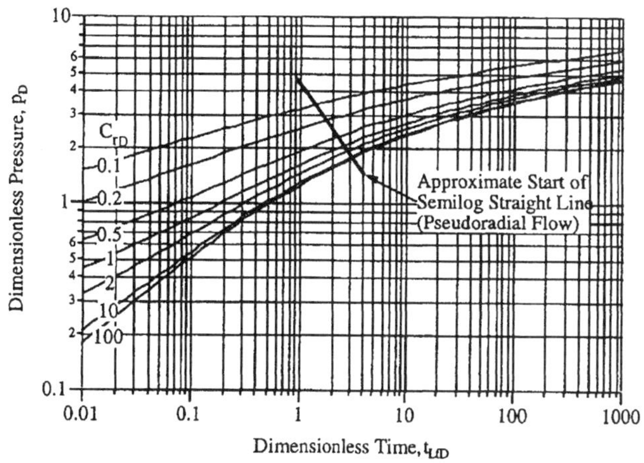

Fig. 6.3—Dimensionless pressure/time plot showing start of pseudoradial-flow period for finite-conductivity vertical fractures (after Cinco-Ley *et al.*[7]).

TABLE 6.1—DURATION OF FLOW PERIODS IN A HYDRAULICALLY FRACTURED WELL

Case	L_f, (ft)	k, (md)
1	100	1
2	100	0.01
3	1,000	0.01

Solution. The end of the linear-flow regime occurs at a dimensionless time of $t_{L_fD} \approx 0.016$ or, with Eq. 6.2,

$$t \approx \frac{\phi\mu c_t L_f^2 t_{L_fD}}{0.0002637k} = \frac{(0.15)(0.03)(1 \times 10^{-4})L_f^2(0.016)}{0.0002637k}$$

$$= \frac{2.73 \times 10^{-5} L_f^2}{k}.$$

Similarly, the time to reach pseudoradial flow is $t_{L_fD} \approx 3$, or

$$t \approx \frac{\phi\mu c_t L_f^2 t_{L_fD}}{0.0002637k} = \frac{(0.015)(0.03)(1 \times 10^{-4})L_f^2(3)}{0.0002637k}$$

TABLE 6.2—RESULTS OF LINEAR FLOW				
Case	L_f (ft)	k (md)	Time to End of Linear Flow (hours)	Time to Start of Pseudoradial Flow (hours)
1	100	1	0.273	51.2
2	100	0.01	27.3	5,120**
3	1,000	0.01	2,730*	512,000***

*114 days; **213 days; ***58 years

$$= \frac{5.12 \times 10^{-5} L_f^2}{k}.$$

Table 6.2 summarizes the results.

6.3 Flow Geometry and Depth of Investigation of a Vertically Fractured Well

Fluid flow in a vertically fractured well has been described with elliptical geometry.[8-12] The equation for an ellipse with its major axis along the x axis and minor axis along the y axis is

$$\frac{x^2}{a^2} + \frac{y^2}{b^2} = 1, \quad (6.6)$$

where the endpoints of the major and minor axes are $(\pm a, 0)$ and $(0, \pm b)$, respectively. The foci of the ellipse are $\pm c$ where $c^2 = a^2 - b^2$. In terms of a well with a single vertical fracture with two wings of equal length, L_f, the relation becomes $L_f^2 = a^2 - b^2$ where L_f is the focal length of the ellipse. **Fig. 6.4** shows the elliptical geometry of a vertically fractured well.

Hale and Evers[11] defined a depth of investigation for a vertically fractured well. They based their definition on a definition of dimensionless time at a distance b, the length of the minor axis:

$$t_{bD} = \frac{0.0002637kt}{\phi \mu c_t b^2}. \quad (6.7)$$

Solving for the length of the minor axis,

$$b = \left(\frac{0.0002637kt}{\phi \mu c_t t_{bD}} \right)^{1/2}. \quad (6.8)$$

If we assume that pseudosteady-state flow exists out to distance b at dimensionless time $t_{bD} = 1/\pi$ as in linear systems, Eq. 6.8 becomes

$$b = 0.02878 \left(\frac{kt}{\phi \mu c_t} \right)^{1/2}, \quad (6.9)$$

which represents the depth of investigation in a direction perpendicular to the fracture at time t for a vertically fractured well. In gas wells, the terms μ and c_t should be $\bar{\mu}$ and $\bar{c}_t$, evaluated at average drainage-area pressure, $\bar{p}$.

The elliptical pattern of the propagating pressure transient can be fully described in terms of the lengths of the major axis, a, the minor axis, b, and the focus, L_f. With the estimate of b from Eq. 6.9 and an estimate of L_f obtained by one of the methods described later, we can estimate the length of the major axis from

$$a = \sqrt{L_f^2 + b^2}. \quad (6.10)$$

Given values of a and b, we can calculate the depth of investigation at a particular time, t, in any direction from the fracture by use of Eq. 6.6. Furthermore, the area, A, enclosed by the ellipse at time t (i.e., the area of the reservoir sampled by the pressure transient) is given by

$$A = \pi ab. \quad (6.11)$$

The coefficient 0.02878 in Eq. 6.9 is strictly correct only for highly conductive fractures ($C_{rD} \geq 100$). As C_{rD} becomes smaller, the ratio a/b also becomes smaller. The lower bound of a/b is 1 (i.e., a circle) as C_{rD} approaches 0.

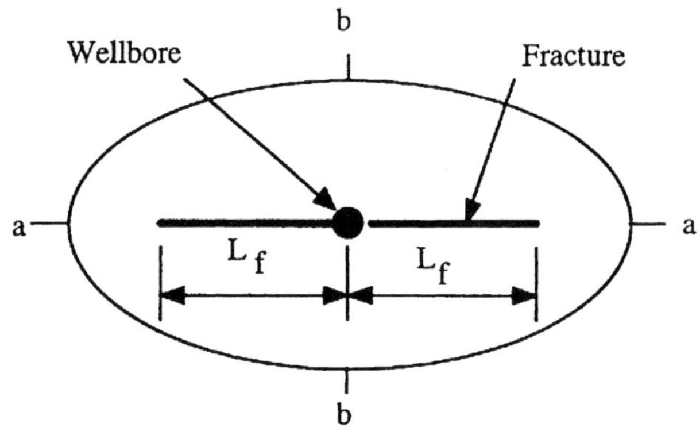

Fig. 6.4—Elliptical flow pattern around a vertically fractured well.

6.4 Specialized Methods for Post-Fracture Well-Test Analysis

Generally, the objectives of post-fracture pressure-transient-test analysis are to assess the success of the fracture treatment and to estimate the fracture half-length, fracture conductivity, and formation permeability. In this section, we discuss three specialized methods for analyzing post-fracture pressure transient tests—pseudoradial flow, bilinear flow, and linear flow. We include examples with these analysis methods in the following section.

6.4.1 Pseudoradial-Flow Method. The pseudoradial-flow method[3] applies when a short, highly conductive fracture is created in a high-permeability formation so that pseudoradial flow develops in a short time. The time required to achieve pseudoradial flow for an infinitely conductive fracture ($C_{rD} \geqq 100$) in either a flow test or a pressure-buildup test is estimated by

$$t_{L_f D} = \frac{0.0002637kt}{\phi \mu c_t L_f^2} \approx 3. \quad (6.12)$$

The beginning of pseudoradial flow is characterized by the flattening of the pressure derivative on a log-log plot and by the start of a straight line on a semilog plot. Therefore, when the pseudoradial-flow regime is reached, conventional semilog analysis, described in Chaps. 2 and 3, can be used to calculate permeability and skin factor. For a highly conductive fracture, skin factor is related to fracture half-length by[13]

$$L_f \approx 2r_w e^{-s}. \quad (6.13)$$

Table 6.3 summarizes working equations for pseudoradial-flow analysis for five cases: (1) slightly compressible liquid, (2) gas with adjusted pressure and adjusted time as plotting functions, (3) gas with real pressure and real time as plotting functions, (4) gas with pressure squared and real time as plotting functions, and (5) gas with pseudopressure and real time as plotting functions. We recommend the following procedure for analyzing test data from the pseudoradial-flow regime.

1. For a drawdown test, make a semilog plot of p_{wf} vs. log t. For a buildup test, make a semilog plot of p_{ws} vs. the Horner time ratio. Use adjusted pressure for gas wells; for shut-in periods, use adjusted time.
2. Determine the position and slope, m, of the semilog straight line.
3. With m, calculate values of k and s (or s' for a gas well) with the appropriate equations from Table 6.3.
4. Calculate the fracture half-length, L_f, with Eq. 6.13.

The pseudoradial-flow method has the following limitations that make it seldom applicable in practice.[3]

1. The conditions that are most favorable for the occurrence of pseudoradial flow are short, highly conductive fractures in high-permeability formations. These formations, however, are rarely fractured. The most common application of hydraulic fractures,

TABLE 6.3—SUMMARY OF WORKING EQUATIONS: RADIAL- OR PSEUDORADIAL FLOW CASE

	Case	
	Oil	Gas, With Adjusted Variables
Flow Test		
Semilog graph variables	p_{wf} vs. t	$p_{a,wf}$ vs. t
Permeability from slope, m, of semilog straight line	$k = \dfrac{162.6 q_o B_o \mu_o}{mh}$	$k = \dfrac{162.6 q_g \bar{B}_g \bar{\mu}_g}{mh}$
Skin factor calculation	$s = 1.151\left[\dfrac{\bar{p} - p_{1hr}}{m} - \log\left(\dfrac{k}{\phi \mu_o c_t r_w^2}\right) + 3.23\right]$	$s' = 1.151\left[\dfrac{\bar{p} - p_{1hr}}{m} - \log\left(\dfrac{k}{\phi \bar{\mu}_g \bar{c}_t r_w^2}\right) + 3.23\right]$
Buildup Test		
Semilog graph variables	p_{ws} vs. $(t_p + \Delta t)/\Delta t$	$p_{a,ws}$ vs. $(t_p + \Delta t_a)/\Delta t_a$
Permeability from slope, m, of semilog straight line	$k = \dfrac{162.6 q_o B_o \mu_o}{mh}$	$k = \dfrac{162.6 q_g \bar{B}_g \bar{\mu}_g}{mh}$
Skin factor calculation	$s = 1.151\left[\dfrac{p_{1hr} - p_{wf}}{m} - \log\left(\dfrac{k}{\phi \mu_o c_t r_w^2}\right) + 3.23\right]$	$s' = 1.151\left[\dfrac{p_{1hr} - p_{wf}}{m} - \log\left(\dfrac{k}{\phi \bar{\mu}_g \bar{c}_t r_w^2}\right) + 3.23\right]$
Definition of $p_{MBH,D}$	$\dfrac{kh(p^* - \bar{p})}{70.6 q_o B_o \mu_o}$	$\dfrac{kh(p_a^* - \bar{p}_a)}{70.6 q_g \bar{B}_g \bar{\mu}_g}$

TABLE 6.3—SUMMARY OF WORKING EQUATIONS: RADIAL- OR PSEUDORADIAL-FLOW CASE (continued)

	Case	
	Gas, With Pressure and Time	Gas, With Pressure-Squared and Time
Flow Test		
Semilog graph variables	p_{wf} vs. t	p_{wf}^2 vs. t
Permeability from slope, m, of semilog straight line	$k = \dfrac{162.6 q_g \bar{B}_g \bar{\mu}_g}{mh}$	$k = \dfrac{1,637 q_g T \bar{z} \bar{\mu}_g}{mh}$
Skin factor calculation	$s' = 1.151\left[\dfrac{\bar{p} - p_{1hr}}{m} - \log\left(\dfrac{k}{\phi \bar{\mu}_g \bar{c}_t r_w^2}\right) + 3.23\right]$	$s' = 1.151\left[\dfrac{\bar{p} - p_{1hr}^2}{m} - \log\left(\dfrac{k}{\phi \bar{\mu}_g \bar{c}_t r_w^2}\right) + 3.23\right]$
Buildup Test		
Semilog graph variables	p_{ws} vs. $(t_p + \Delta t)/\Delta t$	p_{ws}^2 vs. $(t_p + \Delta t)/\Delta t$
Permeability from slope, m, of semilog straight line	$k = \dfrac{162.6 q_g \bar{B}_g \bar{\mu}_g}{mh}$	$k = \dfrac{1637 q_g T \bar{z} \bar{\mu}_g}{mh}$
Skin factor calculation	$s' = 1.151\left[\dfrac{p_{1hr} - p_{wf}}{m} - \log\left(\dfrac{k}{\phi \bar{\mu}_g \bar{c}_t r_w^2}\right) + 3.23\right]$	$s' = 1.151\left[\dfrac{p_{1hr} - p_{wf}^2}{m} - \log\left(\dfrac{k}{\phi \bar{\mu}_g \bar{c}_t r_w^2}\right) + 3.23\right]$
Definition of $p_{MBH,D}$	$\dfrac{kh(p^* - \bar{p})}{70.6 q_g \bar{B}_g \bar{\mu}_g}$	$\dfrac{kh(p^{*2} - \bar{p}^2)}{711 q_g T \bar{z} \bar{\mu}_g}$

wells with long fractures in low-permeability formations, requires impractically long test times to reach pseudoradial flow.

2. The second limitation of the pseudoradial-flow method is that, for gas wells, the apparent skin factor, s', calculated from test data often affected by non-Darcy flow.

3. The pseudoradial-flow method applies only to highly conductive ($C_{rD} \geqq 100$) fractures. For lower-conductivity fractures, fracture lengths calculated with the skin factor (Eq. 6.13) will be too low. **Fig. 6.5**, which relates effective wellbore radius, r_{wa}, to C_{rD}, is useful in estimating L_f if both s and C_{rD} can be estimated.

6.4.2 Bilinear-Flow Method. The bilinear-flow method[3] applies to test data obtained during the bilinear-flow regime in wells with finite-conductivity vertical fractures. Bilinear flow is indicated by a quarter-slope line on a log-log graph of $(p_i - p_{wf})$ vs. t for a constant-rate flow test, or $(p_{ws} - p_{wf})$ vs. Δt_e for a buildup test, with Agarwal's[14] equivalent time function, $\Delta t_e = \Delta t/(1 + \Delta t/t_p)$. The log-log plot of the pressure derivative will also have a quarter slope during this same time period. In gas-well analysis, real time should be used for flow periods and adjusted time (pseudotime normalized by use of $\bar{\mu}$ and $\bar{c}_t$ evaluated at $\bar{p}$) for shut-in periods.

During bilinear flow,

$$p_D = \frac{1.38}{\sqrt{C_{rD}}} t_{L_fD}^{1/4} \quad \text{(6.14)}$$

$$\text{and } t_{L_fD} \frac{dp_D}{dt_{L_fD}} = \frac{0.345}{\sqrt{C_{rD}}} t_{L_fD}^{1/4}. \quad \text{(6.15)}$$

We recommend the following procedure for analyzing test data obtained in the bilinear-flow regime.

1. For a constant-rate flow test, plot p_{wf} vs. $t^{1/4}$ on Cartesian coordinate paper. For a buildup test, plot p_{ws} vs. $\Delta t_e^{1/4}$ (or $\Delta t_{ae}^{1/4}$ for a gas well test).

2. Determine the slope, m_B, of the linear region of the plot.

3. From independent knowledge of k (e.g., from a prefracture well test), estimate the fracture conductivity, $w_f k_f$, with m_B and the relationship

$$w_f k_f = \left(\frac{44.1 qB\mu}{hm_B}\right)^2 \left(\frac{1}{\phi \mu c_t k}\right)^{0.5}, \quad \text{(6.16)}$$

where $\bar{\mu}$ and $\bar{c}_t$ evaluated at $\bar{p}$ are used for a gas-well test.

TABLE 6.3—SUMMARY OF WORKING EQUATIONS: RADIAL- OR PSEUDORADIAL-FLOW CASE (continued)

	Gas, With Pseudopressure and Time
Flow Test	
Semilog graph variables	p_p vs. t
Permeability from slope, m, of semilog straight line	$k = \dfrac{1{,}637 q_g T}{mh}$
Skin factor calculation	$s' = 1.151\left[\dfrac{\bar{p}_p - p_{p,1hr}}{m} - \log\left(\dfrac{k}{\phi \bar{\mu}_g \bar{c}_t r_w^2}\right) + 3.23\right]$
Buildup Test	
Semilog graph variables	p_p vs. $(t_p + \Delta t)/\Delta t$
Permeability from slope, m, of semilog straight line	$k = \dfrac{1{,}637 q_g T}{mh}$
Skin factor calculation	$s' = 1.151\left[\dfrac{p_{p,1hr} - p_{p,wf}}{m} - \log\left(\dfrac{k}{\phi \bar{\mu}_g \bar{c}_t r_w^2}\right) + 3.23\right]$
Definition of $p_{MBH,D}$	$\dfrac{kh\left(p_p^* - \bar{p}_p\right)}{711\, q_g T}$

The bilinear-flow analysis method has the following important limitations.

1. Fracture half-length, L_f, cannot be estimated with this analysis technique.
2. In wells with low-conductivity fractures, wellbore storage frequently distorts early test data for a sufficient length of time so that the quarter-slope line characteristic of bilinear flow may not appear on a log-log plot of test data.
3. The greatest limitation is that an independent estimate of k is required, suggesting that prefracture well tests should be conducted before fracturing the well to obtain independent estimates of formation properties.

6.4.3 Linear-Flow Method. The linear-flow method[3] applies to test data obtained during formation-linear flow in wells with high-conductivity fractures ($C_{fD} \geqq 100$). After wellbore-storage effects have ended, formation-linear flow occurs up to a dimensionless time of $t_{L_fD} \approx 0.016$. Formation-linear flow is modeled by

$$p_D = \left(\pi t_{L_fD}\right)^{1/2} \quad \text{(6.17)}$$

so that

$$\log p_D = \frac{1}{2}\log\left(t_{L_fD}\right) + \frac{1}{2}\log(\pi), \quad \text{(6.18)}$$

which means that a log-log plot of pressure change vs. time will have a slope of one-half. Note that the pressure derivative is

$$t_{L_fD}\frac{dp_D}{dt_{L_fD}} = \frac{1}{2}\left(\pi t_{L_fD}\right)^{1/2} \quad \text{(6.19)}$$

so that

$$\log\left(t_{L_fD}\frac{dp_D}{dt_{L_fD}}\right) = \frac{1}{2}\log\left(t_{L_fD}\right) + \frac{1}{2}\log(\pi), \quad \text{(6.20)}$$

which indicates that a log-log plot of the derivative against time will have a slope of one-half during the same time period.

According to Economides,[2] Agarwal's[14] equivalent time function,

$$\Delta t_e = \frac{\Delta t}{1 + \Delta t_p / t_p}, \quad \text{(6.21)}$$

is used in place of t in the definition of dimensionless time in Eq. 6.3 when analyzing a pressure-buildup test. However, the definition of Δt_e given by Eq. 6.21 applies rigorously only for radial flow in an infinite-acting reservoir. Under the conditions when linear flow is the only flow pattern occurring during the test, a more appropriate form of the equivalent time function is

$$\sqrt{\Delta t_{eL}} = \sqrt{t_p} + \sqrt{\Delta t} - \sqrt{t_p + \Delta t}. \quad \text{(6.22)}$$

Note that test conditions where only linear flow occurs are rare, and, consequently, Eq. 6.22 is infrequently used for well-test analysis. **Table 6.4** summarizes working equations for linear-flow analysis for five cases: (1) slightly compressible liquid, (2) gas with adjusted pressure and adjusted time as plotting functions, (3) gas with real pressure and real time as plotting functions, (4) gas with pressure squared and real time as plotting functions, and (5) gas with pseudopressure and real time as plotting functions. Note that the pressure-buildup equations in Table 6.4 are presented in terms of Eq. 6.21.

We recommend the following procedure for analyzing test data obtained in the linear-flow regime.

1. For a constant-rate flow test, plot p_{wf} vs. $t^{1/2}$ on Cartesian coordinate paper. For a buildup test, plot p_{ws} vs. $\Delta t_e^{1/2}$ (or $\Delta t_{ae}^{1/2}$ for a gas well test).
2. Determine the slope, m_L, of the linear region of the plot.
3. From independent knowledge of k (e.g., from a prefracture well test), estimate the fracture half-length, L_f, with the appropriate equation for $\sqrt{kL_f}$ from Table 6.4.

The linear-flow analysis method also has limitations.

1. The method applies only for fractures with high conductivities. Strictly speaking, linear flow occurs for the condition of uniform flux into a fracture (same flow rate from the formation per unit cross-sectional area of the fracture at all points along the fracture)

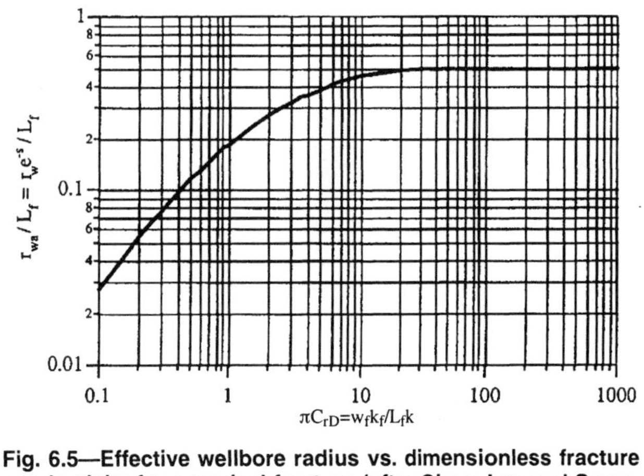

Fig. 6.5—Effective wellbore radius vs. dimensionless fracture conductivity for a vertical fracture (after Cinco-Ley and Samaniego-V.[1]).

TABLE 6.4—SUMMARY OF WORKING EQUATIONS: LINEAR FLOW			
	Oil	Gas, With Adjusted Variables	Gas, With Pressure and Time
Flow Test			
Cartesian-coordinate graph variables	p_{wf} vs. $\sqrt{t}$	$p_{a,wf}$ vs. $\sqrt{t}$	p_{wf} vs. $\sqrt{t}$
$\sqrt{k}L_f$ from slope, m_L, of straight line	$\sqrt{k}L_f = \dfrac{4.064qB}{m_L h}\left(\dfrac{\mu_o}{\phi c_t}\right)^{1/2}$	$\sqrt{k}L_f = \dfrac{4.064q_g\bar{B}_g}{m_L h}\left(\dfrac{\bar{\mu}_g}{\phi \bar{c}_t}\right)^{1/2}$	$\sqrt{k}L_f = \dfrac{4.064q_g\bar{B}_g}{m_L h}\left(\dfrac{\bar{\mu}_g}{\phi \bar{c}_t}\right)^{1/2}$
Buildup Test			
Cartesian-coordinate graph variables	p_{ws} vs. $\sqrt{\Delta t_e}$	$p_{a,ws}$ vs. $\sqrt{\Delta t_{ae}}$	p_{ws} vs. $\sqrt{\Delta t_e}$
$\sqrt{k}L_f$ from slope, m_L, of straight line	$\sqrt{k}L_f = \dfrac{4.064qB}{m_L h}\left(\dfrac{\mu_o}{\phi c_t}\right)^{1/2}$	$\sqrt{k}L_f = \dfrac{4.064q_g\bar{B}_g}{m_L h}\left(\dfrac{\bar{\mu}_g}{\phi \bar{c}_t}\right)^{1/2}$	$\sqrt{k}L_f = \dfrac{4.064q_g\bar{B}_g}{m_L h}\left(\dfrac{\bar{\mu}_g}{\phi \bar{c}_t}\right)^{1/2}$

TABLE 6.4—SUMMARY OF WORKING EQUATIONS: LINEAR FLOW (Continued)		
	Gas, With Pressure Squared and Time	Gas, With Pseudopressure and Time
Flow Test		
Cartesian-coordinate graph variables	p_{wf}^2 vs. $\sqrt{t}$	p_p vs. $\sqrt{t}$
$\sqrt{k}L_f$ from slope, m_L, of straight line	$\sqrt{k}L_f = \dfrac{40.93q_g T\bar{z}}{m_L h}\left(\dfrac{\bar{\mu}_g}{\phi \bar{c}_t}\right)^{1/2}$	$\sqrt{k}L_f = \dfrac{40.93q_g T}{m_L h}\left(\dfrac{1}{\phi \bar{\mu}_g \bar{c}_t}\right)^{1/2}$
Buildup Test		
Cartesian-coordinate graph variables	p_{ws}^2 vs. $\sqrt{\Delta t_e}$	p_p vs. $\sqrt{\Delta t_e}$
$\sqrt{k}L_f$ from slope, m_L, of straight line	$\sqrt{k}L_f = \dfrac{40.93q_g T\bar{z}}{m_L h}\left(\dfrac{\bar{\mu}_g}{\phi \bar{c}_t}\right)^{1/2}$	$\sqrt{k}L_f = \dfrac{40.93q_g T}{m_L h}\left(\dfrac{1}{\phi \bar{\mu}_g \bar{c}_t}\right)^{1/2}$

Note: $\Delta t_e = \dfrac{\Delta t}{(1 + \Delta t/t_p)}$

rather than for infinite fracture conductivity. Therefore, only very early test data ($t_{LfD} \leqq 0.016$) exhibit linear flow in a high-conductivity fracture.

2. Some or all of these early data may be distorted by wellbore storage, further limiting the amount of linear-flow data available for analysis.

3. Like the bilinear-flow method, estimating fracture half-length requires an independent estimate of permeability, k, which suggests that a prefracture well test should be conducted.

6.5 Post-Fracture Well-Test Analysis With Type Curves

We discussed well-test analysis with type curves for homogeneous-acting reservoirs in Chap. 4. In this section, we introduce four type curves developed specifically for analyzing well tests from hydraulically fractured wells. Because type-curve methods are more general than the specialized analysis methods, they span the entire range of flow regimes and include the intervening transition regions. We illustrate type-curve analysis techniques with the following type curves: (1) the Gringarten-Ramey-Raghavan[15] type curve, (2) the Cinco *et al.*[7] type curve, (3) the Agarwal *et al.*[5] type curve, and (4) the Barker-Ramey type curve.[16] **Table 6.5** summarizes interpretation of match points and parameters for type-curve plots for both oil and gas-well tests.

6.5.1 Gringarten-Ramey-Raghavan[15] Type Curve. The Gringarten-Ramey-Raghavan[15] type curve (**Fig. 6.6**) is useful for post-fracture analysis of data from a constant-rate flow test. The type curve is a graph of solutions to flow equations modeling a vertical hydraulic fracture in a finite reservoir under the following assumptions: (1) the fracture is infinitely conductive, (2) the well is centered in a square drainage area with no-flow boundaries, (3) the fracture has two equal-length wings, and (4) wellbore-storage effects are ignored.

We recommend the following procedure for analyzing test data with the Gringarten-Ramey-Raghavan type curve. Although presented for well tests in oil wells, the procedure is also valid for gas well tests when the appropriate adjusted plotting functions are used.

1. Plot the pressure change, Δp, and pressure derivative, $t\Delta p'$, vs. t for a constant-rate flow test or Δp and $\Delta t_e \Delta p'$ vs Δt_e for a buildup test on tracing paper or log-log paper with the same grid size as the type curve.

2. Perform a qualitative type-curve analysis. The purpose of this preliminary analysis is to obtain an initial match and to use the type curves to identify any flow regimes characteristic of infinite-conductivity vertical fractures.

If the type-curve match indicates early data points with $t_{L_fD} < 0.016$ and a slope equal to one-half for both pressure change and pressure derivative, the linear-flow pattern may be present, and we can analyze the data with the technique presented previously.

If the type-curve match indicates a flattening derivative and several data points with $t_{L_fD} > 3$, pseudoradial flow has occurred and we can analyze the data with the technique presented previously. The pseudoradial-flow period in a hydraulically fractured well is similar to radial flow in a homogeneous-acting reservoir, so the pressure derivative during this time period will form a horizontal line.

If boundary effects occur during the test, the test data will deviate upward from the infinite-acting curve defined by $L_e/L_f = \infty$ and will match one of the curves for a finite reservoir characterized by the parameter $L_e/L_f = \infty$.

3. If any flow regimes characteristic of infinite-conductivity vertical fractures are identified, perform a post-fracture analysis with the specialized techniques discussed previously.

If the preliminary type-curve match indicates data points for $t_{L_fD} < 0.016$, plot p_{wf} vs. $t^{1/2}$ on Cartesian coordinate paper for a constant-rate flow test or p_{ws} vs. $\Delta t_e^{1/2}$ for a buildup test. A straight

TABLE 6.5—SUMMARY OF INTERPRETATION PROCEDURES

Case	Interpretation of Unit Slope Line	Definition of C_D	Interpretation of Pressure Match Point
	Slightly Compressible Fluids (Oil)		
Δp vs. Δt	$C_D = \frac{0.0372 q_o B_o}{\phi h c_t r_w^2}\left(\frac{\Delta t}{\Delta p}\right)_{USL}$	$C_D = \frac{22.92 A_{wb}}{\rho_{wb} \phi c_t h r_w^2}$	$k = \frac{141.2 q_o B_o \mu_o}{h}\left(\frac{p_D}{\Delta p}\right)_{MP}$
	Compressible Fluids (Gas)		
Δp vs. Δt	$C_D = \frac{0.0372 q_g \bar{B}_g}{\phi h \bar{c}_t r_w^2}\left(\frac{\Delta t}{\Delta p}\right)_{USL}$	$C_D = \frac{0.8936 V_{wb} c_{wb}}{\phi h \bar{c}_t r_w^2}$	$k = \frac{141.2 q_g \bar{\mu}_g \bar{B}_g}{h}\left(\frac{p_D}{\Delta p}\right)_{MP}$
Δp_a vs. Δt_a	$C_D = \frac{0.0372 q_g \bar{B}_g}{\phi h \bar{c}_t r_w^2}\left(\frac{\Delta t}{\Delta p}\right)_{USL}$	$C_D = \frac{0.8936 V_{wb} T}{\phi h r_w^2 T_{wb}}$	$k = \frac{141.2 q_g \bar{\mu}_g \bar{B}_g}{h}\left(\frac{p_D}{\Delta p}\right)_{MP}$
Δp^2 vs. Δt	$C_D = \frac{0.375 q_g T \bar{z}}{\phi h \bar{c}_t r_w^2}\left(\frac{\Delta t}{\Delta (p^2)}\right)_{USL}$	$C_D = \frac{0.8936 V_{wb} c_{wb}}{\phi \bar{c}_t h r_w^2}$	$k = \frac{1422 q_g T \bar{z} \bar{\mu}_g}{h}\left[\frac{p_D}{(\Delta p^2)}\right]_{MP}$
Δp_p vs. Δt	$C_D = \frac{0.375 q_g T}{\phi h \bar{c}_t r_w^2}\left(\frac{\Delta t}{\Delta p_p}\right)_{USL}$	$C_D = \frac{0.8936 V_{wb} c_{wb}}{\phi \bar{c}_t h r_w^2}$	$k = \frac{1422 q_g T}{h}\left[\frac{p_D}{(\Delta p)}\right]_{MP}$

For drawdown-test analysis, the plotting functions are defined as $\Delta p = \bar{p} - p_{wf}$ (or equivalent) and $\Delta t = t$ = flow time (or equivalent). For buildup-test analysis, the plotting functions are defined as $\Delta p = p_{ws} - p_{wf}$ ($\Delta t = 0$) (or equivalent) and $\Delta t = \Delta t_e = \Delta t/(1 + \Delta t/t_p)$ (or equivalent).

TABLE 6.5—SUMMARY OF INTERPRETATION PROCEDURES (continued)

Case	Interpretation Time Match Point in Gringarten Type Curve	Interpretation of Time Match Point in Fractured Well Type Curves
	Slightly Compressible Fluids (Oil)	
Δp vs. Δt	$C_D = \frac{0.0002637 k}{\phi \mu_o c_t r_w^2}\left(\frac{\Delta t}{t_D/C_D}\right)_{MP}$	$L_f = \left[\left(\frac{0.0002637 k}{\phi \mu_o c_t}\right)\left(\frac{\Delta t}{t_{L_f D}}\right)_{MP}\right]^{1/2}$
	Compressible Fluids (Gas)	
Δp vs. Δt	$C_D = \frac{0.0002637 k}{\phi \bar{\mu}_g \bar{c}_t r_w^2}\left(\frac{\Delta t}{t_D/C_D}\right)_{MP}$	$L_f = \left[\left(\frac{0.0002637 k}{\phi \bar{\mu}_g \bar{c}_t}\right)\left(\frac{\Delta t}{t_{L_f D}}\right)_{MP}\right]^{1/2}$
Δp_a vs. Δt_a	$C_D = \frac{0.0002637 k}{\phi \bar{\mu}_g \bar{c}_t r_w^2}\left(\frac{\Delta t}{t_D/C_D}\right)_{MP}$	$L_f = \left[\left(\frac{0.0002637 k}{\phi \bar{\mu}_g \bar{c}_t}\right)\left(\frac{\Delta t}{t_{L_f D}}\right)_{MP}\right]^{1/2}$
$\Delta(p^2)$ vs. Δt	$C_D = \frac{0.0002637 k}{\phi \bar{\mu}_g \bar{c}_t r_w^2}\left(\frac{\Delta t}{t_D/C_D}\right)_{MP}$	$L_f = \left[\left(\frac{0.0002637 k}{\phi \bar{\mu}_g \bar{c}_t}\right)\left(\frac{\Delta t}{t_{L_f D}}\right)_{MP}\right]^{1/2}$
Δp_p vs. Δt	$C_D = \frac{0.0002637 k}{\phi \bar{\mu}_g \bar{c}_t r_w^2}\left(\frac{\Delta t}{t_D/C_D}\right)_{MP}$	$L_f = \left[\left(\frac{0.0002637 k}{\phi \bar{\mu}_g \bar{c}_t}\right)\left(\frac{\Delta t}{t_{L_f D}}\right)_{MP}\right]^{1/2}$

For drawdown-test analysis, the plotting functions are defined as $\Delta p = p_i - p_{wf}$ (or equivalent) and $\Delta t = t$ = flow time (or equivalent) For buildup-test analysis, the plotting functions are defined as $\Delta p = p_{ws} - p_{wf}$ ($\Delta t = 0$) (or equivalent) and $\Delta t = \Delta t_e = \Delta t/(1 + \Delta t/t_p)$ (or equivalent).

line should appear with a slope inversely proportional to $\sqrt{k}L_f$. If permeability can be estimated from a pseudoradial-flow analysis or if a prefracture value is available, we can estimate fracture half-length, L_f.

If the type-curve match indicates a flattening of the derivative and several data points with $t_{L_f D} > 3$, plot p_{wf} vs. log t for a constant-rate flow test or p_{ws} vs. log $(t_p + \Delta t)/\Delta t$ for a buildup test. If a semilog straight line appears, we can make a unique estimate of effective permeability.

If boundary effects occur during the test and if we have an estimate of the drainage area shape and relative location of the well in this area, we can estimate drainage area with the parameter L_e/L_f and the estimate of fracture half-length from the linear-flow analysis.

4. Next, perform a quantitative type-curve analysis to either confirm the results from the specialized analysis techniques or to obtain estimates when these analysis techniques are not possible.

If a prefracture permeability estimate is available, or, with the permeability estimate from the pseudoradial flow analysis, precalculate a pressure match point. Use a convenient and arbitrary value of dimensionless pressure, p_D, to precalculate Δp.

$$(\Delta p)_{MP} = \frac{141.2 qB\mu}{kh}(p_D)_{MP} \quad \text{. (6.23)}$$

Compare the type-curve match with the preliminary match obtained in Step 2. If the matches differ greatly, then repeat Steps 2 and 3 until a consistent match is obtained.

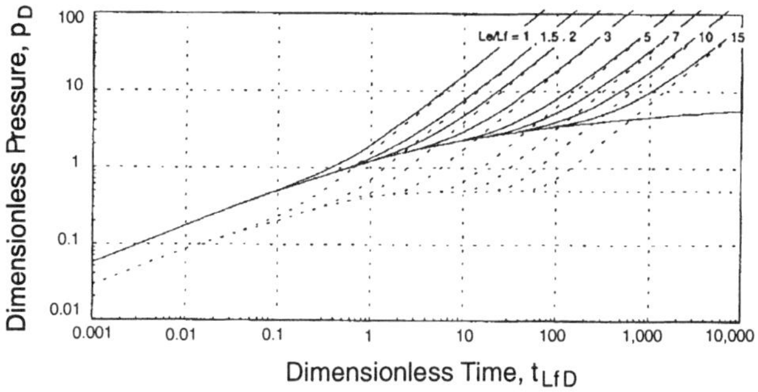

Fig. 6.6—Gringarten-Ramey-Raghavan[15] type curve for a vertically fractured well in the center of a closed square, infinite-conductivity fracture, no wellbore storage.

If a value of permeability is not available, choose a pressure match point $(p_D, \Delta p)_{MP}$ from the qualitative type-curve match and estimate formation permeability from

$$k = \frac{141.2qB\mu}{h}\left(\frac{p_D}{\Delta p}\right)_{MP} \quad \text{.......................} \quad (6.24)$$

Choose a time match point $(\Delta t, t_{L_fD})_{MP}$ and estimate fracture half-length,

$$L_f = \left[\frac{0.0002637k}{\phi\mu c_t}\left(\frac{\Delta t}{t_{L_fD}}\right)_{MP}\right]^{1/2} \quad \text{...............} \quad (6.25)$$

The value of fracture half-length computed with the time match point should be consistent with the value obtained from the linear-flow analysis. If the values are not consistent, repeat Steps 2 and 3.

Example 6.2—Analysis of a Post-Fracture, Constant-Rate Flow Test With Boundary Effects. A constant-rate drawdown test was run in a gas well following a fracture treatment. **Table 6.6** summarizes pressure-drawdown data, while other known data are summarized next. Assume that wellbore-storage effects are negligible. Determine formation permeability, skin factor, and fracture half-length. Use the adjusted pressure and time plotting functions introduced in Chaps. 3 and 4.

q_g = 3,000 Mscf/D
h = 60 ft
r_w = 0.25 ft

TABLE 6.6—PRESSURE DRAWDOWN TEST DATA, EXAMPLE 6.2

t (hours)	p_{wf} (psi)	t (hours)	p_{wf} (psi)
0.001410	4,993.1	56.275	3,977.9
0.002813	4,990.2	89.236	3,792.9
0.005613	4,986.2	141.91	3,588.9
0.011199	4,980.5	225.78	3,369.2
0.028131	4,969.1	360.02	3,135.3
0.070662	4,951.0	455.26	3,016.8
0.17750	4,922.4	576.26	2,892.6
0.44586	4,877.0	730.43	2,766.7
1.1200	4,806.2	927.60	2,637.6
1.7751	4,758.3	1,180.7	2,502.5
3.5421	4,668.5	1,507.4	2,355.7
5.6144	4,594.6	1,932.0	2,188.4
8.8995	4,507.3	2,488.9	1,984.4
11.205	4,458.4	3,227.3	1,720.3
17.764	4,347.7	4,221.8	1,347.9
22.368	4,284.9	5,588.5	683.8
35.473	4,142.8		

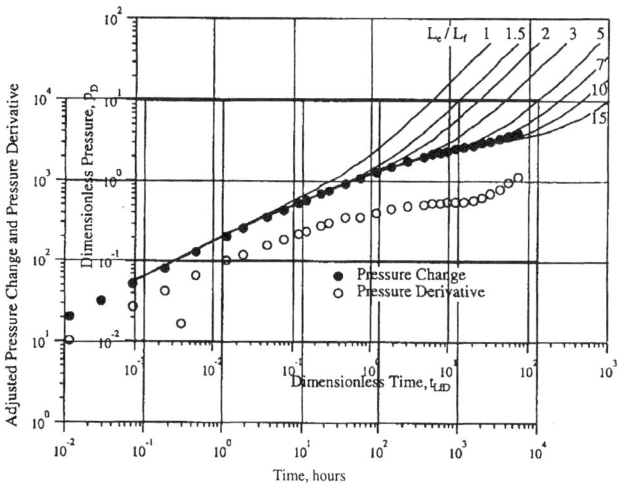

Fig. 6.7—Qualitative type-curve analysis with the Gringarten-Ramey-Raghavan[15] model, Example 6.2.

T = 110°F
γ_g = 0.65
$\bar{z}$ = 0.9910
$p_i = \bar{p}$ = 5,000 psi
$p_{a,1hr}$ = 5,239 psi
$\bar{p}_a$ = 3,974 psi
$\bar{c}_t$ = 2.084 10^{-4} psi^{-1}
$\bar{B}_g$ = 0.7085 RB/Mscf
μ_g = 0.01961 cp
ϕ = 0.10
S_w = 0.35

Solution. To facilitate post-fracture well-test analysis, we have developed worksheets similar to those discussed in Chap. 4. The Appendix gives blank worksheets and worksheets for Example 6.2.

Qualitative Type-Curve Analysis.

1. Plot the adjusted pressure change, Δp_a, and pressure derivative, $t\Delta p'_a$, vs. t on log-log paper with the same grid size as the type curve (**Fig. 6.7**). **Table 6.7** summarizes plotting functions.

2. From a preliminary type-curve analysis using the curve for $L_e/L_f \approx 10$, we observe the following. Formation-linear flow, which ends at $t_{L_fD} = 0.016$, appears to exist for $0.0014 \leqq t \leqq 1.775$ hours. Pseudoradial flow, which begins at $t_{L_fD} = 3$, appears to be present for $141.91 \leqq t \leqq 1{,}932.0$ hours. In addition, the pressure derivative is stable (unchanging) in this time region. Boundary effects are exhibited for $t > 1{,}932.0$ hours, indicating a finite reservoir with $L_e/L_f \approx 10$. Note that the pressure derivative increases rapidly in this region. Because the linear- and pseudoradial-flow regimes have been identified tentatively, we should attempt to analyze the data in these regimes and obtain an estimate of formation permeability and fracture half-length.

Pseudoradial-Flow Analysis.

1. First, we plot the adjusted pressure, p_a, vs. the log of flow time (**Fig. 6.8**). The slope of the straight line through the points during the time period $141.91 \leqq t \leqq 1{,}932.0$ hours is $m = 1{,}284.5$ psi/cycle.

2. With the slope, the formation permeability (Table 6.3) is

$$k = \frac{162.6q_g\bar{B}_g\mu}{mh} = \frac{(162.6)(3,000)(0.7085)(0.01961)}{(1,284.5)(60)}$$

$$= 0.088 \text{ md.}$$

3. From Table 6.3, the appropriate equation for skin factor, s', is

$$s' = 1.151\left[\frac{p_{ai} - p_{a,1hr}}{m} - \log\left(\frac{k}{\phi\overline{\mu c}_t r_w^2}\right) + 3.23\right]$$

$$= 1.151\left\{\frac{3,974 - 5,239}{1,284.5}\right.$$

TABLE 6.7—PLOTTING FUNCTIONS FOR WELL-TEST ANALYSIS, EXAMPLE 6.2

$p_{wf,a}$ (psi)	Δp_a (psi)	$t\Delta p'_a$ (psi)	Square Root Time Function (hours$^{1/2}$)	$p_{wf,a}$ (psi)	Δp_a (psi)	$t\Delta p'_a$ (psi)	Square Root Time Function (hours$^{1/2}$)
3,966.6	7.1113		0.03755	3,082.1	891.57	347.83	5.9559
3,963.6	10.102	5.1497	0.05304	2,909.8	1,063.9	396.05	7.5017
3,959.5	14.226	7.2395	0.07492	2,716.4	1,257.4	439.35	9.4512
3,953.6	20.104	10.332	0.10583	2,503.3	1,470.4	476.02	11.913
3,941.8	31.859	16.513	0.16772	2,274.9	1,698.8	504.21	15.026
3,923.2	50.522	26.161	0.26582	2,033.9	1,939.8	520.13	18.974
3,893.7	80.050	41.495	0.42131	1,913.0	2,060.7	525.19	21.337
3,846.7	126.96	65.240	0.66773	1,787.5	2,186.2	527.36	24.005
3,773.5	200.23	98.385	1.0583	1,661.8	2,311.9	533.97	27.026
3,723.8	249.88	118.66	1.3323	1,534.7	2,439.0	542.66	30.457
3,630.6	343.10	154.08	1.8820	1,404.1	2,569.6	564.34	34.361
3,553.8	419.95	182.11	2.3695	1,265.5	2,708.2	607.58	38.826
3,462.8	510.88	214.81	2.9832	1,112.2	2,861.5	678.45	43.955
3,411.8	561.87	232.59	3.3474	933.36	3,040.3	782.32	49.888
3,296.3	677.42	274.30	4.2147	718.29	3,255.4	923.50	56.809
3,230.7	743.02	298.10	4.7295	453.14	3,520.6	1,091.8	64.975
3,082.1	891.57	347.83	5.9559	119.69	3,854.0		74.756

$$- \log\left[\frac{0.088}{(0.10)(0.01961)(2.084\times10^{-4})(0.25)^2}\right] + 3.23\Bigg\}$$

$$= -4.94,$$

where $\bar{p}_{a,1\text{hr}}$ = the adjusted pressure on the extrapolated semilog straight line at $t = 1$ hour.

3. For an infinite-conductivity fracture, the fracture half-length is

$$L_f \approx 2r_w e^{-s} = (2)(0.25)\, e^{-(-4.94)} = 70 \text{ ft}.$$

Linear-Flow Analysis.

1. First, we plot adjusted pressure, p_a, vs. the square root of time (**Fig. 6.9**).

2. The slope of the straight line through the data points during the time period $0.0014 \leqq t \leqq 1.7751$ hours is $m_L = 193.9$ psi/$\sqrt{\text{hr}}$. With this slope and with permeability from the pseudoradial flow analysis, we estimate the fracture half-length (Table 6.4) to be

$$L_f = \frac{4.064 q_g \bar{B}_g}{m_L h \sqrt{k}}\left(\frac{\bar{\mu}}{\phi c_t}\right)^{1/2}$$

$$= \frac{(4.064)(3,000)(0.7085)}{(193.9)(60)\sqrt{0.088}}\left[\frac{0.01961}{(0.10)(2.084\times10^{-4})}\right]^{1/2}$$

$$= 77 \text{ ft},$$

which agrees with the estimate of 70 ft from the pseudoradial-flow analysis. Because this estimate is obtained from a straight-line slope of data in a particular flow regime, we have more confidence in its accuracy than in an estimate from type-curve or pseudoradial-flow analysis.

Quantitative Type-Curve Analysis.

1. Because the pseudoradial-flow analysis provided us with an estimate of formation permeability, we can confirm our analysis with type curves. First, we precalculate a pressure match point (**Fig. 6.10**) with an arbitrary value of $p_D = 1$.

The pressure match point is

$$(\Delta p_a)_{MP} = \frac{141.2 q_g \bar{B}_g \bar{\mu}}{kh}(p_D)_{MP}$$

$$= \frac{(141.2)(3,000)(0.7085)(0.0196)}{(0.088)(60)}(1)$$

$$= 1,115 \text{ psi}.$$

2. Maintaining the pressure match point from Step 1, we slide the log-log plot horizontally until a match is obtained. As Fig. 6.10 shows, a good match is obtained at an interpolated value of the parameter $L_e/L_f = 8.5$. This value is approximately equal to the value obtained in the preliminary type-curve analysis. With the data in the

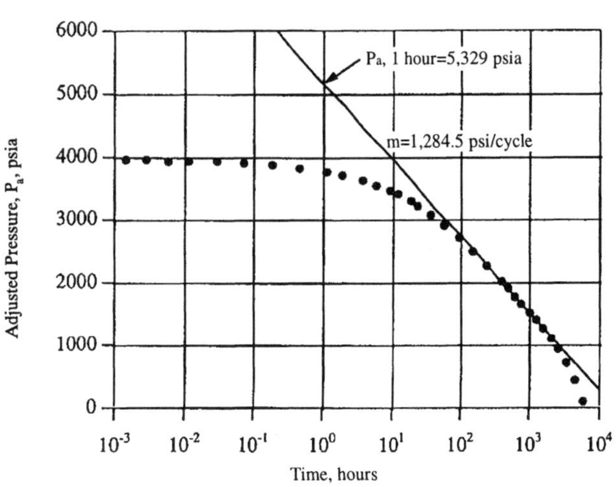

Fig. 6.8—Pseudoradial-flow analysis, Example 6.2.

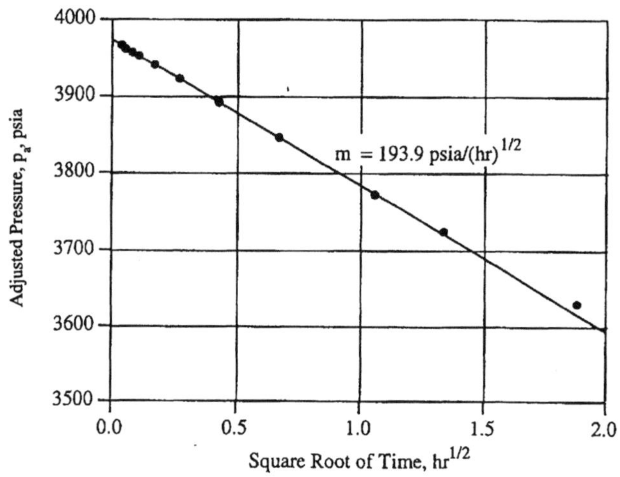

Fig. 6.9—Linear-flow analysis, Example 6.2.

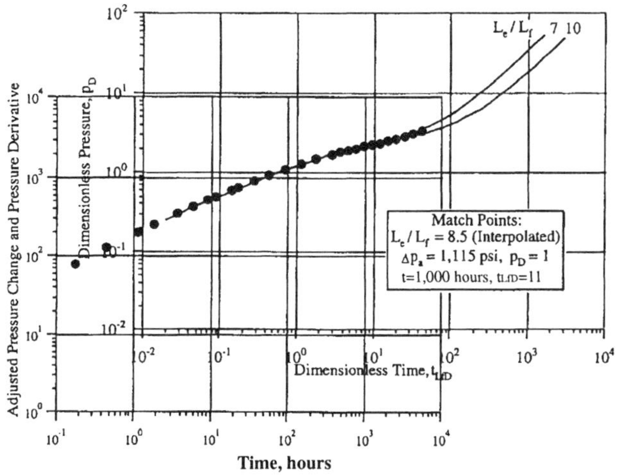

Fig. 6.10—Type-curve match with the Gringarten-Ramey-Raghavan[15] model, Example 6.2.

matched position, we choose the time match point Δt_a = 1,000 hours and $t_{L_fD} = 11$.

Estimate the fracture half-length with the time match point:

$$L_f = \left[\frac{0.0002637k}{\phi \overline{\mu c}_t} \left(\frac{\Delta t_a}{t_{L_fD}} \right)_{MP} \right]^{1/2}$$

$$= \left[\frac{(0.0002637)(0.088)}{(0.10)(0.01961 \times 10^{-4})} \left(\frac{1,000}{11} \right) \right]^{1/2}$$

$$= 72 \text{ ft},$$

which is consistent with the values obtained from both the linear- and pseudoradial-flow analyses. Inconsistency among calculated values of hydraulic fracture half-length from the linear, pseudoradial, and type-curve analyses may require iteration until the values are approximately equal. Because $L_e/L_f = 8.5$ and because the best value of $L_f = 77$ ft, $L_e = 655$ ft.

6.5.2 Cinco-Ley *et al.*[7] Type Curve. The Cinco-Ley *et al.*[7] type curve (**Fig. 6.11**) can be used for post-fracture analysis of data from a constant-rate flow test or a pressure-buildup test. The type curve is a graph of solutions to flow equations modeling a vertical hydraulic fracture in an infinite-acting reservoir under the following assumptions: (1) the fracture has finite conductivity that is uniform throughout the fracture, (2) the fracture has two equal-length wings, and (3) wellbore-storage effects are ignored.

We recommend the following procedure for analyzing test data with the Cinco-Ley *et al.* type curve. Although presented in terms of variables for slightly compressible liquids, the procedure is also applicable for gas-well tests when the appropriate plotting functions are used.

1. Plot the pressure change, Δp, and pressure derivative, $t\Delta p'$, vs. t for a constant-rate flow test or Δp and $\Delta t_e \Delta p'$ vs. Δt_e for a buildup test on tracing paper or log-log paper with the same grid size as the type curve.

2. Perform a preliminary or qualitative type-curve analysis to obtain an initial match and to use the type curves to identify any flow regimes characteristic of finite-conductivity vertical fractures. Find the type curve that best matches the test data. Because of the similar curve shapes, a unique match is difficult to obtain, especially for low-conductivity fractures.

For infinite-conductivity fractures, pseudoradial flow begins at $t_{L_fD} = 3$. Pseudoradial flow begins somewhat sooner for finite-conductivity fractures. In either case, the pressure derivative flattens during this flow regime. If the type-curve match indicates a flat derivative and several data points with $t_{L_fD} > 3$, pseudoradial-flow regime is present and we can analyze these data with the technique presented in Sec. 6.4.

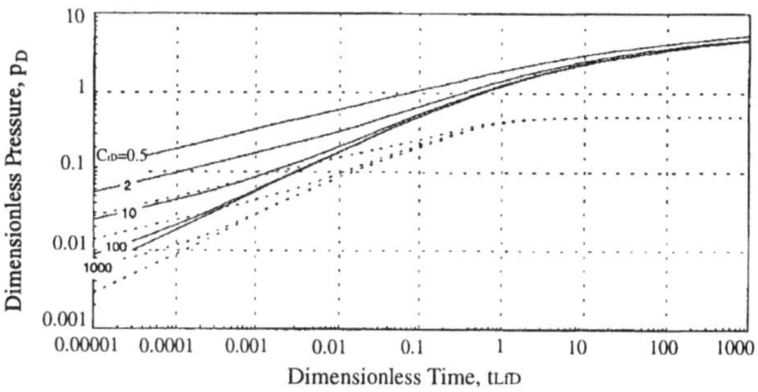

Fig. 6.11—Cinco-Ley *et al.*[7] type curve for a vertically fractured well, finite-conductivity fracture.

Bilinear flow may appear with a finite-conductivity fracture if wellbore-storage effects do not distort the pressure response. If the pressure-change and pressure-derivative curve match indicates early data points with a one-quarter slope, then the bilinear-flow pattern may be present and we can perform bilinear-flow analysis.

If boundary effects occur during the test, the test data will deviate from the type curves, which were developed assuming an infinite-acting reservoir.

3. If any flow regimes characteristic of finite-conductivity vertical fractures are identified, perform a post-fracture analysis with the specialized techniques discussed previously.

If the type-curve match indicates several data points with a flat derivative and with $t_{L_fD} > 3$, plot p_{wf} vs. log t for a constant-rate flow test or p_{ws} vs. log $(t_p + \Delta t)/\Delta t$ for a buildup test. If a semilog straight line appears, a unique estimate of formation permeability can be made.

If the type-curve match indicates a one-quarter slope for both pressure change and pressure derivative, then plot p_{wf} vs. $t^{1/4}$ on Cartesian coordinate paper for a constant-rate flow test or p_{ws} vs. $\Delta t_e^{1/4}$ for a buildup test. We can estimate fracture conductivity (Eq. 6.16) from the slope of the straight line indicative of bilinear flow, but we must have an estimate of formation permeability, either from a prefracture well test or from the pseudoradial-flow analysis.

4. Next, perform a quantitative type-curve analysis. The purpose of this analysis is either to confirm the results from the specialized analysis techniques or to obtain estimates when these analysis techniques are not possible.

If a prefracture permeability estimate is available or with the permeability estimate from the pseudoradial flow analysis precalculate a pressure match point. Use an arbitrary value of dimensionless pressure, p_D, to precalculate Δp.

$$(\Delta p)_{MP} = \frac{141.2qB\mu}{kh} (p_D)_{MP} . \quad \text{(6.23)}$$

Compare the type-curve match with the preliminary match obtained in Step 2. If the matches are significantly different, repeat Steps 2 and 3 until a consistent match is obtained.

If a value of permeability is not available, choose a pressure match point (p_D, Δp) from the qualitative type-curve match and estimate formation permeability with

$$k = \frac{141.2qB\mu}{h} \left(\frac{p_D}{\Delta p} \right)_{MP} . \quad \text{(6.24)}$$

Note that, without an independent estimate of k, unambiguous type-curve matches are difficult to obtain.

Choose a time match point ($\Delta t, t_{L_fD}$) and estimate fracture half-length,

$$L_f = \left[\frac{0.0002637k}{\phi \mu c_t} \left(\frac{\Delta t}{t_{L_fD}} \right)_{MP} \right]^{1/2} . \quad \text{(6.25)}$$

TABLE 6.8—POST-FRACTURE BUILDUP TEST DATA, EXAMPLE 6.3			
Δt (hours)	p_{ws} (psi)	Δt (hours)	p_{ws} (psi)
0.0480	4,673.8	23.129	4,742.6
0.1440	4,677.1	39.257	4,761.6
0.3360	4,680.7	62.537	4,781.0
0.7200	4,685.1	86.537	4,796.0
1.4880	4,691.1	110.54	4,808.7
3.0240	4,698.9	134.54	4,818.6
6.0960	4,709.8	158.54	4,827.9
12.240	4,725.1	168.00	4,831.1

The value of fracture half-length computed with the time match point should be consistent with the value obtained from the linear-flow analysis. If the values do not agree, repeat Steps 2 and 3.

After k and L_f have been determined, the fracture conductivity, $w_f k_f$, can be found with the value of dimensionless fracture conductivity, C_{rD}, from the type-curve match. Fracture conductivity is given by

$$w_f k_f = C_{rD} \pi k L_f \quad \text{(6.26)}$$

The value of fracture conductivity computed with Eq. 6.26 should be comparable with the value estimated from the bilinear-flow analysis; however, the value obtained from the bilinear-flow analysis should be more accurate because method uses the pressure behavior from a Cartesian coordinate plot (rather than a less sensitive log-log plot).

Example 6.3—Post-Fracture Pressure-Buildup Test in a Gas Well With a Finite-Conductivity Vertical Fracture. A pressure buildup test was run following a fracture treatment in a gas well. **Table 6.8** summarizes pressure and time data, while other known data are summarized next. A prefracture pressure-buildup test gave a permeability estimate of $k = 0.1$ md. Determine the fracture half-length, L_f, and fracture conductivity, $w_f k_f$ with the Cinco-Ley *et al.* type curves and assuming a finite-conductivity vertical fracture. In addition, identify and use any specialized analysis techniques to confirm the results.

$\phi =$ 0.20
$h =$ 100 ft
$r_w =$ 0.25 ft
$c_t =$ 1.057×10^{-4} psi^{-1}
$B_g =$ 0.6992 RB/Mscf
$\mu =$ 0.0244 cp
$q_g =$ 1,000 Mscf/D

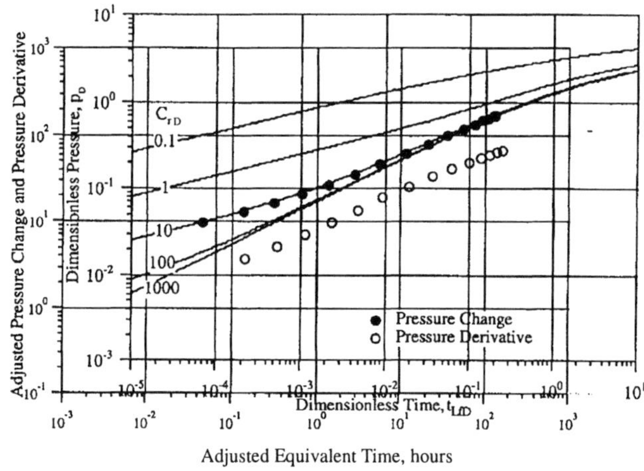

Fig. 6.12—Qualitative type-curve match with the Cinco-Ley *et al.*[7] model, Example 6.3.

$t_p =$ 168 hours
$t_{pa} =$ 355.5 hours
$\bar{z} =$ 1.0136
$\gamma_g =$ 0.68
$T =$ 225°F
$p_i = \bar{p} =$ 5,000 psi
$p_{wf} =$ 4,664 psi
$S_w =$ 0.25

Solution. *Qualitative Type-Curve Analysis.*

1. Plot Δp_a and $\Delta t_{ae} \Delta p'_a$ vs. Δt_{ae} on tracing paper or log-log paper with the same grid size as the type curve (**Fig. 6.12**). **Table 6.9** summarizes the well-test analysis plotting functions.

2. From a preliminary type-curve analysis, we observe the following: for infinite-conductivity fractures, pseudoradial flow begins at $t_{L_f D} \approx 3$ and the pressure derivative flattens. On the basis of the match shown in Fig. 6.12, the pseudoradial-flow regime probably is not present. Bilinear flow, which may evolve with a finite-conductivity fracture, has a characteristic quarter slope on both pressure change and pressure-derivative plots at early times. The preliminary type curve match suggests the first several points may exhibit bilinear flow. No boundary effects appear at the end of the test, indicating an infinite-acting reservoir.

Bilinear-Flow Analysis.

1. First, we plot the adjusted pressure, p_a, vs. the fourth root of time function (**Fig. 6.13**).

TABLE 6.9—PLOTTING FUNCTIONS FOR WELL-TEST ANALYSIS, EXAMPLE 6.3				
Δt_{ae} (hours)	p_a (psi)	Δp_a (psi)	$\Delta t_{ae}\, \Delta p'_a$ (psi)	Fourth Root of Time Function (hours$^{1/4}$)
0.044595	3,098.2	9.4277	—	0.45954
0.13390	3,101.4	12.703	3.6761	0.60492
0.31259	3,105.0	16.275	5.0108	0.74773
0.67002	3,109.4	20.642	6.9953	0.90474
1.3847	3,115.3	26.596	9.5877	1.0848
2.8120	3,123.1	34.346	13.237	1.2949
5.6557	3,133.9	45.179	18.763	1.5421
11.293	3,149.1	60.384	25.048	1.8332
21.103	3,166.5	77.776	32.839	2.1433
35.202	3,185.4	96.683	40.673	2.4358
54.660	3,204.7	115.99	48.228	2.7191
73.666	3,219.6	130.92	53.865	2.9297
91.671	3,232.3	143.57	58.620	3.0943
108.74	3,242.2	153.44	63.746	3.2292
124.94	3,251.4	162.70	65.625	3.3433
131.11	3,254.6	165.89	—	3.3838

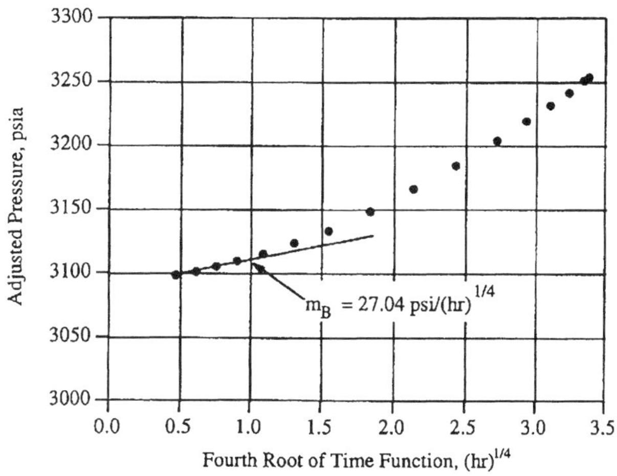

Fig. 6.13—Bilinear-flow analysis, Example 6.3.

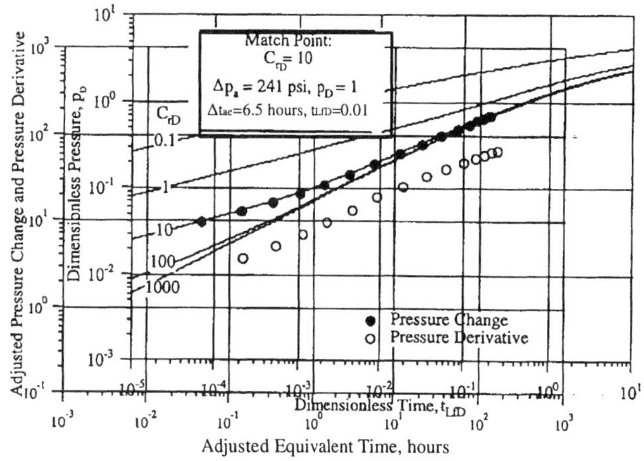

Fig. 6.14—Type-curve match with Cinco-Ley *et al.*[7] model for a finite-conductivity fracture, Example 6.3.

2. The first several data points appear to be linear. The slope of the straight line through the data for $0.0446 \le (\Delta t_{ae})^{1/4} \le 0.67$ hr$^{1/4}$ is $m_B = 27.04$ psi/(hr)$^{1/4}$. From Eq. 6.16, the fracture conductivity is estimated to be

$$w_f k_f = \left(\frac{44.1 qB\mu}{hm_B}\right)^2 \left(\frac{1}{\phi\mu c_t k}\right)^{0.5}$$

$$= \left[\frac{(44.1)(1,000)(0.6992)(0.0244)}{(100)(27.04)}\right]^2$$

$$\times \left[\frac{1}{(0.20)(0.0244)(1.057 \times 10^{-4})(0.1)}\right]^{0.5}$$

$$= 340.9 \text{ md-ft.}$$

Quantitative Type-Curve Analysis.

1. Find the type curve (**Fig. 6.14**) that best matches the test data. For this test, we have an estimate of formation permeability before the fracture treatment, so the pressure match point is precalculated with an arbitrary value of $p_D = 1$.

$$(\Delta p_a)_{MP} = \frac{141.2 q_g \overline{B}_g \overline{\mu}}{kh} (p_D)_{MP}$$

$$= \frac{(141.2)(1,000)(0.6992)(0.0244)}{(0.1)(100)}(1)$$

$$= 241 \text{ psi.}$$

2. Find the "best" match by sliding the log-log plot horizontally while maintaining the pressure match point. We match on a dimensionless fracture conductivity of $C_{rD} = 10$; a time match point is $\Delta t_{ae} = 6.5$ hours and $t_{L_fD} = 0.01$.

From the time match point, the fracture half-length is

$$L_f = \left[\frac{0.0002637k}{\phi\overline{\mu}\overline{c}_t}\left(\frac{\Delta t_{ae}}{t_{L_fD}}\right)_{MP}\right]^{1/2}$$

$$= \left[\frac{(0.0002637)(0.1)}{(0.20)(0.0244 \times 10^{-4})}\left(\frac{6.5}{0.01}\right)\right]^{1/2}$$

$$= 182.3 \text{ ft.}$$

3. The fracture conductivity, $w_f k_f$, can be found with the value of dimensionless fracture conductivity, $C_{rD} = 10$, from the type curve match, $w_f k_f = \pi k C_{rD} L_f = \pi\,(0.10)(10)(182.3) = 572.7$ md-ft, which is fairly consistent with the bilinear analysis, thus confirming the presence of the bilinear-flow regime.

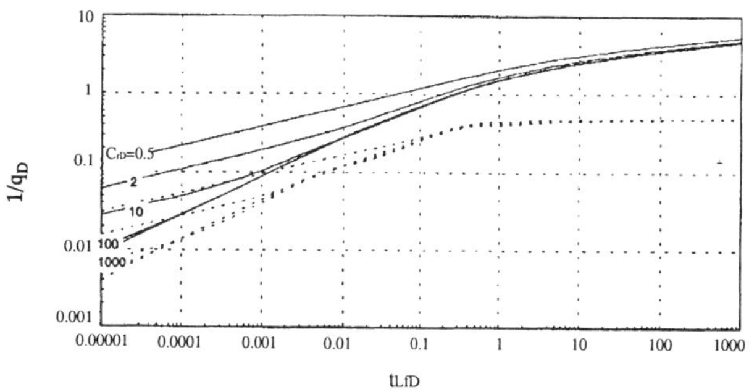

Fig. 6.15—Agarwal *et al.*[5] type curve for a vertically fractured well, finite-conductivity fracture, constant BHP production.

6.5.3 Agarwal *et al.*[5] Type Curve. The Agarwal *et al.*[5] type curve (**Fig. 6.15**) is useful for analyzing flow tests or long-term production data in wells produced at essentially constant BHP. The type curve is a graph of solutions to flow equations modeling a vertical hydraulic fracture in an infinite-acting reservoir under the following assumptions: (1) the fracture has finite conductivity that is uniform throughout the fracture and (2) the fracture has two equal-length wings. When a well produces at constant BHP, wellbore-storage effects (other than wellbore unloading immediately after starting production from a previously shut-in well) are not present, so wellbore storage is not of concern in analyzing test data for this case.

We recommend the following procedure to analyze data from constant BHP well tests with the Agarwal *et al.* type curve. Although presented in terms of variables for a slightly compressible liquid, the procedure is also valid for gas-well test analysis with the appropriate plotting functions:

1. Plot the reciprocal of rate, $1/q$, vs. flowing test time, t, on tracing paper or log-log paper with the same grid size as the type curve.

2. Find the type curve that best matches the test data. The data may fit more than one curve, particularly for lower values of C_{rD}. An unambiguous match cannot be obtained without an independent estimate of permeability. Once the best match is obtained, record the value of the type curve correlating parameter, C_{rD}.

3. If a type-curve match is attempted without an independent estimate of permeability, estimate k with a rate match point $(1/q, 1/q_D)_{MP}$ from the type-curve match,

$$k = \frac{141.2 qB\mu}{h(p_i - p_{wf})}\left(\frac{1/q_D}{1/q}\right)_{MP} \quad \text{.....................} \quad (6.27)$$

4. If an independent estimate of formation permeability is available, a unique type-curve match can be found. First, precalculate a

TABLE 6.10—FLOW-TEST DATA, EXAMPLE 6.4			
t (hours)	q_g (Mscf/D)	t (hours)	q_g (Mscf/D)
1.5102	6,693.7	119.96	834.7
1.9013	6,071.2	151.02	745.5
2.3935	5,493.1	190.13	666.4
3.0133	4,958.9	239.35	596.1
3.7935	4,467.8	301.33	533.7
4.7757	4,018.2	379.35	478.0
6.0123	3,608.3	477.57	428.5
7.5690	3,235.9	601.23	384.3
9.5289	2,898.7	756.90	344.8
11.996	2,594.3	952.89	309.6
15.102	2,320.0	1,199.6	278.4
19.013	2,073.3	1,510.2	250.6
23.935	1,851.9	1,901.3	225.7
30.133	1,653.5	2,393.5	203.6
37.935	1,475.7	3,013.3	183.9
47.757	1,316.8	3,793.5	166.4
60.123	1,174.7	4,775.7	150.8
75.690	1,048.0	6,012.3	136.9
95.289	935.1	7,569.1	124.5

TABLE 6.11—PLOTTING FUNCTIONS, EXAMPLE 6.4			
t (hours)	$1/q_g$ ($\times 10^6$) (Mscf/D)$^{-1}$	t (hours)	$1/q_g$ ($\times 10^6$) (Mscf/D)$^{-1}$
1.5102	149.4	119.96	1,198.1
1.9013	164.7	151.02	1,341.4
2.3935	182.0	190.13	1,500.7
3.0133	201.7	239.35	1,677.6
3.7935	223.8	301.33	1,873.9
4.7757	248.9	379.35	2,091.9
6.0123	277.1	477.57	2,333.8
7.5690	309.0	601.23	2,602.2
9.5289	345.0	756.90	2,899.8
11.996	385.5	952.89	3,229.5
15.102	431.0	1,199.6	3,592.1
19.013	482.3	1,510.2	3,990.3
23.935	540.0	1,901.3	4,431.6
30.133	604.8	2,393.5	4,912.7
37.935	677.6	3,013.3	5,438.5
47.757	759.4	3,793.5	6,010.1
60.123	851.3	4,775.7	6,631.4
75.690	954.2	6,012.3	7,303.7
95.289	1,069.4	7,569.1	8,029.5

rate match point by selecting an arbitrary value for $1/q_D$ and calculating $1/q$,

$$(1/q)_{MP} = \frac{141.2B\mu}{kh(p_i - p_{wf})}(1/q_D)_{MP} \text{ . . . . } (6.28)$$

Align the matching values of $1/q$ and $1/q_D$ and move the test data horizontally to obtain a unique match. Record a time match point $(\Delta t,\ t_{L_fD})_{MP}$ from the match.

5. Estimate fracture half-length, L_f, with Eq. 6.25 and a time match point and the permeability.

$$L_f = \left[\frac{0.0002637k}{\phi\mu c_t}\left(\frac{\Delta t}{t_{L_fD}}\right)_{MP}\right]^{1/2} \text{ . . . . } (6.25)$$

6. After k and L_f have been determined, calculate the fracture conductivity, $w_f k_f$ (Eq. 6.26), with the value of dimensionless fracture conductivity, C_r, from the type-curve match.

$$w_f k_f = C_{rD}\pi k L_f \text{ . . . . } (6.26)$$

Example 6.4 —Analyzing a Post-Fracture Production Test Under Constant Bottomhole Flowing Conditions. A hydraulically fractured gas well produced at a constant flowing BHP with the rate history given in **Table 6.10**. Analysis of a prefracture buildup test indicates $k = 0.01$ md. Estimate the hydraulic fracture half-length and fracture conductivity with the Agarwal *et al.* type curve.

$r_w = 0.25$ ft
$h = 50$ ft
$\phi = 0.10$
$T = 150°F$
$\gamma_g = 0.80$
$\bar{z} = 0.985$
$\overline{B_g} = 1.3639$ RB/Mscf
$\bar{c_t} = 4.5 \times 10^{-4}$ psi^{-1}
$\bar{\mu} = 0.0142$ cp
$p_i = \bar{p} = 2{,}400$ psi
$p_{ai} = 1{,}476$ psi
$p_{wf} = 2{,}000$ psi
$p_{awf} = 1{,}077$ psi

Solution.

1. First, we plot the reciprocal of rate, $1/q_g$, vs. elapsed test time, t, on log-log paper with the same grid size as the type curve. Because this is a flow test, use real time rather than adjusted time. **Table 6.11** summarizes the plotting functions.

2. Because we have a prefracture estimate of the formation permeability, we can precalculate a rate match point and obtain a unique match. For this problem, we arbitrarily select $(1/q_D)_{MP} = 10^{-1}$.

The corresponding reciprocal rate is

$$(1/q)_{MP} = \frac{141.2\overline{B}_g\overline{\mu}}{kh(p_{a,i} - p_{a,wf})}(1/q_D)_{MP}$$

$$= \frac{(141.2)(1.364)(0.0142)}{(0.01)(50)(1{,}476 - 1{,}077)}(10^{-1})$$

$$= 0.00137 \text{ D/Mscf.}$$

Maintaining the rate match point, we obtain a good match with $C_{rD} = 100$ (**Fig. 6.16**). In addition, we select the following time match point: $t = 100$ hours and $t_{L_fD} = 9.75 \times 10^{-4}$.

3. With the time match point, the fracture half-length is

$$L_f = \left[\frac{0.0002637k}{\phi\bar{\mu}\bar{c}_t}\left(\frac{\Delta t}{t_{L_fD}}\right)_{MP}\right]^{1/2}$$

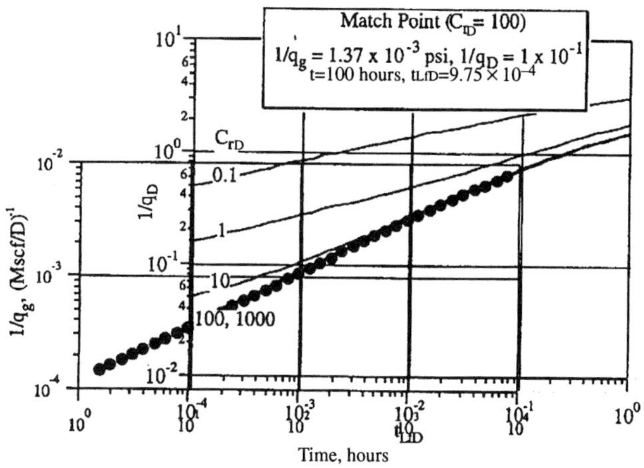

Fig. 6.16—Type-curve match, Example 6.4.

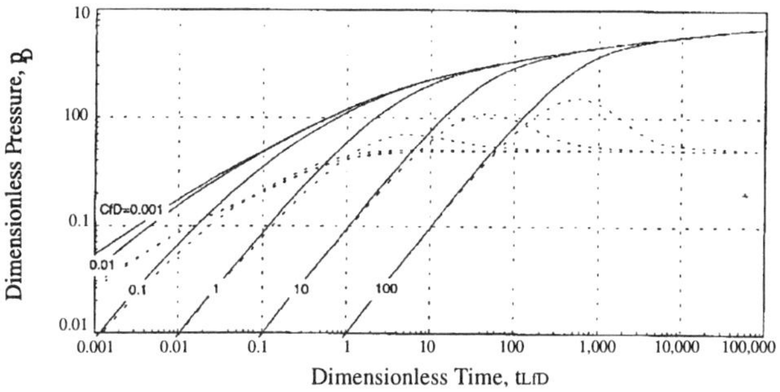

Fig. 6.17—Barker-Ramey type curve for a vertically fractured well with wellbore storage.[16]

$$= \left[\frac{(0.0002637)(0.1)}{(0.10)(0.0142)(4.5\times10^{-4})}\left(\frac{100}{9.75\times10^{-4}}\right)\right]^{1/2}$$

$$= 651 \text{ ft.}$$

4. The fracture conductivity, $w_f k_f$, is $w_f k_f = pC_{rD}kL_f = \pi(100)(0.01)(651) = 2{,}045$ md-ft.

6.5.4 Barker-Ramey Type Curve.[16] The Barker-Ramey type curve (**Fig. 6.17**) includes wellbore-storage effects on a post-fracture constant-rate flow test or pressure-buildup test.[16] The type curve is a graph of solutions to flow equations modeling a vertical hydraulic fracture in an infinite-acting reservoir under the following assumptions: (1) the fracture is infinitely conductive and (2) the fracture has two equal-length wings.

We recommend the following procedure for analyzing test data with the Barker-Ramey type curve. Although presented in terms of variables for a slightly compressible liquid, the procedure also is valid for gas-well-test analysis with the appropriate plotting functions.

1. Plot the pressure change, Δp, and pressure derivative, $t\Delta p'$ vs. t, for a constant-rate flow test or Δp and $\Delta t_e \Delta p'$ vs. Δt_e or a buildup test on tracing paper or log-log paper with the same grid size as the type curve.

2. Perform a preliminary or qualitative type-curve analysis. The purpose of this preliminary analysis is to obtain an initial match and to use the type curves to identify any flow regimes characteristic of infinite-conductivity vertical fractures.

Wellbore-storage effects are indicated for data matched on type curves for $C_{fD} > 0$. Early-time data with a slope significantly steeper than that of any type curve usually indicates wellbore-storage distortion of the test data.

If the type-curve match indicates several data points with $t_{L_fD} > 3$ and if the pressure-derivative plot flattens, the pseudoradial-flow regime is present and we can perform a pseudoradial-flow analysis.

If the type curve match indicates early pressure change and pressure derivative points with slopes of one half for $t_{L_fD} < 0.016$, the linear-flow pattern may be present and we can perform a linear-flow analysis.

If boundary effects occur during the test, then late-time data will deviate from the type curves that were developed assuming an infinite-acting reservoir.

3. If any flow regimes characteristic of infinite-conductivity vertical fractures are identified, perform a post-fracture analysis with the specialized techniques discussed previously.

If the type-curve match indicates several data points with $t_{L_fD} > 3$ and with a flat pressure derivative, then plot p_{wf} vs. log t for a constant-rate flow test or p_{ws} vs. log $(t_p+\Delta t)/\Delta t$ for a buildup test. If a semilog straight line appears, formation permeability can be estimated uniquely.

If the type curve indicates pressure change and pressure derivatives with half-slopes for $t_{L_fD} < 0.016$, plot p_{wf} vs. $\Delta t_e^{1/2}$ on Cartesian coordinate paper for a constant-rate flow test or for a buildup test. A straight line should appear with a slope inversely proportional to $\sqrt{kL_f}$. If we can estimate permeability from a pseudoradial-flow analysis or if a prefracture value is available, we can estimate fracture half-length, L_f.

4. Next, perform a quantitative type-curve analysis. The purpose of this analysis is either to confirm the results from the specialized analysis techniques or to obtain estimates of formation and fracture properties when these analysis techniques are not possible.

If a permeability estimate is available from either a prefracture test or from pseudoradial flow analysis, precalculate a pressure match point. Use a convenient and arbitrary value of dimensionless pressure, p_D, to precalculate Δp.

$$(\Delta p)_{MP} = \frac{141.2qB\mu}{kh}(p_D)_{MP}. \quad\text{(6.23)}$$

Compare the type-curve match with the preliminary match obtained in Step 2. If the matches are significantly different, repeat Steps 2 and 3 until a consistent match is obtained.

If a value of permeability is not available, choose a pressure match point $(p_D, \Delta p)$ from the qualitative type-curve match and estimate formation permeability from

$$k = \frac{141.2qB\mu}{h}\left(\frac{p_D}{\Delta p}\right)_{MP}. \quad\text{(6.24)}$$

Choose a time match point $(\Delta t, t_{L_fD})$ and estimate fracture half-length,

$$L_f = \left[\frac{0.0002637k}{\phi\mu c_t}\left(\frac{\Delta t}{t_{L_fD}}\right)_{MP}\right]^{1/2}. \quad\text{(6.25)}$$

The value of fracture half-length computed with the time match point should be consistent with the value obtained from the linear-flow analysis. If the values do not agree, repeat Steps 2 and 3.

The matching parameter is a dimensionless fracture storage coefficient (analogous to the dimensionless wellbore-storage coefficient), C_{fD}, defined by

$$C_{fD} = \frac{0.8936C}{\phi hc_t L_f^2}. \quad\text{(6.29)}$$

With the estimated fracture half-length and the value of C_{fD} from the type-curve match, estimate the wellbore-storage coefficient, C, from Eq. 6.29. To verify the type-curve analysis if data are available, estimate C_{fD} from wellbore properties shown in Table 6.5.

Estimate a lower bound on the fracture conductivity. For an infinite-conductivity fracture, $C_{rD} \geqq 100$ and $w_f k_f \geqq C_{rD}\pi kL_f \geqq 100\pi kL_f$.

Example 6.5—Analyzing a Post-Fracture Pressure-Buildup Test With Wellbore-Storage Distortion. A pressure-buildup test was run following a fracture treatment in a gas well. **Table 6.12** summarizes pressure and time data, while other known data are given next. With a prefracture permeability estimate of 0.1 md, determine the fracture half-length and conductivity. In addition, verify the type-curve analysis with any applicable specialized analysis techniques. Because this is a gas well, use adjusted pressure and time variables for your analyses.

$r_w = 0.25$ ft
$h = 100$ ft
$\phi = 0.12$
$\gamma_g = 0.65$
$T = 200°F$
$\bar{z} = 0.945$
$\bar{c}_t = 1.506\times10^{-4}$ psi^{-1}
$\bar{B}_g = 0.7855$ RB/Mscf
$\bar{\mu} = 0.0216$ cp
$q_g = 1{,}400$ Mscf/D
$t_p = 1{,}644$ hours
$t_{pa} = 865.6$ hours
$p_{wf} = 3{,}249$ Mscf/D
$p_i = \bar{p} = 4{,}000$ psi
$S_w = 0.25$
$k = 0.1$ md

TABLE 6.12—PRESSURE BUILDUP TEST DATA, EXAMPLE 6.5

t (hours)	p_{ws} (psi)	$p_{a,ws}$ (psi)	t (hours)	p_{ws} (psi)	$p_{a,ws}$ (psi)
0.0041	3,249.1	1,822.0	1.3063	3,274.2	1,845.6
0.0065	3,249.2	1,822.1	1.6446	3,279.0	1,850.2
0.0082	3,249.3	1,822.2	2.0705	3,284.5	1,855.3
0.0131	3,249.4	1,822.3	2.6066	3,290.9	1,861.4
0.0165	3,249.5	1,822.4	3.2815	3,298.3	1,868.4
0.0207	3,249.7	1,822.6	4.1312	3,306.6	1,876.2
0.0261	3,249.8	1,822.7	5.2011	3,316.1	1,885.2
0.0328	3,250.0	1,822.9	6.5478	3,326.6	1,895.1
0.0413	3,250.3	1,823.2	8.2436	3,338.3	1,906.2
0.0520	3,250.6	1,823.5	10.379	3,351.3	1,918.6
0.0655	3,250.9	1,823.7	13.067	3,365.5	1,935.1
0.0824	3,251.4	1,824.2	16.451	3,381.0	1,946.8
0.1038	3,252.0	1,824.8	20.714	3,397.9	1,962.9
0.1306	3,252.7	1,825.4	26.080	3,416.2	1,980.4
0.1645	3,253.5	1,826.2	32.834	3,436.1	1,999.5
0.2070	3,254.5	1,827.1	41.350	3,457.6	2,020.1
0.2606	3,255.7	1,828.2	52.071	3,480.9	2,042.5
0.3281	3,257.2	1,829.7	65.576	3,506.0	2,066.7
0.4131	3,259.0	1,831.3	82.591	3,530.0	2,089.8
0.5200	3,261.1	1,833.3	104.03	3,556.0	2,115.0
0.6547	3,263.6	1,835.7	131.06	3,582.0	2,140.2
0.8242	3,266.6	1,838.5	165.13	3,609.0	2,166.5
1.0376	3,270.1	1,841.8			

Solution. *Qualitative Type-Curve Analysis.*

1. Plot adjusted pressure change, Δp_a, and adjusted pressure derivative, $\Delta t_{ae}\Delta p'_a$, vs. Δt_{ae} on log-log paper with the same grid size as the type curve (**Fig. 6.18**). **Table 6.13** summarizes the plotting functions required for this analysis.

2. On the basis of the preliminary type-curve match with the curve for $C_{fD}=0.1$, we observe that pseudoradial flow, which begins at $t_{L_fD}\approx 3$, does not appear to be present for this test. There are some data for $t_{L_fD}\leqq 0.016$ (i.e., the linear-flow time range); however, the data are all distorted by wellbore storage and we cannot perform linear-flow analysis. No boundary effects are present, indicating an infinite-acting reservoir. Because we cannot perform a linear-flow analysis, we attempt a quantitative type-curve analysis.

Quantitative Type-Curve Analysis.

1. Because we have a permeability estimate, we can precalculate a pressure match point. We arbitrarily choose the dimensionless pressure $p_D=3.0$, and the pressure match point is

$$(\Delta p_a)_{MP} = \frac{141.2 q_g \overline{B}_g \overline{\mu}}{kh}(p_D)_{MP}$$

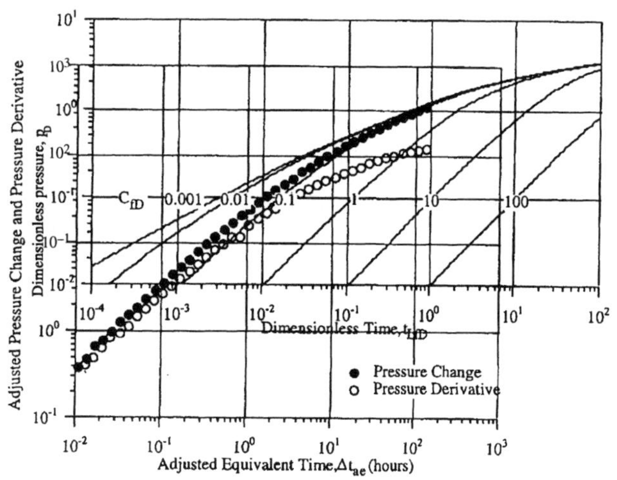

Fig. 6.18—Qualitative type-curve analysis with the Barker-Ramey type curve, Example 6.5.[16]

$$= \frac{(141.2)(1,400)(0.7855)(0.0216)}{(0.1)(100)}(3.0)$$

$$= 1,000 \text{ psi}.$$

With this precalculated match point, the type curve for $C_{fD}=0.06$ matches the data (**Fig. 6.19**). In addition, we obtain the time match point $\Delta t_{ae}=1,000$ hours and $t_{L_fD}=6.25$.

2. The fracture half-length is computed with the time match point.

$$L_f = \left[\frac{0.0002637k}{\phi\overline{\mu}\,\overline{c}_t}\left(\frac{\Delta t_{ae}}{t_{L_fD}}\right)_{MP}\right]^{1/2}$$

$$= \left[\frac{(0.0002637)(0.10)}{(0.12)(0.0216)(1.506\times10^{-4})}\left(\frac{1,000}{6.25}\right)\right]^{1/2}$$

$$= 104 \text{ ft}.$$

3. Compute the wellbore-storage coefficient, C, with the type curve correlating parameter, C_{fD}.

$$C = \frac{\phi h \overline{c}_t L_f^2}{0.8936} C_{fD}$$

$$= \frac{(0.12)(100)(1.506\times10^{-4})(104)^2}{0.8936}(0.06)$$

$$= 1.31 \text{ bbl/psi}.$$

4. For an infinite-conductivity fracture (i.e., $C_{rD}\geqq 100$), the fracture conductivity is estimated to be at least $w_f k_f \geqq \pi C_{rD} k_f = \pi(100)\times(0.1)(104)=3,270$ md-ft.

6.5.5 Recommended Procedure for Type-Curve Analysis of Post-Fracture Well Tests. We recommend the following step-by-step procedure for analysis of post-fracture buildup and constant-rate flow tests. The procedure combines both type-curve and specialized analysis techniques.

TABLE 6.13—LOG-LOG PLOTTING FUNCTIONS FOR WELL-TEST ANALYSIS, EXAMPLE 6.5

Δt_{ae} (hours)	Δp_a (psi)	$\Delta t_{ae}\Delta p'_a$ (psi)	Δt_{ae} (hours)	Δp_a (psi)	$\Delta t_{ae}\Delta p'_a$ (psi)
0.0033	0.0942	—	1.0610	23.694	18.240
0.0053	0.1881	0.2297	1.3367	28.207	21.213
0.0067	0.2824	0.2299	1.6841	33.398	24.597
0.0106	0.3761	0.4081	2.1220	39.443	28.181
0.0133	0.4702	0.4762	2.6741	46.433	32.253
0.0168	0.6582	0.6215	3.3704	54.273	36.446
0.0211	0.7522	0.8166	4.2485	63.247	40.865
0.0266	0.9403	0.9189	5.3558	73.188	45.695
0.0334	1.2224	1.1226	6.7525	84.290	50.552
0.0421	1.5046	1.4287	8.5144	96.624	55.755
0.0530	1.7864	1.7349	10.736	110.11	61.150
0.0667	2.2565	2.1425	13.536	124.88	66.740
0.0840	2.8208	2.6533	17.066	140.99	72.802
0.1058	3.4788	3.1631	21.512	158.48	79.208
0.1332	4.2312	3.7746	27.103	177.52	86.165
0.1677	5.1713	4.5906	34.141	198.15	93.686
0.2111	6.2996	5.6093	42.972	220.53	98.513
0.2659	7.7100	6.7305	54.035	244.72	104.01
0.3348	9.4023	8.0543	67.854	267.88	107.92
0.4216	11.377	9.5812	85.049	293.05	111.26
0.5310	13.727	11.311	106.36	318.27	116.72
0.6687	16.548	11.343	132.59	344.50	—
0.8423	19.839	15.681			

1. Plot pressure change and pressure derivative as a function of elapsed time, either flow time for drawdown tests or shut-in time for buildup tests, on log-log coordinates.

For an oilwell flow test, plot pressure change, Δp, and pressure derivative, $t\Delta p'$, vs. elapsed time since the start of the test, t. For an oilwell buildup test, plot Δp and $\Delta t_e \Delta p'$ vs. Agarwal's equivalent time, Δt_e.

For a gas-well buildup test, plot adjusted pressure change, Δp_a, and adjusted pressure derivative, $\Delta t_{ae}\Delta p'_a$, vs. equivalent adjusted time, Δt_{ae}. For a gas-well flow test, plot adjusted pressure change, Δp_a, and adjusted pressure derivative, $t\Delta p'_a$, vs. real time, t, rather than adjusted time. Table 6.5 summarizes interpretation procdures for type-curve plots.

2. Select the appropriate type curve for analysis. Attempt to match the test data with the Cinco-Ley *et al.*[7] type curve for finite-conductivity fractures. For long fractures in low-permeability formations, this type curve usually is required and often will suffice.

Early-time data with a slope significantly steeper than that of any type curve usually indicate wellbore-storage distortion of the test data. If a significant portion of the data exhibits this pattern, then the Barker-Ramey type curve may provide better test interpretation.[16]

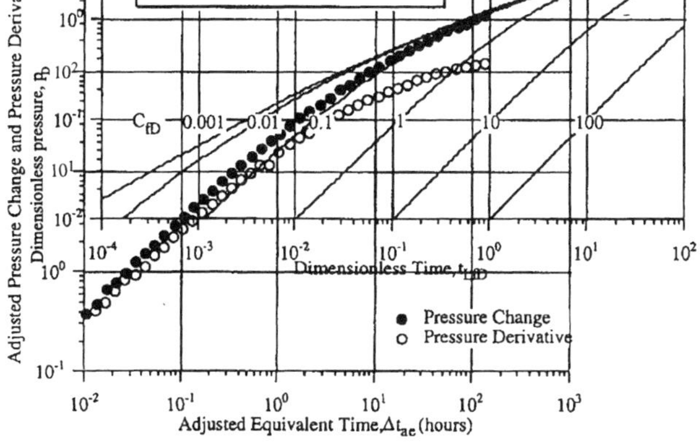

Fig. 6.19—Qualitative type-curve analysis with the Barker-Ramey type curve, Example 6.5.[16]

The Barker-Ramey type curve is limited, however, to infinite-conductivity fractures.

If fracture conductivity is high ($C_{rD} \geqq 10$ or, preferably, 100) and late-time data exhibit a steepening slope, boundary effects may be indicated. The Gringarten-Ramey-Raghavan[15] type curve for finite reservoirs can be used and can confirm the presence of boundary effects. Quantitative application of this type curve also is limited to infinite-conductivity fractures ($C_{rD} \geqq 100$).

If fracture conductivity is very high ($C_{rD} \geqq 100$), then the Bourdet *et al.*[17] type curve (discussed in Chap. 4) for radial flow in a homogeneous reservoir also can be used to analyze the test data. The result is a skin factor from which the fracture half-length can be estimated as $L_f \approx 2r_w e^{-s}$.

The Agarwal *et al.*[5] type curve should be used for a constant BHP flow test or long-term production-data analysis. Satisfactory matches usually can be obtained with this curve as long as the reservoir is infinite-acting.

3. While the test-data plot is in a fitted position on a type curve, determine whether specific flow regimes can be analyzed. Better estimates of formation permeability and fracture half-length can be obtained if specific flow regimes can be identified in the test data, particularly if a permeability estimate is available from a prefracture test.

If some data extend beyond $t_{L_fD} \approx 3$ and the pressure derivative flattens, a straight line indicative of pseudoradial flow may appear on a semilog plot. If the pseudoradial regime is present, permeability can be estimated more acclurately from a semilog plot than from type-curve analysis alone and can be used to obtain a better type-curve fit. A post-fracture estimate of permeability, particularly if it can be obtained from pseudoradial-flow analysis, is likely to be more accurate than a prefracture estimate because more of the formation may be in communication with the wellbore after fracturing. If $C_{rD} \geqq 100$, L_f may be estimated by $L_f \approx 2r_w e^{-s}$.

If a half-slope appears on both pressure-change and pressure-derivative plots at early times ($t_{L_fD} \leqq 0.016$) and $C_{rD} \geqq 100$, a straight line should appear on a $\sqrt{t}$ plot. From this plot, we can estimate $\sqrt{k}L_f$, which can be used to verify or improve estimates from type-curve analysis or, given a prefracture permeability estimate, to obtain a good estimate of fracture half-length.

If a quarter-slope appears on both pressure-change and pressure-derivative plots at early times, a straight line, indicative of bilinear flow, should appear on a $t^{1/4}$ plot. From this plot, $w_f k_f$ can be estimated and used to verify or improve estimates from type-curve analysis.

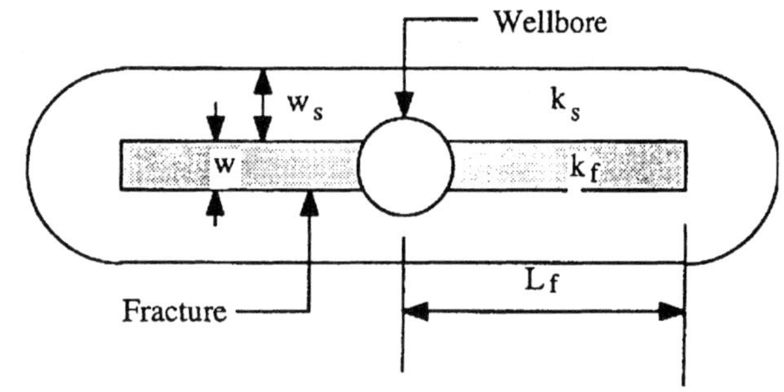

Fig. 6.20—Infinite-conductivity vertical fracture with fluid-loss damage (after Cinco-Ley and Samaniego-V.[4]).

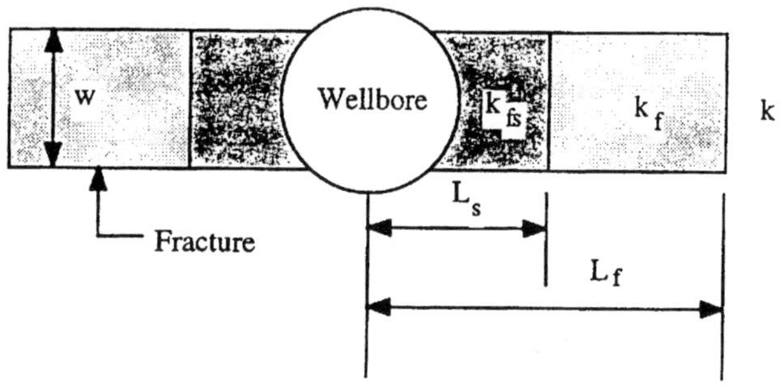

Fig. 6.21—Choked fracture (after Cinco-Ley and Samaniego-V.[4]).

6.5.6 Limitations of Type-Curve Analysis. Because they encompass all flow regimes and intervening transition periods, type curves provide a more general method of test analysis than do methods based on a particular flow regime (bilinear, linear, or pseudoradial). However, type-curve analysis does have some significant limitations.[3]

1. Type-curve matches frequently are not unique because the test data often match more than one curve equally well. This problem is particularly likely to occur with low-conductivity (i.e., $C_{fD} < 10$) fractures. An independent estimate of formation permeability can reduce the ambiguity problem, but to obtain the estimate requires either the achievement of pseudoradial flow during a post-fracture test (unlikely in low-permeability formations with long fractures) or a prefracture test in the same interval open to the well after fracturing.

2. Type curves necessarily focus on key features (such as conductivity range) and may not reflect additional complications, such as wellbore storage, boundary effects, or fracture closure (which results in continuously variable fracture conductivity along the length of a fracture or with continued test time). As a result, they generally are oversimplified models of the tested formation and fracture.

3. Test interpretation may be complicated by variable fracture conductivity. In most gas-well buildup tests, non-Darcy flow in the fracture continues for a long time after shut-in, causing the apparent fracture conductivity to increase continuously throughout much of a test.[18] Fracture conductivity also will change if higher-permeability proppants are injected near the ends of a fracture or if the fracture becomes narrower as it extends laterally into the formation.[19] Because a type curve has a fixed value of fracture conductivity as a parameter, variable fracture conductivity may render type-curve analysis of test data impossible.

4. Unequal-length fracture wings can cause the value of L_f estimated with either the Agarwal *et al.*[5] type curve or the Cinco-Ley *et al.*[7] type curve to be less than the actual fracture half-length.[19] The error occurs when the ratio of the shorter wing length to that of the longer wing is less than 0.3.

6.6 Effects of Fracture and Formation Damage

Cinco-Ley and Samaniego-V.[4,20] suggested that two types of fracture damage can occur during the hydraulic fracturing process: within the fracture adjacent to the wellbore and in the formation around the fracture face. The first type of damage, often described as a "choked" fracture, is thought to be caused by proppant crushing and embedding in the formation. The second type of damage, quantified as a fracture-face skin, is probably caused by fracture-fluid losses into the formation and adverse reactions between the formation and fracture fluids.

The presence of skin damage around the fracture face may complicate post-fracture type-curve analysis of the formation and the fracture.[4,20] The fracture-skin factor is defined as

$$s_{fs} = \frac{\pi w_s}{2L_f}\left(\frac{k}{k_s} - 1\right), \quad \text{(6.30)}$$

where w_s = the extent of fluid-loss damage into the formation (normal to the fracture face) and k_s = the permeability of the damaged zone (**Fig. 6.20**). Cinco-Ley and Samaniego-V.[20] developed type curves for analyzing well tests in hydraulically fractured wells with fluid-loss damage.

The pressure response in a well with a choked fracture is similar to that for a fracture with skin damage.[4] The skin factor representing a choked fracture is given by

$$s_{fs,ch} = \frac{\pi L_s k}{w_f k_{fs}}, \quad \text{(6.31)}$$

where L_s = the length of the choked portion of the fracture, w_f = the fracture width, and k_{fs} = the permeability of the choked portion (**Fig. 6.21**). Cinco-Ley and Samaniego[4] developed type curves for analyzing well tests from hydraulically fractured wells with choked fractures.

6.7 Chapter Summary

In this chapter, we discussed interpretation methods for wells that have been hydraulically fractured. Fracturing is often necessary in low-permeability formations to achieve economic flow rates.

In Sec. 6.2, we discussed the five flow patterns that may occur in hydraulically fractured wells. These five patterns include fracture-linear flow, bilinear flow, formation-linear flow, elliptical flow, and pseudoradial flow. Fracture-linear flow is usually very short-lived and does not play an important role in well-test analysis. Bilinear flow occurs in finite-conductivity fractures as fluid flows from the formation into the fracture, then down the fracture to the wellbore. Formation-linear flow occurs only in high-conductivity fractures where linear flow occurs within the formation to the fracture. Elliptical flow is a transitional flow pattern that occurs between either bilinear or formation-linear flow at early times and pseudoradial flow at late times. The last flow pattern encountered in an infinite acting reservoir is pseudoradial flow, which may occur with fractures of any conductivity. We also presented equations for estimating the beginning and end of each of these flow periods.

In Sec. 6.3, we discussed the flow geometry and depth of investigation of a vertically fractured well. We presented equations that allow us to estimate depth of investigation in any direction from the well and the total area sampled by a well test of a given duration. Although these expressions assume highly conductive fractures, they may be used to get rough estimates that apply to lower-conductivity fractures as well.

In Sec. 6.4, we presented specialized analysis methods that are appropriate for three different flow regimes. These methods allow us to analyze wells that are exhibiting pseudoradial, bilinear, or formation-linear flow. For each of the specialized analysis methods, we presented working equations for oil. We also presented working equations for gas in four different forms: adjusted pressure and adjusted time, pressure and time, pressure-squared and time, and pseudopressure and time. A consistent theme throughout this section and Sec. 6.5 is the need for independent permeability estimates to analyze post-fracture transient tests best. These independent permeability estimates preferably are obtained from analysis of prefracture pressure transient tests.

In Sec. 6.4.1, we discussed pseudoradial-flow analysis. This method is applicable to short, high-conductivity fractures in high-permeability formations, where the pseudoradial-flow pattern is likely to be established during the test. When the pseudoradial-flow method is applicable, it allows the engineer to estimate formation permeability, k, and skin factor, s. If the fracture conductivity is known, the fracture half-length can be estimated from the skin factor. The biggest limitation of the pseudoradial flow method is that pseudoradial flow is seldom achieved in low-permeability formations with long fractures. Other limitations include the fact that the skin factor measured during the test includes an apparent skin resulting from non-Darcy flow and the need to know fracture conductivity to estimate fracture half-length.

In Sec. 6.4.2, we discussed the bilinear-flow method. This method applies to data within the bilinear-flow regime from wells that have finite-conductivity fractures. The bilinear-flow regime may be recognized on a log-log graph of pressure change vs. time change by the appearance of a quarter-slope line. The pressure derivative also follows a quarter-slope line during this flow regime. If the formation permeability, k, is known, we can estimate the fracture conductivity $w_f k_f$ from data obtained during the bilinear-flow regime. The limitations of the bilinear-flow method are that we cannot estimate fracture half-length L_f, that wellbore storage often masks the bilinear-flow regime, and that an independent estimate of formation permeability k is required.

In Sec. 6.4.3, we discussed the linear-flow method. This method applies to test data obtained during formation linear flow in wells with high-conductivity fractures. During the linear-flow period, a log-log graph of pressure change vs. time has a slope of one-half, as will the pressure-derivative graph. If we have an independent estimate of formation permeability k, we can estimate the fracture half-length L_f by use of linear-flow analysis. Limitations of the linear-flow method include the assumption of a high-conductivity fracture, possible distortion of data during the linear-flow period by wellbore storage, and the necessity of an independent estimate of formation permeability.

In Sec. 6.5, we presented four type curves that may be useful in analyzing post-fracture transient well tests. These type curves include the Gringarten-Ramey-Raghavan type curve, the Cinco-Ley *et al.* type curve, the Agarwal *et al.* type curve, and the Barker-Ramey type curve. Analysis with the type curves is complemented by the specialized analysis methods presented in Sec. 6.4.

The Gringarten-Ramey-Raghavan type curve, discussed in Sec. 6.5.1, is a family of solutions to the flow equation with the following assumptions: the fracture has infinite conductivity, the well is centered in a square drainage area with no-flow boundaries, the fracture has two equal-length wings, and wellbore-storage effects are ignored. The correlating parameter for this family of type curves is L_e/L_f. Thus, if sufficient data are available for a complete analysis with the Gringarten-Ramey-Raghavan type curve, we can obtain formation permeability k from the pressure match point, fracture half-length L_f from the time match point, and distance to the boundary L_e from the correlating parameter.

In Sec. 6.5.2, we presented the Cinco-Ley *et al.* type curve. This type curve presents solutions to the flow equation with the following assumptions: the well has a finite-conductivity fracture, the fracture has two equal-length wings, and wellbore storage effects are negligible. For this type curve, the correlating parameter is the dimensionless fracture conductivity, C_{rD}. A complete analysis with this type curve provides estimates of formation permeability k from the pressure match point, fracture half-length L_f from the time match point, and fracture conductivity $w_f k_f$ from the dimensionless fracture conductivity C_{rD}.

The Agarwal *et al.* type curve presented in Sec. 6.5.3 makes essentially the same assumptions as the Cinco-Ley *et al.* type curve. However, this type curve is a solution for production at constant pressure rather than from production at constant rate. The dependent variable is the reciprocal of the dimensionless production rate, $1/q_D$. The correlating parameter is the dimensionless fracture conductivity, C_{rD}. A complete analysis with this type curve provides estimates of formation permeability k from the rate match point, fracture half-length L_f from the time match point, and fracture conductivity $w_f k_f$ from the dimensionless fracture conductivity C_{rD}.

In Sec. 6.5.4, we discussed the Barker-Ramey type curve. This type curve includes the effect of wellbore or fracture storage. Other assumptions for this type curve include infinite-conductivity fracture and equal-length wings. The correlating parameter for this type curve is the dimensionless wellbore-storage coefficient based on fracture half-length, C_{fD}. If sufficient data are available, this type curve can provide estimates of formation permeability k from the pressure match point, fracture half-length L_f from the time match point, and wellbore storage coefficient C from the dimensionless wellbore storage coefficient C_{fD}.

In Sec. 6.5.5, we recommended a procedure for type-curve analysis of post-fracture well tests. This procedure has three main steps: (1) plot pressure change and pressure derivative as functions of elapsed time on log-log coordinates, (2) select the appropriate type curve for analysis, and (3) identify and analyze any specific flow-regime data appearing during the test. The type-curve analysis and the specific-flow-regime analysis must be consistent for a correct analysis.

In Sec. 6.5.6, we discussed the limitations of type-curve analysis, which include the following: (1) it is often difficult to get a unique match; (2) type curves focus on a single key feature and may not reflect all complications affecting the test; (3) the assumption of uniform fracture conductivity is not achieved in practice; and (4) if the fracture has unequal length wings, the fracture half-length estimates may be too high.

In Sec. 6.6, we briefly discussed the effects of fracture and formation damage. Two types of damage may occur. The first, damage to the fracture face, may be caused by fracture-fluid losses into the formation and adverse reactions between fracture and formation fluids. Because of the tremendous surface area of the fracture, the permeability reduction near the fracture must be high for this mechanism to cause significant damage. The second type of damage is often referred to as a "choked" fracture. Here, the fracture conductivity adjacent to the wellbore is decreased because of the crushing of the proppant or to overflushing the fracture treatment. We presented equations for estimating the skin factor resulting from each of these types of damage.

Exercises

1. Estimate the beginning and end of the linear flow period and the time at which pseudoradial flow begins for each of the following cases. Assume that pseudoradial flow begins at $t_{LfD} = 3$. Convert times to successively larger units (days, months, years) until the result can be expressed as a number between 1 and 100. $\phi = 8\%$; $\mu = 0.022$ cp; $C_{rD} = 1000$; $c_t = 2.5 \times 10^{-4}$ psi^{-1}.

DATA FOR EXERCISE 6.1

Case No.	L_f, ft	k, md
1	75	5
2	75	0.5
3	75	0.05
4	225	1.5
5	225	0.15
6	225	0.015
7	675	0.5
8	675	0.05
9	675	0.005

2. Calculate the dimensionless fracture conductivity C_{rD} for each of the following cases. Determine whether each situation is a high conductivity or low conductivity case. For the finite-conductivity cases, estimate when bilinear flow ends. For high-conductivity cases, estimate the beginning and end of the linear flow period. Estimate the time at which pseudoradial flow begins for all cases; assume that pseudoradial flow begins at $t_{LfD} = 3$ regardless of conductivity. Convert times to successively larger units (days, months, years) until the result can be expressed as a number between 1 and 100. $\phi = 12\%$; $\mu = 0.025$ cp; $w_f k_f = 1000$ md-ft; $c_t = 5 \times 10^{-4}$ psi^{-1}.

DATA FOR EXERCISE 6.2

Case No.	L_f, ft	k, md
1	65	15
2	65	1.5
3	65	0.15
4	215	1.5
5	215	0.15
6	215	0.015
7	650	1.5
8	650	0.15
9	650	0.015

3. Compare the dimensionless conductivity for two hydraulically fractured wells, one in an oil reservoir and the other in a gas reservoir. Note that the term kh/μ is the same for both reservoirs. How high would the fracture conductivity $w_f k_f$ have to be in each reservoir to give a dimensionless conductivity C_{rD} of 100? Can you draw any conclusions about the feasibility of creating long, high conductivity hydraulic fractures in oil wells vs. gas wells?

Oil reservoir properties: $\phi = 12\%$; $k = 1.5$ md; $B_o = 1.31$ RB/STB; $p_i = 2400$ psi; $L_f = 500$ ft; $\mu_o = 0.5$ cp; $h = 25$ ft; $w_f k_f = 1000$ md-ft; $c_t = 2.0 \times 10^{-5}$ psi^{-1}.

Gas reservoir properties: $\phi = 12\%$; $k = 0.054$ md; $B_g = 1.31$ RB/STB; $p_i = 2400$ psi; $L_f = 500$ ft; $\mu_g = 0.018$ cp; $h = 25$ ft; $w_f k_f = 1000$ md-ft; $c_t = 4.0 \times 10^{-4}$ psi^{-1}.

4. Given the following formation and fluid properties, estimate formation permeability, fracture half-length, and drainage radius from the drawdown test data for this oil well. $q = 200$ STB/D; $h = 12$ ft; $B = 1.325$ RB/STB; $p_i = 3343.40$ psia; $\phi = 11.8\%$; $c_t = 14.7 \times 10^{-6}$ psi^{-1}; $r_w = 0.25$ ft; $\mu = 0.49$ cp.

PRESSURE-TIME DATA FOR EXERCISE 6.4

Time (hours)	Pressure (psi)	Time (hours)	Pressure (psi)	Time (hours)	Pressure (psi)
0.0000	3343.40	0.219	3031.45	14.76	1904.57
0.0010	3313.75	0.285	2994.99	19.18	1803.26
0.0023	3302.25	0.372	2955.06	24.9	1700.73
0.0040	3291.60	0.484	2911.26	32.4	1597.23
0.0062	3280.78	0.630	2863.12	42.2	1492.94
0.0090	3269.31	0.820	2810.17	54.8	1387.98
0.0128	3256.86	1.067	2751.98	71.2	1282.51
0.0176	3243.17	1.389	2688.23	92.6	1176.82
0.0239	3227.97	1.806	2618.84	120.4	1071.07
0.0320	3211.07	2.35	2543.91	156.5	964.31
0.0426	3192.27	3.05	2463.80	203.5	853.11
0.0564	3171.42	3.97	2378.98	264.5	730.51
0.0743	3148.39	5.16	2290.03	343.9	586.07
0.0976	3123.04	6.71	2197.55	447.0	406.67
0.1279	3095.23	8.73	2102.14	581.1	177.07
0.1673	3064.77	11.35	2004.33	600.0	144.85

5. Given the following formation and fluid properties, estimate formation permeability, fracture half-length, and drainage radius from the drawdown test data for this gas well. $T_f = 147$°F; $h = 11$ ft; $\gamma_g = 0.68$ (air = 1.0); $q = 750$ Mscf/D; $p_i = 2618$ psia; $\phi = 21.3\%$; $c_w = 3.6 \times 10^{-6}$ psi^{-1}; $r_w = 0.35$ ft; $S_w = 26.8\%$; $c_f = 4 \times 10^{-6}$ psi^{-1}.

PRESSURE-TIME DATA FOR EXERCISE 6.5

Time (hours)	Pressure (psi)	Time (hours)	Pressure (psi)	Time (hours)	Pressure (psi)
0.000	2618.00	5.02	2513.82	261.9	2149.74
0.030	2608.34	6.55	2501.31	340.5	2112.75
0.069	2603.95	8.55	2487.60	442.7	2074.82
0.120	2599.87	11.15	2472.53	575.5	2036.02
0.186	2595.71	14.52	2455.92	748.2	1996.43
0.271	2591.29	18.91	2437.59	972.7	1956.14
0.383	2586.49	24.61	2417.37	1265	1915.26
0.528	2581.21	32.02	2395.14	1644	2873.70
0.716	2575.37	41.65	2370.82	2137	1830.63
0.961	2568.89	54.18	2344.43	2778	1783.96
1.279	2561.74	70.46	2316.06	3612	1730.15
1.692	2553.84	91.63	2285.82	4695	1664.20
2.23	2545.17	119.15	2253.90	6104	1579.81
2.93	2535.65	154.93	2220.46	6500	1556.00
3.84	2525.23	201.4	2185.69		

6. Given the following formation and fluid properties, estimate fracture conductivity and fracture half-length (if possible) from the buildup test data. Note that the permeability is already known from a pre-fracture buildup test. $T_f = 145$°F; $h = 44$ ft; $\gamma_g = 0.593$ (air = 1.0); $q = 900$ Mscf/D; $r_w = 0.48$ ft; $\phi = 12.7\%$; $c_w = 3.6 \times 10^{-6}$ psi^{-1}; $t_p = 2880$ hr; $k = 0.012$ md; $S_w = 51.1\%$; $c_f = 4 \times 10^{-6}$ psi^{-1}; $p_{wf} = 2180.46$ psia.

PRESSURE-TIME DATA FOR EXERCISE 6.6

Time (hours)	Pressure (psi)	Time (hours)	Pressure (psi)	Time (hours)	Pressure (psi)
0.000	2180.46	1.940	2253.36	22.7	2408.45
0.040	2196.50	2.37	2260.20	27.3	2429.09
0.088	2201.30	2.88	2267.63	32.8	2451.58
0.146	2205.43	3.50	2275.69	39.4	2476.05
0.215	2209.37	4.24	2284.46	47.3	2502.68
0.298	2213.31	5.12	2294.01	56.8	2531.65
0.397	2217.35	6.19	2304.40	68.2	2563.15
0.517	2221.56	7.47	2315.72	81.8	2597.37
0.660	2226.00	9.00	2328.05	98.2	2634.53
0.832	2230.72	10.84	2341.49	117.9	2674.84
1.038	2235.77	13.05	2356.14	141.6	2718.51
1.286	2241.19	15.70	2372.10	169.9	2765.69
1.583	2247.04	18.88	2389.49	196.0	2805.15

7. Given the following formation and fluid properties, estimate fracture conductivity and fracture half-length (if possible) from the buildup test data. Note that the permeability is already known from a pre-fracture buildup test. $T_f = 154$°F; $h = 147$ ft; $\gamma_g = 0.714$ (air = 1.0); $q = 3950$ Mscf/D; $r_w = 0.23$ ft; $\phi = 6.7\%$; $c_w = 3.6 \times 10^{-6}$ psi^{-1}; $t_p = 3600$ hr; $k = 0.061$ md; $S_w = 55.2\%$; $c_f = 4 \times 10^{-6}$ psi^{-1}; $p_{wf} = 970.82$ psia.

PRESSURE-TIME DATA FOR EXERCISE 6.7					
Time (hours)	Pressure (psi)	Time (hours)	Pressure (psi)	Time (hours)	Pressure (psi)
0.000	970.82	1.277	1136.73	16.143	1274.39
0.030	1037.74	1.626	1146.53	20.209	1291.76
0.068	1052.47	2.063	1156.71	25.291	1310.39
0.114	1063.67	2.609	1167.32	31.644	1330.44
0.173	1073.48	3.291	1178.40	39.585	1352.08
0.246	1082.64	4.143	1190.00	49.511	1375.48
0.338	1091.50	5.209	1202.19	61.919	1400.85
0.452	1100.28	6.541	1215.02	77.428	1428.37
0.595	1109.10	8.207	1228.57	96.815	1458.18
0.774	1118.07	10.288	1242.91	121.049	1490.40
0.998	1127.26	12.890	1258.15	144.000	1517.15

8. Given the following formation and fluid properties, estimate fracture conductivity and fracture half-length (if possible) from the buildup test data. Note that the permeability is already known from a pre-fracture buildup test. $T_f = 211°F$; $h = 157$ ft; $\gamma_g = 0.711$ (air = 1.0); $q = 4750$ Mscf/D; $r_w = 0.42$ ft; $\phi = 8.5\%$; $c_w = 3.6 \times 10^{-6}$ psi^{-1}; $t_p = 720$ hr; $k = 0.092$ md; $S_w = 46.6\%$; $c_f = 4 \times 10^{-6}$ psi^{-1}; $p_{wf} = 2669.89$ psia.

PRESSURE-TIME DATA FOR EXERCISE 6.8					
Time (hours)	Pressure (psi)	Time (hours)	Pressure (psi)	Time (hours)	Pressure (psi)
0.0000	2669.89	0.274	2709.72	8.07	2768.44
0.0025	2682.23	0.345	2712.08	10.09	2775.50
0.0056	2685.00	0.434	2714.57	12.61	2783.23
0.0095	2687.13	0.545	2717.20	15.77	2791.71
0.0144	2689.00	0.684	2719.99	19.71	2801.02
0.0205	2690.76	0.857	2722.95	24.6	2811.24
0.0281	2692.48	1.074	2726.11	30.8	2822.43
0.0377	2694.18	1.345	2729.48	38.5	2834.68
0.0496	2695.90	1.684	2733.10	48.1	2848.02
0.0645	2697.67	2.11	2737.00	60.2	2862.48
0.0831	2699.48	2.64	2741.21	75.2	2878.05
0.1064	2701.36	3.30	2745.77	94.0	2894.67
0.1355	2703.31	4.13	2750.72	117.5	2912.23
0.1719	2705.35	5.16	2756.11	144.0	2928.91
0.217	2707.48	6.45	2762.00		

9. Given the following formation and fluid properties, estimate fracture conductivity and fracture half-length (if possible) from the buildup test data. Note that the permeability is already known from a pre-fracture buildup test. $T_f = 217°F$; $h = 247$ ft; $\gamma_g = 0.722$ (air = 1.0); $q = 6500$ Mscf/D; $r_w = 0.37$ ft; $\phi = 5.3\%$; $c_w = 3.6 \times 10^{-6}$ psi^{-1}; $t_p = 7200$ hr; $k = 0.027$ md; $S_w = 50.1\%$; $c_f = 4 \times 10^{-6}$ psi^{-1}; $p_{wf} = 2602.34$ psia.

PRESSURE-TIME DATA FOR EXERCISE 6.9					
Time (hours)	Pressure (psi)	Time (hours)	Pressure (psi)	Time (hours)	Pressure (psi)
0.0000	2602.34	0.1097	2630.85	3.23	2712.76
0.0010	2610.23	0.1381	2633.18	4.04	2724.53
0.0023	2612.00	0.1736	2635.76	5.04	2737.69
0.0038	2613.37	0.218	2638.62	6.31	2752.40
0.0058	2614.59	0.274	2641.79	7.88	2768.83
0.0082	2615.76	0.343	2645.32	9.86	2787.19
0.0113	2616.91	0.430	2649.24	12.32	2807.71
0.0151	2618.09	0.538	2653.62	15.40	2830.64
0.0198	2619.31	0.674	2658.49	19.26	2856.27
0.0258	2620.60	0.843	2663.93	24.1	2884.91
0.0333	2621.98	1.055	2670.00	30.1	2916.93
0.0426	2623.46	1.320	2676.77	37.6	2952.70
0.0542	2625.07	1.650	2684.34	47.0	2992.62
0.0688	2626.82	2.06	2692.78	58.8	3037.06
0.0869	2628.74	2.58	2702.22	72.0	3081.72

10. Given the following formation and fluid properties, estimate permeability, fracture conductivity, and fracture half-length (if possible) from the production data. $p_i = 1759.3$ psia; $h = 33$ ft; $\gamma_g = 0.599$ (air = 1.0); $p_{wf} = 650$ psia; $T_f = 120°F$; $\phi = 12.2\%$; $c_w = 3.6 \times 10^{-6}$ psi^{-1}; $r_w = 0.45$ ft; $S_w = 52.0\%$; $c_f = 4 \times 10^{-6}$ psi^{-1}.

PRESSURE-TIME DATA FOR EXERCISE 6.10					
Time (hours)	Pressure (psi)	Time (hours)	Pressure (psi)	Time (hours)	Pressure (psi)
3.05	1990.0	98.30	511.6	453.10	319.7
4.80	1625.8	123.90	471.5	483.10	314.6
7.00	1391.9	153.10	438.7	513.10	309.8
9.75	1221.8	183.10	413.9	543.10	305.5
13.20	1088.2	213.10	394.8	573.10	301.5
17.45	978.1	243.10	379.4	603.10	297.8
22.80	884.4	273.10	366.8	633.10	294.3
29.55	802.8	303.10	356.1	663.10	291.1
37.95	730.6	333.10	346.9	693.10	288.1
48.40	666.2	363.10	338.8	714.05	286.1
61.50	608.6	393.10	331.8		
77.85	557.2	423.10	325.4		

11. Given the following formation and fluid properties, estimate permeability, fracture conductivity, and fracture half-length (if possible) from the production data. $p_i = 2725.5$ psia; $h = 175$ ft; $\gamma_g = 0.633$ (air = 1.0); $p_{wf} = 900$ psia; $T_f = 148°F$; $\phi = 5.4\%$; $c_w = 3.6 \times 10^{-6}$ psi^{-1}; $r_w = 0.43$ ft; $S_w = 31.4\%$; $c_f = 4 \times 10^{-6}$ psi^{-1}.

PRESSURE-TIME DATA FOR EXERCISE 6.11					
Time (hours)	Pressure (psi)	Time (hours)	Pressure (psi)	Time (hours)	Pressure (psi)
2.90	4362.9	44.05	1118.7	298.60	477.2
4.50	3504.4	53.85	1012.8	359.35	444.1
6.40	2935.8	65.60	919.8	432.25	413.8
8.65	2519.7	79.70	838.1	519.70	385.8
11.40	2196.6	96.65	766.2	624.65	359.8
14.70	1935.6	117.00	703.0	750.55	335.7
18.65	1718.9	141.40	647.2	901.65	313.3
23.40	1535.1	170.70	597.6	1007.05	300.3
29.10	1376.9	205.85	553.3		
35.90	1239.2	248.00	513.4		

12. Given the following formation and fluid properties, estimate as many fracture and formation properties as possible from the buildup test data. $q = 50$ STB/D; $h = 11$ ft; $B = 1.073$ RB/STB; $t_p = 900$ hr; $\phi = 23.2\%$; $c_t = 9.25 \times 10^{-6}$ psi^{-1}; $p_{wf} = 74.2$ psia; $r_w = 0.25$ ft; $\mu = 1.47$ cp.

PRESSURE-TIME DATA FOR EXERCISE 6.12					
Time (hours)	Pressure (psi)	Time (hours)	Pressure (psi)	Time (hours)	Pressure (psi)
0.000	74.2	0.688	137.00	13.19	376.72
0.0100	75.73	0.870	148.17	16.50	400.14
0.0225	77.56	1.097	160.65	20.6	423.46
0.0381	79.72	1.381	174.47	25.8	446.57
0.0577	82.29	1.736	189.68	32.3	469.39
0.0831	85.33	2.18	206.27	40.3	491.87
0.1126	88.92	2.74	224.22	50.4	513.95
0.1507	93.16	3.43	243.44	63.1	535.58
0.1984	98.12	4.30	263.82	78.8	556.74
0.258	103.90	5.38	285.19	98.6	577.38
0.333	110.59	6.74	307.38	120.0	595.08
0.426	118.28	8.43	330.16		
0.542	127.06	10.55	353.34		

13. Given the following formation and fluid properties, estimate as many fracture and formation properties as possible from the buildup test data. $q = 1000$ STB/D; $h = 38$ ft; $B = 1.243$ RB/STB;

$t_p = 720$ hr; $\phi = 6.4\%$; $c_t = 12.6 \times 10^{-6}$ psi^{-1}; $p_{wf} = 1497.33$ psia; $r_w = 0.31$ ft; $\mu = 0.58$ cp.

PRESSURE-TIME DATA FOR EXERCISE 6.13					
Time (hours)	Pressure (psi)	Time (hours)	Pressure (psi)	Time (hours)	Pressure (psi)
0.000	1497.33	0.831	1739.98	8.57	1994.90
0.025	1513.86	1.064	1770.25	10.74	2014.90
0.056	1531.72	1.355	1799.92	13.45	2034.35
0.095	1551.44	1.719	1828.60	16.84	2053.32
0.144	1573.23	2.17	1856.04	21.1	2071.89
0.205	1597.10	2.74	1882.14	26.4	2090.09
0.282	1622.97	3.45	1906.89	33.0	2107.94
0.377	1650.59	4.34	1930.39	41.3	2125.46
0.496	1679.62	5.45	1952.79	48.0	2137.12
0.645	1709.59	6.84	1974.24		

14. Given the following formation and fluid properties, estimate as many fracture and formation properties as possible from the buildup test data. Note that formation permeability is known from a prefracture buildup test. $T_f = 86°$F; $h = 181$ ft; $\gamma_g = 0.572$ (air = 1.0); $q = 875$ Mscf/D; $r_w = 0.2$ ft; $\phi = 11.0\%$; $c_w = 3.6 \times 10^{-6}$ psi^{-1}; $t_p = 2400$ hr; $k = 0.007$ md; $S_w = 54.2\%$; $c_f = 4 \times 10^{-6}$ psi^{-1}; $p_{wf} = 753.45$ psia.

PRESSURE-TIME DATA FOR EXERCISE 6.14					
Time (hours)	Pressure (psi)	Time (hours)	Pressure (psi)	Time (hours)	Pressure (psi)
0.000	753.45	0.293	770.40	9.16	821.52
0.0030	753.94	0.384	772.60	11.92	829.36
0.0069	754.52	0.502	774.92	15.49	838.17
0.0120	755.22	0.655	777.39	20.1	848.07
0.0186	756.06	0.855	780.05	26.2	859.17
0.0271	757.05	1.115	782.95	34.0	871.61
0.0383	758.20	1.452	786.13	44.3	885.51
0.0527	759.52	1.891	789.65	57.6	901.03
0.0716	761.00	2.46	793.57	74.8	918.28
0.0960	762.64	3.20	797.94	97.3	937.42
0.1279	764.41	4.17	802.85	126.5	958.56
0.1692	766.30	5.42	808.36	144.0	969.80
0.223	768.30	7.05	814.56		

References

1. Cinco-Ley, H. and Samaniego-V., F.: "Transient Pressure Analysis for Fractured Wells," *JPT* (September 1981) 1749.
2. Economides, M.J.: "Post-Treatment Evaluation and Fractured Well Performance," *Reservoir Stimulation*, M.J. Economides and K.G. Nolte (eds.), Schlumberger Educational Services, Houston (1987) Chap. 11, 1–17.
3. Lee, W.J.: "Post-Fracture Formation Evaluation," *Recent Advances in Hydraulic Fracturing,* J.L. Gidley *et al.* (eds.), Monograph Series, SPE, Richardson, Texas (1989) **12**, 317–340 and 416–430.
4. Cinco-Ley, H. and Samaniego-V., F.: "Transient Pressure Analysis: Finite-Conductivity-Fracture Case vs. Damaged-Fracture Case," paper SPE 10179 presented at the 1981 SPE Annual Technical Conference and Exhibition, San Antonio, Texas, 5–7 October.
5. Agarwal, R.G., Carter, R.D., and Pollock, C.B.: "Evaluation and Performance Prediction of Low-Permeability Gas Wells Stimulated by Massive Hydraulic Fracturing," *JPT* (March 1979) 362; *Trans.*, AIME, **267**.
6. Prats, M., Hazebroek, P., and Strickler, W.R.: "Effect of Vertical Fractures on Reservoir Behavior-Compressible Fluid Case," *SPEJ* (June 1962) 87; *Trans.*, AIME, **225**.
7. Cinco-Ley, H., Samaniego-V., F., and Dominguez, N.: "Transient Pressure Behavior for a Well With a Finite-Conductivity Vertical Fracture," *SPEJ* (August 1976) 253.
8. Coats, K.H, Tek, M.R., and Katz, D.L.: "Unsteady-State Liquid Flow Through Porous Media Having Elliptical Boundaries," *Trans.*, AIME (1959) **216**, 460.
9. Kucuk, F. and Brigham, E.W.: "Transient Flow in Elliptical Systems," *SPEJ* (December 1979) 401; *Trans.*, AIME, **267.**
10. Kucuk, F.: "Unsteady-State Water Influx in Elliptical and Anisotropic Reservoir/Aquifer Systems," *SPEJ* (June 1981) 309.
11. Hale, B.W. and Evers, J.F.: "Elliptical Flow Equations for Vertically Fractured Gas Wells," *JPT* (December 1981) 2489.
12. Obut, S.T. and Ertekin, T.: "A Composite System Solution in Elliptical Flow Geometry," *SPEFE* (September 1987) 227.
13. Lee, W.J.: *Well Testing*, Textbook Series, SPE, Richardson, Texas (1977) **1**.
14. Agarwal, R.G.: "A New Method To Account for Producing-Time Effects When Drawdown Type Curves Are Used To Analyze Pressure-Buildup and Other Test Data," paper SPE 9289 presented at the 1980 SPE Annual Technical Conference and Exhibition, Dallas, 21–24 September.
15. Gringarten, A.C., Ramey, H.J. Jr., and Raghavan, R.: "Unsteady-State Pressure Distributions Created by a Well With a Single Infinite-Conductivity Vertical Fracture," *SPEJ* (August 1974) 347; *Trans.*, AIME, **257**.
16. Ramey, H.J. Jr. and Gringarten, A.C.: "Effect of High Volume Vertical Fractures on Geothermal Steam Well Behavior," *Proc.*, Second United Nations Symposium on the Use and Development of Geothermal Energy, San Francisco (1975).
17. Bourdet, D. *et al.*: "A New Set of Type Curves Simplifies Well Test Analysis," *World Oil* (May 1983) 95.
18. Holditch, S.A. and Morse, R.A.: "The Effects of Non-Darcy Flow on the Behavior of Hydraulically Fractured Wells," *JPT* (October 1976) 1169.
19. Bennett, C.O. *et al.*: "Influence of Fracture Heterogeneity and Wing Length on the Response of Vertically Fractured Wells," *SPEJ* (April 1983) 219.
20. Cinco-Ley, H. and Samaniego-V., F.: "Effect of Wellbore Storage and Damage on the Transient Pressure Behavior of Vertically Fractured Wells," paper SPE 6752 presented at the 1977 SPE Annual Technical Conference and Exhibition, Denver, Colorado, 9–12 October.
21. Lee, W.J. and Holditch, S.A.: "Fracture Evaluation With Pressure Transient Testing in Low-Permeability Gas Reservoirs," *JPT* (September 1981) 1776.
22. Holditch, S.A.: "Pretreatment Formation Evaluation," *Recent Advances in Hydraulic Fracturing,* J.L. Gidley *et al.* (eds.), Monograph Series, SPE, Richardson, Texas (1989) **12**, 47.

Chapter 7
Interpretation of Well-Test Data in Naturally Fractured Reservoirs

7.1 Overview

This chapter focuses on the interpretation of well-test data from wells completed in naturally fractured reservoirs. In previous chapters, we developed well-test analysis techniques for homogeneous-acting reservoirs; however, because of the presence of two distinct types of porous media, the assumption of homogeneous behavior is no longer valid in naturally fractured reservoirs. In this chapter, we describe two naturally fractured reservoir models, including the physics governing fluid flow in these reservoirs. We also introduce special semilog and type-curve analysis techniques for well tests in these reservoirs. Finally, we use several examples to illustrate the underlying theoretical concepts and practical applications of the analysis techniques.

7.2 Naturally Fractured Reservoir Models

A characteristic of naturally fractured reservoirs is the presence of two distinct types of porous media called the matrix and fracture. Because of the different fluid-storage and conductivity characteristics of the matrix and fractures, these reservoirs often are called dual-porosity reservoirs. **Fig. 7.1** illustrates a naturally fractured reservoir composed of a rock matrix surrounded by an irregular system of vugs and natural fractures. Fortunately, it has been observed that a real, heterogeneous naturally fractured reservoir has a characteristic behavior that can be interpreted with an equivalent, homogeneous dual-porosity model like that shown in the idealized sketch.

Several models have been proposed to represent the pressure behavior in a naturally fractured reservoir. These models differ conceptually only in the assumptions made to describe fluid flow in the matrix. Most dual-porosity models assume that production from the naturally fractured system goes from the matrix to the fracture and thence to the wellbore (i.e., the matrix does not produce directly into the wellbore). Furthermore, the models assume that the matrix has low permeability but large storage capacity relative to the natural-fracture system, while the fractures have high permeability but low storage capacity. Warren and Root[1] introduced two dual-porosity parameters in addition to the usual single-porosity parameters that can be used to describe dual-porosity reservoirs.

Interporosity flow is the fluid exchange between the two media (i.e., the matrix and fractures) constituting a dual-porosity system. Warren and Root defined the interporosity flow coefficient, λ, as

$$\lambda = \alpha r_w^2 \frac{k_m}{k_f}, \qquad (7.1)$$

where k_m = permeability of the matrix, k_f = permeability of the natural fractures, and α = parameter characteristic of the system geometry.

The interporosity flow coefficient is a measure of how easily fluid flows from the matrix to the fractures. The parameter α is defined by[2]

$$\alpha = \frac{4j(j+2)}{L^2}, \qquad (7.2)$$

where L = a characteristic dimension of a matrix block and j = the number of normal sets of planes limiting the less-permeable medium ($j = 1, 2, 3$). For example, $j = 3$ in the idealized reservoir block model in Fig. 7.1. On the other hand, for the multilayered or "slab" model that **Fig. 7.2** shows, $j = 1$. For the "slab" model, by letting $L = h_m$ (the thickness of an individual matrix block), λ becomes

$$\lambda = 12 r_w^2 \frac{k_m}{k_f h_m^2}. \qquad (7.3)$$

The storativity ratio,[1] ω, is defined by

$$\omega = \frac{(\phi V c_t)_f}{(\phi V c_t)_{f+m}} = \frac{(\phi V c_t)_f}{(\phi V c_t)_f + (\phi V c_t)_m}, \qquad (7.4)$$

where V = ratio of the total volume of one medium to the bulk volume of the total system and ϕ = ratio of the pore volume (PV) of one medium to the total volume of that medium. The subscripts f and $f+m$ refer to the fracture and to the total system (fractures plus matrix), respectively. Consequently, the storativity ratio is a measure of the relative fracture–storage capacity in the reservoir.

Although many models have been developed for naturally fractured reservoirs, we present two common models—pseudosteady-state and transient flow—describing flow in the less-permeable medium, the matrix. Barenblatt *et al.*[4] and subsequent authors[1,5-7] assumed pseudosteady-state flow, while others,[8-13] notably deSwaan,[8] assumed transient flow in the matrix. Intuition suggests that, in a low-permeability matrix, very long times should be required to reach pseudosteady-state and that transient matrix flow should dominate; however, test analysis indicates that pseudosteady-state flow is quite common. A possible explanation of this seeming inconsistency is that matrix flow is almost always transient but can exhibit a pseudosteady-state-like behavior if there is a signifi-

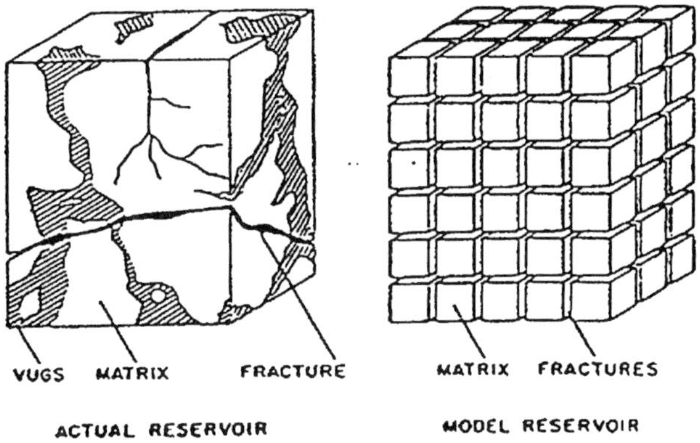

Fig. 7.1—Actual and idealized dual-porosity reservoir model (after Warren and Root[1]).

cant impediment to flow from the less-permeable medium to the more-permeable one (e.g., solution deposits in fissures).

7.3 Pseudosteady-State Matrix Flow Model

The pseudosteady-state flow model assumes that, at a given time, the pressure in the matrix is decreasing at the same rate at all points, and, thus, flow from the matrix to the fracture is proportional to the difference between matrix pressure and pressure in the adjacent fracture. Specifically, this model, which does not allow unsteady-state pressure gradients within the matrix, assumes pseudosteady-state flow conditions are present from the beginning of flow.

Because it assumes a pressure distribution in the matrix that would be reached only after what could be a considerable flow period, the pseudosteady-state flow model obviously is oversimplified. Again, however, this model seems to match a surprising number of field tests. One possible reason is that damage to the face of the matrix could cause the flow from matrix to fracture to be controlled by a sort of choke (the thin, low-permeability damaged zone) and, therefore, is proportional to pressure differences upstream and downstream of the choke. The next two sections present semilog and log-log analysis techniques for well tests in naturally fractured reservoirs exhibiting pseudosteady-state flow characteristics.

7.3.1 Semilog Analysis Technique. The pseudosteady-state matrix flow solution developed by Warren and Root[1] predicts that two parallel straight lines will develop on a semilog graph of test data. Curve A in **Fig. 7.3** represents this characteristic pressure response.

The initial straight line reflects flow in the fracture system only. At this time, the formation is behaving like a homogeneous formation with fluid flow originating only from the fracture system with no contribution from the matrix. Consequently, the slope of the initial semilog straight line is proportional to the permeability-thickness product of the natural-fracture system, just as it is for any homogeneous system. Following a discrete pressure drop in the fracture system, the fluid in the matrix begins to flow into the fracture and a rather flat transition region appears. Finally, the matrix and fracture reach an equilibrium condition and a second straight line appears. At this time, the reservoir again behaves like a homogeneous system, but now the system consists of both the matrix and the fractures. The slope of the second semilog straight line is proportional to the total permeability-thickness product of the matrix/fracture system. Because the permeability of the fractures is much greater than that of the matrix, the slope of the second line is almost identical to that of the initial line.

The shape of a semilog plot of test data from a naturally fractured reservoir is almost never the same as predicted by Warren and Root's model. Wellbore storage almost always obscures the initial straight line and often obscures part of the transition region between the straight lines. Curve B in Fig. 7.3 gives a more common pressure response.

The reservoir permeability-thickness product, kh (actually the kh of the fractures, or $(kh)_f$, because $(kh)_m$ is usually negligible), can be obtained from the slope, m, of the two semilog straight lines. Storativity, ω, can be determined from their vertical displacement, δp. The in-

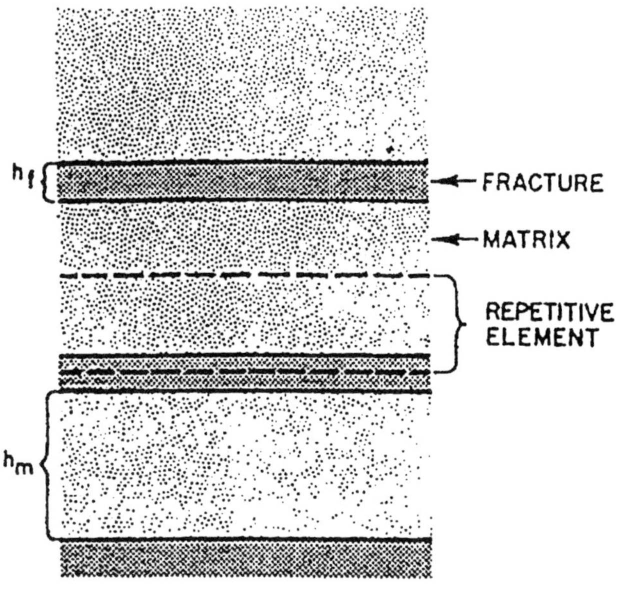

Fig. 7.2—Schematic reservoir with rectangular matrix elements (after Serra *et al.*[3]).

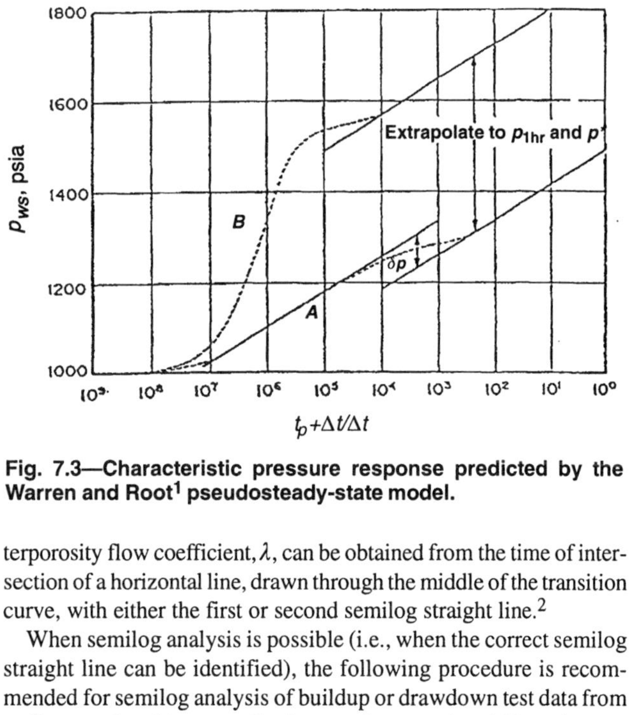

Fig. 7.3—Characteristic pressure response predicted by the Warren and Root[1] pseudosteady-state model.

terporosity flow coefficient, λ, can be obtained from the time of intersection of a horizontal line, drawn through the middle of the transition curve, with either the first or second semilog straight line.[2]

When semilog analysis is possible (i.e., when the correct semilog straight line can be identified), the following procedure is recommended for semilog analysis of buildup or drawdown test data from wells completed in naturally fractured reservoirs. Although presented in variables for slightly compressible fluids (i.e., liquids), the same procedure is applicable to gas well tests when the appropriate variables are used.

1. From the slope of the initial straight line (if present) or final straight line (more likely to be present), determine the permeability-thickness product, kh. In either case, the slope, m, is related to the total kh of the system; total system kh is essentially all in the fractures. The permeability-thickness product is given by

$$(kh)_f = \bar{k}h = \frac{162.6 qB\mu}{m}, \quad (7.5)$$

where $\bar{k} = (kh)_f / h$. Strictly speaking, the slope of the second straight line is related to $[(kh)_f + (kh)_m]$, but $(kh)_m$ ordinarily is much less than $(kh)_f$.

2. If both initial and final straight lines can be identified (or the position of the initial line can at least be approximated) and the pres-

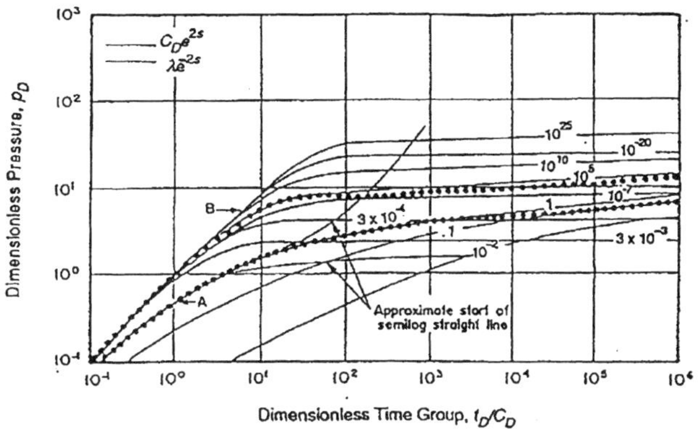

Fig. 7.4—Type curves for pseudosteady-state matrix flow (after Bourdet *et al.*[16]).

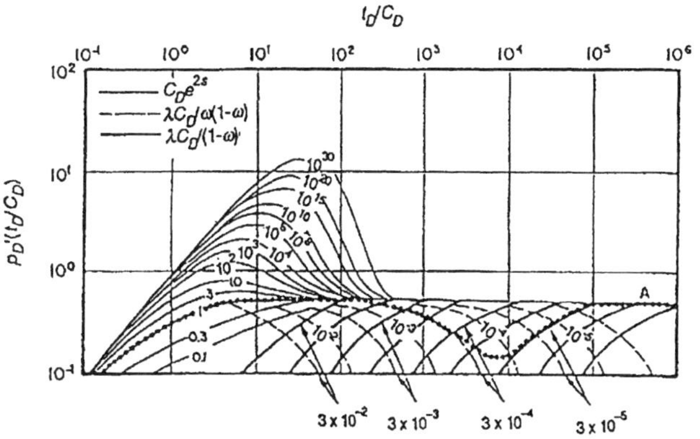

Fig. 7.5—Derivative type curves for pseudosteady-state matrix flow (after Bourdet[16]).

sure difference, δp, established, then the storativity ratio, ω, is calculated from

$$\omega = 10^{-\delta p/m}. \quad (7.6)$$

If the times of intersection of a horizontal line drawn through the midpoint of the transition data with the first and second semilog straight lines are denoted by t_1 and t_2, respectively, the storativity ratio also may be calculated from

$$\omega = \frac{t_1}{t_2}. \quad (7.7)$$

For a buildup test, where the times of intersection of a horizontal line drawn through the midpoint of the transition data with the first and second semilog straight lines are denoted by $[(t_p+\Delta t)/\Delta t]_1$ and $[(t_p+\Delta t)/\Delta t]_2$, respectively, the storativity ratio may be calculated from

$$\omega = \frac{\left[(t_p+\Delta t)/\Delta t\right]_2}{\left[(t_p+\Delta t)/\Delta t\right]_1}. \quad (7.8)$$

3. The interporosity flow coefficient, λ, is calculated[14] for a drawdown test by

$$\lambda = \frac{(\phi V c_t)_f \mu r_w^2}{\gamma \bar{k} t_1} = \frac{(\phi V c_t)_{f+m} \mu r_w^2}{\gamma \bar{k} t_2}, \quad (7.9)$$

or for a buildup test by

$$\lambda = \frac{(\phi V c_t)_f \mu r_w^2}{\gamma \bar{k} t_p}\left(\frac{t_p+\Delta t_1}{\Delta t_1}\right) = \frac{(\phi V c_t)_{f+m} \mu r_w^2}{\gamma \bar{k} t_p}\left(\frac{t_p+\Delta t_2}{\Delta t_2}\right), \quad (7.10)$$

where γ is the exponential of Euler's constant ($\gamma = 1.781$). Because t_1 and t_2 often are approximated, the value of λ obtained by this method may not be very accurate but usually is of the same order of magnitude as the correct value.

The terms $(\phi V)_m$ and $(c_t)_m$ in Eq. 7.10 are obtained by conventional methods. A porosity log usually reads only the matrix porosity (not the fracture porosity) and thus gives ϕ_m, while $(c_t)_m$ is the sum of c_oS_o, c_gS_g, c_wS_w, and c_f. V_m usually can be assumed to be essentially 1.0. From the definition of ω in Eq. 7.4,

$$(\phi V c_t)_f = (\phi V c_t)_m \left(\frac{\omega}{1-\omega}\right). \quad (7.11)$$

4. The second semilog straight line should be extrapolated to $p_{1\text{hr}}$, and the skin factor is

$$s = 1.151\left[\frac{\Delta p_{1\text{hr}}}{m} - \log\left(\frac{\bar{k}}{\phi \mu c_t r_w^2}\right) + 3.23\right], \quad (7.12)$$

where $\Delta p_{1\text{hr}} = (p_i - p_{1\text{hr}})$ for a drawdown test or $[p_{1\text{hr}} - p_{wf}(\Delta t = 0)]$ for a buildup test.

5. The second semilog straight line should be extrapolated to p^* (Fig. 7.3). From p^*, $\bar{p}$ can be found with conventional methods (e.g., the Matthew-Brons-Hazebroek[15] p^* method described in Chap. 2).

7.3.2 Type-Curve Analysis Technique. Particularly because of wellbore-storage distortion, type curves are quite useful for identifying and analyzing dual-porosity systems. **Fig. 7.4** shows an example of the Bourdet *et al.*[16] type curves developed for pseudosteady-state matrix flow. Initially, test data follow a curve for some value of C_De^{2s} where C_D is the dimensionless wellbore-storage coefficient. In the example in Fig. 7.4, the earliest data for Well A follow the curve for $C_De^{2s} = 1$. The data then deviate from the early fit and follow a transition curve characterized by the parameter λe^{-2s}. In Fig. 7.4, the data follow the curve for $\lambda e^{-2s} = 3 \times 10^{-4}$. When equilibrium is reached between the matrix and fracture system, the data then follow another C_De^{2s} curve; in the example, the later data follow the $C_De^{2s} = 0.1$ curve.

At earliest times, the reservoir behaves like a homogeneous reservoir, with all fluid originating from the fracture system. During intermediate times, there is a transition region as the matrix begins to produce into the fractures. At later times, the system again behaves like a homogeneous system, with both matrix and fractures contributing to fluid production.

Fig. 7.5 illustrates derivative type curves for a formation with pseudosteady-state matrix flow.[16] The most notable feature, characteristic of naturally fractured reservoirs, is the dip below the homogeneous reservoir curve. The curves dipping downward are characterized by a parameter $\lambda C_D/\omega(1-\omega)$, while the curves returning to the homogeneous reservoir curves are characterized by the parameter $\lambda C_D/(1-\omega)$. Test data that follow this pattern on the derivative type curve can be interpreted reasonably as identifying a dual-porosity reservoir with pseudosteady-state matrix flow (a theory that needs to be confirmed with geological information and reservoir performance). Pressure and pressure-derivative type curves can be used together for analysis of a dual-porosity reservoir. The pressure-derivative data are especially useful for identifying dual-porosity behavior.

The following procedure is recommended for type-curve analysis of buildup or drawdown test data from wells completed in naturally fractured reservoirs following the pseudosteady-state model. The procedure is presented in terms of variables for slightly compressible fluids; however, the procedure is also applicable for gas-well tests when the appropriate plotting functions are used. The procedure suggests the use of equivalent time, Δt_e, for buildup tests. Equivalent time as we have defined it is rigorously correct only for radial flow in homogeneous-acting formations.

1. Plot pressure-change and pressure-derivative data on log-log graph paper or on tracing paper. Plot $(p_i - p_{wf})$ vs. t for a drawdown test or $[p_{ws} - p_{wf}(\Delta t = 0)]$ vs. Δt_e for a buildup test.

2. Perform a qualitative type-curve analysis. The purpose of this analysis is to identify the various flow regimes characteristic of

dual-porosity behavior, including the early and late semilog straight lines that reflect fracture and fracture/matrix flow, respectively.

If there is a horizontal line on the derivative plot of the test data at the end of the test, overlay this line onto the $(t_D/C_D)\,p_D'=0.5$ line on the type curve. These data represent flow from the fracture/matrix system.

If there is a unit-slope line at early times, fix the horizontal match of the data by overlaying the unit-slope line on the type curve. If no unit-slope line is present, fix the horizontal match by finding the best match possible for the early data on the derivative type curve.

3. If the two straight lines are present and can be identified, perform a semilog analysis with the procedure discussed in Sec. 7.3.1.

4. Perform a quantitative type-curve analysis. The objective of this analysis is either to confirm the results from the semilog analysis or to estimate reservoir properties when no semilog analysis is possible.

A. If an estimate of permeability is available from the semilog analysis, calculate a pressure match point with Eq. 7.13,

$$(\Delta p)_{MP} = \frac{141.2qB\mu}{\bar{k}h}(p_D)_{MP}, \qquad (7.13)$$

where p_D is an arbitrarily chosen value. Record a time match point $(\Delta t,\, t_D/C_D)$.

B. With the type curve in matched position, determine the value of C_De^{2s} characterizing the fit of the earliest data on the pressure type curve. The fit on the derivative type curve should confirm this parameter value. The earliest data are those that appear before the transition region. This value of C_De^{2s}, which we call $(C_De^{2s})_f$, characterizes the fracture system.

C. With the type curve remaining in the same position, determine the value of C_De^{2s}, which we call $(C_De^{2s})_{f+m}$, that characterizes the test data after the transition to total system flow has been completed.

D. Read the value of λe^{-2s} that characterizes the horizontal transition curve crossed by the test data on the pressure type curve at an intermediate time in the transition region. There is a significant uniqueness problem in determining the best-fitting transition curve. For Test B in Fig. 7.4, the transition curve chosen was for $\lambda e^{-2s}=10^{-7}$.

E. Calculate the storativity ratio, ω, from the ratio of the two C_De^{2s} values:

$$\omega = \frac{(C_De^{2s})_{f+m}}{(C_De^{2s})_f}. \qquad (7.14)$$

F. If no semilog analysis is possible, calculate permeability from the pressure match point $(\Delta p,\, p_D)$.

$$\bar{k} = \frac{141.2qB\mu}{\Delta ph}p_D. \qquad (7.15)$$

Record a time match point $(\Delta t,\, t_D/C_D)$.

G. Compute the wellbore-storage coefficient from the time match point $(\Delta t,\, t_D/C_D)$.

$$C_D = \frac{0.0002637\bar{k}}{\phi\mu c_t r_w^2}\left(\frac{\Delta t}{t_D/C_D}\right)_{MP}. \qquad (7.16)$$

For a gas-well-test analysis with adjusted pressure, p_a, and adjusted time, t_a, $\underline{\mu}$ and c_t in Eq. 7.16 are evaluated at average drainage-area pressure, $\bar{p}$.

H. From the value of C_D determined in Step 4G and the value of $(C_De^{2s})_{f+m}$ determined in Step 4C, compute the skin factor, s.

$$s = 0.5\ln\left[\frac{(C_De^{2s})_{f+m}}{C_D}\right]. \qquad (7.17)$$

I. Using the estimate of s from Step 4H and the parameter λe^{-2s} from Step 4D, calculate λ. This value should be consistent with the estimate of λ from semilog analysis. If it is not, then the type-curve match should be refined.

TABLE 7.1—RATE DATA, EXAMPLE 7.1

t (hours)	q (Mscf/D)	t (hours)	q (Mscf/D)
1,296	212.60	991	147.30
1,968	229.80	10,656	181.90
2,688	222.70	11,400	135.80
3,408	206.30	12,048	150.40
4,080	203.70	12,720	148.50
4,824	202.40	13,536	143.40
5,592	198.30	14,256	137.90
6,240	194.80	14,952	153.10
7,032	139.80	15,696	142.60
7,776	177.90	16,440	142.10
8,520	155.30	17,232	149.30
9,240	156.30		

Example 7.1—Analysis of a Buildup Test in a Naturally Fractured Reservoir With Pseudosteady-State Matrix Flow. A pressure-buildup test was obtained from a gas well completed in a naturally fractured reservoir. Wells in this formation usually exhibit pseudosteady-state matrix flow. The test period was preceded by more than one rate. **Tables 7.1 and 7.2** summarize the rate data and pressure-buildup-test data, respectively. Determine permeability; skin factor; the storativity ratio, ω; and the interporosity flow coefficient, λ, with semilog and type-curve analysis techniques.

$r_w = 0.3$ ft
$\phi = 0.0533$
$\mu_g = 0.01118$ cp
$h = 286$ ft
$S_g = 1.0$
$c_t = 0.002658$ psia^{-1}
$A = 43.0$ acres
$B_g = 6.48012$ RB/Mscf
$p_{a\,wf}(\Delta t=0) = 36.67$ psia

Solution. *Calculate t_p.* First, we calculate the effective producing time before buildup test with the principle of superposition,

$$t_p = \frac{\sum\limits_{i=1}^{n=\text{last}}(q_i - q_{i-1})\left(t_n - \sum\limits_{j=0}^{j=i-1} t_j\right)}{q_n}$$

$$= \frac{\begin{array}{l}\times\,(212.6-0)(17{,}232-0)\\ +\,(229.8-212.6)(17{,}232-1{,}296)\\ +\ldots+(149.3-142.1)(17{,}232-16{,}440)\end{array}}{149.3}$$

$$= 19{,}836 \text{ hours.}$$

Construct Data Plots. We then plot adjusted pressure change and pressure derivative vs. equivalent-time function on a log-log plot (**Fig. 7.6**) and adjusted pressure vs. equivalent-time function on a semilog graph (**Fig. 7.7**). **Table 7.3** gives semilog and log-log plotting functions.

Qualitative Type-Curve Analysis. Dual-porosity behavior is evident from a match of both adjusted pressure change and pressure-derivative data (Fig. 7.6) with the Bourdet *et al.*[16] type curve for pseudosteady-state interporosity flow. The derivative data from longest shut-in times ($\Delta t_e > 100$ hours) can be aligned with the $(t_D/C_D)\,p_D'=0.5$ line, whereas earlier derivative data ($0.3 < \Delta t_e < 100$ hours) fall below this line, characteristic of a dual-porosity system. Because the earliest data are trending toward a unit-slope line, we also can establish a tentative horizontal match.

Semilog Analysis. Because we have tentatively identified the second straight line indicative of the naturally fractured system behaving like a homogeneous system, we now attempt a semilog analysis.

TABLE 7.2—PRESSURE-BUILDUP TEST DATA, EXAMPLE 7.1

Δt (hours)	p_{ws} (psia)	Δt (hours)	p_{ws} (psia)	Δt (hours)	p_{ws} (psia)
0	171.75	0.66300	244.32	20.213	268.09
0.016998	174.06	0.74800	247.11	22.695	269.11
0.034000	177.77	0.85000	249.52	25.466	270.19
0.050999	181.26	0.96900	251.46	28.594	271.36
0.068001	184.62	1.0880	252.76	32.113	272.61
0.084999	187.83	1.2240	253.81	36.040	273.92
0.10200	190.90	1.3770	254.62	40.460	275.32
0.13600	196.81	1.5470	255.28	45.424	276.80
0.15300	199.53	1.7510	255.89	51.022	278.35
0.17000	202.15	1.9720	256.41	57.330	280.01
0.18700	204.67	2.2270	256.90	64.385	281.73
0.20400	207.09	2.4990	257.33	72.353	283.56
0.22100	209.41	2.8050	257.74	81.234	285.45
0.23800	211.64	3.1620	258.19	91.194	287.44
0.25500	213.77	3.5530	258.61	102.340	289.48
0.27200	215.83	3.995	259.04	114.960	291.67
0.28900	217.79	4.488	259.49	128.990	293.92
0.32300	221.69	5.049	259.97	144.780	296.29
0.34000	223.41	5.678	260.47	162.460	298.76
0.35700	225.06	6.375	260.99	182.300	301.30
0.37400	226.63	7.157	261.53	204.570	303.84
0.39100	228.12	8.041	262.09	229.550	306.38
0.40800	229.56	9.027	262.69	257.630	308.88
0.42500	230.91	10.047	263.26	289.110	311.40
0.44200	232.21	11.305	263.94	324.470	313.87
0.45900	233.45	12.699	264.66	364.250	316.53
0.47600	234.63	14.280	265.43	408.770	319.25
0.51000	236.93	16.031	266.26	458.680	321.93
0.57800	240.65	18.003	267.14	507.420	324.22

1. On the Horner plot in Fig. 7.7, a straight line with slope $m = 107 - 67 = 40$ psi/cycle is fit through the later data identified from the qualitative type-curve match. Another straight line with the same slope is drawn through the early data.

2. From the second (later-time) semilog straight line, we calculate the permeability-thickness product with Eq. 7.5,

$$(kh)_f = \bar{k}h = \frac{162.6 q_{\text{last}} B_g \mu_g}{m}$$

$$= \frac{(162.6)(149.31)(6.48012)(0.01118)}{40.0}$$

$$= 43.97 \text{ md-ft.}$$

The average fracture permeability is

$$\bar{k} = \frac{\bar{k}h}{h} = \frac{43.97}{286} = 0.154 \text{ md.}$$

3. The vertical displacement between the two parallel lines in Fig. 7.7 is $\delta p = 116 - 51 = 65$ psi.

The storativity ratio is estimated with Eq. 7.6: $\omega = 10^{-\delta p/m} = 10^{-65/40} = 0.02371$.

4. Forcing a horizontal line through the approximate middle of the transition data at $(p_a)_{ws} = 86$ psia in Fig. 7.7, we find that this line intersects the first and second parallel semilog straight lines at Horner time ratios of $\left[(t_p + \Delta t_a)/\Delta t_a\right]_1 = 27{,}938$ and $\left[(t_p + \Delta t_a)/\Delta t_a\right]_2 = 661.2$, respectively. Now we calculate the term $\phi V c_t$ for the matrix, $(\phi V c_t)_m = (0.0533)\ (0.002658) = 1.4167 \times 10^{-4}$.

From Eq. 7.11 the same term for the fracture system is

$$(\phi V c_t)_f = (\phi V c_t)_m \left(\frac{\omega}{1 - \omega}\right)$$

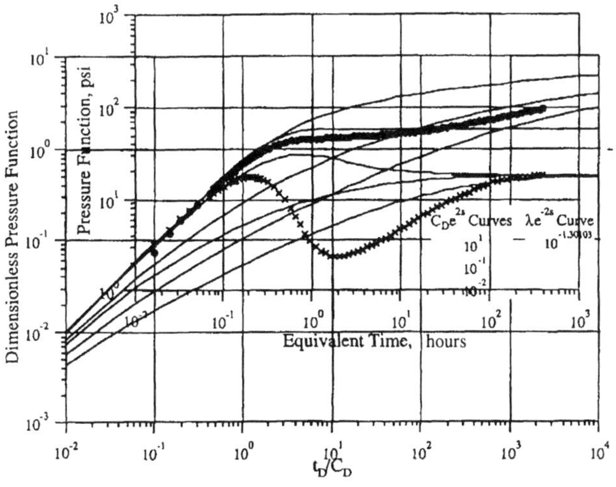

Fig. 7.6—Qualitative type-curve match, Example 7.1.

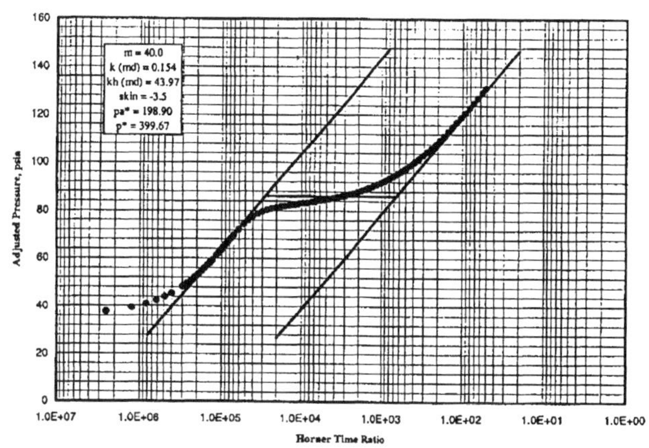

Fig. 7.7—Horner plot, Example 7.1.

TABLE 7.3—SEMILOG AND LOG-LOG PLOTTING FUNCTIONS, EXAMPLE 7.1

Δt_e (hours)	Horner Time Ratio	p_{ws} (psia)	$p_{a,ws}$ (psia)	Δp_a (psi)	$\Delta t_e \frac{d\Delta p_a}{d\Delta t_e}$, psi
0.007856	2,525,013	174.06	37.661	0.994	2.3237
0.015844	1,251,957	177.77	39.291	2.6241	3.2321
0.023987	826,947.9	181.26	40.854	4.1872	4.626
0.032278	614,536.2	184.62	42.382	5.7155	5.9853
0.04071	487,251.3	187.83	43.879	7.2125	7.2615
0.049278	402,532.6	190.9	45.327	8.6606	8.3864
0.066799	296,950.6	196.81	48.19	11.524	10.408
0.075745	261,878.7	199.53	49.53	12.864	11.301
0.084804	233,904.1	202.15	50.489	14.182	12.13
0.093973	211,081.9	204.67	52.124	15.457	12.826
0.10325	192,116.2	207.09	53.371	16.704	13.545
0.11262	176,132.1	209.41	54.575	17.908	14.266
0.12209	162,470.3	211.64	55.749	19.082	14.888
0.13166	150,660.8	213.77	56.882	20.215	15.402
0.14131	140,372.2	215.83	57.986	21.319	15.834
0.15105	131,320.8	217.79	59.052	22.385	16.132
0.17077	116,156.2	221.69	61.19	24.523	16.678
0.18074	109,748.8	223.41	62.147	25.48	16.875
0.19079	103,967.7	225.06	63.064	26.397	17.031
0.20091	98,730.77	226.63	63.953	27.286	17.104
0.21109	93,969.4	228.12	64.805	28.138	17.025
0.22134	89,617.78	229.56	65.619	28.952	16.979
0.23164	85,632.88	230.91	66.402	29.736	16.913
0.24199	81,970.33	232.21	67.157	30.49	16.589
0.2524	78,589.54	233.45	67.874	31.207	16.54
0.26286	75,462.22	234.63	68.561	31.895	16.464
0.28393	69,862.29	236.93	69.919	33.252	15.98
0.32657	60,740.42	240.65	72.14	35.473	14.905
0.38063	52,113.6	244.32	74.361	37.694	13.259
0.43536	45,562.29	247.11	76.079	39.412	11.427
0.50169	39,538.36	249.52	77.573	40.906	9.5998
0.5797	34,217.7	251.46	78.788	42.122	7.7489
0.65818	30,137.65	252.76	79.609	42.942	6.2778
0.74826	26,509.5	253.81	80.274	43.607	5.0686
0.84995	23,337.84	254.62	80.787	44.121	4.1069
0.96323	20,593.21	255.28	81.206	44.539	3.4501
1.0995	18,040.93	255.89	81.599	44.932	2.9531
1.2474	15,901.88	256.41	81.933	45.267	2.648
1.4184	13,984.77	256.9	82.249	45.582	2.4722
1.601	12,389.76	257.33	82.529	45.862	2.3507
1.8068	10,978.53	257.74	82.793	46.126	2.2986
2.0473	9,688.858	258.19	83.078	46.412	2.2832
2.3111	8,582.926	258.61	83.347	46.681	2.327
2.6097	7,600.874	259.04	83.63	46.963	2.4173
2.9433	6,739.374	259.49	83.915	47.249	2.5082
3.3235	5,968.407	259.97	84.228	47.561	2.6358

$$= \left(1.4167 \times 10^{-4}\right)\left(\frac{0.02371}{1 - 0.02371}\right)$$

$$= 3.4406 \times 10^{-6}.$$

The interporosity flow coefficient is estimated with Eq. 7.10 and the intersection of the first semilog line with the horizontal line through the transition data,

$$\lambda = \frac{(\phi V c_t)_f \mu_g r_w^2}{\gamma \bar{k} t_p}\left(\frac{t_p + \Delta t_a}{\Delta t_a}\right)_1$$

$$= \frac{(3.4406 \times 10^{-6})(0.01118)(0.3)^2}{(1.781)(0.154)(19,836)}(27,938)$$

$$= 1.78 \times 10^{-8}.$$

Similarly, using the intersection of the second semilog line with the horizontal line through the transition data,

$$\lambda = \frac{(\phi V c_t)_{f+m} \mu_g r_w^2}{\gamma \bar{k} t_p}\left(\frac{t_p + \Delta t_a}{\Delta t_a}\right)_2$$

$$= \frac{(1.4167 \times 10^{-4})(0.01118)(0.3)^2}{(1.781)(0.154)(19,836)}(661.2)$$

$$= 1.73 \times 10^{-8}.$$

Averaging the two values, which theoretically should be equal, we obtain

$$\lambda = 1.75 \times 10^{-8}.$$

TABLE 7.3—SEMILOG AND LOG-LOG PLOTTING FUNCTIONS, EXAMPLE 7.1 (CONTINUED)					
Δt_e (hours)	Horner Time Ratio	p_{ws} (psia)	$p_{a,ws}$ (psia)	Δp_a (psi)	$\Delta t_e \frac{d\Delta p_a}{d\Delta t_e}$, psi
3.7505	5,288.895	260.47	84.557	47.89	2.7666
4.2245	4,695.467	260.99	84.895	48.228	2.9263
4.7572	4,169.68	261.53	85.248	48.582	3.0626
5.3606	3,700.332	262.09	85.619	48.953	3.2268
6.0349	3,286.881	262.69	86.015	49.348	3.4188
6.7338	2,945.736	263.26	86.386	49.719	3.6172
7.5976	2,610.824	263.94	86.831	50.164	3.8846
8.5569	2,318.129	264.66	87.304	50.637	4.1605
9.6476	2,056.055	265.43	87.814	51.148	4.5003
10.859	1,826.688	266.26	88.369	51.702	4.8366
12.226	1,622.444	267.14	88.959	52.293	5.2031
13.764	1,441.151	268.09	89.598	52.932	5.6083
15.496	1,280.072	269.11	90.276	53.61	6.0201
17.436	1,137.646	270.19	91.005	54.338	6.4496
19.633	1,010.34	271.36	91.779	55.132	6.895
22.114	896.9883	272.61	92.65	55.984	7.3834
24.894	796.8185	273.92	93.547	56.88	7.8526
28.036	707.5189	275.32	94.505	57.839	8.3288
31.579	628.139	276.8	95.534	58.867	8.8366
35.592	557.3163	278.35	96.61	59.944	9.3601
40.136	494.2197	280.01	97.758	61.091	9.8884
45.241	438.4518	281.73	98.98	62.314	10.434
51.036	388.6668	283.56	100.27	63.604	10.985
57.526	344.818	285.45	101.61	64.948	11.598
64.843	305.9081	287.44	103.04	66.375	12.199
73.075	271.4471	289.48	104.51	67.844	12.836
82.44	240.6114	291.67	106.11	69.439	13.525
92.909	213.4992	293.92	107.76	71.091	14.186
104.76	189.3471	296.29	109.5	72.838	14.816
118.1	167.9594	298.76	111.34	74.676	15.299
133.15	148.9748	301.3	113.25	76.582	15.667
150.14	132.1167	303.84	115.18	78.511	15.963
169.28	117.1786	306.38	117.11	80.448	16.093
190.9	103.9078	308.88	119.04	82.374	16.353
215.24	92.15759	311.4	121	84.332	16.691
242.67	81.74064	313.87	122.93	86.266	17.018
273.63	72.49205	316.53	125.03	88.361	17.314
308.38	64.32324	319.25	127.19	90.524	17.638
347.45	57.09023	321.93	129.35	92.683	17.72
385.68	51.43124	324.22	131.19	94.528	17.966

5. Extrapolating the second semilog straight line in Fig. 7.7 to a shut-in adjusted time of 1 hour, we obtain $p_{a\,1hr} = 27$ psia. The skin factor is

$$s = 1.151\left[\frac{p_{a1hr} - p_{awf}(\Delta t = 0)}{m} - \log\left(\frac{\bar{k}}{\phi\mu_g c_t r_w^2}\right) + 3.23\right]$$

$$= 1.151\left\{\frac{27.0 - 36.67}{40.0}\right.$$

$$\left. - \log\left[\frac{0.154}{(0.0533)(0.01118)(0.002658)(0.3)^2}\right] + 3.23\right\}$$

$$= -3.505.$$

Quantitative Type-Curve Analysis. The purpose of this final analysis is to use type curves to confirm the results from the semilog analysis. **Fig. 7.8** illustrates the type-curve match.

1. We have an estimate of $\bar{k}h = 0.154$ md-ft from the semilog analysis, so we precalculate a pressure match point with an arbitrary value of $p_D = 10$,

$$(\Delta p)_{MP} = \frac{141.2 q_{last} B_g \mu_g}{\bar{k}h}(p_D)_{MP}$$

$$= \frac{(141.2)(149.3)(6.48012)(0.01118)}{43.97}(10)$$

$$= 343.4 \text{ psi}.$$

2. The match also forces the time derivative data to overlay $(t_D/C_D)\,p_D' = 0.5$ and $(C_De^{2s})_{f+m} = 10^{-1.4}$ (interpolating to a better fit). The early-time data are fit to the curve for $(C_De^{2s})_f = 10^1$, while our best estimate of the matching parameter for the transition data is $\lambda e^{-2s} = 10^{-1.30}$. Finally, we choose a time match point; $\Delta t_e = 6.0$ hours and $t_D/C_D = 1.0$.

3. The storativity ratio is

$$\omega = \frac{(C_De^{2s})_{f+m}}{(C_De^{2s})_f} = \frac{10^{-1.4}}{10^1} = 0.00398.$$

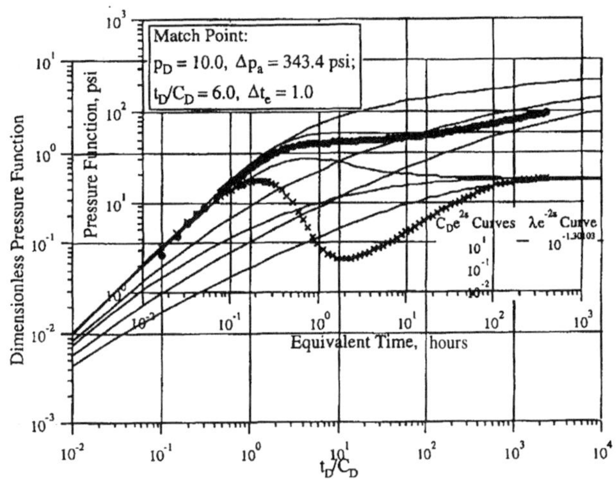

Fig. 7.8—Quantitative type-curve match, Example 7.1.

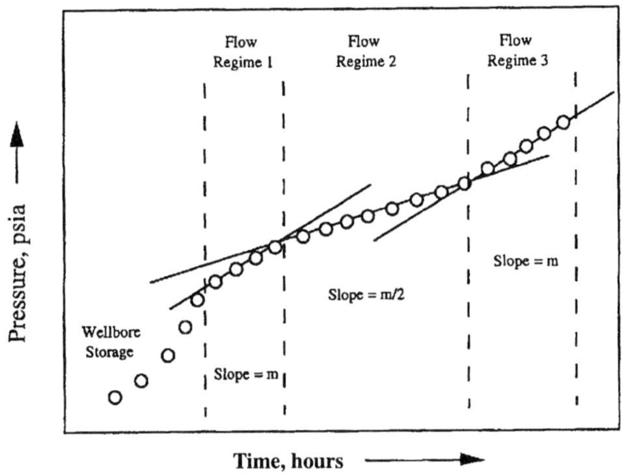

Fig. 7.9—Characteristic flow regimes in a dual-porosity system with transient matrix flow.

4. The dimensionless wellbore-storage coefficient is computed from the time match point.

$$C_D = \frac{0.0002637\bar{k}}{\phi\mu c_t r_w^2}\left(\frac{\Delta t}{t_D/C_D}\right)_{MP}$$

$$= \frac{0.0002637(0.154)}{0.0533(0.01118)(0.002658)(0.3)^2}\left(\frac{1}{6}\right)$$

$$= 47.48.$$

5. The skin factor is calculated with Eq. 7.17,

$$s = 0.5\ln\frac{(C_De^{2s})_{f+m}}{C_D} = 0.5\ln\left(\frac{10^{-1.4}}{47.48}\right)$$

$$= -3.542,$$

which agrees with the estimate of $s = -3.505$ from semilog analysis.

6. The interporosity flow coefficient, λ, is calculated with $\lambda e^{-2s} = 10^{-1.30}$ from the type-curve match and $s = -3.542$ from Step 5, $\lambda e^{-2s} = 10^{-1.30}$ or

$$\lambda = \frac{10^{-1.30}}{e^{-2s}} = \frac{0.05}{e^{-2(-3.542)}} = 4.19 \times 10^{-5}.$$

The estimates of ω and λ are in poor agreement with the values from semilog analysis. The lack of closer agreement is related to the fact that the early-time test data do not overlay the 0.5 line of the derivative type curve. This suggests that the initial straight line identified in the semilog analysis is not correct and hence the estimates of ω and λ are not correct.

7.4 Transient Matrix Flow Model

The more probable flow regime in the matrix is unsteady-state or transient flow (i.e., flow in which an increasing pressure drawdown starts at the matrix/fracture interface and moves farther into the matrix with increasing time). Only at late times should pseudosteady-state be achieved, although a matrix with a thin, low-permeability damaged zone at the fracture face may behave as predicted by the pseudosteady-state matrix flow model even though the flow in the matrix is actually unsteady-state.

A semilog graph of test data for a formation with transient matrix flow has a characteristic shape different from that for pseudosteady-state flow in the matrix. Three distinct flow regimes have been identified that are characteristic of dual-porosity reservoir behavior with transient matrix flow. **Fig. 7.9** illustrates these on a semilog graph. We label these as Flow Regimes 1, 2, and 3.

Flow Regime 1 occurs at early times when all production comes from the fractures. Flow Regime 2 occurs when production from the matrix into the fracture begins and continues until the matrix-to-fracture transfer reaches equilibrium. This equilibrium point marks the beginning of Flow Regime 3, during which total system flow, from matrix to fracture to wellbore, is dominant. The same three flow regimes appear when there is pseudosteady-state matrix flow. The duration and shape of the transition flow regimes, however, is considerably different for the two matrix flow models.

Serra *et al.*[13] observed that pressures from each of these flow regimes will plot as straight lines on conventional semilog graphs. Flow Regimes 1 and 3, which correspond to the classic early- and late-time semilog straight-line periods, respectively, have the same slope.[1,7] Flow Regime 2 is an intermediate transitional period between the first and third flow regimes. The semilog straight line of Flow Regime 2 has a slope of approximately one-half that of Flow Regimes 1 and 3. If all or any two of these regimes can be identified, then a complete analysis is possible with semilog methods alone. Certain nonideal conditions, however, may make this analysis difficult to apply.

Flow Regime 1 often is distorted or even totally obscured by wellbore storage, which often makes this flow regime difficult to identify. Flow Regime 2, the transition region, also may be obscured by wellbore storage. Flow Regime 3 sometimes requires a long flow period followed by a long shut-in time to be observed, especially in low-permeability formations. Furthermore, boundary effects may appear before Flow Regime 3 is fully developed. Secs. 7.4.1 and 7.4.2 present semilog and log-log plotting analysis techniques for well tests from dual-porosity reservoirs exhibiting transient matrix flow.

7.4.1 Semilog Analysis Techniques. Serra *et al.*[3] presented a semilog method for analyzing well-test data in dual-porosity reservoirs exhibiting transient matrix flow. They found that the existence of the transition region, Flow Regime 2, and either Flow Regime 1 or Flow Regime 3 is sufficient to obtain a complete analysis of drawdown- or buildup-test data. Further, they assumed unsteady-state flow in the matrix, no wellbore storage, and rectangular matrix-block geometry (Fig. 7.2). On the basis of our experience, the rectangular matrix-block geometry is adequate, although different assumed geometries can lead to slightly different interpretation results.

The major weakness of the Serra *et al.* method is that it assumes no wellbore storage. In many cases, Flow Regimes 1 and 2 are partially or even totally obscured by wellbore storage, making analysis by their method impossible or difficult. Despite this limitation, the Serra *et al.* method has great practical value when used in conjunction with type-curve methods. The following discussion presents calculation procedures for application of the Serra *et al.* method. These calculations apply to both buildup- and drawdown-test data. Although presented for well-test analysis of slightly compressible liquids, this procedure is also applicable to gas-well-test analysis when adjusted time and adjusted pressure variables are used, with gas properties evaluated at average drainage-area pressure, $\bar{p}$.

Semilog Analysis Based on Flow Regimes 1 and 2. This case is rarely observed because of wellbore storage. If these flow regimes are present, however, the following procedure can be used to analyze the test data.

1. Estimate $k_f h_{ft} = \bar{k}h$ from the slope of the semilog straight line (m or m^*, where $m^* = m/2$),

$$k_f h_{ft} \approx \bar{k}h = \frac{162.6qB\mu}{m} = \frac{81.29qB\mu}{m^*}, \quad (7.18)$$

where h_{ft} = net fracture thickness.

2. Estimate the skin factor, s, in one of two ways.

A. First, assuming a value of $n^2 k_m \phi_m c_{mt}$ (which will be improved in Step 7 as part of an iterative procedure),

$$s = 0.5756\left\{\frac{\Delta p^*_{1hr}}{m^*} - \log\left[\frac{(k_f h_{ft})^2}{n^2 k_m \phi_m c_{mt} \mu r_w^4}\right] + 3.729\right\}, \quad (7.19)$$

where n = number of fractures (equal to net matrix thickness, h_{mt}, divided by individual matrix-block thickness, h_m, or net fracture thickness, h_{ft}, divided by individual fracture thickness, h_f); k_m = matrix permeability, md; ϕ_m = matrix porosity, fraction; c_{mt} = total matrix compressibility, psi^{-1}; and k_f = fracture permeability, md.

B. Alternatively, assuming a value of $\phi_f h_{ft} c_{ft}$ (which will be checked in Step 6 as part of an iterative process),

$$s = 1.151\left\{\frac{\Delta p_{1hr}}{m} - \log\left[\frac{(k_f h_{ft})}{\phi_f h_{ft} c_{ft} \mu r_w^2}\right] + 3.23\right\}, \quad (7.20)$$

where ϕ_f = fracture porosity, fraction, and c_{ft} = fracture system total compressibility, psi^{-1}.

3. Calculate p_{wD} with:

$$p_{wD} = \frac{k_f h_{ft} \Delta p}{141.2qB\mu}. \quad (7.21)$$

4. Plot p_{wD} vs. Δt on tracing paper or on semilog graph paper with the same scale as in **Fig. 7.10.**

5. Align the vertical axis of the data plot with the vertical axis of Fig. 7.10. Move the data plot in the horizontal direction only to obtain a time match point. Record values of the parameters λ', ω', t_1^*, t_{1D}, and (t_D, Δt) at the match point, where t_1^* is the time (hours) at the point of intersection of Flow Regimes 1 and 2 and t_{1D} is the dimensionless time at this point, defined by

$$t_{1D} = \frac{0.0002637kt_1^*}{\phi\mu c_t r_w^2}. \quad (7.22)$$

6. Estimate $\phi_f c_{ft} h_{ft}$ from the time match point,

$$\phi_f c_{ft} h_{ft} = \frac{2.637 \times 10^{-4} k_f h_{ft}}{\mu r_w^2}\left(\frac{\Delta t}{t_D}\right)_{MP}. \quad (7.23)$$

If the value does not agree with that in Step 2B, use this updated value, return to Step 2B, and repeat until there is no further change in the value of $\phi_f c_{ft} h_{ft}$.

7. Estimate $n^2 k_m \phi_m c_{mt}$,

$$n^2 k_m \phi_m c_{mt} = \frac{532.3\mu(\phi_f c_{ft} h_{ft})^2}{t_1^*}. \quad (7.24)$$

If the value does not agree with that in Step 2A, use this updated value, return to Step 2A, and repeat until there is no further change in the value of $n^2 k_m \phi_m c_{mt}$.

8. Assume a value of $\phi_m c_{mt}$ and estimate k_m/h_m^2,

$$\frac{k_m}{h_m^2} = \frac{(n^2 k_m \phi_m c_{mt})}{h^2 \phi_m c_{mt}}. \quad (7.25)$$

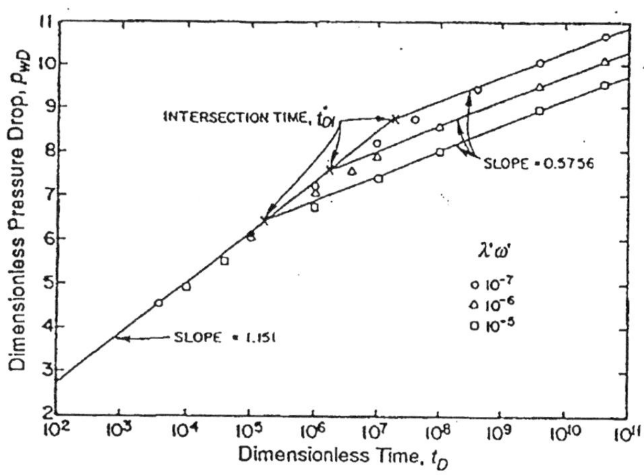

Fig. 7.10—Estimating ω' and λ' with data from Flow Regimes 1 and 2 (after Serra *et al.*[3]).

9. Assume that $h_{mt} = h$ and estimate λ' and ω'.

$$\omega' = \frac{\phi_m c_{mt} h}{\phi_f c_f h_{ft}} \quad (7.26)$$

$$\text{and } \lambda' = 12\frac{k_m}{h_m^2}\frac{h}{k_f h_{ft}} r_w^2. \quad (7.27)$$

10. Estimate ω and λ.

$$\omega = \frac{1}{1+\omega'} \quad (7.28)$$

$$\text{and } \lambda \approx \lambda'. \quad (7.29)$$

Semilog Analysis Based on Flow Regimes 2 and 3. Because of wellbore-storage effects, Flow Regimes 2 and 3 are more commonly observed in practice. The following procedure can be used to analyze test data under these conditions.

1. Estimate $k_f h_{ft} = \bar{k}h$ from the slope of the semilog straight line (m or m^*, where $m^* = m/2$) with Eq. 7.18.

2. Assume a value of $\phi_m c_{mt} \mu$ and calculate k_m/h_m^2.

$$\frac{k_m}{h_m^2} = \frac{532.3\phi_m c_{mt}\mu}{\Delta t^*}, \quad (7.30)$$

where Δt^* = time at which Flow Regimes 2 and 3 intersect.

3. Estimate λ'.

$$\lambda' = 12\frac{k_m h_{mt}}{k_f h_{ft}}\frac{r_w^2}{h_m^2} \approx 12\frac{k_m}{h_m^2}\frac{h}{k_f h_{ft}} r_w^2, \quad (7.31)$$

where $h \approx h_{mt}$.

4. Estimate the time, t_{b2} (or Δt_{b2} for buildup), at which Flow Regime 2 begins and calculate

$$\phi_f c_{ft} h_{ft} = 8.33 \times 10^{-4}\left(\frac{k_f h_{ft} \phi_m c_{mt} h \lambda' t_{b2}}{\mu r_w^2}\right)^{1/2}. \quad (7.32)$$

5. Estimate ω', ω, and λ.

$$\omega' = \frac{\phi_m c_{mt} h_{mt}}{\phi_f c_{ft} h_{ft}}, \quad (7.33)$$

$$\omega = \frac{1}{1+\omega'}, \quad (7.34)$$

$$\text{and } \lambda \approx \lambda', \quad (7.35)$$

assuming $h \approx h_{mt}$.

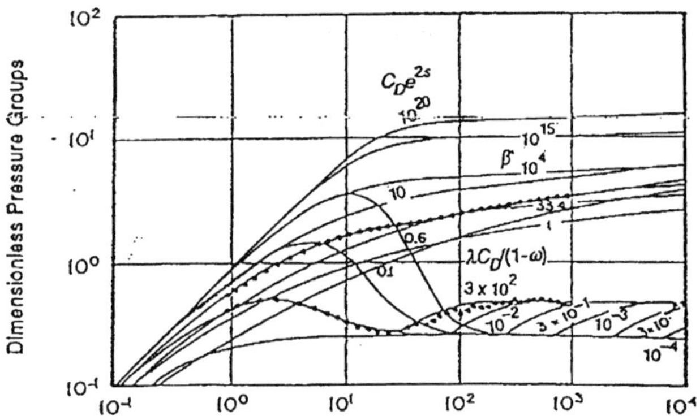

Fig. 7.11—Type curves for transient matrix flow (after Bourdet *et al.*[17]).

6. Estimate the skin factor, s.

$$s = 1.151\left[\frac{\Delta p_{1hr}}{m} - \log\left(\frac{k_f h_{ft}}{\phi_m c_{mt} h \mu r_w^2}\right) + 3.23\right]. \quad \ldots (7.36)$$

For a flow test, $\Delta p_{1hr} = p_i - p_{1hr}$. For a buildup test, $\Delta p_{1hr} = p_{1hr} - p_{wf}\,(\Delta t = 0)$.

Both of these semilog methods require some knowledge of or assumptions about matrix and fracture properties. The advantage of semilog analysis is the unique interpretation possible when the correct semilog straight lines are identified. The disadvantages are that (1) wellbore storage can distort data to the extent that application of semilog analysis is impossible and (2) the test duration is sometimes insufficient to reach Flow Regime 3 (especially in low-permeability formations), or boundary effects may occur before Flow Regime 3 develops.

7.4.2 Type-Curve Analysis Technique. Bourdet *et al.*[17] presented type curves for analyzing well tests in dual-porosity reservoirs that include the effects of wellbore storage and unsteady-state flow in the matrix. The type curves are useful supplements to the Serra *et al.* semilog analysis. **Fig. 7.11** gives an example of the pressure and pressure-derivative type curves for transient matrix flow. Early (fracture-dominated) data are fit on the pressure type curve by a C_De^{2s} value indicative of homogeneous behavior. Data in the transition region are fit by curves characterized by a parameter β'. Finally, data in the homogeneous-acting, fracture-plus-matrix flow regime are fit by another C_De^{2s} curve.

On the derivative type curve, early data also are fit by a derivative curve reflecting homogeneous behavior. If wellbore-storage distortion ceases before the transition region begins (unlikely but possible), the derivative data will be horizontal and should be aligned with the $(t_D/C_D)\,p_D' = 0.5$ curve. However, if the transition region is present (recall that its semilog slope is half that of the middle-time straight line), the derivative curve will flatten and should be aligned with the $(t_D/C_D)\,p_D' = 0.25$ curve. After wellbore distortion has ceased and before boundary effects have appeared, the homogeneous (fracture-plus-matrix) data should be horizontal on the derivative type curve and should be aligned with the $(t_D/C_D)\,p_D' = 0.5$ curve.

We recommend the following procedure for using the Bourdet transient or unsteady-state matrix-flow type curves. For analyzing gas-well-test data, use the appropriate adjusted pressure and time plotting functions presented in Chaps. 3 and 4. In addition, evaluate the gas properties at the average reservoir pressure.

1. Plot data as Δp vs. Δt on tracing paper or on log-log graph paper with the same scale as the type curves.

2. Next, perform a qualitative type-curve analysis. The purpose of this step is to identify various flow regimes to determine whether semilog analysis is possible. Overlay the log-log graph on the type curves and move the graph horizontally and vertically until a good match is obtained on the basis of the following criteria: A unit-slope line on the test data should overlay that line on the derivative type curve. Similarly, horizontal lines should overlay the $(t_D/C_D)\,p_D' = 0.25$ line (transition region) or the $(t_D/C_D)\,p_D' = 0.5$ line (homogeneous-acting region) on the derivative type curve.

3. If the two straight lines are present and can be identified, perform a semilog analysis with the procedures discussed in the previous sections.

4. Perform a quantitative type-curve analysis. The objective of this analysis is to confirm the results from the semilog analysis or to estimate reservoir properties when no semilog analysis is possible.

If an estimate of permeability is available from the semilog analysis, calculate a pressure match with Eq. 7.13.

$$(\Delta p)_{MP} = \frac{141.2qB\mu}{\bar{k}h}(p_D)_{MP}, \quad \ldots (7.13)$$

where p_D = an arbitrarily chosen value. If no estimate of permeability is available, use the match from the qualitative type-curve analysis. Obtain a time match point $(\Delta t, t_D/C_D)$.

If no semilog analysis is possible, match the data with type curves and obtain pressure and time match points, $(\Delta p, p_D)$ and $(\Delta t, t_D/C_D)$, respectively. Compute the average system permeability with the pressure match point.

$$\bar{k} = \frac{141.2qB\mu}{h}\left(\frac{p_D}{\Delta p}\right)_{MP}. \quad \ldots (7.37)$$

Calculate $(C_D)_{f+m}$ from the time match point $(\Delta t, t_D/C_D)$,

$$(C_D)_{f+m} = \frac{0.0002637k}{\phi_m c_{mt} \mu r_w^2}\left(\frac{\Delta t_e}{t_D/C_D}\right)_{MP}. \quad \ldots (7.38)$$

Calculate the skin factor, s.

$$s = 0.5\ln\left(\frac{C_De^{2s}}{C_D}\right)_{f+m}. \quad \ldots (7.39)$$

Calculate the interporosity coefficient, λ.

$$\lambda = 1.8914\frac{(C_De^{2s})_{f+m}}{(\beta')_{MP}e^{-2s}}. \quad \ldots (7.40)$$

Calculate the storativity ratio, ω.

$$\omega = \frac{(C_De^{2s})_{f+m}}{(C_De^{2s})_f}. \quad \ldots (7.41)$$

If $(C_De^{2s})_f$ cannot be determined uniquely, then

$$\omega \leqq \frac{(C_De^{2s})_{f+m}}{(C_De^{2s})_f}. \quad \ldots (7.42)$$

Although this procedure can be used to analyze tests that have wellbore-storage distortion, it does require the existence of Flow Regime 3 for a good interpretation. For best results, the type curves and the semilog analysis methods should be applied simultaneously until a consistent interpretation can be found with both approaches. Some tests may exhibit behavior that can be interpreted equally well as single- or dual-porosity. In these cases, additional data from such sources as cores, logs, or additional well tests are needed to determine which interpretation is more likely.

Example 7.2—Analysis of a Pressure-Buildup Test in Naturally Fractured Reservoir With Transient Matrix Flow. A pressure-buildup test was simulated for an oil well in a naturally fractured reservoir with transient matrix flow behavior. Well and reservoir data are summarized next, and **Table 7.4** gives the buildup-test data. Determine permeability, skin factor, storativity ratio, and interporosity flow coefficient with the semilog and type-curve analysis techniques.

q_g = 333 STB/D
t_p = 2,000 hours
S_o = 0.67
r_w = 0.33 ft
h = 33 ft

TABLE 7.4—PRESSURE BUILDUP TEST DATA, EXAMPLE 7.2

t (hours)	p_{wf} (psi)	t (hours)	p_{wf} (psi)
0.001	2,697.7	0.7	2,820.1
0.002	2,703.0	0.8	2,820.9
0.003	2,708.0	0.9	2,821.5
0.004	2,712.6	1.0	2,822.1
0.005	2,717.0	2.0	2,825.7
0.006	2,721.2	3.0	2,827.8
0.007	2,725.1	4.0	2,829.2
0.008	2,728.8	5.0	2,830.3
0.009	2,732.3	6.0	2,831.2
0.01	2,735.6	7.0	2,831.9
0.02	2,760.5	8.0	2,832.5
0.03	2,775.8	9.0	2,833.1
0.04	2,785.5	10.0	2,833.6
0.05	2,792.0	20.0	2,837.0
0.06	2,796.4	30.0	2,839.4
0.07	2,799.6	40.0	2,841.3
0.08	2,802.0	90.0	2,847.6
0.09	2,803.8	100.0	2,848.5
0.1	2,805.3	200.0	2,854.3
0.2	2,812.0	300.0	2,857.6
0.3	2,814.9	400.0	2,859.8
0.4	2,816.8	900.0	2,865.4
0.5	2,818.2	1,000.0	2,866.1
0.6	2,819.2		

$\phi = 0.0633$
$B_o = 1.3$ RB/STB
$c_t = 2.52\ 10^{-5}$ psi^{-1}
$\mu_o = 1.3$ cp
$S_w = 0.33$
$c_{mt} = 4.0\ 10^{-6}$ psi^{-1}
$p_i = \bar{p} = 2{,}876$ psi
$p_{wf}(\Delta t = 0) = 2{,}692.05$ psi

Solution. *Construct Data Plots.* Because the simulated well test is an oilwell buildup test, we plot pressure change and pressure derivative vs. equivalent time on log-log paper (**Fig. 7.12**) and pressure vs. Horner time ratio (**Fig. 7.13**) on semilog paper. **Table 7.5** gives type-curve plotting functions, while **Table 7.6** gives semilog plotting functions.

Qualitative Type-Curve Analysis. The objective of this initial qualitative type-curve analysis is to determine whether Flow Regimes 1, 2, and 3 can be identified.

Match both pressure change and pressure-derivative data (Fig. 7.12) with the Bourdet *et al.*[17] type curves. Dual-porosity behavior is evident from the match of the test data with the type curve for transient flow in the matrix. Specifically, the trend of the derivative data for $\Delta t_e \geq 150$ hours is horizontal and can be aligned on the $(t_D/C_D)\,p_D' = 0.5$ line, while the trend of the earlier derivative data for $4 \leqq \Delta t_e \leqq 15$ hours is flat and matches the $(t_D/C_D)\,p_D' = 0.25$ line. This behavior is characteristic of a dual-porosity system. With the earliest data trending toward a unit-slope line, we also can establish a reasonable horizontal match. The type-curve match, in particular the derivative type-curve match, suggests that Flow Regimes 2 and 3 are present.

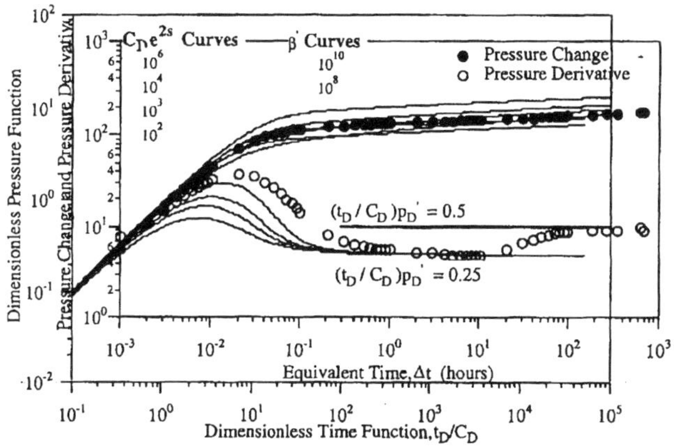

Fig. 7.12—Qualitative type-curve match, Example 7.2.

TABLE 7.5—PLOTTING FUNCTIONS FOR TYPE CURVE ANALYSIS, EXAMPLE 7.2

Δt_e, (hours)	Δp, (psi)	$\Delta t_e \Delta p'$	Δt_e, (hours)	Δp, (psi)	$\Delta t_e \Delta p'$
0.001	5.65	7.65	0.6998	128.05	5.63
0.002	10.95	10.60	0.7997	128.85	5.58
0.003	15.95	14.67	0.8996	129.45	5.54
0.004	20.55	18.49	0.9995	130.05	5.47
0.005	24.95	21.51	1.9980	133.65	5.19
0.006	29.15	24.29	2.9955	135.75	5.06
0.007	33.05	26.68	3.9920	137.15	4.99
0.008	36.75	28.32	4.9875	138.25	4.82
0.009	40.25	30.05	5.9821	139.15	4.82
0.01	43.55	31.64	6.9756	139.85	4.78
0.02	68.45	37.07	7.9681	140.45	4.78
0.03	83.75	35.07	8.9597	141.05	4.78
0.0399	93.45	30.27	9.9502	141.55	4.84
0.0499	99.95	26.21	19.802	144.95	5.61
0.0599	104.35	22.57	29.557	147.35	6.43
0.0699	107.55	19.38	39.216	149.25	7.11
0.0799	109.95	17.76	86.120	155.55	7.76
0.0899	111.75	15.56	95.240	156.45	8.01
0.0999	113.25	13.84	181.82	162.25	8.67
0.1998	119.95	8.08	260.87	165.55	9.10
0.2998	122.85	6.85	333.33	167.75	9.04
0.3999	124.75	6.29	620.69	173.35	9.72
0.4999	126.15	5.97	666.67	174.05	9.09
0.5998	127.15	5.80			

TABLE 7.6—PLOTTING FUNCTIONS FOR HORNER SEMILOG ANALYSIS, EXAMPLE 7.2

p, (psi)	Horner Time Ratio	p, (psi)	Horner Time Ratio
2,697.7	2,000,000	2,820.1	2,858.1
2,703.0	1,000,000	2,820.9	2,501.0
2,708.0	666,667	2,821.5	2,223.2
2,712.6	500,000	2,822.1	2,001.0
2,717.0	400,000	2,825.7	1,001.0
2,721.2	333,333	2,827.8	667.67
2,725.1	285,720	2,829.2	601.00
2,728.8	250,000	2,830.3	401.00
2,732.3	222,222	2,831.2	334.33
2,735.6	200,000	2,831.9	286.71
2,760.5	100,000	2,832.5	251.00
2,775.8	66,668	2,833.1	223.22
2,785.5	50,001	2,833.6	201.00
2,792.0	40,001	2,837.0	101.00
2,796.4	33,334	2,839.4	67.67
2,799.6	28,572	2,841.3	51.00
2,802.0	25,001	2,847.6	23.22
2,803.8	22,223	2,848.5	21.00
2,805.3	20,001	2,854.3	11.00
2,812.0	10,001	2,857.6	7.67
2,814.9	6667.7	2,859.8	6.00
2,816.8	5001.0	2,865.4	3.22
2,818.2	4001.0	2,866.1	3.00
2,819.2	3334.3		

Semilog Analysis Based on Flow Regimes 2 and 3. Because we have tentatively identified Flow Regimes 2 and 3, we now attempt a semilog analysis.

1. A final straight line in Fig. 7.13 with slope $m = 20.8$ psi/cycle is a reasonable fit of the later data (i.e., Flow Regime 3). We can force a line with slope $m^* = m/2 = 10.4$ psi/cycle through the earlier data that fall on the $(t_D/C_D)\,p_D' = 0.25$ line on the derivative type curve.

The two semilog lines intersect at a Horner time ratio of

$$\frac{\Delta t^* + t_p}{\Delta t^*} = 59.8,$$

or the point of intersection is

$$\Delta t^* = \frac{t_p}{59.8 - 1} = \frac{2,000}{58.8} = 34 \text{ hours.}$$

Although the beginning of Flow Regime 2 is unclear because of wellbore-storage distortion of the test data, we assume it begins at

$$\frac{\Delta t_{b2} + t_p}{\Delta t_{b2}} = 2,200$$

$$\text{or } \Delta t_{b2} = \frac{2,000}{2,200 - 1} = 0.91 \text{ hour.}$$

2. Determine $k_f h_{ft} \approx \overline{k}h$ from the slope of the semilog straight line.

$$k_f h_{ft} \approx \overline{k}h = \frac{162.6 q_o B_o \mu_o}{m}$$

$$= \frac{(162.6)(333)(1.3)(1.3)}{20.8}$$

$$= 4,399.3 \text{ md-ft}$$

or, for $h = 33$ ft,

$$\overline{k} = 4,399.3/33 = 133.3 \text{ md.}$$

3. Calculate k_m/h_m^2. From the available data, we have $\phi_m = 0.0633$, $c_{mt} = 4.0 \times 10^{-6}$ psia^{-1}, and $\mu_o = 1.3$ cp. Then

$$\frac{k_m}{h_m^2} = \frac{532.3 \phi_m c_{mt} \mu_o}{\Delta t^*}.$$

$$= \frac{532.3(0.0633)(4.0 \times 10^{-6})(1.3)}{34}$$

$$= 5.066 \times 10^{-6} \text{ md/ft}^2.$$

4. Estimate λ'.

$$\lambda' \approx 12 \frac{k_m}{h_m^2} \frac{h}{k_f h_{ft}} r_w^2$$

$$= \frac{(12)(5.066 \times 10^{-6})(33)(0.33)^2}{4,399.3}$$

$$= 5 \times 10^{-8}.$$

5. From time Δt_{b2} when Flow Regime 2 begins, estimate $\phi_f c_{ft} h_{ft}$.

$$\phi_f c_{ft} h_{ft} = 8.33 \times 10^{-4} \left(\frac{k_f h_{ft} \phi_m c_{mt} h \lambda' \Delta t_{b2}}{\mu_o r_w^2} \right)^{1/2}$$

$$= (8.33 \times 10^{-4})$$

$$\times \left[\frac{(4399.3)(0.0633)(4 \times 10^{-6})(33)(5 \times 10^{-8})(0.91)}{(1.3)(0.33)^2} \right]^{1/2}$$

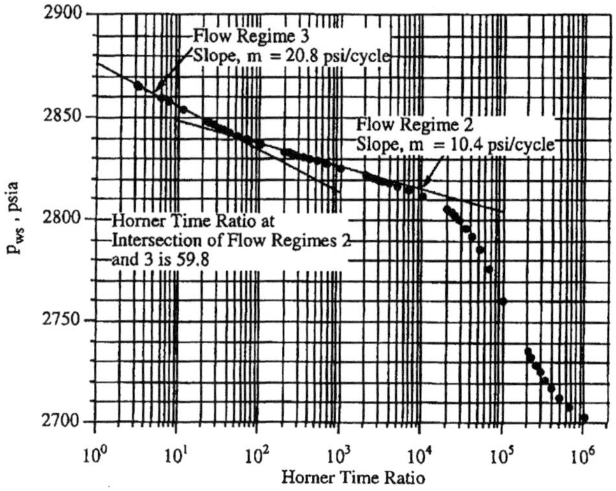

Fig. 7.13—Horner plot, Example 7.2.

$$= 9.1 \times 10^{-8} \text{ ft/psi.}$$

6. Estimate ω', ω, and λ, where $h \approx h_{mt}$.

$$\omega' = \frac{\phi_m c_{mt} h_{mt}}{\phi_f c_{ft} h_{ft}} = \frac{(0.0633)(4 \times 10^{-6})(33)}{9.1 \times 10^{-8}} = 91.8.$$

The parameter ω is

$$\omega = \frac{1}{1 + \omega'} = \frac{1}{1 + 91.8} = 0.0108.$$

Then, $\lambda \approx \lambda' = 5 \times 10^{-8}$.

7. To calculate skin factor, we need $p_{1\text{hr}}$ on the semilog straight line of Flow Regime 3. At $\Delta t = 1$ hour, the adjusted Horner time ratio is $(t_p + \Delta t)/\Delta t = (2,000 + 1)/1 = 2,001$. At this value in Fig. 7.13, we obtain $p_{1\text{hr}} = 2,807.4$ psia. The skin factor is

$$s = 1.151 \left[\frac{p_{1\text{hr}} - p_{wf}(\Delta t = 0)}{m} - \log\left(\frac{k_f h_{ft}}{\phi c_t h \mu_o r_w^2} \right) + 3.23 \right]$$

$$= 1.151 \left\{ \left(\frac{2,807.4 - 2,692.05}{20.8} \right) \right.$$

$$\left. - \log\left[\frac{4,399.3}{(0.033)(2.52 \times 10^{-5})(33)(1.3)(0.33)^2} \right] + 3.23 \right\}$$

$$= 0.$$

Quantitative Type-Curve Analysis. The purpose of this final analysis is to use type curves to confirm the results from the semilog analysis.

1. Using the log-log plot in **Fig. 7.14**, find the match points and matching parameters. In this case, we have an estimate of $\overline{k}h$ from semilog analysis, so we precalculate a pressure match point. We arbitrarily choose $p_D = 1$ and calculate

$$(\Delta p)_{MP} = \frac{141.2 q_o B_o \mu_o}{\overline{k}h} (p_D)_{MP}$$

$$= \frac{(141.2)(333)(1.3)(1.3)(1)}{(4399.3)}$$

$$= 18.1 \text{ psi.}$$

The match also forces the late-time derivative data to overlay $(t_D/C_D)\,p_D' = 0.5$ and the early derivative data to overlay $(t_D/C_D)\,p_D' = 0.25$. The type-curve correlating parameters were $(C_D e^{2s})_f = 10^4$ (early data), while parameter $(C_D e^{2s})_{f+m} = 10^2$ (late-

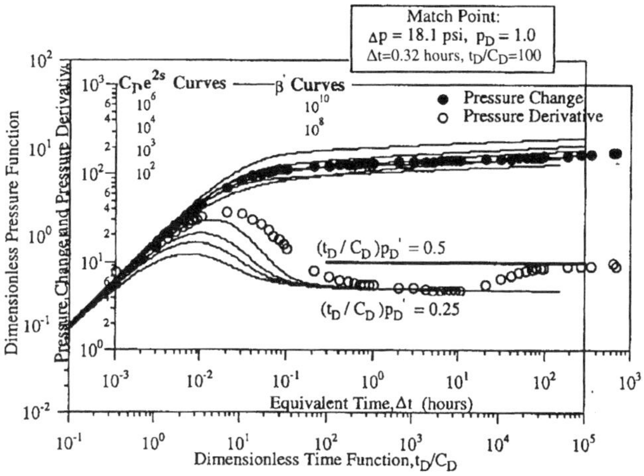

Fig. 7.14—Quantitative type-curve match, Example 7.2.

time data). β' is estimated to be 10^{10}. A time match point is $\Delta t_e = 0.32$ hour and $t_D/C_D = 100$.

2. Calculate $(C_D)_{f+m}$ from the time match point.

$$(C_D)_{f+m} = \frac{0.0002637k}{\phi c_t \mu_o r_w^2}\left(\frac{\Delta t_e}{t_D/C_D}\right)_{MP}$$

$$= \frac{(0.0002637)(133.3)}{(0.0633)(2.52 \times 10^{-5})(1.3)(0.33)^2}\left(\frac{0.32}{100}\right)$$

$$= 3,138.$$

3. Compute the skin factor.

$$s = 0.5\ln\left(\frac{C_D e^{2s}}{C_D}\right)_{f+m} = 0.5\ln\left(\frac{1,000}{3138}\right) = -0.57,$$

which agrees with $s \cong 0$ from the semilog analysis.

4. Calculate λ with Eq. 7.40.

$$\lambda = \frac{1.8914(C_D e^{2s})_{f+m}}{(\beta')_{MP} e^{-2s}} = \frac{(1.8914)(1,000)}{(10^{10})e^{-2(-0.57)}} = 6.0 \times 10^{-8},$$

which agrees with $\lambda' = 5.0 \times 10^{-8}$ from semilog analysis.

5. Calculate ω with Eq. 7.41.

$$\omega \approx \frac{(C_D e^{2s})_{f+m}}{(C_D e^{2s})_f} = \frac{10^2}{10^4} = 1.0 \times 10^{-2},$$

which agrees very well with $\omega = 0.0108$ from semilog analysis.

7.5 Chapter Summary

In this chapter, we introduced interpretation methods for naturally fractured reservoirs. Naturally fractured reservoirs have two distinct porosity systems: the fracture system and the matrix. Typically, most of the PV of the reservoir lies in the matrix, while most or all of the flow capacity is provided by the fracture system.

In Sec. 7.2, we discussed two conceptual models of a naturally fractured or dual-porosity reservoir. The pseudosteady-state model assumes that flow from the matrix to the fracture system occurs under pseudosteady-state conditions; that is, that the flow rate between the matrix and the fracture is proportional to the difference between the pressure in the fracture and the average pressure in the matrix. The transient dual-porosity model assumes that a transient flow equation is needed to describe flow from the matrix to the fracture system. Both pseudosteady-state and transient dual-porosity models are characterized by two properties: the interporosity flow coefficient, λ, which describes the degree of communication between the two porosity systems, and the storativity ratio, ω, which is the fraction of the total PV compressibility product provided by the fracture system. For a reservoir where both porosity systems have the same compressibility, ω is simply the fraction of the total PV attributable to the fracture system.

In Sec. 7.3, we discussed methods for analyzing tests in pseudosteady-state dual-porosity reservoirs. Although neglecting transient flow in the matrix would appear to be an oversimplification of the behavior of the system, this model does match a surprising number of field cases.

Sec. 7.3.1 discusses semilog analysis methods for pseudosteady-state dual-porosity systems. Ideally, a pseudosteady-state dual-porosity system exhibits two parallel, straight lines on a semilog graph. The initial straight line represents flow in the fracture system only, before the matrix begins to respond. As the matrix begins to provide fluid to the fracture system, a rather flat transition region appears. The final straight line represents total system flow. Wellbore storage almost always obscures the first semilog straight line and frequently obscures part of the transition region as well.

If either straight line can be identified with semilog analysis, the permeability-thickness product can be determined. If both straight lines can be identified, the storativity ratio, ω, and the interporosity flow coefficient, λ, also can be estimated. The second straight line is used to estimate skin factor, s, and extrapolated pressure, p^*.

In Sec. 7.3.2, we discussed the Bourdet *et al.*[16] type curves for pseudosteady-state dual-porosity systems. These type curves account for wellbore storage and skin and allow tests where the first semilog straight line is obscured by wellbore storage to be analyzed. On a log-log graph, as the well is coming out of wellbore storage, the early pressure data again represent flow in the fracture system only. These data are identified by the parameter group $(C_D e^{2s})_f$ characterizing the fracture system only. The data within the transition region are characterized by the parameter group λe^{-2s}. Once the matrix begins to contribute fully to fluid flow, the system again behaves like a homogeneous system. The data within this region are identified by the parameter group $(C_D e^{2s})_{f+m}$, characterizing the total system.

Next, we presented a recommended procedure for analysis of a pseudosteady-state dual-porosity system, combining both semilog and type-curve analysis. If a match is possible, and the parameter groups can be estimated from the match, the test can be analyzed, even if semilog analysis is not possible. The storativity ratio is ω obtained from the two groups $(C_D e^{2s})_f$ and $(C_D e^{2s})_{f+m}$. The permeability is calculated from the pressure match point. The time match point is used to calculate the wellbore-storage coefficient, C_D. The skin factor, s, is obtained from the wellbore-storage coefficient and the group $(C_D e^{2s})_{f+m}$. Finally, the interporosity flow coefficient is obtained from s and the group λe^{-2s}.

An example analysis of a buildup test in a naturally fractured reservoir with pseudosteady-state matrix flow completes Sec. 7.3.2.

Sec. 7.4 discusses methods for analyzing tests in dual-porosity reservoirs that exhibit transient flow in the matrix. The ideal semilog response from a transient dual-porosity system has two straight lines, just as the response from a pseudosteady-state system. The early straight line (Flow Regime 1) represents flow in the fractures, and the late straight line (Flow Regime 3) represents total system response. The major difference between the responses of the two dual-porosity models lies in the transition region, Flow Regime 2. In the transient dual-porosity model, the data within the transition region follow a third straight line, which has a slope one-half that of the first and last straight lines. Wellbore storage often obscures Flow Regimes 1 and 2. Flow Regime 3 may be obscured by boundary effects or insufficient producing or shut-in time.

We present semilog analysis methods appropriate for these reservoirs in Sec. 7.4.1. Semilog analysis may be based on either Flow Regimes 1 and 2 or Flow Regimes 2 and 3. Analysis with either of these methods requires some knowledge of or assumptions about matrix and fracture properties.

If Flow Regimes 1 and 2 are present, the permeability, k, may be obtained from the slope of either straight line. The skin factor, s, is then obtained with an iterative process. Finally, storativity ratio and interporosity flow coefficient, λ, are estimated. If Flow Regimes 2 and 3 are present, k may be obtained from the slope of either straight line. λ is estimated from the time at which Flow Regimes 2 and 3 intersect, while ω is estimated from λ and the time at which Flow Regime 2 begins. Finally, s is estimated from the straight line for Flow Regime 3.

In Sec. 7.4.2, we discussed type curves for transient dual-porosity systems. These type curves account for wellbore storage and skin. As with the pseudosteady-state dual-porosity type curves, the early and late data are identified by the parameter groups $(C_De^{2s})_f$ and $(C_De^{2s})_{f+m}$, respectively. Data within the transition region are characterized by a new parameter group, β'. Assuming that wellbore storage has ended, the derivative of the early and late data should fall on the horizontal line corresponding to $(t_D/C_D)\,p_D'$. The derivative of the transition region data should fall on a line corresponding to $(t_D/C_D)\,p_D'=0.25$.

Interpretation of the match points and matching parameters with the transient dual-porosity type curves follows very closely that with the pseudosteady-state dual-porosity type curves, except that λ is obtained from the parameter group β'.

Finally, we illustrated both semilog and type-curve analysis of transient dual-porosity reservoirs with an example problem.

Exercises

1. Given the following formation and fluid properties, estimate formation permeability, skin factor, λ, and ω from the buildup test data. $q=125$ STB/D; $h=17$ ft; $B=1.054$ RB/STB; $t_p=1200$ hr; $\phi=13.0\%$; $c_t=7.19\times10^{-6}$ psi^{-1}; $p_{wf}=211.20$ psia; $r_w=0.30$ ft; $\mu=1.72$ cp.

PRESSURE-TIME DATA FOR EXERCISE 7.1

Time (hours)	Pressure (psi)	Time (hours)	Pressure (psi)	Time (hours)	Pressure (psi)
0.0	211.20	0.219	456.20	14.76	462.03
0.0010	390.73	0.285	456.65	19.18	463.40
0.0023	404.32	0.372	456.90	24.94	465.08
0.0040	413.00	0.484	457.03	32.42	467.10
0.0062	419.73	0.630	457.11	42.15	469.49
0.0090	425.39	0.820	457.18	54.80	472.24
0.0128	430.36	1.067	457.27	71.24	475.32
0.0176	434.81	1.389	457.39	92.61	478.66
0.0239	438.82	1.806	457.55	120.39	482.17
0.0320	442.43	2.35	457.75	156.51	485.76
0.0426	445.66	3.05	458.01	203.5	489.32
0.0564	448.48	3.97	458.35	264.5	492.79
0.0743	450.87	5.16	458.78	343.9	496.11
0.0976	452.84	6.71	459.33	447.0	499.25
0.1279	454.36	8.73	460.03	581.1	502.16
0.1673	455.46	11.35	460.92	720.0	504.37

2. Given the following formation and fluid properties, estimate formation permeability, skin factor, λ, and ω from the buildup test data. Compare your results from those from Exercise 7.1. $q=125$ STB/D; $h=17$ ft; $B=1.054$ RB/STB; $t_p=1200$ hr; $\phi=13.0\%$; $c_t=7.19\times10^{-6}$ psi^{-1}; $p_{wf}=211.20$ psia; $r_w=0.30$ ft; $\mu=1.72$ cp.

PRESSURE-TIME DATA FOR EXERCISE 7.2

Time (hours)	Pressure (psi)	Time (hours)	Pressure (psi)	Time (hours)	Pressure (psi)
0.0	211.20	0.219	341.53	14.76	461.85
0.0010	212.07	0.285	365.13	19.18	463.23
0.0023	213.19	0.372	388.74	24.94	464.92
0.0040	214.64	0.484	410.60	32.42	466.95
0.0062	216.50	0.630	428.91	42.15	469.35
0.0090	218.90	0.820	442.40	54.80	472.11
0.0128	221.98	1.067	450.83	71.24	475.21
0.0176	225.91	1.389	455.12	92.61	478.57
0.0239	230.92	1.806	456.85	120.39	482.11
0.0320	237.26	2.35	457.47	156.51	485.71
0.0426	245.22	3.05	457.80	203.5	489.29
0.0564	255.11	3.97	458.15	264.5	492.77
0.0743	267.26	5.16	458.58	343.9	496.09
0.0976	281.94	6.71	459.14	447.0	499.23
0.1279	299.31	8.73	459.84	581.1	502.15
0.1673	319.31	11.35	460.73	720.0	504.36

3. Given the following formation and fluid properties, estimate formation permeability, skin factor, λ, and ω from the buildup test data. $q=20$ STB/D; $h=12$ ft; $B=1.108$ RB/STB; $t_p=1600$ hr; $\phi=17.8\%$; $c_t=10.4\times10^{-6}$ psi^{-1}; $p_{wf}=385.68$ psia; $r_w=0.37$ ft; $\mu=1.16$ cp.

PRESSURE-TIME DATA FOR EXERCISE 7.3

Time (hours)	Pressure (psi)	Time (hours)	Pressure (psi)	Time (hours)	Pressure (psi)
0.0	385.68	0.836	940.37	43.6	1118.22
0.0050	392.84	1.092	992.84	56.7	1120.41
0.0115	402.01	1.425	1033.80	73.8	1123.10
0.0200	413.72	1.858	1062.77	95.9	1126.35
0.0309	428.57	2.42	1081.58	124.7	1130.21
0.0452	447.31	3.15	1093.23	162.1	1134.69
0.0638	470.76	4.10	1100.51	210.8	1139.72
0.0879	499.77	5.34	1105.24	274.0	1145.21
0.1193	535.20	6.94	1108.37	356.2	1150.98
0.1601	577.71	9.03	1110.43	463.0	1156.83
0.213	627.61	11.74	1111.83	602.0	1162.55
0.282	684.56	15.27	1112.89	782.6	1167.97
0.372	747.29	19.86	1113.90	1017.3	1172.98
0.488	813.35	25.8	1115.04	1200.0	1175.88
0.640	879.15	33.6	1116.45		

4. Given the following formation and fluid properties, estimate formation permeability, skin factor, λ, and ω from the buildup test data. $T_f=140$°F; $h=72$ ft; $\gamma_g=0.636$ (air $=1.0$); $q=2800$ Mscf/D; $r_w=0.24$ ft; $\phi=9.9\%$; $c_w=3.6\times10^{-6}$ psi^{-1}; $t_p=288$ hr; $S_w=34.0\%$; $c_f=4\times10^{-6}$ psi^{-1}; $p_{wf}=1442.19$ psia.

PRESSURE-TIME DATA FOR EXERCISE 7.4

Time (hours)	Pressure (psi)	Time (hours)	Pressure (psi)	Time (hours)	Pressure (psi)
0.0	1442.19	0.344	4044.08	6.60	4268.68
0.0050	1552.50	0.435	4077.73	8.25	4300.91
0.0113	1684.60	0.548	4091.44	10.32	4335.25
0.0191	1841.77	0.691	4098.95	12.90	4371.02
0.0288	2027.06	0.868	4106.07	16.14	4407.55
0.0410	2242.80	1.090	4114.46	20.2	4444.24
0.0563	2489.58	1.368	4124.59	25.2	4480.67
0.0754	2764.60	1.715	4136.80	31.5	4516.55
0.0992	3058.59	2.15	4151.42	39.4	4551.68
0.1290	3352.58	2.69	4168.75	49.3	4585.91
0.1663	3618.93	3.37	4189.06	60.0	4615.19
0.213	3829.12	4.22	4212.49		
0.271	3968.46	5.27	4239.09		

5. Given the following formation and fluid properties, estimate formation permeability, skin factor, λ, and ω from the buildup test data. $T_f=110$°F; $h=236$ ft; $\gamma_g=0.695$ (air $=1.0$); $q=9500$ Mscf/D; $r_w=0.21$ ft; $\phi=9.0\%$; $c_w=3.6\times10^{-6}$ psi^{-1}; $t_p=3600$ hr; $S_w=56.8\%$; $c_f=4\times10^{-6}$ psi^{-1}; $p_{wf}=1515.89$ psia.

PRESSURE-TIME DATA FOR EXERCISE 7.5

Time (hours)	Pressure (psi)	Time (hours)	Pressure (psi)	Time (hours)	Pressure (psi)
0.0	1515.89	0.1673	1875.03	8.73	1974.61
0.0010	1543.26	0.219	1890.26	11.35	1981.35
0.0023	1569.76	0.285	1903.97	14.76	1989.53
0.0040	1596.61	0.372	1916.05	19.18	1999.29
0.0062	1623.85	0.484	1926.38	24.9	2010.66
0.0090	1651.21	0.630	1934.92	32.4	2023.58
0.0128	1678.35	0.820	1941.67	42.2	2037.85
0.0176	1704.90	1.067	1946.81	54.8	2053.16
0.0239	1730.56	1.389	1950.63	71.2	2069.13
0.0320	1755.10	1.806	1953.53	92.6	2085.45
0.0426	1778.39	2.35	1955.99	120.4	2101.85
0.0564	1800.36	3.05	1958.45	156.5	2118.15
0.0743	1821.02	3.97	1961.28	196.0	2131.98
0.0976	1840.36	5.16	1964.77		
0.1279	1858.37	6.71	1969.14		

6. Given the following formation and fluid properties, estimate formation permeability, skin factor, λ, and ω from the buildup test data. Compare your results from those from Exercise 7.1. $q = 125$ STB/D; $h = 17$ ft; $B = 1.054$ RB/STB; $t_p = 1200$ hr; $\phi = 13.0\%$; $c_t = 7.19 \times 10^{-6}$ psi^{-1}; $p_{wf} = 211.20$ psia; $r_w = 0.30$ ft; $\mu = 1.72$ cp.

PRESSURE-TIME DATA FOR EXERCISE 7.6

Time (hours)	Pressure (psi)	Time (hours)	Pressure (psi)	Time (hours)	Pressure (psi)
0.0	211.20	0.219	423.71	14.76	461.60
0.0010	352.90	0.285	426.44	19.18	463.71
0.0023	365.64	0.372	429.10	24.9	465.83
0.0040	373.73	0.484	431.69	32.4	468.03
0.0062	379.99	0.630	434.22	42.2	470.39
0.0090	385.27	0.820	436.70	54.8	472.98
0.0128	389.93	1.067	439.13	71.2	475.85
0.0176	394.18	1.389	441.51	92.6	478.99
0.0239	398.13	1.806	443.86	120.4	482.35
0.0320	401.83	2.35	446.17	156.5	485.83
0.0426	405.35	3.05	448.46	203.5	489.35
0.0564	408.70	3.97	450.71	264.5	492.79
0.0743	411.92	5.16	452.94	343.9	496.11
0.0976	415.02	6.71	455.14	447.0	499.24
0.1279	418.01	8.73	457.32	581.1	502.16
0.1673	420.91	11.35	459.47	720.0	504.37

7. Given the following formation and fluid properties, estimate formation permeability, skin factor, λ, and ω from the buildup test data. Compare your results from those from Exercises 7.1, 7.2, and 7.6. $q = 125$ STB/D; $h = 17$ ft; $B = 1.054$ RB/STB; $t_p = 1200$ hr; $\phi = 13.0\%$; $c_t = 7.19 \times 10^{-6}$ psi^{-1}; $p_{wf} = 211.20$ psia; $r_w = 0.30$ ft; $\mu = 1.72$ cp.

PRESSURE-TIME DATA FOR EXERCISE 7.7

Time (hours)	Pressure (psi)	Time (hours)	Pressure (psi)	Time (hours)	Pressure (psi)
0.0	211.20	0.219	333.31	14.76	461.28
0.0010	212.07	0.285	353.87	19.18	463.47
0.0023	213.19	0.372	374.07	24.9	465.65
0.0040	214.63	0.484	392.61	32.4	467.88
0.0062	216.48	0.630	408.33	42.2	470.25
0.0090	218.86	0.820	420.59	54.8	472.85
0.0128	221.91	1.067	429.47	71.2	475.73
0.0176	225.78	1.389	435.67	92.6	478.90
0.0239	230.70	1.806	440.13	120.4	482.28
0.0320	236.88	2.35	443.63	156.5	485.79
0.0426	244.58	3.05	446.64	203.5	489.32
0.0564	254.08	3.97	449.38	264.5	492.77
0.0743	265.62	5.16	451.95	343.9	496.09
0.0976	279.38	6.71	454.40	447.0	499.23
0.1279	295.42	8.73	456.76	581.1	502.15
0.1673	313.55	11.35	459.05	720.0	504.36

8. Given the following formation and fluid properties, estimate formation permeability, skin factor, λ, and ω from the buildup test data. $q = 425$ STB/D; $h = 12$ ft; $B = 1.161$ RB/STB; $t_p = 2400$ hr; $\phi = 21.7\%$; $c_t = 11.5 \times 10^{-6}$ psi^{-1}; $p_{wf} = 1372.92$ psia; $r_w = 0.39$ ft; $\mu = 0.89$ cp.

PRESSURE-TIME DATA FOR EXERCISE 7.8

Time (hours)	Pressure (psi)	Time (hours)	Pressure (psi)	Time (hours)	Pressure (psi)
0.0	1372.92	0.320	1800.25	12.91	1959.35
0.0025	1389.41	0.418	1816.22	16.79	1973.70
0.0058	1409.40	0.546	1829.89	21.83	1989.48
0.0100	1433.40	0.713	1842.13	28.38	2006.42
0.0155	1461.74	0.929	1853.49	36.89	2024.16
0.0226	1494.48	1.210	1864.30	47.96	2042.33
0.0319	1531.29	1.575	1874.75	62.35	2060.65
0.0440	1571.26	2.05	1884.94	81.06	2078.93
0.0596	1612.91	2.67	1894.93	105.38	2097.07
0.0800	1654.24	3.47	1904.75	136.99	2115.00
0.1065	1693.08	4.52	1914.51	178.09	2132.66
0.1410	1727.64	5.87	1924.47	196.00	2139.03
0.1858	1756.91	7.64	1934.99		
0.244	1780.86	9.93	1946.50		

9. Given the following formation and fluid properties, estimate formation permeability, skin factor, λ, and ω from the buildup test data. Is it realistic to run a buildup test this long? Why or why not? Is it possible to estimate λ or ω from a shorter test? $T_f = 139$°F; $h = 76$ ft; $\gamma_g = 0.668$ (air = 1.0); $q = 1000$ Mscf/D; $r_w = 0.28$ ft; $\phi = 8.8\%$; $c_w = 3.6 \times 10^{-6}$ psi^{-1}; $t_p = 7200$ hr; $S_w = 25.8\%$; $c_f = 4 \times 10^{-6}$ psi^{-1}; $p_{wf} = 2308.07$ psia.

PRESSURE-TIME DATA FOR EXERCISE 7.9

Time (hours)	Pressure (psi)	Time (hours)	Pressure (psi)	Time (hours)	Pressure (psi)
0.0	2308.07	2.77	2712.86	139.8	3001.72
0.0200	2326.04	3.76	2737.36	188.7	3023.20
0.0470	2347.75	5.09	2760.76	255	3044.58
0.0835	2373.69	6.90	2783.46	344	3066.34
0.1327	2403.99	9.33	2805.71	464	3089.48
0.1991	2438.34	12.62	2827.71	627	3115.14
0.289	2475.87	17.06	2849.56	846	3144.16
0.410	2515.14	23.0	2871.33	1143	3176.53
0.573	2554.38	31.1	2893.08	1543	3211.37
0.794	2591.94	42.0	2914.83	2083	3247.20
1.092	2626.66	56.8	2936.58	2812	3282.41
1.494	2658.15	76.7	2958.33	3600	3309.96
2.04	2686.68	103.5	2980.07		

10. Given the following formation and fluid properties, estimate formation permeability, skin factor, λ, and ω from the buildup test data. $T_f = 109$°F; $h = 190$ ft; $\gamma_g = 0.794$ (air = 1.0); $q = 12800$ Mscf/D; $r_w = 0.34$ ft; $\phi = 13.5\%$; $c_w = 3.6 \times 10^{-6}$ psi^{-1}; $t_p = 3600$ hr; $S_w = 55.6\%$; $c_f = 4 \times 10^{-6}$ psi^{-1}; $p_{wf} = 1005.81$ psia.

PRESSURE-TIME DATA FOR EXERCISE 7.10

Time (hours)	Pressure (psi)	Time (hours)	Pressure (psi)	Time (hours)	Pressure (psi)
0.0	1005.81	0.1880	1509.21	12.74	1744.55
0.0010	1044.35	0.255	1528.76	17.20	1761.13
0.0024	1085.41	0.345	1547.44	23.2	1779.38
0.0042	1129.46	0.467	1565.47	31.4	1800.08
0.0066	1175.60	0.631	1582.99	42.3	1823.56
0.0100	1222.36	0.853	1600.10	57.1	1849.58
0.0144	1267.97	1.152	1616.88	77.1	1877.46
0.0205	1310.82	1.557	1633.39	104.1	1906.36
0.0287	1349.77	2.10	1649.68	140.6	1935.57
0.0397	1384.41	2.84	1665.79	189.8	1964.62
0.0546	1414.95	3.83	1681.74	256.2	1993.23
0.0747	1442.00	5.18	1697.56	345.9	2021.24
0.1018	1466.30	6.99	1713.23	360.0	2024.92
0.1385	1488.52	9.44	1728.80		

References

1. Warren, J.E. and Root, P.J.: "The Behavior of Naturally Fractured Reservoirs," *SPEJ* (September 1963) 245; *Trans.*, AIME, **228.**

2. Gringarten, A.C.: "Interpretation of Tests in Fissured and Multilayered Reservoirs With Double-Porosity Behavior: Theory and Practice," *JPT* (April 1984) 549; *Trans.*, AIME, **277.**
3. Serra, K.V., Reynolds, A.C., and Raghavan, R.: "New Pressure Transient Analysis Methods for Naturally Fractured Reservoirs," *JPT* (December 1983) 2271; *Trans.*, AIME, **275.**
4. Barenblatt, G.E., Zheltov, I.P., and Kochina, I.N.: "Basic Concepts in the Theory of Homogeneous Liquids in Fissured Rocks," *J. Appl. Math. Mech.* (1960) **24**, 852–864.
5. Odeh, A.S.: "Unsteady-State Behavior of Naturally Fractured Reservoirs," *SPEJ* (March 1965) 60; *Trans.*, AIME, **234**.
6. Kazemi, H., Seth, M.S., and Thomas, G.W.: "The Interpretation of Interference Tests in Naturally Fractured Reservoirs With Uniform Fracture Distribution," *SPEJ* (December 1969) 463; *Trans.*, AIME, **246**.
7. Mavor, M.J. and Cinco, H.: "Transient Pressure Behavior of Naturally Fractured Reservoirs," paper SPE 7977 presented at the 1979 SPE California Regional Meeting, Ventura, California,18–20 April.
8. deSwaan, A.: "Analytical Solutions for Determining Naturally Fractured Reservoir Properties by Well Testing," *SPEJ* (June 1976) 117; *Trans.*, AIME, **261.**
9. Najurieta, H.L.: "A Theory for Pressure Transient Analysis in Naturally Fractured Reservoirs," *JPT* (July 1980) 1241; *Trans.*, AIME, **269**.
10. Kazemi, H.: "Pressure Transient Analysis of Naturally Fractured Reservoirs With Uniform Fracture Distribution," *SPEJ* (December 1969) 451; *Trans.*, AIME, **246**.
11. Boulton, N.S. and Streltsova, T.D.: "Unsteady Flow to a Pumped Well in a Fissured Water-Bearing Formation," *J. Hydrol.* (1977) **35**, 257.
12. Cinco, H. and Samaniego, F.: "Pressure Transient Analysis for Naturally Fractured Reservoirs," paper SPE 11026 presented at the 1982 SPE Annual Technical Conference and Exhibition, New Orleans, 26–29 September.
13. Streltsova, T.D.: "Well Pressure Behavior of a Naturally Fractured Reservoir," *SPEJ* (October 1983) 769; *Trans.*, AIME, **275.**
14. Bourdet, D. and Gringarten, A.C.: "Determination of Fissured Volume and Block Size in Fractured Reservoirs by Type-Curve Analysis," paper SPE 9293 presented at the 1980 SPE Annual Technical Conference and Exhibition, Dallas, 21–24 September.
15. Matthews, C.S., Brons, F., and Hazebroek, P.: "A Method for Determination of Average Pressure in a Bounded Reservoir," *Trans.*, AIME (1954) **201**, 182.
16. Bourdet, D. *et al.*: "Interpreting Well Tests in Fractured Reservoirs," *World Oil* (October 1983) 77–87.
17. Bourdet, D. *et al.*: "New Type Curves Aid Analysis of Fissured Zone Well Tests," *World Oil* (April 1984) 111–224.

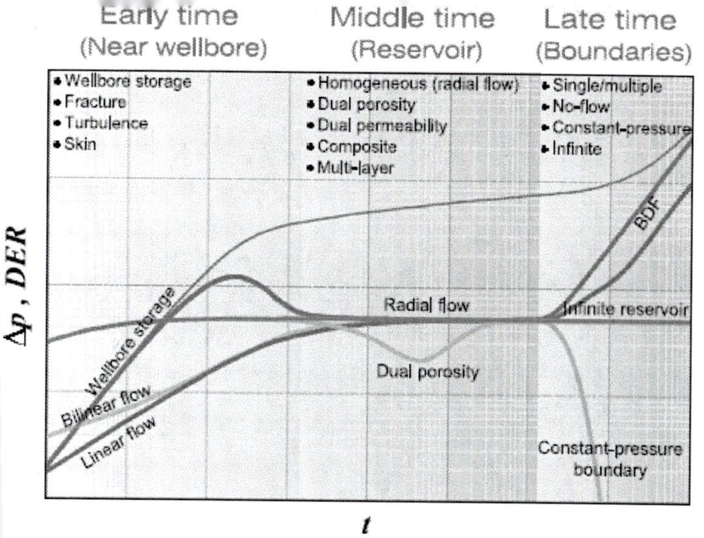

STORAGE

Storage Flow Regime is equivalent to emptying a tank

$$c = \frac{-1}{V}\frac{\Delta V}{\Delta p} = \frac{-1}{V}\frac{qt}{\Delta p}$$

Early time: Wellbore Storage (WBS)

- Pressure and derivative have unit slope

Late time: Boundary-Dominated Flow (BDF)

- Also known as:
  - Pseudo-steady state
  - Stabilized
  - Tank-type behavior
- Applies to Drawdown only – **NOT Buildup**
- Derivative unit slope occurs much earlier than pressure unit slope

Flow regime		Derivative Slope	Time Function*	Result
Wellbore storage		1	t	C, C_D
Bilinear flow		1/4	$\sqrt[4]{t}$	$k^{1/4}\sqrt{k_f w}$
Linear flow	fracture	1/2	$\sqrt{t}$	$x_f\sqrt{k}$
	horiz. well			$L_e\sqrt{k_y}$
Spherical flow		-1/2	$1/\sqrt{t}$	$\sqrt[3]{k_x k_y k_z}$
Vertical radial flow in horizontal wells		0	$\log(t)$	$L_e\sqrt{k_y k_z}$
Radial flow (∞-acting)		0	$\log(t)$	kh
Linear flow – Channel		1/2	$\sqrt{t}$	$W\sqrt{k}$
Boundary-Dominated Flow		1	t	V_p

** x-axis of specialized plots.*

Wellbore Storage – Unit Slope

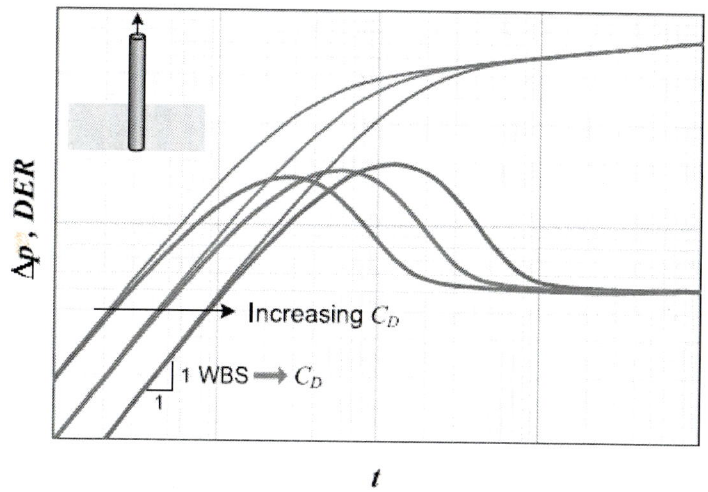

Boundaries

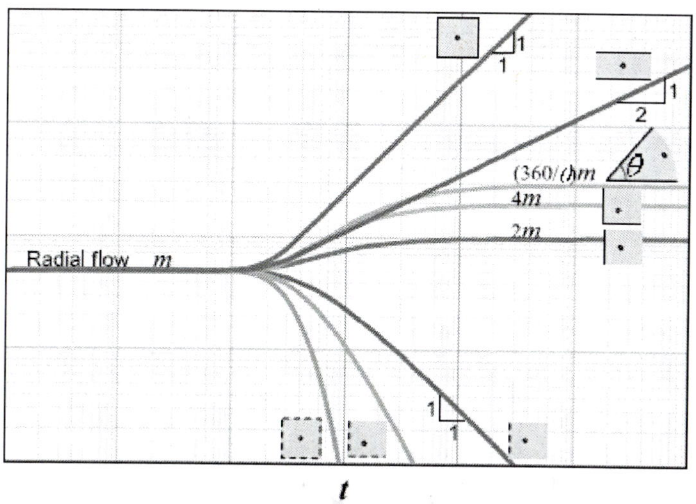

Reservoir Storage (BDF) – Unit Slope

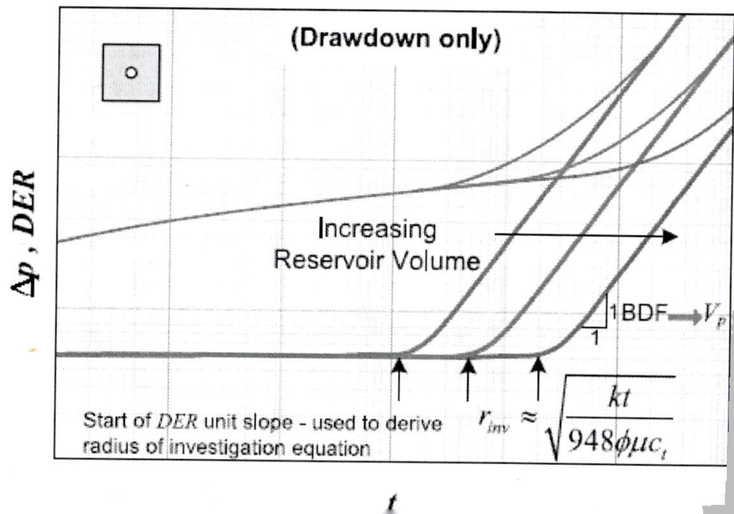

Chapter 8
Drillstem Testing and Analysis

8.1 Overview

Petroleum engineers have used the drillstem test (DST) as a formation evaluation method for many years. Originally used to identify reservoir fluids, the DST also has become an important method for estimating reservoir pressure and well potential. If sufficient fluid is produced into the drillpipe or captured in the sample chamber, the fluid content of a reservoir from a DST of the test tool can be established. In addition, well potential can be determined from information on pressure and flow rate as functions of time. In this chapter, we describe conventional DST analysis techniques, present guidelines for test design and monitoring, discuss the closed-chamber DST and the impulse-testing method, and present recommended pressure transient analysis techniques. Examples illustrate all analysis techniques.

8.2 Conventional DST

The most common application of a DST is to obtain information about a zone during the drilling phase but before the completion phase. Although the DST also can be used to monitor reservoir and well conditions in developed productive zones, this is a less common application.[1-3] DST can be viewed as a temporary well completion whose purpose is to obtain some or all of the following information: (1) identification of reservoir fluid; (2) an indication of well productivity; and (3) pressure transient data to estimate formation permeability, skin factor, and static reservoir pressure.

The following is a common procedure for running a conventional DST.

1. A DST tool that includes packers and a tester valve is attached to the end of the drillstring and run into the mud-filled wellbore to the zone to be tested.

2. The packers on test tool are set, isolating the interval of interest from the mud column in the annulus. Opening the tester valve imposes a sudden pressure drop at the formation face, causing formation fluid to flow into the drillpipe and increasing the liquid level in the wellbore. For wells in which the liquid level does not reach the surface during a flow period, the DST typically displays a decreasing flow rate throughout the flow period. When the tester valve is closed for a buildup period, the wellbore-storage coefficient decreases by as much as two orders of magnitude from a relatively large value for a rising liquid level to a relatively small value for fluid compression. Pressure is recorded continuously during the test.

3. A DST commonly consists of two flow periods and two shut-in periods. The initial flow period is a brief (5 minutes or less) production period whose purpose is to draw down the pressure slightly near the wellbore, permitting any mud-filtrate-invaded zone to bleed back to or below static reservoir pressure.

The initial shut-in period, which should be at least 1 hour long, allows the formation pressure to build up to true static formation pressure and, if its duration is long enough for wellbore-storage effects to end, provides buildup data for initial estimates of reservoir properties.

The final flow period (30 minutes to several hours) must provide a reservoir fluid sample in the test chamber and must draw down the pressure as far out into the formation as possible to generate a pressure transient whose radius of investigation reaches beyond any altered zone around the wellbore.

The final shut-in period (one to two times as long as the final flow period) provides pressure transient data to estimate reservoir properties. The extrapolated shut-in pressures (estimates of initial reservoir pressure) from the initial and final shut-in periods should be in agreement. If the second extrapolated pressure is significantly smaller than the first, reservoir depletion may be indicated.

Fig. 8.1 shows a typical pressure chart recorded during a DST. The chart usually is an oxide-coated metal with the pressure reading indicated by a scratch on the coating material made by a stylus attached to the pressure gauge. The points, lines, and curves on the chart have the following meanings.

1. The baseline represents the surface pressure.
2. Line 1 shows the increasing hydrostatic pressure of the mud column as the tool is lowered into the hole.
3. p_{ihm} is the initial hydrostatic pressure of the mud column when the tool reaches the interval to be tested. When the tester valve is opened, the pressure drops immediately from p_{ihm} to the initial flowing pressure during the first flow period, p_{if1}.
4. Line 2 is the pressure response during the initial flow period. As fluid flows into the drillpipe, the liquid level in the pipe rises and causes pressure at the tool to increase from p_{if1} to the final flowing pressure during the first flow period, p_{ff1}.
5. Curve 3 is the initial shut-in period. When the tester valve is closed, pressure builds up to the initial shut-in pressure, p_{isi}.
6. When the tester valve is reopened for the final flow period, pressure drops immediately from the initial shut-in pressure, p_{isi}, to the initial flowing pressure during the second flow period, p_{if2}.
7. Line 4 is the pressure response during the final flow period. As in the initial flow period, pressure increases from p_{if2} to p_{ff2}, the final flowing pressure in the second flow period, as the liquid level in the pipe rises.
8. Curve 5 is the final shut-in period. When the tester valve is reclosed, pressure builds up to the final shut-in pressure, p_{fsi}.

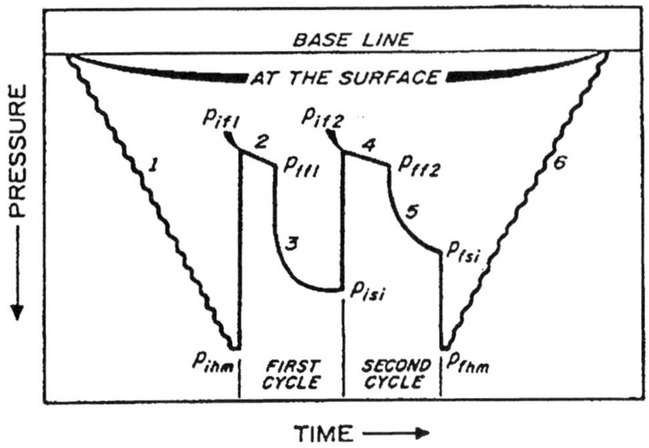

Fig. 8.1—Typical DST pressure chart (after Earlougher[2]).

9. When the packers are pulled loose after the end of the test, pressure returns to the final hydrostatic pressure of the mud column, p_{fhm}.

10. Line 6 shows the decreasing hydrostatic pressure of the mud column as the tool is pulled out of the hole.

DST tools usually include two or more clock-driven, bourdon-tube recording pressure gauges, one or two packers, and a set of flow valves. **Fig. 8.2** illustrates the sequence of events in a conventional DST for a typical Halliburton tool assembly, which occurs as follows.

1. Running into hole (Fig. 8.2a). Closed-in pressure (CIP) valve open, tester valve closed, bypass ports open.
2. Flowing formation (Fig. 8.2b). CIP valve open, tester valve open, bypass ports closed; formation fluid flowing into drillstem (may or may not reach surface).
3. Formation closed in (Fig. 8.2c). CIP valve closed, tester valve open, bypass ports closed; cessation of flow from formation; pressure buildup in formation.
4. Equalizing pressure (Fig. 8.2d). CIP valve closed, tester valve closed (sample of formation fluid trapped between valves), bypass ports open; pressure equalized across packer.
5. Reverse circulating (Fig. 8.2e). CIP valve closed; mud pumped down annulus to displace produced fluids up drillstring for measurement at surface.
6. Pulling out of hole (Fig. 8.2f). CIP valve closed; mud in drillstring bled into annulus through open valve.

8.3 Conventional DST Design

Appropriate equipment and hole-preparation procedures are required to obtain DST data that are sufficiently accurate for formation evaluation. We adapted the following recommended guidelines from Erdle.[5]

1. Length of test zone. The length of the zone to be tested should be minimized. The shorter the zone, the less mud that must be displaced out of the test tool during the first flow period, thus increasing the likelihood of recovering reservoir fluid. Wellbore-storage effects during shut-in periods are shorter lived for a shorter test zone.

2. Cushions. A cushion is a volume of liquid placed in the tool to help control high-pressure formations and to minimize slugging during the test. The use of cushions, either liquid or gas, in the string during a DST should be minimized. The presence of cushions in the pipe can inhibit the recovery of reservoir fluids because of the added hydrostatic pressure of the cushions and can interfere with the description of recovered fluid samples. Cushions should be run during a DST only for the following reasons: to prevent pipe collapse during deep tests or when mud weights are high, to control differential pressure across the packers at the start of the initial flow period (for normal bottomhole temperatures, the pressure differential should not exceed 5,000 psi for openhole tests or 7,500 psi for cased-hole tests), to prevent high differential pressures at the formation face when testing unconsolidated formations or gravel-pack completions (the pressure differential should be less than 400 psi to prevent formation sand production, gravel-pack plugging, or hydraulic-fracture proppant production), and to protect pipe from corrosive gases by running a slug of inhibitor.

3. Test tools. Tools that provide a large sealing area and that can take advantage of high operating forces with minimal rotation of the string should be used. Rotationally operated tools are less reliable than those operated by reciprocating motion of the pipe. If used at all, rotational tools should be used only for shallow (less than

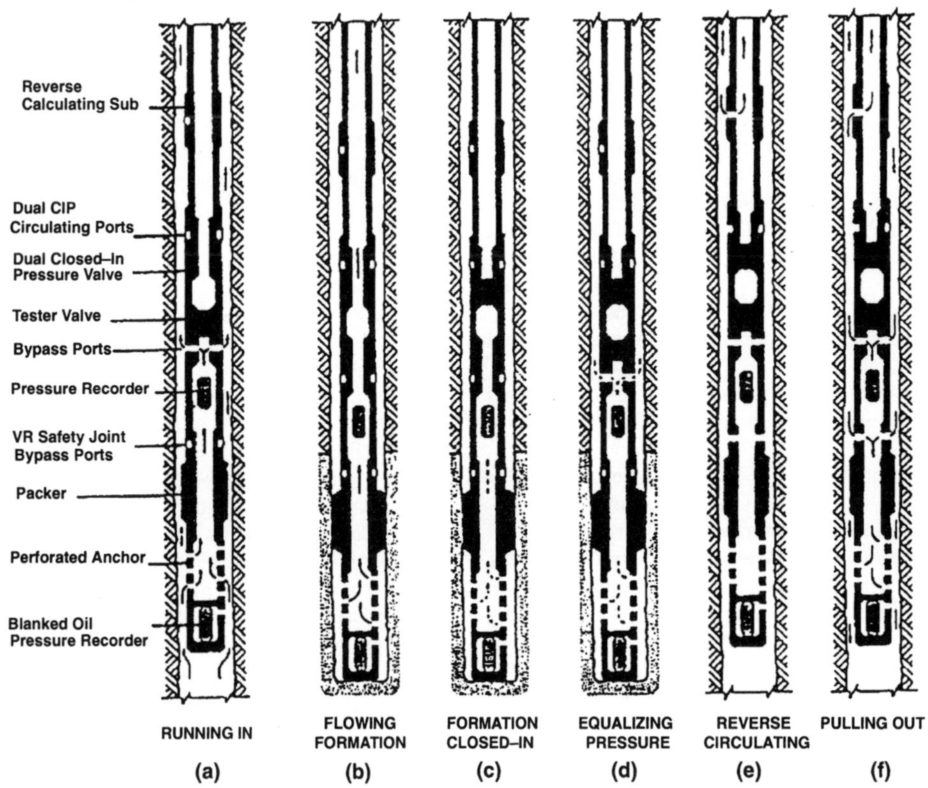

Fig. 8.2—DST tool operation for openhole formation test (after Edwards and Winn[4]).

LINEAR FLOW

Early Time ⟶ fracture

- Infinite conductivity fracture
- Pressure and derivative have slope of ½ (Separated by a factor of 2)
- Equivalent to negative skin

$$r_{wa} = \frac{x_f}{2} \Rightarrow s = \ln\left(\frac{2r_w}{x_f}\right)$$

- Fracture effectiveness may be reduced by:
  - Skin on fracture face
  - Choke skin
  - Finite conductivity within the fracture

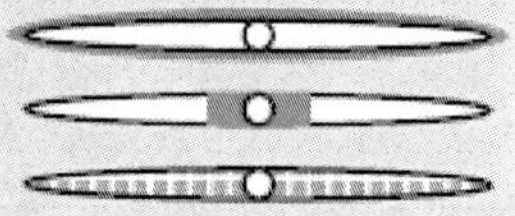

Late Time ⟶ channel (parallel boundaries)

- Derivative has slope of ½

BILINEAR FLOW

- Finite conductivity fracture
- Dimensionless Fracture Conductivity $FCD = \frac{k_f w}{k x_f}$
- Bilinear flow regime precedes linear flow
- $FCD > 100$ ⟶ infinite conductivity fracture

SPHERICAL FLOW

- Partial Penetration or limited perforations
- Slope of derivative is -½
- Component of total skin
- Magnitude of skin depends on vertical permeability and perforation interval
- Radial flow from completed interval may be observed before spherical flow

Reservoir Linear Flow (Fracture) – Half Slope

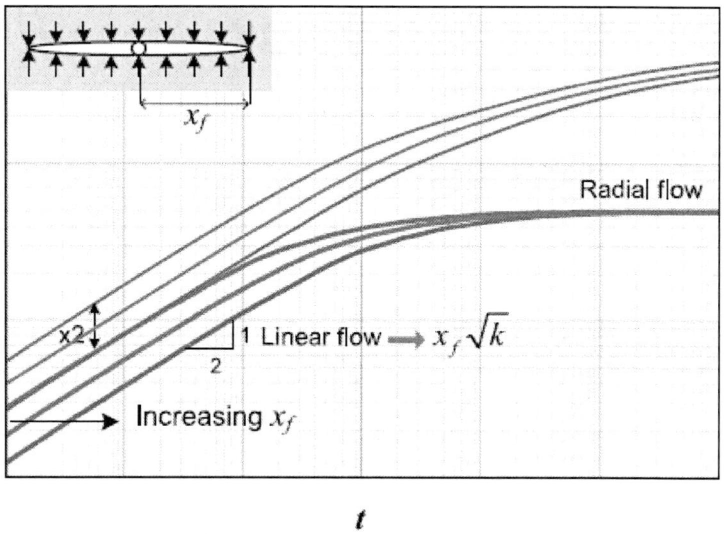

Bilinear Flow – Quarter Slope (Finite Conductivity Fracture)

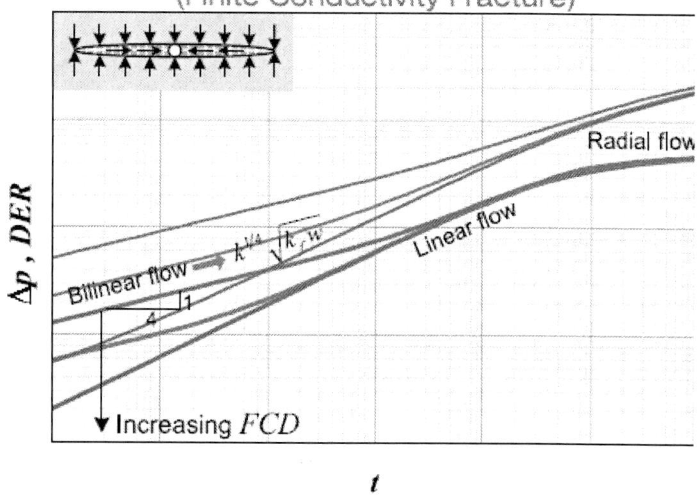

Reservoir Linear Flow (Channel) – Half Slope

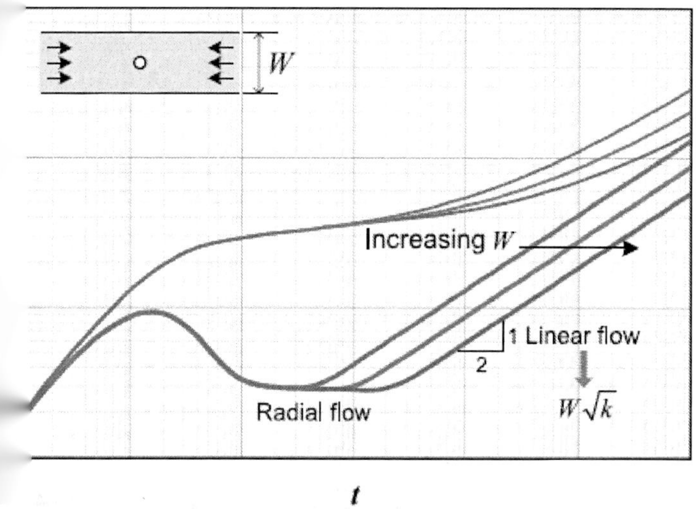

Spherical Flow – Negative Half Slope (Partial Penetration)

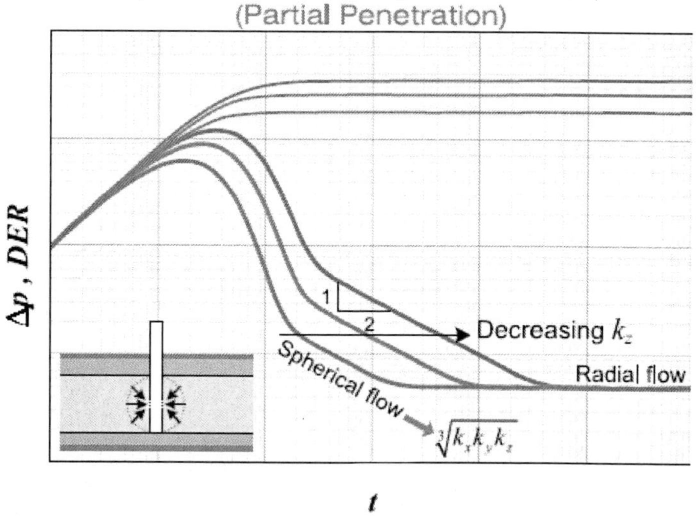

2,000-ft depth) openhole DST's. Pressure-operated tools have less operating force available for opening and closing and so require more maintenance than tools operated by reciprocation. Sleeve-type valves have more potential sealing area than full-bore valves and so are more reliable.

4. Packers. Packers with short elements are preferred to packers with long elements. When the packer is set, long rubber elements deform just as short elements do, providing the same contact area with the borehole wall. Furthermore, long-element packers may allow more rubber extrusion around the bottom of the packer.

5. Hole conditions. To help ensure the success of an openhole DST, the hole should be clean and a low-water-loss, low-viscosity mud should be used. American Petroleum Inst. (API) water loss[6] should be no more than 10 cm^3 to minimize mudcake thickness. Marsh-funnel viscosity[6] should be 80 seconds or less to minimize swabbing effects. Before a cased-hole test, a scraper should be run to remove residual cement that could damage the test tool or interfere with packer sealing.

6. Packer seats. Before an openhole test, a caliper log should be run to help identify an appropriate packer seat. The seat should be opposite a consolidated sandstone or an unfractured carbonate. A cased-hole packer seat should be located away from sections of overlapping liner and casing and from squeezed-off sections.

7. Bottomhole-pressure (BHP) and temperature recorders. At least three recorders should be used. Two gauges are positioned at the bottom of the tool to measure flowing and shut-in pressures during the test and hydrostatic mud-column pressure while running in or out of the hole. A third gauge is positioned above the tester valve to measure pressure inside the drillstring. This recorder can detect leaks in the tool or drillstring during shut-in or while running in or out of the hole, calculate liquid-fillup rates in the pipe, and provide data for slug-test analysis of pipe fillup during flow periods.

8. Isolation of tested interval. The number of packer seats and the exposure of the formation to drilling fluids should be minimized. The longer the sandface is exposed to drilling fluids, the worse washing out and mud-filtrate invasion will be.

An on-bottom test that uses conventional hard-rubber packers to isolate an interval at the bottom of the wellbore is the best technique for an openhole DST. This method has the following advantages: (1) only one packer seat is required, (2) mud-filtrate invasion is minimized because the test is usually done shortly after the test interval has been drilled, and (3) the test interval is short. In practice, two packers may be set at the top of the interval to be tested as a safeguard against losing the packer seat during the test.

If the wellbore extends below the bottom of the zone to be tested, a straddle test that uses conventional packers to isolate the interval from above and below is the second-best technique for openhole drillstem testing. Because two packer seats are required, this method may produce less reliable test data than the technique requiring only one packer seat. In practice, four packers (two above the interval to be tested and two below) may be set to reduce the chance of losing a packer seat.

Replacing conventional hard-rubber packers with inflatable packers in either of these two configurations may reduce the quality of the DST data significantly. If the hole is washed out, however, inflatable packers may be required. Under such circumstances, the formation usually has been exposed to drilling fluid for a long period of time. Skin damage caused by mud-filtrate invasion may be high, and pressure transient analysis may not produce as good results as when wellbore conditions are more favorable.

Although reservoir conditions may dictate the durations of test periods in specific situations, we offer the following general guidelines[5] for designing the durations of the flow and shut-in periods of a DST.

1. Determine the total on-bottom time available for the test.

2. Subtract the durations of the initial flow and shut-in periods from the total time. The inital flow period should be 5 minutes or less. The initial shut-in period should be at least 1 hour.

3. If the interval to be tested is expected to be homogeneous (single-layer, single-porosity), then the final flow period should be one-third of the remaining on-bottom test time. The final shut-in period should be two-thirds of the remaining time. This allocation is intended to maximize the probability of obtaining good pressure-buildup data.

4. If the interval to be tested is expected to be heterogeneous (multilayer or dual-porosity), then the final flow and final shut-in periods should each be allocated half of the remaining on-bottom test time. The relatively longer final flow period in this case is important for detecting heterogeneities.

Example 8.1—Conventional DST Design Without a Water Cushion. A development well[5] has been drilled in a consolidated sandstone formation that is thought to be relatively homogeneous acting. Total well depth is 5,300 ft. From log analysis, the interval 5,120 to 5,150 ft has been identified tentatively as a productive zone. Casing has not yet been set in this zone. The drillpipe is 3.5 in. (2.992 in. inside diameter), 9.50-lbm/ft internal-upset pipe with a collapse rating of 10,040 psi. In addition, the wellbore is filled with 9-lbm/gal mud. Temperature at the midpoint of the interval (5,135 ft) is 120°F, and pressure at the midpoint is 2,390 psia. Assuming that no water cushion is required, design a DST to evaluate the interval. The objectives of the DST are to identify formation fluids and to obtain pressure transient data for formation evaluation.

Solution. Erdle's[5] design guidelines can be used to design the DST. To maximize the likelihood of recovering reservoir fluids and to minimize wellbore-storage effects during the test, the test is limited as nearly as possible to the 30-ft interval. The drilling mud in the hole is a low-water-loss (10 cm^3 or less) mud to minimize mudcake thickness. To minimize swabbing effects, the Marsh-funnel viscosity does not exceed 80 seconds.

Assuming that no water cushion is required in this case and with a design safety factor of 1.125 for collapse, the collapse pressure owing to the mud column on the outside of the pipe is

$$p_{\text{collapse}} = \frac{(g_w)(W_m)(D_c)(F_d)}{W_w}$$

$$= \frac{(0.433)(9)(5,150)(1.125)}{8.33}$$

$$= 2,711 \text{ psi},$$

where g_w = water gradient, psi/ft; W_m = mud weight, lbm/gal; D_c = depth, ft; F_d = design factor; and W_w = water weight, lbm/gal. The collapse rating of 10,040 psi is more than adequate in this case, so the test can be run with empty drillpipe.

A straddle design is used for the test. The test string includes two conventional packers to be positioned immediately above and below the interval to be tested. Satisfactory packer seats are available in this vicinity, the sand is consolidated, and the caliper log indicates no washouts in the neighborhood of the interval. The test tool has three pressure recorders and is operated by reciprocating motion of the drillstring.

Because of the objectives of the DST, the test is designed with two flow and two shut-in periods. Experience with this formation suggests that a total test time of 10 hours should be adequate to achieve the test objectives. The initial flow period is 5 minutes, and the initial shut-in period is 90 minutes. On the basis of a previous evaluation of this formation, the reservoir is thought to be homogeneous (single-layer, single-porosity). Therefore, the final flow period is 168 minutes (one-third of the remaining 505 minutes), and the final shut-in period is 337 minutes (two-thirds of the remaining time).

Example 8.2—Conventional DST Design With a Water Cushion. A development well[5] has been drilled in a carbonate formation that is probably naturally fractured (i.e., a dual-porosity reservoir). Total depth is 10,050 ft. The interval 9,950 to 10,020 ft has been identified from logs as a potentially productive zone. Casing has not yet been set in this zone. The drillpipe is 4.5-in. (3.958 in. inside diameter), 13.75-lbm/ft internal-upset pipe with a collapse rating of 7,200 psi. The wellbore is filled with 12.4-lbm/gal mud. Temperature and pressure at the midpoint of the interval (9,985 ft) are 180°F and 4,500 psia, respectively. Assuming a water cushion is required, design a DST to evaluate the interval. The objectives of the DST are to identify formation fluids and to obtain pressure transient data for formation evaluation.

Solution. Erdle's[5] design guidelines can be used to design the DST. The test is limited as nearly as possible to the interval to maximize the likelihood of recovering reservoir fluids and to minimize wellbore-storage effects during the test. The drilling mud in the hole is a low-water-loss (10 cm^3 or less) mud to minimize mudcake thickness. To minimize swabbing effects, the Marsh-funnel viscosity does not exceed 80 seconds.

A water cushion is needed to avoid collapse of the drillpipe in this case. By use of a design safety factor of 1.125 for collapse, the collapse pressure owing to the mud column on the outside of the pipe is

$$p_{\text{collapse}} = \frac{(g_w)(W_m)(D_c)(F_d)}{W_w}$$

$$= \frac{(0.433)(12.4)(10,020)(1.125)}{8.33}$$

$$= 7,266 \text{ psi},$$

which exceeds the collapse rating of 7,200 psi for this pipe. To offset the hydrostatic pressure and reduce the net collapsing pressure on the pipe below 7,200 psi, a 158-ft cushion of water is run in the drillpipe. At a water gradient of 0.433 psi/ft, a 158-ft water column inside the pipe creates an offsetting hydrostatic pressure of 68 psi to reduce the net collapsing pressure to 7,200 psi.

Because the bottom of the interval to be tested is very near the bottom of the wellbore, an on-bottom design is used. The caliper log indicates numerous washouts in the wellbore, limiting the availability of good packer seats. According to the log, the nearest acceptable site for a packer seat is 50 ft above the top of the interval. If a conventional packer were to be placed at this point, the total tested interval would be 150 ft. With such a long interval, a substantial amount of mud would have to be displaced out of the test tool, possibly preventing the collection of a good sample of formation fluids. Furthermore, wellbore-storage effects could last too long for satisfactory pressure transient data to be obtained. Therefore, an inflatable packer is used to isolate the interval below 9,950 ft. The test tool has three pressure recorders and is operated by reciprocating motion of the drillstring.

Because of the objectives of the DST, the test is designed with two flow periods and two shut-in periods. Experience with this formation suggests that a total test time of 12 hours should be adequate to achieve the test objectives. The initial flow period is 5 minutes, and the initial shut-in period is 90 minutes. Because the reservoir is naturally fractured (dual-porosity), the final flow and shut-in periods each is 312.5 minutes (one-half of the remaining 625 minutes). The relatively longer final flow period, compared with that for a homogeneous reservoir, is necessary for detecting heterogeneities.

8.4 DST-Monitoring Procedures

Monitoring a DST is essential to ensure that the desired information is obtained from the test and that safe operating conditions are maintained during the test.[5] Three monitoring systems are currently in use: (1) the traditional bubble-bucket system, (2) the closed-chamber system, and (3) the surface-pressure-readout system.

The bubble-bucket system involves observing bubbles of air displaced out of the drillstring into a bucket of water during a flow period of a DST. The flow rate of the air is related qualitatively to reservoir-flow capacity and reservoir-fluid type. The lack of quantitative measurements commonly results in erroneous interpretations with this monitoring system.

The closed-chamber system is an inexpensive monitoring method that provides quantitative measurements of fluid influx from the formation throughout a test. Flow rates can be estimated as the test progresses. Sec. 8.6 discusses this test method.

The surface-pressure-readout system is the most sophisticated and expensive method of monitoring a DST. BHP data are conducted to the surface electronically, providing continuous real-time data on the reservoir pressure response throughout the test.

The nature and volume of the fluid produced into the drillstring during the DST are not known with certainty until the fluid is recovered and analyzed at the surface. The fluid sample, trapped between the closed-in pressure valve and the tester valve, is recovered by reversing out.

Proper analysis of fluid samples is extremely important. Gas recovery from the sample chamber should be measured to ensure correct identification of reservoir fluids, especially when stable, foamy emulsions are recovered. Compositions and chemical properties of samples recovered from the sample chamber and the drillstring should be evaluated after emulsions have been broken. Chemical analysis should be done on water-cushion samples taken before and after the DST.

8.5 DST Analysis Techniques

Conventional semilog and type-curve methods[1-5,7-9] are frequently used to analyze pressure transient data from DST's. The Horner method, discussed in Chaps. 2 and 3, is the most common analysis technique for buildup data, but the underlying assumptions often do not accurately describe the physical processes occurring during a DST before the time when produced fluids reach the surface. The Horner method assumes that the well is produced at a constant rate before shut-in. The solution to the diffusivity equation for a constant production rate gives a declining flowing pressure with time. Most DST's, however, exhibit a declining flow rate and increasing flowing pressure during the flow periods. Furthermore, produced fluids do not reach the surface during the flow period of most DST's. As a result, the Horner method may lead to incorrect interpretations of DST buildup data, including false indications of reservoir heterogeneities or depletion.

Conventional type-curve matching of DST data may also give misleading or inconclusive results when the well does not flow to the surface during the test.[5] Guidelines for determining the end of wellbore-storage effects from type curves (i.e., the line indicating the start of the semilog straight line shown on some type curves) do not apply to DST data and may indicate erroneously that the correct semilog straight line is not present. In addition, a changing wellbore-storage coefficient during flow periods may render type-curve matching[9,10-18] impossible.

The Horner method and conventional type-curve matching are adequate for analyzing a DST only when the well flows to the surface and when the rate remains constant. Refer to Chaps. 2 and 4 for use of the Horner semilog and log-log type-curve methods, respectively, for analyzing DST data. For those DST's in which the well does not flow to the surface, other analysis techniques, such as those developed by Ramey *et al.*[9] and Peres *et al.*,[7] should be considered.

8.5.1 Ramey *et al.*[9] Type Curves. Ramey *et al.* developed semilog and log-log type curves for analyzing data from DST's in which the flow does not reach the surface. The type curves are also applicable to slug tests in underpressured reservoirs. Slug testing involves the imposition of an instantaneous change in pressure in a well and the measurement of the resulting change in pressure as a function of time. This change in pressure is created by either injecting a specific volume of fluid (i.e., a slug) into the well or withdrawing a slug from the well. Slug testing is most suitable for wells that do not flow to the surface or for wells that are capable of flowing to the surface but are restrained mechanically from reaching the surface. A DST with a short flow period is similar to and can be modeled like a slug test.

The type curves in **Figs. 8.3 to 8.5** are solutions to the diffusivity equation modeling the pressure behavior in a well with a rising liquid level in the wellbore during the test. The rising liquid level results from accumulation of fluid in the wellbore following an instantaneous pressure change at the sandface. In addition, Ramey *et al.* include steady-state skin effects.

The type curves are presented in terms of a dimensionless pressure ratio, p_{DR}, and a dimensionless time-plotting function, t_D/C_D. The dimensionless pressure ratio is defined by

$$p_{DR} = \frac{p_i - p_{wf}(t)}{p_i - p_o}, \quad \text{(8.1)}$$

RADIAL FLOW (INFINITE-ACTING)

- Derivative has a slope of zero
- Used to obtain permeability

SKIN

- $\Delta p|_{skin}$ = difference between ideal and measured flowing pressure
- $s = \Delta p|_{skin}$ expressed in dimensionless form
- Skin $s = \left(\frac{k}{k_{skin}} - 1\right) \ln \frac{r_{skin}}{r_w}$
- Apparent wellbore radius $r_{wa} = r_w e^{-s}$; $s = \ln \frac{r_w}{r_{wa}}$
- Total skin, $s' = s_{damage} + s_{turbulence} + s_{partial\ penetration} + s_{geometry} + s_{fracture} + \ldots$

Permeability

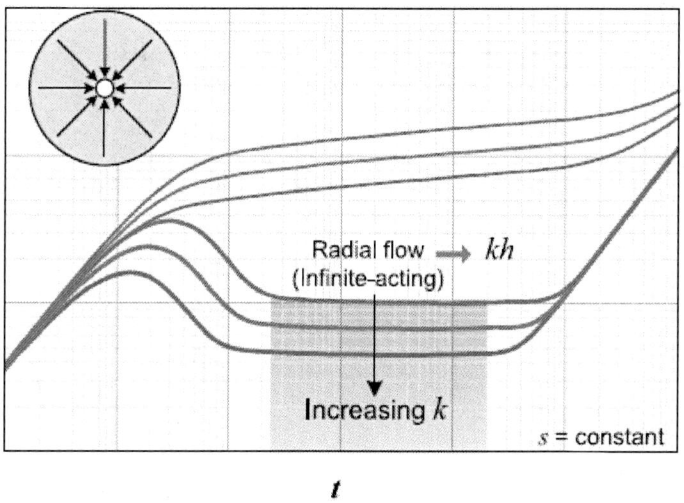

Skin

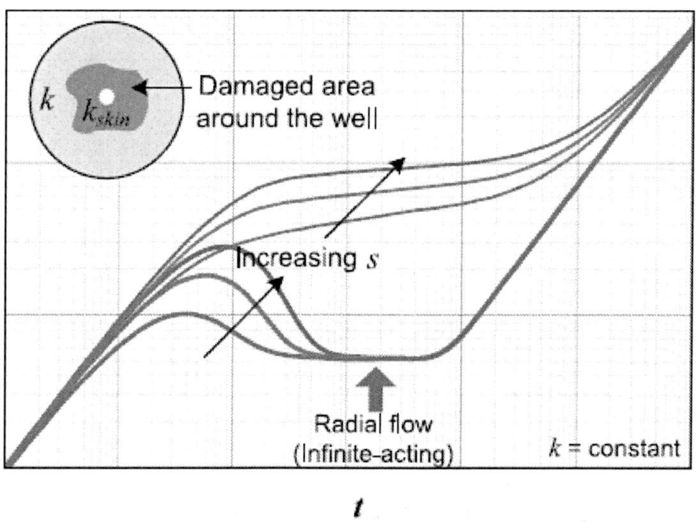

HORIZONTAL WELL FLOW REGIMES

- Vertical radial flow → vertical permeability, $\sqrt{k_y k_z}$ skin around wellbore, s_{damage}
- Linear flow → k_y or effective wellbore length, L_e
- Once linear flow is reached, horizontal well is similar to vertical fractured well + $s_{convergence}$
- Horizontal radial flow → horizontal permeability, $\sqrt{k_x k_y}$, and skin equivalent to vertical well, $s_{effective}$

Horizontal Well

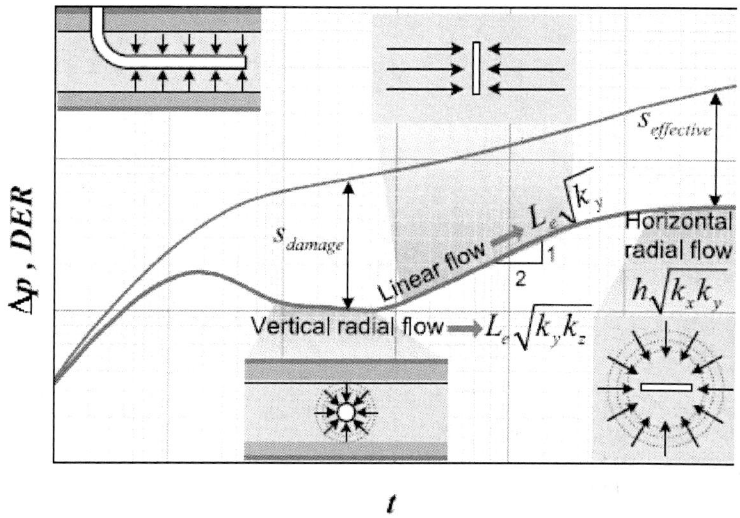

Effective Well Length

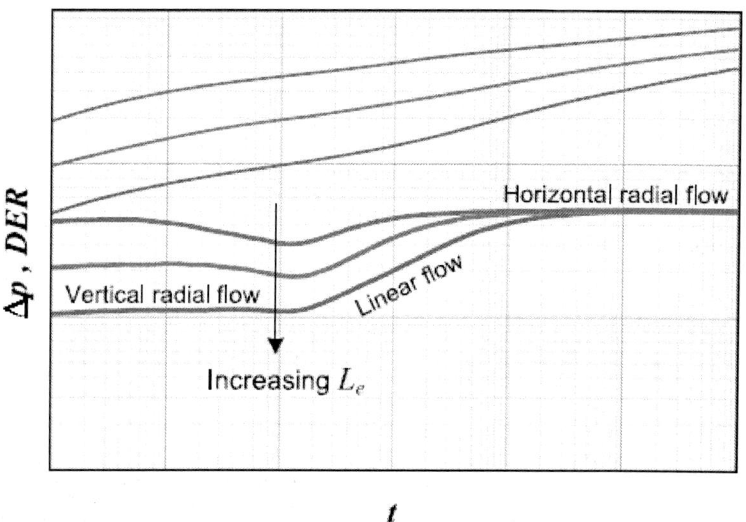

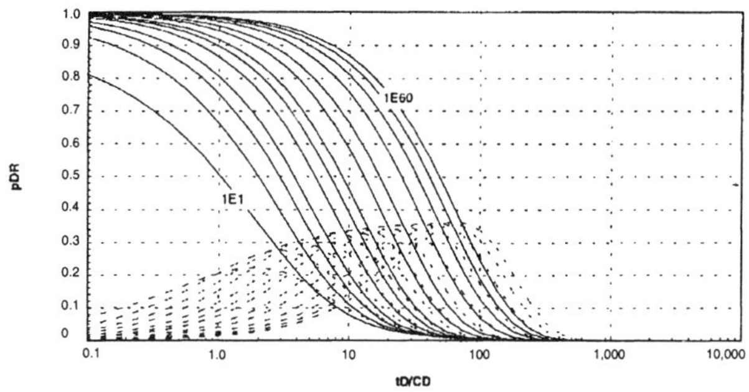

Fig. 8.3—Ramey *et al.*[9] semilog type curve for analysis of early- and late-time DST flow-period data. Reprinted with permission from the Canadian Inst. of Mining, Metallurgy, and Petroleum.

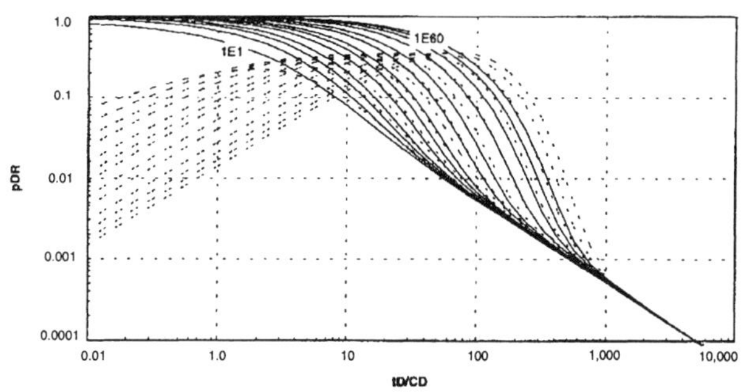

Fig. 8.4—Ramey *et al.*[9] log-log type curve for analysis of late-time DST flow-period data. Reprinted with permission from the Canadian Inst. of Mining, Metallurgy, and Petroleum.

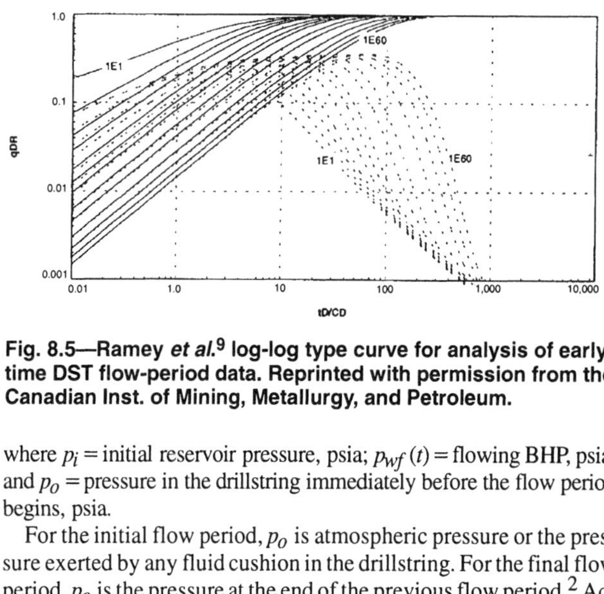

Fig. 8.5—Ramey *et al.*[9] log-log type curve for analysis of early-time DST flow-period data. Reprinted with permission from the Canadian Inst. of Mining, Metallurgy, and Petroleum.

where p_i = initial reservoir pressure, psia; $p_{wf}(t)$ = flowing BHP, psia; and p_o = pressure in the drillstring immediately before the flow period begins, psia.

For the initial flow period, p_o is atmospheric pressure or the pressure exerted by any fluid cushion in the drillstring. For the final flow period, p_o is the pressure at the end of the previous flow period.[2] According to Ramey and Agarwal,[15] Eq. 8.1 also describes the fluid unloading rate in the drillstring. Alternatively, the wellbore unloading rate is

$$p_{DR} = C_D \frac{dp_{wD}}{dt_D}, \quad (8.2)$$

where p_{wD} = dimensionless wellbore pressure,

$$p_{wD} = \frac{kh(p_i - p_{wf})}{141.2qB\mu}. \quad (8.3)$$

In addition, dimensionless time is defined by

$$t_D = \frac{0.0002637kt}{\phi \mu c_t r_w^2} \quad (8.4)$$

and the dimensionless wellbore-storage coefficient is

$$C_D = \frac{0.8936C}{\phi c_t h r_w^2}, \quad (8.5)$$

where the wellbore-storage coefficient C (in bbl/psi) owing to a changing fluid level in the wellbore is

$$C = \frac{25.65A_{wb}}{\rho_{wb}}. \quad (8.6)$$

In Eq. 8.6, A_{wb} = cross-sectional area of the wellbore or drillstring and ρ_{wb} = density of the fluid in the wellbore or drillstring. Similar to the Gringarten *et al.*[18] type curves, Ramey *et al.*[9] plot their type curves as functions of the correlating parameter $C_D e^{2s}$, where C_D = dimensionless wellbore-storage coefficient defined by Eq. 8.5 and s = dimensionless skin factor.

Ramey and Agarwal[15] also define a dimensionless sandface rate as

$$q_{DR} = 1 - p_{DR} = 1 - C_D \frac{dp_{wD}}{dt_D}, \quad (8.7)$$

which is often used as a plotting function to provide better resolution of very early DST or slug-test data (Fig. 8.5).

Type-Curve-Analysis Procedure. The following type-curve-analysis procedure, which is similar to that described in Chap. 4 for other type curves, is adapted from Earlougher[2] for analyzing DST flow data or slug-test data with the Ramey *et al.*[9] type curves.

1. Prepare semilog and log-log plots of p_{DR} vs. t and a log-log plot of $q_{DR} = 1 - p_{DR}$ vs. t. Make the plots either on tracing paper or on semilog and log-log paper the same size as the type curves.
2. Overlay the test-data plots on the appropriate type curve (Figs. 8.3 through 8.5) and find the type curve that nearly fits all the plotted test data. Fig. 8.3 provides better resolution of both early- and late-time data, while Figs. 8.4 and 8.5 are better for analyzing either late-time or early-time data, respectively.

Note that p_{DR} and q_{DR} have values between zero and one and are independent of flow rate and formation properties. Consequently, we simply fit equal values of p_{DR} or q_{DR} from the type curves and test data and slide the test plots horizontally only until a match is obtained. To reduce the ambiguity (i.e., uniqueness) problems associated with type-curve analysis, we also recommend matching p_{DR} and q_{DR} simultaneously until a consistent match is obtained.

3. From the matches, record the value of the correlating parameter $C_D e^{2s}$. The value of this parameter from both p_{DR} and q_{DR} matches should be equal.
4. With the test-data plots still fitted to the type curves, select a convenient time match point, $(t, t_D/C_D)_{MP}$.
5. Compute the wellbore-storage coefficient with Eq. 8.6.
6. Compute the dimensionless wellbore-storage coefficient, C_D, defined by Eq. 8.5.
7. Calculate permeability with the time match point from Step 4 and the wellbore-storage coefficient from Step 5. Note that time, t, is measured in hours in Eq. 8.8.

$$k = \frac{3,390\mu C}{h}\left(\frac{t_D/C_D}{t}\right)_{MP}. \quad (8.8)$$

8. Estimate the skin factor from the type-curve correlating parameter, $C_D e^{2s}$, and the dimensionless wellbore-storage coefficient, C_D.

$$s = 0.5\ln\left[\frac{\phi c_t r_w^2 (C_D e^{2s})_{MP}}{0.8936C}\right] = 0.5\ln\left[\frac{(C_D e^{2s})_{MP}}{C_D}\right]. \quad (8.9)$$

Ramey *et al.*'s[9] type curves were developed for homogeneous-acting, single-layered reservoirs, but similar type curves are available for double-porosity (i.e., naturally fractured) reservoirs,[19] hydraulically fractured wells,[20] and multilayer formations.[21] Example 8.3 demonstrates the analysis of a flow period of a DST with the Ramey type curves.

Example 8.3—Analysis of DST Flow Data With the Ramey Type Curves. A DST consisting of a 5-hour flow period followed by a

t (hours)	p_{wf} (psi)	t (hours)	p_{wf} (psi)
0.0506	354.6	0.675	1,204.6
0.0675	383.2	0.90	1,438.3
0.090	420.6	1.20	1,706.4
0.120	469.3	1.60	2,000.8
0.160	532.2	2.13	2,305.8
0.213	613.1	2.85	2,598.8
0.285	716.2	3.79	2,854.1
0.379	846.3	5.06	3,050.7
0.506	1,007.8		

TABLE 8.1—PRESSURE AND TIME DATA FROM THE FLOW PERIOD OF A DST, EXAMPLE 8.3

buildup period of 6 hours was conducted in an oil well. The pressure and time data, given in **Table 8.1,** were taken during the flow period of the test. Use the Ramey *et al.*[9] type curves to analyze the data and estimate the permeability, skin factor, and dimensionless wellbore-storage coefficient.

$p_i = 3{,}315.2$ psi
$p_o = 263.1$ psi
$\phi = 0.13$
$r_w = 0.365$ ft
$h = 23$ ft
$C = 0.01609$ bbl/psi
$\mu = 0.43$ cp
$c_t = 14.7 \times 10^{-6}$ psi^{-1}

Solution.

1. Prepare semilog and log-log plots of dimensionless pressure, p_{DR}, vs. t and a log-log plot of dimensionless rate, $q_{DR} = 1 - p_{DR}$, vs. t. Make the plots either on tracing paper or on semilog and log-log paper with the same size log cycles as the type curves. **Table 8.2** summarizes the plotting functions.

For example, at a flowing time of $t = 0.0506$ hours, p_{DR} and q_{DR} are

$$p_{DR} = \frac{p_i - p_{wf}(t)}{p_i - p_o}$$

$$= \frac{3,315.2 - 354.6}{3,315.2 - 263.1}$$

$$= 0.9700$$

and $q_{DR} = 1.0 - p_{DR} = 1.0 - 0.9700 = 0.0300.$

At a flowing time of 0.0675 hours,

$$t'_{pDR} = -\frac{p_{DR1} - p_{DR2}}{\ln t_1 - \ln t_2}$$

$$= -\frac{0.9700 - 0.9484}{[\ln(0.0506) - \ln(0.090)]}$$

$$= -0.0376.$$

2. Overlay the test-data plot on the appropriate type curve and find the type curve that nearly fits all the plotted test data. Note that, because p_{DR} and q_{DR} have values between 0.0 and 1.0 on both the type curves and data plots, we simply align the vertical axes and slide the data plot horizontally until a match is found. From **Figs. 8.6 through 8.8,** we obtain a good match for C_De^{2s} between 10^{10} and 10^{15}.

3. Interpolating between the type curves, the type-curve-correlating parameter is $C_De^{2s} = 3 \times 10^{12}$.

4. From the match, we select a time match point, $(t, t_D/C_D)_{MP}$; $t = 1.2$ hours and $t_D/C_D = 10.0$.

5. Compute the dimensionless wellbore-storage coefficient, C_D, defined by Eq. 8.6.

$$C_D = \frac{0.8936C}{\phi c_t h r_w^2}$$

$$= \frac{(0.8936)(0.01609)}{(0.13)(14.7 \times 10^{-6})(23)(0.354)^2}$$

$$= 2,610.$$

6. Calculate permeability with the time match point and the wellbore-storage coefficient from Step 5.

$$k = \frac{3,390\mu C}{h}\left(\frac{t_D/C_D}{t}\right)_{MP}$$

$$= \frac{(3,390)(0.43)(0.01609)}{23}\left(\frac{10.0}{1.2}\right)$$

$$= 8.50 \text{ md}.$$

7. Estimate the skin factor with the type-curve correlating parameter, C_De^{2s}, from Step 3 and the dimensionless wellbore-storage coefficient, C_D, computed in Step 5.

$$s = 0.5\ln\left[\frac{(C_De^{2s})_{MP}}{C_D}\right] = 0.5\ln\left(\frac{3 \times 10^{12}}{2,610}\right) = 10.4.$$

8.5.2 Peres *et al.*[7] Method. Peres *et al.* provide a method for analyzing pressure-buildup data from a DST when produced liquids do not reach the surface during a flow period. The method accounts for producing time and variable flow rate during the flow period. Similar to the Horner graphical method presented in Chaps. 2 and 3, a straight-line plot is used to obtain permeability, k; skin factor, s; and initial reservoir pressure, p_i. The following definitions, which are given in oilfield units, are used to develop the Peres *et al.* method.

The drawdown wellbore-storage coefficient, C_F (RB/psi), is defined by

$$C_F - = \frac{25.65A_{wb}}{\rho_{wb}} = \frac{25.65(\pi r_p^2)}{\rho_{wb}} \quad \text{(8.10)}$$

TABLE 8.2—DIMENSIONLESS-PRESSURE AND RATE-PLOTTING FUNCTIONS, EXAMPLE 8.3

t (hours)	p_{DR}	q_{DR}	t'_{pDR}	t (hours)	p_{DR}	q_{DR}	t'_{pDR}
0.0506	0.9700	0.0300		0.675	0.6915	0.3085	0.2449
0.0675	0.9607	0.0393	0.0376	0.90	0.6150	0.3850	0.2857
0.090	0.9484	0.0516	0.0490	1.20	0.5271	0.4729	0.3203
0.120	0.9324	0.0676	0.0635	1.60	0.4307	0.5693	0.3423
0.160	0.9118	0.0882	0.0821	2.13	0.3307	0.6693	0.3394
0.213	0.8853	0.1147	0.1045	2.85	0.2347	0.7653	0.3118
0.285	0.8515	0.1485	0.1326	3.79	0.1511	0.8489	0.2579
0.379	0.8089	0.1911	0.1664	5.06	0.0867	0.9133	
0.506	0.7560	0.2440	0.2034				

- A specialized plot is a plot of the pressure data on a time axis that is specific to a particular flow regime:

Flow Regime	X-Axis
Radial flow (horizontal)	$\log(t)$
Radial flow (vertical)	$\log(t)$
Linear flow (fracture)	$\sqrt{t}$
Linear flow (channel)	$\sqrt{t}$
Bilinear flow	$\sqrt[4]{t}$
Spherical flow	$1/\sqrt{t}$
Wellbore storage (afterflow)	t
Boundary-dominated flow	t

- It exhibits a straight line during that flow regime
- The slope of the line gives the result of interest

Radial Flow - Dual Porosity

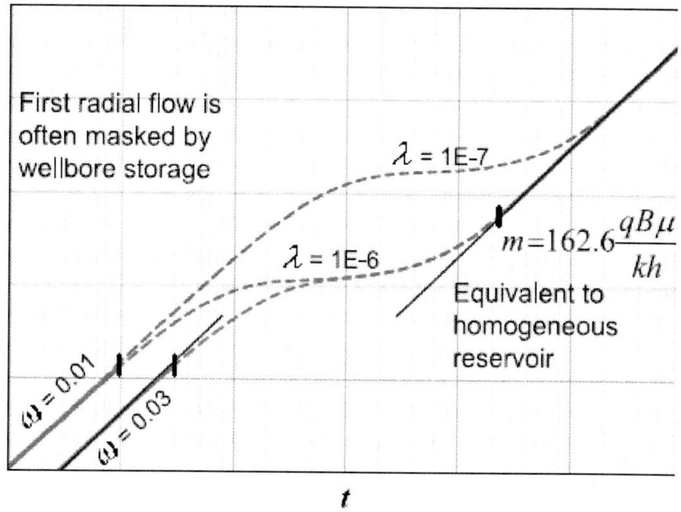

Radial Flow

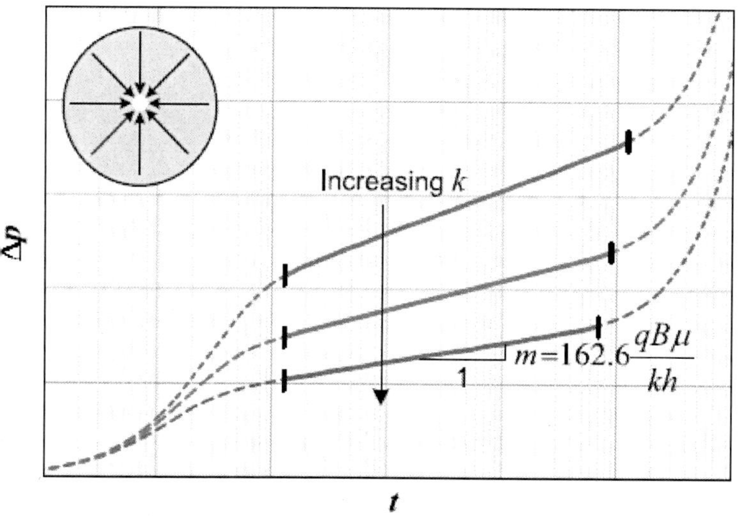

Reservoir Linear Flow - Fracture or Channel

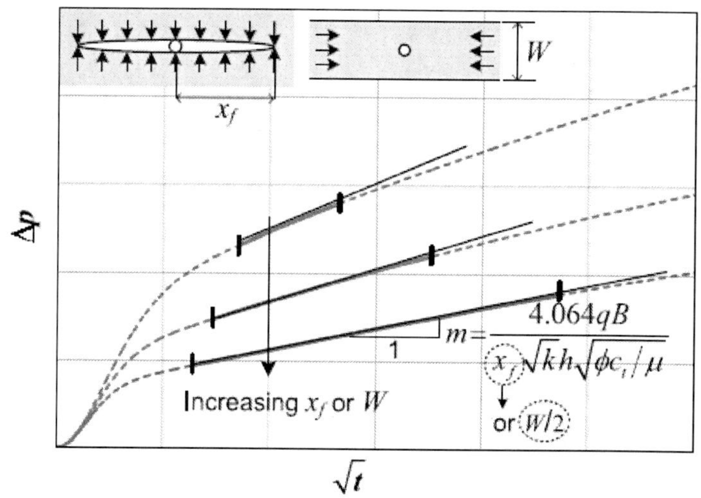

Radial Flow - Horizontal Well

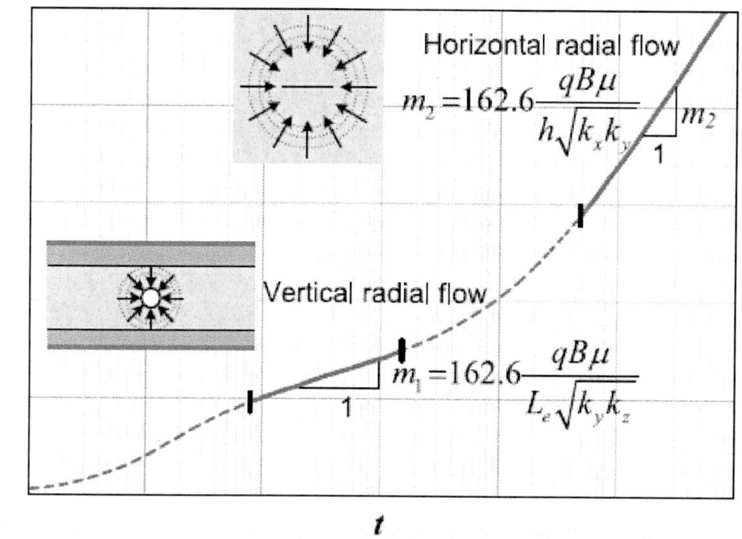

Spherical Flow

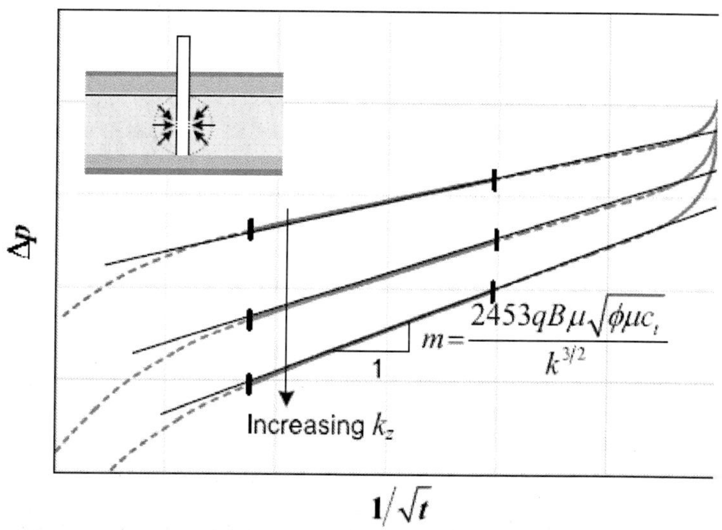

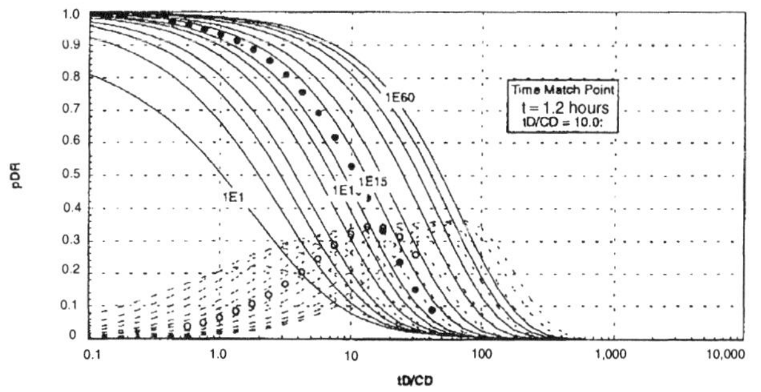

Fig. 8.6—Semilog type-curve match of early- and late-time data with the Ramey *et al.*[9] type curves, Example 8.3.

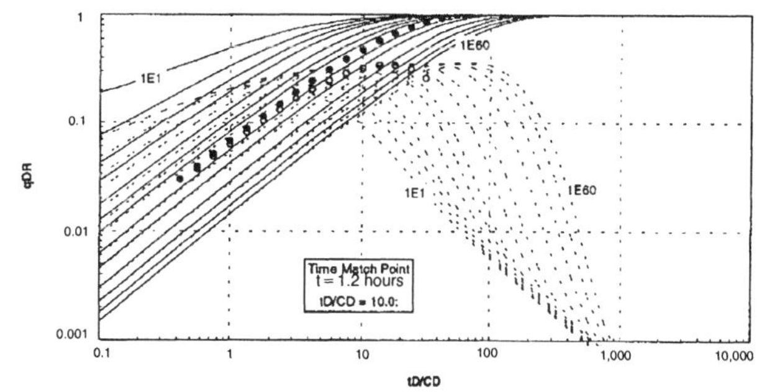

Fig. 8.8—Log-log type-curve match of early-time data with the Ramey *et al.*[9] type curves, Example 8.3.

for a rising liquid level in the wellbore, where r_p = inside diameter of the drillpipe, in., and ρ_{wb} = density of the fluid produced into the drillpipe, lbm/ft^3.

Similarly, the buildup wellbore-storage coefficient is defined by

$$C_S = V_{wb}\, c_{wb}\ , \quad (8.11)$$

where V_{wb} = volume of the wellbore during the buildup period, bbl, and c_{wb} = isothermal compressibility of the fluid in the wellbore, psi^{-1}.

The pressure response during a buildup period in a DST is given by

$$p_{ws}(\Delta t) = p_i - m_m t_m\ , \quad (8.12)$$

where p_{ws} = bottomhole shut-in pressure, psia; p_i = initial reservoir pressure, psia; Δt = elapsed time since shutting in the well, hours; t_m = multirate time function defined by Eq. 8.14, hours; and m_m = slope of straight line through data.

The producing time interval $[0, t_p]$ is partitioned (with an arbitrary value of n) as

$$0 = t_{0,p} < t_{1,p} < t_{2,p} < \cdots < t_{n,p} < t_{n+1,p} = t_p\ , \quad (8.13)$$

where t_p = the producing time during the flow period of the test. The multirate time function, t_m, in Eq. 8.12 is defined by

$$t_m = \sum_{j=0}^{n} \left\{ \left(\frac{t_p}{t_p + \Delta t - t_{j,p}} \right) \left[\frac{p_{wf}(t_{j+1,p}) - p_{wf}(t_{j,p})}{p_{wf}(t_p) - p_o} \right] \right\}, \quad (8.14)$$

where p_{wf} = bottomhole flowing pressure, psia, and p_o = initial BHP at the start of the test, psia. Note that the multirate time function is dimensionless.

Similar to conventional pressure-buildup tests discussed in Chap. 2, we can estimate the formation permeability with the slope of the straight line, m_m,

$$k = \frac{141.2\mu}{m_m h} \frac{24C_F\left[p_{wf}(t_p) - p_o\right]}{2t_p}. \quad (8.15)$$

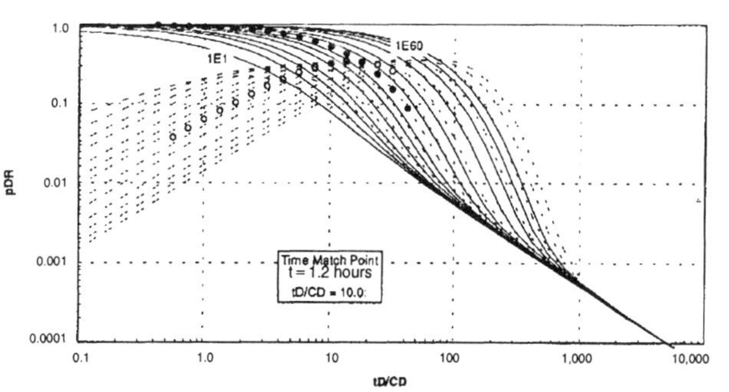

Fig. 8.7—Log-log type-curve match of late-time data with the Ramey *et al.*[9] type curves, Example 8.3.

Extrapolating the straight line of Eq. 8.12 to $t_m = 0$ gives the initial reservoir pressure, p_i. Permeability, k, is obtained from Eq. 8.15. **Fig. 8.9** illustrates this graphical analysis technique.

The skin factor, s, is given by

$$s = 1.151\left\{ \left[\frac{p_i - \overline{p}_f}{m_m \ln 10} - \log\left(\frac{k}{\phi\mu c_t r_w^2}\right) + 3.23 \right] - \left\{ \sum_{j=0}^{n} \left[\log\left(t_p - t_{j,p}\right) \frac{p_{wf}(t_{j+1,p}) - p_{wf}(t_{j,p})}{p_{wf}(t_p) - p_o} \right] \right\} \right\}, \quad (8.16)$$

where $\overline{p}_f$ is the average flowing pressure during the flow period and is defined by

$$\overline{p}_f = \frac{1}{t_p} \int_0^{t_p} p_{wf}(\tau)\mathrm{d}\tau. \quad (8.17)$$

The integration in Eq. 8.17 can be performed with enough accuracy for most field applications with the trapezoidal rule. Partioning the producing-time interval as in Eq. 8.13 and integrating Eq. 8.17 by the trapezoidal rule gives

$$\overline{p}_f = \frac{1}{t_p} \sum_{j=0}^{n} \left[\frac{p_{wf}(t_{j+1,p}) + p_{wf}(t_{j,p})}{2} \left(t_{j+1,p} - t_j\right) \right]$$

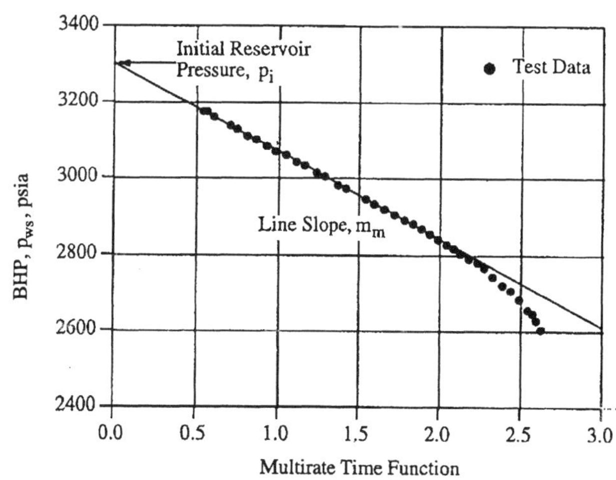

Fig. 8.9—Graphical analysis technique for pressure-buildup data from DST.

TABLE 8.3—DST DATA, EXAMPLE 8.4

Flow Period

Elapsed Test Time (hours)	Flowing BHP, p_{wf} (psi)	Elapsed Test Time (hours)	Flowing BHP, p_{wf} (psi)
0.00	263.10	2.64	1,170.4
0.24	475.80	2.88	1,221.6
0.48	520.45	3.12	1,278.0
0.72	584.78	3.36	1,331.9
0.96	684.57	3.60	1,377.8
1.20	765.97	3.84	1,421.1
1.44	852.63	4.08	1,467.1
1.68	927.47	4.32	1,513.1
1.92	999.68	4.56	1,556.4
2.16	1,060.1	4.8576	1,606.3
2.40	1,107.3	—	—

Shut-In Period

Elapsed Test Time (hours)	Shut-in BHP, p_{ws} (psi)	Buildup Time, Δt (hours)	Elapsed Test Time (hours)	Shut-In BHP, p_{ws} (psi)	Buildup Time, Δt (hours)
4.8768	2,602.8	0.0192	5.5200	2,917.9	0.6624
4.8864	2,626.5	0.0288	5.6064	2,931.1	0.7488
4.8960	2,643.5	0.0384	5.6928	2,944.2	0.8352
4.9056	2,654.0	0.0480	5.9136	2,973.1	1.0560
4.9248	2,682.9	0.0672	6.0096	2,984.9	1.1520
4.9440	2,703.9	0.0864	6.2016	3,005.9	1.3440
4.9632	2,721.0	0.1056	6.3456	3,017.7	1.4880
4.9920	2,743.3	0.1344	6.5376	3,032.2	1.6800
5.0208	2,764.3	0.1632	6.7296	3,045.3	1.8720
5.0400	2,777.4	0.1824	6.9696	3,059.7	2.1120
5.0688	2,790.6	0.2112	7.2096	3,072.9	2.3520
5.0976	2,805.0	0.2400	7.4976	3,087.3	2.6400
5.1264	2,816.8	0.2688	7.8336	3,103.1	2.9760
5.1552	2,827.3	0.2976	8.1696	3,114.9	3.3120
5.1936	2,841.8	0.3360	8.6016	3,129.3	3.7440
5.2320	2,854.9	0.3744	9.0336	3,142.5	4.1760
5.2800	2,868.0	0.4224	10.1376	3,166.1	5.2800
5.3280	2,879.9	0.4704	10.7616	3,176.6	5.9040
5.3856	2,893.0	0.5280	11.0400	3,179.2	6.1824
5.4528	2,906.1	0.5952			

$$= \frac{1}{t_p}\left[\frac{p_{wf}(t_{1,p}) + p_{wf}(t_{0,p})}{2}(t_{1,p} - t_{0,p})\right.$$

$$+ \frac{p_{wf}(t_{2,p}) + p_{wf}(t_{1,p})}{2}(t_{2,p} - t_{1,p}) + \ldots$$

$$\left.+ \frac{p_{wf}(t_p) + p_{wf}(t_{n,p})}{2}(t_p - t_{n,p})\right]. \quad (8.18)$$

The following procedure is recommended for analyzing DST's with the Peres *et al.*[7] method.

Analysis Procedure—Peres et al. Method.

1. To analyze a DST with Peres *et al.*'s[7] method, we must prepare a plot of shut-in BHP, p_{ws}, vs. the multirate time function, t_m, defined by Eq. 8.14.

$$t_m = \sum_{j=0}^{n}\left\{\left(\frac{t_p}{t_p + \Delta t - t_{j,p}}\right)\left[\frac{p_{wf}(t_{j+1,p}) + p_{wf}(t_{j,p})}{p_{wf}(t_p) - p_o}\right]\right\}. \quad (8.14)$$

In the absence of wellbore-storage effects, the data should lie on a straight line. Consequently, any deviations from a straight line during early shut-in times suggest wellbore-storage effects.

2. Draw the best-fit line through the data and compute the slope, m_m,

$$m_m = \left|\frac{p_{ws2} - p_{ws1}}{t_{m2} - t_{m1}}\right|. \quad (8.19)$$

For subsequent calculations, use the absolute value of the slope.

3. Using the slope from Step 2, calculate the formation permeability with Eq. 8.15.

$$k = \frac{141.2\mu}{m_m h}\frac{24C_F[p_{wf}(t_p) - p_o]}{2t_p}. \quad (8.15)$$

4. Estimate the initial reservoir pressure, p_i, from the extrapolation of the straight line to $t_m = 0$.

5. Compute the skin factor with Eq. 8.16,

$$s = 1.151\left\{\frac{p_i - \overline{p}_f}{m_m \ln 10} - \log\left(\frac{k}{\phi\mu c_t r_w^2}\right) + 3.23\right.$$

TABLE 8.4—PRESSURE AND TIME-PLOTTING FUNCTIONS, EXAMPLE 8.4

Flow Period

Elapsed Test Time (hours)	p_{wf} (psi)	Index, j	Flowing Time, $t_{j,p}$ (hours)	Elapsed Test Time (hours)	p_{wf} (psi)	Index, j	Flowing Time, $t_{j,p}$ (hours)
0.00	263.10	0	0.00	2.64	1,170.4	11	2.64
0.24	475.80	1	2.64	2.88	1,221.6	12	2.88
0.48	520.45	2	2.88	3.12	1,278.0	13	3.12
0.72	584.78	3	3.12	3.36	1,331.9	14	3.36
0.96	684.57	4	3.36	3.60	1,377.8	15	3.60
1.20	765.97	5	3.60	3.84	1,421.1	16	3.84
1.44	852.63	6	3.84	4.08	1,467.1	17	4.08
1.68	927.47	7	4.08	4.32	1,513.1	18	4.32
1.92	999.68	8	4.32	4.56	1,556.4	19	4.56
2.16	1,060.1	9	4.56	4.8576	1,606.3	20	4.8576
2.40	1,107.3	10	4.8576				

Shut-In Period

Elapsed Test Time, (hours)	p_{ws} (psi)	Δt (hours)	t_m	Elapsed Test Time, (hours)	p_{ws} (psi)	Δt (hours)	t_m
4.8768	2,602.8	0.0192	2.6300	5.5200	2,917.9	0.6624	1.6616
4.8864	2,626.5	0.0288	2.5991	5.6064	2,931.1	0.7488	1.5974
4.8960	2,643.5	0.0384	2.5695	5.6928	2,944.2	0.8352	1.5394
4.9056	2,54.0	0.0480	2.5410	5.9136	2,973.1	1.0560	1.4129
4.9248	2,682.9	0.0672	2.4871	6.0096	2,984.9	1.1520	1.3656
4.9440	2,703.9	0.0864	2.4371	6.2016	3,005.9	1.3440	1.2816
4.9632	2,721.0	0.1056	2.3903	6.3456	3,017.7	1.4880	1.2263
4.9920	2,743.3	0.1344	2.3255	6.5376	3,032.2	1.6800	1.1606
5.0208	2,764.3	0.1632	2.2661	6.7296	3,045.3	1.8720	1.1026
5.0400	2,777.4	0.1824	2.2292	6.9696	3,059.7	2.1120	1.0387
5.0688	2,790.6	0.2112	2.1774	7.2096	3,072.9	2.3520	0.9826
5.0976	2,805.0	0.2400	2.1292	7.4976	3,087.3	2.6400	0.9235
5.1264	2,816.8	0.2688	2.0842	7.8336	3,103.1	2.9760	0.8636
5.1552	2,827.3	0.2976	2.0420	8.1696	3,114.9	3.3120	0.8116
5.1936	2,841.8	0.3360	1.9897	8.6016	3,129.3	3.7440	0.7538
5.2320	2,854.9	0.3744	1.9412	9.0336	3,142.5	4.1760	0.7040
5.2800	2,868.0	0.4224	1.8852	10.1376	3,166.1	5.2800	0.6034
5.3280	2,879.9	0.4704	1.8337	10.7616	3,176.6	5.9040	0.5587
5.3856	2,893.0	0.5280	1.7770	11.0400	3,179.2	6.1824	0.5408
5.4528	2,906.1	0.5952	1.7166				

$$- \sum_{j=0}^{n}\left[\log(t_p - t_{j,p})\frac{p_{wf}(t_{j+1,p}) - p_{wf}(t_{j,p})}{p_{wf}(t_p) - p_o}\right]\Bigg\}, \quad \text{(8.16)}$$

where $\overline{p_f}$ is calculated with Eq. 8.18.

Example 8.4—DST Analysis With the Peres *et al.* Method. A DST[7] consisting of one 5-hour flow period and one 6-hour shut-in period was conducted. **Table 8.3** gives drawdown and buildup data from the test. Other known data are summarized next. Determine formation permeability, skin factor, and initial reservoir pressure with the Peres *et al.*[7] method.

$p_i =$ 3,315.2 psi
$p_o =$ 263.1 psi
$\phi =$ 0.13
$r_w =$ 0.365 ft
$h =$ 23 ft
$C_F =$ 0.01609 bbl/psi
$\mu =$ 0.43 cp
$c_t = 14.7 \times 10^{-6}$ psi^{-1}

Solution.

1. Calculate the multirate time function defined by Eq. 8.14. Producing time during the flow period is partioned (Eq. 8.13), giving values of $t_{1,p} = 0.24$ hours, $t_{2,p} = 0.48$ hours, . . . , $t_{19,p} = t_{n,p} = 4.56$ hours, and $t_{n+1,p} = t_p = 4.8576$ hours. The multirate time function, t_m, is calculated for each shut-in time, Δt, with $n = 19$ (**Table 8.4**). For example, the value of t_m at $\Delta t = 0.0192$ hours is

$$t_m = \sum_{j=0}^{n}\left\{\left(\frac{t_p}{t_p + \Delta t - t_{j,p}}\right)\left[\frac{p_{wf}(t_{j+1,p}) + p_{wf}(t_{j,p})}{p_{wf}(t_p) - p_o}\right]\right\}.$$

$$= \left(\frac{4.8576}{4.8576 + 0.0192 - 0}\right)\left(\frac{475.8 - 263.1}{1606.3 - 263.1}\right)$$

$$+ \left(\frac{4.8576}{4.8576 + 0.0192 - 0.24}\right)\left(\frac{520.45 - 475.8}{1606.3 - 263.1}\right) + \ldots$$

$$+ \left(\frac{4.8576}{4.8576 + 0.0192 - 4.56}\right)\left(\frac{1606.3 - 1556.4}{1606.3 - 263.1}\right)$$

$$= 2.63.$$

2. Reservoir properties are found from a plot of shut-in BHP, p_{ws}, as a function of multirate time, t_m (**Fig. 8.10**). Wellbore-storage effects during the buildup period cease at about $t_m = 2.2$.

3. Draw a straight line through the data points in Fig. 8.10. This line has a slope of $m_m = 231.5$ psi. By rearranging Eq. 8.15, we get the formation permeability.

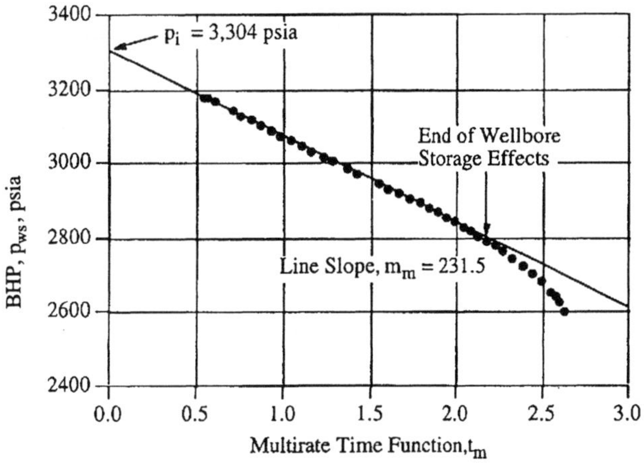

Fig. 8.10—Pressure and multirate time plot, Example 8.4.

$$k = \frac{141.2\mu}{m_m h}\,\frac{24C_F\left[p_{wf}(t_p) - p_o\right]}{2t_p} \quad \ldots\ldots\ldots\ldots\ldots\ldots \quad (8.15)$$

$$= \left[\frac{(141.2)(0.43)}{(231.5)(23)}\right]\left[\frac{(24)(0.01609)(1,606.3 - 263.1)}{2(4.8576)}\right].$$

4. Extrapolating the straight line to $t_m = 0$ in Fig. 8.7, the initial reservoir pressure is estimated to be $p_i = 3,304$ psi.

5. Finally, we can calculate the skin factor. First, the average flowing pressure is given by Eq. 8.18,

$$\overline{p_f} = \frac{1}{4.8576}\left[\left(\frac{475.8 + 263.1}{2}\right)(0.24 - 0)\right.$$

$$+ \left(\frac{520.45 + 475.8}{2}\right)(0.48 - 0.24) + \ldots$$

$$\left. + \left(\frac{1606.3 + 1556.4}{2}\right)(4.8576 - 4.56)\right]$$

$$= 1,069 \text{ psia}.$$

From Eq. 8.16, the skin factor is

$$s = 1.151\left\{\frac{p_i - \overline{p_f}}{m_m \ln 10} - \log\left(\frac{k}{\phi\mu c_t r_w^2}\right) + 3.23\right.$$

$$\left. - \sum_{j=0}^{n}\left[\log\left(t_p - t_{j,p}\right)\frac{p_{wf}\left(t_{j+1,p}\right) - p_{wf}\left(t_{j,p}\right)}{p_{wf}\left(t_p\right) - p_o}\right]\right\}$$

$$s = 1.151\left\{\frac{3,304 - 1,069}{231.5\ln(10)}\right.$$

$$- \log\left[\frac{0.609}{(0.13)(0.43)(1.45 \times 10^{-5})(0.354)^2}\right] + 3.23$$

$$- \left[\log(4.8576 - 0)\left(\frac{475.8 - 263.1}{1,606.3 - 263.1}\right)\right.$$

$$+ \log(4.8576 - 0.24)\left(\frac{520.45 - 475.8}{1606.3 - 263.1}\right) + \ldots$$

$$\left.\left. + \log(4.8576 - 4.56)\left(\frac{1,606.3 - 1,556.4}{1,606.3 - 263.1}\right)\right]\right\} = 0.28,$$

which agrees with $s = 0.3$ from Example 8.3.

8.6 Closed-Chamber DST

The closed-chamber DST[22] is a well-testing method that permits monitoring of the initial flow period of a test to identify formation fluids and provide good field estimates of gas and liquid flow rates during the initial flow period. If the remainder of the DST is to be conventional, information about the initial flow period from the closed-chamber test can be used to establish minimum times for the final flow and buildup periods and to preset surface equipment for the remainder of the test. A closed-chamber test differs from a conventional DST in that the well is closed in at the surface during flow periods and is open at the surface when the well is shut in at the formation. Fluid influx into the test tool is monitored throughout the test, allowing flow rates to be estimated as the test progresses. Flow rates and recoveries can be confirmed upon test completion.

With accurate measurement of both downhole and surface pressures, closed-chamber tests also can be analyzed to obtain estimates of formation properties. Kabir *et al.*[23] present a method to estimate flow rates of both gas and liquid with surface and bottomhole pressure gauges. Their model accounts for the effects of gas and liquid compression, friction, and mass transfer. With accurate estimates of flow rates of both phases, they were able to obtain accurate estimates of formation permeability.

A set of equations describing pressures and flow rates can be derived from a mass balance of gas and liquids in the drillpipe during a closed-chamber DST run with empty drillpipe. When the tester valve is opened, gas and liquids enter the drillpipe. Gas is vented from the pipe but liquids do not reach the surface. If the rate at which gas is vented differs from the gas-influx rate, then the mass of gas in the system will change. Under the given conditions, the mass balance for single-phase gas flow is described by change of the mass in the system divided by change in time equals mass rate in minus mass rate out.

The mass in the system at any time t is pMV/RTz, the mass rate in is $p_{sc}Mq_{in}/RT_{sc}$, and the mass rate out is $p_{sc}Mq_{out}/RT_{sc}$. V is the gas volume inside the chamber (drillpipe). If the chamber is initially empty, then V is initially equal to the drillpipe volume. Assuming Tz to be constant, the mass balance becomes

$$q_{in} - q_{out} = \frac{T_{sc}\left(V\mathrm{d}p/\mathrm{d}t + p\mathrm{d}V/\mathrm{d}t\right)}{p_{sc}Tz}. \quad \ldots\ldots\ldots\ldots\ldots \quad (8.20)$$

While the assumption of constant Tz is adequate for identifying reservoir fluids and for obtaining field estimates of flow rates, it can cause errors in calculated flow rates, as shown by Kabir *et al.*[23]

For a rising liquid level in the drillpipe during the flow period of the DST, V decreases as liquid enters the pipe. Thus, the term dV/dt is negative, and $-dV/dt$ represents the rate at which liquid enters the drillpipe. For a standard temperature of 60°F (520°R) and a base pressure of 14.65 psia, Eq. 8.20 becomes

$$q_{in} - q_{out} = \frac{286\left(Vdp/dt + pdV/dt\right)}{Tz}, \quad \ldots\ldots\ldots\ldots\ldots \quad (8.21)$$

where flow rate is in Mscf/D, volume is in barrels, and time is in minutes. Useful modeling equations can be derived from Eq. 8.21 for the following four cases.

1. Surface valve closed, tester valve open, only gas produced: $q_{out} = 0$, $\mathrm{d}V/\mathrm{d}t = 0$. Under these conditions, Eq. 8.21 reduces to

$$q_{in} = \frac{286V}{zT}\frac{\mathrm{d}p}{\mathrm{d}t}, \quad \ldots\ldots\ldots\ldots\ldots\ldots\ldots\ldots \quad (8.22)$$

which gives the flow rate of clean gas. In Eq. 8.22, T is the average flowing temperature of the gas in the chamber.

2. Surface valve closed, tester valve open, only water produced: $q_{out} = 0$, $q_{in} = 0$ and Eq. 8.21 reduces to

TABLE 8.5—WATER PRODUCTION AS A PERCENTAGE OF INITIAL CHAMBER VOLUME (AFTER ALEXANDER[22]).

Surface Pressure (psig)	Gas-Free Production (%)	Water Production With Complete Solution-Gas Breakout, %								
		A: 600 B: 1.05	1,000 1.58	1,500 2.10	2,000 2.50	3,000 3.29	4,000 3.95	5,000 4.48	6,000 4.87	8,000 5.53
0	0	0	0	0	0	0	0	0	0	0
1	6	3	3	2	2	2	1	1	1	1
2	11	6	5	4	4	3	3	2	2	2
3	16	9	7	6	6	5	4	4	3	3
4	21	12	10	8	7	6	5	5	4	4
5	25	14	12	10	9	7	6	6	5	5
6	28	17	14	12	10	9	8	7	7	6
7	32	19	16	13	12	10	9	8	8	7
8	35	21	17	15	13	11	10	9	8	8
9	37	23	19	17	15	12	11	10	9	9
10	40	25	21	18	16	14	12	11	10	9
15	50	33	28	25	23	19	17	16	15	14
20	57	40	35	31	28	24	22	20	19	17
25	62	45	40	35	33	28	26	24	22	21
30	67	50	44	40	37	32	29	27	26	24
35	70	54	48	43	40	36	32	30	29	27
40	73	57	51	47	44	39	35	33	32	29
45	75	60	54	50	47	42	38	36	34	32
50	77	62	57	52	49	44	41	38	37	34
60	80	67	61	57	54	49	45	43	41	38
80	84	73	68	64	61	56	52	50	48	45
100	87	77	73	69	66	61	58	55	54	51
150	91	83	80	77	74	70	67	65	63	61
200	93	87	84	81	80	76	73	71	70	68

A: Reservoir pressure, psia
B: Gas-water ratio, bbl/bbl

$$V\frac{dp}{dt} = -\overline{p_c}\frac{dV}{dt}, \qquad (8.23)$$

which relates the rate of water influx to the change in pressure. The average chamber pressure, $\overline{p_c}$, is related to surface pressure by

$$\overline{p_c} = \frac{p\left(e^{\gamma_g L/53.34Tz} - 1\right)}{\gamma_g L/53.34Tz}, \qquad (8.24)$$

where γ_g = gas-specific gravity (air = 1); L = height of the gas column in feet; and 53.34 is the universal gas constant, ft-lbf/lbm°R.

3. Surface valve open, tester valve closed, gas vented: $q_{in} = 0$, $dV/dt = 0$. Under these conditions, Eq. 8.21 reduces to

$$-q_{out} = \frac{286V}{zT}\frac{dp}{dt}. \qquad (8.25)$$

The volume of gas in the drillpipe at a particular time can be calculated by measuring q_{out} and the rate of pressure change, dp/dt. The volume of liquid in the drillpipe is equal to the difference between the drillpipe volume and the gas volume.

4. Surface valve closed, tester valve open, gas and liquid produced: $q_{out} = 0$, $dV/dt \neq 0$, and Eq. 8.21 reduces to

$$q_{in} = \frac{286}{zT}\left(V\frac{dp}{dt} + \overline{p_c}\frac{dV}{dt}\right). \qquad (8.26)$$

The gas volume, V, is determined at the end of the flow period. The rate dV/dt is calculated as an average rate during the flow period (implicitly assuming a constant rate during the flow period even though the flow rate during a DST declines with time). The use of Eq. 8.26 is limited to final calculations after the test has been completed and subsurface-pressure charts have been analyzed.

When only liquid (water or mud) is produced into the drillpipe during a flow period, the fraction of drillpipe fillup is

$$\frac{V_p}{V_c} = \frac{p_2 - p_1}{p_2 + 14.65}, \qquad (8.27)$$

where p_1 = surface pressure at the beginning of the flow period, psig; p_2 = surface pressure at the end of the flow period, psig; V_c = initial chamber volume, bbl; and V_p = volume of liquid produced into the chamber, bbl.

Table 8.5 gives the gas-free chamber fillup as a percentage for a range of surface pressures with $p_1 = 0$ psig.

If gas comes out of solution, then the fillup fraction is reduced from its gas-free value. Assuming that all solution gas is liberated, Eq. 8.18 becomes

$$\frac{V_p}{V_c} = \frac{p_2 - p_1}{\left[(p_1 + 14.65)R_s + (p_2 + 14.65)\right]}, \qquad (8.28)$$

where R_s = solution gas/liquid ratio (bbl/bbl). Table 8.5 gives values of V_p/V_c as a percentage with complete liberation of solution gas for a range of BHP's. For example, as Table 8.5 shows, in a well with gas-free liquid flowing into the sample chamber at a surface pressure of p_2 = 15 psig (initial surface pressure p_1 = 0 psig), the liquid influx during the flow period is equal to 50% of the original chamber volume by Eq. 8.27. If reservoir pressure is 2,000 psi and all the solution gas is liberated, then liquid influx is reduced to 23%. A recovery of less than 23% would indicate the presence of free gas. For an initial surface pressure, p_1, greater than zero, the values in Table 8.5 would differ in accordance with Eqs. 8.27 and 8.28.

By use of Eqs. 8.22 and 8.23, the range of the rates of change in surface pressure (dp/dt) that can be expected during a closed-chamber DST can be estimated as part of pretest planning. The highest and lowest expected rates of surface-pressure change are associated with production only of gas and only of water, respectively. If the rate of change in surface pressure is higher than the lowest expected rate, then the presence of oil or gas is likely. The following procedure is recommended for estimating the maximum expected rate of surface-pressure change if only gas is produced during a DST. The observed rate of change may be considerably less than the estimated value because of well damage, formation drawdown, or tool configuration.

TABLE 8.6—ORIFICE COEFFICIENTS FOR CRITICAL FLOW PROVER[24]

Orifice Size (in.)		Meter Coefficient, (Mscf-D)/psi	
Fraction	Decimal	2 in.	4 in.
1/16	0.0625	0.0848	
3/32	0.09375	0.1867	
1/8	0.125	0.3506	
3/16	0.1875	0.8052	
7/32	0.21875	1.1112	
¼	0.25	1.4390	1.387
5/16	0.3125	2.2130	
3/8	0.375	3.1484	3.117
7/16	0.4375	4.5123	
½	0.50	5.6647	5.576
5/8	0.625	8.5694	8.686
¾	0.750	12.5147	12.448
7/8	0.875	17.2112	16.927
1	1.0	22.6311	22.052
1 1/8	1.125	28.9803	27.778
1¼	1.250	36.5871	34.300
1 3/8	1.375	44.9506	41.295
1½	1.5	55.5750	49.208
1¾	1.75		67.220
2	2.0		88.811
2¼	2.25		113.851
2½	2.5		142.787
2¾	2.75		176.787
3	3.0		217.241

Analysis Procedure—Closed-Chamber Testing With Only Gas Production.

1. Specify the sizes of the orifice and meter to be used. The smallest orifice that will not become plugged should be used.
2. Estimate BHP with the best available data for the particular area.
3. Calculate the sample chamber volume in barrels.
4. Estimate bottomhole temperature and gas-compressibility factor.
5. Calculate the maximum gas flow rate, q_g, with the equation for a critical flow prover[24]:

$$q_g = C_m p_u F_{tf} F_g F_{pv}, \quad \text{(8.29)}$$

where C_m = orifice coefficient for critical flow prover (**Table 8.6**); p_u = upstream pressure, psia, equal to BHP; F_{tf} = flowing temperature factor = $\sqrt{519.67/T_m}$; T_m = temperature of gas flowing through meter, °R, equal to bottomhole temperature; F_g = gravity factor = $\sqrt{0.6/\gamma_g}$; and F_{pv} = gas-supercompressibility factor = $\sqrt{1/z}$.

6. Calculate the maximum expected rate of surface-pressure change (psi/min) from Eq. 8.22, rearranging and solving for dp/dt. The estimated average flowing temperature, rather than bottomhole temperature, is used in this calculation.

A similar procedure can be followed to estimate the minimum expected rate of surface-pressure change if only water is produced during a DST. An observed rate of change significantly higher than the estimated value is a good indication that oil or gas is present. A lower observed rate does not preclude the possibility of a very low rate of oil or gas production.

Analysis Procedure—Closed-Chamber Testing With Only Water Production.

1. Determine the size of the smallest choke that can be used without plugging.
2. Estimate BHP with the best available data for the particular area.
3. Calculate the sample chamber volume in barrels.
4. Calculate the maximum water flow rate with the equation for liquid flow through an orifice[22]:

$$-dV/dt = q_w = 0.35d^2\sqrt{p}, \quad \text{(8.30)}$$

where d = orifice diameter, in., and p = BHP, psig.

5. Calculate the minimum expected rate of surface-pressure change (psi/min) from Eq. 8.23, rearranging and solving for dp/dt. Assume that surface pressure (psia) is equal to the average pressure in the chamber.

Example 8.5—Pretest Planning Calculations—Closed-Chamber Testing. A well[22] is to be tested at a depth of 10,000 ft with drillpipe with a capacity of 0.0108 bbl/ft. Other known data are summarized next. Estimate the range of the rates of change in surface pressure that can be expected during a DST.

Local pressure gradient: p_{grad} = 0.45 psi/ft.
Local temperature gradient: T_{grad} = 1.6°F/100 ft.
Local mean surface temperature: T_{surf} = 60°F.
Atmospheric pressure: p_{atm} = 14.7 psia.
Local gas-specific gravity: γ_g = 0.65.
Local gas-compressibility factor: z = 0.9.

Solution. A 2-in. meter with a 3/8 in. orifice is used to measure flow rates during the DST. With the local pressure gradient of 0.45 psi/ft for wells in this area, BHP is estimated to be $p = (p_{grad})(L_c) = (0.45)(10{,}000) = 4{,}500$ psig, or in terms of inches of water, $h_w = p/(0.0361\,\gamma_w) = 4{,}500/[0.0361(1)] = 1.247 \times 10^5$ in., where γ_w = specific gravity of water and L_c = chamber length. The volume of the chamber is

$$V = (S_c)(L_c)$$

$$= (0.0108)(10{,}000)$$

$$= 108 \text{ bbl},$$

where S_c = chamber capacity.

The maximum rate of change in surface pressure during the initial flow period of the test will occur if only gas is produced. From the local temperature gradient of 1.6°F/100 ft, the bottomhole temperature is estimated to be

$$T = T_{surf} + (T_{grad})(\text{depth}) = 60 + (1.6/100)/(10{,}000) = 220°\text{F}.$$

From Eq. 8.29 and Table 8.6, the maximum gas flow rate is

$$C_m = 3.1484,$$

$$F_{tf} = \sqrt{519.67/(220 + 460)} = 0.874,$$

$$F_g = \sqrt{0.6/0.651} = 0.961,$$

$$F_{pv} = \sqrt{1/0.94} = 1.054,$$

and $q_g = (3.1484)(4500 + 14.7)(0.874)(0.961)(1.054) = 12{,}583$ Mscf/D.

The maximum expected rate of surface-pressure change (psi/min) is found by rearranging Eq. 8.22 and solving for dp/dt. In the following calculation, the average flowing temperature is estimated with the local temperature gradient at the midpoint of the chamber (5,000 ft), giving

$$T = 60 + (1.6/100)/(5{,}000) = 140°\text{F}.$$

With this average flowing temperature, dp/dt is

$$dp/dt = \frac{q_{in}Tz}{286V}$$

$$= \frac{(12{,}583)(140 + 460)(0.9)}{(286)(108)}$$

$$= 220 \text{ psi/min}.$$

TABLE 8.7—CLOSED-CHAMBER DST DATA, EXAMPLE 8.6

Initial Flow Period		
Cumulative Flow Time (minutes)	Manometer Pressure (in. of water)	Surface Pressure (psig)
0	0.0	0.0000
1	1.2	0.0433
2	2.3	0.0830
3	2.5	0.0903
4	2.6	0.0939
5	3.6	0.1300
6	3.8	0.1372
7	3.9	0.1408
8	4.0	0.1444
9	4.1	0.1480
10	4.2	0.1516

Final Flow Period		
Cumulative Flow Time (minutes)	Surface Pressure (psig)	Pressure-Buildup Rate, $\Delta p/\Delta t$ (psi/min)
0	0.0	
5	2.4	0.48
10	3.6	0.24
15	4.5	0.18
20	5.6	0.22
25	7.0	0.28
30	8.0	0.20
35	9.3	0.26
40	10.5	0.24
45	11.6	0.22
50	13.2	0.32
55	14.5	0.26
60	16.0	0.30

The minimum rate of change in surface pressure will occur if only water is produced during the flow period. By use of Eq. 8.30, the maximum water flow rate is calculated to be

$$-\,dV/dt = q_w = 0.35(3/8)^2\sqrt{4500} = 3.30\text{ bbl/min}.$$

The average chamber pressure is

$$\bar{p} = \frac{p\left(e^{\gamma_g L/53.34\,\bar{T}z} - 1\right)}{\gamma_g L/53.34\bar{T}z}$$

$$= \frac{(14.7)\left\{e^{[(0.65)(10,000)]/[(53.34)(140+460)(0.9)]} - 1\right\}}{[(0.65)(10,000)]/[(53.34)(140 + 460)(0.9)]}$$

$$= 16.49\text{ psia},$$

where $\bar{T}$ = average flowing temperature. Rearranging Eq. 8.23, the rate of change in surface pressure if only water is produced during the initial flow period of the DST is

$$dp/dt = \frac{-\,\bar{p}dV/dt}{V} = \frac{(16.49)(3.30)}{108} = 0.504\text{ psi/min}.$$

Any gas coming out of solution would increase this rate slightly; therefore, for purposes of preplanning estimation the rate may simply be doubled to 1 psi/min.

In summary, the rate of change in surface pressure during the initial flow period of the DST is expected not to exceed 234 psi/min. If observed pressure changes are significantly more than 1 psi/min, then either oil or gas probably is present.

Closed-Chamber Test Procedure. The following procedure,[22] which is based on field experience, is recommended for closed-chamber DST's.

1. Initial flow period. Like the pretest planning procedure, calculate the maximum and minimum expected rates of surface-pressure change if gas or water, respectively, is encountered during the test. Connect a water manometer to the pressure-gauge manifold. Open the tool with only the bubble hose open at the surface. When bubbles appear in the bucket, shut off the bubble hose and begin timing with a stopwatch. Observe and record pressures every minute. Terminate the initial flow period after 5 minutes or when pressure reaches a predetermined safe operating level. Record the duration of the initial flow period.

2. Initial shut-in period. Shut in the well for at least 1 hour. Wait until surface pressure stablizes. If surface pressure is greater than 15 psig, install a critical-flow prover with a $^1/_{16}$-in. orifice plate on the blowdown line. Vent the critical-flow prover to the atmosphere and record chamber pressures for a few minutes. The volume of free gas and hence the volume of liquid produced then can be calculated with Eqs. 8.25 and 8.29 as described. Review the test objectives and plan the remainder of the test (additional flow and shut-in periods) accordingly.

Example 8.6—Analysis of a Closed-Chamber DST. A closed-chamber DST consisting of two flow periods and two shut-in periods was run in a newly drilled well.[22] The objectives of the test were to identify formation fluids and to obtain flow-rate data to be used in determining the necessary shut-in time required to obtain satisfactory pressure transient data. The total liquid recovered after the final flow period was 1.8 bbl of filtrate and muddy water. The blowdown consisted of only air, while the initial chamber volume was 49.7 bbl. **Table 8.7** gives time and pressure data for the two flow periods; the durations of the flow and shut-in periods are summarized next. With the flow-period data, identify the formation fluids and, if gas was produced during the test, determine the flow rate: initial flow period = 10 minutes; initial shut-in period = 90 minutes; final-flow period = 60 minutes; final shut-in period = 120 minutes.

Solution.

1. Pretest planning calculations similar to those illustrated in Example 8.5 indicated that only liquid entered the drillpipe during the initial flow period. Rearranging Eq. 8.27, the liquid influx into the chamber during the initial flow period is estimated to be

$$V_p = \frac{V_c(p_2 - p_1)}{(p_2 + 14.65)}$$

$$= \frac{(49.7)(0.1516 - 0)}{(0.1516 + 14.65)}$$

$$= 0.51\text{ bbl}.$$

2. We solve for the volume of gas in the chamber after the final-flow period, expressed at standard conditions, to determine how much gas entered the chamber during the final flow period. At the beginning of the final flow period, the chamber volume was equal to the initial volume less the liquid influx during the initial flow period; i.e., 49.7 − 0.5 = 49.2 bbl. Given the total recovery of 1.8 bbl, 1.3 bbl of liquid was produced during the final flow period. The volumes of gas produced during the initial flow period can be calculated from the real gas law, expressed in the form

$$p_{sc}V_{2,sc}/T_{sc}z_{sc} = p_2V_2/T_2z_2.$$

At a standard temperature of 60°F (520°R) and an initial pressure of p_1 = 14.65 psia, the number of standard cubic feet in the drillpipe at the end of the test can be calculated. Assuming the average chamber temperature to be 90°F, then the values of z_{sc} and z_2 are 1.0 and 0.98, respectively, and the initial volume of gas in the chamber, expressed at standard conditions, is

$$V_{2,sc} = \frac{p_2V_2T_{sc}z_{sc}}{p_{sc}T_2z_2}$$

$$= \frac{(16 + 14.65)(49.7 - 1.8)(60 + 460)(1)}{(14.65)(90 + 460)(0.98)}$$

$= 96.7$ bbl.

The drillpipe contained $49.7 - 0.51 = 49.2$ bbl of liquid at the end of the initial flow period or at the beginning of the final flow period. Thus, the volume of gas produced into the chamber during the final flow period was $96.7 - 49.2 = 47.5$ bbl or 267 scf. Because this volume of gas was produced during a period of 1 hour, the average flow rate is equal to

$$\overline{q}_g = \frac{(267)(24)}{(1,000)} = 6.4 \text{ Mscf/D}.$$

From Eq. 8.26 and assuming that the average chamber pressure is equal to the surface pressure, the instantaneous flow rate at the end of the final flow period is estimated to be

$$q_{in} = \frac{286\left(V\mathrm{d}p/\mathrm{d}t + \overline{p}_c \mathrm{d}V/\mathrm{d}t\right)}{Tz}$$

$$= \left(286\left\{(49.7 - 1.8)(0.30) + (16 + 14.65) \times [(0 - 1.8)/(60 - 0)]\right\}\right)/[(90 + 460)(0.98)]$$

$$= 7.1 \text{ Mscf/D},$$

where the rate of surface-pressure change, $\mathrm{d}p/\mathrm{d}t$, at the end of the final flow period is equal to 0.30 psi/min from Table 8.7. Had the test been run as a conventional DST, the gas production might have been missed.

8.7 Impulse Testing

Impulse testing[25,26] is a method suitable for testing wells that do not flow to the surface during a test or for which extended flow is undesirable. The formation is subjected to an impulse generated by a short flow or injection period, followed by a buildup or falloff period. Accurate measurements of the pressure response with time and of the total quantity of fluid produced or injected are required. Pressure-change and pressure-derivative type curves[10-12] are used to interpret the pressure response. The impulse method permits a low-cost evaluation of the reservoir and wellbore condition before final completion.

The impulse-testing technique method is particularly useful for analyzing buildup data following a short flow period (no more than a few minutes). The initial flow period of a DST is short enough to be considered an impulse. During a tubing-conveyed perforating operation, perforating when the well is underbalanced results in a rapid influx of fluid into the tubing, followed by shut-in after a short time. A backsurge operation also creates a high flow rate into the wellbore for a short period. All three situations are well-suited to analysis by the impulse-testing method.

When a formation is subjected to an instantaneous impulse, the resulting pressure response over time is given by Green's functions.[27] The functions are the time derivatives of the dimensionless pressure solutions developed as type curves.[28-31] The pressure response, not its derivative, will match the pressure-derivative curve. In practice, the impulse rate is not instantaneous, so buildup or falloff data will not follow the pressure-derivative response until the duration of the impulse, t_p, becomes small compared with the shut-in time, t. For this condition or when the impulse is instantaneous, the pressure response during the shut-in period is given by[25]

$$(p_i - p)\Delta t = \frac{3388.8 Q_t \mu}{kh} p_D', \quad \text{(8.31)}$$

where Q_t = the total quantity of fluid (reservoir barrels) produced or injected. The pressure response during the impulse (flow) period is given by

$$(p_i - p)t_p = \frac{3388.8 Q_t \mu}{kh} p_D. \quad \text{(8.32)}$$

Therefore, a log-log plot of the pressure/time group ($\Delta p t_p$, psi-hours) as a function of time during the impulse period will follow the pressure-change type-curve response, p_D, while the pressure/time group ($\Delta p \Delta t$, psi-hours) during the subsequent shut-in period will follow the pressure-derivative type-curve response, p_D'.

The behavioral model of a reservoir will be apparent in the pressure response during the impulse period (flow or injection) and the subsequent shut-in period (buildup or falloff). The pressure response during the shut-in period will exhibit the derivative behavior characteristic of a particular model. Chap. 4 discusses the pressure-derivative characteristics of various reservoir models.

Theoretically, an impulse generates a signal that contains all the information characterizing a reservoir. In practice, however, the resolution capabilities of the pressure gauge limit how much of the signal can be read. The higher the transmissibility (kh/μ) of a reservoir, the larger the quantity of flowing fluid required to create a given amplitude of pressure response. Hence, impulse testing may not work well in very-high-transmissibility reservoirs.

Example 8.7—Analysis of an Impulse Test. Fig. 8.11 shows an impulse-test analysis[25] of a tubing-conveyed perforating operation in an unconsolidated sand. The well was perforated while underbalanced, allowed to flow 17 bbl in 9 minutes, and then shut in. The following reservoir and fluid properties were known:

$c_t = 12.6 \times 10^{-6}$ psi^{-1}
$\mu = 0.39$ cp
$h = 35$ ft
$\phi = 0.25$
$r_w = 0.29$ ft
$p_i = 5,074$ psia
$C = 0.13$ bbl/psi

Solution.

1. The following type-curve correlating parameter and pressure and time match points were obtained from the match shown in Fig. 8.11:

Correlating parameter: $C_D e^{2s} = 1.5 \times 10^4$.

Pressure match point: $\left(\frac{p_D'}{\Delta p \Delta t}\right)_{MP} = 0.44.$

Time match point: $\left(\frac{t_D/C_D}{\Delta t}\right)_{MP} = 54.$

2. If we rearrange Eq. 8.31, we can calculate permeability from the type-curve analysis:

$$k = \frac{3,388.8 Q_t \mu}{h}\left(\frac{p_D'}{\Delta p \Delta t}\right)_{MP}$$

$$= \frac{(3,388.8)(17)(0.39)}{35}(0.44)$$

$$= 282 \text{ md}.$$

3. Next, calculate the dimensionless wellbore-storage coefficient, C_D, with Eq. 8.6.

$$C_D = \frac{0.8936C}{\phi c_t h r_w^2}$$

$$= \frac{(0.8936)(0.13)}{(0.25)(0.39)(12.6 \times 10^{-6})(0.29)^2}$$

$$= 1.1 \times 10^6.$$

4. The skin factor is estimated with the type-curve correlating parameter, $C_D e^{2s}$, and the dimensionless wellbore-storage coefficient, C_D.

$$s = 0.5 \ln\left[\frac{(C_D e^{2s})_{MP}}{C_D}\right]$$

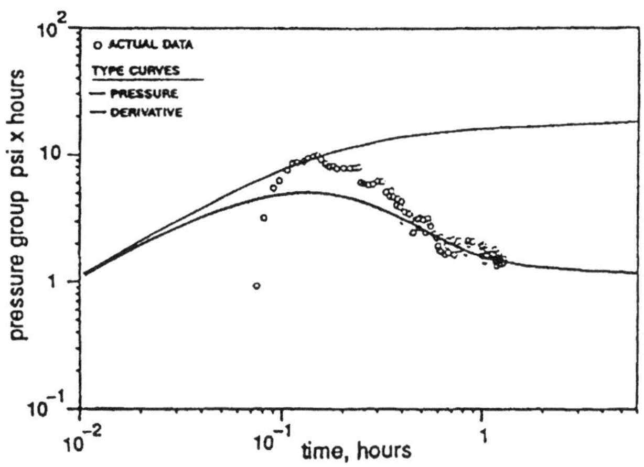

Fig. 8.11—Impulse analysis, tubing-conveyed perforating operation (after Ayoub *et al.*[25]).

$$= 0.5 \ln\left(\frac{1.5 \times 10^4}{1.1 \times 10^6}\right)$$

$$= -2.1.$$

The negative skin factor indicates effective perforation cleanup of drilling fluids. In addition, a possible improvement in permeability was obtained near the wellbore.

8.8 Chapter Summary

In this chapter, we discussed DST design, monitoring, and analysis procedures.

Sec. 8.2 describes a conventional DST. DST's are run to obtain one or more of the following: (1) identification of reservoir fluid; (2) estimate of well productivity; and (3) estimates of formation permeability, skin factor, and static reservoir pressure. A conventional DST consists of two flow periods and two shut-in periods. The DST tool is first run into the hole. The valves are then open for a short flow period, usually about 5 minutes. At the end of the first flow period, the valves are closed for a buildup test, which should last 1 hour or more to obtain a good estimate of true static formation pressure. A second flow period and shut-in period are then obtained. The second flow period typically lasts from 30 minutes to several hours. The final shut-in period typically lasts one to two times as long as the final flow period. After the final shut-in period, the DST tool is pulled out of the hole and the test chart recovered and analyzed.

Sec. 8.3 discusses guidelines for designing a conventional DST. Mechanical considerations include the height of the test zone, the use of a cushion, choice of test tools and packers, hole conditions, availability of packer seats, number of pressure and temperature gauges, and isolation of the tested interval. We also presented guidelines for determining test duration. Finally, we illustrated the use of these guidelines with two example problems.

Sec. 8.4 outlines procedures for monitoring a DST. Three methods currently in use include (in increasing order of sophistication and expense) the bubble-bucket system, the closed-chamber system, and the surface-pressure-readout system.

Sec. 8.5 presents techniques for analyzing DST's. Analysis methods, such as Horner analysis and conventional type-curve analysis may be used to analyze DST's where fluids reach the surface during the flow period and the flow rate remains constant. Unfortunately, fluids do not usually reach the surface during a typical DST. Instead, a typical test exhibits a declining flow rate and increasing flow pressure during the flow periods as the hydrostatic head of the fluid column in the drillstring increases.

In Sec. 8.5.1, we discuss the use of the Ramey *et al.* type curves for DST's in which flow does not reach the surface, or "slug" tests. This method allows us to analyze the pressure data obtained during the flow period of a DST. These type curves are based on solutions to the diffusivity equation that gives the pressure reponse to the instantaneous production or injection of a specific volume of fluid at time zero. The type curves also include the effects of wellbore storage and skin factor. The solutions are presented in terms of a dimensionless pressure ratio as a function of the dimensionless time group t_D/C_D. The matching parameter is C_De^{2s}. The dimensionless pressure ratio is calculated directly from field pressure data. A match is then obtained by sliding the field data horizontally until the best match is found. We then illustrated the use of the Ramey type curves through analysis of an example DST.

Sec. 8.5.2 presents the Peres *et al.* method for analyzing DST's. This method provides a way to analyze the pressure data obtained during the shut-in period. The Peres *et al.* method takes into account producing time and variable flow rate during the flow period through the use of a multirate time function. The Peres *et al.* method provides estimates of formation permeability k, skin factor s, and initial reservoir pressure p_i. We illustrated the Peres *et al.* method with an example.

In Sec. 8.6, we discuss the closed-chamber DST. This is a method of monitoring the initial flow period of a DST to indentify produced fluids and to provide estimates of gas and liquid flow rates during the initial flow period. Information gained during the closed-chamber test can be used to establish minimum times for the remaining flow and shut-in periods. The closed-chamber DST is based on a mass balance of fluids in the wellbore. We can estimate the maximum pressure change to be expected assuming that either gas only or water only is produced. If the pressure rises faster than predicted for water production only, then either oil or gas is flowing into the wellbore. After the test is completed, we can estimate the flow rate from the pressures measured during the test. We illustrate both pre- and post-calculations for closed-chamber DST analysis with examples.

Sec. 8.7 describes impulse testing. In an impulse test, the formation is subjected to a short flow or injection period followed by a much longer shut-in period. To use the impulse-test method, the quantity of fluid produced or injected must be known. The pressure response is analyzed with conventional pressure- and pressure-derivative type curves. However, the pressure response during the shutin period is matched to the derivative type curve. To close the section, we illustrate the impulse-test-analysis method with an example problem.

Exercises

1. A development well has been drilled in an existing field. The productive interval is a naturally fractured limestone at 4,300–4,370 ft. Total depth is 4,500 ft. Casing has not yet been set. The drillpipe is 4½ in. (3.958 in. ID) 13.75 lb/ft internal upset pipe with a collapse rating of 7,200 psi. The wellbore is filled with 9.2 ppg mud. Temperature at the midpoint of the interval is 135°F, and pressure at the midpoint of interval is 1975 psi. Assuming no water cushion is required, design a 12 hour drillstem test to evaluate the interval. The objectives of the drillstem test are to identify formation fluids and to obtain pressure transient data for formation evaluation.

2. A development well has been drilled in an existing field. The productive interval is a homogeneous sandstone at 6,170–6,230 ft. Total depth is 6,475 ft. Casing has not yet been set. The drillpipe is 3½ in. (2.992 in. ID) 9.50 lb/ft internal upset pipe with a collapse rating of 10,040 psi. The wellbore is filled with 13 ppg mud. Temperature at the midpoint of the interval is 183°F, and pressure at the midpoint of interval is 4,650 psi. Assuming no water cushion is required, design a 6 hour drillstem test to evaluate the interval. The objectives of the drillstem test are to identify formation fluids and to obtain pressure transient data for formation evaluation.

3. A development well has been drilled in an existing field. The productive interval is a naturally fractured carbonate at 11,780–11,840 ft. Total depth is 11,950 ft. Casing has not yet been set. The drillpipe is 4½ in. (3.958 in. ID) 13.75 lb/ft internal upset pipe with a collapse rating of 7,200 psi. The wellbore is filled with 10.5 ppg mud. Temperature at the midpoint of the interval is 210°F, and pressure at the midpoint of interval is 5110 psi. Determine whether or not a water cushion is required, then design a 12 hour drillstem test to evaluate the interval. The objectives of the drillstem

test are to identify formation fluids and to obtain pressure transient data for formation evaluation.

4. It is often desired to keep the reservoir fluid above the bubble point pressure during a test so that flow occurs under single-phase conditions. Assuming that the bubble point pressure in Exercise 2 is 3767 psi, design a drillstem test to evaluate the interval while using a water cushion to maintain pressure above the bubble point pressure.

5. A drillstem test was conducted with a 4-hr flow period followed by a 6-hour buildup period. Use the Ramey, *et al.* type curves to analyze the flow period data given in the table below. Estimate permeability and skin factor. $p_i = 5321.66$ psi; $h = 18$ ft; $c_t = 14.1 \times 10^{-6}$ psi^{-1}; $p_o = 2475.93$ psi; $\phi = 12.4\%$; $\mu = 0.46$ cp; $r_w = 0.33$ ft; $C = 0.0722$ bbl/psi.

PRESSURE-TIME DATA FOR EXERCISE 8.5

Time (hours)	Pressure (psi)	Time (hours)	Pressure (psi)	Time (hours)	Pressure (psi)
0.0	2475.93	0.0838	2663.05	0.772	3656.01
0.0120	2507.07	0.1106	2716.00	1.019	3895.77
0.0158	2516.30	0.1460	2783.00	1.346	4153.01
0.0209	2528.22	0.193	2867.16	1.776	4414.47
0.0276	2543.60	0.254	2971.89	2.34	4662.56
0.0364	2563.39	0.336	3100.69	3.09	4876.02
0.0481	2588.82	0.443	3256.66	4.00	5027.82
0.0635	2621.42	0.585	3441.81		

6. A drillstem test was conducted with a 5-hr flow period followed by a 5-hour buildup period. Use the Ramey, *et al.* type curves to analyze the flow period data given in the table below. Estimate permeability and skin factor. $p_i = 2342.3$ psi; $h = 25$ ft; $c_t = 12.4 \times 10^{-6}$ psi^{-1}; $p_o = 576.43$ psi; $\phi = 21.1\%$; $\mu = 0.76$ cp; $r_w = 0.22$ ft; $C = 0.068$ bbl/psi.

PRESSURE-TIME DATA FOR EXERCISE 8.6

Time (hours)	Pressure (psi)	Time (hours)	Pressure (psi)	Time (hours)	Pressure (psi)
0.0	576.43	0.0645	639.15	0.785	1170.46
0.0070	583.70	0.0852	658.18	1.036	1309.11
0.0092	585.97	0.1124	682.80	1.368	1465.35
0.0122	588.94	0.1484	714.50	1.805	1633.56
0.0161	592.83	0.1959	755.09	2.38	1804.33
0.0213	597.90	0.259	806.71	3.15	1965.04
0.0281	604.54	0.341	871.72	4.15	2102.34
0.0370	613.20	0.451	952.62	5.00	2173.03
0.0489	624.47	0.595	1051.67		

7. Use the Peres, *et al.* method to analyze the buildup period data for the test in Exercise 5. The buildup data are given in the table below. Estimate permeability and skin factor. Compare your results to those from Exercise 5. $p_i = 5321.66$ psi; $h = 18$ ft; $c_t = 14.1 \times 10^{-6}$ psi^{-1}; $p_o = 2475.93$ psi; $\phi = 12.4\%$; $\mu = 0.46$ cp; $r_w = 0.33$ ft; $C_F = 0.0822$ bbl/psi.

PRESSURE-TIME DATA FOR EXERCISE 8.7

Time (hours)	Pressure (psi)	Time (hours)	Pressure (psi)	Time (hours)	Pressure (psi)
4.00100	5057.85	4.0542	5203.58	4.674	5247.86
4.00225	5085.78	4.0688	5207.40	4.843	5252.73
4.00381	5111.07	4.0869	5211.12	5.055	5257.76
4.00577	5132.88	4.1097	5214.81	5.319	5262.91
4.00821	5150.78	4.1381	5218.52	5.650	5268.13
4.01126	5164.82	4.1736	5222.30	6.06	5273.35
4.01507	5175.53	4.218	5226.18	6.58	5278.50
4.01984	5183.66	4.274	5230.18	7.23	5283.50
4.0258	5189.98	4.343	5234.33	8.03	5288.27
4.0333	5195.12	4.430	5238.66	9.04	5292.75
4.0426	5199.55	4.538	5243.16	10.00	5295.98

8. Use the Peres, *et al.* method to analyze the buildup period data for the test in Exercise 6. The buildup data are given in the table for this exercise. Estimate permeability and skin factor. Compare your results to those from Exercise 6. $p_i = 2342.3$ psi; $h = 25$ ft; $c_t = 12.4 \times 10^{-6}$ psi^{-1}; $p_o = 576.43$ psi; $\phi = 21.1\%$; $\mu = 0.76$ cp; $r_w = 0.22$ ft; $C_F = 0.068$ bbl/psi.

PRESSURE-TIME DATA FOR EXERCISE 8.8

Time (hours)	Pressure (psi)	Time (hours)	Pressure (psi)	Time (hours)	Pressure (psi)
5.00100	2207.18	5.0542	2308.65	5.674	2320.65
5.00225	2236.86	5.0688	2309.66	5.843	2321.98
5.00381	2260.71	5.0869	2310.67	6.055	2323.34
5.00577	2278.16	5.1097	2311.67	6.319	2324.75
5.00821	2289.80	5.1381	2312.69	6.650	2326.19
5.01126	2296.96	5.1736	2313.72	7.06	2327.64
5.01507	2301.16	5.218	2314.77	7.58	2329.09
5.01984	2303.66	5.274	2315.86	8.23	2330.52
5.0258	2305.29	5.343	2316.99	9.03	2331.90
5.0333	2306.53	5.430	2318.16	10.00	2333.16
5.0426	2307.62	5.538	2319.38		

9. A well is to be tested at a depth of 8209 ft using drillpipe with a capacity of 0.0087 bbl/ft. Other known data are summarized below. Estimate the range of the rates of change in surface pressure that can be expected during a drillstem test. A 2-in. meter with a 5/16 in. orifice will be used to measure flow rates. $p_{atm} = 14.7$ psi; $t_{surf} = 69°F$; $\gamma_g = 0.83$ (air = 1.0); $p_{grad} = 0.72$ psi/ft; $t_{grad} = 1.12°F/100$ ft; $z = 1.08$.

10. A well is to be tested at a depth of 9145 ft using drillpipe with a capacity of 0.0152 bbl/ft. Other known data are summarized below. Estimate the range of the rates of change in surface pressure that can be expected during a drillstem test. A 2-in. meter with a 3/16 in. orifice will be used to measure flow rates. $p_{atm} = 14.7$ psi; $t_{surf} = 54°F$; $\gamma_g = 0.62$ (air = 1.0); $p_{grad} = 0.57$ psi/ft; $t_{grad} = 1.31°F/100$ ft; $z = 1.02$.

11. The following pressures were recorded for the well in Exercise 9 during the first flow period. Is there significant influx of gas or oil during the test?

PRESSURE-TIME DATA FOR EXERCISE 8.11

Time (min)	Pressure (psig)	Time (min)	Pressure (psig)	Time (min)	Pressure (psig)
0	0	6	114	12	228
3	57	9	171	15	285

12. No liquids were recovered during the test in Exercise 11. Estimate the average gas flow rate during the flow period.

13. The following pressures were recorded for the well in Exercise 10 during the two flow periods. Total fluid recovery was 16.7 bbl. Estimate the fluid produced during the first flow period, then find the average gas flow rate during the final flow period.

Initial flow period:

PRESSURE-TIME DATA FOR EXERCISE 8.13

Time (min)	Pressure (psig)	Time (min).	Pressure (psig)	Time (min)	Pressure (psig)
0	0.00	4	0.15	8	0.30
2	0.07	6	0.22	10	0.37

Final flow period:

PRESSURE-TIME DATA FOR EXERCISE 8.13

Time (min)	Pressure (psig)	Time (min)	Pressure (psig)	Time (min)	Pressure (psig)
0	0.00	24	6.48	48	12.96
12	3.24	36	9.72	60	16.20

14. An impulse test was conducted on a new well using a tubing-conveyed perforation gun. The well was perforated underbalanced, and allowed to flow 40 bbl in 6 minutes, then shut in. Analyze the pressure data given below to estimate formation permeability and

skin factor. $p_i = 2930.1$ psi; $h = 13$ ft; $c_t = 10.7 \times 10^{-6}$ psi^{-1}; $\phi = 22.8\%$; $\mu = 0.90$ cp; $r_w = 0.43$ ft; $C = 0.000522$ bbl/psi.

PRESSURE-TIME DATA FOR EXERCISE 8.14					
Time (hours)	Pressure (psi)	Time (hours)	Pressure (psi)	Time (hours)	Pressure (psi)
0.0010	2892.50	0.0405	1978.43	0.222	2833.91
0.0021	2852.35	0.0456	1914.77	0.245	2864.36
0.0033	2809.56	0.0512	1853.41	0.271	2884.64
0.0046	2764.08	0.0573	1794.95	0.299	2897.99
0.0061	2715.88	0.0640	1739.96	0.330	2906.70
0.0077	2664.99	0.0714	1688.91	0.364	2912.41
0.0095	2611.43	0.0795	1642.15	0.401	2916.18
0.0114	2555.30	0.0885	1599.91	0.443	2918.73
0.0136	2496.74	0.0983	1562.29	0.488	2920.52
0.0159	2435.95	0.1092	1838.87	0.538	2921.83
0.0185	2373.18	0.1211	2115.77	0.592	2922.85
0.0214	2308.77	0.1342	2335.77	0.653	2923.68
0.0245	2243.11	0.1486	2505.33	0.719	2924.37
0.0280	2176.69	0.1645	2632.01	0.792	2924.97
0.0318	2110.03	0.1819	2723.95	0.872	2925.50
0.0360	2043.74	0.201	2788.97	0.960	2925.97

References

1. Matthews, C.S. and Russell, D.G.: *Pressure Buildup and Flow Tests in Wells*, Monograph Series, SPE, Richardson, Texas (1977) **1,** 84–91.
2. Earlougher, R.C. Jr.: *Advances in Well Test Analysis*, Monograph Series, SPE, Richardson, Texas (1977) **5,** 90–104.
3. Lee, W.J.: *Well Testing*, Textbook Series, SPE, Richardson, Texas (1977) **1,** 97–98.
4. Edwards, A.G. and Winn, R.H.: "A Summary of Modern Tools and Techniques Used in Drill Stem Testing," Halliburton Services, Duncan, Oklahoma (May 1974) 6.
5. Erdle, J.C.: "Current Drillstem Testing Practices: Design, Conduct and Interpretation," paper SPE 13182 presented at the 1984 SPE Annual Technical Conference and Exhibition, Houston, 16–19 September.
6. Gatlin, C.: *Petroleum Engineering*, Prentice-Hall Inc., Englewood Cliffs, New Jersey (1960) 71–74.
7. Peres, A., Onur, M., and Reynolds, A.: "New Methods for the Analysis of Drillstem Test Data," paper SPE 19843 presented at the 1989 SPE Annual Technical Conference and Exhibition, San Antonio, 8–11 October.
8. Correa, A.C. and Ramey, H.J. Jr.: "A Method for Pressure Buildup Analysis of Drillstem Tests," paper SPE 16802 presented at the 1987 SPE Annual Technical Conference and Exhibition, Dallas, 27–30 September.
9. Ramey, H.J. Jr., Agarwal, R., and Martin, I.: "Analysis of 'Slug Test' or DST Flow Period Data," *J. Cdn. Pet. Tech.* (July–September 1975) 37.
10. Bourdet, D. *et al.*: "A New Set of Type Curves Simplifies Well Test Analysis," *World Oil* (May 1983) 95.
11. Bourdet, D. *et al.*: "Interpreting Well Tests in Fractured Reservoirs," *World Oil* (October 1983) 77.
12. Bourdet, D. *et al.*: "New Type-Curves for Tests of Fissured Formations," *World Oil* (April 1984).
13. van Poollen, H.K. and Weber, J.B.: "Data Analysis for High Influx Wells," paper SPE 3017 presented at the 1970 SPE Annual Fall Meeting, Houston, 4–7 October.
14. Kohlhaas, C.A.: "A Method for Analyzing Pressures Measured During Drillstem-Test Flow Periods," *JPT* (October 1972) 1278; *Trans.*, AIME, **253.**
15. Ramey, H.J. Jr. and Agarwal, R.G.: "Annulus Unloading Rates as Influenced by Wellbore Storage and Skin Effect," *SPEJ* (October 1972) 453; *Trans.*, AIME, **253.**
16. Onur, M. and Reynolds, A.C.: "A New Approach for Constructing Type Curves for Well Test Analysis," *SPEFE* (March 1988) 197.
17. Peres, A., Onur, M., and Reynolds, A.C.: "A New General Pressure Analysis Procedure for Slug Tests," paper SPE 18801 presented at the 1989 SPE California Regional Meeting, Bakersfield, California, 5–7 April.
18. Gringarten, A.C. *et al.*: "A Comparison Between Different Skin and Wellbore Storage Type Curves for Early-Time Transient Analysis," paper SPE 8205 presented at the 1979 SPE Annual Technical Conference and Exhibition, Las Vegas, Nevada, 23–26 September.
19. Grader, A.S. and Ramey, H.J. Jr.: "Slug-Test Analysis in Double-Porosity Reservoirs," *SPEFE* (June 1988) 329.
20. Rushing, J.A. *et al.*: "Analysis of Slug Test Data From Hydraulically Fractured Coalbed Methane Wells," paper SPE 21492 presented at the 1991 SPE Gas Technology Symposium, Houston, 23–25 January.
21. Karsaki, K., Long, J.C.S., and Witherspoon, P.A.: "Analytical Models of Slug Tests," *Water Resources Research J.* (1988) **24,** No. 1, 115.
22. Alexander, L.G.: "Theory and Practice of the Closed-Chamber Drillstem Test Method," *JPT* (December 1977) 1539.
23. Kabir, C.S., Badry, R.A., and Hasan, A.R.: "Two-Phase Gas/Liquid Flow Rate Estimation During Closed-Chamber Testing," *J. Cdn. Pet. Tech.* (March–April 1991) 102.
24. Jennings, J.W. *et al.*: "Deliverability Testing of Natural Gas Wells, State of Texas," Texas Railroad Commission report, Petroleum Engineering Dept., Texas A&M U., College Station, Texas (August 1989) 13–14, 118.
25. Ayoub, J.A., Bourdet, D.P., and Chauvel, Y.L.: "Impulse Testing," *SPEFE* (September 1988) 534.
26. Simmons, J.F.: "Interpretation of Underbalanced Surge Pressure Data by Rate/Time Convolution," paper SPE 19843 presented at the 1986 SPE Annual Technical Conference and Exhibition, New Orleans, 5–8 October.
27. Carslaw, H.S. and Jaeger, J.C.: *Conduction of Heats in Solids*, second edition, Oxford U. Press, New York City (1959).
28. Johnson, C.R., Greenkorn, R.A., and Woods, E.G.: "Pulse Testing: A New Method for Describing Reservoir Flow Properties Between Wells," *JPT* (December 1966) 1599; *Trans.*, AIME, **237.**
29. Houze, O.P., Horne, P., and Ramey, H.J. Jr.: "Pressure-Transient Response of an Infinite-Conductivity Vertical Fracture in a Reservoir With Double-Porosity Behavior," *SPEFE* (September 1988) 510.
30. Cooper, H.H. Jr., Bredehoeft, J.D., and Papadopulos, I.S.: "Response of a Finite-Diameter Well to an Instantaneous Charge of Water," *Water Resources Research*, (1967) **3,** No. 1, 263.
31. Erdle, J.C., Upchurch, J.M., and Warren, D.A.: "Early Fluid Entry Determination: Key to Safe, Optimum Drillstem Testing," paper SPE 6884 presented at the 1977 SPE Annual Technical Conference and Exhibition, Denver, Colorado, 9–12 October.

Chapter 9
Injection-Well Testing

9.1 Overview

In many reservoirs, especially in mature waterflooded fields, the injection wells may be as numerous as the producing wells. Consequently, testing injection wells is particularly important for efficient planning and operation of both secondary and tertiary recovery projects. Satisfactory injection performance over a long period of time and prompt detection of increasing wellbore damage are important to the economics of any recovery project. This chapter begins with a discussion of injectivity and falloff testing in liquid-filled, unit-mobility-ratio reservoirs. Then, we address methods for determining average reservoir pressure and for analyzing composite systems in nonunit-mobility-ratio reservoirs. Finally, we discuss well testing and analysis with the step-rate method. We base much of the material discussed in this chapter on the material from Ref. 1.

9.2 Injectivity Testing in a Liquid-Filled Reservoir: Unit-Mobility-Ratio Reservoir Conditions

Craig[2] defines the mobility ratio during an injection process as the ratio of the mobility of the displacing fluid (i.e., the injected fluid) to the fluid being displaced. In equation form, the mobility ratio for a water injection into an oil reservoir is

$$M = \frac{(k/\mu)_w}{(k/\mu)_o}. \quad (9.1)$$

The value of the mobility ratio often is used as an indicator of the displacement efficiency. For example, in a waterflood project in an oil reservoir, a mobility ratio of less than one suggests an efficient, "piston-like" oil-displacement process, while mobility ratios greater than one may result in inefficient displacement.

When the mobility ratio between the injected and in-situ fluids is near unity, the analysis techniques for injection tests are similar to those developed for production tests. The unit-mobility-ratio approximation applies not only in mature waterfloods that initially had mobility ratios significantly different from unity, but also early in the life of tertiary recovery projects when so little fluid has been injected that it appears only as a skin effect. When the unit-mobility-ratio condition is satisfied, injection-well testing for liquid-filled systems is analogous to a drawdown test, while shutting in an injection well results in a pressure falloff that is analogous to a pressure-buildup test. Therefore, the equations for production-well testing apply to injection-well testing as long as proper sign conventions are observed.

Injectivity testing is pressure transient testing during injection into a well and is analogous to drawdown testing for both constant and variable injection rates.[1] For results of the analysis methods presented in this section to be valid, the mobility ratio of the injected and in-situ fluids must be near unity. If the mobility ratio is not unity but the radius of investigation has not exceeded the radius of the injected-fluid bank, the effective permeability and skin factor (but not static drainage-area pressure) in the inner zone still can be determined.

Fig. 9.1 depicts an idealized injectivity test. Before the test, the well is shut in to allow pressure to stabilize at the initial reservoir pressure, p_i. The test begins at time $t=0$ with a constant injection rate, q, where by convention q is negative. If the injection rate varies significantly during the test, a variable-rate analysis method similar to that discussed in Chap. 2 must be used. Because unit-mobility-ratio injection-well testing is analogous to production-well testing, analysis methods for drawdown and multirate testing can be applied directly to injection-well testing.

For the constant-rate injectivity test illustrated in Fig. 9.1, the injection bottomhole pressure (BHP), p_{wf}, can be modeled by the logarithmic approximation to the *Ei*-function solution (discussed in Chap. 2),

$$p_{wf} = p_i - \frac{162.6(-q)B\mu}{kh}\left[\log\left(\frac{kt}{\phi\mu c_t r_w^2}\right) - 3.23 + 0.869s\right]. \quad (9.2)$$

Eq. 9.2 describes the variation of BHP as a function of time during a constant-rate injection test. Like the analysis of pressure-drawdown tests, a plot of BHP, p_{wf}, vs. the logarithm of injection time, t, forms a straight line from which the slope, m, allows us to estimate formation permeability, k, and skin factor, s.

Like the semilog analysis technique for constant-rate drawdown tests presented in Chaps. 2 and 3, the effective permeability in the drainage area of the well is

$$k = -\frac{162.6(-q)B\mu}{mh}. \quad (9.3)$$

We also can develop an expression for the skin factor, s, by combining Eqs. 9.2 and 9.3,

$$s = 1.151\left[\left(\frac{p_{1\text{hr}} - p_i}{m}\right) - \log\left(\frac{kt}{\phi\mu_w c_t r_w^2}\right) + 3.23\right], \quad (9.4)$$

where $p_{1\text{hr}}$ = BHP at an injection time of 1 hour and necessarily lies either on the semilog straight line or its extrapolation. **Fig. 9.2** illus-

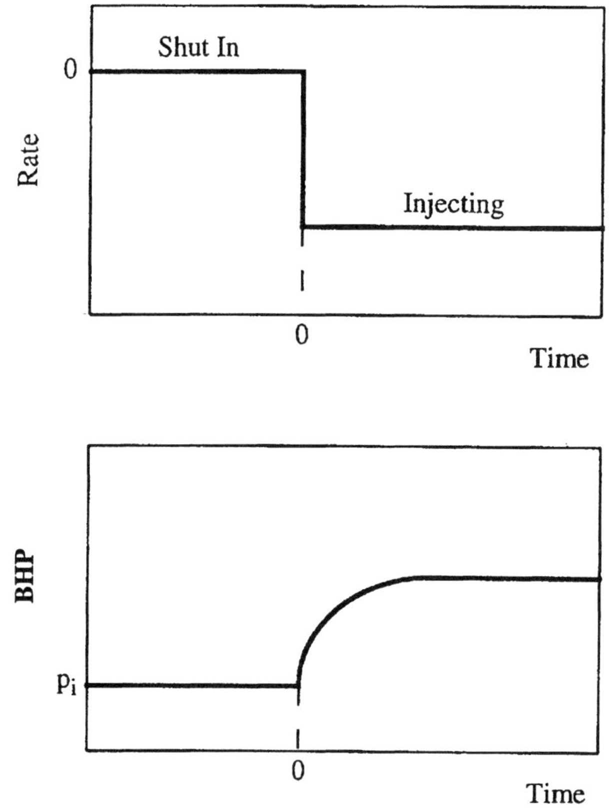

Fig. 9.1—Idealized injectivity test (after Earlougher[1]).

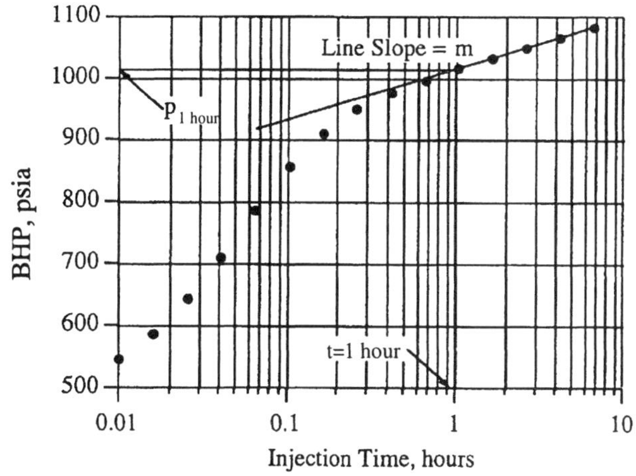

Fig. 9.2—Graphical semilog analysis technique for constant-rate injection test data.

trates the graphical semilog analysis technique for constant-rate injection tests.

Similar to drawdown testing, wellbore storage may have a significant effect on the pressure response during an injection-well test. In some cases, the reservoir pressure may be low enough so that a free liquid surface exists in a shut-in well. In that case, the wellbore-storage coefficient is relatively large and is given by

$$C = \frac{25.65 A_{wb}}{\rho_{wb}}, \quad (9.5)$$

where A_{wb} = area of the wellbore, ft^2, and ρ_{wb} = density of the fluid in the wellbore, lbm/ft^3.

Wellbore storage causes the BHP response to fall below the extrapolated semilog straight line during the duration of storage (Fig. 9.2). The time to the end of wellbore storage, t_{wbs} (hours), or the beginning of the semilog straight line is estimated from

$$t_{wbs} \cong \frac{(200,000 + 12,000s)C}{(kh/\mu)}. \quad (9.6)$$

Example 9.1—Analyzing an Injectivity Test in a Liquid-Filled Reservoir: Unit-Mobility-Ratio Reservoir Conditions. A constant-rate water-injectivity well test was conducted in a waterflooded reservoir. Before the test, all other wells in the reservoir had been shut in for several weeks and the reservoir pressure had stabilized. **Table 9.1** summarizes injectivity-test data. Assuming a unit-mobility ratio, determine k and s.

$q = -100$ STB/D
$p_i = 449$ psia
$r_w = 0.25$ ft
$\rho_{wb} = 62.4$ lbm/ft^3
$h = 16$ ft
$\phi = 0.15$
$c_t = 7.7\ 10^{-6}$ psi^{-1}
$\mu = 1.0$ cp
$B = 1.0$ RB/STB

TABLE 9.1—CONSTANT-RATE-INJECTION TEST DATA, EXAMPLE 9.1

Time (hours)	Pressure (psia)	Time (hours)	Pressure (psia)
0.0100	546.30	0.4120	976.40
0.0159	587.30	0.6560	997.80
0.0253	642.30	1.040	1,016.5
0.0403	711.00	1.660	1,034.2
0.0642	785.60	2.640	1,051.2
0.1020	855.20	4.210	1,068.0
0.1630	910.20	6.700	1,084.5
0.2590	949.10		

Solution. For an injectivity test with unit-mobility-ratio conditions, we follow a systematic procedure similar to that described in Chaps. 2 and 4 for a constant-rate drawdown test.

Construct Data Plots. The first step in our analysis is to construct appropriate data plots. **Fig. 9.3** shows a log-log plot of the pressure change and pressure derivative vs. injection time, while **Fig. 9.4** gives a semilog plot. **Table 9.2** summarizes both log-log and semilog plotting functions.

Qualitative Type-Curve Analysis. Similar to the analysis of a constant-rate drawdown test, we first perform a qualitative type-curve analysis. The purpose of this analysis is to obtain an initial match and to use type curves to identify early-, middle-, and late-time flow regimes.

With the Bourdet *et al.*[3] type curve for radial flow with wellbore storage and skin, a good match is the $C_D e^{2s} = 1 \times 10^5$ curve (interpolating in Fig. 9.3). The pressure-derivative data become horizontal at

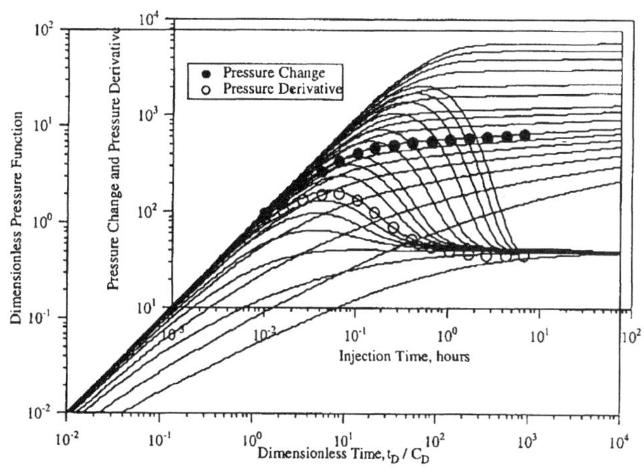

Fig. 9.3—Qualitative type-curve analysis with the Bourdet *et al.*[3] type curve, Example 9.1.

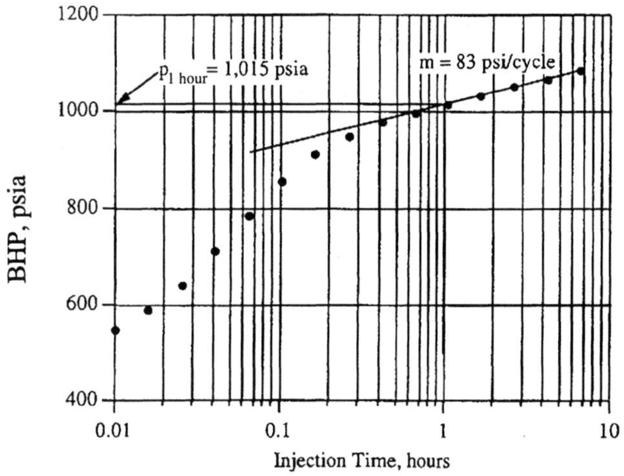

Fig. 9.4—Semilog analysis, Example 9.1.

the end of the test and can be matched with $p_D'(t_D/C_D) = 0.5$, which indicates the semilog straight line begins around 0.5 hour.

In addition, the pressure-derivative data at the end of the test do not deviate from the type curve, indicating that the reservoir is infinite-acting throughout the test. Because injectivity-test data are available during the infinite-acting time period, a semilog analysis is possible.

Semilog Analysis.

1. The semilog straight line in Fig. 9.4 begins at 0.53 hour and continues for the remainder of the test. The semilog straight line, which is consistent with the results of the preliminary type-curve analysis, has a slope

$$m = \frac{p_{ws2} - p_{ws1}}{\log t_2 - \log t_1} = \frac{1{,}084.5 - 1{,}034.2}{\log(6.70) - \log(1.66)} = 83.0 \text{ psi/cycle.}$$

2. The permeability is estimated with the slope of the semilog line,

$$k = -\frac{162.6qB\mu}{mh} = -\frac{(162.6)(-100)(1.0)(1.0)}{(83.0)(16)} = 12.2 \text{ md.}$$

3. The skin factor is

$$s = 1.151\left[\left(\frac{p_{1\text{hr}} - p_i}{m}\right) - \log\left(\frac{k}{\phi\mu c_t r_w^2}\right) + 3.23\right]$$

$$= 1.151\left\{\left(\frac{1{,}015 - 449}{83.0}\right)\right.$$

$$\left. - \log\left[\frac{12.2}{(0.15)(1.0)(7.7 \times 10^{-6})(0.25)^2}\right] + 3.23\right\}$$

$$= 2.1,$$

where $p_{1\text{hr}}$ = pressure corresponding to $\Delta t = 1$ hour from the semilog straight line, and p_i = BHP before the beginning of the test ($\Delta t = 0$) and is equal to the initial drainage-area pressure.

Quantitative Type-Curve Analysis. We now perform a quantitative type-curve analysis (**Fig. 9.5**) to confirm the results of the semilog analysis or, if necessary, to refine the semilog analysis to obtain consistency between results from the two analyses.

1. Using $k = 12.2$ md from the semilog analysis, precalculate a pressure-match point. Choosing a convenient value of $p_D = 10$, we calculate the corresponding pressure change, Δp,

$$(\Delta p)_{MP} = -\frac{141.2(-q)B\mu}{kh}(p_D)_{MP}$$

TABLE 9.2—SEMILOG AND LOG-LOG PLOTTING FUNCTIONS FOR ANALYZING INJECTIVITY-TEST DATA, EXAMPLE 9.1

Time (hours)	Pressure (psia)	Pressure Change (psi)	Pressure Derivative
0.0100	546.3	97.4	88.41
0.0159	587.3	138.4	103.4
0.0253	642.3	193.4	132.9
0.0403	711.0	262.1	153.9
0.0642	785.6	336.7	155.3
0.1020	855.2	406.3	133.93
0.1630	910.2	461.3	100.6
0.2590	949.1	500.2	71.4
0.4120	976.4	527.5	52.4
0.6560	997.8	548.9	43.3
1.040	1,016.5	567.6	39.2
1.660	1,034.2	585.3	37.2
2.640	1,051.2	602.3	36.3
4.210	1,068.0	619.1	35.8
6.700	1,084.5	635.6	35.5

$$= -\frac{(141.2)(-100)(1.0)(1.0)}{(12.2)(16)} = 723 \text{ psi.}$$

2. The type-curve match is found by aligning the precalculated pressure-match point and sliding the pressure change and pressure-derivative curves horizontally. With interpolation, the best fit is obtained with the type curve defined by the correlating parameter $C_De^{2s} = 2.0 \times 10^4$, which is slightly different from the qualitative type-curve analysis.

3. We select the following time-match point.

$\Delta t = 0.8$ hour and $t_D/C_D = 100$.

4. By use of the time-match point, the dimensionless wellbore-storage coefficient is

$$C_D = \frac{0.0002637k}{\phi r_w^2 c_t \mu}\left(\frac{\Delta t}{t_D/C_D}\right)_{MP}$$

$$= \frac{(0.0002637)(12.2)}{(0.15)(0.25)^2(7.7 \times 10^{-6})(1.0)}\left(\frac{1}{125.6}\right)$$

$$= 355.$$

5. The skin factor is calculated with C_D from Step 4 and the correlating parameter, $C_De^{2s} = 2.0 \times 10^4$, from the type-curve match.

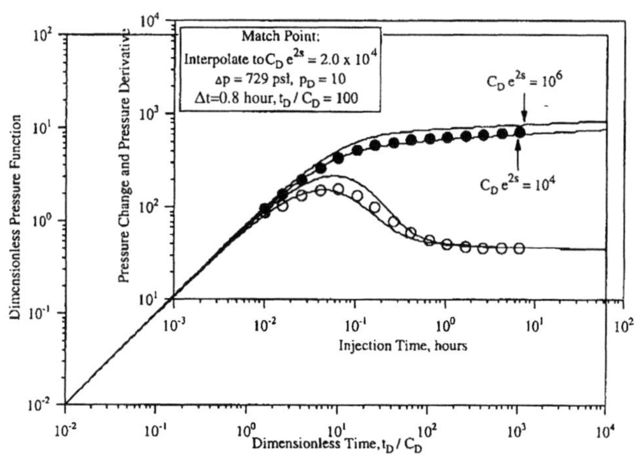

Fig. 9.5—Qualitative type-curve analysis with the Bourdet *et al.*[3] type curve, Example 9.1.

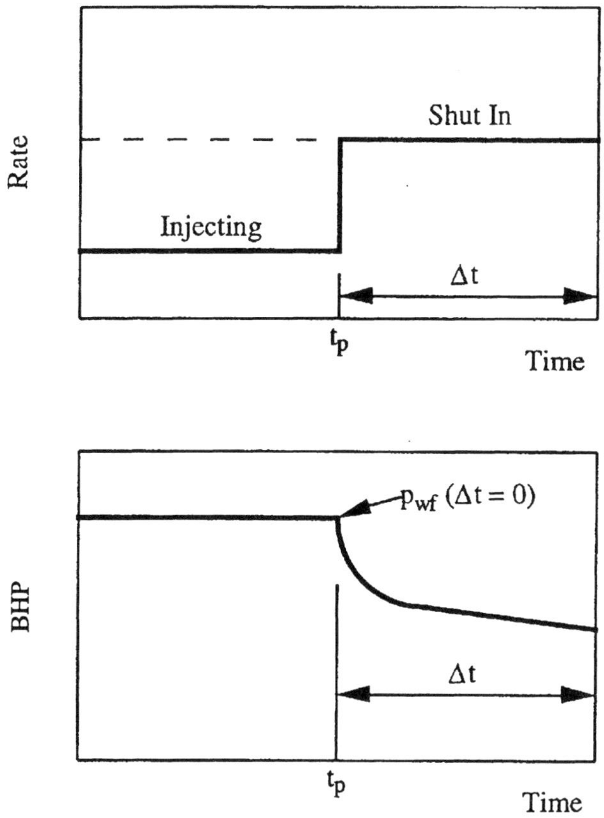

Fig. 9.6—Idealized falloff test (after Earlougher[1]).

$$s = 0.5\ln\left(\frac{C_D e^{2s}}{C_D}\right) = 0.5\ln\left(\frac{2.0 \times 10^4}{355}\right) = 2.0,$$

which agrees with $s = 2.1$ from the semilog analysis, thus confirming our semilog analysis.

6. We can estimate the radius of investigation achieved at various times before and during the injectivity test.

The radius of investigation, $r_{i,\text{start}}$, at the beginning of the semilog straight line is

$$r_{i,\text{start}} = \left(\frac{kt}{948\phi\mu c_t}\right)^{1/2} = \left[\frac{(12.2)(0.53)}{(948)(0.15)(1.0)(7.7 \times 10^{-6})}\right]^{1/2}$$

$$= 77 \text{ ft.}$$

The radius of investigation, $r_{i,\text{end}}$, at the end of the injectivity test is

$$r_{i,\text{end}} = \left(\frac{kt}{948\phi\mu c_t}\right)^{1/2} = \left[\frac{(12.2)(6.7)}{(948)(0.15)(1.0)(7.7 \times 10^{-6})}\right]^{1/2}$$

$$= 273 \text{ ft.}$$

7. We now determine whether the assumption of unit-mobility-ratio conditions, which was the basis of our analysis, applies. A volumetric balance provides an estimate of the distance to the front of the water bank, r_{wb}. The volume of water injected, W_i, is

$$W_i = \frac{\pi r_{wb}^2 h\phi\Delta S_w}{5.615}.$$

The distance to the water bank, r_{wb}, is

$$r_{wb} = \left(\frac{5.615 W_i}{\pi h\phi\Delta S_w}\right)^{1/2}.$$

Coreflooding tests suggest that $\Delta S_w = 0.4$. If the waterflooding has been under way for 2 years, the volume of injected water is

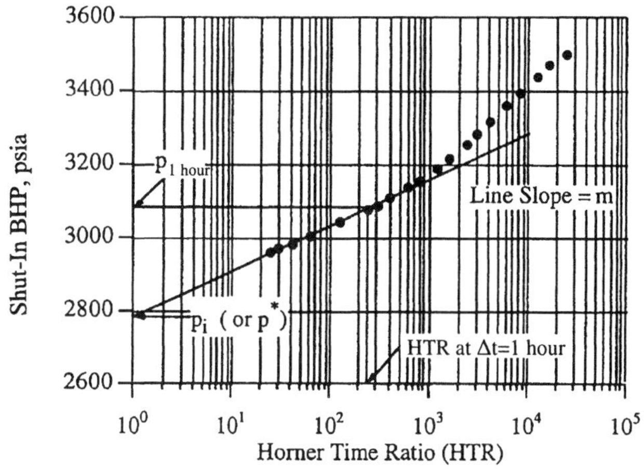

Fig. 9.7—Horner[4] semilog analysis of a pressure-falloff test.

$W_i = (100 \text{ STB/D})(1.0 \text{ RB/STB})(2 \text{ years})(365 \text{ D/yr}) = 73{,}000 \text{ RB}.$

The distance to the water bank is

$$r_{wb} = \left(\frac{5.615 W_i}{\pi h\phi\Delta S_w}\right)^{1/2} = \left[\frac{(5.615)(73{,}000)}{\pi(16)(0.15)(0.4)}\right]^{1/2} = 369 \text{ ft.}$$

Because the distance to the front of the water bank is greater than the radius of investigation achieved during the test, we are justified in assuming the unit-mobility-ratio conditions. In addition, the results of the analysis indicated that no boundary effects had distorted the data at the end of the test.

Note that, if the other wells in the reservoir had been active, then interference effects may have distorted the data at the end of the test, making semilog analysis incorrect or impossible.

9.3 Falloff Testing in a Liquid-Filled Reservoir: Unit-Mobility-Ratio Reservoir Conditions

Falloff testing in an injection well is analogous to pressure-buildup testing in a production well.[1] An idealized falloff test, which **Fig. 9.6** shows, is conducted by injecting at a constant rate before shut-in at time t_p. Pressure data taken immediately before and during the shut-in period are analyzed with the methods developed for pressure-buildup tests.

For both infinite-acting and finite reservoirs, the pressure-falloff behavior can be modeled by

$$p_{ws} = p^* - m\log\left(\frac{t_p + \Delta t}{\Delta t}\right). \qquad (9.7)$$

As **Fig. 9.7** shows, a Horner[4] plot of shut-in BHP, p_{ws}, vs. the logarithm of the Horner time ratio, $(t_p + \Delta t)/\Delta t$, should exhibit a straight line with intercept p^* at infinite shut-in time (i.e., at a Horner time of 1 and with slope m). The formation permeability is computed from the slope of this line.

$$k = -\frac{162.6(-q)B\mu}{mh}. \qquad (9.3)$$

The extrapolated pressure, p^*, is equivalent to the initial pressure, p_i, for an infinite-acting reservoir with little or no pressure depletion. As in pressure-buildup testing, if the injection rate varies before the falloff test, the equivalent injection time[4] can be approximated with a pseudoproduction time,[4] t_p,

$$t_p = \frac{24V_p}{q}, \qquad (9.8)$$

where V_p = cumulative volume injected since the last pressure equalization (not cumulative injection since the well was put on injection) and q = constant rate just before shut-in. If t_p is greater than about twice t_{pss}, the time to reach pseudosteady state, t_{pss} should be used rather than t_p.[5,6]

TABLE 9.3—PRESSURE-FALLOFF-TEST DATA, EXAMPLE 9.2			
Time (hours)	Pressure (psia)	Time (hours)	Pressure (psia)
0.01	3,500	0.3004	3,155
0.015	3,470	0.4007	3,137
0.02	3,442	0.6015	3,112
0.03	3,397	0.8027	3,091
0.04	3,362	1.0042	3,075
0.06	3,317	2.0168	3,042
0.08	3,282	4.0678	3,007
0.10	3,257	6.1538	2,982
0.1501	3,216	8.2759	2,972
0.2002	3,187	10.435	2,962

We also must ensure that the duration of the injection period before the falloff test exceeds the duration of wellbore storage during injection. The minimum injection time is estimated with

$$t_{p,\min} = \frac{\phi \mu c_t r_w^2}{0.0002637k}(120 + 7s)C_D. \quad \ldots\ldots\ldots\ldots\ldots \quad (9.9)$$

To analyze a pressure-falloff test in a liquid-filled reservoir with unit-mobility-ratio conditions, we can use the semilog and type-curve analysis procedures presented in Chaps. 2 and 4 for pressure-buildup tests. Example 9.2 illustrates these procedures.

Example 9.2—Falloff-Test Analysis in a Liquid-Filled Reservoir: Unit-Mobility-Ratio Reservoir Conditions. Following an injection period of 240 hours, a falloff test was conducted on a brine-disposal well. Using the data summarized below and in **Table 9.3**, determine k and s.

t_p = 240 hours
q_w = −807 STB/D
$\bar{p}$ = 2,788 psia
μ_w = 1.0 cp
B_w = 1.0 RB/STB
c_t = 1.18×10^{-5} psi^{-1}
ϕ = 0.25
h = 28 ft
r_w = 0.4 ft
depth = 4,819 ft
A = 20 acres = 871,200 ft^2

Solution. *Construct Data Plots.* Like the analysis procedures presented in previous chapters, we first construct the appropriate data plots. **Fig. 9.8** gives a log-log plot of pressure change and pressure derivative vs. Agarwal's[7] equivalent time, while **Fig. 9.9** shows a Horner semilog plot. **Table 9.4** summarizes semilog and log-log plotting functions.

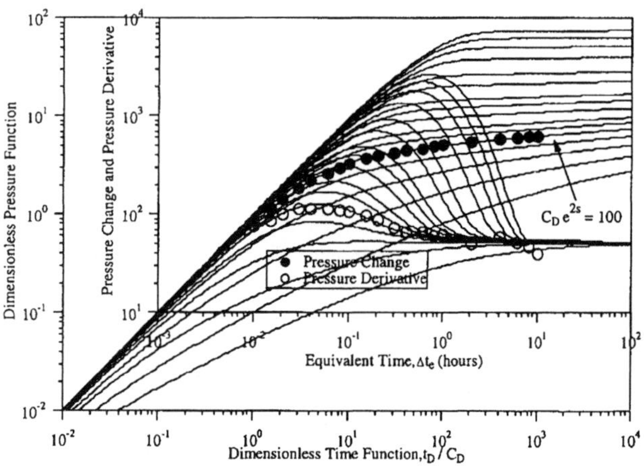

Fig. 9.8—Qualitative type-curve analysis with the Bourdet *et al.*[3] model, Example 9.2.

TABLE 9.4—SEMILOG AND LOG-LOG PLOTTING FUNCTIONS, EXAMPLE 9.2				
Equivalent Time (hours)	Pressure Change (psi)	Pressure Derivative	Pressure (psia)	Horner Time Ratio
0.010	82	73.99	3,500	24,001.0
0.015	112	85.56	3,470	16,001.0
0.020	140	100.91	3,442	12,001.0
0.030	185	112.6	3,397	8,001.0
0.040	220	112.6	3,362	6,001.0
0.060	265	113.9	3,317	4,001.0
0.080	300	109.9	3,282	3,001.0
0.100	325	108.3	3,257	2,401.0
0.150	366	96.2	3,216	1,600.0
0.200	395	87.0	3,187	1,200.0
0.300	427	72.7	3,155	799.9
0.400	445	65.5	3,137	600.0
0.600	470	66.4	3,112	400.0
0.800	491	60.8	3,091	299.9
1.00	507	61.9	3,075	240.0
2.00	540	49.1	3,042	120.0
4.00	575	57.5	3,007	60.0
6.00	600	51.7	2,982	40.0
8.00	610	46.2	2,972	30.0
10.0	620	39.2	2,962	24.0

Qualitative Type-Curve Analysis. A qualitative type-curve analysis (Fig. 9.8) that uses the Bourdet *et al.*[3] type curve for radial flow with wellbore storage and skin indicates a good match is with the $C_De^{2s} = 10^2$ curve. The trend of the pressure-derivative data suggests that the semilog straight line begins at approximately 0.7 hour. Boundary effects appear at about 8 hours, indicated by the deviation of the pressure-derivative data.

Horner Semilog Analysis.

1. Fig. 9.9 shows the Horner analysis. The beginning of the semilog straight line occurs at a Horner time of 412, which corresponds to an equivalent shut-in time $\Delta t_e = 0.58$ hour. Boundary effects are indicated by the deviation from the semilog straight line at a Horner time ratio of between 30 and 40, or between equivalent shut-in times of 6 and 8 hours. The slope of the semilog straight line is

$$m = \frac{2,982 - 3,091}{\log(40) - \log(299.99)} = 124.7 \text{ psi/cycle}.$$

2. The permeability is calculated with the slope of the semilog straight line.

$$k = -\frac{162.6(-q)B\mu}{mh} = -\frac{(162.6)(-807)(1.0)(1.0)}{(124.7)(28)}$$

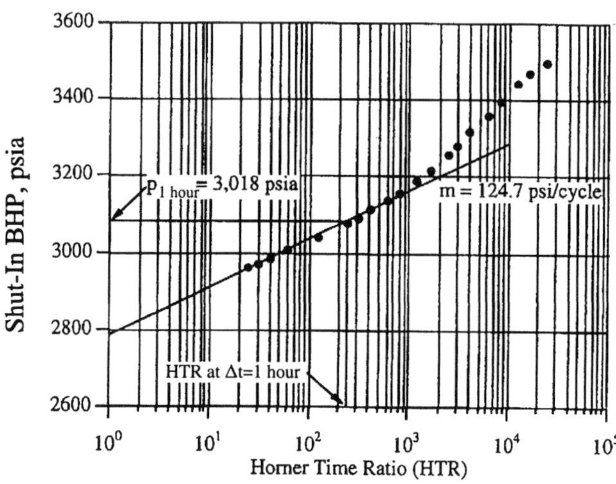

Fig. 9.9—Horner semilog analysis, Example 9.2.

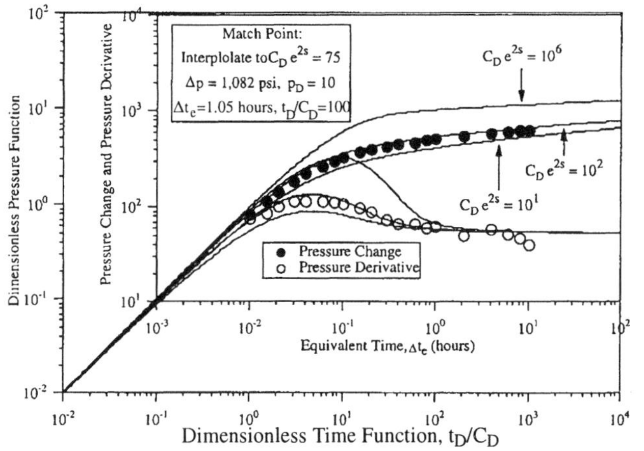

Fig. 9.10—Qualitative type-curve analysis with the Bourdet *et al.*[3] model, Example 9.2.

$= 37.6$ md.

3. The skin factor is

$$s = 1.151\left[\left(\frac{p_{wf} - p_{1\text{hr}}}{m}\right) - \log\left(\frac{k}{\phi\mu c_t r_w^2}\right) + 3.23\right]$$

$$= 1.151\left\{\left(\frac{3,582 - 3,081}{124.7}\right)\right.$$

$$\left. - \log\left[\frac{37.6}{(0.25)(1.0)(1.18\times10^{-5})(0.4)^2}\right] + 3.23\right\}$$

$$= -0.75,$$

where $p_{1\text{hr}}$ = pressure corresponding to $\Delta t = 1$ hour from the semilog straight line, and p_{wf} = BHP just before the beginning of the test ($\Delta t = 0$).

Quantitative Type-Curve Analysis.

1. Choosing a convenient value of $p_D = 10$, we precalculate the corresponding pressure change.

$$(\Delta p)_{MP} = -\frac{141.2(-q)B\mu}{kh}(p_D)_{MP}$$

$$= -\frac{(141.2)(-807)(1.0)}{(37.6)(28)}(10)$$

$$= 1,082 \text{ psi}.$$

2. The type-curve match (**Fig. 9.10**) is obtained by aligning the precalculated pressure-match point and sliding the pressure-change and pressure-derivative data horizontally while honoring the pressure-match point. A good match is obtained with the interpolated type curve described by the correlating parameter $C_De^{2s} = 75$, which is slightly different from the qualitative curve match.

3. With the type-curve and test data in place, we select the following time-match point:

$\Delta t_e = 1.05$ hours and $t_D/C_D = 100$.

4. Using the time match point, calculate the dimensionless wellbore-storage coefficient.

$$C_D = \frac{0.0002637k}{\phi r_w^2 c_t\mu}\left(\frac{\Delta t_e}{t_D/C_D}\right)_{MP}$$

$$= \frac{(0.0002637)(37.6)}{(0.25)(0.4)^2(1.18\times10^{-5})(1.0)}\left(\frac{1.05}{100}\right)$$

$$= 221.$$

5. The skin factor is calculated with C_D from Step 4 and the correlating parameter, $C_De^{2s} = 75$, from the type-curve match.

$$s = 0.5\ln\left(\frac{C_De^{2s}}{C_D}\right) = 0.5\ln\left(\frac{75}{221}\right) = -0.54.$$

The skin factor calculated from the type-curve match agrees with the value from the semilog analysis, confirming the validity of our results.

6. We must ensure that the duration of the injection period before the test exceeded the duration of wellbore storage during the injection period. From Eq. 9.9, the minimum injection time is

$$t_{p,\min} = \frac{\phi\mu c_t r_w^2}{0.0002637k}(120 + 7s)C_D$$

$$= \frac{(0.25)(1.0)(1.8\times10^{-5})(0.4)^2}{0.0002637(37.6)}$$

$$\times\{[120 + (7)(-0.74)](221)\}$$

$$= 1.2 \text{ hours},$$

which is less than the injection time before the test, $t_p = 240$ hours. Thus, the test results are valid. If $t_{p,\min}$ had exceeded t_p, we would have had to run the test again with a longer pretest injection period.

7. We can estimate the radius of investigation achieved at various times before and during the falloff test.

The radius of investigation achieved during the injection period before the start of the falloff test is

$$r_{i,inj} = \left(\frac{kt_p}{948\phi\mu_w c_t}\right)^{½} = \left[\frac{(37.6)(240)}{948(0.25)(1.0)(1.18\times10^{-5})}\right]^{½}$$

$$= 1,799 \text{ ft}.$$

At the beginning of the semilog straight line,

$$r_{i,\text{start}} = \left(\frac{kt_p}{948\phi\mu_w c_t}\right)^{½} = \left[\frac{(37.6)(0.583)}{948(0.25)(1.0)(1.18\times10^{-5})}\right]^{½}$$

$$= 89 \text{ ft}.$$

Taking the end of the semilog straight line to be 6.2 hours (the time when the data begin to deviate from the straight line), the radius of investigation at the end of the semilog straight line is

$$r_{i,\text{end}} = \left(\frac{kt_p}{948\phi\mu_w c_t}\right)^{½} = \left[\frac{(37.6)(6.2)}{948(0.25)(1.0)(1.18\times10^{-5})}\right]^{½}$$

$$= 289 \text{ ft}.$$

At the end of the falloff test,

$$r_{i,\text{final}} = \left(\frac{kt_p}{948\phi\mu_w c_t}\right)^{½} = \left[\frac{(37.6)(10.435)}{948(0.25)(1.0)(1.18\times10^{-5})}\right]^{½}$$

$$= 375 \text{ ft}.$$

Because the well is a brine-disposal well, not an oil well, the mobility of the injected fluid is the same as the fluid in the formation. Therefore, because the radius of investigation achieved during the falloff test (375 ft) is less than that achieved during the injection period (1,799 ft), we are confident that the last two data points of the test represent reservoir boundaries or heterogeneities. If $r_{i,inj}$ were less than $r_{i,\text{end}}$, the interpretation of these data as representing boundaries or

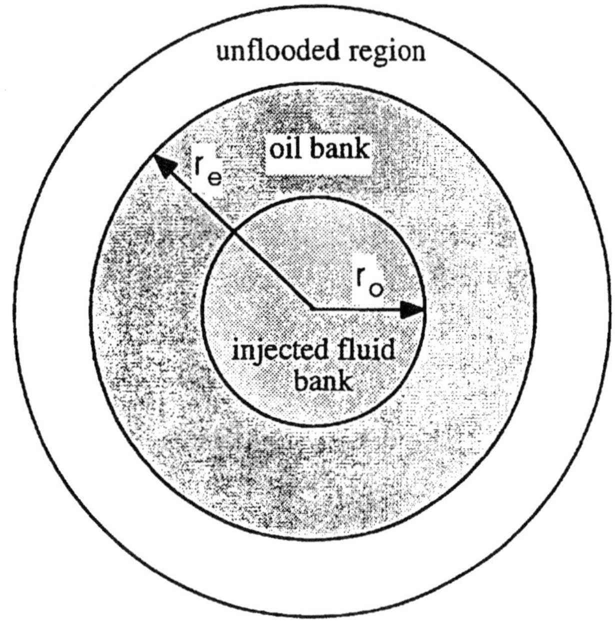

Fig. 9.11—Schematic diagram of composite reservoir system.

heterogeneities would be highly questionable because the falloff during the test cannot sample reservoir properties beyond the distance reached by the pressure transient during injection.

9.4 Estimating Average Drainage-Area Pressure

In a liquid-filled reservoir of uniform mobility and $\phi c_t h$, p^* is obtained by extrapolating the straight-line portion of the Horner plot from a falloff test to $(t_p + \Delta t)/\Delta t = 1$. In new wells or wells with short injection times (infinite-acting), $p^* \approx p_i$. For a finite reservoir, however, average drainage-area pressure is obtained by correcting p^* with the Matthews *et al.*[8] method. Average drainage-area pressure, $\bar{p}$, is[5,6,8]

$$p = p^* - \frac{m p_{DMBH}(t_{pDA})}{2.3025}, \quad (9.10)$$

where the dimensionless producing time is defined by

$$t_{pDA} = \frac{0.0002637 k t_p}{\phi \mu c_t A}. \quad (9.11)$$

Eq. 9.10 should be applied judiciously with an injectivity test to estimate average pressure in the drainage area of a well. According to Earlougher,[1] Eq. 9.10 is strictly applicable only under the following conditions: (1) a homogeneous-acting, single-layer reservoir with a single-phase, slightly compressible liquid; (2) no large contrasts in values of $\phi c_t h$ of the injected and original fluid; and (3) no hydraulic fractures.

9.5 Composite-System-Test Analysis for Nonunit-Mobility-Ratio Reservoir Conditions

When the unit-mobility-ratio assumption is not satisfied, the problem becomes one of a composite system. The analogy between production-well testing and injection-well testing is then less complete than in the unit-mobility-ratio case. Analysis must consider the relative sizes of the injected-fluid (usually water) bank, the in-situ-fluid (oil) bank, and possibly a third bank with a mobile gas saturation. In general, analysis is possible only when the radius of the outer bank is greater than 10 times the radius of the inner bank.[9,10]

Some injection wells have the fluid distribution that **Fig. 9.11** shows, in which the locations of the banks move as fluid is injected. The radius of the water bank is r_o, and the radius of the oil bank is r_e. Such a system is called a composite system.[1,9] When the mobility ratio is not unity, the pressure behavior cannot be analyzed with the falloff-test analysis described in Sec. 9.3. However, two analysis methods are applicable for injection-well tests in nonunit-mobility-ratio reservoirs—the Hazebroek *et al.*[10,11] and Merrill *et al.*[12] methods.

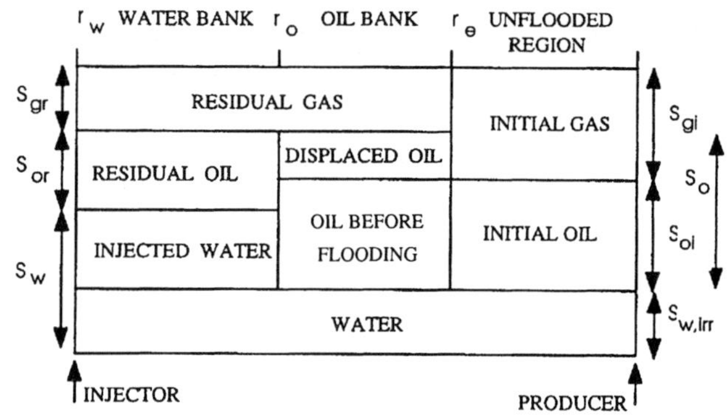

Fig. 9.12—Fluid saturations in a composite reservoir system (after Hazebroek *et al.*[10]).

9.5.1 Hazebroek *et al.*[10] Method. The method of Hazebroek *et al.*[10] applies to falloff-test analysis in injection wells before gas fillup. Their model was developed on the basis of the following assumptions: (1) a significant gas saturation lies ahead of the oil bank, (2) the pressure at the outer edge of the oil bank is constant and dominated by the pressure in the gas phase, and (3) the flow of water into the formation after shut-in is negligible. Although the second assumption is not strictly correct, the method can be used to estimate permeability, skin factor, and average drainage-area pressure from falloff-test analysis. The third assumption implies that the method should be applied to the latter part of the falloff curve, when afterflow distortion has ceased. Furthermore, the model assumes that only gas is produced during the fillup period, although oil may be mobile in the reservoir.[2,10]

Fig. 9.12 illustrates a model of the saturation distribution during injection. In the water bank, injected water has displaced gas and oil down to their residual saturations, S_{gr} and S_{or}, respectively. Injection has built up the oil saturation in the oil bank to a saturation, S_o, greater than the preflooding (initial) saturation, S_{oi}. Only water is mobile in the water bank, with saturation S_w, while only oil is mobile in the oil bank. In the unflooded region, all fluid saturations are at their initial values of S_{gi}, S_{oi}, and $S_{w,irr}$. The simple model is a good approximation for reservoirs producing oils with viscosities less than 50 cp.

Falloff BHP is modeled by

$$p_{ws} = p_e + b_1 e^{-\beta_1 \Delta t}, \quad (9.12)$$

where p_e = pressure at the outer edge of the oil bank at shut-in time Δt. Eq. 9.12 indicates that a plot of $\log(p_{ws} - p_e)$ vs. Δt should be linear, with intercept β_1 and slope $b_1/2.303$. The intercept b_1 is found by extrapolating the straight line to $\Delta t = 0$. The correct value of p_e is determined graphically by plotting $\log(p_{ws} - p_e)$ vs. Δt with trial values of p_e until a straight line is obtained at later shut-in times. If the trial value of p_e is too high, the data curve downward. Conversely, if the trial value of p_e is too low, the data curve upward. **Fig. 9.13** illustrates this trial-and-error process.

The permeability-thickness product in the water bank is related to b_1 by

$$k_w h = \frac{q \mu_w}{b_1}(2F), \quad (9.13)$$

where **Figs. 9.14 through 9.16** give the function F as a function of the mobility ratio, $M = (k/\mu)_1/(k/\mu)_2$, the ratio of the total compressibilities in the oil and water banks, $\gamma = c_{to}/c_{tw}$, and the ratio of the inner and outer radii of the oil bank, r_{oD}.

We now develop an equation to calculate r_{oD}. We start by noting that the pore volume (PV) of the oil bank is

$$V_{po} = \phi \pi h (r_e^2 - r_o^2), \quad (9.14)$$

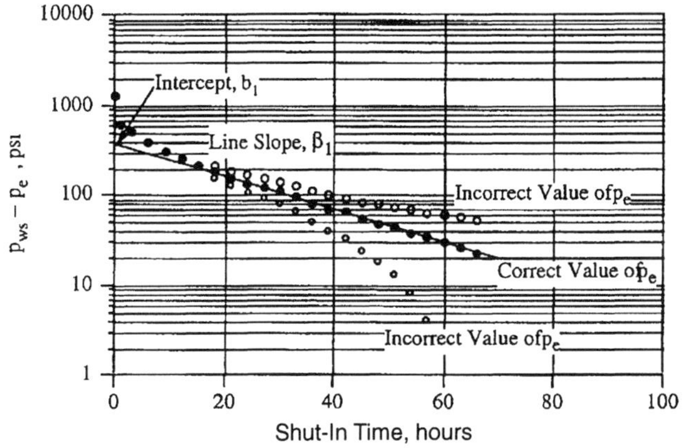

Fig. 9.13—Determining p_e from pressure-falloff-test data (after Hazebroek *et al.*[10]).

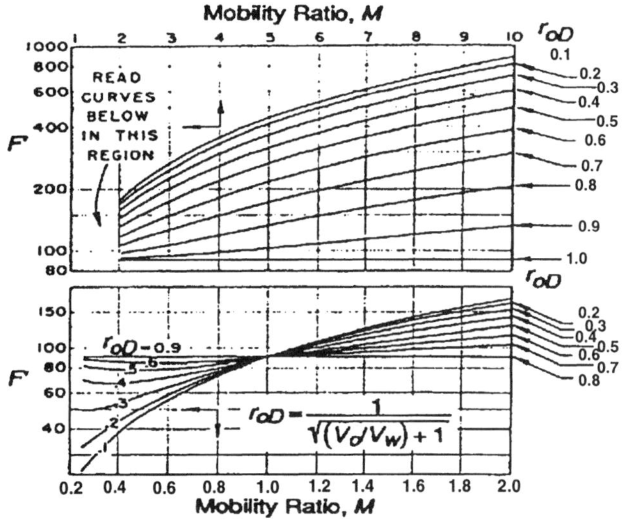

Fig. 9.14—Function F for calculating kh, $\gamma = 1$ (after Hazebroek *et al.*[10]).

and the PV of the water bank is

$$V_{pw} = \phi\pi h\left(r_o^2 - r_w^2\right) \approx \phi\pi h r_o^2 \quad (9.15)$$

for r_w much less than r_o. Combining Eqs. 9.14 and 9.15,

$$\frac{V_{po}}{V_{pw}} = \frac{r_e^2 - r_o^2}{r_o^2} = \frac{r_e^2}{r_o^2} - 1 \quad (9.16)$$

$$\text{or } \frac{r_e^2}{r_o^2} = \frac{V_{po}}{V_{pw}} + 1. \quad (9.17)$$

Defining the dimensionless radius r_{oD} as

$$r_{oD} = \frac{r_o}{r_e} \quad (9.18)$$

and combining Eqs. 9.17 and 9.18, r_{oD} can be expressed in terms of PV's,

$$r_{oD} = \frac{1}{\sqrt{\frac{V_{po}}{V_{pw}} + 1}}. \quad (9.19)$$

Before fillup, the moveable-oil volume in the oil bank, V_o, equals the PV of the oil bank, V_{po}, times the oil saturation, S_o, less the residual oil saturation. Because the oil saturation is at the residual value, S_{or}, in the water bank (i.e., no moveable oil in the water bank), the moveable-oil volume, V_o, equals the moveable-oil volume in both the water and oil banks. Furthermore, the moveable-oil volume contained in the combined PV of the oil and water banks, $(V_{po} + V_{pw})$, equals the combined PV times the initial oil saturation less the residual oil saturation; i.e.,

$$\begin{aligned} V_o &= V_{po}(S_o - S_{or}) \\ &= V_{po}(S_o - S_{or}) + V_{pw}(S_o - S_{or}) \\ &= \left(V_{po} + V_{pw}\right)(S_{oi} - S_{or}). \end{aligned} \quad (9.20)$$

Because only gas is produced before fillup, the volume occupied by the injected water, V_w, throughout the combined PV of the oil and water banks, $(V_{po} + V_{pw})$, equals the combined PV times the initial gas saturation less the residual gas saturation; i.e.,

$$V_w = \left(V_{po} + V_{pw}\right)\left(S_{gi} - S_{gr}\right). \quad (9.21)$$

Therefore, the ratio of moveable-oil volume to injected-water volume is

$$\frac{V_o}{V_w} = \frac{\left(V_{po} + V_{pw}\right)(S_{oi} - S_{or})}{\left(V_{po} + V_{pw}\right)\left(S_{gi} - S_{gr}\right)} = \frac{S_{oi} - S_{or}}{S_{gi} - S_{gr}}. \quad (9.22)$$

Because the saturations of the immobile fluids (S_{or}, S_{gr}, and $S_{w,irr}$) are the same in the oil and water banks, the moveable-oil saturation in the oil bank, $S_{om} = S_o - S_{or}$, must be equal to the injected-water saturation in the water bank, $S_{w,inj} = S_w - S_{w,irr}$. The moveable-oil volume, V_o, equals S_{om} times the oil-bank PV, V_{po}. Similarly, the injected-water volume, V_w, equals $S_{w,inj}$ times the water-bank PV, V_{pw}. Therefore, the ratio of the moveable-oil volume to injected-water volume equals the ratio of the PV's of the oil and water banks; i.e.,

$$\frac{V_o}{V_w} = \frac{V_{po}}{V_{pw}}. \quad (9.23)$$

Combining Eqs. 9.22 and 9.23 with Eq. 9.19, we obtain the desired working equation for r_{oD} in terms of oil and gas saturations:

$$r_{oD} = \frac{1}{\sqrt{\frac{S_{oi} - S_{or}}{S_{gi} - S_{gr}} + 1}}. \quad (9.24)$$

Where no curves are given for $r_{oD} = 0$ in Figs. 9.14 to 9.16, the curves for $r_{oD} = 0.1$ may be used instead. Plotted values of F are such that $k_w h$ is obtained in md-ft when q_w is in STB/D, m_w is in cp, and b_1 is in psi. For a given mobility ratio and a value of $k_w h$, the permeability-thickness product for the oil bank is calculated as follows:

$$k_o h = \frac{\mu_o}{\mu_w}\frac{k_w h}{M}, \quad (9.25)$$

giving an estimate of permeability in the oil bank for a known value of h. The outer radius of the oil bank, r_e, is

$$r_e = \sqrt{\frac{5.615 W_i}{\pi\phi h\left(S_{gi} - S_{gr}\right)}}, \quad (9.26)$$

where the wellbore radius, r_w, is negligibly small compared with r_e. The skin factor, s, then can be obtained from the relation

$$s + \ln\frac{r_e}{r_w} = 0.00708\frac{(p_w - p_e)k_w h}{q\mu_w} - \frac{M-1}{2}\ln\left(\frac{V_o}{V_w} + 1\right). \quad (9.27)$$

If unit-mobility-ratio analysis is applied to a reservoir in which the mobility ratio is greater than one, as in the case of viscous oil, the skin factor may be underestimated when $s > 0$ or overestimated when $s < 0$.

An important assumption of this analysis method is that afterflow is negligible. Generally, this assumption is valid under two conditions. In the first condition, the surface pressure decreases slowly and the well remains filled with liquid for long shut-in times. This

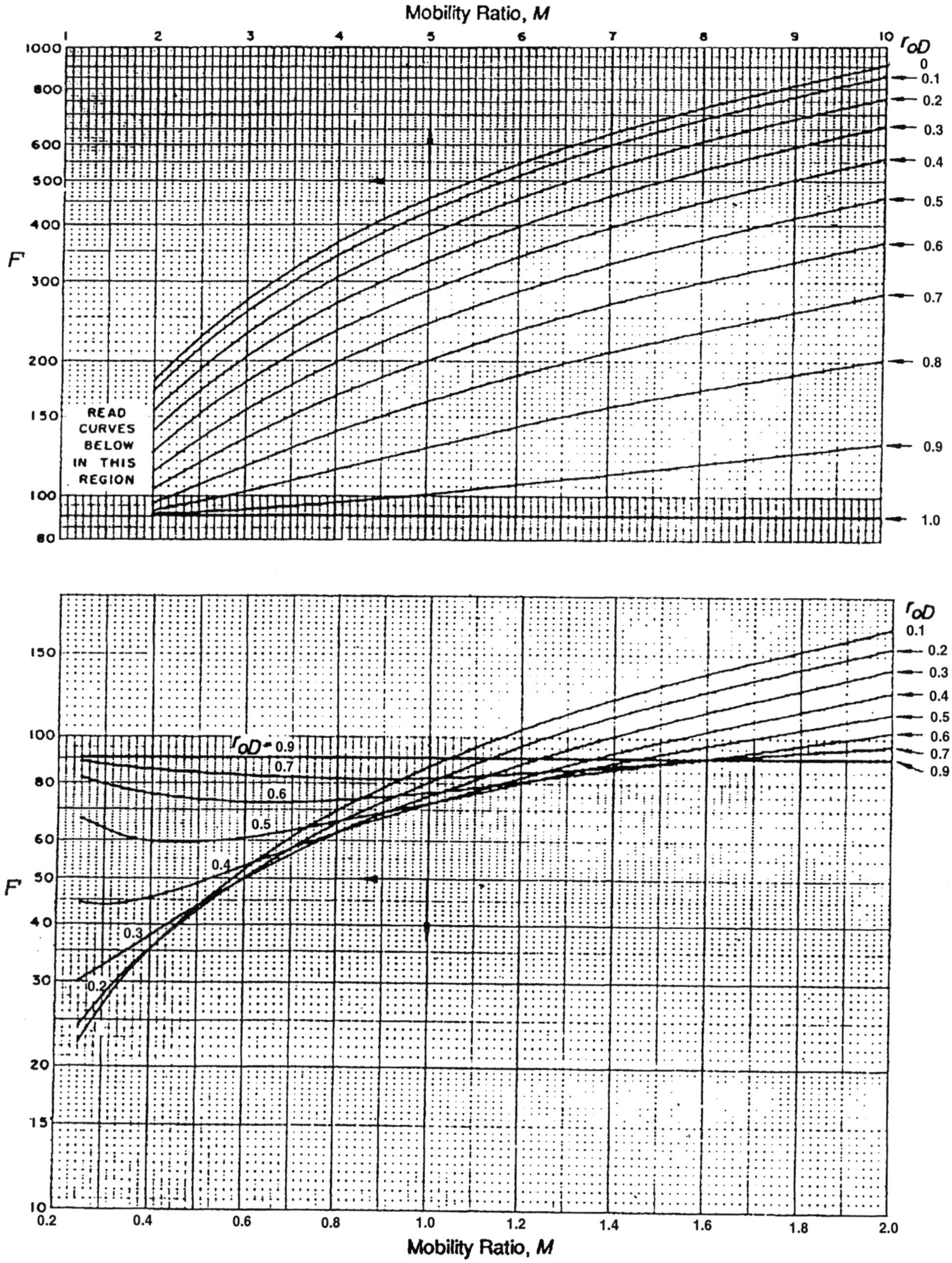

Fig. 9.15—Function *F* for calculating *kh*, $\gamma = 2$ (after Hazebroek *et al.*[10]).

situation occurs with high reservoir pressures, with afterflow into the formation resulting only from expansion of the water in the wellbore as pressure decreases. For this case, a definitive constant related to the wellbore-storage characteristics is

$$C_1 = 0.0538 \frac{d_t^2 \beta_1 b_1 c_w (p_w - p_t)}{q \rho_w}, \quad \ldots\ldots\ldots\ldots\ldots\ldots \quad (9.28)$$

where p_t = surface pressure and ρ_w = density of water (lbm/ft^3).

In the second condition, the surface pressure drops to zero shortly after shut-in, after which the fluid level in the wellbore starts to fall. In this case, the volume of afterflow into the formation at any time practically would be equal to the volume between the top of the well and the water level in the wellbore, with comparatively negligible expansion of the water. For this case,

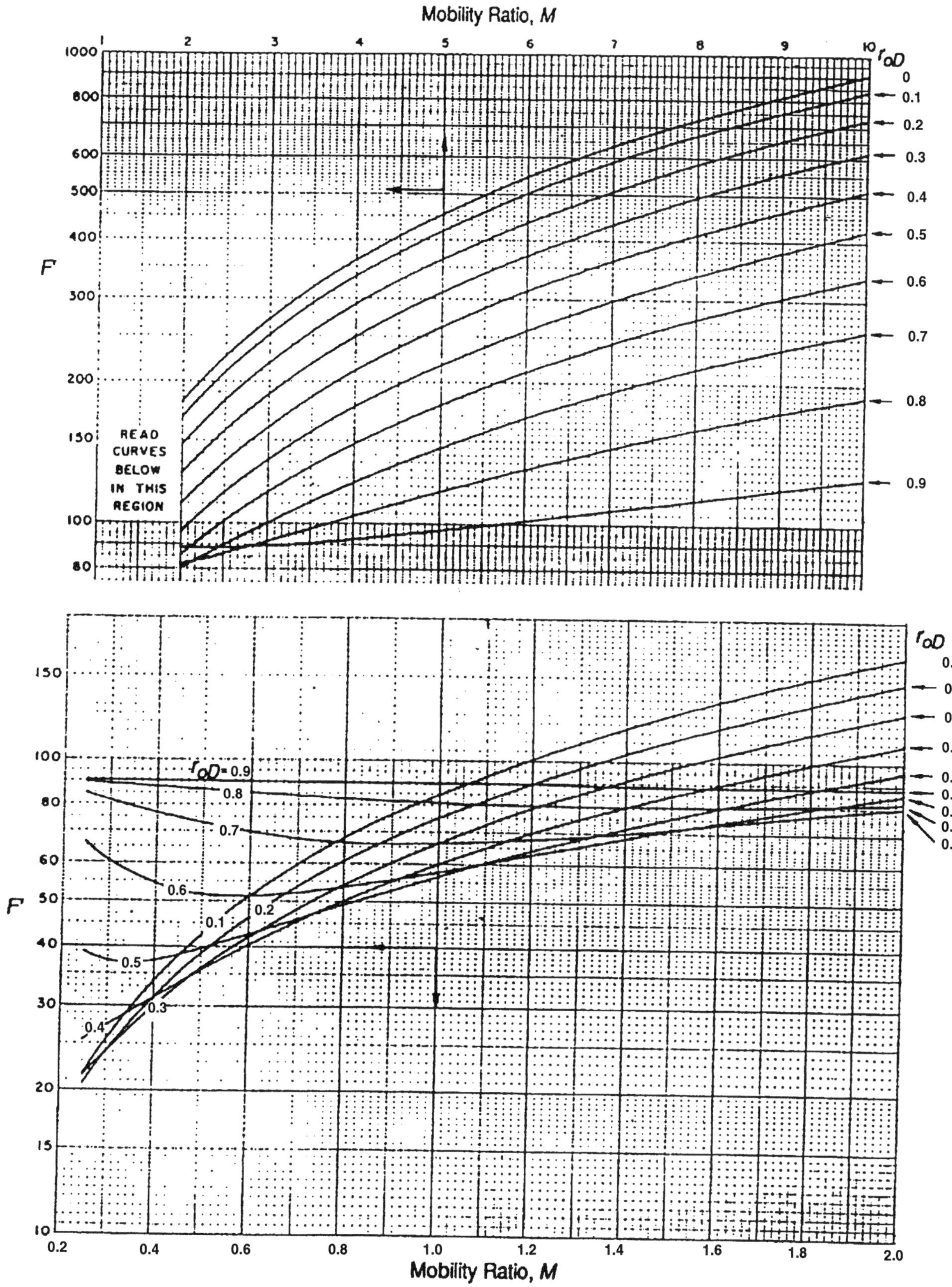

Fig. 9.16—Function *F* for calculating *kh*, $\gamma = 4$ (after Hazebroek *et al.*[10]).

$$C_1 = 3.3571 \frac{d_i^2 \beta_1 b_1}{q \rho_w} . \quad \text{(9.29)}$$

A small value of C_1 in either case indicates negligible afterflow for each unit drop in pressure. Except at large shut-in times, nonunit-mobility-ratio analysis should not be applied when $C_1 \geqq 0.1$.

Hazebroek *et al.*[10] concluded that permeability and skin factor for the zone near the well can be estimated equally well by their method or by the Horner method. The Horner method gives the properties of the fluid bank within the radius of investigation reached during the test; however, if that radius of investigation exceeds the boundary between the inner and outer banks, the results will be incorrect.

TABLE 9.5—PRESSURE-FALLOFF-TEST DATA, EXAMPLE 9.3

Shut-In Time (hours)	Pressure (psia)	$p_{ws}-p_e$ (psi)
0.0	598.0	566.0
1.0	512.0	480.0
3.0	407.0	375.0
5.0	332.0	300.0
7.0	282.0	250.0
9.0	249.0	217.0
11.0	227.0	195.0
13.0	208.0	176.0
15.0	190.0	158.0
17.0	175.0	143.0
19.0	161.0	129.0
21.0	148.0	116.0

When the distance to the boundary between the two banks is small compared with the radius of investigation, the permeability of the outer region can be obtained by pressure transient analysis and the skin factor reflects the presence of the inner bank.

For small injected volumes, the Hazebroek *et al.* method is preferred for estimating permeability and skin factor in the inner zone when the important assumption of constant pressure at the outer edge of the oil bank is satisfied. For systems with large injection volumes, the pressure transient generally will not encounter the gas-saturated region during a test, so normal analysis methods should provide equally reliable results.

Example 9.3—Falloff Analysis in a Composite System With the Hazebroek *et al.* Method. A pressure-falloff analysis[10] was performed in a reservoir under waterflood. The static pressure, $p_e=32$ psi, was determined following a lengthy shut-in period. The flowing pressure immediately before shut-in, p_{wf}, was 598 psi. The k_w/k_o ratio was found to be 0.3 from laboratory measurements of core permeability to water at the water saturation in the water zone and core permeability to oil at the water saturation in the oil zone. The injection rate during the test was $q=1{,}020$ STB/D for 21 hours, while the cumulative water injected in the well was $W_i=6{,}097$ bbl. The inside diameter of the tubing was $d_t=3.548$ in. The wellhead pressure fell to zero soon after shut-in. **Table 9.5** summarizes pressure-change and time data for the test. Other known data are summarized next. Determine the effective permeabilities to water and oil and the skin factor.

$h=$ 45 ft
$\rho_w=$ 62.4 lbm/ft^3
$\phi=$ 0.3
$\mu_w=$ 0.9 cp
$\mu_o=$ 12.0 cp
$S_g=$ 0.12
$S_o=$ 0.56
$S_{or}=$ 0.20
$S_{gr}=$ 0.0
$c_{to}=4.0\times10^{-6}$ psi^{-1} (oil zone)
$c_{tw}=3.0\times10^{-6}$ psi^{-1} (injected-water zone)

Solution.

1. First, prepare a semilog plot (**Fig. 9.17**) of the pressure difference, $p_{ws}-p_e$, vs. shut-in time, Δt. Table 9.5 gives the plotting functions.

The intercept is

$b_1=347$ psi at shut-in time $\Delta t=0$.

The variable β_1 is

$$\beta_1 = (2.303)(\text{slope}) = 2.303\left[\frac{\log(347)-\log(116)}{0-21}\right]$$

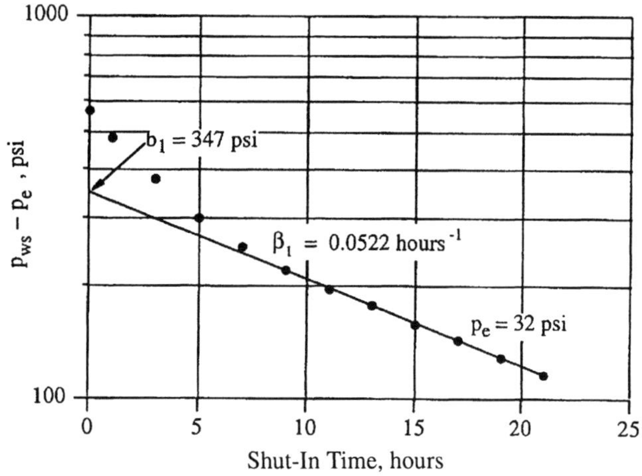

Fig. 9.17—Pressure-falloff curve, Example 9.3 (after Matthews and Russell[11]).

$$= 0.0522 \text{ hour}^{-1}.$$

2. To ensure that the nonunit-mobility-ratio method applies, we check the value of C_1 for this case. C_1 is determined with Eq. 9.28,

$$C_1 = 3.3571\frac{d_t^2\beta_1 b_1}{q\rho_w}$$

$$= 3.3571\frac{(3.548)^2(0.0522)(347)}{(1{,}020)(62.4)}$$

$$= 0.012.$$

Because C_1 is small (less than 0.1), afterflow is limited. Therefore, the nonunit-mobility-ratio method applies. Note that the signs of β_1 and q need not be taken into account in these calculations.

3. From the definition of mobility ratio given by Eq. 9.1, in this reservoir

$$M = \frac{k_w\mu_o}{k_o\mu_w} = \frac{(0.3)(12)}{0.9} = 4.$$

4. The ratio of the inner and outer radii of the oil bank is

$$r_{oD} = \frac{1}{\sqrt{\dfrac{S_{oi}-S_{or}}{S_{gi}-S_{gr}}+1}} = \frac{1}{\sqrt{\dfrac{0.56-0.20}{0.12-0}+1}} = 0.5.$$

5. The ratio of total compressibilities in the oil and water banks is

$$\gamma = \frac{c_{to}}{c_{tw}} = \frac{4.0\times10^{-6}}{3.0\times10^{-6}} \approx 1.$$

6. Reading $F=220$ from Fig. 9.14 at $M=4$ and $\gamma=1$, we calculate the permeability-thickness product for the water phase,

$$k_wh = \frac{q\mu_w}{b_1}(2F) = \frac{(1{,}020)(0.9)}{347}(2)(220) = 1{,}164 \text{ md-ft}.$$

The effective permeability to water, k_w , in the water zone is

$$k_w = \frac{k_wh}{h} = \frac{1{,}164 \text{ md-ft}}{45 \text{ ft}} = 26 \text{ md}.$$

7. The permeability-thickness product in the oil bank is

$$k_oh = \frac{\mu_o}{\mu_w}\frac{k_wh}{M} + \frac{(12)(1{,}164)}{(0.9)(4)} = 3{,}880 \text{ md-ft}.$$

The effective permeability to oil in the oil zone, k_o, is

$$k_o = \frac{k_oh}{h} = \frac{3{,}880 \text{ md-ft}}{45 \text{ ft}} = 86 \text{ md}.$$

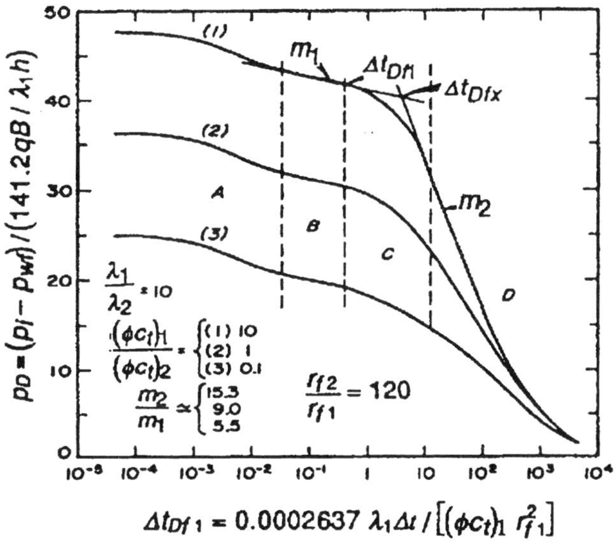

Fig. 9.18—Pressure-falloff behavior in a two-zone composite system (after Merrill *et al.*[12]).

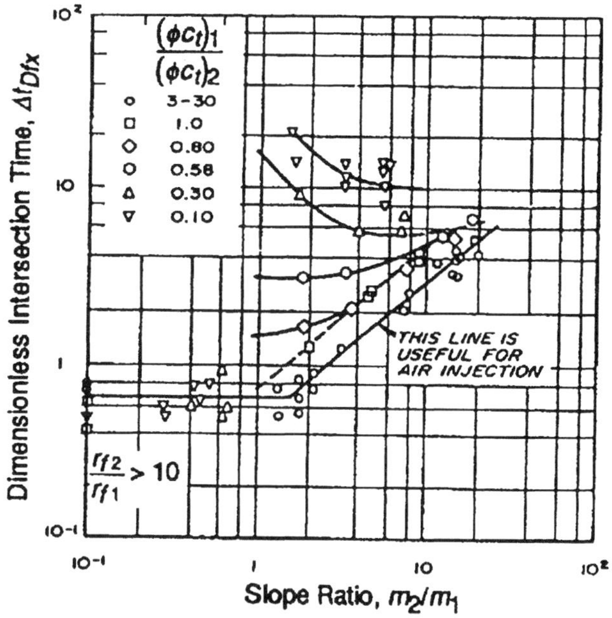

Fig. 9.19—Correlation of dimensionless intersection time for pressure-falloff data in a two-zone composite reservoir (after Merrill *et al.*[12]).

8. The outer radius of the reservoir is

$$r_e = \sqrt{\frac{5.615W_i}{\pi\phi h(S_g - S_{gr})}} = \sqrt{\frac{(5.615)(6,097)}{\pi(0.3)(45)(0.12 - 0)}} = 82 \text{ ft.}$$

9. The skin factor is

$$s = 0.00708\frac{(p_{wf} - p_e)k_w h}{q\mu_w} - \frac{M-1}{2}\ln\left(\frac{V_o}{V_w} + 1\right) - \ln\frac{r_e}{r_w}$$

$$= \left[\frac{0.00708(598 - 32)(1,164)}{(1,020)(0.9)}\right]$$

$$- \left(\frac{4-1}{2}\right)\ln(3 + 1) - \ln\left(\frac{82}{1}\right)$$

$$= -1.4.$$

9.5.2 Merrill *et al.* Method. For liquid-filled systems, the method of Merrill *et al.*[12] may give more accurate results than the Hazebroek *et al.* method because it avoids the assumption of piston-like displacement. **Fig. 9.18** illustrates expected injection-well falloff behavior in a two-bank, liquid-filled system. The three falloff curves apply for different ratios of porosity-compressibility product, ϕc_t, between the first and second bank, with a mobility ratio of $M = \lambda_1/\lambda_2 = (k/\mu)_1/(k/\mu)_2$. Sec. A is dominated by wellbore-storage effects. Sec. B includes a semilog straight line that provides information about the injected-fluid bank, while Sec. C is a transition region in which the oil bank begins to influence the pressure-falloff behavior. Sec. D includes a second semilog straight line whose slope is determined by properties of the injected-fluid bank and the oil bank.

Merrill *et al.*[12] provide methods for estimating the location of the front of the injected-fluid bank, r_{f1}, and the permeabilities of the two fluid banks in a two-zone system. Their data were derived from computer simulations in which the ratio of the distance to the displaced-fluid bank to the distance to the injected-fluid bank is $r_{f2}/r_{f1} > 50$. In practice, their technique provides acceptable results as long as $r_{f2}/r_{f1} > 10$. Unlike the Hazebroek *et al.* method, the Merrill *et al.* technique does not require *a priori* knowledge of the mobility ratio, although an estimate of the ϕc_t ratio is required.

The Merrill *et al.* approach provides two methods for estimating the distance to the front of the injected-fluid bank, r_{f1}, from a plot of p_{ws} vs. log Δt. The first method uses the extrapolated intersection time of the two semilog straight lines on the Horner plot, Δt_{fx}, to obtain

$$r_{f1} = \sqrt{\frac{0.0002637(k/\mu)_1}{(\phi c_t)_1}\frac{\Delta t_{fx}}{\Delta t_{Dx}}}. \quad \text{(9.30)}$$

Fig. 9.19 shows a correlation of Δt_{fDx} with the ratio of the slopes of the two semilog straight lines, m_2/m_1, and the ratio of the porosity/compressibility products in the two fluid banks, $(\phi c_t)_1/(\phi c_t)_2$.

The second method uses the point of deviation of the observed pressure data from the first semilog straight line, Δt^*_{f1}, to obtain

$$r_{f1} = \sqrt{\frac{0.0002637(k/\mu)_1}{(\phi c_t)_1}\frac{\Delta t^*_{f1}}{\Delta t^*_{fD1}}}. \quad \text{(9.31)}$$

The range of Δt^*_{fD1} is 0.13 to 1.39, with a mean value of 0.389 used in Fig. 9.18. The range and mean of Δt^*_{fD1} agree with an interpretive rule of thumb that the slope of the injected-fluid bank normally is valid up to a time equivalent to $\Delta t_{fD1} \approx 0.25$. Because Δt^*_{fD1} does not correlate well with slope and specific-storage (ϕc_t) ratios, the second method for estimating r_{f1} should be used only when insufficient data are available to estimate Δt_{fx} for the first method.

The permeability in the injected-fluid bank, k_1, and the skin factor, s, can be estimated from the slope, m_1, of the first semilog straight line. If $r_{f2} > 10r_{f1}$, the mobility in the second zone can be estimated from

$$\left(\frac{k}{\mu}\right)_2 = \frac{(k/\mu)_1}{(\lambda_1/\lambda_2)}, \quad \text{(9.32)}$$

where the term λ_1/λ_2 is obtained from **Fig. 9.20.**

If semilog straight lines appear in both zones and if the specific storage capacity, ϕc_t, can be estimated for both zones, the mobility or permeability in each zone can be found. The mobility or permeability in the second zone cannot be obtained simply on the basis of the slope for the second zone, m_2. When $r_{f2} < 10r_{f1}$, reservoir simulation is required to find mobility in the second zone.

The maximum wellbore-storage coefficient that still permits the first semilog straight line to be observed is

$$C \leqq 5.9 \times 10^{-7}\left(\frac{k}{\mu}\right)_1 h\Delta t^*_{f1}e^{-0.14s}, \quad \text{(9.33)}$$

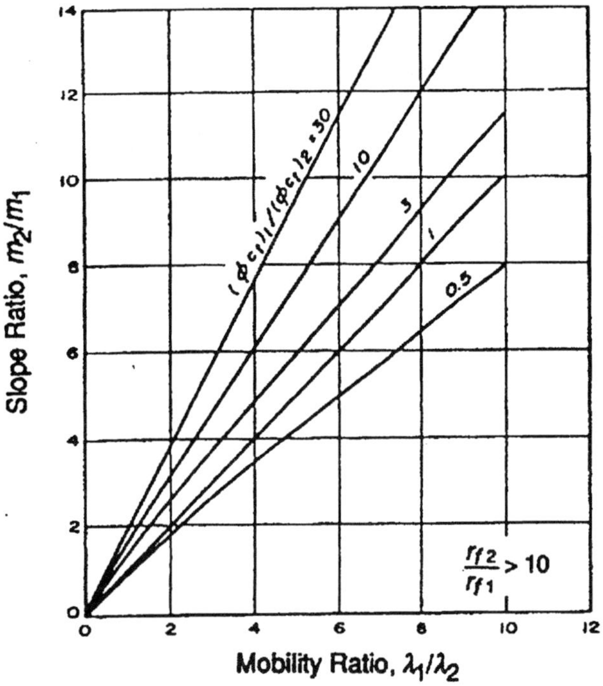

Fig. 9.20—Correlation of mobility ratio with slope ratio and specific-storage ratio (after Merrill *et al.*[12]).

which allows for about one cycle of the first semilog straight line between the cessation of afterflow and initial deviation from the first line as second-bank effects begin to affect the pressure response. This criterion is seldom met, especially if the boundaries of the injected-fluid bank and the oil bank are relatively close to the injection well.

In a nonunit-mobility-ratio reservoir, the injectivity tends to change as water is injected.[12] During the early stages of injection, the changing injectivity appears as a changing skin factor. After enough fluid has been injected to form a significant water bank around the well, the mobility of that bank can be found from well testing. The skin factor calculated from the first-bank slope, m_1, does not change unless the injected fluid is actually damaging or stimulating the formation.

Example 9.4—Falloff Analysis in a Composite System With the Merrill *et al.* Method. Following a waterflood in an oil reservoir, a falloff test was conducted.[12] The radius of investigation achieved during the test encountered both the water bank surrounding the wellbore and the oil bank between the injected water zone and the unaffected region of the reservoir. Determine the effective permeability in each zone and find the skin factor with the Merrill *et al.* method.

Pressure-time data are given in **Table 9.6.**

$t_p =$ 100 hours
$q_w =$ −400 STB/D
$p_i =$ 1,008 psia
$r_w =$ 0.25 ft
$\phi =$ 0.20
$p_{wf} =$ 1,314.7 psi
$\mu_w =$ 1.0 cp
$\mu_o =$ 1.0 cp
$h =$ 20 ft
$B_o =$ 1.2 RB/STB
$B_w =$ 1.0 RB/STB
$S_{w1} =$ 0.75
$S_{w2} =$ 0.30 (irreducible)
$S_{o2} =$ 0.70
$S_{o1} =$ 0.25 (residual)

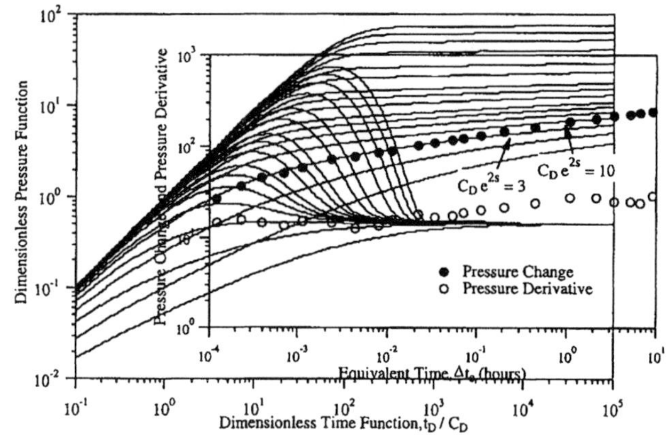

Fig. 9.21—Qualitative type-curve analysis of the early pressure data by use of the Bourdet *et al.*[3] model, Example 9.4.

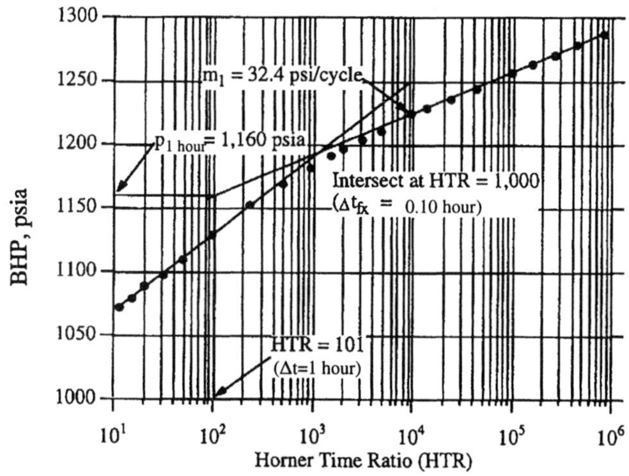

Fig. 9.22—Horner[4] semilog analysis, Example 9.4

$c_{t1} = 6.9 \times 10^{-6}$ psi^{-1}
$c_{t2} = 7.7 \times 10^{-6}$ psi^{-1}

Solution. *Construct Data Plots.* As a first step, we construct the appropriate data plots. **Fig. 9.21** gives a log-log plot of pressure change and pressure derivative vs. Agarwal's[7] equivalent time, and **Fig. 9.22** shows a Horner semilog plot. **Table 9.7** presents semilog and log-log plotting functions.

Qualitative Type-Curve Analysis. A qualitative type-curve analysis (Fig. 9.21) indicates that semilog straight lines exist in both the water and oil banks. Using the Bourdet *et al.*[3] type curve for radial flow with

TABLE 9.6—FALLOFF-TEST DATA, EXAMPLE 9.4

Time (hours)	Pressure (psia)	Time (hours)	Pressure (psia)
0.00012	1,288.1	0.0521	1,196.6
0.00023	1,278.8	0.0687	1,191.1
0.00038	1,270.3	0.1104	1,181.8
0.00067	1,263.7	0.2036	1,169.0
0.0011	1,257.1	0.4399	1,152.4
0.0023	1,244.1	1.0678	1,129.6
0.0042	1,236.6	2.1267	1,110.4
0.0076	1,229.0	3.4116	1,097.7
0.0107	1,224.4	5.2687	1,088.5
0.0214	1,211.4	7.0717	1,078.5
0.0330	1,204.0	9.7019	1,072.1

TABLE 9.7—SEMILOG AND LOG-LOG PLOTTING FUNCTIONS, EXAMPLE 9.4

Equivalent Time (hours)	Pressure Change (psi)	Pressure Derivative	Horner Time Ratio	Pressure (psia)
0.00012	26.6	14.661	826,930	1,288.1
0.00023	35.9	15.679	438,500	1,278.8
0.00038	44.4	14.277	262,010	1,270.3
0.00067	51.0	13.300	150,270	1,263.7
0.0011	57.6	15.179	95,239.	1,257.1
0.0023	70.6	14.232	43,123	1,244.1
0.0042	78.1	12.682	23,777	1,236.6
0.0076	85.7	13.319	13,111	1,229.0
0.0107	90.3	15.320	9,355.5	1,224.4
0.0214	103.3	17.691	4,682.6	1,211.4
0.0330	110.7	16.636	3,030.4	1,204.0
0.0520	118.1	17.556	1,921.9	1,196.6
0.0686	123.6	18.829	1,457.5	1,191.1
0.1103	132.9	20.188	907.04	1,181.8
0.2032	145.7	21.235	492.11	1,169.0
0.4380	162.3	23.608	228.32	1,152.4
1.06	185.1	27.250	94.650	1,129.6
2.08	204.3	27.882	48.021	1,110.4
3.30	217.0	24.701	30.312	1,097.7
5.01	226.2	24.920	19.980	1,088.5
6.60	236.2	23.621	15.141	1,078.5
8.84	242.6	28.808	11.307	1,072.1

wellbore storage and skin indicates that (in the water bank that the pressure transient encounters first during the test) the early pressure change and pressure-derivative data appear to match the $C_De^{2s}=3$ curve, with the semilog straight line in this zone existing from about 0.004 to 0.01 hour. In the oil bank encountered later during the test, the pressure-change and derivative data match best with the $C_De^{2s}=10$ curve, with the semilog straight line in this zone appearing to extend from about 0.4 hour through the end of the test (9.7 hours).

Horner Semilog Analysis. Fig. 9.22 shows the fits of the semilog straight lines in the two zones. The straight line at larger Horner time ratio values (earlier times) represents the water bank, while the straight line at smaller Horner time ratio values (later times) represents the oil bank.

1. The semilog straight line in the water bank occurs from approximately 0.004 to 0.010 hour and is consistent with the preliminary type-curve match. The slope of the first line is $m_1=32.4$ psi/cycle.

2. We estimate permeability in the water bank to be

$$k_1 = -\frac{162.6q_wB_w\mu_w}{m_1h} = -\frac{(162.6)(-400)(1.0)(1.0)}{(32.4)(20)}$$

$$= 100 \text{ md.}$$

3. The skin factor, which is estimated for the water zone only, is

$$s = 1.151\left[\left(\frac{p_{wf}-p_{1\text{hr}}}{m_1}\right) - \log\left(\frac{k_1}{\phi\mu_wc_tr_w^2}\right) + 3.23\right]$$

$$= 1.151\left\{\left(\frac{1,314.7-1,160}{32.4}\right)\right.$$

$$\left. -\log\left[\frac{100}{(0.20)(1.0)(6.9\times10^{-6})(0.25)^2}\right] + 3.23\right\}$$

$$= -1.2,$$

where $p_{1\text{hr}}=1,160$ psia is obtained from the extrapolation of the line drawn through the early pressure data (i.e., the water bank).

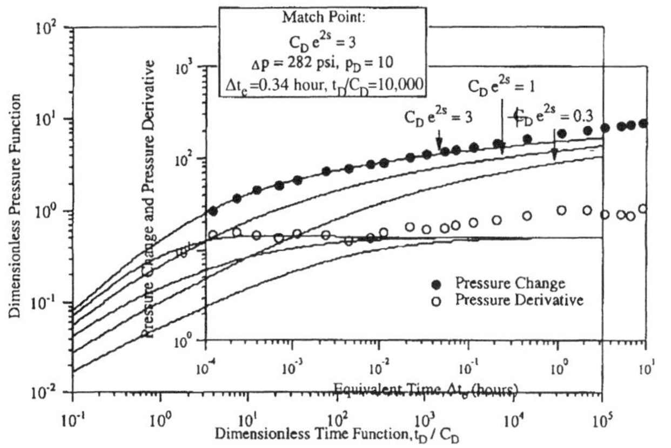

Fig. 9.23—Quantitative type-curve analysis of the early pressure data with the Bourdet *et al.*[3] model, Example 9.4.

Quantitative Type-Curve Analysis. We now perform a quantitative type-curve analysis (**Fig. 9.23**).

1. Precalculate a pressure-match point with k_1 from the semilog analysis in the water bank. Choosing a convenient value of $p_D=10$, the corresponding pressure change is

$$(\Delta p)_{MP} = -\frac{141.2(-q_w)B_w\mu_w}{kh}(p_D)_{MP}$$

$$= -\frac{(141.2)(-400)(1.0)(1.0)}{(100)(20)}(10)$$

$$= 282 \text{ psi.}$$

2. The type-curve match is obtained by aligning the precalculated pressure-match point from Step 1 and sliding the pressure-change and -derivative data horizontally to honor the pressure-match point Δp and p_D. A good match is obtained with the type curve defined by the correlating parameter $C_De^{2s}=3$.

3. We select a time-match point

$\Delta t_e=0.34$ hour and $t_D/C_D=10{,}000$.

4. The dimensionless wellbore-storage coefficient is

$$C_D = \frac{0.0002637k_1}{\phi r_w^2c_t\mu_w}\left(\frac{\Delta t_e}{t_D/C_D}\right)_{MP}$$

$$= \frac{(0.0002637)(100)}{(0.20)(0.25)^2(6.9\times10^{-6})(1.0)}\left(\frac{0.34}{10,000}\right)$$

$$= 10.4.$$

5. The skin factor is calculated with C_D from Step 4 and the correlating parameter, $C_De^{2s}=3$, from the type-curve match.

$$s = 0.5\ln\left(\frac{C_De^{2s}}{C_D}\right) = 0.5\ln\left(\frac{3}{10.4}\right) = -0.62,$$

which agrees with the value from the semilog analysis.

6. We now proceed with the composite-system-test analysis and calculate the following quantities:

$m_2/m_1=(61.1)/(32.4)=1.89$

$(\phi c_t)_1=(0.20)(6.9\times10^{-6})=1.38\times10^{-6}$ psi^{-1}

$(\phi c_t)_2=(0.20)(7.7\times10^{-6})=1.54\times10^{-6}$ psi^{-1}

$(\phi c_t)_1/(\phi c_t)_2=(1.38\times10^{-6})/(1.54\times10^{-6})=0.90.$

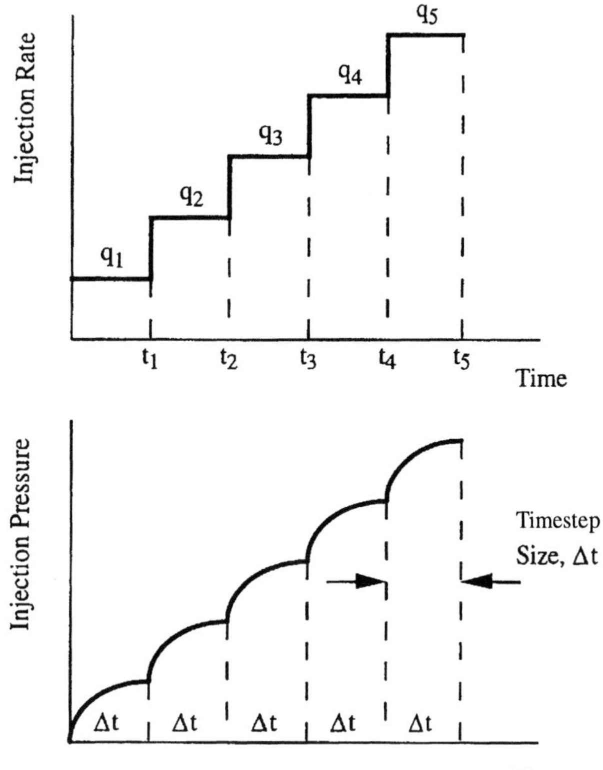

Fig. 9.24—Typical step-rate injectivity-test procedure.[15]

The mobility of the water zone is

$$\lambda_1 = \left(\frac{k}{\mu}\right)_1 = -\frac{162.6 q_w B_w \mu_w}{m_1 h \mu_w} = -\frac{162.6(-400)(1.0)}{(32.4)(20)}$$

$$= 100 \text{ md/cp.}$$

To estimate $\lambda_2 = (k/\mu)_2$, we enter Fig. 9.20 with the values of $m_2/m_1 = 1.89$ and $(\phi c_t)_1/(\phi c_t)_2 = 0.90$ and read $\lambda_1/\lambda_2 = 2.0$. Thus,

$$\lambda_2 = \frac{\lambda_1}{\lambda_1/\lambda_2} = \frac{100}{2.0} = 50 \text{ md/cp}$$

and $k_2 = \mu_o \lambda_2 = 1.0(50) = 50$ md.

For a composite system, we determine the effective permeability in the oil zone, k_2, as illustrated. Had we attempted to use the slope of the second semilog straight line, we would have calculated erroneously

$$k_2 = -\frac{162.6 q_o B_o \mu_o}{m_2 h} = -\frac{162.6(-400)(1.2)(1.0)}{(61.1)(20)}$$

$$= 63.8 \text{ md.}$$

Note that the value of 63.8 md reflects properties of both the water and oil zones.

7. The location of the front of the water bank, r_{f1}, is estimated from the time, Δt_{fx}, at which the two semilog straight lines intersect. From the semilog plot of the falloff data, the two semilog straight lines intersect at a Horner time ratio of 1,000, which corresponds to a time of $\Delta t_{fx} = 0.10$ hour. Using Fig. 9.19 with $m_2/m_1 = 1.89$ and $(\phi c_t)_1/(\phi c_t)_2 = 0.90$, we get $\Delta t_{fDx} = 1.4$. Then from Eq. 9.30,

$$r_{f1} = \sqrt{\frac{0.0002637(k/\mu)_1}{(\phi c_t)_1}\frac{\Delta t^*_{f1}}{\Delta t^*_{fD1}}}$$

$$= \sqrt{\frac{(0.0002637)(100)(0.10)}{(1.38 \times 10^{-6})(1.4)}}$$

$$= 37 \text{ ft.}$$

9.6 Step-Rate Testing

A step-rate injectivity test normally is used to estimate the formation or fracture-parting pressure, which is the injection pressure that will initiate a fracture in an unfractured well or extend an existing fracture in a fractured well.[13] The parting pressure determined from a step-rate test is equivalent to the fracture extension/propagation pressure in the hydraulic-fracturing literature and generally is greater than the closure pressure required to initiate the opening of an existing fracture.[14] Determination of the parting pressure is important because injection above the parting pressure can result in premature breakthrough, poor sweep, reduced oil recovery, and loss of expensive injection fluids through uncontrolled fracture growth. **Fig. 9.24** illustrates the step-rate-test procedure.[15]

The well either is shut in or stabilized at a reduced, constant injection rate before the beginning of the test. If the well is shut in, the shut-in period should be sufficiently long so that BHP is near the static reservoir pressure. If the injection rate is stabilized at a reduced and constant rate, the stabilization period should be long enough to achieve steady-state or pseudosteady-state flow. The step-rate injectivity test that follows the shut-in or stabilization period is simple, inexpensive, and fast.

The test consists of a series of constant-rate injections of equal duration, with injection rates increasing in a stepwise manner. In relatively-low-permeability formations ($k < 5$ md), each injection rate should last approximately 1 hour. In formations with permeability exceeding 10 md, 30-minute injection times are adequate. As few as four rates may be used for conventional or multirate analysis, but normally from six to eight rates are preferred. Under appropriate conditions, a two-step rate test[16] can be used.

9.6.1 Conventional Analysis Technique. The conventional analysis method[15] consists of plotting injection pressure at the end of each rate against injection rate. BHP is preferable to surface pressure; however, surface pressure can be used if it is positive throughout the test and friction effects are not significant. This method assumes steady-state Darcy flow into an injection well and is based on the steady-state radial-flow equation,

$$p_{wf} = p_e - \frac{141.2 q B \mu}{kh}\left(\ln\frac{r_e}{r_w} + s\right), \quad \ldots\ldots\ldots\ldots\ldots (9.34)$$

where r = radius of drainage, ft; p_e = pressure at the radius of drainage, psia; and p_{wf} = injection BHP, psia.

If steady-state flow is achieved at each injection rate, r_e and p_e are constant and the injection BHP, p_{wf}, at the end of each constant-rate injection period is a linear function of the corresponding injection rate, q. If steady-state flow is not achieved during each injection timestep, r_e can be replaced by r_d, a time-dependent drainage radius representing the radius beyond which no appreciable amount of fluid is drained at a particular flowing time.[17] If the timestep size is constant for all injection rates during the test, p_{wf} still is linearly related to q at pressures below the parting pressure because the $\ln(r_d/r_w)$ term is insensitive to small changes in r_d during the test.

The conventional analysis technique requires that the pressure at the end of each injection period or timestep be plotted as a function of the corresponding injection rate. Theoretically, the plot should have two straight-line segments. **Fig. 9.25** illustrates the pressure behavior of a hypothetical step-rate test.

The pressure corresponding to the point where the two lines intersect is interpreted as the parting pressure. Note that this point can also indicate the breakdown pressure of the cement bond. However, when the cement bond fails, the slope of the second straight line in Fig. 9.25 usually continues below the parting pressure as the injection rate is decreased. The parting pressure may vary depending on fluid-saturation conditions in the formation and long-term variations in reservoir pressure level with time.

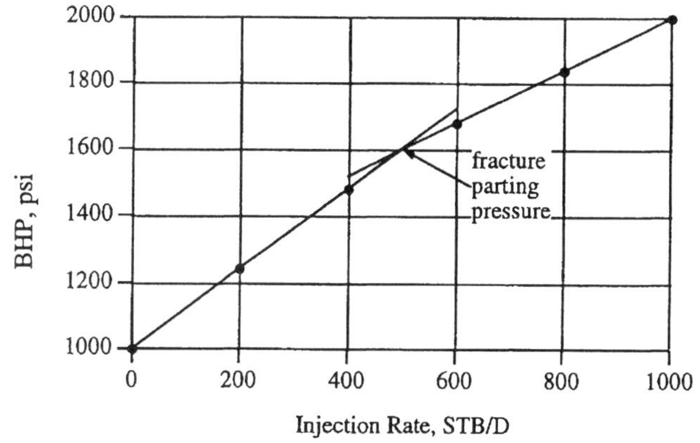

Fig. 9.25—Pressure behavior of step-rate injectivity test.

9.6.2 Multirate Analysis Technique. If pressures are recorded continuously and injection rates are measured accurately, a multirate pressure transient technique can be applied to the step-rate test.[15] The multirate analysis is based on the principle of superposition and assumes transient radial flow into the reservoir during each constant-rate-injection period. A plot is made of the Odeh-Jones[18] pressure and time-plotting functions,

$$\frac{p_i - p_{wf}}{q_n} \text{ vs. } \sum_{j=1}^{n} \frac{q_j - q_{j-1}}{q_n} \log\left(t_n - t_{j-1}\right),$$

where rates are negative for injection.

Alternatively, Agarwal's[7] multirate equivalent time can be applied instead of the Odeh-Jones plotting functions. To use Agarwal's equivalent-time analysis, a plot is made of

$$\frac{p_{wf_n} - p_{wf_{n-1}}\left(t_{n-1}\right)}{q_{n-1} - q_n} \text{ vs.}$$

$$\Delta t \prod_{j=1}^{n-1} \left(\frac{t_{n-1} - t_{j-1}}{\Delta t + t_{n-1} - t_{j-1}} \right)^{\left(q_j - q_{j-1}\right)/\left(q_{n-1} - q_n\right)}.$$

For either the Odeh-Jones plot or the Agarwal equivalent-time plot, the test data plotted in this fashion produce a single straight line of slope m' and intercept b' at pressures less than the parting pressure, where

$$m' = \frac{162.6\mu B}{kh} \quad \text{(9.35)}$$

$$\text{and } b' = \frac{162.6\mu B}{kh}\left(\log \frac{k}{\phi \mu c_t r_w^2} - 3.23 + 0.869s \right). \quad \text{(9.36)}$$

Skin factor, s, and formation flow capacity, kh, or formation permeability, k, can be estimated from this line at pressures below the parting pressure. At pressures higher than parting pressure, flow is no longer radial and test data no longer lie on the straight line.

9.6.3 Step-Rate Testing for Fractured Wells. For a hydraulically fractured well, the assumption of transient radial flow during the preparting steps is invalid unless the timesteps are sufficiently long and the transient is no longer affected by the fracture (i.e., pseudoradial flow). If the well is fractured, the parting pressure may not be evident either on the conventional pressure-rate plot or from multirate analysis based on transient radial flow. The multirate analysis method must be modified for a fractured well with the linear-flow equation[1,15] given in Chap. 6

$$p_i - p_{wf} = \frac{4.064qB}{L_f h} \sqrt{\frac{\mu t}{\phi c_t k}}. \quad \text{(9.37)}$$

Superposing,

$$\frac{p_i - p_{wf}}{q_n} = \frac{4.064qB}{L_f h} \sqrt{\frac{\mu t}{\phi c_t k}} \sum_{j=1}^{n} \frac{q_j - q_{j-1}}{q_n} \sqrt{t_n - t_{j-1}}. \quad \text{(9.38)}$$

The resulting Odeh-Jones plot prepared with Eqs. 9.37 and 9.38 is similar to that for radial flow. At pressures higher than parting pressure, the data diverge from the preparting data. As before, Agarwal's multirate equivalent-time approach can be used rather than the Odeh-Jones time function.

9.6.4 Two-Step Rate Test. A two-step rate test[16] can be used when pressures are measured continuously during the test. Agarwal's multirate equivalent-time method is used to analyze the test data. The two-step rate test can be run in at least four different modes.

Mode I. Injection has been at a low, stabilized rate before the test. The injection well is then shut in for the first step. The second step consists of a constant injection rate high enough so that parting pressure is exceeded.

Mode II. Injection has been at a low, stabilized rate before the test. The first step of the test consists of a constant injection rate higher than the stabilized rate but low enough to avoid reaching the parting pressure. For the second step, the injection rate is increased to a constant rate high enough so that parting pressure is exceeded.

Mode III. Injection has been at a low, stabilized rate before the test. The injection rate is reduced to a lower constant rate for the first step, then increased to a higher constant rate sufficient to surpass the parting pressure during the second step.

Mode IV. The injection well has been shut in before the test. For the first step, the injection rate is constant and low enough that parting pressure is not reached. For the second step, the injection rate is constant and high enough so that parting pressure is exceeded.

The test data are plotted as an Agarwal multirate equivalent-time plot. The first step provides a baseline pressure transient response of the well/reservoir system below the parting pressure. As long as the parting pressure is not exceeded and wellbore effects are constant during the test, data in the first and second steps coincide on the plot. When the parting pressure is exceeded during the second step, the data deviate from the baseline response and exhibit a decreased slope. The pressure at which the second-step data begin to deviate from the baseline represents the parting pressure.

Design and Analysis Considerations for Two-Step Rate Tests. The first timestep of the test should last beyond the end of wellbore-storage effects.[16] If such a timestep is of impractical length, the duration of the step should be at least long enough for the pressure data to be out of the flow period that is completely dominated by wellbore storage (unit-slope region on log-log plot). For an unfractured well, the first timestep can be set equal to the time required to reach the radial- or pseudoradial-flow regime. For a fractured well, the first timestep can be limited to the end of the linear-flow regime. The appropriate multirate analysis method then can be applied. The duration of the first and second steps of a two-step rate test should be such that the two steps are of similar length on an equivalent-time scale. This design of the timestep lengths ensures that enough baseline data are collected during the first step to permit a comparison with data from the second step. If the relative duration of the two steps is such that the parting pressure is exceeded during the second step at an equivalent time beyond the last data point from the first step, no baseline data correspond to the pressure response in the second step and the parting pressure cannot be determined. For a specific test, the relative duration of the two steps can be determined from the following equations and the proposed injection rates. Agarwal's equivalent-time function for radial flow is

$$\Delta t_{e_{nR}} = \Delta t \prod_{j=1}^{n-1} \left(\frac{t_{n-1} - t_{j-1}}{\Delta t + t_{n-1} - t_{j-1}} \right)^{\left(q_j - q_{j-1}\right)/\left(q_{n-1} - q_n\right)}. \quad \text{(9.39)}$$

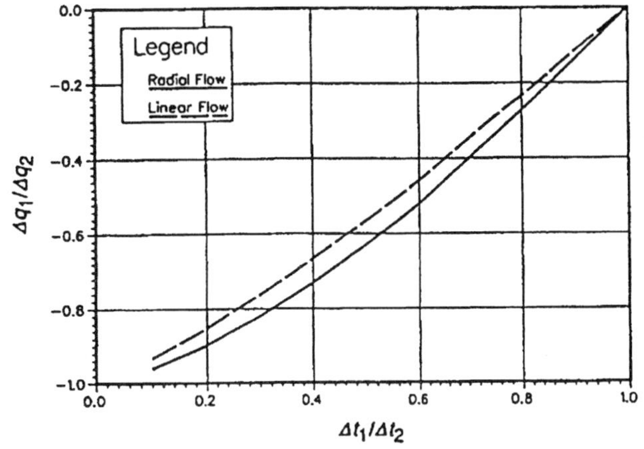

Fig. 9.26—Timestep-size design, Modes I and III (after Singh and Agarwal[16]).

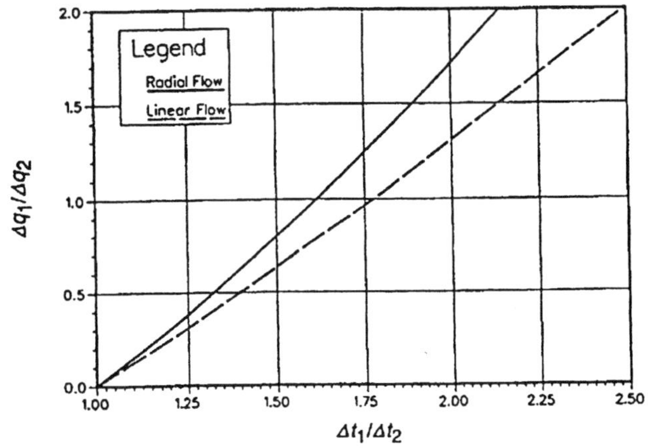

Fig. 9.27—Timestep-size design, Modes II and IV (after Singh and Agarwal[16]).

For linear flow, the equivalent-time function is

$$\Delta t_{e_{nR}} = \Delta t + \sum_{j=1}^{n-1} \times \left[\left(\frac{q_j - q_{j-1}}{q_{n-1} - q_n} \right) \left(\sqrt{t_{n-1} - t_{j-1}} - \sqrt{\Delta t + t_{n-1} - t_{j-1}} \right) \right]. \quad (9.40)$$

The pressure function for either case is

$$\frac{\Delta p}{\Delta q} = \frac{p_{wf_n}(\Delta t) - p_{wf_{n-1}}(t_{n-1})}{q_{n-1} - q_n}. \quad (9.41)$$

Alternatively, **Figs. 9.26 and 9.27** can be used to determine the relative duration of the two steps. Fig. 9.26 gives the ratio of rate increments ($\Delta q_0/\Delta q_1$) as a function of the ratio timestep sizes ($\Delta t_1/\Delta t_2$) for Modes I and III for radial and linear flow. Fig. 9.27 gives similar data for Modes II and IV. For a constant injection rate, q_0, followed by rate increments of $\Delta q_1 = q_1 - q_0$ for time Δt_1 in the first step and $\Delta q_2 = q_2 - q_1$ for Δt_2 in the second step, either Fig. 9.26 or Fig. 9.27 gives the ratio of timesteps ($\Delta t_1/\Delta t_2$) corresponding to $\Delta q_1/\Delta q_2$.

The duration of the second step, Δt_2, can be found from this ratio and the desired duration of the first step, Δt_1. Equivalent times based on Δt_1 and Δt_2 then are similar, ensuring a satisfactory baseline from the first-step data for comparison with the second-step data. For most two-step rate tests, equal timestep size can be used for both steps for Modes I and III. For Modes II and IV, the first step should be twice as long as the second step.

The relationships in Figs. 9.26 and 9.27 are based on the assumption that injection before the test was stabilized at constant rate for a much longer time than the duration of the first step. If the injection rate varies shortly before the test, then the appropriate equivalent time equation for radial or linear flow should be used in conjunction with the historical and planned injection history to calculate the equivalent times for both steps.

An important requirement is that the pretest period and each step of the test be in the same flow regime. If the pretest period is in a different flow regime than the test steps, the data for the two steps may not overlay. The two-step rate test then is inconclusive, and the parting pressure cannot be determined.

In selecting the injection rates to be used in the test, pressures at both the stabilized pretest rate and the rate during the second step must not exceed the parting pressure. This condition can be satisfied for the Modes I and III procedures by setting the stabilized rate well below the last estimated parting pressure and for the Modes II and IV procedures by choosing the pretest and first-step rates on the basis of a previous conventional step-rate test from the same well.

While the objective of the two-step rate test is to exceed the parting pressure during the second step, the injection rate during the second step should not be so high that the parting pressure is reached very early in the second step. If insufficient preparting data are obtained during the second step to compare with the first-step baseline data, the parting pressure may not be apparent. Therefore, the rate during the second step should be high enough that the parting pressure is exceeded during the latter half of the step. This aspect of the test can be designed on the basis of either an analytical or numerical model if good descriptions of the reservoir and well are available or if one has previous experience with the subject well or neighboring wells.

Although errors in pressure or rate data affect the plots of the test data, the parting pressure can still be determined in some cases. Data-measurement errors that may occur during a test include inaccurate corrections to surface pressure when BHP cannot be measured directly, fluctuations in the injection rate during a step, and inaccurate measurement of injection rates.[16]

Testing Considerations for Two-Step Rate Tests. The interpretation of a two-step rate test[16] depends solely on observed changes in the pressure transient response resulting from exceeding the parting pressure during a single step. Therefore, a pressure recorder accurate at both low- and high-flow rates must be used to measure pressures for the test.

The two-step-rate-test analysis is based on the assumptions that rate changes during the two steps of the test are instantaneous and that the rates are constant before and after these instantaneous changes. For the test to provide accurate data for determining the parting pressure, test conditions should satisfy these assumptions as closely as possible. Any rate changes that occur before the start of the test must be measured accurately and accounted for in the superposition analysis.

Accurate test data are obtained only if the data are not influenced by wellbore-storage effects. If wellbore storage changes during the test, the first- and second-step data plots do not overlie one another at pressures below the parting pressure, and the parting pressure cannot be determined. The appropriate mode should be chosen in designing the test so that wellbore storage remains constant during the test.

If an online pressure readout is used during a test and indicates that the parting pressure may not have been exceeded during the second step, a third step at a higher injection rate can be included as part of the test. The data from the resulting three-step rate test are analyzed as for a two-step test.

Example 9.5—Two-Step Rate Test of a Fractured Well With Radial-Flow Analysis. Problem: A two-step rate test was performed on a water-injection well. The well was stabilized at an injection rate

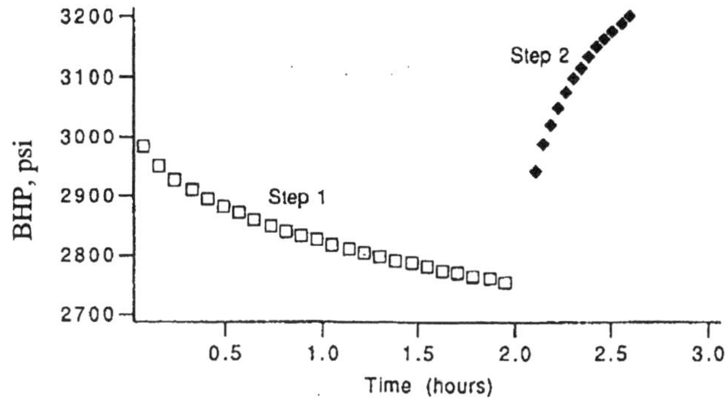

Fig. 9.28—Pressure and time data, Example 9.5.

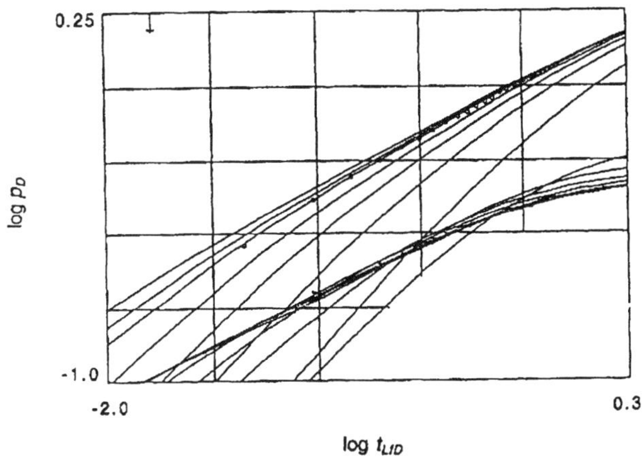

Fig. 9.29—Type-curve match of Step 1 data, Example 9.5

of $q_1 = 800$ STB/D for 1 week ($n = 1$, $t_1 = 168$ hours) before the test. The test consisted of a shut-in period of 2.031 hours (Step 1, $n = 2$, $t_2 = 170$ hours, $q_2 = 0$) followed by a 1-hour injection period (Step 2, $n = 3$) at a rate of $q_3 = 2{,}000$ STB/D. **Fig. 9.28 and Table 9.8** give BHP and time data. Other known data are summarized next. Determine fracture-parting pressure.

$k =$ 4.8 md
$C =$ 0.005 bbl/psi
$s =$ 0
$B =$ 1.0 RB/STB
$r_w =$ 0.3 ft
$\mu =$ 0.8 cp
$h =$ 72.5 ft
$c_t = 3.0 \times 10^{-6}$ psi^{-1}
$L_f =$ 90 ft
$\phi =$ 0.133

Solution. Type-curve analysis of the Step 1 data, shown in **Fig. 9.29,** indicates that the duration of Step 1 is long enough for the data to be beyond the linear-flow regime and into the pseudoradial-flow transition period. Hence, multirate analysis with radial-flow calculations is appropriate for finding the fracture-parting pressure.

Agarwal's radial-flow equivalent time, $\Delta t_{e_{nR}}$, and the pressure function, $\Delta p/\Delta q$, are calculated for each of the two steps of the test ($n = 2$ and $n = 3$), as Table 9.8 shows. For example, the equivalent time at $p_{wf} = 2{,}770.37$ psi in Step 1 ($n = 2$) is

$$\Delta t_{e_{nR}} = \Delta t \prod_{j=1}^{n-1} \left(\frac{t_{n-1} - t_{j-1}}{\Delta t + t_{n-1} - t_{j-1}} \right)^{(q_j - q_{j-1})/(q_{n-1} - q_n)}$$

and $$\Delta t_{e_{2R}} = \Delta t \left(\frac{t_1 - t_0}{\Delta t + t_1 - t_0} \right)^{(q_1 - q_0)/(q_{n-1} - q_n)}$$

$$= (1.7060 - 0)$$

$$\times \left(\frac{168 - 0}{1.7060 - 0 + 168 - 0} \right)^{(-800-0)/(-800-0)}$$

$$= 1.6889 \text{ hours}$$

and the pressure function is

$$\frac{\Delta p}{\Delta q} = \frac{p_{wf_n}(\Delta t) - p_{wf_{n-1}}(t_{n-1})}{q_{n-1} - q_n} = \frac{p_{wf_2} - p_{ref}}{q_1 - q_2}$$

TABLE 9.8—DATA FOR EXAMPLE 9.5

Time (hours)	p_{wf} (psi)	Pressure Derivative	Δt_{enR} (hours)	$\Delta p/\Delta q$ (psi/STB-D)
		Step 1		
0.0000	3,059.000			
0.0812	2,982.254	46.705	0.0812	0.0959
0.1625	2,949.867	51.832	0.1623	0.1364
0.2437	2,927.678	58.752	0.2434	0.1642
0.3250	2,910.036	64.779	0.3244	0.1862
0.4062	2,895.195	69.662	0.4052	0.2048
0.4874	2,882.139	74.016	0.4860	0.2211
0.5687	2,870.504	78.039	0.5668	0.2356
0.6499	2,859.946	81.417	0.6474	0.2488
0.7312	2,850.199	84.258	0.7280	0.2610
0.8124	2,841.110	87.034	0.8085	0.2724
0.8936	2,832.811	89.678	0.8889	0.2827
0.9749	2,825.024	91.862	0.9693	0.2925
1.0561	2,817.653	93.696	1.0495	0.3017
1.1374	2,810.641	95.251	1.1297	0.3104
1.2186	2,803.943	97.007	1.2098	0.3188
1.2998	2,797.561	98.446	1.2898	0.3268
1.3811	2,791.668	100.27	1.3698	0.3342
1.4623	2,786.044	101.79	1.4497	0.3412
1.5436	2,780.634	103.09	1.5296	0.3480
1.6248	2,775.416	104.54	1.6092	0.3545
1.7060	2,770.370	105.83	1.6888	0.3608
1.7873	2,765.482	107.18	1.7685	0.3669
1.8685	2,760.741	108.75	1.8480	0.3728
1.9498	2,756.135	110.25	1.9274	0.3786
2.0310	2,751.656	2.0067	0.3842	
		Step 2		
2.0710	2,879.148	0.0397	0.0637	
2.1110	2,941.543	0.0788	0.0949	
2.1510	2,984.976	0.1173	0.1167	
2.1910	3,019.437	0.1553	0.1339	
2.2310	3,048.677	0.1927	0.1485	
2.2710	3,074.148	0.2296	0.1612	
2.3110	3,095.010	0.2661	0.1717	
2.3510	3,113.920	0.3020	0.1811	
2.3910	3,131.085	0.3375	0.1897	
2.4310	3,148.915	0.3726	0.1986	
2.4710	3,161.659	0.4072	0.2050	
2.5110	3,175.486	0.4414	0.2119	
2.5510	3,188.486	0.4753	0.2184	
2.5910	3,200.615	0.5087	0.2245	
2.6310	3,206.091	0.5417	0.2272	
2.6710	3,213.005	0.5744	0.2307	
2.7110	3,219.421	0.6068	0.2339	
2.7510	3,225.389	0.6388	0.2369	
2.7910	3,230.950	,0.6705	0.2396	
2.8310	3,236.139	0.7018	0.2422	
2.8710	3,240.813	0.7328	0.2446	
2.9110	3,245.045	0.7636	0.2467	
2.9510	3,248.951	0.7940	0.2486	
2.9910	3,252.575	0.8242	0.2505	
3.0310	3,255.941	0.8540	0.2521	

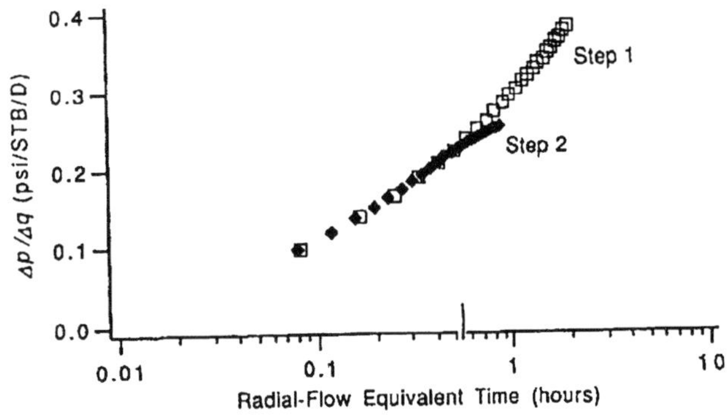

Fig. 9.30—Radial-flow equivalent-time plot, Example 9.5.

$$= \frac{2777.3 - 3057.0}{-800 - 0} = 0.3608 \text{ psi/STB-D}.$$

Similarly, the equivalent time at $p_{wf}=3{,}095.01$ psi in Step 2 ($n=3$) is

$$\Delta t_{e_{3R}} = \Delta t \left(\frac{t_2 - t_0}{\Delta t + t_2 - t_0}\right)^{(q_1-q_0)/(q_2-q_3)}$$

$$\times \left(\frac{t_2 - t_1}{\Delta t + t_2 - t_1}\right)^{(q_2-q_1)/(q_2-q_3)}$$

$$= (2.3110 - 2.03100)$$

$$\times \left(\frac{170.031 - 0}{2.3110 - 2.0310 + 170.031 - 168}\right)^{(-800-0)/[0-(-2000)]}$$

$$\times \left(\frac{170.031 - 168}{2.3110 - 2.0310 + 170.031 - 168}\right)^{[-0(-800)]/[0-(-2000)]}$$

$$= 0.2661 \text{ hour}$$

and the pressure function is

$$\frac{\Delta p}{\Delta q} = \frac{p_{wf_3} - p_{wf_2}}{q_2 - q_3}$$

$$= \frac{3095.01 - 2751.656}{0 - (-2000)}$$

$$= 0.1717 \text{psi/STB-D}.$$

Rates are negative for injection. In the expressions for equivalent time, Δt represents the elapsed time from the beginning of the test. In the pressure function, Δp is the difference between measured p_{wf} and p_{wf} at the beginning of the step.

Fig. 9.30 shows a radial-flow multirate equivalent-time plot of the two-step-rate-test data. Data for the two steps coincide up until an equivalent time of 0.51 hour, after which the Step 2 data deviate from the baseline data of Step 1. From Table 9.8, the pressure that corresponds to the equivalent time of 0.51 hour in Step 2 is 3,201 psi, which is interpreted to be the fracture-parting pressure.

9.7 Chapter Summary

In Chap. 9, we present an overview of pressure transient testing of injection wells.

In Sec. 9.2, we discuss methods for analyzing injectivity tests under unit-mobility-ratio reservoir conditions. Injectivity tests are conducted by injecting into the reservoir at a constant rate and measuring the resulting pressure response. Thus, injectivity tests correspond directly to drawdown tests in producing wells.

Unit-mobility-ratio conditions occur when the pressure transients established during a test encounter only fluid having a single value of mobility. This can occur (1) when the injected fluid and the displaced fluid have the same mobility, as in a brine-disposal well; (2) when the injection period before the test is long enough so that the pressure transients during the test encounter only that region of the reservoir that has been swept by the injection fluid; or (3) when the injection period before the test is short enough so that the region swept by the injected fluid can be modeled by a zone of altered permeability, or skin.

If unit-mobility-ratio reservoir conditions are present, injectivity tests may be analyzed with techniques that are analogous to those used in analyzing drawdown tests. We illustrate the application of both semilog and type-curve analysis methods to injectivity tests through an example.

Just as injectivity tests correspond to drawdown tests in producing wells, injection-well falloff tests correspond to buildup tests in producing wells. In Sec. 9.3, we discuss analysis of falloff tests under unit-mobility-ratio conditions. Again, we can use both Horner semilog-analysis and type-curve-analysis methods that have been developed for producing wells. We illustrate these techniques with an example.

In Sec. 9.4, we briefly discuss the application of the Matthews *et al.* method for estimating average drainage-area pressure to injection wells. In addition to the assumption of unit-mobility-ratio reservoir conditions, application of the Matthews *et al.* method to injection wells requires that the injected and displaced fluids have comparable values of $\phi c_t h$.

In Sec. 9.5, we turn our attention to composite systems (that is, systems where the assumption of unit-mobility-ratio conditions is not valid). In this situation, there are at least two and possibly three different zones with different fluid saturations and mobility ratios.

In Sec. 9.5.1, we discuss the Hazebroek *et al.* method for analyzing falloff tests in injection wells before fillup. This method assumes that (1) a significant gas saturation exists in the reservoir outside the oil bank, (2) the pressure at the outer edge of the oil bank is constant, and (3) wellbore-storage effects are negligible.

The Hazebroek *et al.* method requires the following information: the mobility ratio, the ratio of the total compressibilities in the oil and water banks, and the ratio of the inner and outer radii of the oil bank. The size of the oil bank can be estimated if the fluid saturations in the three zones and the water-injection volume are known.

The first step in applying the Hazebroek *et al.* method is to determine the pressure at the outer edge of the oil bank by a trial-and-error process with a semilog plot of $\log(p_{ws} - p_e)$ vs. Δt. The permeability thickness product $k_w h$ is then determined from the y intercept of the plot with correlations that Hazebroek *et al.* developed. After $k_w h$ is determined, the effective permeability-thickness product for oil in the oil bank, $k_o h$, is estimated from the mobility ratio and the viscosities of the oil and water.

We present an example analysis of a falloff test using the Hazebroek *et al.* method.

In Sec. 9.5.2, we discuss the Merrill *et al.* method for analyzing falloff tests in liquid-filled reservoirs. This method allows the engineer to estimate both the location of the front of the injected-fluid bank and the permeabilities of the two fluid banks. The Merrill *et al.* method is developed for situations where the oil bank is much larger than the water bank, $r_{f2}/r_{f1} > 50$. However, it provides acceptable results when $r_{f2}/r_{f1} > 10$. Unlike the Hazebroek *et al.* method, the Merrill *et al.* method does not require prior knowledge of the mobility ratio.

A conventional Horner plot is used as the basis for the Merrill *et al.* method. When this method is applicable, two semilog straight lines are present, reflecting responses from the two fluid banks. The distance to the front of the injected-fluid bank may be estimated from the time of intersection of the two straight lines or from the time when the observed pressure data deviate from the first straight line. The permeability in the injected-fluid bank and the skin factor may be obtained from the slope of the first semilog straight line. Although the permeability in the oil bank is related to the slope of the second semilog straight line, it may not be computed directly from the slope. Instead, the permeability in the oil bank must be estimated from the permeability in the water bank by use of correlations that Merrill *et al.* developed. We illustrate the use of the Merrill *et al.* method with an example.

In Sec. 9.6, we discuss step-rate testing. The objective of step-rate testing is to determine the parting pressure, which is the pressure at which fractures are created or extended. Care must be taken to avoid injecting at pressures higher than the parting pressure to prevent premature breakthrough, poor sweep efficiency, reduced oil recovery, and loss of expensive injection fluids.

In Secs. 9.6.1 through 9.6.3, we discuss step-rate-test analysis with conventional and multirate analysis methods. The conventional analysis method, discussed in Sec. 9.6.1, requires either steady-state flow be achieved during the test or that equal-length timesteps be used. The parting pressure is determined from the intersection of two straight lines on a plot of injection BHP vs. injection rate. The conventional analysis method uses only a single pressure point at each rate.

In Sec. 9.6.2, we discuss the multirate analysis method. This method requires pressures to be recorded continuously, rather than at the end of each injection period as in the conventional method. The multirate method assumes transient radial flow into the reservoir during each constant-rate-flow period. In Sec. 9.6.3, we extend the multirate method to apply to linear flow to a vertical fracture.

In Sec. 9.6.4, we present a special case of multirate testing, the two-step rate test. In this test, the first rate is selected to be low enough so that the parting pressure is not exceeded during the first injection period, while the second rate is selected so ensure that the parting pressure is exceeded during the second period. The two tests are plotted as pressure difference divided by flow rate vs. the Agarwal multirate equivalent-time function. If the parting pressure is not exceeded in either test, the data from the two tests lie along a single trend. Departure from the trend established by the first injection period indicates that the parting pressure was exceeded during the second injection period. The parting pressure is then obtained directly from the pressure corresponding to the point of departure from the trend. We present an example application of the two-step-rate-test method to determine parting pressure.

Exercises

1. An injection test was conducted in a salt water disposal well. Given the following rock and fluid properties, estimate permeability and skin factor from the injection test data. $q_w = 450$ STB/D; $h = 54$ ft; $B_w = 1.028$ RB/STB; $p_i = 3430$ psia; $\phi = 20.8\%$; $c_t = 6.16 \times 10^{-6}$ psi^{-1}; $r_w = 0.46$ ft; $\mu = 0.65$ cp.

PRESSURE-TIME DATA FOR EXERCISE 9.1

Time (hours)	Pressure (psi)	Time (hours)	Pressure (psi)	Time (hours)	Pressure (psi)
0.0	3430.00	0.831	3623.03	8.57	3767.05
0.0250	3438.91	1.064	3651.97	10.74	3769.27
0.0563	3449.66	1.355	3679.32	13.45	3771.33
0.0953	3462.55	1.719	3703.61	16.84	3773.29
0.1441	3477.85	2.17	3723.68	21.1	3775.18
0.205	3495.79	2.74	3738.96	26.4	3777.03
0.281	3516.50	3.45	3749.69	33.0	3778.85
0.377	3539.98	4.34	3756.76	41.3	3780.63
0.496	3565.98	5.45	3761.34	48.0	3781.83
0.645	3593.96	6.84	3764.54		

2. An injection test was conducted in an injection well in a mature waterflood project. Given the following rock and fluid properties, estimate permeability and skin factor from the injection test data, assuming that unit-mobility ratio conditions apply. $q_w = 1750$ STB/D; $h = 79$ ft; $B_w = 1.02$ RB/STB; $p_i = 2402.90$ psia; $\phi = 20.0\%$; $c_t = 6.08 \times 10^{-6}$ psi^{-1}; $r_w = 0.31$ ft; $\mu = 0.78$ cp.

PRESSURE-TIME DATA FOR EXERCISE 9.2

Time (hours)	Pressure (psi)	Time (hours)	Pressure (psi)	Time (hours)	Pressure (psi)
0.0	2402.90	0.333	2721.91	3.43	2877.79
0.0100	2419.43	0.426	2760.83	4.30	2880.17
0.0225	2439.19	0.542	2794.62	5.38	2882.43
0.0381	2462.58	0.688	2821.75	6.74	2884.62
0.0577	2489.91	0.869	2841.69	8.43	2886.77
0.0821	2521.34	1.079	2855.11	10.55	2888.87
0.1126	2556.76	1.381	2863.54	13.19	2890.95
0.1507	2595.72	1.736	2868.80	16.50	2893.00
0.1984	2637.28	2.18	2872.40	20.6	2895.03
0.258	2679.99	2.74	2875.26	24.0	2896.40

3. An injection-falloff test was conducted in a salt water disposal well. Given the following rock and fluid properties, estimate permeability and skin factor from the injection-falloff test data. $q_w = 625$ STB/D; $h = 18$ ft; $B_w = 1.007$ RB/STB; $t_p = 24$ hr; $\phi = 24.4\%$; $c_t = 6.08 \times 10^{-6}$ psi^{-1}; $p_{wf} = 1927.92$ psia; $r_w = 0.25$ ft; $\mu = 1.10$ cp.

PRESSURE-TIME DATA FOR EXERCISE 9.3

Time (hours)	Pressure (psi)	Time (hours)	Pressure (psi)	Time (hours)	Pressure (psi)
0.0	1927.92	0.01064	1479.11	0.1345	1379.17
0.00025	1887.24	0.01355	1463.46	0.1684	1372.35
0.00056	1842.62	0.01719	1451.06	0.211	1365.61
0.00095	1794.65	0.0217	1440.69	0.264	1358.94
0.00144	1744.52	0.0274	1431.54	0.330	1352.32
0.00205	1693.88	0.0345	1423.12	0.413	1345.75
0.00281	1644.77	0.0434	1415.19	0.516	1339.23
0.00377	1599.31	0.0545	1407.60	0.645	1332.77
0.00496	1559.34	0.0684	1400.26	0.807	1326.35
0.00645	1526.01	0.0857	1393.10	1.000	1320.25
0.00831	1499.52	0.1074	1386.08		

4. An injection-falloff test was conducted in an injection well in a mature waterflood project. Given the following rock and fluid properties, estimate permeability and skin factor from the injection-falloff test data, assuming that unit-mobility ratio conditions apply. $q_w = 600$ STB/D; $h = 15$ ft; $B_w = 1.006$ RB/STB; $t_p = 720$ hr; $\phi = 20.6\%$; $c_t = 5.98 \times 10^{-6}$ psi^{-1}; $p_i = 1135.83$ psia; $r_w = 0.29$ ft; $\mu = 1.32$ cp.

PRESSURE-TIME DATA FOR EXERCISE 9.4

Time (hours)	Pressure (psi)	Time (hours)	Pressure (psi)	Time (hours)	Pressure (psi)
0.0	1135.83	0.0593	930.72	0.502	841.17
0.00230	1091.43	0.0765	919.17	0.629	832.22
0.00518	1058.16	0.0979	908.28	0.789	823.35
0.00877	1030.97	0.1247	897.89	0.988	814.55
0.01326	1008.15	0.1581	887.87	1.238	805.79
0.01888	988.64	0.200	878.15	1.549	797.08
0.0259	971.67	0.252	868.67	1.939	788.40
0.0347	956.65	0.318	859.36	2.000	787.20
0.0456	943.12	0.399	850.21		

5. An injection-falloff test was conducted in an injection well in a waterflood project. Given the following rock and fluid properties, estimate permeability and skin factor from the injection-falloff test data using the Hazebroek-Rainbow-Matthews Method. $p_e = 47$ psi; $q_w = 649$ STB/D; $h = 23$ ft; $\mu_w = 0.75$ cp; $S_g = 0.14$; $S_o = 0.65$; $p_{wf} = 543$ psi; $W_i = 8121$ bbl; $\phi = 23.4\%$; $\mu_o = 6.0$ cp; $S_{gr} = 0.0$; $S_{or} = 0.23$; c_t*(injected water zone)* $= 4 \times 10^{-6}$ psi^{-1}; c_t*(oil*

zone) $= 16 \times 10^{-6}$ psi^{-1}; $k_w/k_o = 0.25$; $d_t = 2.992$ in.; $r_w = 0.33$ ft; $\rho_w = 62.4$ lbm/ft^3.

PRESSURE-TIME DATA FOR EXERCISE 9.5					
Time (hours)	Pressure (psi)	Time (hours)	Pressure (psi)	Time (hours)	Pressure (psi)
1.00	175.80	9.00	69.47	17.00	58.66
2.00	101.75	10.00	67.70	18.00	57.74
3.00	86.33	11.00	66.07	19.00	56.90
4.00	81.31	12.00	64.57	20.00	56.12
5.00	78.27	13.00	63.19	21.00	55.40
6.00	75.75	14.00	61.91	22.00	54.74
7.00	73.48	15.00	60.74	23.00	54.13
8.00	71.39	16.00	59.66	24.00	53.57

6. An injection-falloff test was conducted in an injection well in a waterflood project. Given the following rock and fluid properties, estimate permeability and skin factor from the injection-falloff test data using the Hazebroek-Rainbow-Matthews Method. $p_e = 125$ psi; $q_w = 225$ STB/D; $h = 18$ ft; $\mu_w = 0.75$ cp; $S_g = 0.08$; $S_o = 0.332$; $p_{wf} = 1293$ psi; $W_i = 1920$ bbl; $\phi = 18.5\%$; $\mu_o = 0.6$ cp; $S_{gr} = 0.0$; $S_{or} = 0.19$; c_t(*injected water zone*) $= 4 \times 10^{-6}$ psi^{-1}; c_t(*oil zone*) $= 8 \times 10^{-6}$ psi^{-1}; $k_w/k_o = 0.5$; $d_t = 3.45$ in.; $r_w = 0.25$ ft; $\rho_w = 62.4$ lbm/ft^3.

PRESSURE-TIME DATA FOR EXERCISE 9.6					
Time (hours)	Pressure (psi)	Time (hours)	Pressure (psi)	Time (hours)	Pressure (psi)
1.00	417.18	9.00	276.07	17.00	239.17
2.00	330.50	10.00	270.87	18.00	235.25
3.00	313.06	11.00	265.85	19.00	231.45
4.00	305.19	12.00	261.01	20.00	227.79
5.00	298.80	13.00	256.33	21.00	224.26
6.00	292.80	14.00	251.81	22.00	220.84
7.00	287.02	15.00	247.45	23.00	217.55
8.00	281.45	16.00	243.24	24.00	214.36

7. An injection-falloff test was conducted in an injection well in a waterflood project. Given the following rock and fluid properties, estimate effective permeability in each zone and skin factor from the injection-falloff test data using the Merrill-Kazemi-Gogarty method. $t_p = 720$ hr; $r_w = 0.32$ ft; $\mu_w = 0.75$ cp; $S_{w1} = 0.77$; $S_{o1} = 0.23$; $c_{t1} = 6 \times 10^{-6}$ psi^{-1}; $q_w = 325$ STB/D; $\phi = 18.5\%$; $\mu_o = 6$ cp; $S_{w2} = 0.35$; $S_{o2} = 0.65$; $c_{t2} = 12 \times 10^{-6}$ psi^{-1}; $p_i = 623$ psi; $p_{wf} = 883.3$ psi; $h = 27$ ft; $B_w = 1.0$ bbl/STB.

PRESSURE-TIME DATA FOR EXERCISE 9.7					
Time (hours)	Pressure (psi)	Time (hours)	Pressure (psi)	Time (hours)	Pressure (psi)
0.0	883.30	0.1097	748.43	3.23	711.63
0.00100	789.29	0.1381	746.40	4.03	708.49
0.00225	782.20	0.1736	744.37	5.04	705.28
0.00381	777.61	0.218	742.30	6.31	702.00
0.00577	774.01	0.274	740.19	7.88	698.66
0.00821	770.94	0.343	738.02	9.86	695.26
0.01126	768.19	0.430	735.79	12.32	691.82
0.01507	765.66	0.538	733.47	15.40	688.35
0.01984	763.28	0.674	731.07	19.26	684.84
0.0258	761.00	0.843	728.58	24.1	681.32
0.0333	758.80	1.055	725.99	30.1	677.79
0.0426	756.65	1.319	723.31	37.6	674.27
0.0542	754.56	1.650	720.53	47.0	670.75
0.0688	752.49	2.06	717.65	48.0	670.43
0.0869	750.45	2.58	714.68		

8. An injection-falloff test was conducted in an injection well in a waterflood project. Given the following rock and fluid properties, estimate effective permeability in each zone and skin factor from the injection-falloff test data using the Merrill-Kazemi-Gogarty method. $t_p = 1440$ hr; $r_w = 0.37$ ft; $\mu_w = 0.6$ cp; $S_{w1} = 0.73$; $S_{o1} = 0.27$; $c_{t1} = 7.5 \times 10^{-6}$ psi^{-1}; $q_w = 575$ STB/D; $\phi = 22.3\%$; $\mu_o = 10.3$ cp; $S_{w2} = 0.28$; $S_{o2} = 0.72$; $c_{t2} = 7.5 \times 10^{-6}$ psi^{-1}; $p_i = 1193$ psi; $p_{wf} = 1541.43$ psi; $h = 39$ ft; $B_w = 1.02$ bbl/STB.

PRESSURE-TIME DATA FOR EXERCISE 9.8					
Time (hours)	Pressure (psi)	Time (hours)	Pressure (psi)	Time (hours)	Pressure (psi)
0.0	1541.43	0.1381	1441.18	5.04	1385.73
0.00100	1476.59	0.1736	1439.70	6.31	1379.45
0.00225	1467.80	0.218	1438.10	7.88	1372.92
0.00381	1463.73	0.274	1436.35	9.86	1366.16
0.00577	1460.87	0.343	1434.40	12.32	1359.22
0.00821	1458.54	0.430	1432.20	15.40	1352.12
0.01126	1456.52	0.538	1429.73	19.26	1344.89
0.01507	1454.69	0.674	1426.94	24.1	1337.56
0.01984	1452.99	0.843	1423.82	30.1	1330.16
0.0258	1451.37	1.055	1420.35	37.6	1322.71
0.0333	1449.83	1.319	1416.50	47.0	1315.25
0.0426	1448.33	1.650	1412.29	58.8	1307.78
0.0542	1446.87	2.06	1407.69	73.5	1300.34
0.0688	1445.43	2.58	1402.72	91.8	1292.94
0.0869	1444.02	3.23	1397.39	96.0	1291.48
0.1097	1442.61	4.03	1391.72		

9. An two-step rate test was conducted in a salt water disposal well to determine fracture parting pressure. Given the following rock and fluid properties, estimate formation permeability, skin factor, and fracture parting pressure. $h = 54$ ft; $B_w = 1.028$ bbl/STB; $q_1 = 450$ STB/D; $q_2 = 0$ STB/D; $q_3 = 2700$ STB/D; $\phi = 20.8\%$; $c_t = 6.16 \times 10^{-6}$ psi^{-1}; $t_1 = 288$ hrs; $t_2 = 292$ hrs; $t_3 = 296$ hrs; $r_w = 0.46$ ft; $\mu = 0.65$ cp; $\Delta t_2 = 4$ hrs; $\Delta t_3 = 4$ hrs.

Step 1 pressures (falloff test, $q = 0$ STB/D):

PRESSURE-TIME DATA FOR EXERCISE 9.9, STEP 1					
Time (hours)	Pressure (psi)	Time (hours)	Pressure (psi)	Time (hours)	Pressure (psi)
0.0	3564.21	0.0426	3497.80	0.538	3477.66
0.00100	3547.32	0.0542	3495.75	0.674	3475.95
0.00225	3535.32	0.0688	3493.79	0.843	3474.24
0.00381	3526.48	0.0869	3491.88	1.055	3472.54
0.00577	3519.88	0.1097	3490.02	1.319	3470.84
0.00821	3514.84	0.1381	3488.19	1.650	3469.15
0.01126	3510.85	0.1736	3486.40	2.06	3467.46
0.01507	3507.57	0.218	3484.62	2.58	3465.77
0.01984	3504.75	0.274	3482.86	3.23	3464.09
0.0258	3502.24	0.343	3481.12	4.00	3462.49
0.0333	3499.94	0.430	3479.38		

Step 2 pressures (injection test, $q = 2700$ STB/D):

PRESSURE-TIME DATA FOR EXERCISE 9.9, STEP2					
Time (hours)	Pressure (psi)	Time (hours)	Pressure (psi)	Time (hours)	Pressure (psi)
0.0	3462.49	0.0426	3860.89	0.538	3971.14
0.00100	3563.79	0.0542	3873.13	0.674	3978.78
0.00225	3635.78	0.0688	3884.90	0.843	3986.27
0.00381	3688.82	0.0869	3896.31	1.055	3993.56
0.00577	3728.43	0.1097	3907.44	1.319	4000.60
0.00821	3758.69	0.1381	3918.34	1.650	4007.31
0.01126	3782.62	0.1736	3929.07	2.06	4013.66
0.01507	3802.32	0.218	3939.66	2.58	4019.58
0.01984	3819.22	0.274	3950.12	3.23	4025.05
0.0258	3834.25	0.343	3956.33	4.00	4029.82
0.0333	3848.01	0.430	3963.53		

References

1. Earlougher, R.C. Jr.: *Advances in Well Test Analysis*, Monograph Series, SPE, Richardson, Texas (1977) **5**, 74–89.
2. Craig, F.F. Jr.: *The Reservoir Engineering Aspects of Waterflooding*, Monograph Series, SPE, Richardson, Texas (1971) **3**, 79.
3. Bourdet, D. *et al.*: "A New Set of Type Curves Simplifies Well Test Analysis," *World Oil* (May 1983) 95.

4. Horner, D.R.: "Pressure Buildup in Wells," *Proc.*, Third World Pet. Cong., The Hague, The Netherlands (1951) Sec. II, 503-523; *Pressure Analysis Methods*.
5. Pinson, A.E. Jr.: "Concerning the Value of Producing Time Used in Average Pressure Determinations From Pressure Buildup Analysis," *JPT* (November 1972) 1369.
6. Kazemi, H.: "Determining Average Reservoir Pressure From Pressure Buildup Tests," *SPEJ* (February 1974) 55; *Trans.*, AIME, **257.**
7. Agarwal, R.G.: "A New Method To Account for Producing Time Effects When Drawdown Type Curves Are Used To Analyze Pressure Buildup and Other Test Data," paper SPE 9289 presented at the 1980 SPE Annual Technical Conference and Exhibition, Dallas, 21–24 September.
8. Matthews, C.S., Brons, F., and Hazebroek, P.: "A Method for Determination of Average Pressure in a Bounded Reservoir," *Trans.*, AIME (1954) **201,** 182.
9. Carter, R.D.: "Pressure Behavior of a Limited Circular Composite Reservoir," *SPEJ* (December 1966) 328; *Trans.*, AIME (1966) **237.**
10. Hazebroek, P., Rainbow, H., and Matthews, C.S.: "Pressure Fall-Off in Water Injection Wells," *Trans.*, AIME (1958) **213,** 250.
11. Matthews, C.S. and Russell, D.G.: *Pressure Buildup and Flow Tests in Wells*, Monograph Series, SPE, Richardson, Texas (1977) **1,** 72–83.
12. Merrill, L.S., Kazemi, H., and Gogarty, W.B.: "Pressure Falloff Analysis in Reservoirs With Fluid Banks," *JPT* (July 1974) 809; *Trans.*, AIME (1974) **257.**
13. Hall, H.N.: "How To Analyze Waterflood Injection Performance," *World Oil* (October 1963) 128.
14. Nolte, K.G.: "Principles for Fracture Design Based on Pressure Analysis," *SPEPE* (February 1988) 22; *Trans.*, AIME, **285.**
15. Singh, P.K., Agarwal, R.G., and Krase, L.D.: "Systematic Design and Analysis of Step-Rate Tests To Determine Formation Parting Pressure," paper SPE 16798 presented at the 1987 SPE Annual Technical Conference and Exhibition, Dallas, 27–30 September.
16. Singh, P.K. and Agarwal, R.G.: "Two-Step Rate Test: New Procedure for Determining Formation Parting Pressure," *JPT* (January 1990) 84.
17. Aronofsky, J.S. and Jenkins, R.: "A Simplified Analysis of Unsteady Radial Gas Flow," *Trans.*, AIME (1954) **201,** 149.
18. Odeh, A.S. and Jones, L.G.: "Pressure Drawdown Analysis, Variable-Rate Case," *JPT* (August 1965) 960; *Trans.*, AIME (1965) **234.**

Chapter 10
Interference and Pulse Testing

10.1 Overview

Well tests in which a pressure disturbance is created in one well and the pressure response is measured in one or more wells are called multiple-well tests. Multiple-well tests are designed to determine properties in a region centered along a line connecting pairs of test wells and, therefore, are more sensitive to directional variations of reservoir properties, such as permeability. These tests also are ideal for establishing continuity of the pay zone within a reservoir, estimating the volume of hydrocarbons within the drainage area of the wells, and quantifying reservoir anisotropies and heterogeneities. In this chapter, we discuss the underlying theoretical concepts and practical applications of two common multiple-well tests—interference and pulse tests.

10.2 Interference Tests

An interference test involves creating a significant pressure disturbance, either by producing from or injecting into at least one well, called the active well, and observing the pressure response in at least one other well, called the observation well, located a distance r from the active well (**Fig. 10.1**).[1,2]

The characteristics of the pressure behavior as a function of time reflect the reservoir properties between the active and observation wells. An interference test can be conducted with more than one active well and/or more than one observation well. **Figs. 10.2 and 10.3** depict a typical rate schedule at an active well and the pressure responses at the active and observation wells, respectively.

Note that a time lag exists between the time when a rate change is made at the active well and the time when the pressure transient is seen in the observation well. The area investigated in an interference test, as **Fig. 10.4** illustrates, is defined by the radius of investigation, r_i, which was derived in Chap. 1 and is given by

$$r_i = \sqrt{\frac{kt}{948\phi\mu c_t}}. \quad \text{(10.1)}$$

Recall that the radius of investigation, derived in Chap. 1 and defined by Eq. 10.1, represents the distance that a significant pressure disturbance is propagated (at Time t) by production or injection at a constant rate. As Fig. 10.4 shows, reservoir properties throughout a rectangular area of sides $2r_i$ and $2r_i + r$ influence the pressure response in a test because properties are measured in more of the reservoir than just between wells.

In Secs. 10.2.1 and 10.2.2, we present the underlying theory and practical applications of interference testing in infinite-acting, homogeneous reservoirs. First, we assume isotropic reservoir conditions; then, we investigate the effects of reservoir anisotropy on the interference test response.

10.2.1 Interference Test Analysis in Infinite-Acting, Homogeneous, Isotropic Reservoirs. A homogeneous reservoir is a reservoir in which the formation properties do not depend on location, while an isotropic reservoir is a reservoir in which the properties are independent of direction. In an infinite-acting, homogeneous, isotropic reservoir, the exponential integral solution (i.e., the *Ei*-function solution) of the line-source model describes the pressure behavior at the observation well, or in field units,

$$p_i - p_r = \Delta p = -70.6\frac{q\mu B}{kh}Ei\left(\frac{-948\phi\mu c_t r^2}{kt}\right), \quad \text{(10.2)}$$

where p_r = pressure at the observation well located a distance r from the active well. The pressure drawdown, Δp, at radius r (i.e., at the observation well) is caused by production at rate q at the active well. Because the skin factor of the active well does not affect the drawdown at the observation well, s does not appear in the equation. Wellbore-storage effects at both the active and response wells can be assumed to be negligible for most practical purposes, although these effects should be considered in a rigorous analysis. Ogbe and Brigham[3] discuss wellbore storage in interference testing.

Type-curve analysis is convenient for analyzing pressure data from an interference test with constant-rate production at the active well. We define dimensionless pressure, radius, and time variables, respectively, as

$$p_D = \frac{kh(p_i - p_r)}{141.2qB\mu}, \quad \text{(10.3)}$$

$$r_D = \frac{r}{r_w}, \quad \text{(10.4)}$$

and $$t_D = \frac{0.0002637kt}{\phi\mu c_t r_w^2}. \quad \text{(10.5)}$$

After substituting Eqs. 10.3 through 10.5 into Eq. 10.2, we can write the *Ei*-function solution in dimensionless form as

$$p_D = -\frac{1}{2}Ei\left(\frac{-r_D^2}{4t_D}\right). \quad \text{(10.6)}$$

A log-log plot of p_D vs. t_D/r_D^2 (i.e., the *Ei*-function solution shown in **Fig. 10.5**) is a useful type curve for analyzing interference tests.

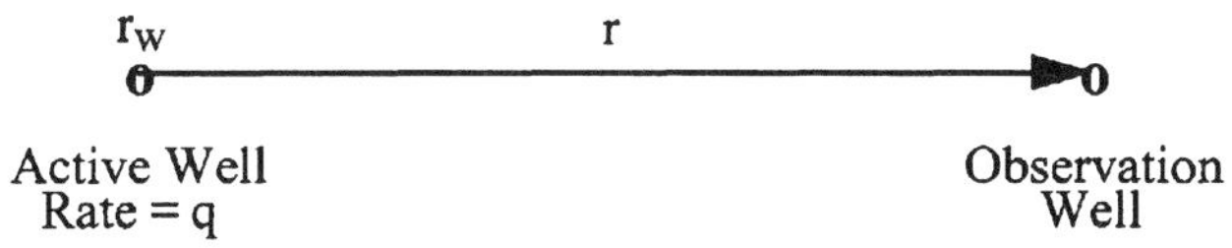

Fig. 10.1—Active and observation wells, interference or pulse test (after Earlougher[2]).

The solution given by Eq. 10.6 does not include wellbore-storage or skin effects.

We recommend the following procedure for analyzing interference tests in infinite-acting, homogeneous, isotropic reservoirs.

1. Plot pressure change, $\Delta p = p_i - p_r$, as a function of t, where p_i = initial reservoir pressure or average reservoir pressure measured at the beginning of the test, psia; p_r = measured bottomhole pressure (BHP) measured during the interference test, psia; and t = elapsed test time, hours. Make the plot either on tracing paper or on log-log graph paper with the same size log cycles as the type curve.
2. Overlay the type curve on the test-data plot. Shift the plotted data both horizontally and vertically until a match is found with the type curve.
3. Select convenient pressure- and time-match points: pressure-match point = $(p_D, \Delta p)_{MP}$ and time-match point = $(t_D/r_D^2, t)_{MP}$.
4. Calculate permeability, k, from the pressure-match point.

$$k = \frac{141.2qB\mu}{h}\left(\frac{p_D}{\Delta p}\right)_{MP}. \qquad (10.7)$$

5. Calculate the porosity/total-compressibility product, ϕc_t, from the time-match point.

$$\phi c_t = \frac{0.0002637k}{\mu r^2}\left(\frac{t}{t_D/r_D^2}\right)_{MP}, \qquad (10.8)$$

where r = distance between the active and observation wells and k = permeability calculated from Step 4. Note that, if we have an estimate of total compressibility, c_t, we can estimate the porosity with the total-compressibility/porosity product, ϕc_t, from Eq. 10.9.

$$\phi = \frac{\phi c_t}{c_t}. \qquad (10.9)$$

Example 10.1—Interference Test Analysis in an Infinite-Acting, Homogeneous, Isotropic Reservoir. Water was injected into a well during a 48-hour interference test. The pressure response at an observation well located 119 ft away from the active well was observed for 148 hours. **Table 10.1** gives time and pressure data from the observation well, along with other well and formation properties. Determine k and ϕ with type-curve analysis.

Solution.

1. First, make a log-log plot of Δp as a function of t (**Fig. 10.6**). To maintain the sign convention that $\Delta p/q$ is always positive, plot Δp as a positive number. **Table 10.2** gives the plotting functions.

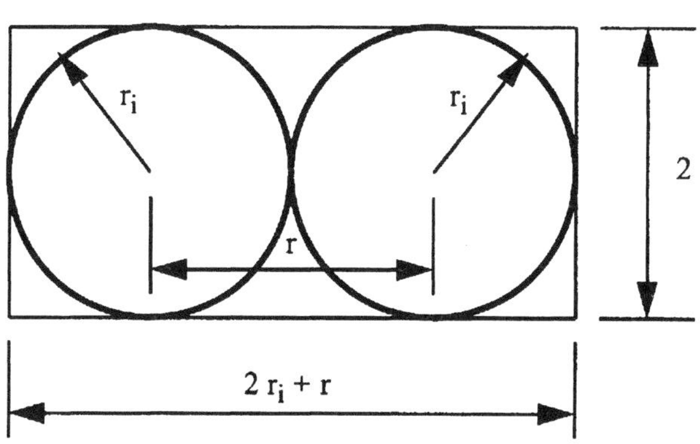

Fig. 10.4—Area investigated during interference test.

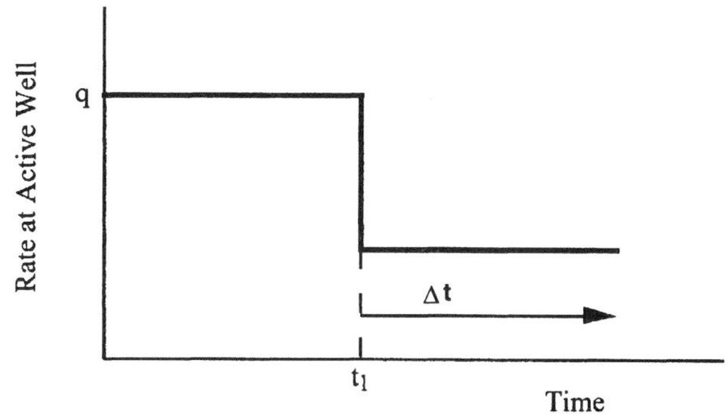

Fig. 10.2—Rate schedule at active well during interference test (after Earlougher[2]).

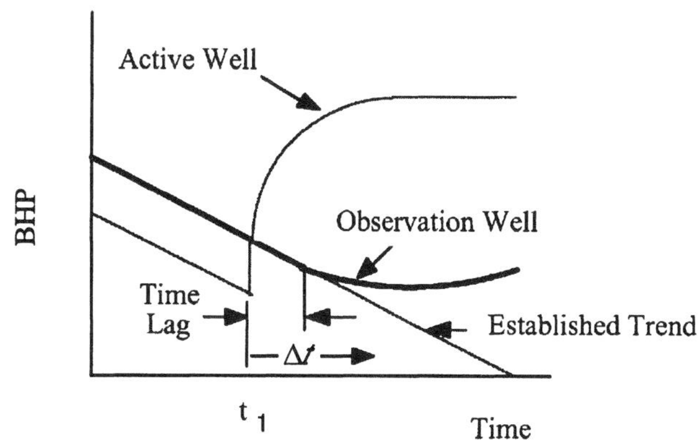

Fig. 10.3—Pressure responses at active and observation wells during interference test (after Earlougher[2]).

2. Matching the data to the *Ei*-function solution (p_D vs. t_D) in Fig. 10.5, select the following pressure- and time-match points.

Pressure-match point: Δp = 105 psi, p_D = 1.0.

Time-match point: t = 12 hours, t_D/r_D^2 = 1.0.

3. From the pressure-match point, the permeability is

$$k = \frac{141.2qB\mu}{h}\left(\frac{p_D}{\Delta p}\right)_{MP} = \frac{141.2(170)(1)(1)}{45}\left(\frac{1.0}{105}\right) = 5.1 \text{ md}.$$

4. From the time-match point, the porosity/total-compressibility product is

$$\phi c_t = \frac{0.0002637k}{\mu r^2}\left(\frac{t}{t_D/r_D^2}\right)_{MP}$$

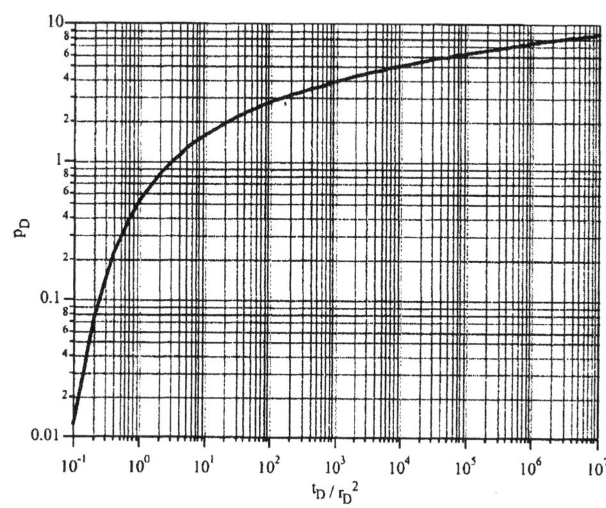

Fig. 10.5—Exponential integral solution.

TABLE 10.1—DATA, EXAMPLE 10.1	
Well and Formation Properties	
μ_w, cp	1.0
B_w, bbl/STB	1.0
c_t, psi^{-1}	12.7×10^{-6}
q, STB/D	−170
p_i, psig	0
h, ft	45
r, ft	119
Depth; ft	2,000
Interference Test Data	
t (hours)	$p_w = p_r$ (psig)
0.0	0
4.3	22
21.6	82
28.2	95
45.0	119
48.0	End

$$= \frac{(0.0002637)(5.1)}{(1)(119)^2}\left(\frac{12}{1.0}\right) = 1.14 \times 10^{-6}\text{psi}^{-1}.$$

Because we have an estimate of total compressibility, c_t, we can also estimate the porosity.

$$\phi = \frac{\phi c_t}{c_t} = \frac{1.14 \times 10^{-6}}{12.7 \times 10^{-6}} = 0.09 \text{ or } 9.0\%.$$

10.2.2 Interference Testing in Infinite-Acting, Homogeneous, Anisotropic Reservoirs. Assuming single-phase flow of a slightly compressible fluid and using the well-coordinate system shown in **Fig. 10.7,** Papadopulos[4] derived the following solution for the pressure response $p(x,y,t)$ at an observation well caused by an active well (producer or injector) located at the origin in a rectangular (x–y) coordinate system:

$$\sqrt{k_{xx}k_{yy} - k_{xy}^2}\left[\frac{h(p_i - p_{xyt})}{141.2qB\mu}\right] =$$

$$-\frac{1}{2}Ei\left[\frac{-\phi\mu c_t}{0.00105t}\left(\frac{k_{xx}y^2 + k_{yy}x^2 - 2k_{xy}xy}{k_{xx}k_{yy} - k_{xy}^2}\right)\right], \quad \text{(10.10)}$$

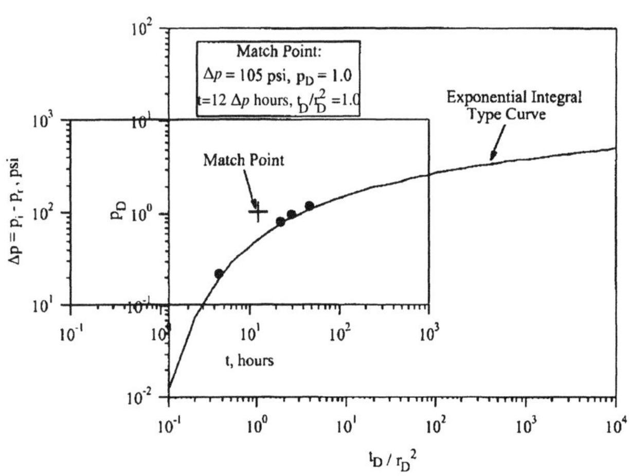

Fig. 10.6—Type-curve match of interference test data, Example 10.1.

TABLE 10.2—PLOTTING FUNCTIONS FOR INTERFERENCE TEST ANALYSIS, EXAMPLE 10.2	
t (hours)	$p_i - p_w$ (psi)
0.0	—
4.3	−22
21.6	−82
28.2	−95
45.0	−119
48.0	—

$$k_{XX} = \frac{1}{2}\left\{\left(k_{xx} + k_{yy}\right) + \left[\left(k_{xx} - k_{yy}\right)^2 + 4k_{xy}^2\right]^{1/2}\right\}, \quad \text{(10.11)}$$

$$k_{YY} = \frac{1}{2}\left\{\left(k_{xx} + k_{yy}\right) - \left[\left(k_{xx} - k_{yy}\right)^2 + 4k_{xy}^2\right]^{1/2}\right\}, \quad \text{(10.12)}$$

$$\text{and } \theta = \arctan\left(\frac{k_{XX} - k_{xx}}{k_{xy}}\right). \quad \text{(10.13)}$$

The variables in Eqs. 10.11 through 10.13 are defined as k_{xx}, k_{yy}, k_{xy} = components of the permeability tensor, md; k_{XX} = maximum principal permeability, md; k_{YY} = minimum principal permeability, md; and θ = angle of orientation.

The principal permeabilities, k_{XX} and k_{YY}, and the orientation, θ, can be determined from interference-test data from three observation wells located on different rays from the active well at the origin. The method requires performing a type-curve match of the test data with the exponential-integral solution in Fig. 10.5. For the anisotropic case, dimensionless pressure and time variables are defined, respectively, as

$$p_D = \sqrt{k_{xx}k_{yy} - k_{xy}^2}\left[\frac{h(p_i - p_{xyt})}{141.2qB\mu}\right] \quad \text{(10.14)}$$

$$\text{and } \frac{t_D}{r_D^2} = \frac{0.0002637t}{\phi\mu c_t}\left(\frac{k_{xx}k_{yy} - k_{xy}^2}{k_{xx}y^2 + k_{yy}x^2 - 2k_{xy}xy}\right). \quad \text{(10.15)}$$

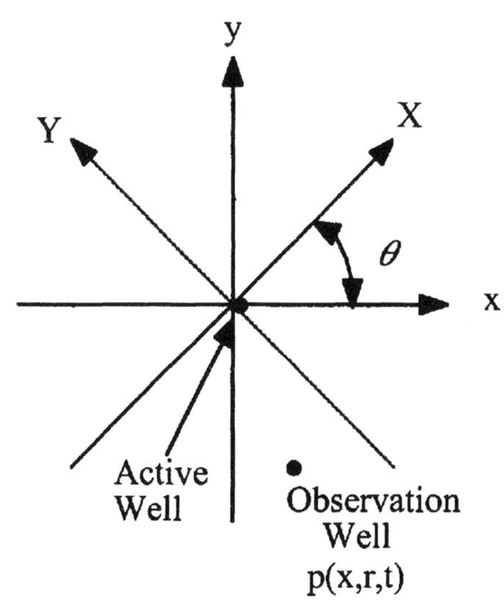

Fig. 10.7—Coordinate system and well locations for interference testing (after Ramey[5]).

Eq. 10.4 defines dimensionless distance, r_D. The rising pressure change, Δp, for each observation well is plotted as a function of time on a single log-log graph of the same dimensions as the type curve. The data are then matched with the type curve. The same pressure-match point (p_D, Δp) is found for all three wells, but three different time-match points are found. Unrealistic parameter estimates (e.g., negative permeabilities) may result if different pressure matches are used. While the alignment of the pressure scale is maintained, the data are shifted horizontally to find an individual time-match point for each well.

The following is a recommended procedure for analyzing interference data from three wells in an anisotropic reservoir.

1. On a single graph, plot the rising pressure change as a function of time for each of the three observation wells.
2. Overlay the type curve (Fig. 10.5) with the test data. Shifting the data plot both vertically and horizontally, find a single pressure-match point (p_D, Δp) that satisfies all three sets of data.
3. Maintaining the pressure-match point and shifting the data plot only horizontally, find a time-match point for each set of data. The result is three time-match points, all with the same value of t but three different values of t_D/r_D^2.
4. Setting the right side of Eq. 10.14 equal to the value of p_D from the pressure-match point, calculate the value of the expression $k_{xx}k_{yy} - k_{xy}{}^2$. The resulting equation contains three unknowns (k_{xx}, k_{yy}, and k_{xy}).
5. Setting the right side of Eq. 10.15 equal to each value of t_D/r_D^2 from the time match points, establish three equations. The unknowns are $\phi\mu c_t$, k_{xx}, k_{yy}, and k_{xy}.
6. Simultaneously solve the four equations (one equation from Step 4 and three equations from Step 5) for the four unknowns, k_{xx}, k_{yy}, k_{xy}, and $\phi\mu c_t$.
7. Calculate k_{XX}, k_{YY}, and θ with Eqs. 10.11, 10.12, and 10.13, respectively, and the variables obtained in Step 6.

In a field example demonstrating this method, Ramey[5] considered an interference test from a water-injection well in an oil reservoir. He applied the solution for single-phase flow to a case of multiphase flow on the basis that the formation was thoroughly watered out with a high-producing water cut and no mobile gas. Thus, the oil and any nonmobile gas in the reservoir would have an appreciable effect only on total system compressibility, c_t. He further discussed the applicability of single-phase-flow analysis to multiphase-flow and composite-reservoir cases.[5]

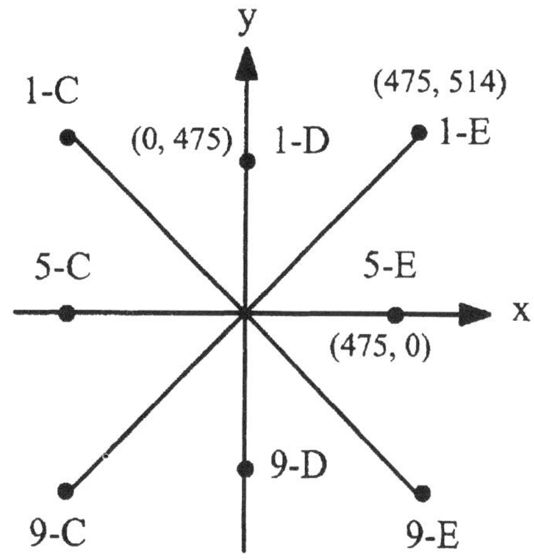

Fig. 10.8—Well pattern, Example 10.2.

Example 10.2—Interference Test Analysis in an Infinite-Acting, Homogeneous, Anisotropic Reservoir. An interference test[5] was run in a single water-injection well and three observation oil wells. The producers have high water cuts and no mobile gas. The three observation wells selected were Wells 1-D, 1-E, and 5-E, with coordinates given in feet (**Fig. 10.8**). **Table 10.3** gives observed pressure and time data. Determine the maximum and minimum principal permeabilities and the orientation.

TABLE 10.3—INTERFERENCE-TEST DATA FOR WELLS 1-D, 1-E, AND 5-E, EXAMPLE 10.2

Well and Formation Properties	
h, ft	25
r_w, ft	0.563
ϕ	0.20
c_t, psi^{-1}	8.0×10^{-6}
B_w, RB/STB	1.0
μ_w, cp	1.0
q_w, STB/D	115
p_i, psi	240
T, °F	72
oil gravity, °API	36

Well 1-D		Well 1-E		Well 5-E	
t (hours)	Δp (psi)	t (hours)	Δp (psi)	t (hours)	Δp (psi)
23.5	6.7	27.5	3	21	4
28.5	7.2	47	5	47	11
51	15	72	11	72	16.3
77	20	95	13	94	21.2
95	25	115	16	115	22
119	24	125	16	122	25
125	23.2	142	13	140	22.3
141	19	192	10	188	19.2
163	18	215	10	210	18
188	14	240	6	285	15
215	12	295	5.8	—	—
265	10	—	—	—	—
290	10	—	—	—	—

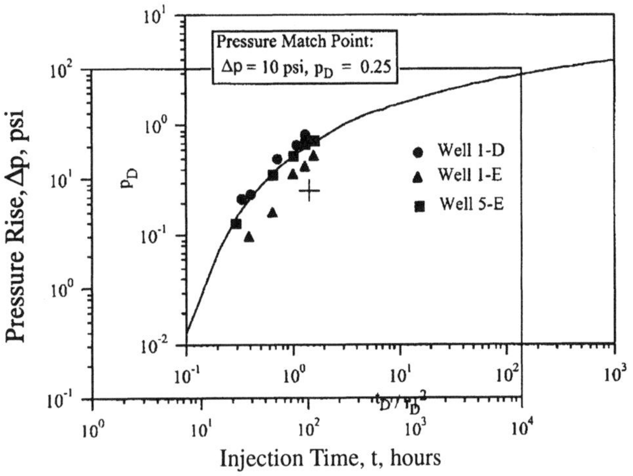

Fig. 10.9—Determining pressure-match point of interference test data, Example 10.2.

Solution.

1. First, plot Δp data from the three observation wells as a function of injection time on a single log-log graph (**Fig. 10.9**).
2. Shifting the data plot both horizontally and vertically, all three sets of data satisfactorily match the exponential-integral-solution type curve at a single pressure-match point of ($\Delta p = 10$ psi, $p_D = 0.25$).
3. Now, select a convenient match time of $t = 100$ hours for all three plots. Maintaining the pressure-match point, shift the data plot horizontally only to obtain a separate match of each data plot with the type curve (see **Figs. 10.10 through 10.12** for Wells 1-D, 1-E, and 5-E, respectively).

Well 1-D: ($t = 100$ hours, $t_D/r_D^2 = 1.40$).
Well 1-E: ($t = 100$ hours, $t_D/r_D^2 = 0.70$).
Well 5-E: ($t = 100$ hours, $t_D/r_D^2 = 1.10$).

4. From the pressure-match point and Eq. 10.14, calculate

$$p_D = \sqrt{k_{xx}k_{yy} - k_{xy}^2}\left[\frac{h(p_i - p_{xyt})}{141.2qB\mu}\right]$$

or $$\sqrt{k_{xx}k_{yy} - k_{xy}^2} = \frac{141.2qB\mu}{h(p_i - p_{xyt})}p_D$$

$$= \frac{(141.2)(115)(1.0)(1.0)}{(25)(10)}(0.25) = 16.24$$

and $k_{xx}k_{yy} - k_{xy}^2 = 263.7$ md^2.

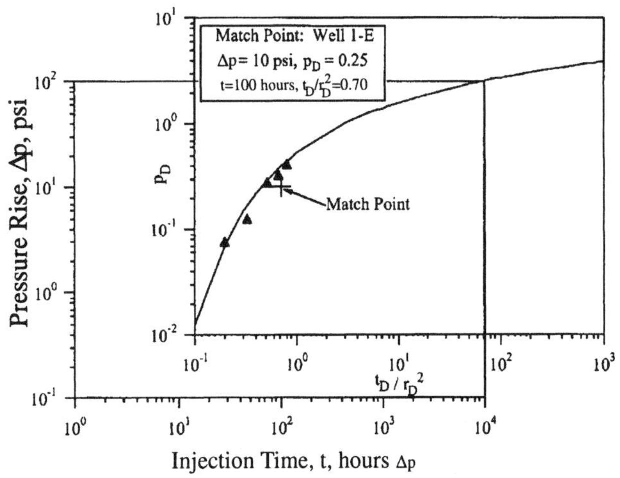

Fig. 10.11—Type-curve analysis for Well 1-E, Example 10.2.

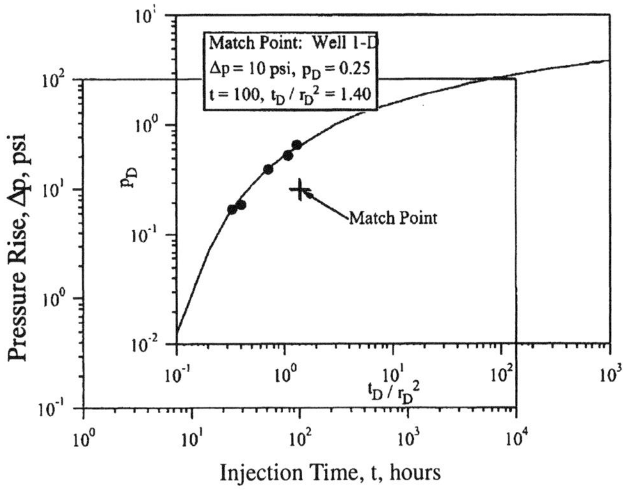

Fig. 10.10—Type-curve analysis for Well 1-D, Example 10.2.

5. From the time-match points (all of which have $t = 100$ hours) and Eq. 10.15, calculate the following for each well.

Well 1-D ($x = 0$, $y = 475$ ft),

$$\frac{t_D}{r_D^2} = \frac{0.0002637(t)}{\phi\mu c_t}\left(\frac{k_{xx}k_{yy} - k_{xy}^2}{k_{xx}y^2 + k_{yy}x^2 - 2k_{xy}xy}\right)$$

$$1.40 =$$

$$\frac{0.0002637(100)}{\phi\mu c_t}\left[\frac{263.7}{k_{xx}(475)^2 + k_{yy}(0)^2 - 2k_{xy}(0)(475)}\right]$$

$$1.40 = \frac{3.082 \times 10^{-5}}{\phi\mu c_t k_{xx}}.$$

Well 1-E ($x = 475$, $y = 514$ ft),

$$\frac{t_D}{r_D^2} = \frac{0.0002637t}{\phi\mu c_t}\left(\frac{k_{xx}k_{yy} - k_{xy}^2}{k_{xx}y^2 + k_{yy}x^2 - 2k_{xy}xy}\right)$$

$$0.70 =$$

$$\frac{0.0002637(100)}{\phi\mu c_t}\left[\frac{263.7}{k_{xx}(514)^2 + k_{yy}(475)^2 - 2k_{xy}(475)(514)}\right]$$

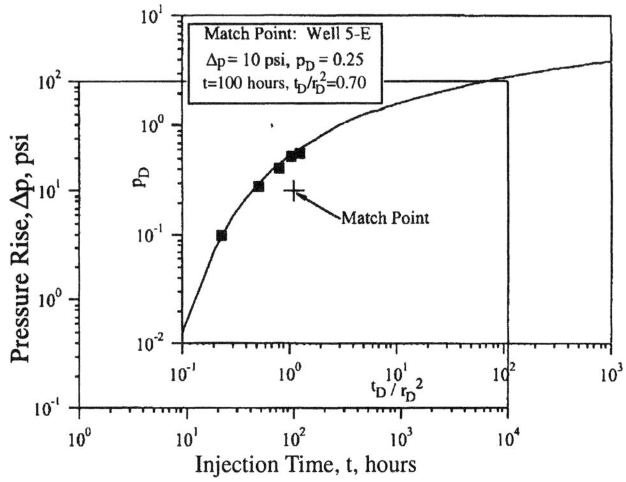

Fig. 10.12—Type-curve analysis for Well 5-E, Example 10.2.

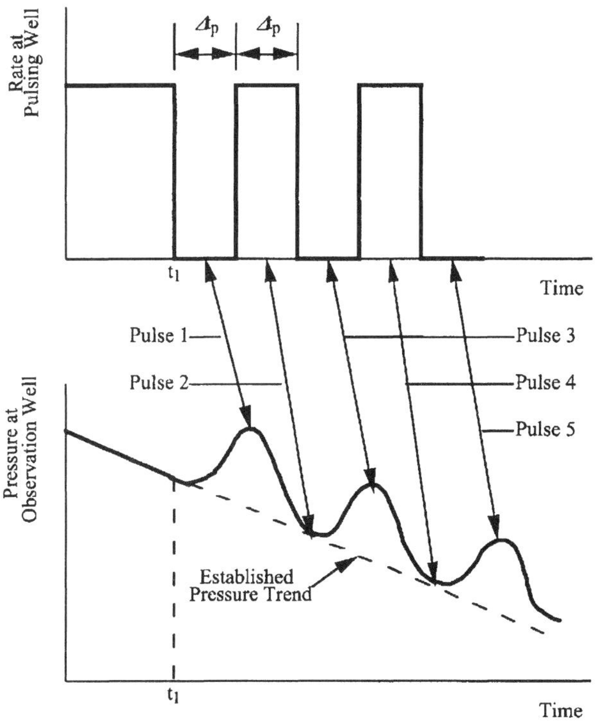

Fig. 10.13—Rate history at active well and pressure response at observation well for a pulse test (after Earlougher[2]).

$$0.70 = \frac{1}{\phi\mu c_t}$$

$$\times \left\{ 6.954 / \left[\left(2.642 \times 10^5\right)k_{xx} + \left(2.256 \times 10^5\right)k_{yy} \right.\right.$$

$$\left.\left. - \left(4.833 \times 10^5\right)k_{xy} \right] \right\}.$$

Well 5-E ($x = 475$, $y = 0$),

$$\frac{t_D}{r_D^2} = \frac{0.0002637t}{\phi\mu c_t} \left(\frac{k_{xx}k_{yy} - k_{xy}^2}{k_{xx}y^2 + k_{yy}x^2 - 2k_{xy}xy} \right)$$

$$1.10 = \frac{0.0002637(100)}{\phi\mu c_t}$$

$$\times \left[\frac{263.7}{k_{xx}(0)^2 + k_{yy}(475)^2 - 2k_{xy}(475)(0)} \right]$$

$$1.10 = \frac{3.082 \times 10^{-5}}{\phi\mu c_t k_{yy}}.$$

6. Simultaneous solution of these four equations (one equation from Step 4 and three equations from Step 5) yields $k_{xx} = 14.6$ md, $k_{yy} = 18.6$ md, $k_{xy} = 3.0$ md, and $\phi\mu c_t = 1.50\ 10^{-6}$ cp/psi.

7. Using Eqs. 10.11, 10.12, and 10.13, calculate the principal permeabilities and the angle of orientation as follows:

The maximum principal permeability, k_{XX}, is

$$k_{XX} = \frac{1}{2} \left\{ \left(k_{xx} + k_{yy}\right) + \left[\left(k_{xx} - k_{yy}\right)^2 + 4k_{xy}^2 \right]^{1/2} \right\}$$

$$= \frac{1}{2} \left\{ (14.6 + 18.6) + \left[(14.6 - 18.6)^2 + 4(3.0)^2 \right]^{1/2} \right\}$$

$$= 20.2 \text{ md}.$$

The minimum principal permeability, k_{YY}, is

$$k_{YY} = \frac{1}{2} \left\{ \left(k_{xx} + k_{yy}\right) - \left[\left(k_{xx} - k_{yy}\right)^2 + 4k_{xy}^2 \right]^{1/2} \right\}$$

$$= \frac{1}{2} \left\{ (14.6 + 18.6) - \left[(14.6 - 18.6)^2 + 4(3.0)^2 \right]^{1/2} \right\}$$

$$= 13.0 \text{ md}.$$

The orientation of the principal axes is

$$\theta = \arctan\left(\frac{k_{XX} - k_{xx}}{k_{xy}} \right)$$

$$= \arctan\left(\frac{20.2 - 14.6}{3.0} \right)$$

$$= 61.8°.$$

10.3 Pulse Tests

In a pulse test,[1,2] a coded signal is sent from an active well (producer or injector) to a shut-in observation well. **Fig. 10.13** illustrates an example rate schedule at the active well and the resulting pressure response at the observation well. The signal is generated by producing from or injecting into the active well, shutting in the well, producing again, shutting in again, and repeating this sequence in a regular pattern. The well is pulsed so that we can identify readily the effect of the active well on the pressure response at the observation well amid the noise of the reservoir pressure trend. With short disturbances to established production, we can determine communication between wells and establish reservoir properties. Highly sensitive pressure gauges are used to detect these small disturbances, which may be on the order of 0.01 psi. Like interference tests, pulse tests can quantify directional reservoir properties, such as permeability.

Pulse tests have certain advantages over conventional interference tests. Pulse tests have fewer problems with reservoir noise and reservoir-pressure trends that cause errors in measuring pressure responses at observation wells. Further, pulse-test analysis is based on simple infinite-acting reservoir equations, which are adequate for most pulse tests. When interference tests are analyzed, however, boundary effects frequently must be taken into account. Secs. 10.3.1 and 10.3.2 discuss pulse testing for two reservoir conditions—isotropic and anisotropic properties. Both conditions assume infinite-acting, homogeneous reservoirs.

10.3.1 Pulse-Test Analysis in Infinite-Acting, Homogeneous, Isotropic Reservoirs. Before discussing pulse-test analysis, we define the variables that will be used.[6,7] **Fig. 10.14** illustrates some of these. By convention, the ratio of the pressure change to change in flow rate, $\Delta p/\Delta q$, is always positive. The time lag, t_L, is the time between the end of a pulse and the pressure peak caused by the pulse. For example, t_{L1} in Fig. 10.14 is the time between the end of the first shut-in period and the pressure peak caused by Pulse 1.

The amplitude of the pressure response, Δp, is the vertical distance between two parallel lines, one tangential to two adjacent peaks (or valleys) on either side of the pulse to be analyzed, and the other parallel to the first line and tangential to the intervening valley (or peak) that is being measured. In Fig. 10.14, for example, Δp_4 is the maximum pressure change corresponding to Pulse 4, which is a flow period.

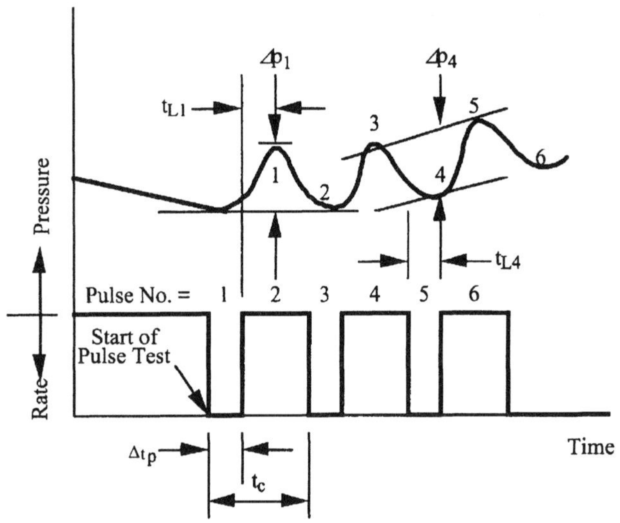

Fig. 10.14—Pulse-rate test rate and pressure history showing definition of time lag and response amplitude (after Earlougher[2]).

TABLE 10.4—APPLICATION OF PULSE-TEST CHARTS	
Pulse Analyzed	Applicable Figures
First normal pulse (first pulse)	10.15, 10.16
First inverted pulse (second pulse)	10.17, 10.18
Later normal pulses (pulses 3, 5, 7, . . .)	10.19, 10.20
Later inverted pulses (pulses 4, 6, 8, . . .)	10.21, 10.22

F' denotes the ratio of the length of the first pulse, Δt_p, to the cycle length (pulse plus shut-in), Δt_c, and is given by

$$F' = \frac{\Delta t_p}{\Delta t_c}. \qquad (10.16)$$

All times are in hours or, in the case of ratios, in consistent units. Eq. 10.4 defines the dimensionless distance between the active well and the observation well, r_D.

The dimensionless time lag, t_{LD}, is

$$t_{LD} = \frac{0.0002637kt_L}{\phi\mu c_t r_w^2}, \qquad (10.17)$$

and the dimensionless pressure-response amplitude, Δp_D, is

$$\Delta p_D = \frac{kh\Delta p}{141.2qB\mu}. \qquad (10.18)$$

Special charts[6,8] are used to analyze pulse tests. These charts are based on simulations of the pressure response at an observation well

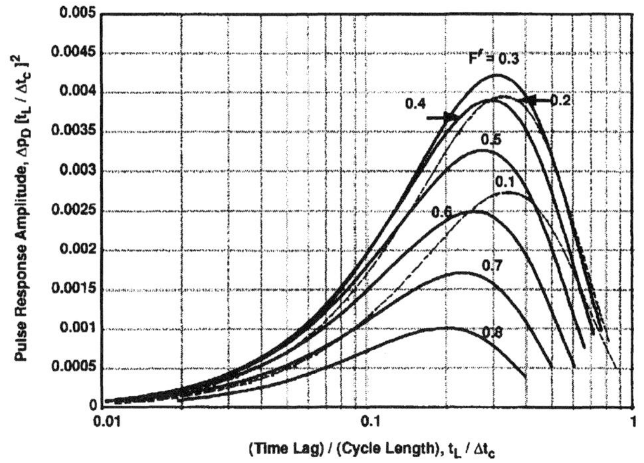

Fig. 10.15—Relation between time lag and response amplitude for first normal pulse (Pulse 1) (after Ogbe[9]).

by use of superposition of the *Ei*-function solution to the diffusivity equation to model the rate changes in the pulsing sequence. The charts apply as shown in **Table 10.4** and are given in **Figs. 10.15 through 10.22.**

Ogbe[9] noted that the charts originally presented by Kamal and Brigham[6] contained some inaccuracies. Figs. 10.15 through 10.22 show a complete set of corrected charts. Ogbe[9] recommends that the corrected charts be used in place of those originally presented by Kamal and Brigham,[6] who designated pulses as "odd" or "even." This terminology has led to much confusion. We adopt the notation used by Ogbe, with the first, third, fifth, ... pulses denoted as "normal" and the second, fourth, sixth ... pulses denoted as "inverted."

We can determine k and ϕc_t with the charts in Figs. 10.15 through 10.22 and the following procedure for pulse-test analysis in infinite-acting, homogeneous, isotropic reservoirs.

1. Plot the test data on Cartesian-coordinate graph paper. Draw parallel lines tangentially to the pulse to be analyzed, and determine the amplitude of the pressure response, Δp, and the lag time, t_L.

2. Calculate the variable F' from Eq. 10.16. Calculate the ratio of the time lag t_L to the cycle length Δt_c, $[t_L/\Delta t_c]$.

3. With the appropriate chart (Fig. 10.15, 10.17, 10.19, or 10.21), read the dimensionless-response-amplitude correlating parameter $p_D[t_L/\Delta t_c]^2$ corresponding to F' and $[t_L/\Delta t_c]$. Calculate Δp_D from

$$\Delta p_D = \frac{\Delta p_D (t_L/\Delta t_c)^2}{(t_L/\Delta t_c)^2}. \qquad (10.19)$$

4. Calculate the formation permeability, k, with the pressure response amplitude, Δp, from Step 1 and the dimensionless pressure-response amplitude, Δp_D, from Step 3.

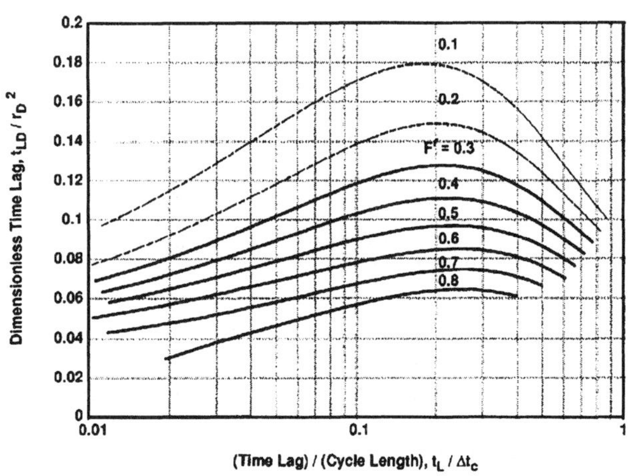

Fig. 10.16—Relation between time lag and cycle length for first normal pulse (Pulse 1) (after Ogbe[9]).

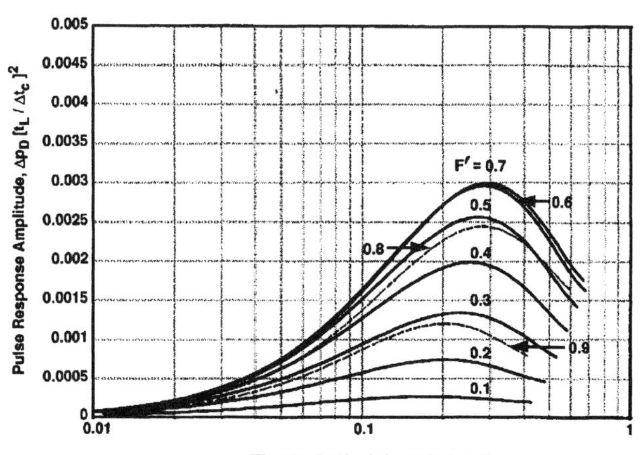

Fig. 10.17—Relation between time lag and response amplitude for first inverted pulse (Pulse 2) (after Ogbe[9]).

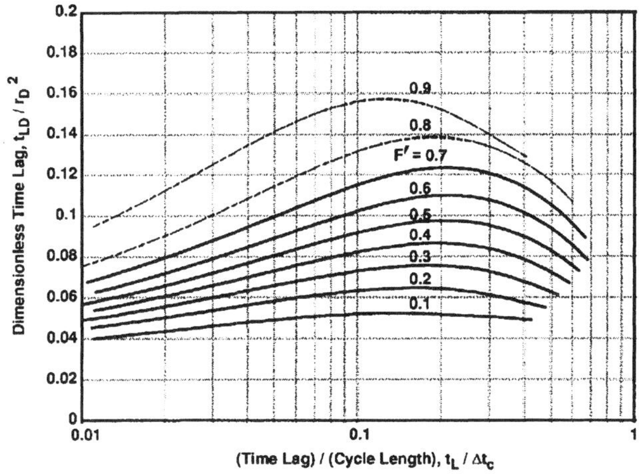

Fig. 10.18—Relation between time lag and cycle length for first inverted pulse (Pulse 2) (after Ogbe[9]).

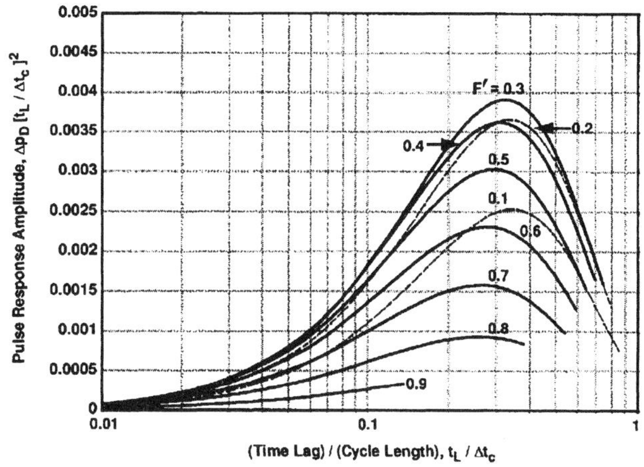

Fig. 10.19—Relation between time lag and response amplitude for all normal pulses after the first pulse (Pulses 3, 5, 7, . . .).

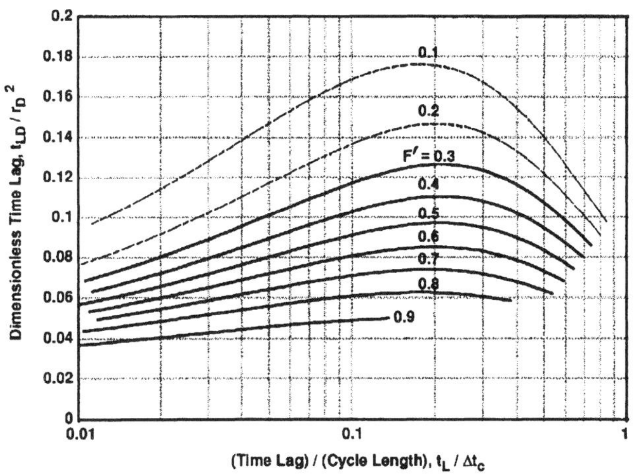

Fig. 10.20—Relation between time lag and cycle length for all normal pulses after the first pulse (Pulses 3, 5, 7, . . .).

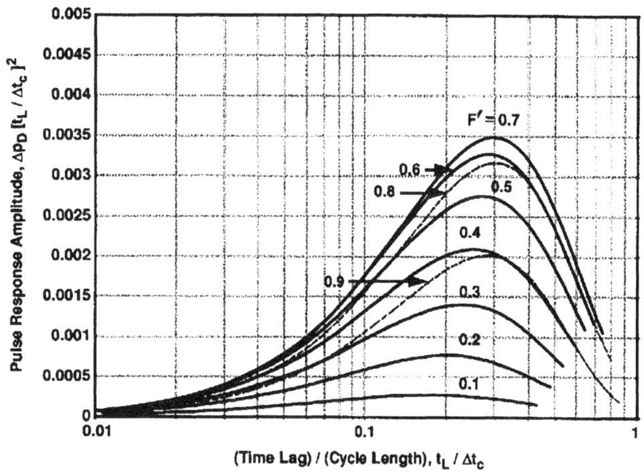

Fig. 10.21—Relation between time lag and response amplitude for all inverted pulses after the first pulse (Pulses 4, 6, 8, . . .).

$$k = \frac{141.2qB\mu\Delta p_D}{h\Delta p}. \qquad (10.20)$$

5. With the appropriate chart (Figs. 10.16, 10.18, 10.20, or 10.22), read the dimensionless-time-lag correlating parameter t_{LD}/r_D^2 corresponding to F' and $[t_L/\Delta t_c]$. Calculate the dimensionless time lag as

$$t_{LD} = \left(t_{LD}/r_D^2\right)r_D^2 .$$

6. Calculate ϕc_t with

$$\phi c_t = \frac{0.0002637kt_L}{\mu r_w^2 t_{LD}}. \qquad (10.21)$$

Although Eqs. 10.20 and 10.21 indicate that k and ϕc_t can be determined by analyzing a single pulse, all available data should be analyzed to minimize the chances of error. This method does not consider wellbore-storage effects, which can be important in the test analysis if the distance between the active and observation wells is short or if the duration of wellbore storage is long. Further, skin effects at the wells are not modeled.

Pulse tests with wellbore storage may be analyzed with the method presented by Ogbe[9] and Ogbe and Brigham.[8]

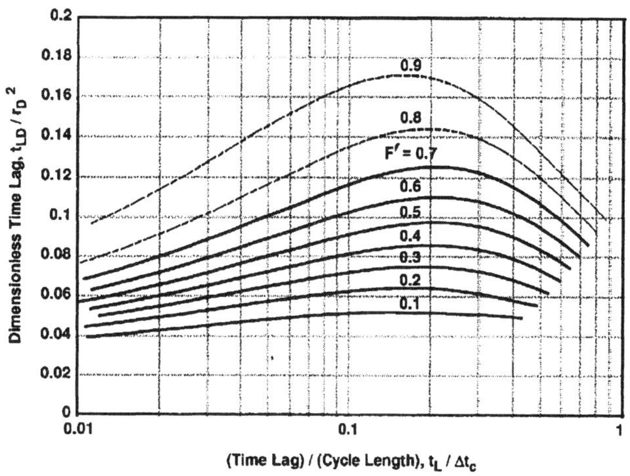

Fig. 10.22—Relation between time lag and cycle length for all inverted pulses after the first pulse (Pulses 4, 6, 8, . . .).

Example 10.3—Pulse-Test Analysis for an Infinite-Acting, Homogeneous, Isotropic Reservoir. A pulse test was conducted between two wells 492 ft apart in a reservoir. One of the wells was produced for 72 minutes, shut in for 72 minutes, produced for 72 minutes, and so on. The pressure was monitored in the observation well as a function of time; **Table 10.5** presents the tabulated test data and other known data. Calculate the permeability and porosity for the region influenced by the test. Analyze the fourth pulse, i.e., the second pulse in which the pressure increases (the second inverted pulse).

Solution.

1. First, plot the test data on Cartesian coordinate paper (**Fig. 10.23**). Draw parallel lines tangentially about the pressure response

TABLE 10.5—DATA FOR EXAMPLE 10.3

Known Data	
μ, cp	1
B, RB/STB	1.009
c_t, psi^{-1}	3.0×10^{-6}
h, ft	20
q, STB/D	300
r, ft	492
r_w, ft	0.25

Pressure and Time Data

t (minutes)	p_{wf} (psi)	t (minutes)	p_{wf} (psi)	t (minutes)	p_{wf} (psia)	t (minutes)	p_{wf} (psi)
0.0	2,990.00	80.6	2,987.22	216.0	2,984.84	308.5	2,984.45
1.0	2,990.00	85.4	2,987.07	219.4	2,984.70	319.2	2,984.24
5.5	2,989.99	92.5	2,986.86	224.6	2,984.52	335.2	2,983.79
8.6	2,989.98	103.2	2,986.70	229.4	2,984.38	360.0	2,982.95
13.4	2,989.92	119.2	2,986.63	236.4	2,984.25	365.5	2,982.76
20.5	2,989.77	144.0	2,986.70	247.2	2,984.17	368.6	2,982.67
31.2	2,989.42	149.5	2,986.72	263.2	2,984.21	373.4	2,982.56
47.2	2,988.76	157.4	2,986.69	288.0	2,984.44	380.5	2,982.45
72.0	2,987.60	164.5	2,986.59	293.5	2,984.49	391.2	2,982.41
74.0	2,987.51	175.2	2,986.32	296.6	2,984.51	407.2	2,982.51
77.5	2,987.35	191.2	2,985.79	301.4	2,984.51	432.0	2,982.83

corresponding to the fourth pulse. The peak occurs at 0.2125 day, or 0.0125 day after the end of the pulse. Thus, the time lag for the fourth pulse, t_L, is equal to 0.0125 day or 0.3 hour. The plot also shows that the vertical distance between the parallel lines is 1.1 psi, so the amplitude of the pressure response is $\Delta p = 1.1$ psi.

2. The pulse length, Δt_p, is given as 72 minutes. From the given data, the cycle length, Δt_c, is equal to the production time (pulse length) plus the shut-in time (i.e., $\Delta t_c = 72 + 72 = 144$ minutes or 2.4 hours). Thus

$$F' = \frac{\Delta t_p}{\Delta t_c} = 0.5.$$

Now, we calculate

$$\frac{t_L}{\Delta t_c} = \frac{0.3}{2.4} = 0.125.$$

3. From Fig. 10.21,

$\Delta p_D [t_L/\Delta t_c]^2 = 1.93 \times 10^{-3}$.

Calculating Δp_D,

$$\Delta p_D = \frac{\Delta p_D (t_L/\Delta t_c)^2}{(t_L/\Delta t_c)^2} = \frac{1.93 \times 10^{-3}}{(0.125)^2} = 0.1235.$$

4. Next, calculate the permeability from

$$k = \frac{141.2 q B \mu \Delta p_D}{h \Delta p} \quad \ldots\ldots\ldots\ldots\ldots\ldots\ldots\ldots\ldots \quad (10.20)$$

$$= \frac{(141.2)(300)(1.009)(1.0)(0.1235)}{(20)(1.1)} = 240 \text{ md}.$$

5. From Fig. 10.22, $t_{LD}/r_D^2 = 9.38 \times 10^{-2}$. Calculating t_{LD},

$$t_{LD} = (t_{LD}/r_D^2) r_D^2 = (t_{LD}/r_D^2)\left(\frac{r}{r_w}\right)^2$$

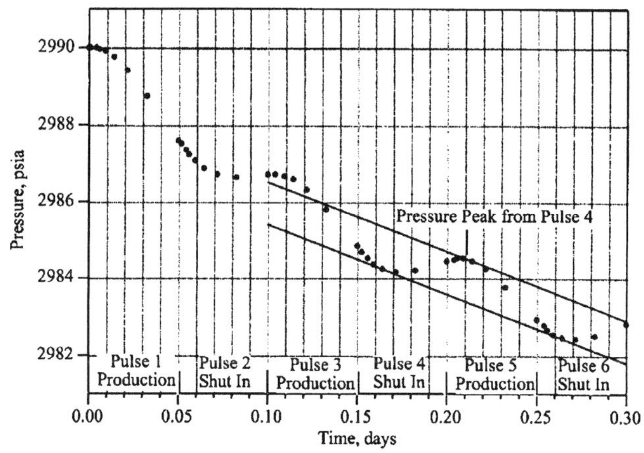

Fig. 10.23—Pulse-test analysis, Example 10.3.

$$= (9.38 \times 10^{-2})\left(\frac{492}{0.25}\right)^2 = 363{,}000.$$

6. Calculate ϕc_t from

$$\phi c_t = \frac{0.0002637 k t_L}{\mu r_w^2 t_{LD}} = \frac{(0.0002637)(240)(0.3)}{(1.0)(0.25)^2(363{,}000)}$$

$$= 8.37 \times 10^{-7} \text{psi}^{-1}.$$

Calculating the porosity ϕ,

$$\phi = \frac{\phi c_t}{c_t} = \frac{8.37 \times 10^{-7}}{3 \times 10^{-6}} = 0.279 = 27.9\%.$$

10.3.2 Pulse-Test Analysis in Infinite-Acting, Homogeneous, Anisotropic Reservoirs. As is the case with interference tests, the analysis of pulse tests in homogeneous, anisotropic reservoirs is similar to that in isotropic reservoirs.[10] For the anisotropic case, Eqs. 10.17 and 10.18 are replaced, respectively, by

$$\Delta t_{cD} = \frac{h\Delta t_c}{948\mu}\left(\frac{k_{xx}k_{yy} - k_{xy}^2}{k_{xx}y^2 + k_{yy}x^2 - 2k_{xy}xy}\right) \quad \text{......... (10.22)}$$

$$\text{and } \Delta p_D = \frac{h\Delta p}{70.6qB\mu}\sqrt{k_{xx}k_{yy} - k_{xy}^2}\,. \quad \text{.............. (10.23)}$$

The maximum principal permeability, k_{XX}, the minimum principal permeability, k_{YY}, and the angle of orientation, θ, are calculated with Eqs. 10.11, 10.12, and 10.13, respectively.

10.4 Recommendations for Multiple-Well Testing

As a general approach to the design and analysis of interference and pulse tests, Kamal[10,11] recommends that the analyst begin with the simplest reservoir model and progress to more complex models if necessary. The simplest case is that of a homogeneous, isotropic reservoir. Porosity, permeability, and thickness are constant throughout the reservoir. Design and analysis of interference and pulse tests are relatively simple for this case. If realistic reservoir properties cannot be obtained with this model, then a more complex model is necessary.

A homogeneous, anisotropic reservoir model should be used next. In this model, porosity and thickness are uniform throughout the reservoir but permeability varies with direction. Analysis is somewhat more complicated than for the simpler case but is still tractable with hand calculations. In type-curve matching, the time-match points are different because permeability varies with direction. If the model is appropriate for the reservoir, then the pressure-match point is the same for all observation wells. If a single pressure-match point cannot be found, this model is inadequate.

Heterogeneous reservoir models should be used when less complex models do not model the reservoir adequately. In a heterogeneous reservoir model, porosity, permeability, and thickness vary spatially. Reservoir simulation is necessary in this case.

The following procedure may be helpful in selecting an appropriate model for a reservoir[11]. If the criteria for validity of a model for a particular reservoir are not met, the next model should be tried.

1. Homogeneous, Isotropic Model. Estimates of reservoir parameters from tests in different wells in the reservoir are in agreement. Estimates of parameters from single-well and multiple-well tests agree. Analysis techniques for interference and pulse tests are described earlier.

2. Homogeneous, Anisotropic Model. Pressure-match point in type-curve analysis is the same for all observation wells. Permeabilities estimated from multiple-well tests are consistent with permeability estimated from single-well tests. Analysis techniques for interference and pulse tests are described earlier.

3. Heterogeneous, 2D Model. Calculated and observed pressures are in reasonable agreement. Reservoir simulation is required.

4. Heterogeneous, 2D Layered or 3D Model. Calculated and observed pressures are in reasonable agreement. Reservoir simulation is required.

10.5 Chapter Summary

In Chap. 10, we discuss pressure transient tests in which the withdrawal or injection of fluid at one well creates a pressure disturbance at another well. Such multiple-well tests can be used to investigate directional variations in permeability, pay-zone continuity, and pore volume in the region affected during the test. Multiple-well tests include both interference and pulse tests.

Sec. 10.2 introduces interference testing. In an interference test, the production rate at one well (the active well) is changed at time t_1. The resulting pressure response at an offset well (the observation well) is measured as a function of elapsed time since the rate change.

Sec. 10.2.1 presents an analysis method for interference tests in infinite-acting, homogeneous, isotropic reservoirs. The first step in the analysis is to plot the pressure change at the observation well since the rate change vs. the elapsed time since the rate change on a log-log scale. The data are then matched to the exponential integral type curve. The final step is to calculate the formation permeability from the pressure-match point and the porosity/total compressibility product from the time-match point. An example problem illustrates this procedure.

Sec. 10.2.2 presents an analysis method for interference tests in anisotropic reservoirs. This method requires that pressures be measured at three or more observation wells. The first step in the analysis is to prepare a log-log plot of pressure change vs. elapsed time for each of the three observation wells. The three data sets are then matched to the exponential integral type curve with the same pressure-match point but different time-match points for the three data sets. Interpreting the pressure- and time-match points results in four equations in four unknowns, which may be solved to obtain the principal permeabilities and the porosity/compressibility product. An example problem from the petroleum literature illustrates this procedure.

Sec. 10.3 discusses the use of pulse tests. A pulse test is conducted by cycling the active well through a series of production and shut-in periods in a regular pattern. The resulting pressure pulses make it easy to distinguish the effect of the active well from the effects of other producing wells.

Sec. 10.3.1 presents an analysis method for isotropic reservoirs. The amplitude and time lag for each pulse may be analyzed with charts presented in this section. Separate charts are required for the first normal pulse, the first inverted pulse, all normal pulses after the first, and all inverted pulses after the first. As with an interference test, permeability and porosity/compressibility product can be estimated from the pulse-test data. An example analysis illustrates this procedure.

Sec. 10.3.2 briefly discusses the analysis of pulse tests in anisotropic reservoirs, pointing out how to modify the analysis procedure for isotropic reservoirs to analyze anisotropic reservoirs.

Sec. 10.4 presents general recommendations for analyzing multiple-well tests.

Exercises

1. An interference test was conducted in an oil reservoir. Given the following reservoir rock and fluid properties, estimate permeability and total compressibility from the interference test data. $p_i = 1658.0$ psia; $B_o = 1.21$ RB/STB; $h = 17$ ft; $q = 775$ STB/D; $\mu = 0.632$ cp; $\phi = 16.5\%$; $r_w = 0.25$ ft; $r = 573$ ft.

PRESSURE-TIME DATA, EXERCISE 10.1

Time (hours)	Pressure (psi)	Time (hours)	Pressure (psi)	Time (hours)	Pressure (psi)
0.00	1658.00	4.77	1653.22	28.42	1620.65
1.00	1657.99	5.96	1650.89	35.53	1614.89
1.25	1657.95	7.45	1647.99	44.41	1608.92
1.56	1657.83	9.31	1644.55	55.51	1602.78
1.95	1657.58	11.64	1640.60	69.39	1596.49
2.44	1657.09	14.55	1636.17	86.74	1590.08
3.05	1656.26	18.19	1631.34	108.42	1583.57
3.81	1654.99	22.74	1626.14		

2. An interference test was conducted in an oil reservoir. Given the following reservoir rock and fluid properties, estimate permeability and total compressibility from the interference test data. $p_i = 2291.7$ psia; $B_o = 1.14$ RB/STB; $h = 32$ ft; $q = 315$ STB/D; $\mu = 1.31$ cp; $\phi = 18.3\%$; $r_w = 0.31$ ft; $r = 660$ ft.

PRESSURE-TIME DATA, EXERCISE 10.2					
Time (hours)	Pressure (psi)	Time (hours)	Pressure (psi)	Time (hours)	Pressure (psi)
0.00	2291.70	7.45	2291.69	69.39	2280.15
1.00	2291.70	9.31	2291.66	86.74	2276.68
1.25	2291.70	11.64	2291.57	108.42	2272.79
1.56	2291.70	14.55	2291.37	135.53	2268.52
1.95	2291.70	18.19	2290.96	169.41	2263.93
2.44	2291.70	22.74	2290.27	211.76	2259.06
3.05	2291.70	28.42	2289.19	264.70	2253.96
3.81	2291.70	35.53	2287.67	330.87	2248.66
4.77	2291.70	44.41	2285.66	413.59	2243.20
5.96	2291.70	55.51	2283.15	516.99	2237.61

3. An interference test was conducted in an anisotropic reservoir. Given the following reservoir rock and fluid properties, find the principal permeabilities, the orientation of the maximum principal permeability, and total compressibility from the interference test data. $p_i = 1561.7$ psia; $B_o = 1.27$ RB/STB; $h = 21$ ft; $q = 450$ STB/D; $\mu = 0.276$ cp; $\phi = 21.7\%$; $r_w = 0.36$ ft. *Well 1 coordinates (ft): (0,660); Well 2 coordinates (ft): (660,660); Well 3 coordinates (ft): (660,0).*

PRESSURE-TIME DATA, EXERCISE 10.3			
Time (hours)	Well 1 Pressure (psi)	Well 2 Pressure (psi)	Well 3 Pressure (psi)
0.00	1561.70	1561.70	1561.70
1.00	1561.70	1561.70	1561.67
1.25	1561.68	1561.70	1561.63
1.56	1561.65	1561.68	1561.53
1.95	1561.58	1561.65	1561.35
2.44	1561.45	1561.57	1561.07
3.05	1561.23	1561.43	1560.66
3.81	1560.88	1561.19	1560.10
4.77	1560.40	1560.84	1559.39
5.96	1559.77	1560.34	1558.54
7.45	1558.99	1559.69	1557.53
9.31	1558.06	1558.89	1556.40
11.64	1556.99	1557.95	1555.15
14.55	1555.80	1556.87	1553.80
18.19	1554.50	1555.66	1552.35
22.74	1553.10	1554.35	1550.84
28.42	1551.62	1552.94	1549.26
35.53	1550.07	1551.45	1547.62
44.41	1548.46	1549.89	1545.95
55.51	1546.80	1548.28	1544.24
69.39	1545.11	1546.62	1542.50
86.74	1543.38	1544.92	1540.73
108.42	1541.63	1543.19	1538.95
135.53	1539.86	1541.44	1537.15

4. An interference test was conducted in an anisotropic reservoir. Given the following reservoir rock and fluid properties, find the principal permeabilities, the orientation of the maximum principal permeability, and total compressibility from the interference test data. $p_i = 2732.3$ psia; $B_o = 1.08$ RB/STB; $h = 35$ ft; $q = 780$ STB/D; $\mu = 0.83$ cp; $\phi = 18.2\%$; $r_w = 0.27$ ft. *Well 1 coordinates (ft): (350.0,606.2); Well 2 coordinates (ft): (375.0, −649.5); Well 3 coordinates (ft): (−720.0,0.0).*

PRESSURE-TIME DATA, EXERCISE 10.4			
Time (hours)	Well 1 Pressure (psi)	Well 2 Pressure (psi)	Well 3 Pressure (psi)
0.00	2732.30	2732.30	2732.30
1.00	2732.30	2732.30	2732.30
1.25	2732.30	2732.30	2732.30
1.56	2732.30	2732.30	2732.28
1.95	2732.30	2732.30	2732.24
2.44	2732.30	2732.30	2732.13
3.05	2732.28	2732.30	2731.92
3.81	2732.24	2732.30	2731.54
4.77	2732.14	2732.29	2730.95
5.96	2731.93	2732.27	2730.09
7.45	2731.56	2732.20	2728.95
9.31	2730.97	2732.05	2727.52
11.64	2730.13	2731.76	2725.78
14.55	2729.00	2731.29	2723.77
18.19	2727.57	2730.58	2721.50
22.74	2725.84	2729.59	2719.01
28.42	2723.84	2728.31	2716.31
35.53	2721.58	2726.73	2713.45
44.41	2719.09	2724.86	2710.44
55.51	2716.40	2722.73	2707.31
69.39	2713.55	2720.35	2704.08
86.74	2710.54	2717.75	2700.77
108.42	2707.42	2714.98	2697.40
135.53	2704.19	2712.04	2693.97
169.41	2700.89	2708.97	2690.49
211.76	2697.51	2705.80	2686.98

5. A pulse test was conducted on the wells in Exercise 1 to confirm the results from the interference test. Given the following reservoir rock and fluid properties, estimate permeability and total compressibility from the pulse test. Compare your results with those from Exercise 1. $p_i = 1658.0$ psia; $B_o = 1.21$ RB/STB; $h = 17$ ft; $q = 375$ STB/D; $\mu = 0.632$ cp; $\phi = 16.5\%$; $r_w = 0.25$ ft; $r = 573$ ft.

PRESSURE-TIME DATA, EXERCISE 10.5					
Time (hours)	Pressure (psi)	Time (hours)	Pressure (psi)	Time (hours)	Pressure (psi)
0.00	1658.00	13.50	1650.55	27.00	1646.61
0.50	1658.00	14.00	1649.85	27.50	1646.90
1.00	1657.87	14.50	1649.16	28.00	1647.21
1.50	1657.47	15.00	1648.61	28.50	1647.51
2.00	1656.84	15.50	1648.37	29.00	1647.81
2.50	1656.09	16.00	1648.36	29.50	1648.10
3.00	1655.29	16.50	1648.50	30.00	1648.37
3.50	1654.46	17.00	1648.71	30.50	1648.64
4.00	1653.65	17.50	1648.97	31.00	1648.77
4.50	1652.85	18.00	1649.23	31.50	1648.60
5.00	1652.19	18.50	1649.50	32.00	1648.20
5.50	1651.85	19.00	1649.77	32.50	1647.67
6.00	1651.74	19.50	1650.03	33.00	1647.07
6.50	1651.79	20.00	1650.27	33.50	1646.45
7.00	1651.92	20.50	1650.51	34.00	1645.82
7.50	1652.09	21.00	1650.61	34.50	1645.20
8.00	1652.28	21.50	1650.41	35.00	1644.72
8.50	1652.48	22.00	1649.99	35.50	1644.53
9.00	1652.67	22.50	1649.43	36.00	1644.59
9.50	1652.86	23.00	1648.81	36.50	1644.79
10.00	1653.05	23.50	1648.16	37.00	1645.06
10.50	1653.22	24.00	1647.51	37.50	1645.37
11.00	1653.26	24.50	1646.86	38.00	1645.69
11.50	1653.01	25.00	1646.36	38.50	1646.01
12.00	1652.53	25.50	1646.15	39.00	1646.33
12.50	1651.92	26.00	1646.19	39.50	1646.63
13.00	1651.25	26.50	1646.36	40.00	1646.93

6. A pulse test was conducted on the wells in Exercise 2 to confirm the results from the interference test. Given the following reservoir rock and fluid properties, estimate permeability and total compressibility from the pulse test. Compare your results with those from Exercise 2. $p_i = 2291.7$ psia; $B_o = 1.14$ RB/STB; $h = 32$ ft; $q = 450$ STB/D; $\mu = 1.31$ cp; $\phi = 18.3\%$; $r_w = 0.31$ ft; $r = 660$ ft.

PRESSURE-TIME DATA, EXERCISE 10.6

Time (hours)	Pressure (psi)	Time (hours)	Pressure (psi)	Time (hours)	Pressure (psi)
0.00	2291.70	54.00	2287.40	108.00	2282.20
2.00	2291.70	56.00	2287.14	110.00	2282.02
4.00	2291.70	58.00	2286.82	112.00	2281.88
6.00	2291.70	60.00	2286.47	114.00	2281.78
8.00	2291.68	62.00	2286.10	116.00	2281.71
10.00	2291.61	64.00	2285.76	118.00	2281.67
12.00	2291.49	66.00	2285.46	120.00	2281.66
14.00	2291.29	68.00	2285.20	122.00	2281.66
16.00	2291.02	70.00	2284.99	124.00	2281.68
18.00	2290.68	72.00	2284.82	126.00	2281.71
20.00	2290.31	74.00	2284.68	128.00	2281.74
22.00	2289.92	76.00	2284.58	130.00	2281.73
24.00	2289.55	78.00	2284.51	132.00	2281.67
26.00	2289.21	80.00	2284.47	134.00	2281.54
28.00	2288.92	82.00	2284.44	136.00	2281.35
30.00	2288.68	84.00	2284.43	138.00	2281.09
32.00	2288.47	86.00	2284.43	140.00	2280.80
34.00	2288.30	88.00	2284.43	142.00	2280.49
36.00	2288.16	90.00	2284.40	144.00	2280.21
38.00	2288.05	92.00	2284.31	146.00	2279.97
40.00	2287.97	94.00	2284.15	148.00	2279.77
42.00	2287.91	96.00	2283.93	150.00	2279.61
44.00	2287.86	98.00	2283.65	152.00	2279.49
46.00	2287.83	100.00	2283.33	154.00	2279.41
48.00	2287.79	102.00	2283.00	156.00	2279.37
50.00	2287.71	104.00	2282.70	158.00	2279.35
52.00	2287.59	106.00	2282.43	160.00	2279.35

References

1. Matthews, C.S. and Russell, D.G.: *Pressure Buildup and Flow Tests in Wells*, Monograph Series, SPE, Richardson, Texas (1967) **1**.
2. Earlougher, R.C. Jr.: *Advances in Well Test Analysis*, Monograph Series, SPE, Richardson, Texas (1977) **5**.
3. Ogbe, D.O. and Brigham, W.E.: "A Correlation for Interference Testing With Wellbore-Storage and Skin Effects," *SPEFE* (September 1989) 391.
4. Papadopulos, I.S.: "Nonsteady Flow to a Well in an Infinite Anisotropic Aquifer," 1965 Symposium of the Intl. Assn. Sci. Hyd., Dubrovinik, Yugoslavia.
5. Ramey, H.J. Jr.: "Interference Analysis for Anisotropic Formations—A Case History," *JPT* (October 1975) 121; *Trans.*, AIME, **259.**
6. Kamal, M. and Brigham, W.E.: "Pulse-Testing Response for Unequal Pulse and Shut-In Periods," *SPEJ* (October 1975) 399; *Trans.*, AIME, **259.**
7. Kamal, M. and Brigham, W.E.: "Design and Analysis of Pulse Tests With Unequal Pulse and Shut-In Periods," *JPT* (February 1976) 205; *Trans.*, AIME, **261.**
8. Ogbe, O.D. and Brigham, W.: "Pulse Testing with Wellbore Storage and Skin Effects," *SPEFE* (March 1987) 29; *Trans.*, AIME, **283.**
9. Ogbe, O.D.: "Pulse Testing in the Presence of Wellbore Storage and Skin Effects," PhD dissertation, Stanford U., Stanford, California (January 1984).
10. Kamal, M.: "Interference and Pulse Testing—A Review," *JPT* (December 1983) 2257; *Trans.*, AIME, **275.**
11. Kamal, M.: "The Use of Pressure Transients To Describe Reservoir Heterogeneity," *JPT* (August 1979) 1060; *Trans.*, AIME, **267.**

Chapter 11
Design and Implementation of Well Tests

11.1 Overview

This chapter presents fundamental concepts for designing and implementing oil- and gaswell tests, both pressure transient and well-deliverability tests. Included in these design concepts are recommendations for selecting the appropriate well test to achieve the desired test objectives, estimating pretest formation properties, selecting the appropriate flow-rate schedule and sequence for the test, and choosing the test duration required to sample a desired reservoir volume and/or to reach stabilized flow conditions. Several examples illustrate the design concepts.

11.2 Types and Purposes of Well Tests

Oil- and gaswell tests can be grouped into two general categories on the basis of their primary function.[1] The first category, called pressure transient tests, includes those tests designed to quantify important reservoir rock and fluid properties (such as permeability, porosity, and average reservoir pressure) and to locate and identify reservoir heterogeneities (such as sealing faults, natural fractures, and layers). The second category of well tests, deliverability tests, includes tests that are designed to evaluate a well's production potential under specific reservoir conditions.

11.2.1 Pressure Transient Tests. Pressure transient tests refer to those tests where we generate and measure pressure changes with time. These tests allow us to evaluate not only near-wellbore conditions but also the in-situ reservoir properties beyond the region that has been affected by drilling and completion operations. In addition, these tests can characterize important formation features needed to design the optimal depletion plan for the reservoir, including pressure in the drainage area of tested wells, possible presence and approximate location of flow barriers (e.g., sealing faults), and the dominant properties of reservoir heterogeneities (e.g., fracture and matrix properties in a naturally fractured reservoir or individual layer properties). Pressure transient tests can be grouped into two subcategories: single-well and multiwell tests.

Single-Well Tests. As the name implies, single-well tests involve only one well where the pressure response is measured following a rate change. From the measured pressure response, we can characterize average properties in a portion or all of the drainage area of the tested well. A common single-well test, a pressure buildup test, is conducted by first stabilizing a producing well at some fixed rate, placing a bottomhole-pressure (BHP) measuring device in the well, and shutting in the well. Following shut-in, the BHP builds up as a function of time; this rate of pressure buildup is used to estimate well and formation properties. From a pressure-buildup test, we can estimate average reservoir pressure and permeability in the drainage area of the well and the properties of the region immediately adjacent to the wellbore.

Another common pressure transient test, a drawdown or flow test, is conducted by producing a well at a known and constant rate while measuring changes in BHP as a function of time. Drawdown tests are designed primarily to quantify reservoir-flow characteristics, including permeability and skin factor. In addition, when the pressure transient is affected by outer reservoir boundaries, drawdown tests can be used to establish the outer limits of the reservoir and to estimate the hydrocarbon volume in the well's drainage area. These specific drawdown tests are called reservoir-limits tests. When economic considerations require a minimum loss of production time, pressure-drawdown tests also can be used to estimate the deliverability of a well and, if conducted and analyzed properly, are viable alternatives to deliverability tests.[1]

A pressure-falloff test is similar to a pressure-buildup test except that it is conducted on an injection well. Following stabilization at some injection rate, the well is shut in. BHP, which then begins to decline, is measured as a function of time. An alternative test for an injection well is an injectivity test where we inject into a well at a measured rate and measure pressure as it increases with time. Injectivity testing is analogous to pressure-drawdown testing.

Multiwell Tests. When the flow rate is changed in one well and the pressure response is measured in one or more other wells, the test is called a multiwell test. Multiwell tests are designed to determine properties in a region centered along a line connecting pairs of test wells; therefore, they are more sensitive to directional variations of reservoir properties, such as permeability. In addition, these tests can determine the presence or lack of communication between two points in the reservoir. The basic concept in a multiwell test is either to produce from or inject into one well (the active well) and observe the pressure response in one or more offset wells (observation wells). From these data, we can estimate both permeability and porosity in the drainage area of the wells and can quantify some reservoir anisotropies. For example, multiwell tests can be used to determine the orientation of natural fractures and to quantify the ratio of the porosity/compressibility products of the matrix and fracture systems.[2]

Interference and pulse tests are two common multiwell tests. In interference testing, the active well is produced at a measured, constant rate throughout the test. Other wells in the field must be shut in so that any observed pressure response can be attributed to the active well only. In pulse tests, the active well produces and then is shut in, returned to production, and shut in again. This production/shut-in sequence, which is repeated but with production or shut-in periods rarely exceeding more than a few hours, produces a pressure response in the observation wells that usually can be interpreted unambiguously even when other wells in the field continue to produce.

11.2.2 Deliverability Tests. Deliverability tests are flow tests designed to measure the production capabilities of a well under specific reservoir conditions. Although these tests are used primarily for gas wells, deliverability testing also is applicable to oil wells. Unlike most pressure transient tests, some deliverability tests require stabilized flowing conditions for proper analysis. A common productivity indicator obtained from gaswell deliverability tests is the absolute open flow (AOF) potential. Another application of deliverability testing is for generating a reservoir inflow performance relationship (IPR) or gas-backpressure curve. The IPR, which describes the relationship between surface production rates and bottomhole flowing pressures (BHFP's), is used to design surface production facilities.

Common deliverability tests include flow-after-flow, single-point, isochronal, and modified isochronal tests. These tests are briefly reviewed here for completeness of the discussion of well-test design and implementation.

Flow-After-Flow Tests. Flow-after-flow tests, which sometimes are referred to as gas-backpressure or four-point tests, are conducted by producing the well at a series of different stabilized flow rates and measuring the stabilized BHFP at the sandface. Each different flow rate is established in succession either without or with a very short intermediate shut-in period. Typically, conventional flow-after-flow tests are conducted with a sequence of increasing flow rates; however, if stabilized flow rates are attained, the rate sequence does not affect the test.[1] The requirement that the flowing periods be continued until stabilization is sometimes a major limitation of the flow-after-flow test, especially in low-permeability formations that take long times to reach stabilized flowing conditions.

Single-Point Tests. A single-point test is an attempt to overcome the limitation of long test times required for the flow-after-flow tests in low-permeability formations. Single-point tests are conducted by flowing the well at a single rate until the BHFP is stabilized. Single-point tests are particularly appropriate when a well's deliverability characteristics are being updated, as required by many regulatory agencies. One limitation of this type of test, however, is that analysis of the results requires prior knowledge of the well's deliverability behavior, either from previous testing on the well or from correlations with other wells producing in the same field under similar conditions. In these tests, we must also ensure that the well has flowed long enough to be out of wellbore storage and in the pseudosteady-state-flow regime.

Isochronal Tests. The isochronal test also was developed to shorten test times in wells that take long periods to stabilize. Specifically, the isochronal test is a series of single-point tests developed to estimate stabilized deliverability characteristics without actually flowing the well for the time required to achieve stabilized conditions. The isochronal test is conducted by alternately producing the well, then shutting in the well and allowing it to build up to the average reservoir pressure before the beginning of the next production period. Pressures are measured at several time increments during each flow period. The times at which the pressures are measured should be the same relative to the beginning of each flow period. For example, BHFP's could be measured at 0.5, 1.0, 1.5, and 2.0 hours after the beginning of each flow period. Because less time is required to build up to essentially initial pressure after short flow periods than to reach stabilized flow in a flow-after-flow test, the isochronal test is more practical for low-permeability formations. Although not required for analyzing the test, a final, stabilized flow point usually is obtained at the end of the test.

Modified Isochronal Tests. The time to build up to the average reservoir pressure before flowing for a certain period of time may still be impractical, even after short flow periods in an isochronal test. Consequently, the isochronal test was modified to shorten test times further. The objective of the modified isochronal test is to obtain the same data as in an isochronal test without using the sometimes lengthy shut-in periods required to reach the average reservoir pressure in the well's drainage area. The modified isochronal test is conducted like an isochronal test except that the shut-in periods are of equal duration and should equal or exceed the length of the flow period. Because the well often does not build up to average reservoir pressure after each flow period, the shut-in sandface pressures recorded immediately before each flow period are used in the test analysis instead of the average reservoir pressure. Consequently, the modified isochronal test is less accurate than the isochronal test. It should be noted that, as the duration of the shut-in periods increases, the accuracy of the modified isochronal test also increases. Again, a final stabilized flow point often is obtained at the end of the test but is not required for analyzing the test data.

11.3 General Test-Design Considerations

If properly designed and implemented, a well test can provide much useful information about both individual wells and the reservoir. Common applications of oil- and gaswell tests include identifying and locating important reservoir heterogeneities, such as sealing faults, reservoir layering, or natural fractures; calculating the resource in place and estimating the reserves for various stages of reservoir depletion; selecting optimal field development and production strategies; and estimating future deliverability for design of surface production and processing equipment. Other applications include determining the nature of the formation fluids and obtaining fluid samples for laboratory analysis.

Over the years, a variety of well tests have been developed by the petroleum industry. Many of these tests, however, are appropriate only under specific testing conditions, and each test yields different information about the well and/or the reservoir. Therefore, one of the first steps in well-test design is to identify the test objectives clearly. Once the test objectives have been identified, we may then proceed with the test design, including selecting a particular well test to achieve the desired objectives, developing procedures to implement the test safely and economically, selecting test equipment required to obtain the proper data, and collecting and analyzing the data.

In general, the goals of a well test are not only to obtain sufficient data to meet the stated objectives but also to accomplish these tasks in a timely and inexpensive manner. Satisfying these general goals requires that the design engineer recognize the various well-testing environments and understand how these environments affect the well-test design and implementation. This section summarizes some of the testing environments commonly encountered in well–test design and discusses how these various situations affect the test design.

11.3.1 Type and Status of Well. When deciding the type of well test to use, the foremost design considerations are the type and status of the well.[3] The type of well refers to its major function, such as development/exploration or producer/injector, while the well's status indicates whether the well is active or shut-in. Further, depending on the type and status of the well and the information required, we may select either a pressure transient test or a deliverability test.

Development vs. Exploration Wells. The major difference between testing development and exploration wells is the type of information needed. In a developed field where the general geology and reservoir behavior are known, development wells are drilled either to increase reserves or to accelerate production. Consequently, testing of development wells concentrates more on obtaining specific properties of the individual well rather than the reservoir. For example, a pressure buildup test may be run to estimate the current average reservoir pressure in the well's drainage area, the amount of skin damage from a recent workover, or the properties required for stimulation design and evaluation.

Conversely, for an exploration well drilled in an area where no or few wells exist, the tests must yield more information about the reservoir in a large region around the well rather than just near the individual well. More emphasis is placed on evaluating deliverability, estimating reserves, and obtaining geologic information. For example, in a wildcat well in a new reservoir where little or no data are available, a drillstem test may be run with the stated purposes of obtaining fluid samples for laboratory analysis, estimating well productivity, and calculating the formation permeability and initial static reservoir pressure.

Producing vs. Injection Wells. The major differences in testing these wells are the type of test and the equipment needed. To estimate reservoir properties, a pressure drawdown or buildup test typically is conducted, while a four-point or single-point test is used to estimate future well deliverability in a producing well. Conversely, reservoir properties can be estimated from an injection or falloff test in an injection well. In general, because of the equipment required to perform an injection test, testing injection wells is more expensive than testing producers.

Gas-storage fields are one of the few examples of reservoirs developed by wells that alternate between injection and production, depending on the time of the year. From late spring to early fall, gas is injected into the wells, while from late fall through the winter and early spring, the wells are produced to satisfy the short-term demand for gas associated with the colder weather. Other applications of gas-injection wells include pressure maintenance projects in gas-condensate reservoirs and CO_2 displacement programs in oil reservoirs.

Shallow vs. Deep Wells. The major difference between these testing environments is the type of equipment required to implement the test. The volume of fluids stored in the wellbore distorts the early-time pressure response and controls the duration of wellbore storage, especially in deep wells with large wellbore volumes. If wellbore-storage effects are not minimized or if the test is not continued beyond the end of the wellbore-storage period, the test data are difficult, if not impossible, to analyze with conventional methods. To minimize wellbore-storage distortion and to keep well tests within reasonable and economical lengths of time, it may be necessary to run tubing, packers, and bottomhole shut-in devices. Although this additional equipment greatly reduces or even eliminates wellbore-storage effects, it also increases the cost and time required to perform the test.

Another consideration is the higher reservoir pressure and temperature encountered in deep wells. In deep, high-pressure gas wells, equipment must be specified that can accommodate safely the highest pressure and flow rate expected during the test. For example, threaded pipe connections on flow pipe and lubricators should be used only at lower reservoir pressures, while at higher pressures, welded connections, integral connections, and/or flanges should be used to connect lubricators to the wellhead. In general, equipment designed to sustain higher pressures is also more costly.

In addition, because temperature increases with depth, temperature-related tool failures occur more frequently in deeper wells and must be incorporated into the design process. A very common temperature-related problem is failure of lubricants in mechanical tools. Care should be taken to lubricate all moving tool parts with temperature oils and greases. Another common temperature-related problem is with battery packs in electric gauges with bottomhole memories. Generally, to prevent tool failures, high-temperature batteries must be used when the bottomhole temperature exceeds 300°F.

Stimulated vs. Unstimulated Wells. The pressure transient data used to calculate permeability and skin factor must be taken from the middle-time region or data representing radial flow in an infinite-acting reservoir. These data are no longer affected by wellbore storage and are representative of the effective reservoir permeability that is unaltered by stimulation or damage. Many wells, particularly those in low-permeability gas-bearing formations, require hydraulic fracturing to be economically viable producers. In these stimulated wells, the pressure transient often is affected by the hydraulic fracture long after wellbore storage ceases. In low-permeability wells that have been hydraulically fractured, pressure transient data representing the radial-flow period may not appear for many months, or even years.

Because of these long testing times, we generally recommend an initial well test before the fracture treatment. A reasonably short prefracture test yields data that can be used to determine the formation permeability. Then, a post-fracture test can be used to assess the success of the fracture treatment and to evaluate the fractured well's performance. If the formation permeability has been estimated previously, the post-fracture test can be terminated when wellbore-storage effects end, thus considerably shortening the post-fracture testing time and reducing the cost of the test.

Another consideration in testing hydraulically fractured wells is proper selection of surface equipment. Often, sand used for the proppant is displaced and produced during the flowback period of the test. Because sand is very abrasive, it can destroy chokes, valves, and flow lines. If sand production is expected to be a problem, sand catchers and knockouts should be placed upstream of the production measuring equipment.

11.3.2 Effects of Reservoir Properties. ***Low Permeability vs. High Permeability Formations.*** Formation properties also affect the test design. Principal among these properties is permeability, which dictates the test flow rate as well as the duration of the test. When selecting the test flow rate and flow times, several criteria must be satisfied. First, the test must be maintained for a sufficient length of time to obtain data beyond near-wellbore effects, such as the period of wellbore storage distortion, formation damage, or stimulation. In addition, the test must be run long enough to reach the desired radius of investigation and to evaluate a representative volume of the formation. Often, in low-permeability reservoirs, the flowing time required to satisfy both these criteria is prohibitive, especially when flaring gas to the atmosphere.

The duration of the wellbore-storage period depends on several wellbore parameters (e.g., wellbore volume and fluid compressibility) and reservoir parameters (e.g., porosity, permeability, net-pay thickness, and fluid properties). Although permeability does affect the duration of wellbore-storage distortion, it is not the only parameter that must be considered in well-test design. For example, depending on well depth and reservoir pressure, low-permeability wells can exhibit wellbore storage effects lasting several weeks. However, a reservoir with equally low permeability and the same reservoir pressure may be encountered at a shallower depth. Because of the smaller wellbore volume, the duration of wellbore storage could be much shorter than in the deeper well. The engineer should, however, estimate the duration and severity of wellbore storage effects for each well-test design.

Although the reservoir properties cannot be changed to eliminate or reduce wellbore-storage effects, the wellbore can be altered mechanically to change its properties. For example, running a bottomhole packer on a string of tubing can reduce the effective wellbore volume by a factor of three or four and, consequently, reduce the duration of wellbore-storage effects. Further reductions in the wellbore-storage period can be achieved by running bottomhole valves in the tubing string, which allows the well to be shut in just above the sandface rather than at the surface. Note that all these special tools are expensive. Therefore, the design engineer must compare the cost of longer test times vs. the cost of running tubing packers and bottomhole shut-in valves.

Similar to the duration of wellbore-storage effects, the time to reach a desired radius of investigation in a reservoir increases with decreasing permeability. However, in this case the flowing time cannot be changed by changing the wellbore configuration. Additional test time results in increased costs, especially when production losses become significant during the test period.

When selecting the test flow rate, it is necessary to create a measurable drawdown at the sandface and in the reservoir during the test. The magnitude of the pressure drawdown depends on the reservoir properties and the rate at which fluid flows in the reservoir. Because the reservoir properties are fixed, the pressure drawdown can be controlled by varying only the flow rate. The difference between the sandface and reservoir pressure varies directly with the change in flow rate and inversely with permeability. For higher-permeability wells, a given flow rate yields a smaller pressure drawdown than with the same rate in lower-permeability wells. Therefore, to increase the pressure drawdown, a higher rate must be used.

Note that higher rates require larger separators and meter runs at the surface. In addition, higher flow rates result in wasting more gas through venting and flaring, unless the well is connected to a pipeline. Therefore, in terms of cost per day, the test cost generally increases in higher-permeability wells. The increased cost is offset somewhat, however, by the fact that the total test time is reduced significantly in high-permeability wells. Finally, higher rates can create larger pressure drawdowns, which may result in undue retrograde condensation in gas-condensate reservoirs or formation sloughing in unconsolidated sandstones. Both these conditions may damage the reservoir immediately adjacent to the wellbore.

Single vs. Multiple Zones. Again, the primary difference under these conditions is the type of equipment used to conduct the test. Generally, the test objectives and techniques are the same. In a single zone, a drawdown and a buildup test are typically run to determine reservoir properties, such as permeability; skin factor; or, in hydraulically fractured wells, fracture half-length. During the test, the producing zone communicates with the wellbore through an openhole interval or perforated casing.

In wells intersecting multiple zones, the same drawdown and buildup tests used in the single zone are used. The testing procedure, however, must be modified to isolate the tested zone from the other producing zones surrounding it. The equipment used to isolate individual zones varies, depending on the isolation method. For example, when isolating the zone from below, either a wireline bridge plug or a mechanical bridge plug can be used. After placing the plug between zones, the wireline or tubing is pulled out of the hole. The test is conducted by use of bottomhole gauges placed in the well on a wireline.

When isolating the zone from above, a packer and tubing are used, with the packer set between the test zone and the producing zones above. The test is conducted with gauges on a wireline run through the tubing. Alternatively, the gauges could be placed in a bomb carrier mounted below a bridge plug that is set between the test zone and zones above. To isolate the test zone above and below, set a bridge plug between the test zone and the zones below. In addition, use a packer on tubing to isolate the test zone from the zones above. Pressure gauges are run on a wireline through the tubing.

11.3.3 Safety and Environmental Considerations. ***Sweet vs. Sour and Corrosive Gases.*** The type of gas in the reservoir is a major design consideration, especially when implementing safety procedures in the actual test and when incorporating corrosion protection of the equipment. Sweet gases refer to hydrocarbon mixtures with little or no impurities, while sour and corrosive gases contain large quantities of nonhydrocarbon gases, such as hydrogen sulfide (H_2S) and carbon dioxide (CO_2).

When testing in environments containing H_2S, all industry-standard safety procedures must be followed. Generally, H_2S is recognized by its characteristic foul odor. Prolonged exposure to low concentrations tends to affect the olfactory nerves, thereby dulling the sense of smell. When high concentrations are present, however, death may occur before the odor is detected. Recommended operating and testing procedures are provided in American Petroleum Inst. (API) *RP55*.[4] Metal embrittlement and fatigue also can be problems when testing wells with H_2S. New surface equipment has been known to fail during tests in H_2S wells. Therefore, it is very important to use equipment specifically designed to accommodate H_2S. In addition, all personnel at the testing site must be trained in H_2S safety procedures.

Wire pitting and wireline failure can occur in wells where other corrosive gases, such as CO_2, are present. Wire metallurgies are designed to accommodate varying environments, so to prevent costly fishing jobs related to wireline failures and lost tools, corrosion-resistant materials should be used when operating in environments with corrosive gases. Again, this special equipment is more costly but must be included in the design process.

Environmental Concerns. Finally, the design engineer must be aware of environmental hazards, especially when flaring large quantities of natural gas to the atmosphere. Ideally, the test is designed to minimize the volume of gas flared. Fortunately, the volume of gases released to the atmosphere can be minimized when the well is connected to a pipeline. When designing any well test, however, the engineer should be aware of and adhere to all environmental regulations.

11.3.4 Gauge Considerations. In this section, we discuss several factors that affect gauge design and selection and the advantages and disadvantages of several gauges in widespread use today.

The first pressure gauges used in the petroleum industry were mechanical, clock-driven gauges. Over the years, a wide variety of gauges have been developed, with significant improvements in resolution and accuracy.

Gauges can be divided into two types: mechanical and electronic. Mechanical gauges consist of a Bourdon tube pressure-sensing element, a stylus, and a clock-driven cylinder. As the clock rotates the cylinder, the stylus records the pressure on a metal foil wrapped around the cylinder. Electronic gauges consist of a battery-power supply, a pressure transducer, electronics for interpreting the signal produced by the transducer, and a digital memory. Many electronic gauges also provide for surface readout so that the test may be monitored and extended if necessary to ensure adequate data are taken.

The principal factors affecting gauge design and selection are maximum pressure range, maximum temperature range, resolution, accuracy, drift, stabilization time, and cost.

The operating range for a gauge is determined by the operating ranges for its transducer, power source, and recording mechanism. In general, clock-driven mechanical gauges can operate at higher temperatures and pressures than battery-driven electronic gauges.

Resolution is the smallest pressure change that can be detected by the gauge; accuracy is the maximum error in the pressure measured by the gauge. In applications, such as pressure-derivative type-curve analysis and interference and pulse-test analysis, high resolution is more important than high accuracy. Accuracy becomes more important when comparing tests run with different gauges, as in estimating average reservoir pressure from individual well measurements or material-balance calculations or flow-rate estimation with multiple gauges for closed-chamber testing.

Drift is a change in gauge response over time that is not attributable to an actual pressure change. Drift is typically highest following a step pressure change. Stabilization time is the time required for the gauge to reach a stable pressure reading following a step change in either pressure or temperature. Stabilization time governs the ability of the gauge to track rapid changes of pressure with time.

Commonly used gauges include mechanical gauges, conventional and sapphire strain gauges, capacitance gauges, and standard and compensated quartz gauges. **Table 11.1** summarizes the operating characteristics of these gauges.[5,6]

Mechanical gauges were described earlier.

Strain gauges use a strain-sensitive resistor placed on a pressure-sensitive substrate. Conventional strain gauges use a metal membrane substrate. As the pressure changes, the physical dimensions of the membrane change, thereby changing the resistance. Because of the characteristics of the metal membrane, conventional strain gauges are subject to hysteresis effects. Sapphire strain gauges replace the metal membrane substrate with a box constructed from a sapphire crystal. The sapphire crystal responds more quickly and is more stable than conventional strain gauges.[5]

Capacitance gauges measure changes in capacitance caused by pressure. Relatively large plates are required to achieve an acceptable signal-to-noise ratio. These gauges are sensitive to mechanical vibrations and orientation.[5]

Quartz gauges measure the change in resonant frequency of a quartz crystal with pressure (1.5 Hz/psi).[5] Unfortunately, the resonant frequency of a quartz crystal is also quite sensitive to temperature (15 Hz/°C). Thus, a viable quartz gauge must account for changes in temperature. One approach is to use a separate temperature measurement and calculate the pressure; another approach is to measure the beat frequency between two quartz crystals—one exposed to the pressure to be measured, the other contained in a vacuum chamber. Both these approaches suffer from thermal time lag. The compensated quartz gauge[5] measures the change in frequency of two vibrational modes of a single quartz crystal. Because the two modes differ in their relative sensitivity to temperature, the pressure can be determined very accurately. Because both temperature and pressure are measured with a single quartz crystal, there is very little thermal lag.

TABLE 11.1—COMPARISON OF PRESSURE-GAUGE PERFORMANCE (adapted from Ref. 6)

	Gauge Type					
	Mechanical	Metal Strain	Capacitance	Sapphire	Standard Quartz	Compensated Quartz
Advantages	Reliable Simple Rugged	Improved resolution Fast pressure response Rugged and small	Greater stability Lower power requirement	Improved stability Improved accuracy Fast pressure and temperature response Good resolution Very rugged	Best resolution Best stability Best accuracy	Best resolution Best stability Best accuracy Best dynamics Higher pressure range
Disadvantages	Poor resolution Poor stability Poor accuracy Tedious to read	Moderate stability Moderate accuracy Moderate temperature-response time	Poor dynamic response Slower sampling Temperature and vibration sensitivity	Moderate temperature sensitivity	Poor dynamic response High temperature sensitivity Limited pressure range Cost	More electronics Cost
			Pressure, psi			
Maximum range						
Pressure, psi	20,000	20,000	20,000	20,000	11,000	15,000
Temperature, °F	392	347	347	347	347	347
Resolution (full scale and sampling time) from analog chart						
At 20,000 psi and 1 second	−5	0.50	0.20	0.20	NA	NA
At 15,000 psi and 1 second	−2	0.20	0.15	0.10	NA	0.003
At 10,000 psi and 10 seconds	−2	0.10	0.05	0.05	0.007	0.001
Accuracy						
At 20,000 psi	−20	18	NA	6	NA	NA
At 15,000 psi	−15	12	−12	5	NA	(0.01% of reading +2.0 psi)
At 10,000 psi	−10	10	−10	4	(0.25% of reading +0.05 psi)	(0.01% of reading +20 psi)
Drift, 10,000 psi and 302°F						
First day	−5	<2 to 10	<3	<3	<2	<0.2
First 4 days	−10	<3 to 12	<5	<5	0.2 (18 days)	0.2 (7 days)
Long term, per week	−5	<2 to 4	1 to 7	<1	<0.1	<0.1
			Stabilization Time			
After a 5,000-kpsi step	10 minutes*	30 seconds	8 minutes	20 seconds*	6 minutes	Always within 1 psi
After a 50°F step	10 minutes*	10 minutes	40 minutes	10 minutes	25 minutes	Stable within 25 seconds
			Relative Cost			
	Low	Low	Medium-high	Low-medium	High	High

*Estimated.
NA = not applicable.

11.4 Pressure Transient Test Design

In this section, we address specific design procedures for oil- and gaswell pressure transient tests. Although our discussions are limited to pressure-buildup and -drawdown tests, these same concepts generally can be extended to other pressure transient tests, such as pressure-falloff or injectivity tests. Included in these discussions are pretest estimates of reservoir properties, appropriate test flow rates, and test-duration requirements.

11.4.1 Pretest Estimates of Reservoir Properties. Many aspects of pressure transient test design depend on accurate estimates of permeability and skin factor. Although estimates of other rock and fluid properties are required, permeability and skin factor estimates usually are the most difficult because their estimation is frequently a test objective. Possible means of obtaining permeability estimates include laboratory analysis of core data, results from other well tests in the formation, productivity tests (interpreted by use of the pseudosteady-state radial-flow equation with an estimated skin factor), or, for a well in an infinite-acting reservoir, the unsteady-state flow equation (also with an estimated skin factor).

This section offers suggestions for obtaining pretest estimates of both skin factor and permeability. As shown previously, transient flow of gas is modeled most rigorously in terms of either pseudopressure and pseudotime or adjusted-pressure and -time variables. However, given the numerous approximations of properties that must be made in well-test design, we feel that the design equations formulated in terms of ordinary pressure and time are not only simpler to use but also are sufficiently accurate.

Skin Factor. The skin factor is a dimensionless quantity that is used to quantify the additional pressure difference from a zone of altered permeability in the formation immediately adjacent to the wellbore. Because of drilling and/or completion operations, this near-wellbore zone can be damaged, resulting in a permeability that is lower than the unaltered, in-situ formation permeability. Under these conditions, the skin factor is a positive quantity. Larg-

TABLE 11.2—SKIN FACTOR ESTIMATES FOR VARIOUS WELL STIMULATION AND COMPLETION STRATEGIES

Type of Stimulation or Completion	Skin Factor (s)
Natural completion	0
Small acid treatment	–1.0
Intermediate acid treatment	–2.0
Large acid or small fracture treatment	–3.0
Intermediate fracture treatment	–4.0
Large fracture treatment in low-permeability reservoir	–6.0
Very large fracture treatment in low-permeability reservoir	–8.0

er positive values of skin factors indicate greater reductions in near-wellbore permeability. Conversely, if the formation around the wellbore is stimulated, such as by acidizing or fracturing, the skin factor is negative.

Skin factor can be estimated in various ways. An estimate can be obtained from other pressure transient tests in the same formation and in wells with similar completions, or it can be approximated on the basis of the well's completion and/or stimulation technique. For low-permeability wells that do not flow at measurable rates before either a breakdown or an acid treatment, the skin factor can be approximated by $s=-1$ for a breakdown treatment (e.g., using nitrogen or KCl water) or $s=-2$ for acidization. **Table 11.2** provides estimates of skin factors for various completion and stimulation strategies and is based on a modification and updating of Ref. 7.

Because of non-Darcy effects (i.e., turbulence and inertial effects that cannot be modeled by Darcy's equation) characteristic of gas flow, an additional pressure drop similar to the skin effect in the formation may occur immediately adjacent to the wellbore. Unlike the skin effect caused by alterations in the near-wellbore permeability, the non-Darcy effect is not constant but varies directly with flow rate. Consequently, the skin factor estimated from a gaswell test is an apparent skin factor, s', which is defined as

$$s' = s + Dq_g \quad \text{(11.1)}$$

where s = true skin factor, which is sometimes called mechanical skin, resulting from the zone of altered permeability around the wellbore, and Dq_g = additional, rate-dependent skin factor. In Eq. 11.1, D is the turbulent or non-Darcy flow constant given by[8]

$$D = \frac{2.715\times10^{-15}\beta k_g M p_{sc}}{h r_w T_{sc}\mu_{g,wf}}. \quad \text{(11.2)}$$

The turbulence parameter is estimated with the correlation[9]

$$\beta = 1.88\times10^{10}k^{-1.47}\phi^{-0.53}. \quad \text{(11.3)}$$

Although D is not really constant, the estimate given by Eq. 11.2 is adequate for well-test-design purposes.

Formation Permeability. As stated previously, possible sources of permeability estimates include laboratory analysis of core samples from the well to be tested or from adjacent wells in the same formation. Well tests in adjacent wells are other sources. Alternatively, if productivity tests have been conducted on the well to be tested, the pseudosteady-state-flow equation can be used to estimate permeability, or

$$k = \frac{141.2qB\mu}{h(\bar{p}-p_{wf})}\left[\ln\left(\frac{r_e}{r_w}\right)-\frac{3}{4}+s\right]. \quad \text{(11.4)}$$

When a flowing gas well has stabilized, permeability can be estimated by

$$k = \frac{141.2q_g\bar{B}_g\bar{\mu}_g}{h(\bar{p}-p_{wf})}\left[\ln\left(\frac{r_e}{r_w}\right)-\frac{3}{4}+s'\right] \quad \text{(11.5)}$$

where s' = skin factor defined by Eq. 11.1 and q_g = gas flow rate in Mscf/D. The gas formation volume factor, $\bar{B}_g$, in RB/Mscf, is evaluated at average reservoir pressure, or

$$\bar{B}_g = \frac{5.04T\bar{z}}{\bar{p}}. \quad \text{(11.6)}$$

Note that the length of time to reach stabilization varies, depending on the formation permeability and the stage of development in the field. In new wells in undeveloped fields and most low-permeability gas wells, the well may not reach stabilized flowing conditions for several weeks or even months. In this case, the permeability is estimated with the unsteady-state-flow equation. Eq. 11.7 is the unsteady-state-flow equation in a form that is applicable when the flow rates are changing smoothly. For an oil well,

$$\frac{p_i-p_{wf}}{q} = \frac{70.6B\mu}{kh}\left[\ln\left(\frac{kt}{1{,}688\phi\mu c_t r_w^2}\right)-2s\right] \quad \text{(11.7)}$$

or for a gas well,

$$\frac{p_i-p_{wf}}{q_g} = \frac{70.6\bar{B}_g\bar{\mu}_g}{kh}\left[\ln\left(\frac{kt}{1{,}688\phi\bar{\mu}_g\bar{c}_t r_w^2}\right)-2s'\right]. \quad \text{(11.8)}$$

Again, the gas properties in Eq. 11.8 are evaluated at the average drainage-area pressure for wells in developed fields; in new fields, they are evaluated at the initial reservoir pressure. To solve Eqs. 11.7 or 11.8 for the formation permeability, we use the flow rate from the longest-duration flow period during which a smoothly changing rate was maintained. Effective producing time, t, in hours is equal to cumulative production divided by the chosen flow rate. Further, we define an effective transient radius of drainage, r_d,[10] as

$$r_d = \left(\frac{kt}{377\phi\mu c_t}\right)^{1/2} \quad \text{(11.9)}$$

or, for a gas well,

$$r_d = \left(\frac{kt}{377\phi\bar{\mu}_g\bar{c}_t}\right)^{1/2} \quad \text{(11.10)}$$

where $\bar{\mu}_g$ and $\bar{c}_t$ = average values of viscosity and compressibility, respectively, evaluated at current average pressure in the well's drainage area. Incorporating Eq. 11.9 into Eq. 11.7 yields

$$k = \frac{141.2qB\mu}{h(p_i-p_{wf})}\left[\ln\left(\frac{r_d}{r_w}\right)-0.75+s\right]. \quad \text{(11.11)}$$

Similarly, combining Eqs. 11.8 and 11.10 for a gas well yields

$$k = \frac{141.2q_g\bar{B}_g\bar{\mu}_g}{h(p_i-p_{wf})}\left[\ln\left(\frac{r_d}{r_w}\right)-0.75+s'\right]. \quad \text{(11.12)}$$

Eq. 11.11 or 11.12 can be solved iteratively for permeability with the following recommended calculation procedure. Note that, if the flow data are distorted by wellbore storage, this procedure overestimates permeability. During wellbore storage, most of the surface production is simply fluid being unloaded from the well, so the sandface rate is actually less than the flow rate measured at the surface.

Iterative Procedure for Estimating Permeability.

1. Make an initial estimate of k, and with this value, calculate r_d. Use Eq. 11.9 for an oil well and use Eq. 11.10 for a gas well. The initial estimate of k may be no more than a guess.
2. Using the value of r_d from Step 1, solve Eq. 11.11 or 11.12 for k. Compare this calculated value of k with the initial guess. If the values agree, stop; otherwise, continue with the iterative process.
3. Using the estimate of k from Step 2, calculate an improved estimate of r_d.
4. With this improved estimate of r_d, solve Eq. 11.11 or 11.12 again for k.
5. Repeat Steps 3 and 4 until the value of k converges. Generally, convergence occurs in two or three calculations.

Example 11.1—Pretest Permeability Estimate. In preliminary testing,[11] a gas well in a low-permeability formation was produced for 20 hours at a final BHP, p_{wf}, of 400 psia. At the end of the production period, the rate, q_g, was 110 Mscf/D, while cumulative production was 110 Mscf. Before testing, the formation was broken down with KCl water, so assume $s' = -1.0$. With the well and formation data given below, estimate the permeability for designing a well test. Because the well is in a low-permeability formation, estimate k assuming unsteady-state-flow conditions.

$\phi = 0.118$
$h = 6$ ft
$r_w = 0.365$ ft
$p_{wf} = 400$ psia
$p_i = 3{,}200$ psia
$s' = -1.0$
$V_{wb} = 15$ bbl
$c_{wb} = 2.9 \times 10^{-4}$ psi^{-1}
$\underline{c}_t = 2.0 \times 10^{-4}$ psi^{-1}
$\overline{B}_g = 1.5$ RB/Mscf
$\mu_g = 0.015$ cp
$q_g = 110$ Mscf/D
$G_p = 110$ Mscf

Solution.

1. Calculate the effective producing time, t.

$$t = \frac{24 G_p}{q_g} \quad \text{(11.13)}$$

$$= \frac{(24)(110 \text{ Mscf})}{110 \text{ MMscf/D}} = 24 \text{ hours}$$

2. As a first guess, we assume $k = 0.1$ md. Using Eq. 11.10, we estimate r_d.

$$r_d = \left(\frac{kt}{377\phi\overline{\mu}_g\overline{c}_t}\right)^{1/2}$$

$$= \left[\frac{(0.1)(24)}{(377)(0.118)(0.015)(2.0 \times 10^{-4})}\right]^{1/2}$$

$$= 134 \text{ ft.}$$

3. Using Eq. 11.12, solve for k where the value of r_d is obtained from Step 2.

$$k = \frac{141.2 q_g \overline{B}_g \overline{\mu}_g}{h(p_i - p_{wf})}\left[\ln\left(\frac{r_d}{r_w}\right) - 0.75 + s'\right]$$

$$= \frac{(141.2)(110)(1.5)(0.015)}{(6)(3,200 - 400)}\left[\ln\left(\frac{134}{0.365}\right) - 0.75 - 1.0\right].$$

$$= 0.0864 \text{ md.}$$

4. Because the permeability calculated in Step 3 is different from our initial estimate, we must calculate a new value of r_d using the most recent value of k from Step 3.

$$r_d = \left(\frac{kt}{377\phi\overline{\mu}_g\overline{c}_t}\right)^{1/2}$$

$$= \left[\frac{(0.0864)(24)}{(377)(0.118)(0.015)(2.0 \times 10^{-4})}\right]^{1/2}$$

$$= 125 \text{ ft.}$$

5. With the new value of r_d from Step 4, refine the estimate of k.

$$k = \frac{141.2 q_g \overline{B}_g \overline{\mu}}{h(p_i - p_{wf})}\left[\ln\left(\frac{r_d}{r_w}\right) - 0.75 + s'\right]$$

$$= \frac{(141.2)(110)(1.5)(0.015)}{(6)(3,200 - 400)}\left[\ln\left(\frac{125}{0.365}\right) - 0.75 - 1.0\right].$$

$$= 0.0850 \text{ md.}$$

6. Although the permeability calculated in Step 5 is closer to the value calculated in Step 3, we should iterate once more. With the new value of k, the new value of r_d is

$$r_d = \left(\frac{kt}{377\phi\overline{\mu}_g\overline{c}_t}\right)^{1/2}$$

$$= \left[\frac{(0.0850)(24)}{(377)(0.118)(0.015)(2.0 \times 10^{-4})}\right]^{1/2}$$

$$= 124 \text{ ft.}$$

7. Finally, the latest value of k is

$$k = \frac{141.2 q_g \overline{B}_g \overline{\mu}}{h(p_i - p_{wf})}\left[\ln\left(\frac{r_d}{r_w}\right) - 0.75 + s'\right]$$

$$= \frac{(141.2)(110)(1.5)(0.015)}{(6)(3,200 - 400)}\left[\ln\left(\frac{124}{0.365}\right) - 0.75 - 1.0\right]$$

$$= 0.0848 \text{ md,}$$

which is very close to the value calculated in Step 5. Thus, we have converged to the correct value.

11.4.2 Estimating Test Duration. ***Homogeneous-Acting Reservoirs.*** As an absolute minimum, the flow period must be sufficiently long so that the pressure transient moves beyond the wellbore-storage-distortion period. In terms of dimensionless variables, the duration of wellbore storage during a drawdown or injection test in a well with skin factor $s \geqq -3.5$ is estimated by the empirical fit to the Agarwal *et al.* [12] type curve,

$$t_D = (60 + 3.5s)C_D \quad \text{(11.14a)}$$

where, for t in hours, the dimensionless time is

$$t_D = \frac{0.0002637kt}{\phi\mu c_t r_w^2}. \quad \text{(11.14b)}$$

For gas wells, the gas properties in Eq. 11.14 are evaluated at the average reservoir pressure. Eq. 11.14a predicts the duration of wellbore unloading for a well produced at a constant surface flow rate, but the relationship also is accurate for well-test design when the flow rates vary smoothly.[11] The dimensionless wellbore storage coefficient, C_D, in Eq. 11.13 is defined by

$$C_D = \frac{0.8936C}{\phi c_t h r_w^2}, \quad \text{(11.15)}$$

where C = wellbore-storage coefficient in units of bbl/psi. When both gas and liquid phases exist in the wellbore, the wellbore storage coefficient for a changing gas/liquid interface is

$$C = 25.65\frac{A_{wb}}{\rho_{wb}} \quad \text{(11.16)}$$

where A_{wb} = wellbore area, ft^2, and ρ_{wb} = density of fluids in the wellbore, lbm/ft^3.

When the wellbore is filled with a single-phase fluid, the wellbore storage coefficient is defined by

$$C = c_{wb}V_{wb} \quad (11.17)$$

where c_{wb} = compressibility of wellbore fluid, psi^{-1}, and V_{wb} = volume of fluid in wellbore, bbl.

In terms of dimensional variables, the duration of wellbore storage, t_{wbs}, in hours, for a flow test, the flow period before shut-in for a buildup test, or an injection test is estimated by[3]

$$t_{wbs} = \frac{(200,000 + 12,000s)C}{kh/\mu}. \quad (11.18)$$

Eq. 11.18 often is modified[13] to model the duration of wellbore storage distortion for pressure buildup tests; however, for simplicity, we use the same equation for the design of both drawdown and buildup tests. Alternatively, if a well's productivity index has been measured or can be estimated, t_{wbs} is given by

$$t_{wbs} = \frac{200C}{JB} \quad (11.19)$$

where J = productivity index in STB/D-psi.

If a flow time equal to or greater than the duration of the wellbore-storage period is not attained, then neither the flow-period data nor data from a buildup test following the flow period can be analyzed correctly with semilog plotting techniques. Further, conventional type-curve analysis of these data also is impossible. Under these conditions, the well test should be run again with a longer flow period before shut-in.

If a feasible test length, t_{max} for a drawdown test or Δt_{max} for a buildup test, is greater than four times the duration of wellbore storage (i.e., $t_{max} > 4t_{wbs}$ for a drawdown or $\Delta t_{max} > 4\Delta t_{wbs}$ for a buildup test) and occurs before boundary effects appear, then surface shut-in is possible. For a buildup test with a low probability of reaching the semilog straight line, a bottomhole shut-in device should be considered. With this device, wellbore-storage distortion of the test data should be negligible. For a flow test, there is no simple way to limit the wellbore volume to be unloaded and thus reduce the duration of wellbore-storage distortion.

Semilog analysis techniques are inherently more accurate than type-curve methods and, consequently, are preferred for analyzing pressure transient test data. Semilog analysis techniques, however, require at least half a log cycle of middle-time data. This criterion suggests that the time to reach the end of the middle-time or radial-flow region during either a drawdown or a buildup test should be at least three to four times longer than the time to reach the end of wellbore storage. The end of the middle-time region in a homogeneous-acting reservoir can be estimated for either a pressure-drawdown or -buildup test, respectively, as

$$t_{end} = \frac{948\phi\mu c_t L^2}{k} \quad (11.20a)$$

or

$$\Delta t_{end} = \frac{948\phi\mu c_t L^2}{k} \quad (11.20b)$$

where L is the distance from the well to the closest no-flow boundary. Eq. 11.20 is also valid for gaswell tests when average properties are used. The no-flow boundaries may be either natural reservoir boundaries (i.e., sealing faults, low-permeability barriers, etc.) or artificial boundaries created by adjacent producing wells. For a well centered in a drainage area having a radius r_e, Eq. 11.20b can be rewritten for a buildup test as

$$\Delta t_{end} = \frac{237\phi\mu c_t r_e^2}{k}. \quad (11.21)$$

For a drawdown test, however, Eq. 11.20a should be used with r_e replacing L. Similar expressions can be written for gas wells by replacing the fluid properties in Eqs. 11.20 and 11.21 with average values.

Hydraulically Fractured Wells. Like well-test design in homogeneous-acting reservoirs, the engineer also requires guidelines for identifying both the end of wellbore storage distortion and the onset of boundary effects in a hydraulically fractured well. This section provides suggestions for estimating these flow periods for both infinite- and finite-conductivity vertical fractures.

Infinite-Conductivity Vertical Fractures. Although limited in theory to infinite-conductivity (i.e., $C_r \geqq 100$) fractures in infinite-acting reservoirs, the Barker-Ramey[14] type curves (Fig. 6.15, Chap. 6) provide a convenient method for predicting the end of wellbore-storage effects. Analysis of their type curves indicates that, regardless of fracture half-length, L_f, the dimensionless time at the end of wellbore-storage distortion of the test data is given by

$$t_{L_fD} = 15C_{L_fD} \quad (11.22a)$$

or, more simply,

$$t_D = 15C_D, \quad (11.22b)$$

where the dimensionless time based on the fracture half-length, L_f, is

$$t_{L_fD} = \frac{0.0002637kt}{\phi\mu c_t L_f^2} \quad (11.23)$$

and the wellbore-storage coefficient in terms of fracture half-length is

$$C_{L_fD} = \frac{0.8936C}{\phi c_t h L_f^2}. \quad (11.24)$$

The dimensionless fracture conductivity, C_r, is defined by

$$C_r = \frac{w_f k_f}{\pi k L_f}. \quad (11.25)$$

The start of the middle-time region in a test of a hydraulically fractured well usually is the beginning of pseudoradial flow, which is the point at which the transient moves beyond the influence of the fracture and is no longer affected by the fracture properties. The larger the value of C_r, the later the development of an essentially radial drainage pattern. For infinitely conductive fractures in a well centered in a square drainage area,[15] pseudoradial flow begins approximately when

$$t_{L_fD} \geqq 3. \quad (11.26)$$

In dimensional form, the onset of the pseudoradial-flow period, t_{prf}, (in hours) is

$$t_{prf} \approx \frac{11,400\phi\mu c_t L_f^2}{k}. \quad (11.27)$$

In tests of hydraulically fractured wells, $^1/_3$ to $^1/_2$ log cycle of data in pseudoradial flow is desirable so that more-precise semilog analysis can be used rather than type-curve analysis. This objective is idealistic, however, and rarely is achieved in practice, especially in low-permeability formations with long fractures. As a general rule, the pseudoradial-flow period has not begun in a fractured well test at the time that wellbore-storage distortion has ended.

If the fracture half-length is large relative to the drainage area, then boundary effects may quickly mask the infinite-acting pseudoradial flow regime. For infinite-conductivity fractures in wells centered in square reservoirs, the type curves (Fig. 6.6, Chapter 6) developed by Gringarten *et al.*[16] provide a method to estimate the time at which boundary effects will appear,

$$t_{end} = \frac{948\phi\mu c_t L_e^2}{k} \quad (11.20)$$

where examination of these curves indicates that L_e is the distance from the well to the side of the square reservoir.

Finite-Conductivity Vertical Fractures. Finite-conductivity vertical fractures are characterized by measureable pressure drops in the fracture and, therefore, exhibit unique pressure responses. Because of the number of flow regimes exhibited by finite-conductivity vertical fractures, simple equations for estimating the duration of the wellbore-storage period are difficult to obtain. Type curves[17] for finite-conductivity vertical fractures that include wellbore storage effects have been developed; however, we recommend the Barker-Ramey type curves (Fig. 6.15) for estimating the time when wellbore-storage effects end. Although their type curves[14] are not theoretically correct for analyzing well-test data from finite-conductivity vertical fractures, we have found them to be sufficiently accurate for well-test-design purposes, especially for estimating the duration of wellbore-storage effects. Analysis of these type curves indicates that, regardless of fracture half-length, the duration, t_D, of wellbore storage is given by

$$t_D = 15C_D . \qquad (11.22)$$

Depending on the value of the dimensionless fracture conductivity and the duration of wellbore storage, several intermediate flow periods—fracture linear, bilinear, and formation linear—can occur before the onset of pseudoradial flow (Fig. 6.1, Chap. 6).[18] Of these flow regimes, the fracture linear flow occurs first but is very short-lived and usually is masked by wellbore-storage effects. Because of its extremely short duration, this flow period is of no practical use for most well-test analyses.

The next possible flow period is bilinear flow, which occurs only in finite-conductivity fractures as fluid in the surrounding formation flows linearly into the fracture. Depending on the duration of wellbore-storage effects, the bilinear-flow period may last for a significant time. In addition, formation linear flow occurs only in infinite-conductivity fractures. Between the end of formation linear flow and the beginning of pseudoradial-flow patterns, a transitional flow regime occurs that is characterized by an elliptical flow pattern.

For a well with a finite-conductivity vertical fracture in an infinite-acting reservoir, the type curves (Fig. 6.3, Chap. 6) developed by Cinco-Ley *et al.*[19] provide estimates of the time when pseudoradial flow begins. Note that the type curves in Fig. 6.3 do not include the effects of wellbore storage and thus may underestimate the beginning of the semilog straight line. Again, these curves are sufficiently accurate for designing well tests.

Naturally Fractured Reservoirs. In well tests in naturally fractured reservoirs, early-time data are necessary for estimating the properties of the fracture and matrix system. Often, these early-time data are masked by the effects of wellbore storage. Mavor and Cinco-Ley[20] suggest that, if the dimensionless wellbore-storage coefficient, C_D, is

$$C_D > \frac{\omega(1-\omega)}{36\lambda(60+3.5s)}, \qquad (11.28)$$

then wellbore-storage effects distort the early-time data. Under these conditions, the design engineer should consider using a bottomhole shut-in device. For a pressure-buildup test, this device should reduce the effects of wellbore storage. For a flow test, there is no simple way to limit the wellbore volume to be unloaded and thus reduce the duration of wellbore storage distortion.

Mavor and Cinco-Ley[20] also suggest that the duration of wellbore-storage effects in a naturally fractured reservoir can be estimated by use of the same equation developed for a homogeneous-acting reservoir,

$$t_D = (60 + 3.5s)C_D, \qquad (11.29)$$

where the dimensionless time, t_D, is defined by

$$t_D = \frac{0.0002637k_f t}{\left(\phi_m c_m + \phi_f c_f\right)\mu r_w^2}, \qquad (11.30)$$

where Subscripts m and f = matrix and fracture, respectively. Bourdet *et al.*[21,22] developed type curves (Figs. 7.3 and 7.9 for pseudosteady and transient matrix flow, respectively, Chap. 7) that can be used to estimate the beginning of the semilog straight line under a variety of conditions of wellbore storage and skin factor. Note that these type curves and Eq. 11.30 both require an estimate of fracture permeability, k_f, which often is unknown.

11.4.3 Radius of Investigation. The radius-of-investigation concept is of both quantitative and qualitative value in well-test design. Radius of investigation, r_i, refers to the distance that a pressure transient has moved into the formation following a rate change in a well. The distance traveled by the pressure transient is a function not only of the formation rock and fluid properties but also of the time elapsed since the rate change. To evaluate a representative volume of the reservoir or to identify a suspected reservoir heterogeneity, a well test must be designed that achieves the desired radius of investigation.

For well tests designed to analyze productivity problems, a radius of investigation of $50 \leqq r_i \leqq 200$ ft during a flow test or during the flow period before shut-in for a buildup test should be achieved. In a buildup test, a similar radius of investigation during the shut-in period is desirable. In well tests designed for sampling conditions at the reservoir boundaries, a radius of investigation approximately equal to r_e should be achieved during both flow and shut-in periods. For identifying an important heterogeneity, such as a sealing fault, a radius of investigation of at least four times the expected distance to the heterogeneity should be attained.

The radius of investigation[23] for a well in a homogeneous-acting formation and during pseudoradial flow in a hydraulically fractured well is given by

$$r_i = \left(\frac{kt}{948\phi\mu c_t}\right)^{1/2} \qquad (11.31a)$$

$$\text{or } t = \frac{948\phi\mu c_t r_i^2}{k}. \qquad (11.31b)$$

Comparing Eqs. 11.21 and 11.31b shows that the time when boundary effects begin to be seen in the pressure response at the wellbore is one-fourth of the time required for the radius of investigation to reach r_e during a pressure-buildup test.

11.4.4 Flow-Rate Requirements. ***Constant-Rate Testing.*** Most well-test analysis techniques that we have discussed were developed assuming constant production rates during the entire test. In a drawdown test or during the flow period before a pressure-buildup test, a constant rate should be maintained throughout the test whenever possible. Often, constant or near-constant rates can be achieved by use of an adjustable choke at the wellhead.

For flow before a pressure-buildup test and for a well with established production characteristics, the historical producing rate should be maintained essentially constant for at least the expected duration of the shut-in period. Unfortunately, maintaining constant rates during a drawdown test or during the flow period before a pressure-buildup test is difficult to achieve, especially early in the life of a gas well. When a constant rate cannot be maintained, the instantaneous rates should be measured continuously throughout a flow test and variable-rate analysis techniques should be used.

Whenever possible, the pressure gauge should be placed into the hole with the well flowing. If this cannot be done, after the gauge is placed in the hole, the same constant rate established before the well was shut in should be re-established and maintained for at least eight times the duration of the shut-in period.

Initiating Flow in Low-Permeability Wells. Wells in low-permeability formations, especially natural gas reservoirs, are tested infrequently before fracturing because they do not flow at measurable rates, if at all, without stimulation. Experience indicates that a prefracture breakdown with acid, KCl water, or nitrogen usually enables a low-permeability well to flow at rates adequate for prefracture testing.[11]

Given the importance of prefracture tests, these breakdown treatments are a good investment in formation evaluation. Bostic and Graham[24] reported success in achieving flow in the Cotton Valley formation (permeabilities ranging from 0.001 to 0.01 md) with 2% KCl water breakdown. Similar success has been experienced in the tight gas sands in the Travis Peak formation.[25] Holgate *et al.*[26] suggest nitrogen breakdown treatments, which have proved successful in initiating flow in wells completed in the lower-pressure Eastern Devonian gas shale reservoirs.

Liquid Loading. During either a drawdown test or the flow period before a buildup test in a gas well that produces significant quantities of liquids, the minimum flow rate should equal that required to lift the liquids continuously from the wellbore. Liquid loading or ac-

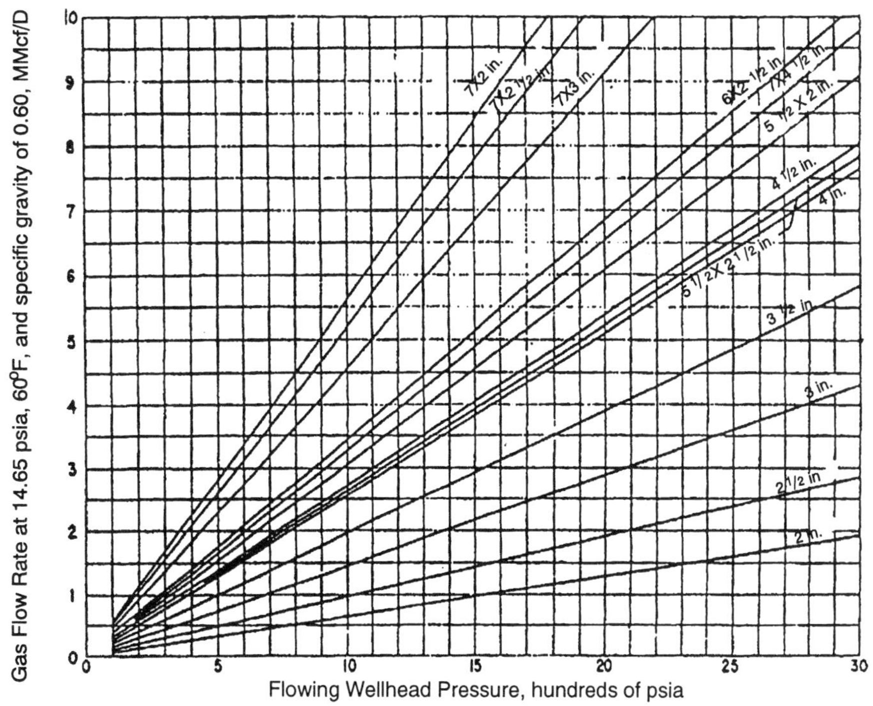

Fig. 11.1—Minimum gas flow rate required to unload liquids from a gas well at a flowing wellhead pressure of 100 psia (from Ref. 27).

cumulation occurs when the gas phase cannot provide adequate energy for the continuous removal of liquids in the wellbore. When liquids are produced with the gas, the pressure drops in the tubing increase not only because of the additional weight of the liquid in the gas stream but also because of the accumulation of the liquids in the bottom of the wellbore. As a result, these liquids impose an additional backpressure on the formation and can affect the test results significantly.

Ideally, the gas velocity should be maintained at some minimum value that is sufficient to eliminate or greatly reduce liquid holdup in the wellbore, thereby continuously removing liquids. When the gas rate falls below a rate that is inadequate to continuously remove the liquids, however, the liquid accumulates in the wellbore. Duggan[27] and Turner *et al.*[28] developed simple methods for estimating the minimum gas flow rate for continuous removal of liquids in the wellbore.

Duggan's[27] Method. In 1961, Duggan developed a method for estimating the gas flow rate required to keep a gas well continuously unloaded. The method is based on the assumption that a minimum wellhead velocity of 5 ft/sec is required to lift liquids continuously in the wellbore. The method is empirical because the velocity of 5 ft/sec was selected after observing the flowing performance of a number of wells under a wide range of operating conditions of liquid loading. **Fig. 11.1** presents the results of Duggan's study as a simple chart.

We emphasize that Duggan's method is empirical and that several field studies subsequent to his work have suggested higher gas velocities may be required. For example, Libson and Henry[29] indicated that gas velocities greater than 15 ft/sec are required in many deep gas wells in southwest Texas. In low-pressure gas wells in the Hugoton field, Smith[30] found that velocities of 5 to 10 ft/sec were necessary to remove condensates, while removal of water required velocities ranging from 10 to 20 ft/sec. These additional results suggest that Duggan's method should be applied judiciously.

Turner et al.[28] Method. Turner *et al.* proposed two physical models for predicting when a gas well will experience liquid loadup: (1) liquid film movement occurring along the walls of the pipe and (2) liquid droplets entrained in a high-velocity gas. Using theoretical particle- and drop-breakup fluid mechanics, they calculated the minimum gas velocity required to remove either water or condensates continuously from the wellbore. On the basis of comparisons between theoretical predictions and field results, they concluded that the controlling mechanism for continuous removal of liquids is the movement of the entrained liquid droplets.

They defined the minimum gas velocity that moves the largest liquid drops that can exist in a gas stream as

$$u_t = 20.4\frac{\Gamma^{1/4}\left(\rho_L - \rho_g\right)^{1/4}}{\rho_g^{1/2}} \quad \text{(11.32)}$$

where u_t = mimimum terminal velocity of free-falling liquid particles, ft/sec; Γ = gas-liquid interfacial tension, dynes/cm; ρ_g = density of gas phase, lbm/ft^3; and ρ_L = density of liquid phase, lbm/ft^3.

However, fluid properties, especially interfacial tensions, are usually not known. Therefore, by assuming typical values for the fluid properties, Turner *et al.* developed the following equations for the minimum velocities of the gas phase required to lift water and condensate, respectively.

$$u_{g-w} = \frac{5.62\left(67 - 0.0031 p_{tf}\right)^{1/4}}{\left(0.0031 p_{tf}\right)^{1/2}} \quad \text{(11.33)}$$

$$u_{g-c} = \frac{4.02\left(45 - 0.0031 p_{tf}\right)^{1/4}}{\left(0.0031 p_{tf}\right)^{1/2}}. \quad \text{(11.34)}$$

Eqs. 11.33 and 11.34 were developed with the assumed fluid properties given in **Table 11.3.**

11.4.5 General Design Procedures—Single-Well Tests. In this section, we offer general procedures for designing single-well pressure buildup tests in both unstimulated and stimulated wells. Note that these procedures are suggestions, and that each test design is unique and will require different considerations depending on the testing objectives and environment.

TABLE 11.3—FLUID PROPERTIES USED IN EQS. 11.33 AND 11.34

Fluid Property	Assumed Value
γ_g	0.6
T, °F	120
ρ_w, lbm/ft³	67
ρ_c, lbm/ft³	45
Γ_{g-w}, dynes/cm	60
Γ_{g-c}, dynes/cm	20

Prefracture Pressure Drawdown and Buildup Tests. The following procedure, which is based on the theoretical and operational principles described in previous sections, should prove adequate for designing drawdown and buildup tests in a well that has not been stimulated, either by hydraulic fracturing or acidization. If a fracture treatment is planned, the procedure applies to the design of a prefracture buildup test. In all equations presented in the following procedure, the gas properties are evaluated at the current average reservoir pressure, $\bar{p}$, for gaswell design.

Test-Design Procedure.

1. Estimate well and reservoir properties required for the test design.

a. Select the flow rate for the drawdown period or for the flow time before shut-in. Using the methods given in Sec. 11.4.5, ensure that this rate is sufficient to lift liquids continuously from the wellbore for gas wells producing liquids.

b. Estimate p_i (or $\bar{p}$) and p_{wf} ($\Delta t=0$). Possible sources include data from a previous drillstem test or pressure surveys from active wells completed in the same formation.

c. Estimate fluid properties (i.e., B, μ, and c_t for oil wells or $\bar{B}_g$, $\bar{\mu}_g$, and $\bar{c}_t$ for gas wells) at $\bar{p}$. Generally, in the absence of laboratory data, correlations are adequate for estimating the fluid properties.

d. Estimate k, h, and ϕ. Permeability can be estimated by use of the suggestions given in Sec. 11.4.1. The best sources for estimating net-pay thickness are well logs, while porosity estimates can be obtained from either cores or well logs.

e. Estimate productivity index,

$$J = \frac{q}{\left(\bar{p} - p_{wf}\right)} \quad \text{.......... (11.35)}$$

or, alternatively, the skin factor.

f. Estimate the wellbore-storage coefficient, C. For wells with both gas and liquid (i.e., condensate and/or water) production, Eq. 11.16 gives the wellbore-storage coefficient for a liquid/gas interface in the wellbore; if only a single fluid phase is produced, estimate C with Eq. 11.17. Average wellbore temperature is the arithmetic average of surface and bottomhole temperatures. For single-phase gas in the wellbore, we can estimate $c_{wb} \approx c_g$ at $\bar{p}$ and average wellbore temperature.

g. Estimate the drainage radius, r_e, as the well spacing from known field development patterns or the distance, L, to a nearby reservoir no-flow boundary. Distances to natural no-flow boundaries may be estimated with geologic data.

2. Estimate the duration of wellbore-storage distortion, t_{wbs}, during the production and/or shut-in period.

a. If the productivity index is unknown and kh/μ and s estimates are available, use Eq. 11.18.

$$t_{wbs} = \frac{(200{,}000 + 12{,}000s)C}{kh/\mu}, \quad s \geqq -3.5. \quad \text{....... (11.18)}$$

b. If the productivity index is known, use Eq. 11.19.

$$t_{wbs} \approx \frac{200C}{JB}. \quad \text{.......... (11.19)}$$

Note that, if the well is stimulated (negative skin factor), using $s=0$ in Eq. 11.18 leads to a more-conservative test design.

3. Estimate the flowing or shut-in time required to achieve the design radius of investigation,

$$t_{\text{min}} \text{ or } \Delta t_{\text{min}} = \frac{948\phi\mu c_t r_i^2}{k}. \quad \text{.......... (11.31)}$$

Guidelines for selecting the design radius of investigation include the following.

a. For evaluating near-wellbore conditions, achieve $r_i=200$ ft to sample a representative portion of the reservoir.

b. If $r_i=200$ ft cannot be attained in a reasonable time, then a smaller r_i can be selected. This smaller r_i, however, must be several times the estimated depth of damage or stimulation, r_s (e.g., $r_i \geqq 5r_s$, with r_s calculated by use of estimated properties in the altered zone near the wellbore rather than in the formation).

c. Sample the entire reservoir, $r_i=r_e$, where r_e is estimated from Step 1g.

d. To confirm the presence of a flow barrier an estimated distance L from the well, use $r_i > 4L$.

4. Estimate the flowing or shut-in time, t_{end}, when boundary effects might appear.

a. For a well a distance L from the nearest boundary,

$$t_{\text{end}} \text{ or } \Delta t_{\text{end}} = \frac{948\phi\mu c_t L^2}{k}. \quad \text{.......... (11.20)}$$

For a drawdown test, r_e should replace L in Eq. 11.20.

b. For a well centered in a circular drainage area, the shut-in time is

$$\Delta t_{\text{end}} \cong \frac{237\phi\mu c_t r_e^2}{k}. \quad \text{.......... (11.21)}$$

5. Select a flow time for the drawdown test or shut-in time for the buildup test that is the greater of $4t_{wbs}$ and t_{min}.

6. An important component in designing a well test is selecting a pressure gauge with the proper sensitivity for the test. The following is a recommended procedure for estimating gauge sensitivity.

a. Estimate the slope, m, of the semilog straight line expected in the flow or buildup test.

$$m = \frac{162.6qB\mu}{kh}. \quad \text{.......... (11.36)}$$

The pressure change between the time at the end of the straight line and 90% of that time is

$$\Delta p = m\log(1/0.90) = 0.0458m. \quad \text{.......... (11.37)}$$

b. Choose a pressure gauge sensitive enough to respond to the expected pressure change during the test.

c. Estimate the maximum pressure to be encountered in the test (e.g., in a new reservoir, $p_{\text{max}}=p_i$).

d. Choose a pressure gauge range so that the maximum test pressure falls between 60 and 80% of the upper limit of the gauge.

e. If possible, choose the clock on a self-contained gauge so that most of the chart is used, but also so that the gauge is run only once during the test. Tandem gauges are essential. An excellent alternative to the conventional mechanical gauge is the memory-type gauge.

To facilitate well-test design, we developed worksheets that incorporate many of the suggestions listed in the design procedure. **Appendix K** shows these worksheets, which can be used for either oil- or gaswell-test design.

Example 11.2—Design of a Prefracture Pressure-Buildup Test for a Gas Well. This example of well-test design is reproduced from Ref. 11. We wish to run a pressure buildup test in a gas well before fracturing. The objective is to design a test that provides estimates of formation permeability and initial reservoir pressure. To minimize flaring gas, the operator would prefer to produce the well for 1 day or less and would prefer to limit the shut-in period to less than 3 days. On the basis of existing development patterns in the field, the drainage area of the well is estimated to be 640 acres. The well is currently producing 100 Mscf/D at a BHFP of 400 psia and has the following parameters.

$\phi = 0.118$

p_i = 3,200 psia
p_{wf} = 400 psia
r_w = 0.365 ft
h = 6 ft
k_g = 0.0848 md
V_{wb} = 15 bbl
c_{wb} = 2.9×10^{-4} psi^{-1}
$\overline{c}_t$ = 2.0×10^{-4} psi^{-1}
$\overline{B}_g$ = 1.5 RB/Mscf
$\overline{\mu}_g$ = 0.015 cp
A = 640 acres

Solution.

1. Estimate gas and formation properties.

a. Estimates of pressures and gas properties are provided above. Note that the gas properties $\overline{B}_g$, $\overline{c}_t$, and $\overline{\mu}$ are evaluated at $\overline{p}$. We choose to set $s' = 0$ for conservative test design.

b. Assuming only gas in the wellbore, with Eq. 11.17, the wellbore-storage coefficient is

$$C = c_{wb}V_{wb} = (2.9 \times 10^{-4})(15) = 0.00435 \text{ bbl/psi}.$$

c. Produce the well at $p_{wf} = 400$ psia and assume that the current well-production rate, $q_g = 100$ Mscf/D, can be maintained until the end of the flow period.

d. For 640-acre spacing and assuming a circular drainage area, the external drainage radius, r_e, is

$$r_e = \sqrt{A/\pi} \quad \text{.................................... (11.38)}$$

$$= \sqrt{(640)(43,560)\pi}$$

$$= 2,979 \text{ ft}.$$

2. From Eq. 11.18, the duration of wellbore-storage distortion is

$$t_{wbs} = \frac{(200,000 + 12,000s')C}{k_g h/\overline{\mu}}$$

$$= \frac{[200,000 + (12,000)(0)](0.00435)}{(0.0848)(6)/(0.015)}$$

$$= 26 \text{ hours}.$$

3. With Eq. 11.20, the shut-in time required to investigate a minimum distance of 200 ft into the reservoir is

$$\Delta t_{min} \approx \frac{948\phi\overline{\mu}_g\overline{c}_t r_i^2}{k_g}$$

$$= \frac{(948)(0.118)(0.015)(2.0 \times 10^{-4})(200)^2}{0.0848}$$

$$= 158 \text{ hours}.$$

4. The shut-in time, Δt_{end}, at which boundary effects appear for a 640-acre drainage area,

$$\Delta t_{end} \approx \frac{237\phi\overline{\mu}\overline{c}_t r_e^2}{k_g}$$

$$= \frac{(237)(0.118)(0.015)(2.0 \times 10^{-4})(2,979)^2}{0.0848}$$

$$= 8,780 \text{ hours}.$$

5. The flow time should be the greater of $4t_{wbs} = (4)(26$ hours$) = 104$ hours and $\Delta t_{min} = 158$ hours. Therefore, the well should be flowed for 7 days because $\Delta t_{min} = 158$ hours. With this flow period, approximately 200 ft of formation is investigated. A rate of 100 Mscf/D is maintained until the end of this period.

6. Similarly, a shut-in time of 7 days should be used.

7. Estimate the pressure gauge sensitivity required for the test.

a. Estimate the slope, m, of the semilog straight line expected in the flow or buildup test with Eq. 11.36.

$$m = \frac{162.6q_g\overline{B}_g\overline{\mu}_g}{kh}$$

$$= \frac{(162.6)(100)(1.5)(0.015)}{(0.0848)(6)}$$

$$= 719 \text{ psi/cycle}.$$

From Eq. 11.xx, the pressure change between the time at the end of the straight line and 90% of that time is

$$\Delta p = m\log(1/0.90)$$

$$= 0.0458m$$

$$= (0.0458)(719) = 33 \text{ psi}.$$

b. Choose a pressure gauge sensitive enough to respond to the expected pressure change during the test. An ordinary Bourdon tube gauge is sufficiently sensitive for this test. However, any gauge with greater sensitivity is acceptable.

c. Estimate the maximum pressure to be encountered in the test (e.g., in a new reservoir, $p_{max} = p_i$). For this test, we anticipate that $p_i = 3,200$ psia.

d. Choose a pressure gauge range so that the maximum test pressure falls between 60 and 80% of the upper limit of the gauge, or 4,000 and 5,300 psi.

e. Choose a 120-hour clock for the pressure gauge.

In summary, the operator cannot test as was originally proposed (flow period of 1 day or less and buildup period of less than 3 days) and obtain a proper test. Instead, the operator should flow the well for seven days at $p_{wf} = 400$ psia, and shut in the well for 7 days. During both the flow period before shut-in and the pressure-buildup test, wellbore-storage distortion should not be a problem. In addition, boundary effects will not be encountered.

Post-Fracture Pressure Drawdown and Buildup Tests. The procedure outlined next provides recommendations for designing post-fracture buildup tests. As with the prefracture test design, a pretest estimate of effective permeability to the produced phase must be available. For gaswell-test design, use gas properties evaluated at $\overline{p}$ for all equations.

Test-Design Procedure.

1. Estimate well and reservoir properties required in the test design.

a. Select the flow rate for the drawdown period or for the flow time before shut-in. Using the methods given in Sec. 11.4.5, ensure that this rate is sufficient to lift liquids continuously from the wellbore.

b. Estimate p_i (or $\overline{p}$) and p_{wf} ($\Delta t = 0$). Possible sources include data from a previous drillstem test or pressure surveys from active wells completed in the same formation.

c. Estimate fluid properties (i.e., B, μ, and c_t for oil wells or $\overline{B}_g$, $\overline{\mu}_g$, and $\overline{c}_t$ for gas wells) at $\overline{p}$. Generally, in the absence of laboratory data, correlations are adequate for estimating the properties.

d. Estimate k, h, and ϕ. Permeability can be estimated with the suggestions given in Sec. 11.4.1. The best sources for estimating net-pay thickness are well logs, while porosity estimates can be obtained from either cores or well logs.

e. Estimate the wellbore storage coefficient, C. For wells with both gas and liquid (i.e., condensate and/or water) production, Eq. 11.16 gives the wellbore-storage coefficient for a liquid/gas interface in the wellbore; if only a single fluid phase is produced, estimate C with Eq. 11.17. Average wellbore temperature is the arithmetic average of surface and bottomhole temperatures. For single-phase gas in the wellbore, estimate $c_{wb} \approx c_g$ at $\overline{p}$ and average wellbore temperature.

f. Estimate the fracture half-length, L_f, and the perpendicular distance, L_e, to the side of a square drainage area where the well is assumed to be completed. In addition, estimate the distance L to any single boundary nearby.

2. Estimate the duration of wellbore-storage distortion, t_{wbs}.

a. Calculate the dimensionless wellbore-storage coefficient, C_D.

$$C_D = \frac{0.8936C}{\phi c_t h r_w^2} . \quad \text{(11.15)}$$

b. Determine the dimensionless time, t_D, at which wellbore storage effects end. For infinite-conductivity fractures ($C_r \geqq 100$),

$$t_D \approx 15C_D . \quad \text{(11.22b)}$$

For the purposes of well-test design, Eq. 11.22 is an adequate approximation for finite-conductivity fractures.

c. Using the definition of dimensionless time,

$$t_{wbs} = \frac{\phi \mu c_t r_w^2 t_D}{0.0002637k} . \quad \text{(11.39)}$$

3. Estimate the time, t_{prf}, required to reach pseudoradial flow. For infinite- and finite-conductivity fractures, the beginning of the semilog straight line in dimensional form is approximated by

$$t_{prf} \approx \frac{11,400\phi \mu c_t L_f^2}{k} . \quad \text{(11.27)}$$

4. Estimate the time, t_{end}, when boundary effects will appear.

a. For a well centered in a symmetrical drainage area with distance L_e from well to boundary,

$$t_{\text{end}} \approx \frac{948\phi \mu c_t L_e^2}{k} . \quad \text{(11.40)}$$

b. For a well much nearer one boundary than others, estimate t_{end} from

$$t_{\text{end}} \approx \frac{948\phi \mu c_t L^2}{k} . \quad \text{(11.41)}$$

5. Select a flow time for the test. The flow time must exceed $4t_{wbs}$. In addition, the flow time should at least equal the lesser of $4t_{prf}$ and the maximum flow time possible for the test. Finally, the well should flow at least long enough to clean up.

6. Select the shut-in time for the buildup test. The shut-in time must exceed $4t_{wbs}$. If impractical, then consider using a bottomhole shut-in device. In addition, the shut-in time should equal the lesser of $4t_{prf}$ or the maximum shut-in time, $\Delta t_{\max}$, possible for the test (which should be less than or equal to the maximum flow time possible).

7. An important component in designing a well test is selecting a pressure gauge with the proper sensitivity for the test. A recommended procedure for estimating the gauge sensitivity is given in the procedure for designing prefracture well tests.

Example 11.3—Design of a Post-Fracture Pressure-Buildup Test for a Gas Well. This example post-fracture pressure buildup test design is reproduced from Ref. 11. The gas well currently produces 1,500 Mscf/D at a BHFP of 400 psia. The longest possible flow period before the buildup test is 1 month, and the longest possible buildup test is 2 weeks. Design a pressure-buildup test to evaluate the success of the fracture treatment for a well with the following parameters.

ϕ = 0.118
p_i = 3,200 psia
p_{wf} = 400 psia
r_w = 0.365 ft
h = 6 ft
k_g = 0.0848 md
V_{wb} = 15 bbl
c_{wb} = 2.9×10^{-4} psi^{-1}
$\underline{c_t}$ = 2.0×10^{-4} psi^{-1}
$\overline{B}_g$ = 1.5 RB/Mscf
$\overline{\mu}_g$ = 0.015 cp
q_g = 1,500 Mscf/D
L_e = 2,640 ft
L_f = 200 ft

Solution.

1. With Eq. 11.17, the wellbore-storage coefficient (single phase flowing in the wellbore) is

$$C = c_{wb}V_{wb} = (2.9 \times 10^{-4})(15) = 0.00435 \text{ bbl/psi}.$$

2. Estimate the duration of wellbore-storage distortion.

a. Calculate the dimensionless wellbore storage coefficient, C_D.

$$C_D = \frac{0.8936C}{\phi \overline{c}_t h r_w^2}$$

$$= \frac{(0.8936)(0.00435)}{(0.118)(2.0 \times 10^{-4})(6)(0.365)^2} = 206.$$

b. With Eq. 11.23, determine the dimensionless time, t_D, at which wellbore storage effects end.

$$t_D \approx 15C_D$$

$$= (15)(206) = 3,090.$$

c. Using the definition of dimensionless time, calculate t_{wbs}.

$$t_{wbs} = \frac{\phi \overline{\mu}_g \overline{c}_t r_w^2 t_D}{0.0002637k_g} \quad \text{(11.42)}$$

$$= \frac{(0.118)(0.015)(2.0 \times 10^{-4})(0.365)^2(3,090)}{(0.0002637)(0.0848)}$$

$$= 6.5 \text{ hours}.$$

3. Estimate the time, t_{prf}, required to reach pseudoradial flow with Eq. 11.27. For infinite-conductivity fractures, the onset of pseudoradial flow occurs at

$$t_{prf} \approx \frac{11,400\phi \overline{\mu}_g \overline{c}_t L_f^2}{k_g}$$

$$= \frac{11,400(0.118)(0.015)(2.0 \times 10^{-4})(200)^2}{0.0848}$$

$$= 1,904 \text{ hours or 79 days}.$$

This flow time is longer than the longest flow period possible (1 month) and longest shut-in period possible (2 weeks). Therefore, pseudoradial flow is not reached during this test and semilog analysis of the pressure transient data is not possible.

4. With Eq. 11.20, the time to boundary effects, either during the flow period or the shut-in period, is

$$t_{\text{end}} \text{ or } \Delta t_{\text{end}} \approx \frac{948\phi \overline{\mu}_g \overline{c}_t L_e^2}{k_g} .$$

$$\approx \frac{(948)(0.118)(0.015)(2.0 \times 10^{-4})(2,640)^2}{0.0848}$$

$$\approx 27,600 \text{ hours or 1,150 days}.$$

Thus, boundary effects are not likely to distort data in this test.

5. The flow time for this test is the maximum possible, or 1 month.

6. The shut-in time for this test is the maximum possible, or 2 weeks.

7. Estimate the pressure-gauge sensitivity required for the test.

a. Estimate the slope, m, of the semilog straight line expected in the flow or buildup test with Eq. 11.36.

$$m = \frac{162.6q_g \overline{B}_g \overline{\mu}_g}{kh} \quad \text{(11.36)}$$

$$= \frac{(162.6)(1,500)(1.5)(0.015)}{(0.0848)(6)}$$

$$= 10,800 \text{ psi/cycle.}$$

Eq. 11.37 gives the pressure change between the time at the end of the straight line and 90% of that time as

$$\Delta p = m \log(1/0.90) \quad \text{(11.37)}$$

$$= 0.0458m$$

$$= (0.0458)(10,800) = 495 \text{ psi.}$$

b. Choose a pressure gauge sensitive enough to respond to the expected pressure change during the test. An ordinary Bourdon tube gauge is sufficiently sensitive for this test. However, any gauge with greater sensitivity is acceptable.

c. Estimate the maximum pressure to be encountered in the test (e.g., in a new reservoir, $p_{max} = p_i$). For this test, we anticipate $p_i = 3{,}200$ psia.

d. Choose a pressure-gauge range so that the maximum test pressure falls between 60 and 80% of the upper limit of the gauge, or 4,000 and 5,300 psi.

e. Choose a 180-hour clock for the pressure gauge.

In summary, before shut-in, the operator should produce the well for 1 month at $p_{wf} = 400$ psia and $q_g = 1{,}500$ Mscf/D. In addition, the pressure-buildup test should last for 2 weeks. During both the flow test and buildup test, wellbore-storage effects last for approximately 6 hours. Neither boundary effects nor pseudoradial flow is achieved during the test; therefore, semilog analysis of the data is not possible.

11.4.6 General Design Procedures—Multiwell Tests. ***Interference-Test Design.*** In designing an interference test, the objective is to obtain a sufficiently large pressure drawdown at the observation well within an acceptable period of time. As a general guideline, a pressure drawdown, Δp, of several psi with a test duration of no more than about 1 day is necessary. Although pressure gauges are sensitive enough to detect a pressure drawdown of a fraction of 1 psi, we cannot be certain whether such a small pressure response is the result of production or injection at the active well or simply caused by reservoir "noise" from other wells in the reservoir.

Test-Design Procedure.

1. Estimate well and reservoir properties required for the test design.

a. Select the flow rate for the interference test. Using the methods given in Sec. 11.4.5, ensure that this rate is sufficient to lift liquids continuously from the wellbore for gas wells producing liquids.

b. Estimate p_i (or $\bar{p}$) and p_{wf} ($\Delta t = 0$). Possible sources include data from a previous drillstem test or pressure surveys from active wells completed in the same formation.

c. Estimate fluid properties (i.e., B, μ, and c_t for oil wells or $\bar{B}_g$, $\bar{\mu}_g$, and $\bar{c}_t$ for gas wells) at $\bar{p}$. Generally, in the absence of laboratory data, correlations are adequate for estimating the fluid properties.

d. Estimate k, h, and ϕ. Permeability can be estimated with the suggestions given in Sec. 11.4.1. The best sources for estimating net-pay thickness are well logs, while porosity estimates can be obtained from either cores or well logs.

e. For interference testing, estimate the distances, r, from the active well to the observation wells.

2. Estimate the flowing time required to achieve the design radius of investigation, r_i. For interference testing, r_i is the distance from the active well to the observation well, i.e., $r_i = r$.

$$t_{min} = \frac{948\phi\mu c_t r_i^2}{k}. \quad \text{(11.31)}$$

3. Using the estimated well and reservoir properties, calculate the following dimensionless variables.

$$r_D = \frac{r}{r_w} \quad \text{(11.43)}$$

and

$$t_D = \frac{0.0002637kt}{\phi\mu c_t r_w^2}, \quad \text{(11.44)}$$

where t = desired flowing time. This flowing time should equal or exceed t_{min}, calculated in Step 2.

4. Using either Eq. 11.45 or Fig. 10.5 (the exponential integral solution, Chap. 10), calculate the dimensionless pressure, p_D.

$$p_D = -\frac{1}{2}\text{Ei}\left(\frac{-r_D^2}{4t_D}\right). \quad \text{(11.45)}$$

5. Using the definition of dimensionless pressure given by Eq. 11.46,

$$p_D = \frac{kh(p_i - p_r)}{141.2qB\mu}, \quad \text{(11.46)}$$

calculate the pressure drawdown, Δp,

$$p_i - p_r = \Delta p = \frac{141.2qB\mu p_D}{kh}. \quad \text{(11.47)}$$

If Δp is less than several psi for a 24-hour production period, an interference test may not be effective in this reservoir. In this case, we may consider running a pulse test where the allowable pressure drawdown can be much smaller for a given test duration.

6. An important component in designing a well test is selecting a pressure gauge with the proper sensitivity for the test. Use the recommended procedure given for single-well tests for estimating the gauge sensitivity.

Example 11.4—Interference-Test Design. We want to conduct an interference test using one active well and one observation well located 600 ft away from the active well. Estimate the expected pressure drawdown at the observation well after 24 hours of production. Determine whether an interference test is appropriate for this reservoir.

k = 50 md
q = 100 STB/D
h = 25 ft
μ = 1.0 cp
B = 1.0 RB/STB
c_t = 3.0×10^{-6} psi^{-1}
r_w = 0.3 ft
r = 600 ft
ϕ = 0.20

Solution. Because estimates of the well and reservoir properties are available, the design is started by calculating the minimum flowing time required for the radius of investigation to reach the observation well 600 ft from the active well.

1. With Eq. 11.31, estimate the flowing time required to achieve the design radius of investigation, r_i. For interference testing, r_i is the distance from the active well to the observation well.

$$t_{min} = \frac{948\phi\mu c_t r_i^2}{k}$$

$$= \frac{(948)(0.20)(1.0)(3.0 \times 10^{-6})(600)^2}{50}$$

$$= 4.1 \text{ hours.}$$

Therefore, if the test is conducted for 24 hours, the pressure transient will be detected at the observation well.

2. Using Eqs. 11.43 and 11.44 and the estimated well and reservoir parameters, calculate the dimensionless radius and time.

$$r_D = \frac{r}{r_w}$$

$$= \frac{600}{0.3} = 2{,}000,$$

$$t_D = \frac{0.0002637kt}{\phi\mu c_t r_w^2}$$

$$= \frac{0.0002637(50)(24)}{(0.20)(1)(3 \times 10^{-6})(0.3)^2} = 5.86 \times 10^6,$$

$$\text{and } \frac{t_D}{r_D^2} = \frac{5.86 \times 10^6}{(2{,}000)^2} = 1.46.$$

3. From the type curve in Fig. 10.5, we read a value of

$p_D = 0.68$ at $t_D/r_D^2 = 1.46$.

Alternatively, we can use tables of Ei functions in Chap. 1 with a value of

$$\frac{r_D^2}{4t_D} = \frac{(2{,}000)^2}{4(5.86 \times 10^6)} = 0.17$$

from which, with Eq. 11.45, we calculate

$$p_D = -\frac{1}{2}\mathrm{Ei}\left(\frac{-r_D^2}{4t_D}\right) = -\frac{1}{2}\mathrm{Ei}(-0.17) = \frac{1}{2}(1.358) = 0.68.$$

4. Using Eq. 10.3 with $p_D = 0.68$, the pressure drawdown at the observation well after $t = 24$ hours is

$$\Delta p = \frac{141.2qB\mu p_D}{kh}$$

$$= \frac{141.2(100)(1)(1)(0.68)}{(50)(25)} = 7.7 \text{ psi.}$$

Our estimate of $\Delta p = 7.7$ psi at $t = 24$ hours meets the criterion that Δp be at least several psi during a test not exceeding 1 day. Therefore, we can run an interference test in this reservoir and be somewhat confident that the pressure data we obtain can be distinguished from reservoir noise. If some of the parameters had been different, e.g., smaller k or larger r, Δp might have been too small at $t = 24$ hours for the interference-test data to be useful. In that case, a pulse test should be run.

Pulse-Test Design. Designing a pulse test[31] requires that the pulse times and the expected pressure response be determined. Estimation of the expected pressure response before running a pulse test is important so that the range and sensitivity of the pressure gauge and the test duration can be properly specified.

Test-Design Procedure.[31]

1. Select the pulse ratio, F'. If a specific ratio is most convenient for a particular application, then that value may be used. Otherwise, the ratio should be selected on the basis of whether the normal or inverted pulses will be analyzed. The choice of normal or inverted is arbitrary. To maximize the amplitude of the pressure response, Δp, to the normal pulses, F' should be 0.3 (Figs. 10.15 and 10.19). To maximize the amplitude of the inverted pulses, F' should be 0.7 (Figs. 10.17 and 10.21). In no case should F' be below 0.2 or above 0.8.

2. Estimate the dimensionless lag time ratio, t_{LD}/r_D^2, for normal pulses by reading the value of $t_L/\Delta t_c$ at the peak amplitude, $\Delta p_D[t_L/\Delta t_c]^2$ for the value of F'chosen from Fig. 10.15 or 10.19 (normal pulse) or 10.17 or 10.21 (inverted pulse) and reading the corresponding value of t_{LD}/r_D^2 from Figs. 10.16 or 10.20 (normal pulse) or 10.18 or 10.22 (inverted pulse).

3. Using the best available estimates of k, ϕ, h, μ, and c_t, together with $\Delta p_D[t_L/\Delta t_c]^2$ and t_{LD}/r_D^2 from the appropriate figure, estimate the cycle length, Δt_c, and the pulse amplitude, Δp, respectively, with

$$\Delta t_c = \frac{3792\phi\mu c_t r^2}{k}\left(\frac{t_{LD}}{r_D^2}\right)\left(\frac{\Delta t_c}{t_L}\right) \quad \text{.......... (11.48)}$$

$$\Delta p = \frac{70.6qB\mu}{kh}\left[\Delta p_D\left(\frac{t_L}{\Delta t_c}\right)^2\right]\left(\frac{\Delta t_c}{t_L}\right)^2. \quad \text{.......... (11.49)}$$

4. Using F' and Δt_c, estimate the pulsing period, Δt_p, and the shut-in period, Δt_s.

$$\Delta t_p = F'\Delta t_c \quad \text{.......... (11.50)}$$

$$\text{and } \Delta t_s = \Delta t_p - \Delta t_c. \quad \text{.......... (11.51)}$$

Example 11.5—Pulse-Test Design for Analysis of First Inverted Pulse. Design a pulse test to analyze the first inverted pulse for the wells and reservoir with the estimated properties given here. Determine the pulse length and the gauge sensitivity required for a pulse test that will provide better estimates of k and ϕc_t.

r = 660 ft
h = 25 ft
ϕ = 0.18
μ = 3 cp
B = 1.1 RB/STB
c_t = 1×10^{-5} psi^{-1}
k = 200 md
q = 100 B/D

Solution.

1. Select $F' = 0.7$ to maximize the pulse-response amplitude in the first inverted pulse. (We could just as well choose to analyze the normal pulses.) From Fig. 10.17 the maximum amplitude, $\Delta p_D(t_L/\Delta t_c)^2 = 0.003$ at $t_L/\Delta t_c = 0.3$ for the first inverted pulse. Fig. 10.21 gives similar (but not identical) results for subsequent inverted pulses.

2. For $F' = 0.7$ and $t_L/\Delta t_c = 0.3$, Fig. 10.18 gives $t_{LD}/r_D^2 = 0.12$ for the first inverted pulse. Figs. 10.22 gives a similar result for subsequent inverted pulses.

3. From Steps 1 and 2 and the given information, we calculate Δt_c from Eq. 11.48.

$$\Delta t_c = \frac{3792\phi\mu c_t r^2}{k}\left(\frac{t_{LD}}{r_D^2}\right)\left(\frac{\Delta t_c}{t_L}\right)$$

$$= \frac{(3792)(0.18)(3)(1 \times 10^{-5})(660)^2}{200}\frac{(0.12)}{(0.3)} = 17.84 \text{ hr.}$$

From Fig. 11.49, we calculate Δp.

$$\Delta p = \frac{70.6qB\mu}{kh}\left[\Delta p_D\left(\frac{t_L}{\Delta t_c}\right)^2\right]\left(\frac{\Delta t_c}{t_L}\right)^2$$

$$= \frac{(70.6)(100)(1.1)(3)}{(200)(25)}[0.003]\left(\frac{1}{0.3}\right)^2 = 0.155 \text{ psi.}$$

4. From Eq. 11.50, the pulsing period is estimated to be

$$\Delta t_p = F'\Delta t_c = (0.7)(17.84) = 12.49 \text{ hours.}$$

Then, the shut-in period is

$$\Delta t_s = \Delta t_c - \Delta t_p = 17.84 - 12.49 = 5.35 \text{ hours.}$$

TABLE 11.4—EFFECTS OF PERMEABILITY AND DRAINAGE AREA ON STABILIZATION TIME

k (md)	A (acres)	t_s Hours	equivalent
0.01	40	25,953	3 years
0.01	640	415,242	47 years
0.1	40	2,595	108 days
0.1	640	41,524	4.7 years
1.0	40	259.5	10.8 days
1.0	640	4,152.4	173 days
10.0	40	25.95	1.1 days
10.0	640	415.2	17.3 days

In summary, the test design is as follows: (1) produce the well for $\Delta t_p = 12.5$ hours, (2) shut in the well for $\Delta t_s = 5.4$ hours, and (3) use a pressure gauge with resolution of 0.01 psi (to permit detection of Δp to two decimal places).

11.5 Deliverability-Test Design

Many of the design considerations presented for pressure transient tests in Sec. 11.4 also are applicable to deliverability testing. Unlike pressure transient tests, however, most analysis techniques for deliverability tests require data obtained under stabilized flowing conditions. Therefore, a critical component in designing well-deliverability tests is estimating the time required to reach such flowing conditions. In this section, we discuss well-stabilization time and introduce the concept of drainage radius for designing deliverability tests. In addition, we discuss selection of the type of test and the flow requirements for each test. Because deliverability testing is more common in gas wells than in oil wells, the discussions focus on designing deliverability tests for gas wells.

11.5.1 Stabilization Time. The analysis techniques for conventional flow-after-flow tests require stabilized flowing conditions. Although isochronal and modified isochronal tests were developed to circumvent the long stabilized flow times required in low-permeability reservoirs, these tests may require a single, stabilized flow period at the end of the test. Stabilization time is defined as the time when the flowing pressure is no longer changing or is no longer changing significantly. Physically, stabilized flow can be interpreted as the time when the pressure transient is affected by all surrounding no-flow boundaries, either a natural reservoir boundaries or artificial boundaries created by active wells surrounding the tested well.

In high-permeability reservoirs, the time to reach stabilized flowing conditions is reasonably short; however, in low-permeability formations, the stabilization time can be months or even years, depending on the formation permeability. For example, consider a well producing a gas with a gravity of 0.6 from a formation at 210°F and an average pressure of 3,500 psia ($\bar{c}_g = 2.468 \times 10^{-4}$ psia^{-1} and $\bar{\mu}_g = 0.02$ cp), with a porosity of 10%. **Table 11.4** clearly shows that several days, if not years, may be required to reach stabilized flow in low-permeability formations, while high-permeability reservoirs stabilize very quickly.

Therefore, it is imperative that the design engineer estimate the time to reach stabilized flow before designing the test. The duration of the stabilization time can affect not only the validity of the data gathered during the test but also the type of deliverability test selected. The stabilization time, t_s, for a well centered in a circular drainage area is estimated with

$$t_s = \frac{948\phi\bar{\mu}_g\bar{c}_t r_e^2}{k_g}, \qquad (11.52)$$

where t_s = stabilization time, hr; ϕ = total porosity, fraction; $\bar{\mu}_g$ = gas viscosity evaluated at $\bar{p}$, cp; $\bar{c}_t$ = total compressibility evaluated at $\bar{p}$, psi^{-1}; r_e = distance to no-flow boundary, ft; and k_g = effective permeability to gas, md.

The average gas properties are evaluated at the average pressure in the drainage area of the well.

Similar to the design process for pressure transient tests, estimating the stabilization time for deliverability tests requires pretest estimates of both formation and fluid properties. If the average reservoir pressure at the time of the test is known or can be estimated, gas properties usually can be obtained with correlations. In addition, the distance to the no-flow boundaries can be estimated from geologic data or known locations of adjacent producing wells. Finally, Sec. 11.4 discusses methods for estimating formation permeability.

The criterion for identifying the beginning of stabilized flowing conditions varies within the natural gas industry. For example, the natural gas regulatory agency in Texas, the Railroad Commission of Texas,[32] defines stabilized flow when two consecutive pressure readings over a 15-minute period agree within 0.1 psi. Similarly, the Interstate Oil Compact Commission,[33] defines stabilized flow as a constant flowing wellhead pressure or static column wellhead pressure and rate of flow for a period of at least 15 minutes.

Rather than basing a definition on a prescribed rate of pressure decline, a better criterion for establishing the beginning of stabilized flow is based on the radius of investigation, or r_i, concept.[1] The radius of investigation, which is defined as the point beyond which pressure drawdown is negligible, is a measure of the distance a pressure transient has traveled into a formation following a rate change in a well and represents the depth to which the formation properties are being investigated at any time during a test. The approximate position of the radius of investigation is estimated with the relation

$$r_i = \sqrt{\frac{k_g t}{948\phi\bar{\mu}_g\bar{c}_t}} \qquad (11.53)$$

where r_i = radius of investigation, ft; ϕ = total porosity, fraction; $\bar{\mu}_g$ = gas viscosity evaluated at $\bar{p}$, cp; $\bar{c}_t$ = total compressibility evaluated at $\bar{p}$, psi^{-1}; t = time since last rate change, hr; k_g = formation permeability, md; and $\bar{p}$ = current average drainage area pressure, psi.

Stabilized flowing conditions occur when the radius of investigation equals or exceeds the drainage boundary of the well, i.e., $r_i \geqq r_e$. As long as the radius of investigation is less than the distance to the well's drainage boundary, i.e., $r_i < r_e$, stabilization has not been attained and the pressure behavior is transient. Therefore, knowing the distance to the drainage boundaries allows us to estimate the time required to reach stabilized flowing conditions and to select the appropriate deliverability test.

11.5.2 Estimating Deliverability-Test Duration. Analysis techniques for deliverability tests may require data obtained under stabilized flowing conditions. Therefore, the flow periods for the conventional four-point test and the single-point test and the extended flow points for the isochronal and modified isochronal tests must equal or exceed the stabilization time estimated by Eq. 11.52. The duration of the isochronal periods for the isochronal and modified isochronal tests, however, is determined by different criteria. Like the flow periods in pressure transient tests, the isochronal flow periods must exceed the time period during which wellbore storage distorts the data. The duration of wellbore storage can be estimated using Eq. 11.18,

$$t_{wbs} = \frac{(200{,}000 + 12{,}000s)C}{kh/\bar{\mu}_g}, \qquad (11.18)$$

where C = wellbore-storage coefficient defined by Eq. 11.16 or 11.17. Ref. 1 suggests a flow time equal to four times the duration of wellbore-storage effects.

Yet another consideration in selecting the duration of the isochronal test periods is that the pressure transient must move beyond the altered zone near the wellbore. This altered zone can be either damage caused by drilling and/or completion operations or stimulation resulting from acidizing or hydraulic fracturing. For test-design purposes, a radius of investigation, r_i, of $50 \leqq r_i \leqq 200$ ft is recommended. With Eq. 11.31b, the flowing time required to reach the design radius of investigation is

$$t_{min} = \frac{948\phi\overline{\mu}_g\overline{c}_t r_i^2}{k_g}. \quad (11.54)$$

If, however, the well is hydraulically fractured, the test must be designed so that the pressure transient moves beyond the region affected by the fracture and pseudoradial flow is achieved. In other words, the dimensionless time based on fracture half-length must be $\geqq 3$. If pseudoradial flow is not attained, the analysis techniques discussed previously for hydraulically fractured wells are no longer valid. In summary, the flow time for the isochronal periods should be the greater value of $4t_{wbs}$ and t_{min}.

11.5.3 Flow-Rate Requirements. ***Constant-Rate Production.*** Conventional deliverability-test-analysis techniques are based on constant production rates during each flow period. Therefore, the test should be designed to maintain constant or near-constant rates. In practice, constant production rates are difficult, if not impossible, to attain during a flow test, especially in low-permeability reservoirs that take long times to reach stabilized flow. Work by Winestock and Colpitts[34] indicates, however, that analysis techniques based on constant-rate production are valid provided the actual rates are declining slightly and vary smoothly and continuously with time.

Sequence of Flow Rates. Typically, deliverability tests are conducted with a sequence of increasing flow rates; i.e., each succeeding rate is higher than the previous rate.[1] If truly stabilized flowing conditions are attained following each rate change during a conventional four-point test, the rate sequence does not affect interpretation of the test. Either an increasing or decreasing rate sequence provides the correct deliverability relationship. To maintain accuracy during an isochronal or modified isochronal test, however, an increasing rate sequence is recommended.

In addition, the timing of the extended flow period for the isochronal and modified isochronal tests is immaterial. If a well is already producing, the extended flow period can be run at the beginning of the test. Otherwise, the extended flow period can be at the end of the test. It should be noted, however, that, if the extended test is conducted at the beginning, the well must be shut in for a length of time required to reach stabilized conditions before the beginning of the isochronal flow periods. This shut-in period may be prohibitively long in low-permeability reservoirs.

A final consideration in the selection of the rate sequence is the presence of liquids in the wellbore and the formation of hydrates. If liquid holdup in the production string is a problem, a decreasing rate sequence is preferred. Sec. 11.4.5 discusses methods for predicting the minimum flow rate required to lift liquids continuously in the wellbore. Similarly, if hydrate formation is probable, a decreasing rate sequence is recommended for a conventional four-point test.

Minimum and Maximum Flow Rates. Several factors must be considered when selecting the flow rates for testing the well.[1] The minimum flow rate should equal or exceed the minimum value required to lift liquids continuously from the wellbore, especially when liquid slugging problems are possible. Sec. 11.4.5 discusses methods for estimating the minimum liquid unloading rate. Another consideration is that the minimum flow rate must be adequate to maintain a wellhead temperature above the point where hydrate formation becomes a problem.

If hydrate formation and liquid accumulations in the wellbore do not affect the flow-rate selection, Ref. 1 suggests choosing minimum and maximum rates that produce pressure drops at the well that equal 5 and 25%, respectively, of the shut-in pressure. Alternatively, if the well's absolute open flow (AOF) potential is known or can be estimated from correlation with similar wells, another criterion is minimum and maximum rates equaling 10 and 75%, respectively, of the AOF. If estimates of reservoir rock and fluid properties can be estimated, the AOF can be calculated with

$$q_{AOF} = \frac{k_g h\overline{p}}{141.2\mu_g B_g\left[\ln\left(\frac{r_e}{r_w}\right) - \frac{3}{4} + s'\right]}. \quad (11.55)$$

The maximum rate should not create excessive pressure drawdowns that might damage the well from water coning or from formation sloughing. Further, the maximum flow rate should not create a pressure drawdown that would cause retrograde condensation in the reservoir, especially in the area adjacent to the wellbore. Finally, if the well is new and has not been connected to a pipeline or production facilities, the maximum flow rate should be chosen to minimize the quantity of gas flared to the atmosphere.

11.5.4 Selection of Deliverability Test. The most important consideration when selecting the deliverability test is the time required for stabilization.[1] If the stabilization time is on the order of several hours, a conventional four-point or backpressure test is appropriate. If the stabilization time greatly exceeds several hours, either an isochronal or modified isochronal test should be considered. The isochronal test is more accurate than the modified isochronal test and, therefore, should be used when greater accuracy is required.

When a well's deliverability characteristics are known or can be estimated from previous well tests, a single-point test is suitable. Single-point tests are especially appropriate when these deliverability characteristics are being updated, as required by many regulatory agencies. A convenient time to conduct a single-point test is before shut-in for a pressure survey of the field and when the well to be tested is stabilized. Under these conditions, only measurements of the flow rate and flowing pressure are required.

11.5.5 General Deliverability Test Design Procedures. This section offers general procedures for designing deliverability tests in both unstimulated and stimulated gas wells. Note that these procedures are suggestions and that each test design is unique and requires different considerations, depending on the testing objectives and environment. In all equations presented in the following procedure, the gas properties are evaluated at the current average reservoir pressure, $\overline{p}$.

Test-Design Procedure.

1. Estimate well and reservoir properties required for the test design.

a. Estimate the initial reservoir pressure, p_i, or the current average reservoir value, $\overline{p}$. Possible sources include data from a previous drillstem test or pressure surveys from active wells completed in the same formation.

b. Estimate gas properties (i.e., $\overline{B}_g, \overline{\mu}_g, \overline{c}_t$) at $\overline{p}$. Generally, correlations are adequate for estimating the gas properties.

c. Estimate k, h, and ϕ. Permeability can be estimated with the suggestions given in Sec. 11.4.1. The best sources for estimating net-pay thickness are well logs, while porosity estimates can be obtained from either cores or well logs.

d. Estimate the wellbore-storage coefficient, C. For gas, estimate C with Eq. 11.17. Average wellbore temperature is the arithmetic average of surface and bottomhole temperatures. For single-phase gas in the wellbore, $c_{wb} \approx c_g$ can be estimated at $\overline{p}$ and average wellbore temperature.

e. Estimate the drainage radius, r_e, as the well spacing from known field development patterns or the distance, L, to a nearby reservoir no-flow boundary. Distances to natural no-flow boundaries may be estimated with geologic data. In addition, if the well is hydraulically fractured, estimate the fracture half-length, L_f.

2. Estimate the stabilization time, t_s, with Eq. 11.52.

$$t_s = \frac{948\phi\overline{\mu}_g\overline{c}_t r_e^2}{k_g}. \quad (11.52)$$

3. On the basis of the length of time required to reach stabilized flowing conditions, select the type of deliverability test. In general, if the stabilization time is on the order of several hours, a conventional four-point test is selected. If the stabilization time greatly exceeds a few hours, an isochronal or modified isochronal test should be considered. When a well's deliverability characteristics are

known or can be estimated from previous well tests, choose a single-point test.

4. Select the duration of the flow periods for the deliverability test. For conventional four-point tests, single-point tests, and the extended flow points in isochronal and modified isochronal tests, the flow time must equal or exceed the stabilization time calculated in Step 2, or $t \geqq t_s$. For the flow periods during isochronal or modified isochronal tests, select the greater value of $4t_{wbs}$ and t_{min}.

a. The duration of wellbore-storage effects, t_{wbs}, is estimated with Eq. 11.18.

$$t_{wbs} = \frac{(200,000 + 12,000s')C}{kh/\overline{\mu}_g}, \quad s' \geqq -3.5. \quad \text{......} (11.18)$$

b. Estimate the flowing time required to achieve the design radius of investigation, $50 \leqq r_i \leqq 200$ ft with Eq. 11.31.

$$t_{min} = \frac{948\phi\overline{\mu}_g\overline{c}_t r_i^2}{k_g}. \quad \text{......} (11.20)$$

If the well is hydraulically fractured with an estimated fracture half-length, L_f, the dimensionless flowing time should exceed the dimensionless time required to reach pseudoradial flow,

$$t_{L_fD} \geqq 3, \quad \text{......} (11.26)$$

or, in dimensional form, the time to reach pseudoradial flow, t_{prf}, in hours, is

$$t_{prf} \approx \frac{11,400\phi\mu_g c_t L_f^2}{k_g}. \quad \text{......} (11.27)$$

5. Calculate the minimum and maximum flow rates. In general, the rates should be selected to reduce or eliminate hydrate formation and liquid-loading problems. The minimum and maximum rates can be estimated as the rates that produce pressure drops at the well equalling 5 and 25%, respectively, of the average shut-in pressure.

$$q_{g_{min}} = \frac{k_g h(\Delta p_{min})}{141.2\mu_g B_g\left[ln\left(\frac{r_e}{r_w}\right) - \frac{3}{4} + s'\right]}, \quad \text{......} (11.56a)$$

where $\Delta p_{min} = 0.05\overline{p}$,

and

$$q_{g_{max}} = \frac{k_g h(\Delta p_{max})}{141.2\mu_g B_g\left[ln\left(\frac{r_e}{r_w}\right) - \frac{3}{4} + s'\right]}, \quad \text{......} (11.56b)$$

where $\Delta p_{max} = 0.25\overline{p}$.

Example 11.6—Designing a Gaswell Deliverability Test. Using the following data, design a suitable deliverability test to assess the production potential of a new gas well. Note that the well is not yet connected to a pipeline.

ϕ = 0.118
$p_i = \overline{p}$ = 3,200 psia
A = 640 acres
r_w = 0.365 ft
h = 6 ft
k_g = 10 md
c_{wb} = 2.9×10^{-4} psi^{-1}
$\overline{c}_t$ = 2.0×10^{-4} psi^{-1}
$\overline{\mu}_g$ = 0.015 cp
V_{wb} = 15 bbl
$\overline{B}_g$ = 1.5 RB/Mscf

Solution.

1. First, estimate gas and formation properties.

a. Estimates of pressures and gas properties are provided. Note that the gas properties, $\overline{B}_g$, $\overline{c}_t$, and $\overline{\mu}_g$, are evaluated at $\overline{p}$. We choose to set $s' = 0$ for conservative test design.

b. From Eq. 11.17, the wellbore storage coefficient is

$$C = c_{wb}V_{wb}$$

$$= (2.9 \times 10^{-4})(15) = 0.00435 \text{ bbl/psi}.$$

c. For 640-acre spacing and assuming a circular drainage area, the external drainage radius, r_e, is

$$r_e = \sqrt{\frac{A}{\pi}} \text{ ft}$$

$$= \sqrt{\frac{(640 \text{ acres})(43,560 \text{ ft}^2/\text{acre})}{\pi}} = 2,979 \text{ ft}.$$

2. Estimate the stabilization time, t_s.

$$t_s = \frac{948\phi\overline{\mu}_g\overline{c}_t r_e^2}{k_g}$$

$$= \frac{(948)(0.118)(0.015)(2.0 \times 10^{-4})(2,979)^2}{10}$$

$$= 298 \text{ hours or 13 days}.$$

3. Because of the long stabilization time, conduct a modified isochronal test.

4. Select the duration of the flow periods for the deliverability test. For the flow periods during modified isochronal tests, select the greater value of $4t_{wbs}$ and t_{min}.

a. Assuming $s' = 0$, the duration of wellbore-storage effects, t_{wbs}, is estimated with

$$t_{wbs} = \frac{(200,000 + 12,000s')C}{k_g h/\overline{\mu}_g}$$

$$= \frac{[200,000 + (12,000)(0)](0.00435)}{(10)(6)/(0.015)}$$

$$= 0.22 \text{ hours}.$$

b. Estimate the flowing time required to achieve the design radius of investigation, $50 \leqq r_i \leqq 200$ ft. For this design, use $r_i = 200$ ft.

$$t_{min} = \frac{948\phi\overline{\mu}_g\overline{c}_t r_i^2}{k_g}$$

$$= \frac{(948)(0.118)(0.015)(2.0 \times 10^{-4})(200)^2}{10}$$

$$= 1.34 \text{ hours}.$$

Now, select the greater value of $4t_{wbs}$ and t_{min}, or choose a flow time for the isochronal periods equal to $t = 1.34$ or $t = 1.5$ hr.

5. Calculate the minimum and maximum flow in rounding up rates. The minimum and maximum flow rates are estimated as follows:

$$q_{g_{min}} = \frac{k_g h(\Delta p_{min})}{141.2\overline{\mu}_g B_g\left[ln\left(\frac{r_e}{r_w}\right) - \frac{3}{4} + s'\right]} \quad \text{......} (11.56a)$$

where $\Delta p_{min} = 0.05\overline{p} = (0.05)(3,200) = 160$ psi

and

$$q_{g_{max}} = \frac{k_g h(\Delta p_{max})}{141.2\overline{\mu}_g B_g\left[ln\left(\frac{r_e}{r_w}\right) - \frac{3}{4} + s'\right]} \quad \text{......} (11.56b)$$

where Δp_{max}

$= 0.25\bar{p} = (0.25)(3,200) = 800$ psi.

Therefore, the minimum and maximum testing rates, respectively, are

$$q_{g_{min}} = \frac{k_g h(\Delta p_{min})}{141.2\bar{\mu}_g \bar{B}_g\left[ln\left(\frac{r_e}{r_w}\right) - \frac{3}{4} + s'\right]}$$

$$= \frac{(10)(6)(160)}{(141.2)(0.015)(1.5)\ln\left[\left(\frac{2,979}{0.365}\right) - \frac{3}{4} + 0\right]}$$

$= 366$ Mscf/D.

$$\text{and } q_{g_{max}} = \frac{k_g h(\Delta p_{max})}{141.2\bar{\mu}_g \bar{B}_g\left[ln\left(\frac{r_e}{r_w}\right) - \frac{3}{4} + s'\right]}$$

$$= \frac{(10)(6)(800)}{(141.2)(0.015)(1.5)\left[\ln\left(\frac{2,979}{0.365}\right) - \frac{3}{4} + 0\right]}$$

$= 1,830$ Mscf/D.

In summary, test the gas well using a modified isochronal test with flow periods of 1.5 hours. A suitable flow schedule is 366, 854, 1,342, and 1,830 Mscf/D. During each flow period, we recommend measuring the flowing pressure at 0.25, 0.5, 1.0, and 1.5 hours.

11.6 Chapter Summary

Sec. 11.1 presents an overview of well test design.

Sec. 11.2 discusses the types and purposes of well tests, including pressure transient tests and deliverability tests.

Sec. 11.2.1 covers pressure transient tests. Pressure transient tests include single-well tests, where a rate change is induced in a well and the corresponding pressure response is measured in the same well, and multiwell tests, where a rate change is induced in one well and the pressure response is monitored in one or more adjacent wells. The primary purpose of a pressure transient test is to provide estimates of formation properties, such as permeability and skin factor. The most common types of single-well tests are buildup tests and flow tests. Buildup tests are often used to estimate average drainage-area pressure, while flow tests may be used to estimate reservoir pore volume. For injection wells, tests include falloff tests (analogous to buildup tests on production wells) and injection tests (analogous to drawdown tests in producing wells). Multiwell tests are designed to estimate formation properties between two wells and to verify the presence or lack of communication between offset wells. Multiwell tests include interference tests, where the active well is produced at a constant rate, and pulse tests, where the active well is cycled off and on in a regular pattern. Pulse tests have the advantage that they can be run without shutting in the field, unlike interference tests.

Sec. 11.2.2 discusses deliverability tests. Objectives of a deliverability test are to determine the AOF potential or to generate a reservoir IPR or gas backpressure curve. The four basic types of deliverability tests include: (1) flow-after-flow or four-point tests, where the well is produced to stabilization at each of several flow rates; (2) single-point tests, where the well is produced to stabilization at a single rate; (3) isochronal tests, where the well is produced at a sequence of different flow rates, with equal-length flow periods and the well is shut in long enough between flow periods to return to the shut-in pressure at the beginning of the first test, and (4) modified isochronal tests, where the shut-in periods are also of equal length, preferably longer than the flow periods.

Sec. 11.3 reviews general design considerations. The first step in well-test design is to identify the test objectives clearly. Once the objectives have been identified, the appropriate test can be selected and designed to achieve the objectives quickly and economically.

Sec. 11.3.1 addresses considerations relating to the well type and its status: development vs. exploration wells, production vs. injection wells, shallow vs. deep wells, and stimulated vs. unstimulated wells. In exploration wells, we are often interested in the reservoir as a whole, while in a development well, we are more interested in the properties near the well. Injection or falloff tests are used to estimate reservoir properties in injection wells, corresponding to drawdown or buildup tests in producing wells. Injection tests are usually more expensive to run than tests on producing wells because of the cost of the additional equipment. In addition, deep wells are often more expensive to test than shallow wells because of the need to minimize wellbore-storage duration. Further, higher pressures and temperatures are likely to be encountered in deep wells, requiring more expensive equipment. Because of the time required to reach pseudoradial flow in wells that have been hydraulically fractured, a prefracture test should be run before the fracture treatment to obtain a permeability estimate. This allows post-fracture tests to be significantly shorter and still provide good estimates of fracture properties.

Sec. 11.3.2 discusses the effects of reservoir properties on well test design: low-permeability vs. high-permeability formations and single-zone vs. multiple-zone reservoirs. In general, low-permeability formations require longer tests to reach the end of wellbore storage and to sample a reasonable distance into the formation than do high-permeability formations. High-permeability reservoirs require higher flow rates to achieve a given pressure drawdown than low-permeability reservoirs. Reservoirs with multiple zones may be tested in the same manner as single zones, provided each zone is isolated and tested in turn.

Sec. 11.3.3 discusses safety and environmental considerations, such as sweet vs. sour and corrosive gases. In the presence of H_2S, the well-test design must follow industry-standard safety precautions. If corrosive gases are present, the design must specify corrosion-resistant equipment to reduce the risk of wireline failure. In testing in areas where gas must be flared, the test design should minimize the amount of gas released. All relevant environmental regulations must be followed in designing any test.

Sec. 11.3.4 discusses the advantages and disadvantages of a variety of types of pressure gauges in common use today.

Sec. 11.4 presents a method for designing pressure transient tests, including techniques for obtaining pretest estimates of reservoir properties, estimating appropriate test flow rates, and determining test duration.

Sec. 11.4.1 discusses methods for obtaining pretest estimates of reservoir properties, including skin factor and formation permeability. It emphasizes that good pretest estimates are necessary to design a pressure transient test properly. Skin factor often can be estimated from the type of completion. Skin factors range from -8 to -3 for hydraulically fractured wells and from -3 to -1 for acidized wells. A natural completion usually is assumed to have a skin factor of 0. Skin factors also may be estimated from offset wells with the same type of completion. For gas wells, non-Darcy flow effects can be estimated through correlations. Permeability can be estimated from core samples or from offset wells. Permeability also may be estimated from productivity tests or from the pseudosteady-state-flow equation. If the reservoir is still infinite-acting, the one-point method may be used to estimate formation permeability from a constant-rate flow test.

Sec. 11.4.2 discusses methods to estimate test duration for homogeneous reservoirs, naturally fractured reservoirs, and reservoirs with vertically fractured wells. Test duration must be long enough for wellbore storage to end. If semilog analysis methods are to be used, the test should be designed to last three to four times longer than wellbore storage. The time to the end of wellbore storage may be estimated from the wellbore-storage coefficient C and either formation permeability, k, or productivity index, J. Wellbore storage may be minimized with a downhole shut-in. The end of the middle-time region may be estimated from the drainage radius or from the distance to the nearest no-flow boundary.

Sec. 11.4.3 applies the concept of radius of investigation to estimate how much of the formation can be sampled with a given test duration and, conversely, the test duration necessary to sample the reservoir to a given distance.

Sec. 11.4.4 discusses flow-rate requirements. The objective in selecting a flow rate is to find a rate that is high enough to produce a significant pressure response yet low enough that the well can produce at the given rate for the duration of the flow period. Other considerations include the need for the rate to be high enough to continuously unload liquids (for gas wells), the need to maintain a wellbore pressure above the bubblepoint (for undersatured oil wells), and the need to initiate flow in low-permeability formations. For producing wells, the recent production rate should be maintained for a period at least as long as the expected shut-in period immediately before the buildup test.

Sec. 11.4.5 presents general design procedures for single-well tests, for both conventionally completed vertical wells and for hydraulically fractured wells. Examples illustrate the design procedures.

Sec. 11.4.6 presents procedures for designing interference tests and pulse tests; the procedures are illustrated with examples.

Sec. 11.5 covers deliverability-test design. Considerations in designing deliverability tests include time to stabilization, type of test, and flow rate or rates.

Sec. 11.5.1 addresses the time to stabilization. Eq. 11.52 shows that stabilized flow is attained for a circular drainage area when $t_s = 948\phi\bar{\mu}_g\bar{c}_t r_e^2/k_g$. We point out that the time required to stabilization can be months or even years for low-permeability gas reservoirs.

Sec. 11.5.2 discusses estimation of deliverability-test duration. For conventional four-point and single-point tests and for the extended flow points for isochronal and modified isochronal tests, the test duration must be such that stabilized flow is achieved. For isochronal and modified isochronal tests, the test duration can be much shorter, subject to the following limitations. The first requirement for an isochronal test is that the flow periods must be long enough that wellbore-storage effects have ended by the end of each period. Further, the periods must be long enough that a significant portion of the reservoir has been sampled. Finally, for hydraulically fractured wells, the test duration must be long enough that pseudoradial flow is achieved by the end of each period.

Sec. 11.5.3 discusses flow-rate requirements for deliverability tests. Although analysis methods are based on the assumption of constant production rate during each flow period, these methods may also be applied to tests where the rate is declining slowly and continuously with time during each flow period. For conventional tests where stabilized flow is achieved during each flow period, the rate sequence does not affect the test interpretation. For isochronal and modified isochronal tests, however, an increasing rate sequence is recommended. The minimum flow rate must be high enough to remove liquids from the wellbore and to prevent hydrates from forming. The maximum rate must be low enough not to damage the well. For wells where gas is flared, the maximum rate should be chosen to minimize the amount of flared gas.

Sec. 11.5.4 addresses the choice of deliverability test. For high-permeability formations where stabilization can be achieved rapidly, conventional four-point or single-point tests are adequate. If the permeability is low enough that stabilization times prohibit the use of four-point tests, isochronal or modified isochronal tests are recommended.

Sec. 11.5.5 presents a general design procedure for deliverability tests and illustrates the procedure with an example.

Exercises

1. A preliminary flow test was conducted on an oil well. Estimate the formation permeability from the flow test data, assuming skin factors of −3, 0, 5, and 10. $p_i = 3942.90$ psia; $h = 33$ ft; $B_o = 1.341$ RB/STB; $p_{wf} = 3256.11$ psia; $q = 165$ STB/D; $\phi = 20.8\%$; $c_t = 12.1 \times 10^{-6}$ psi^{-1}; $t_p = 6$ hr; $r_w = 0.33$ ft; $\mu = 0.632$ cp.

2. A preliminary flow test was conducted on a gas well. Estimate the formation permeability from the flow test data, assuming skin factors of −3, 0, 5, and 10. $T_f = 300$°F; $h = 70$ ft; $\gamma_g = 0.711$ (air = 1.0); $B_{gi} = 0.643$ RB/Mscf; $q = 250$ Mscf/D; $p_i = 6936.90$ psia; $\phi = 10.7\%$; $c_w = 3.6 \times 10^{-6}$ psi^{-1}; $c_{ti} = 3.98 \times 10^{-5}$ psi^{-1}; $t_p = 196$ hr; $r_w = 0.5$ ft; $S_w = 60.2\%$; $c_f = 4 \times 10^{-6}$ psi^{-1}; $\mu_{gi} = 0.031$ cp; $p_{wf} = 4970.07$ psia.

3. Design a pressure buildup test for an oil reservoir with the following properties. Assume skin factor is 0. $p_i = 3942.90$ psia; $h = 33$ ft; $B_o = 1.341$ RB/STB; $V_{wb} = 98$ bbl; $k = 5$ md; $\phi = 20.8\%$; $c_t = 12.1 \times 10^{-6}$ psi^{-1}; $t_p = 6$ hr; $c_{wb} = 9.51 \times 10^{-6}$ psi^{-1}; $r_w = 0.33$ ft; $\mu = 0.632$ cp.

4. Design a pressure buildup test for an oil well with the following properties. Assume skin factor is 5. Calculate the wellbore storage coefficient based on a changing liquid level. Choose a flow rate that will keep the flowing bottomhole pressure above the bubble point pressure. $p_i = 2468.9$ psia; $h = 38$ ft; $B_o = 1.272$ RB/STB; $p_b = 1861.5$ psi; $k = 18$ md; $\phi = 14.0\%$; $c_t = 14.9 \times 10^{-6}$ psi^{-1}; $A_{wb} = 0.1235$ ft^2; $r_w = 0.28$ ft; $\mu = 0.53$ cp; $\rho_{wb} = 44.11$ lbm/ft^3.

5. Design a prefacture pressure buildup test for a gas well with the following properties. Permeability is estimated to be 0.05 md from a preliminary flow test. Assume a skin factor of 5. $T_f = 300$°F; $h = 70$ ft; $\gamma_g = 0.711$ (air = 1.0); $B_{gi} = 0.643$ RB/Mscf; $V_{wb} = 287.4$ bbl; $p_i = 6936.90$ psia; $\phi = 10.7\%$; $c_w = 3.6 \times 10^{-6}$ psi^{-1}; $c_{ti} = 3.98 \times 10^{-5}$ psi^{-1}; $c_{wb} = 84.5 \times 10^{-6}$ psi^{-1}; $r_w = 0.5$ ft; $S_w = 60.2\%$; $c_f = 4 \times 10^{-6}$ psi^{-1}; $\mu_{gi} = 0.031$ cp.

6. Design a postfacture pressure buildup test for a gas well with the following properties. Permeability is estimated to be 0.046 md from a prefracture buildup test. The objective of the test is to estimate fracture conductivity and fracture half-length. $T_f = 300$°F; $h = 70$ ft; $\gamma_g = 0.711$ (air = 1.0); $B_{gi} = 0.643$ RB/Mscf; $V_{wb} = 287.4$ bbl; $L_f = 650$ ft; $p_i = 6936.90$ psia; $\phi = 10.7\%$; $c_w = 3.6 \times 10^{-6}$ psi^{-1}; $c_{ti} = 3.98 \times 10^{-5}$ psi^{-1}; $c_{wb} = 84.5 \times 10^{-6}$ psi^{-1}; $w_f k_f = 2200$ md-ft; $r_w = 0.5$ ft; $S_w = 60.2\%$; $c_f = 4 \times 10^{-6}$ psi^{-1}; $\mu_{gi} = 0.031$ cp.

7. Design an interference test using one active well and one observation well. Rock and fluid properties are given below. Pressure at the active well should remain above 3000 psi. $p_i = 3942.90$ psia; $h = 33$ ft; $B_o = 1.341$ RB/STB; $k = 5$ md; $\phi = 20.8\%$; $c_t = 12.1 \times 10^{-6}$ psi^{-1}; $r_o = 750$ ft; $r_w = 0.33$ ft; $\mu = 0.632$ cp.

8. Design an interference test using one active well and one observation well. Rock and fluid properties are given below. Pressure at the active well should remain above the bubble point pressure. $p_i = 2468.9$ psia; $h = 38$ ft; $B_o = 1.272$ RB/STB; $p_b = 1861.5$ psi; $k = 18$ md; $\phi = 14.0\%$; $c_t = 14.9 \times 10^{-6}$ psi^{-1}; $r_o = 750$ ft; $r_w = 0.28$ ft; $\mu = 0.53$ cp.

9. Design a pulse test for a reservoir with the rock and fluid properties given in Exercise 7.

10. Design a pulse test for a reservoir with the rock and fluid properties given in Exercise 8.

11. Design a deliverability test for a gas well with the following properties. $T_f = 164$°F; $h = 22$ ft; $\gamma_g = 0.742$ (air = 1.0); $B_{gi} = 0.499$ RB/Mscf; $r_w = 0.25$ ft; $A = 760$ acres; $p_i = 7954.8$ psia; $\phi = 13.1\%$; $c_w = 3.6 \times 10^{-6}$ psi^{-1}; $c_{ti} = 4.02 \times 10^{-5}$ psi^{-1}; $V_{wb} = 201.9$ bbl; $k = 36.2$ md; $S_w = 24.9\%$; $c_f = 4 \times 10^{-6}$ psi^{-1}; $\mu_{gi} = 0.042$ cp; $c_{wb} = 47.1 \times 10^{-6}$ psi^{-1}.

12. Design a deliverability test for a gas well with the following properties. $T_f = 100$°F; $h = 87$ ft; $\gamma_g = 0.747$ (air = 1.0); $B_{gi} = 2.171$ RB/Mscf; $r_w = 0.46$ ft; $A = 133$ acres; $p_i = 1072.30$ psia; $\phi = 13.3\%$; $c_w = 3.6 \times 10^{-6}$ psi^{-1}; $c_{ti} = 9.49 \times 10^{-5}$ psi^{-1}; $V_{wb} = 66.8$ bbl; $k = 21.6$ md; $S_w = 14.2\%$; $c_f = 4 \times 10^{-6}$ psi^{-1}; $\mu_{gi} = 0.014$ cp; $c_{wb} = 1.1 \times 10^{-3}$ psi^{-1}.

References

1. *Theory and Practice of the Testing of Gas Wells*, third edition, Energy Resources and Conservation Board, Calgary (1978).
2. Kamal, M.M.: "Interference and Pulse Testing—A Review," *JPT* (December 1983) 2257.
3. Earlougher, R.C. Jr.: *Advances in Well Test Analysis*, Monograph Series, SPE, Richardson, Texas (1977) **5**.
4. *RP55, Conducting Oil and Gas Production Operations Involving Hydrogen Sulfide*, API, Dallas (1981).

5. Vella, M. *et al.*: "The Nuts and Bolts of Well Testing," *Oilfield Review* (April 1992) 14.
6. Ehlig-Economides, C.A., Hegeman, P., and Clark, G.: "Three Key Elements Necessary for Successful Testing," *Oil & Gas J.* (25 July 1994) 84.
7. Riley, H.G.: "A Short Cut to Stabilized Gas Well Productivity," *JPT* (May 1970) 537.
8. Fligelman, H. *et al.*: "Pressure-Drawdown Test Analysis of a Gas Well—Application of New Correlations," *SPEFE* (September 1989) 406.
9. Jones, S.C.: "Using the Inertial Coefficient, *B*, To Characterize Heterogeneity in Reservoir Rock," paper SPE 16949 presented at the 1987 SPE Annual Technical Conference and Exhibition, Dallas, 27–30 September.
10. Aronofsky, J.S. and Jenkins, R.: "A Simplified Analysis of Unsteady Radial Gas Flow," *Trans.*, AIME (1954) **201**, 149.
11. Lee, W.J.: "Pressure-Transient Test Design in Tight Gas Formations," *JPT* (October 1987) 1185.
12. Agarwal, R.G., Al-Hussainy, R., and Ramey, H.J. Jr.: "An Investigation of Wellbore Storage and Skin Effect in Unsteady Liquid Flow: I. Analytical Treatment," *SPEJ* (September 1970) 279; *Trans.,* AIME, **249**.
13. Chen, H.K. and Brigham, W.E.: "Pressure Buildup for a Well With Storage and Skin in a Closed Square," *JPT* (January 1978) 141; *Trans.*, AIME, **265.**
14. Ramey, H.J. Jr. and Gringarten, A.C.: "Effect of High Volume Vertical Fractures on Geothermal Steam Well Behavior," *Proc.*, Second United Nations Symposium on the Use and Development of Geothermal Energy, San Francisco (May 1975).
15. Gringarten, A.C., Ramey, H.J. Jr., and Raghavan, R.: "Applied Pressure Analysis for Fractured Wells," *JPT* (July 1975) 887; *Trans.*, AIME, **259.**
16. Gringarten, A.C., Ramey, H.J. Jr., and Raghavan, R.: "Unsteady-State Pressure Distribution Created by a Well With a Single Infinite-Conductivity Vertical Fracture," *SPEJ* (August 1974) 347; *Trans.*, AIME, **257.**
17. Cinco-Ley, H. and Samaniego-V., F.: "Effects of Wellbore Storage and Damage on the Transient Pressure Behavior of Vertically Fractured Wells," paper SPE 6752 presented at the 1977 SPE Annual Technical Conference and Exhibition, Denver, Colorado, 9–12 October.
18. Cinco-Ley, H., Samaniego-V., F., and Dominguez, N.: "Transient Pressure Behavior for a Well With a Finite-Conductivity Vertical Fracture," *SPEJ* (August 1978) 253.
19. Cinco-Ley, H. and Samaniego-V., F.: "Transient Pressure Analysis for Fractured Wells," *JPT* (September 1981) 1749.
20. Mavor, M.J. and Cinco-Ley, H.: "Transient Pressure Behavior of Naturally Fractured Reservoirs," paper SPE 7977 presented at 1979 SPE California Regional Meeting, Ventura, California, 18–20 April.
21. Bourdet, D. *et al.*: "Interpreting Well Tests in Fractured Reservoirs," *World Oil* (October 1983).
22. Bourdet, D. *et al.*: "New Type Curves Aid Analysis of Fissured Zone Well Tests," *World Oil* (April 1984).
23. Lee, W.J.: *Well Testing*, Textbook Series, SPE, Richardson, Texas (1982) **1.**
24. Bostic, J.N. and Graham, J.A.: "Prefracturing Pressure Transient Testing: East Texas Cotton Valley Tight Gas Play," paper SPE 7941 presented at the 1979 SPE Symposium on Low Permeability Gas Reservoirs, Denver, Colorado, 20–22 May.
25. Holditch, S.A. *et al.*: "The GRI Staged Field Experiment," *SPEFE* (September 1988) 519.
26. Holgate, K.E., Lancaster, D.E., and Lee, W.J.: "Analysis of Drillstem Test Data in Devonian Shale Reservoirs," paper SPE 15925 presented at the 1986 SPE Eastern Regional Meeting, Columbus, Ohio, 12–14 November.
27. Duggan, J.O.: "Estimating Flow Rates Required To Keep Gas Wells Unloaded," *JPT* (December 1961) 1173.
28. Turner, R.G., Hubbard, M.G., and Dukler, A.E.: "Analysis and Prediction of Minimum Flow Rate for the Continuous Removal of Liquids From Gas Wells," *JPT* (November 1969) 1475; *Trans.*, AIME, **246**.
29. Libson, T.N. and Henry, J.R.: "Case Histories: Identification of and Remedial Action for Liquid Loading in Gas Wells—Intermediate Shelf Gas Play," *JPT* (April 1980) 685.
30. Smith, R.V.: *Practical Natural Gas Engineering*, PennWell Publishing Co., Tulsa, Oklahoma (1983) 205.
31. Kamal, M. and Brigham, W.E.: "Design and Analysis of Pulse Tests With Unequal Pulse and Shut-In Periods," *JPT* (February 1976) 205; *Trans.* AIME, **261**.
32. *Back Pressure Tests for Natural Gas Wells*, Railroad Commission of Texas, Reprint Edition, Austin, Texas (1972).
33. *Manual of Back-Pressure Testing of Gas Wells*, Interstate Oil Compact Commission, Oklahoma City, Oklahoma (1970).
34. Winestock, A.G. and Colpitts, G.P.: "Advances in Estimating Gas Well Deliverability," *J. Cdn. Pet. Tech.* (1965) **4**, 111.

Chapter 12
Horizontal Well Analysis

12.1 Overview

In this chapter, we will review methods to analyze pressure-transient tests in horizontal wells and illustrate application of the results of transient test analysis in estimating productivity.

Productivity estimates in horizontal wells are subject to more uncertainty than comparable estimates in vertical wells. Further, it is much more difficult to interpret well-test data from a horizontal well than from a vertical well. The major problem with a horizontal well is that the flow geometry is 3D. We no longer have the radial symmetry that is usually present in a vertical well. Several flow regimes can potentially occur in a horizontal well test, and we need to consider these in analyzing test data from horizontal wells.

Wellbore storage effects can be much more significant in a horizontal well than in vertical wells, and horizontal wells will commonly exhibit partial penetration and end effects that complicate interpretation.

In vertical wells, we are accustomed to dealing with variables such as average permeability, net vertical thickness, and skin. In horizontal wells, we need more detail. Not only is vertical thickness important, but the horizontal dimensions of the reservoir, relative to the horizontal wellbore, need to be known.

12.2 Steps in Evaluating Horizontal Well-Test Data

To evaluate pressure-transient data from a horizontal well, we expect to follow three basic steps. First, we must identify the specific flow regimes in the test data. Second, we need to be able to apply the proper analytical and graphical procedures to the data. Finally, we need to be able to evaluate the uniqueness and sensitivity of the results to properties we have derived from our analysis or simply assumed.

12.2.1 Identify Flow Regimes. Typically, when engineers evaluate completion data from a vertical wellbore, the techniques they use center on a single flow regime, such as infinite-acting radial flow, which is also known as the "middle-time region." However, a pressure-transient test in a horizontal well can involve as many as five major and distinct regimes that need to be identified. These regimes may or may not occur in a given test and may or may not be obscured by wellbore storage effects, end effects, or transition effects between flow regimes.

12.2.2 Apply the Proper Procedures. Each flow regime can be modeled by an equation that can be used to estimate important reservoir properties. At best, only groups of analytical parameters can be determined directly from equations. Therefore, it is imperative that we apply the proper analytical and graphical procedures to the data. In many cases, when solving for specific parameters, the application of these analytical expressions may involve a complex iterative procedure.

12.2.3 Evaluate Uniqueness and Sensitivity. Experience indicates that results of horizontal well-test analysis are seldom unique, so it is important that we evaluate the uniqueness and sensitivity of the results to assumed properties. Simulation of the test using properties that have been determined from the test can confirm that at least the analysis is consistent with the test data. A simulator can also determine whether other sets of formation properties will also lead to a fit of the data.

12.3 Horizontal Well Flow Regimes

Let us now take a look at the five different flow regimes that could occur in a horizontal well. We calculate different formation properties from the data in each of these different flow regimes.

We need to remember that any flow regime may be absent from a plot of test data because of geometry, wellbore storage, or other factors. In addition, the fact that they *can* appear does not mean that they *do* appear.

The five different flow regimes that can occur are (1) early radial, (2) hemiradial, (3) early linear, (4) late pseudoradial, and (5) late linear.

Before we can understand these flow regimes, we need to become acquainted with some nomenclature that we use to describe well and reservoir geometry for a horizontal well. **Fig. 12.1** shows a horizontal well with length, L_w, located within a reservoir that is assumed to be a rectangular parallelepiped, or a "box reservoir" drainage area. We will assume in this discussion that the axes of the coordinate system coincide with the direction of principal permeability. Later, we will state modifications to be used when this statement is not true. We will also assume for now that the well produces over its entire length, L_w.

The axes for this box are the usual x-, y-, and z-axes. Notice that the x-axis is measured along the bottom edge of the reservoir, going from left to right in the direction perpendicular to the well. The y-axis lies along the axis from front to back of the reservoir, parallel to the wellbore. The z-axis is oriented in the direction of reservoir thickness.

The total width of the reservoir, that is, perpendicular to the wellbore, is a_H; the total length in a direction parallel to the wellbore is b_H; and the total height of the reservoir is the net pay thickness, h. Notice that we have parameters for the distance from the well to the

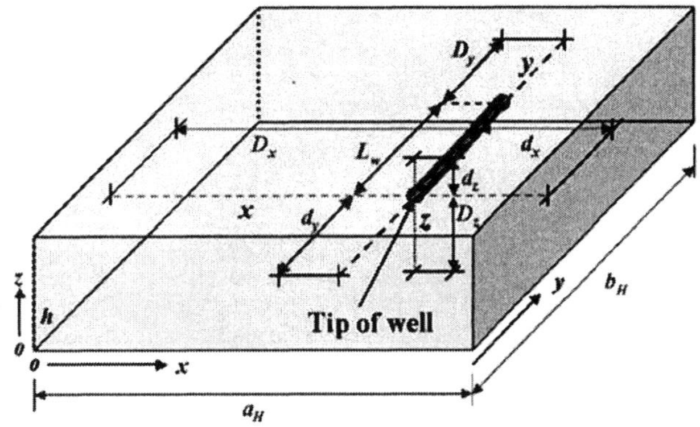

Fig. 12.1—Well and reservoir geometry for a horizontal well.

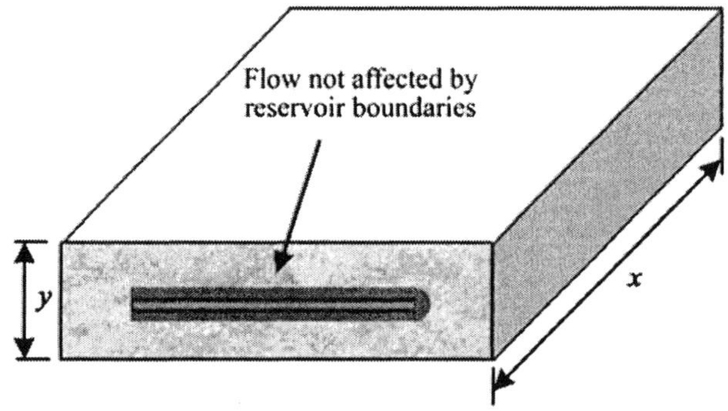

Fig. 12.2—Early-radial flow is not influenced by reservoir boundaries.

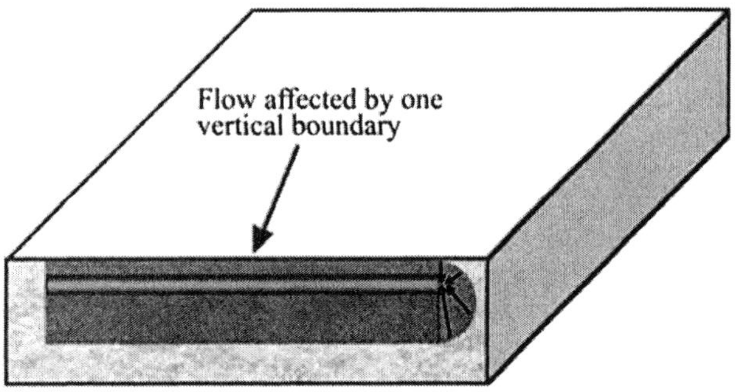

Fig. 12.3—Hemiradial flow occurs when the effects of one vertical boundary are felt much sooner than the effects of the other.

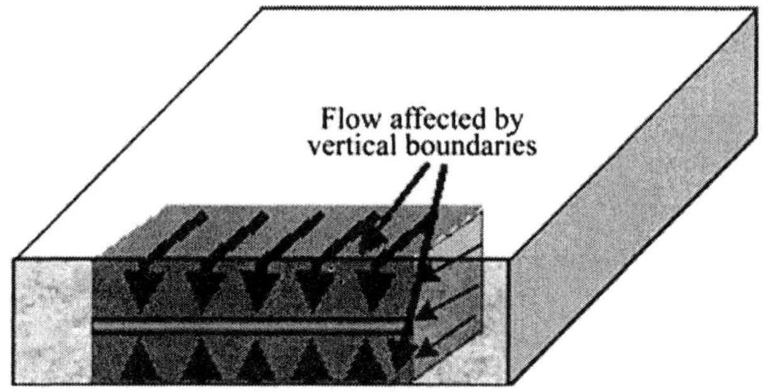

Fig. 12.4—Early-linear flow may develop after both vertical boundaries are encountered.

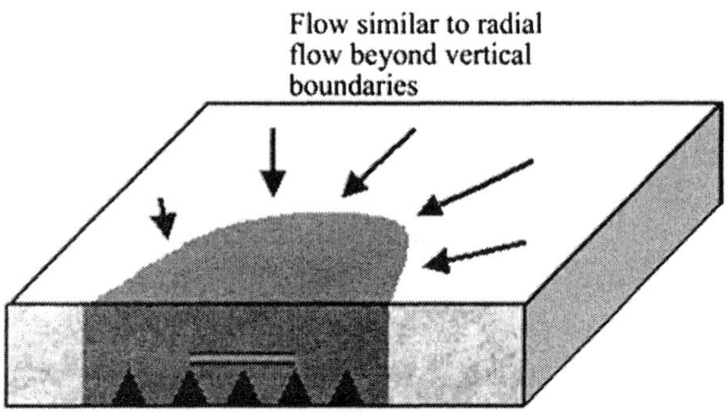

Fig. 12.5—Late-pseudoradial flow begins after flow enters the wellbore from beyond the ends of the well.

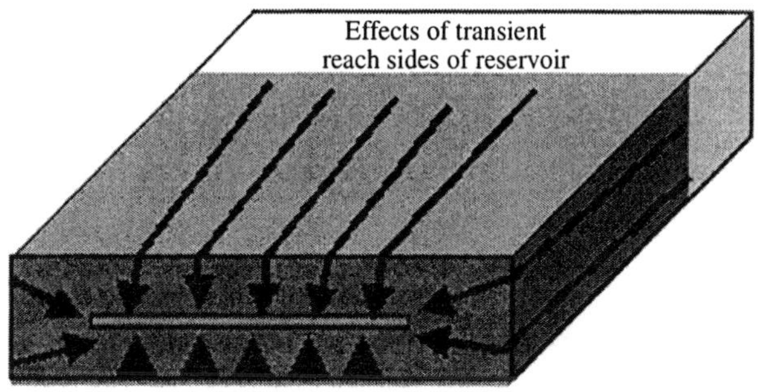

Fig. 12.6—Late-linear flow includes flow from the drainage volume perpendicular to the well.

various borders. Along the axis of the well, the shortest distance from the end of the well to a boundary is d_y; the longest distance from the other end of the well to the boundary is D_y. And in the vertical direction, the shortest distance to a vertical boundary is d_z, while the longest distance to a vertical boundary is D_z.

Given this reservoir geometry, we are now in a position to consider the flow regimes that occur in a horizontal well. We will examine each of the various regimes individually, beginning with early radial flow.

12.3.1 Characteristics of Flow Regimes. Consider a well producing at constant rate. The early-radial flow regime occurs before the area drained or the pressure transient caused by this production encounters either of the boundaries of the reservoir. So a radial flow pattern penetrates out into the reservoir, as shown in **Fig. 12.2.** However, this flow pattern is likely to be elliptical, moving further into the reservoir at a given time in the higher-permeability x-direction than in the lower-permeability z-direction. This phenomenon causes no significant complications in our analysis.

When the wellbore is much nearer one vertical boundary than the other, we may have another flow regime, called half-radial or hemiradial flow **(Fig. 12.3).** Hemiradial flow can occur immediately following the early-radial flow regime if the well is much nearer one of the vertical boundaries than the other.

Eventually, the area affected by the production will include the entire thickness of the reservoir. When that happens, a linear flow pattern may develop, as shown in **Fig. 12.4.**

Once the flow pattern has encountered both the top and the bottom of the reservoir, fluid will begin to flow linearly into the well, as illustrated in Fig. 12.4.

Eventually, flow will begin to come into the wellbore from beyond the ends of the well. Up until the time that these end effects become important, we will continue in early-linear flow. Once end effects become important and after we move through a transition period, we move into a later-pseudoradial flow regime, as illustrated in **Fig. 12.5.**

This flow regime continues until the area affected by the production reaches one of the sides of the reservoir. Once the area affected is the entire width of the reservoir (that is, the pressure transient has reached both sides of the reservoir), we again have a linear flow regime, the late-linear flow regime **(Fig. 12.6).**

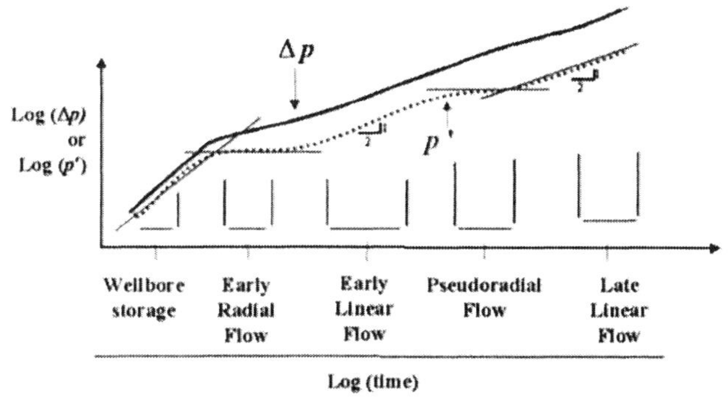

Fig. 12.7—Idealized diagnostic plot identifies all normally expected flow regimes.

12.4 Identifying Flow Regimes in Horizontal Wells

In a test, all these flow regimes can be identified on a diagnostic log-log plot of the pressure change, Δp, and pressure derivative, p', against the logarithm of time **(Fig. 12.7).**

A unit-slope line appears during wellbore storage; a horizontal derivative appears during early-radial flow and then, later, in pseudoradial flow; and a half-slope line appears in early-linear flow and then in late-linear flow. (These half-slope lines appear on the derivative but not on the pressure-change curves.) This does not imply that all these flow regimes will appear in any given test; in fact, that would be rare. But these are the shapes that identify those flow regimes that may appear in the test we are analyzing.

The shapes that may appear in a drawdown test (which is the basis of Fig. 12.7) may not appear in a buildup test because of the complex superposition of flow regimes. For example, a test would have to be in linear flow both at time $(t_p + \Delta t)$ and at time Δt to ensure the appearance of a derivative with a half-slope. This is highly unlikely. The best way to solve the problem is to ensure that a buildup test on a horizontal well is run with a producing time, t_p, much greater than the maximum shut-in time in the test (i.e., $t_p > 10\ \Delta t_{max}$).

We now explore quantitative analysis of the different flow regimes that may appear on a diagnostic plot. **Table 12.1** summarizes the working equations for permeability, skin, and start and end of each of the recognized flow regimes. We must note that different investigators have found different equations for the start and end of various flow regimes, especially linear flow regimes. This results partly from a difference of assumptions about flow into the wellbore. Uniform-flux or infinite-conductivity models are common; neither is rigorously applicable in practice.[1] In this section, we state the equation for duration of flow regimes derived by Odeh and Babu.[2] This model assumes uniform flux into the wellbore.

12.4.1 Early-Radial Flow. We begin with the early-radial flow regime, which is similar to the radial flow period in a vertical well **(Fig. 12.8).** The governing equation for this flow regime is

$$p_i - p_{wf} = \frac{162.6 qB\mu}{\sqrt{k_z k_x}\, L_w}\left[\log_{10}\left(\frac{\sqrt{k_x k_z}\, t}{\phi \mu c_t r_w^2}\right) - 3.227 + 0.868 s_d\right]. \quad (12.1)$$

Data for this period may be masked by wellbore-storage effects, but when present, they may be analyzed on a semilog plot.

The early-radial flow regime may in theory start at time zero, in absence of wellbore storage effects. The end of the early-radial flow regime may occur either when the transient reaches a vertical boundary or when flow comes from beyond the end of the wellbore. The end of the period is the smaller of these two values.

Eq. 12.2[2] says that the period must end when the transient reaches the nearest boundary, d_z, from the well. Note that this equation includes the permeability in the vertical direction:

$$t_{Eerf} = \frac{1{,}800 d_z^2 \phi \mu c_t}{k_z}, \quad (12.2)$$

TABLE 12.1—EQUATIONS FOR ANALYZING FLOW IN HORIZONTAL WELLS WITH UNIFORM FLUX

Wellbore Storage

$$t_{Ewbs} = \frac{(4{,}000 + 240 s_d)C}{\sqrt{k_x k_z}\, L_w/\mu} \quad (1)$$

$$C = V_{wb} c_{wb} \quad (2)$$

$$C = \frac{25.65 A_{wb}}{\rho_{wb} \cos\theta} \quad (3)$$

$$C = qBt/24\Delta p \quad (4)$$

Early Radial Flow

$$\sqrt{k_x k_z} = \frac{162.6 qB\mu}{|m_{erf}| L_w} \quad (5)$$

$$s_d = 1.151\left[\frac{\Delta p_{1hr}}{|m_{erf}|} - \log\left(\frac{\sqrt{k_x k_z}}{\phi \mu c_t r_w^2}\right) + 3.23\right] \quad (6)$$

$$t_{Eerf} = 1{,}800 d_z^2 \phi \mu c_t / k_z \quad (7)$$

$$t_{Eerf} = 125 L_w^2 \phi \mu c_t / k_y \quad (8)$$

Hemiradial Flow

$$\sqrt{k_x k_z} = 2\left(\frac{162.6 qB\mu}{|m_{hrf}| L_w}\right) \quad (9)$$

$$s_d = 2.303\left[\frac{\Delta p_{1hr}}{|m_{hrf}|} - \log\left(\frac{\sqrt{k_x k_z}}{\phi \mu c_t r_w^2}\right) + 3.23\right] \quad (10)$$

$$t_{Ehrf} = 1{,}800 D_z^2 \phi \mu c_t / k_z \quad (11a)$$

$$t_{Ehrf} = 125 L_w^2 \phi \mu c_t / k_y \quad (11b)$$

Early Linear Flow

$$\sqrt{k_x} = \frac{8.13 qB}{|m_{elf}| L_w h}\sqrt{\frac{\mu}{\phi c_t}} \quad (12)$$

$$s_d = \frac{L_w \sqrt{k_x k_z}}{141.2 qB\mu}\Delta p_{t=0} - s_c \quad (13a)$$

$$\text{where } s_c = \ln\left(\frac{h}{r_w}\right) + 0.25\ln\left(\frac{k_x}{k_z}\right) - \ln\left[\sin\left(\frac{\pi d_z}{h}\right)\right] - 1.838 \quad (13b)$$

$$t_{Self} = \frac{1{,}800 D_z^2 \phi \mu c_t}{k_z} \quad (14a)$$

$$t_{Eelf} = 160 L_w^2 \phi \mu c_t / k_y \quad (14b)$$

The radial flow regime will also end when flow from beyond the end of the wellbore becomes important; that time is given by Eq. 12.3:

$$t_{Eerf} = \frac{125 L_w^2 \phi \mu c_t}{k_y}. \quad (12.3)$$

L_w is the completed length of the well, and k_y is the permeability in the direction parallel to the wellbore. It is helpful to check the expected duration of the early-radial flow regime after we have estimated the parameters necessary to make these calculations.

Eq. 12.1 suggests that we identify possible radial flow on the diagnostic plot and then plot bottomhole flowing pressure during

TABLE 12.1—EQUATIONS FOR ANALYZING FLOW IN HORIZONTAL WELLS WITH UNIFORM FLUX (continued)

Late Pseudoradial Flow (requires $L_w/b_H < 0.45$)

$$\sqrt{k_x k_y} = \frac{162.6qB\mu}{|m_{\text{prf}}|h} \tag{15}$$

$$s_d = 1.151\sqrt{\frac{k_z}{k_y}\frac{L_w}{h}}\left[\frac{\Delta p_{1\text{hr}}}{|m_{\text{prf}}|} - \log\left(\frac{k_y}{\phi\mu c_t r_w^2}\right) + 1.76\right] - s_c \tag{16}$$

Note that s_c is defined by Eq. 13b

$$t_{\text{Sprf}} = \frac{1,480L_w^2\phi\mu c_t}{k_y} \tag{17}$$

$$t_{\text{Eprf}} = \frac{2,000\phi\mu c_t(L_w/4 + d_y)^2}{k_y} \tag{18}$$

$$t_{\text{Eprf}} = \frac{1,650\phi\mu c_t d_x^2}{k_x} \tag{19}$$

Late-Linear Flow

$$\sqrt{k_x} = \frac{8.128qB}{|m_{\text{llf}}|b_H h}\sqrt{\frac{\mu}{\phi c_t}} \tag{20}$$

$$s_t = \frac{L_w}{b_H}\left(\frac{b_H\sqrt{k_x k_z}\Delta p_{t=0}}{141.2qB\mu} - s_p - s_c\right) \tag{21}$$

where s_p, the partial penetration skin is given in Eqs. 25 through 35.

$$t_{\text{Sllf}} = \frac{4,800\phi\mu c_t(D_y + L_w/4)^2}{k_y} \tag{22}$$

$$\text{or } t_{\text{Sllf}} = \frac{(1,800)\phi\mu c_t D_z^2}{k_z} \tag{23}$$

$$t_{\text{Ellf}} = \frac{1,650\phi\mu c_t d_x^2}{k_x} \tag{24}$$

Calculating Partial Penetration Skin in Late-Linear Flow and Productivity Equations

$$p_{xyz} = \left(\frac{b_H}{L_w} - 1\right)\left\{\ln\left(\frac{h}{r_w}\right) + 0.25\ln\left(\frac{k_x}{k_z}\right) - \ln\left[\sin\left(\frac{\pi d_z}{h}\right) - 1.838\right]\right\} \tag{25}$$

$$\text{Case 1, } \frac{a_H}{\sqrt{k_x}} > \frac{0.75b_H}{\sqrt{k_y}} >> \frac{0.75h}{\sqrt{k_z}} \tag{26}$$

$$s_p = p_{xyz} + p'_{xy} \tag{27}$$

$$p'_{xy} = \frac{2b_H^2}{L_w h}\sqrt{k_z/k_y}\left\{F\left(\frac{L_w}{2b_H}\right) + 0.5\left[F\left(\frac{4y_m + L_w}{2b_H}\right) - F\left(\frac{4y_m - L_w}{2b_H}\right)\right]\right\} \tag{28}$$

$$\text{where } y_m = d_y + \frac{L_w}{2} \tag{29}$$

$$F(u) = -u\left[0.145 + \ln(u) - 0.137(u)^2\right], u < 1 \tag{30}$$

$$F(u) = (2 - u)\left[0.145 + \ln(2 - u) - 0.137(2 - u)^2\right], u > 1 \tag{31}$$

$$\text{Case 2, } \frac{b_H}{\sqrt{k_y}} > \frac{1.33a_H}{\sqrt{k_x}} >> \frac{0.75h}{\sqrt{k_z}} \tag{32}$$

$$s_p = p_{xyz} + p_y + p_{xy} \tag{33}$$

$$p_y = \frac{6.28b_H^2\sqrt{k_x k_z}}{a_H h \; k_y}\left[\left(\frac{1}{3} - \frac{y_m}{b_H} + \frac{y_m^2}{b_H^2}\right) + \frac{L_w}{24b_H}\left(\frac{L_w}{b_H} - 3\right)\right] \tag{34}$$

$$p_{xy} = \left(\frac{b_H}{L_w} - 1\right)\left(\frac{6.28a_H}{h}\sqrt{\frac{k_z}{k_x}}\right)\left(\frac{1}{3} - \frac{d_x}{a_H} + \frac{d_x^2}{a_H^2}\right),\ d_x \geq 0.25a_H \tag{35}$$

TABLE 12.1—EQUATIONS FOR ANALYZING FLOW IN HORIZONTAL WELLS WITH UNIFORM FLUX (continued)

Productivity and Productivity Index (Uniform Flux Solutions)

$$q = \frac{0.00708 b_H \sqrt{k_x k_z}(\bar{p} - p_{wf})}{B\mu\left[\ln\left(\frac{A^{1/2}}{r_w}\right) + \ln C_H - 0.75 + s_p + \left(\frac{b_H}{L_w}\right)s_d\right]} \quad (36)$$

$$J = \frac{q}{(\bar{p} - p_{wf})} = \frac{0.00708 b_H \sqrt{k_x k_z}}{B\mu\left[\ln\left(\frac{C_H A^{1/2}}{r_w}\right) - 0.75 + s_p + \left(\frac{b_H}{L_w}\right)s_d\right]} \quad (37)$$

$$\ln C_H = 6.28\frac{a_H}{h}\sqrt{\frac{k_z}{k_x}}\left[\frac{1}{3} - \frac{d_x}{a_H} + \left(\frac{d_x}{a_H}\right)^2\right] - \ln\left(\sin\frac{\pi d_z}{h}\right) - 0.5\ln\left[\left(\frac{a_H}{h}\right)\sqrt{\frac{k_z}{k_x}}\right] - 1.088 \quad (38)$$

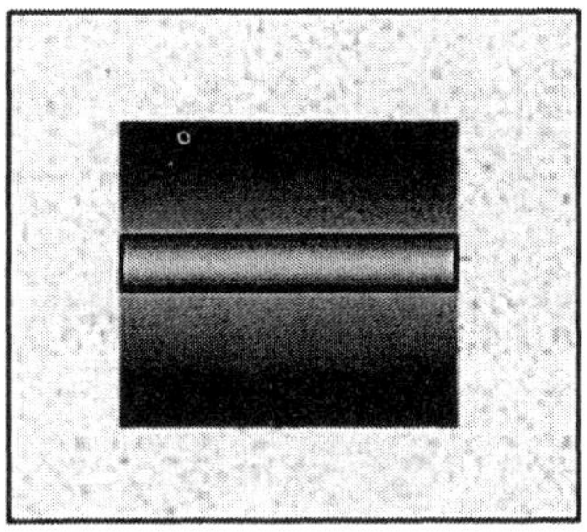

Fig. 12.8—Early-radial flow is similar to radial flow near vertical wells.

the appropriate time range against time on semilog coordinates. The slope of the straight line that results is

$$|m_{erf}| = \frac{162.6 qB\mu}{L_w\sqrt{k_x k_z}}. \quad \text{.......... (12.4)}$$

The group $\sqrt{k_x k_z}$ can be found from the slope, m_{erf}:

$$\sqrt{k_x k_z} = \frac{162.6 qB\mu}{|m_{erf}|L_w}, \quad \text{.......... (12.5)}$$

but notice that to make this calculation, we must know the effective completed length of the well. This is not necessarily the same as the perforated or the completed length of the well; some sections of the well may not produce at all.

The equation for calculating the altered permeability skin, s_d, for early-radial flow is

$$s_d = 1.151\left[\frac{\Delta p_{1hr}}{|m_{erf}|} - \log\left(\frac{\sqrt{k_x k_z}\,t}{\phi\mu c_t r_w^2}\right) + 3.23\right]. \quad \text{.... (12.6)}$$

If we are analyzing a buildup test rather than a constant-rate flow test, we plot the Horner time ratio or equivalent time on the horizontal axis of our semilog plot and then plot shut-in or equivalent time on the vertical axis. Note that this plotting is correct only if $(t_p + \Delta t)$ and Δt appear in this time period simultaneously; that is, we must be in radial flow at both time $(t_p + \Delta t)$ and time Δt. That is unlikely; we may be in the radial-flow regime at time Δt, but we are likely to be in a different flow regime at time $(t_p + \Delta t)$.

Example Well Erf-1. For drawdown test data from Well Erf-1,[2] the diagnostic plot (**Fig. 12.9a**) leads us to believe that the data from approximately 0.2 to 2.0 hours may be in early-radial flow. The following information is available for this well:

q = 800 STB/D
μ = 1 cp
B = 1.25 RB/STB
r_w = 0.25 ft
ϕ = 0.2.
c_t = 15×10^{-6} psi^{-1}
Centered in box-shaped drainage area.
h = 200 ft, a_H = 4,000 ft, and b_H = 2,000 ft.
Lw = 1,000 ft

From analysis of data in the early-linear flow regime, k_x = 200 md.

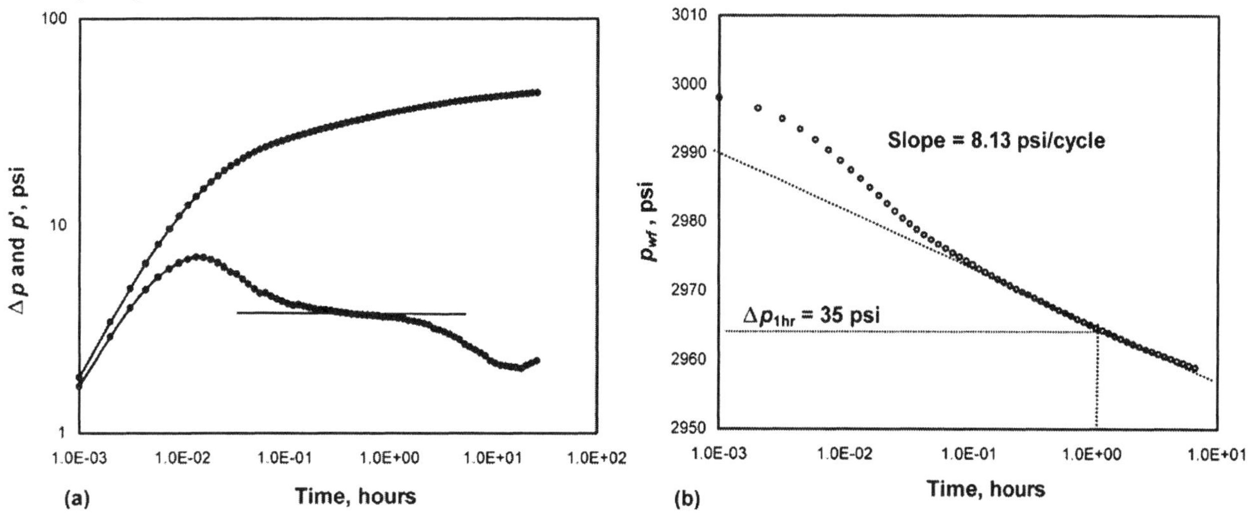

Fig. 12.9—(a) Diagnostic plot confirms early-radial flow in well Erf-1; (b) early-radial flow is indicated by a semilog straight line.

TABLE 12.2—PRESSURE DATA FOR WELL ERF-1

t, hours	p_{wf}, psi	$\Delta p=(p_i-p_{wf})$, psi	p', psi	t, hours	p_{wf}, psi	$\Delta p=(p_i-p_{wf})$, psi	p', psi
0.000	3000.000	0.000	0.000	0.485	2967.646	32.354	3.738
0.001	2998.134	1.866	1.685	0.545	2967.218	32.782	3.724
0.002	2996.572	3.428	2.915	0.611	2966.793	33.207	3.705
0.003	2995.041	4.959	4.012	0.685	2966.371	33.629	3.690
0.004	2993.473	6.527	4.901	0.768	2965.951	34.049	3.676
0.006	2991.933	8.067	5.634	0.861	2965.532	34.468	3.635
0.007	2990.422	9.578	6.197	0.965	2965.117	34.883	3.628
0.009	2988.950	11.050	6.590	1.082	2964.706	35.294	3.617
0.011	2987.528	12.472	6.821	1.213	2964.297	35.703	3.589
0.013	2986.272	13.728	6.983	1.359	2963.892	36.108	3.556
0.016	2984.981	15.019	6.951	1.523	2963.490	36.510	3.486
0.019	2983.760	16.240	6.808	1.707	2963.093	36.907	3.459
0.022	2982.615	17.385	6.578	1.913	2962.701	37.299	3.415
0.025	2981.545	18.455	6.285	2.143	2962.317	37.683	3.374
0.029	2980.551	19.449	5.956	2.401	2961.939	38.061	3.321
0.033	2979.726	20.274	5.833	2.690	2961.568	38.432	3.188
0.038	2978.870	21.130	5.518	3.014	2961.206	38.794	3.137
0.044	2978.076	21.924	5.222	3.376	2960.853	39.147	3.076
0.050	2977.337	22.663	4.950	3.782	2960.510	39.490	2.999
0.057	2976.647	23.353	4.718	4.237	2960.177	39.823	2.912
0.064	2976.046	23.954	4.730	4.746	2959.855	40.145	2.822
0.073	2975.427	24.573	4.564	5.317	2959.542	40.458	2.678
0.083	2974.837	25.163	4.421	5.956	2959.241	40.759	2.604
0.094	2974.272	25.728	4.305	6.671	2958.951	41.049	2.522
0.106	2973.727	26.273	4.209	7.473	2958.672	41.328	2.436
0.119	2973.200	26.800	4.120	8.370	2958.403	41.597	2.356
0.134	2972.702	27.298	4.162	9.376	2958.143	41.857	2.230
0.151	2972.202	27.798	4.099	10.502	2957.891	42.109	2.174
0.170	2971.712	28.288	4.039	11.763	2957.647	42.353	2.126
0.192	2971.233	28.767	3.980	13.175	2957.409	42.591	2.109
0.216	2970.763	29.237	3.926	14.757	2957.174	42.826	2.086
0.242	2970.303	29.697	3.912	16.529	2956.941	43.059	2.071
0.272	2969.847	30.153	3.887	18.513	2956.706	43.294	2.048
0.306	2969.397	30.603	3.852	20.736	2956.469	43.531	2.120
0.343	2968.953	31.047	3.818	23.225	2956.226	43.774	2.177
0.386	2968.514	31.486	3.788	26.013	2955.976	44.024	2.231
0.433	2968.078	31.922	3.744				

In addition, we have pressure data up to 26 hours, as shown in **Table 12.2:**

We plot p_{wf} vs. t on semilog coordinates (**Fig. 12.9b**). The plot results in a straight line with a slope of 8.13 psi/cycle. In Fig. 12.9b, note that at $t = 1.9$ hours, the points begin to deviate from the straight line, just as we expect from our calculations for flow-regime durations that follow. We see that the pressure change at 1 hour is 35 psia ($p_{1\text{hr}} = 2{,}965$ psia).

Using the slope of 8.13 and Eq. 12.5,

$$\sqrt{k_x k_z} = \frac{162.6 q B \mu}{|m_{\text{erf}}| L_w}$$

$$= \frac{162.6(800)(1.25)(1)}{|8.13|1000},$$

$$= 20.0 \text{ md.}$$

and $k_x\ k_z = 400$ md. Thus, because $k_x = 200$ md, $k_z = 2$ md.

Using the value of 35 for $\Delta p_{1\text{hr}}$ from Fig. 12.9b, we can calculate damage skin from Eq. 12.6:

$$s_d = 1.151 * \left[\frac{\Delta p_{1\text{hr}}}{|m_{\text{erf}}|} - \log\left(\frac{k_x k_z}{\phi \mu c_t r_w^2}\right) + 3.23\right]$$

$$= 1.151 * \left\{\left(\frac{35}{8.13}\right) - \log\left[\frac{20.0}{(0.2)(1)(15 \times 10^{-6})(0.25)^2}\right] + 3.23\right\}$$

$$= -0.57.$$

The start of the early-radial flow regime is controlled by wellbore storage, which appears to have vanished at times earlier than 0.2 hours in this example. The end of the early-radial flow regime is expected at the lesser of the two values derived from Eqs. 12.2 and 12.3. For a centered well, $d_z = h/2 = 100$ ft, and Eq. 12.2 gives

$$t_{E\text{erf}} = 1{,}800 d_z^2 \phi \mu c_t / k_z$$

$$= (1{,}800)(100)^2(0.2)(1)\left(1.5 \times 10^{-5}\right)/2$$

$$= 27 \text{ hr.}$$

Assuming $k_y = k_x = 200$ md, Eq. 12.8 gives

$$t_{E\text{erf}} = 125 L_w^2 \phi \mu c_t / k_y$$

$$= (125)(1000)^2(0.2)(1)\left(1.5 \times 10^{-5}\right)/200$$

$$= 1.875 \text{ hr.}$$

Thus, we expect the early-radial flow regime to end at approximately 1.875 hours, which is the smaller value. This is consistent with observed test data.

12.4.2 Hemiradial Flow.

The hemiradial flow period (**Fig. 12.10**) will occur only when the well is close to one of the vertical boundaries (either the upper or the lower boundary) and is analogous to a vertical well near a fault. The governing equation is[1]

$$p_i - p_{wf} = \frac{325.2 q B \mu}{\sqrt{k_z k_x} L_w}\left[\log_{10}\left(\frac{\sqrt{k_x k_z}\, t}{\phi \mu c_t r_w^2}\right) - 3.227 + 0.868 s_d\right]. \quad (12.7)$$

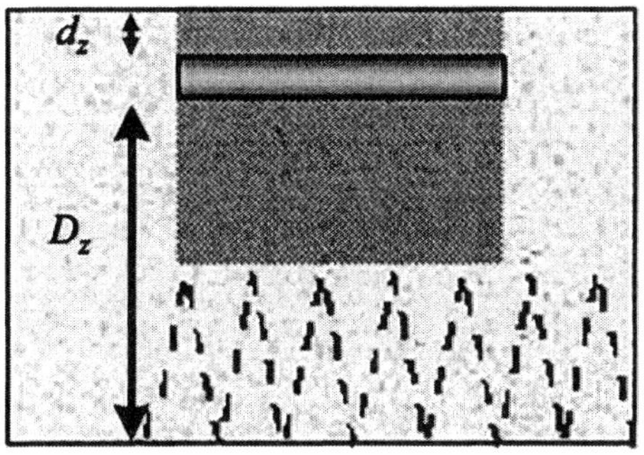

Fig. 12.10—Hemiradial flow can begin when the effects of the nearer vertical boundary are felt.

We identify hemiradial flow on the diagnostic plot with a horizontal derivative. For data that appear to fall into this flow regime, we can make a semilog plot; if a straight line appears, we have more confidence that we have identified radial flow. Consistency checks in the analysis coupled with a well survey will be required to distinguish hemiradial flow from early-radial flow.

The time range in which the analysis for hemiradial flow is valid begins after the closest vertical boundary, d_z, affects the data and before the farthest boundary, D_z, affects them. In the absence of wellbore storage, the start of hemiradial flow is given by

$$t_{Shrf} \cong \frac{1,800 d_z^2 \phi \mu c_t}{k_z}. \quad (12.8)$$

Notice that the start of the hemiradial flow regime involves the shortest distance to a vertical boundary (Fig. 12.10) and the permeability in the vertical direction. However, wellbore storage will most likely determine the actual start of hemiradial flow.

The end of hemiradial flow occurs when the data feel effects of the farther vertical boundary or the ends of the wellbore **(Fig. 12.11).** It is the minimum of the time calculated with either Eq. 12.9 or Eq. 12.10. If the hemiradial flow regime ends when we reach the farthest vertical boundary, it depends on the distance D_z and the vertical permeability, k_z:

$$t_{Ehrf} = \frac{1,800 D_z^2 \phi \mu c_t}{k_z}. \quad (12.9)$$

The appearance of end effects—flow from beyond the ends of the wellbore—causes the end of the hemiradial flow regime to appear; the end of the flow regime occurs when

$$t_{Ehrf} = \frac{125 L_w^2 \phi \mu c_t}{k_y}. \quad (12.10)$$

Note that the completed length of the well, L_w, and the permeability, k_y, parallel to the wellbore appear in this equation. These parameters determine when enough flow has come from beyond the end of the wellbore to distort the radial flow pattern that appeared earlier.

The equation,

$$|m_{hrf}| = \frac{325.2 qB\mu}{L_w \sqrt{k_x k_z}}, \quad (12.11)$$

gives the slope of the semilog straight line for semiradial flow, m_{hrf}. Note that the multiplier, 325.2, is twice the multiplier for early-radial flow.

The equation to estimate the damage-skin factor is also similar to that for radial flow but has a multiplier that differs by a factor of two.

$$s_d = 2.303\left[\frac{\Delta p_{1hr}}{|m_{hrf}|} - \log\left(\frac{\sqrt{k_x k_z}}{\phi \mu c_t r_w^2}\right) + 3.23\right]. \quad (12.12)$$

The equations relating slope and permeability and the equation for skin are similar in a buildup test to those for a drawdown test. The pressure change in the equation for skin is $[p_{1hr} - p_{wf}(\Delta t = 0)]$. Note that semilog plots of buildup-test data from the hemiradial flow regime cannot be analyzed rigorously with data from a Horner plot unless the pressure data at $(t_p + \Delta t)$ and the data obtained at time Δt are in this flow regime simultaneously. As a practical matter, the hemiradial flow regime is likely to appear clearly in the buildup test only when the producing time is much greater than the shut-in time.

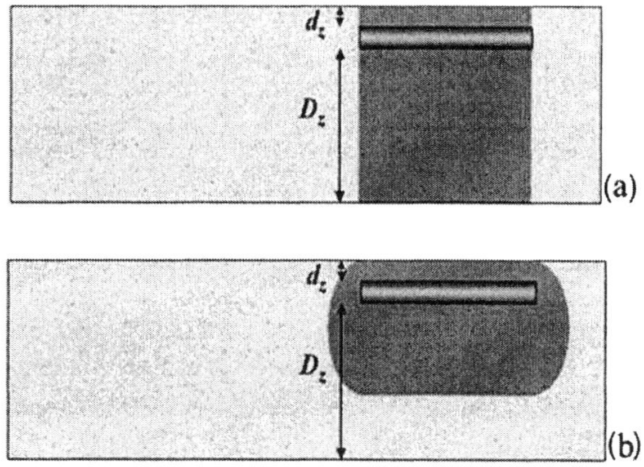

Fig. 12.11—Hemiradial flow ends when (a) the data feel effects of the farther vertical boundary or (b) fluid flows from beyond the ends of the wellbore, whichever comes first.

12.4.3 Early-Linear Flow. The governing equation for early-linear flow is[1]

$$p_i - p_{wf} = \frac{8.128 qB}{L_w}\sqrt{\frac{\mu t}{k_x \phi c_t}} + \frac{141.2 qB\mu}{L_w \sqrt{k_z k_x}}(s_c + s_d). \quad (12.13)$$

The "composite skin," s_c, is discussed later in this section. The start of the early-linear flow regime **(Fig. 12.12)** depends on the farthest distance to a vertical boundary, D_z, and the vertical permeability, k_z.[2]

$$t_{Self} = \frac{1,800 D_z^2 \phi \mu c_t}{k_z}. \quad (12.14)$$

Not until we have reached that farthest vertical boundary can we begin to have a linear flow pattern toward the well. The end of this flow period **(Fig. 12.13)** is

$$t_{Eelf} = \frac{160 L_w^2 \phi \mu c_t}{k_y}. \quad (12.15)$$

Notice that the end depends on the effective completed length of the well, L_w, and on the permeability in the direction parallel to the well. This is the time in which end effects—flow beyond the ends of the well—begin to distort the linear flow pattern significantly.

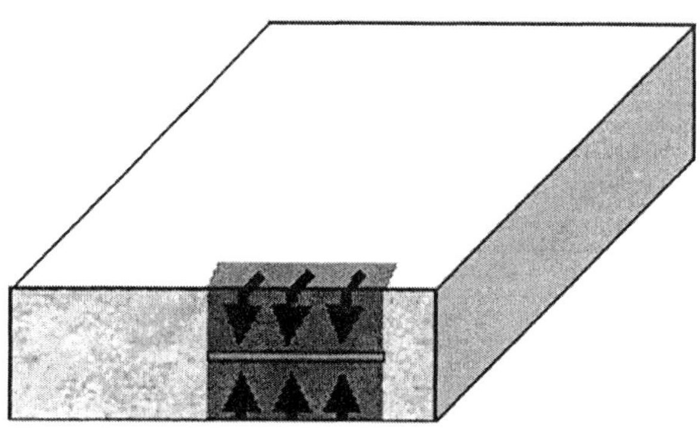

Fig. 12.12—Early-linear flow begins when pressure drawdown is affected by both vertical boundaries of the reservoir, but appreciable flow has not yet occurred from beyond the ends of the wellbore.

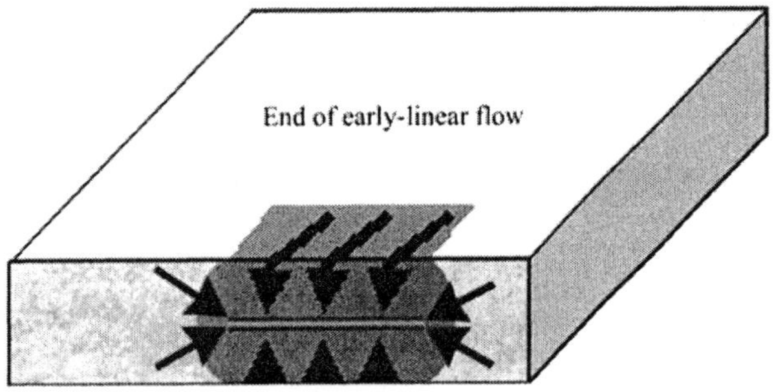

Fig. 12.13—The early-linear flow period ends when fluids flow from beyond the ends of the wellbore.

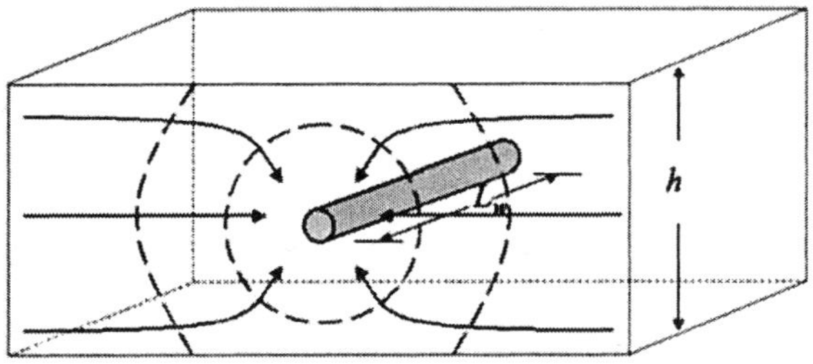

Fig. 12.14—Convergence skin accounts for increased pressure drop as flow deviates from the full vertical thickness of the reservoir into a radial pattern as it enters the well.

We can identify the early-linear flow regime in a drawdown test with a half-slope for the derivative. (Because of the skin effect, the pressure-change curve on the diagnostic plot will only approach a half-slope asymptotically.) For data identified as being in this flow regime, we can plot pressure against the square root of time.

We call the slope of the straight line on such a plot m_{elf}, and from slope m_{elf} we can estimate the square-root of k_x, the horizontal permeability perpendicular to the well:

$$\sqrt{k_x} = \frac{8.128qB}{|m_{elf}|L_w h}\sqrt{\frac{\mu}{\phi c_t}}. \quad (12.16)$$

We can calculate the damage skin from[2]

$$s_d = \frac{L_w\sqrt{k_x k_z}}{141.2qB\mu}\Delta p_{t=0} - s_c. \quad (12.17)$$

This equation includes a convergence skin, s_c, which is[2]

$$s_c = \ln\left(\frac{h}{r_w}\right) + 0.25\ln\left(\frac{k_x}{k_z}\right) - \ln\left[\sin\left(\frac{\pi d_z}{h}\right)\right] - 1.838. \quad (12.18)$$

This convergence skin is an additional pressure drop that acts like a skin effect caused by flow moving throughout the entire formation until it converges down to the small wellbore in the middle of the formation (**Fig. 12.14**). This convergence skin is defined in terms of the ratio of the permeability in the x-direction (which is perpendicular to the wellbore) to the vertical permeability. It also involves the distance to the nearest vertical boundary, d_z, and the net pay thickness, h.

Kuchuk[3] derived a different equation for convergence skin. As a practical matter, the Odeh-Babu and Kuchuk equations lead to the same result.

When we have a single rate preceding shut-in during early-linear flow, the buildup pressure is plotted against $\sqrt{t_p + \Delta t} - \sqrt{\Delta t}$ on a Cartesian plot, which is sometimes called a *tandem-root* plot. The permeability, k_x, is calculated from the slope, m_{elf}, of the plot and Eq. 12.16. Note that k_x has the same relationship to the slope that we had in a drawdown test. Skin for this flow regime is calculated with Eqs. 12.17 and 12.18.

Plots of buildup data from the early-linear flow regime cannot be analyzed rigorously with a plot of p_{ws} vs. $\sqrt{t_p + \Delta t} - \sqrt{\Delta t}$ (that is, the tandem-root plot) unless data at $(t_p + \Delta t)$ and Δt are simultaneously in this flow regime—highly unlikely—or unless t_p is much greater than Δt, in which case we simply plot p_{ws} vs. $\sqrt{\Delta t}$. Little error results from ignoring the $(t_p + \Delta t)$ term, which is essentially constant.

Example Well Elf-2. The diagnostic plot (**Fig. 12.15a**) for a drawdown test from Well Elf-2[2] leads us to believe that it includes data in the early-linear flow regime because the derivative has a half-slope. We also have the following data on this well:

q = 800 STB/D
μ = 1 cp
B = 1.25 RB/STB
r_w = 0.25 ft
ϕ = 0.2.
c_t = 15×10^{-6} psi^{-1}
Centered in box-shaped drainage area 100 ft thick, 4,000 ft long, 4,000 ft wide.
L_w = 2,500 ft
From early-radial flow regime data, $k_x k_z = 8{,}000$ md^2.

In addition, we have the pressure-time data for this well, as shown by **Table 12.3.** Our plot of pressure vs. the square root of time (**Fig. 12.15b**) indicates that we are indeed in early-linear flow. The final point on the straight line is at a time of approximately 10 hours,

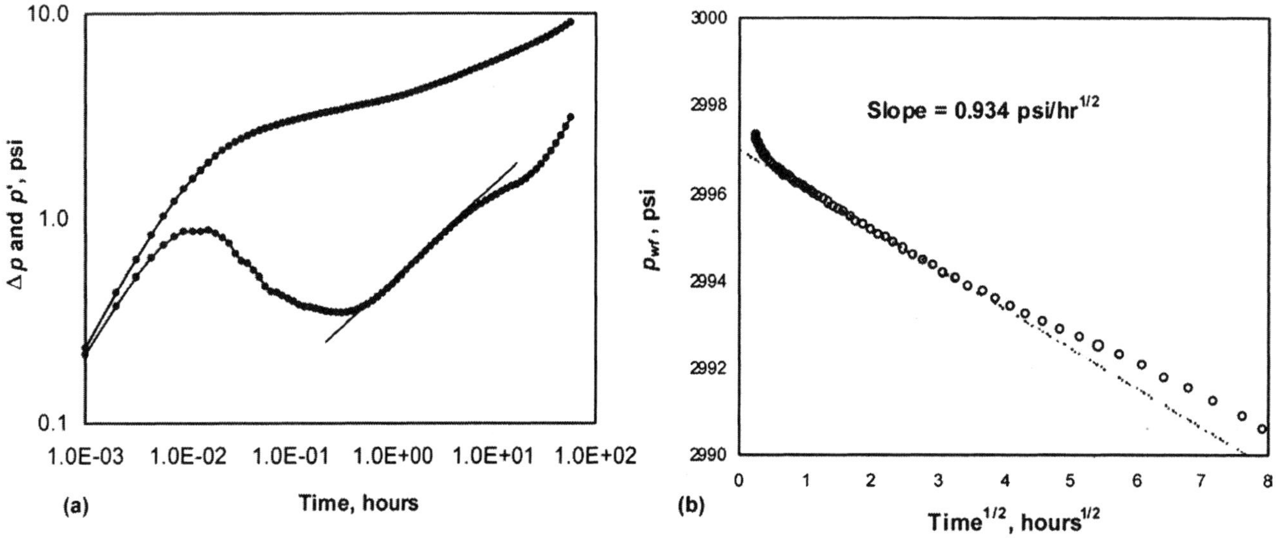

Fig. 12.15—(a) Diagnostic plot confirms early-linear flow in Well Elf-2; (b) early-linear flow is indicated by a straight line on the plot of square root of time.

TABLE 12.3—PRESSURE DATA FOR WELL ELF-2

t, hours	p_{wf}, psi	$\Delta p=(p_i-p_{wf})$, psi	p', psi	t, hours	p_{wf}, psi	$\Delta p=(p_i-p_{wf})$, psi	p', psi
0.000	3000.000	0.000	0.000	0.768	2996.251	3.749	0.436
0.001	2999.766	0.234	0.216	0.861	2996.200	3.800	0.457
0.002	2999.565	0.453	0.376	0.965	2996.146	3.854	0.482
0.003	2999.368	0.632	0.514	1.082	2996.090	3.910	0.508
0.004	2999.169	0.831	0.628	1.213	2996.031	3.969	0.535
0.006	2998.975	1.025	0.718	1.359	2995.968	4.032	0.564
0.007	2998.786	1.214	0.784	1.523	2995.902	4.098	0.591
0.009	2998.605	1.395	0.824	1.707	2995.833	4.167	0.625
0.011	2998.431	1.569	0.841	1.913	2995.760	4.240	0.659
0.013	2998.278	1.722	0.842	2.143	2995.683	4.317	0.694
0.016	2998.126	1.874	0.821	2.401	2995.602	4.398	0.733
0.019	2997.986	2.014	0.785	2.690	2995.517	4.483	0.767
0.022	2997.857	2.143	0.739	3.014	2995.427	4.573	0.811
0.025	2997.740	2.260	0.686	3.376	2995.332	4.668	0.855
0.029	2997.635	2.365	0.630	3.782	2995.233	4.767	0.900
0.033	2997.549	2.451	0.597	4.237	2995.128	4.872	0.945
0.038	2997.463	2.537	0.548	4.746	2995.019	4.981	0.992
0.044	2997.385	2.615	0.505	5.317	2994.904	5.096	1.034
0.050	2997.315	2.685	0.468	5.956	2994.783	5.217	1.083
0.057	2997.250	2.750	0.438	6.671	2994.658	5.342	1.129
0.064	2997.195	2.805	0.433	7.473	2994.527	5.473	1.178
0.073	2997.138	2.862	0.414	8.370	2994.392	5.608	1.223
0.083	2997.085	2.915	0.399	9.376	2994.251	5.749	1.258
0.094	2997.034	2.966	0.387	10.502	2994.105	5.895	1.307
0.106	2996.986	3.014	0.377	11.763	2993.954	6.046	1.356
0.119	2996.939	3.061	0.369	13.175	2993.798	6.202	1.405
0.134	2996.894	3.106	0.372	14.757	2993.637	6.363	1.444
0.151	2996.850	3.150	0.366	16.529	2993.472	6.528	1.477
0.170	2996.806	3.194	0.360	18.513	2993.303	6.697	1.496
0.192	2996.763	3.237	0.356	20.736	2993.129	6.871	1.572
0.216	2996.721	3.279	0.351	23.225	2992.947	7.053	1.641
0.242	2996.680	3.320	0.351	26.013	2992.756	7.244	1.725
0.272	2996.639	3.361	0.350	29.135	2992.555	7.445	1.837
0.306	2996.599	3.401	0.349	32.632	2992.339	7.661	1.968
0.343	2996.558	3.442	0.351	36.549	2992.108	7.892	2.127
0.386	2996.518	3.482	0.355	40.936	2991.856	8.144	2.316
0.433	2996.476	3.524	0.360	45.849	2991.582	8.418	2.538
0.485	2996.434	3.566	0.371	51.351	2991.280	8.720	2.801
0.545	2996.391	3.609	0.384	57.514	2990.946	9.054	3.101
0.611	2996.346	3.654	0.399	62.524	2990.677	9.323	3.234
0.685	2996.300	3.700	0.417				

but there may have been some deviation from the straight line by this time.

From the slope of the straight line, $m_{\text{elf}}=0.934$ psi/hr$^{1/2}$, and Eq. 12.16, we can calculate the permeability in the horizontal plane perpendicular to the wellbore.

$$\sqrt{k_x} = \frac{8.13qB}{|m_{\text{elf}}|L_w h}\sqrt{\frac{\mu}{\phi c_t}}$$

$$= \frac{(8.13)(800)(1.25)}{(0.934)(2,500)(100)}\sqrt{\frac{1}{(0.2)(15)(10^{-6})}}$$

$$= 20.10 \text{ md}^{1/2},$$

or $k_x=(20.1)^2=404$ md.

Analysis of data from the early-radial flow regime indicated that $k_x k_z$ is 8,000 md^2; thus, k_z is approximately 19.8 md.

To calculate s_d, we use Eqs. 12.17 and 12.18, noting that the value for $(p_i-p_{wf})=3.1$ at $t=0$ (that is, $p_0=2{,}996.9$ psia).

$$s_d = \frac{L_w\sqrt{k_x k_z}}{141.2qB\mu}\Delta p_{t=0} - s_c$$

$$= \frac{(2,500)\sqrt{(8,000)}(3.10)}{(141.2)(800)(1.25)(1)} - s_c$$

$$= 4.91 - s_c.$$

$$s_c = \ln\left(\frac{h}{r_w}\right) + 0.25\ln\left(\frac{k_x}{k_z}\right) - \ln\left[\sin\left(\frac{\pi d_z}{h}\right)\right] - 1.838.$$

$$= \ln\left(\frac{100}{0.25}\right) + 0.25\ln\left(\frac{404}{19.8}\right) - \ln\left\{\sin\left[\frac{\pi(50)}{100}\right]\right\} - 1.838.$$

$$= 4.91.$$

Then, $s_d=4.91-4.91=0$.

We can check the expected time range for the early-linear flow regime. Using $D_z=h/2=50$ ft and $k_z=20$ md, we can calculate the beginning of linear flow with Eq. 12.14:

$$t_{\text{Self}} = \frac{(1,800)D_z^2\phi\mu c_t}{k_z}$$

$$= \frac{(1.800)(50)^2(0.2)(1)(1.5\times10^{-5})}{20}$$

$$= 0.675 \text{ hr}.$$

Assuming $k_y=k_x=400$ md, we can use Eq. 12.15 to find the end of early-linear flow.

$$t_{\text{Eelf}} = \frac{160L_w^2\phi\mu c_t}{k_y}$$

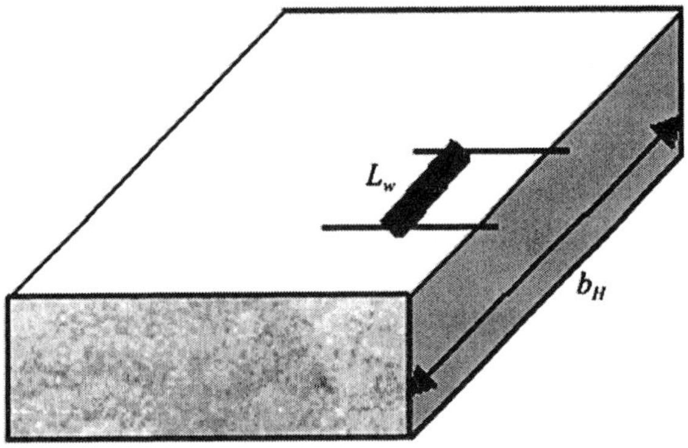

Fig. 12.16—Late-pseudoradial flow occurs when the wellbore is relatively short compared to the length of the drainage area.

$$= \frac{(160)(2,500)^2(0.2)(1)(1.5 \times 10^{-5})}{400}$$

$$= 7.5 \text{ hr.}$$

These limits are reasonably consistent with the time range we analyzed assuming early-linear flow.

12.4.4 Late-Pseudoradial Flow. The governing equation for late-pseudoradial flow is[1]

$$p_i - p_{wf} = \frac{162.6qB\mu}{\sqrt{k_z k_x}\,h}\left[\log_{10}\left(\frac{k_y t}{\phi\mu c_t L_w^2}\right) - 2.303\right]$$

$$\frac{141.2qB\mu}{L_w\sqrt{k_z k_x}}(s_c + s_d). \qquad (12.19)$$

The late-pseudoradial flow period occurs only if[2]

$$\frac{L_w}{b_H} < 0.45. \qquad (12.20)$$

Here, b_H is the dimension of the reservoir parallel to the wellbore. As long as the well is relatively short compared to the length of the drainage area **(Fig. 12.16),** then late-pseudoradial flow can occur.

The start of this flow period occurs when fluid flows from the well beyond the ends of the wellbore **(Fig. 12.17).**

It is approximated with the equation[2]

$$t_{Sprf} = \frac{1,480L_w^2\phi\mu c_t}{k_y}. \qquad (12.21)$$

Notice that this starting time depends on the completed length of the well, L_w, and on the permeability in the direction of the well, k_y.

The end of this period, like others in this section, is approximated by the minimum of the results of two calculations. The first,

$$t_{Eprf} = \frac{2,000\phi\mu c_t(L_w/4 + d_y)^2}{k_y}, \qquad (12.22)$$

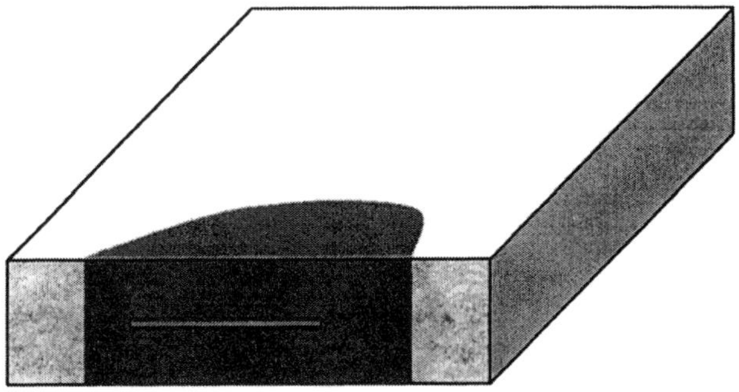

Fig. 12.17—Late-pseudoradial flow begins when fluid flows from the well beyond the ends of the wellbore.

depends on d_y and the length of the wellbore along with k_y, the permeability in the direction parallel to the wellbore. This is the time at which horizontal boundary effects first appear **(Fig. 12.18).**

The other equation gives a time at which the radial flow pattern begins to be distorted depending on the shortest distance, d_x, from the well to a boundary perpendicular to the wellbore, and on k_x, the permeability in that direction.

$$t_{Eprf} = \frac{1,650\phi\mu c_t d_x^2}{k_x}. \qquad (12.23)$$

Whenever boundary effects first appear, whether in a direction that is parallel to the well (Fig. 12.18) or perpendicular to the axis of the well **(Fig. 12.19),** the late-pseudoradial flow period will end.

The diagnostic plot helps us identify the late-pseudoradial flow regime with the characteristic horizontal derivative. For data in the appropriate time range, we prepare a semilog plot of pressure against time for a drawdown test. The slope of this plot will be m_{prf}, and the relationship between that slope and the square root of $k_x k_y$, or the permeabilities in the horizontal plane, is given by

$$\sqrt{k_x k_z} = \frac{162.6qB\mu}{|m_{prf}|h}. \qquad (12.24)$$

The skin equation is similar in form to those that we have seen before:

$$s_d = \left(1.151\sqrt{\frac{k_z}{k_y}\frac{L_w}{h}}\right)\left[\frac{\Delta p_{1hr}}{|m_{prf}|} - \log\left(\frac{k_y}{\phi\mu c_t r_w^2}\right) + 1.76\right] - s_c. \qquad (12.25)$$

Again, the total skin depends on Δp_{1hr}. The convergence skin (Eq. 12.18) is subtracted from the "total" skin to determine the damage skin.

For a buildup test preceded by production at a single rate, we plot pressure against the Horner time ratio on a semilog graph. Permeability is calculated from Eq. 12.26, the same as for a drawdown test. The skin equation is basically the same as for a drawdown test, except that Δp_{1hr} is now $p_{1hr} - p_{wf}$.

To obtain p^*, the extrapolated pressure, we extrapolate the semilog straight line to a Horner time ratio of unity.

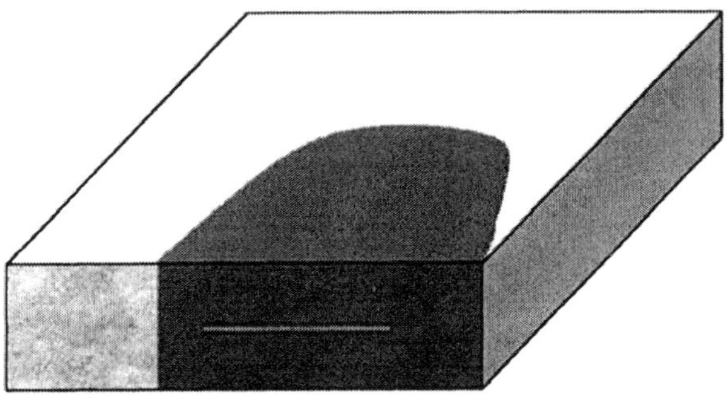

Fig. 12.18—Horizontal boundary effects begin to affect pressures and radial flow pattern ends.

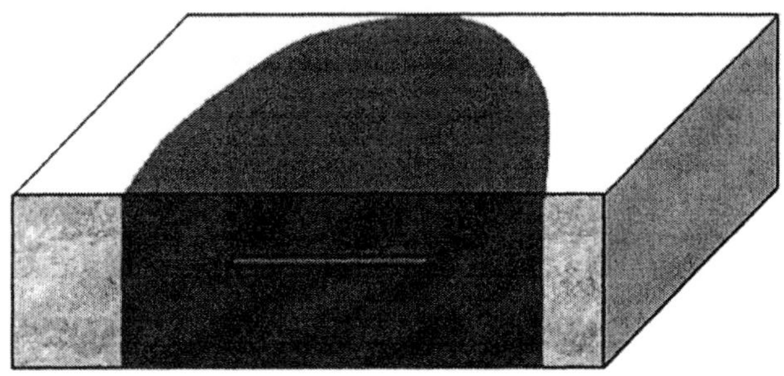

Fig. 12.19—Late-pseudoradial flow effects can end with effects of boundaries perpendicular to the wellbore.

TABLE 12.4—PRESSURE DATA FOR WELL PRF-3

t, hours	p_{wf}, psi	$\Delta p=(p_i-p_{wf})$, psi	p', psi	t, hours	p_{wf}, psi	$\Delta p=(p_i-p_{wf})$, psi	p', psi
0.000	3000.000	0.000	0.000	1.913	2971.031	28.969	2.712
0.001	2996.745	3.255	2.605	2.143	2970.719	29.281	2.801
0.002	2994.507	5.493	3.894	2.401	2970.397	29.603	2.899
0.003	2992.580	7.420	4.745	2.690	2970.063	29.937	2.967
0.004	2990.776	9.224	4.982	3.014	2969.715	30.285	3.112
0.006	2989.243	10.757	4.924	3.376	2969.353	30.647	3.259
0.007	2987.956	12.044	4.700	3.782	2968.976	31.024	3.404
0.009	2986.870	13.130	4.402	4.237	2968.582	31.418	3.547
0.011	2985.939	14.061	4.085	4.746	2968.172	31.828	3.700
0.013	2985.206	14.794	3.980	5.317	2967.744	32.256	3.821
0.016	2984.492	15.508	3.718	5.956	2967.300	32.700	3.981
0.019	2983.856	16.144	3.495	6.671	2966.840	33.160	4.131
0.022	2983.284	16.716	3.311	7.473	2966.365	33.635	4.270
0.025	2982.761	17.239	3.164	8.370	2965.875	34.125	4.398
0.029	2982.278	17.722	3.045	9.376	2965.370	34.630	4.471
0.033	2981.848	18.152	3.085	10.502	2964.852	35.148	4.609
0.038	2981.418	18.582	3.009	11.763	2964.323	35.677	4.731
0.044	2981.009	18.991	2.943	13.175	2963.782	36.218	4.837
0.050	2980.617	19.383	2.882	14.757	2963.231	36.769	4.931
0.057	2980.239	19.761	2.831	16.529	2962.668	37.332	4.966
0.064	2979.878	20.122	2.833	18.514	2962.096	37.904	5.071
0.073	2979.523	20.477	2.798	20.736	2961.515	38.485	5.164
0.083	2979.176	20.824	2.765	23.225	2960.925	39.075	5.270
0.094	2978.838	21.162	2.734	26.013	2960.327	39.673	5.351
0.106	2978.506	21.494	2.702	29.136	2959.721	40.279	5.338
0.119	5978.180	21.820	2.672	32.633	2959.104	40.896	5.438
0.134	2977.859	22.141	2.665	36.550	2958.480	41.520	5.539
0.151	2977.543	22.457	2.654	40.936	2957.848	42.152	5.628
0.170	2977.230	22.770	2.632	45.850	2957.209	42.791	5.708
0.192	2976.921	23.079	2.623	51.353	2956.562	43.438	5.767
0.216	2976.616	23.384	2.601	57.516	2955.906	44.094	5.780
0.242	2976.312	23.688	2.583	64.419	2955.242	44.758	5.876
0.272	2976.011	23.989	2.578	72.150	2954.573	45.427	5.949
0.306	2975.712	24.288	2.572	80.808	2953.901	46.099	5.903
0.343	2975.415	24.585	2.575	90.506	2953.237	46.763	5.879
0.386	2975.120	24.880	2.564	101.368	2952.572	47.428	5.905
0.433	2974.826	25.174	2.525	113.533	2951.902	48.098	5.920
0.485	2974.533	25.467	2.534	127.158	2951.228	48.772	5.953
0.545	2974.242	25.758	2.536	142.418	2950.546	49.454	6.065
0.611	2973.951	26.049	2.544	159.509	2949.853	50.147	6.186
0.685	2973.662	26.338	2.538	178.651	2949.147	50.853	6.312
0.768	2973.374	26.626	2.538	200.090	2948.423	51.577	6.490
0.861	2973.085	26.915	2.512	224.101	2947.676	52.324	6.699
0.965	2972.797	27.203	2.525	250.994	2946.900	53.100	7.007
1.082	2972.509	27.491	2.537	281.114	2946.088	53.912	7.340
1.213	2972.220	27.780	2.558	314.849	2945.231	54.769	7.785
1.359	2971.929	28.071	2.579	352.632	2944.318	55.682	8.318
1.523	2971.635	28.365	2.578	394.949	2943.340	56.660	8.955
1.707	2971.336	28.664	2.640	437.596	2942.385	57.615	8.848

Semilog plots of buildup-test data from the late-pseudoradial flow regime cannot be analyzed rigorously with a Horner plot unless pressures at $(t_p + \Delta t)$ and the pressures obtained at time Δt are simultaneously in the pseudoradial flow regime, which is highly unlikely. However, little error appears if the producing time before shut-in is much greater than the maximum shut-in time achieved in the buildup test.

Example Well Prf-3. The diagnostic plot (Fig. 12.20a) suggests that a constant-rate drawdown test from Well Prf-3[2] includes data in the late-pseudoradial period. The following data are available from the test:

$q =$ 800 STB/D
$\mu =$ 1 cp
$B =$ 1.25 RB/STB
$r_w =$ 0.25 ft
$\phi =$ 0.2
$c_t = 15 \times 10^{-6}$ psi^{-1}
$h =$ 150 ft
$L_w =$ 900 ft
$a_H =$ 5,280 ft
$b_H =$ 5,280 ft

Well centered in drainage volume
$k_x =$ 100 md (from analysis of early-linear flow)
$k_x k_z =$ 1,000 md^2 and $k_z = 10$ md (from analysis of early-radial flow)

The pressure data, p_{wf} vs. time, are given in **Table 12.4.**

A plot of pressure change vs. the logarithm of time **(Fig. 12.20b)** confirms pseudoradial flow. A straight line fits the data from 72 to 142 hours; the slope of the line, m_{prf}, is 15.3 psi/cycle, and $\Delta p_{1hr} = 18.94$ psi (extrapolated; i.e., $p_{1hr} = 2{,}981.06$ psi).

Then, from Eq. 12.24,

$$\sqrt{k_x k_y} = \frac{162.6 qB\mu}{|m_{prf}|h} = \frac{(162.6)(800)(1.25)(1.0)}{(15.3)(150)} = 70.8 \text{ md.}$$

$$\text{Thus, } k_y = \frac{\left(\sqrt{k_x k_y}\right)^2}{k_x} = \frac{(70.8)^2}{100} = 50.2 \text{ md.}$$

From Eqs. 12.25 and 12.18,

$$s_d = 1.151\sqrt{\frac{k_z}{k_y}\frac{L_w}{h}}\left[\frac{\Delta p_{1hr}}{|m_{prf}|} - \log\left(\frac{k_y}{\phi\mu c_t r_w^2}\right) + 1.76\right] - s_c.$$

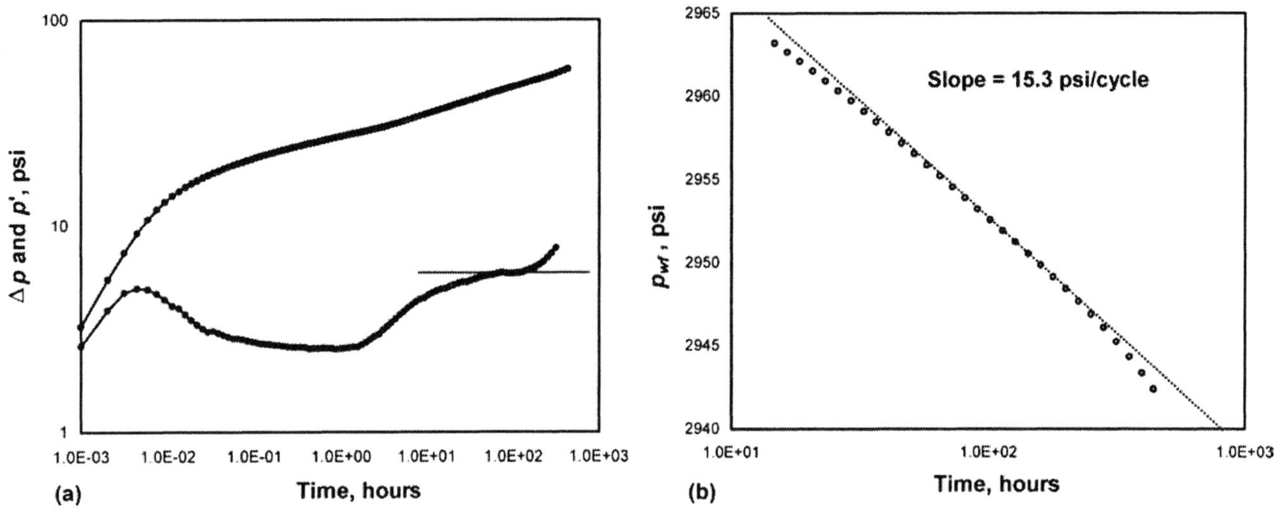

Fig. 12.20—(a) Diagnostic plot confirms pseudoradial flow in Well Prf-3; (b) pseudoradial flow is indicated by a semilog straight line in Well Prf-3.

and

$$s_c = \ln\left(\frac{h}{r_w}\right) + 0.25\ln\left(\frac{k_x}{k_z}\right) - \ln\left[\sin\left(\frac{\pi d_z}{h}\right)\right] - 1.838.$$

Here,

$$s_c = \ln\left(\frac{150}{0.25}\right) + 0.25\ln\left(\frac{100}{10}\right) - \ln\left\{\sin\left[\frac{\pi(75)}{150}\right]\right\} - 1.838.$$

$$= 5.13.$$

$$s_d = 1.151\sqrt{\frac{10}{50.2}\frac{(900)}{(150)}}$$

$$\times\left\{\frac{18.94}{15.3} - \log\left[\frac{50.2}{(0.2)(1)(1.5\times10^{-5})(900)^2}\right] + 1.76\right\} - 5.13$$

$$= 0.057.$$

The pseudoradial flow regime should start at the time given by Eq. 12.21:

$$t_{\text{Sprf}} = \frac{1,480L_w^2\phi\mu c_t}{k_y}$$

$$= \frac{(1,480)(900^2)(0.2)(1)(1.5\times10^{-5})}{50.2} = 71.6\text{ hr.}$$

It should end at a time given by the lesser of values from Eqs. 12.22 and 12.23. From Eq. 12.22,

$$t_{\text{Eprf}} = \frac{2,000\phi\mu c_t(L_w/4 + d_y)^2}{k_y}$$

$$= \frac{(2,000)(0.2)(1)(1.5\times10^{-5})(900/4 + 2,150)^2}{50.2}$$

$$= 674\text{ hr,}$$

where $d_y = \frac{1}{2}(5,200 - 900) = 2,150$ ft for this centered well.

From Eq. 12.23,

$$t_{\text{Eprf}} = \frac{1,650\phi\mu c_t d_x^2}{k_x}$$

$$= \frac{(1,650)(0.2)(1)(1.5\times10^{-5})(2,640)^2}{(100)} = 344\text{ hr.}$$

The smaller of these two values is 344 hours, which is thus the expected end of pseudoradial flow. The data in **Fig. 12.20b** that lie on the straight line span a time range that is generally consistent with the expected duration of the flow regime.

12.4.5 Late-Linear Flow. The governing equation for late-linear flow is[1,2]

$$p_i - p_{wf} = \frac{8.128qB}{b_H h}\sqrt{\frac{\mu t}{k_x\phi c_t}} + \frac{141.2qB\mu}{b_H\sqrt{k_z k_x}}\left(s_p + s_c + \frac{b_H}{L_w}s_d\right). \quad (12.26)$$

The late-linear flow regime starts after the pressure transient has reached the boundaries in the z- and y-directions and the flow behavior with regard to these directions has become pseudosteady-state, as shown by Fig. 12.6.

The start of this time period is the maximum of two equations.[2] The first depends on the time at which we reach the boundary, D_y, beyond the end of the horizontal well. It also depends on the permeability, k_y, in the direction parallel to the wellbore.

$$t_{\text{SIlf}} = \frac{4,800\phi\mu c_t(D_y + L_w/4)^2}{k_y}. \quad (12.27)$$

Another requirement for the start of the late-linear flow regime is the time at which we reach the maximum vertical distance, D_z:

$$t_{\text{SIlf}} = \frac{(1,800)\phi\mu c_t D_z^2}{k_z}. \quad (12.28)$$

Usually, the start of the late-linear flow regime is dictated by the time to reach the boundaries in the y-direction.

The end of this period is given by the equation

$$t_{\text{EIlf}} = \frac{1,650\phi\mu c_t d_x^2}{k_x}. \quad (12.29)$$

The end of the late-linear flow regime depends on reaching the nearest boundary in the direction perpendicular to the wellbore, which is the distance, d_x, away, and on the permeability, k_x, in that direction.

We identify the late-linear flow regime by a half-slope on the derivative in the diagnostic plot of drawdown-test data. (The pressure change may approach a half-slope asymptotically.) For data that appear to be in this flow regime, we plot pressure against the square root of time. From the slope m_{llf} of the plot, we can estimate permeability in the x-direction from

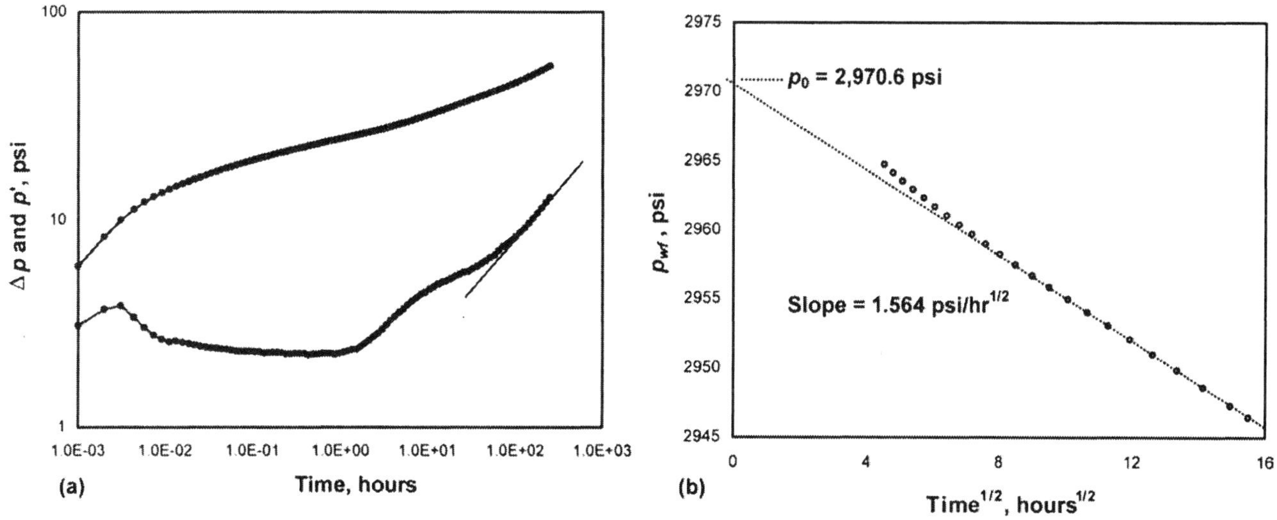

Fig. 12.21—(a) Diagnostic plot confirms late-linear flow in Well Llf-4; (b) late-linear flow is indicated by a straight line on the plot of the square root of time.

$$\sqrt{k_x} = \frac{8.128qB}{|m_{\text{llf}}|b_H h}\sqrt{\frac{\mu}{\phi c_t}}. \qquad (12.30)$$

Alternatively, if we already know k_x from an early-linear flow regime, we can estimate b_H, the length of the drainage area, from

$$b_H = \frac{8.128qB}{|m_{\text{llf}}|h}\sqrt{\frac{\mu}{\phi c_t k_x}}. \qquad (12.31)$$

This late-linear flow regime is the only period that provides the data to calculate the total skin, s, including the partial-penetration skin, s_p, and the convergence skin, s_c. To calculate the damage skin, s_d, we use

$$s_d = \frac{L_w}{b_H}\left[\frac{b_H\sqrt{k_x k_z}(\Delta p_{t=0})}{141.2qB\mu} - s_p - s_c\right]. \qquad (12.32)$$

The total skin depends on $\Delta p_{t=0}$. We have to subtract the partial penetration skin, s_p, and the convergence skin, s_c, from the total skin to get the damage skin. The partial-penetration skin is a complex function that is calculated with Eqs. 25 through 35 in Table 12.1.

For a buildup test, we plot pressure against the Horner time ratio. From the slope, m_{llf}, we can calculate k_x with Eq. 12.30, exactly the same as for drawdown tests. Or, if we already know k_x, we can estimate the length, b_H, of the drainage area with Eq. 12.31. We calculate the damage skin, s_d, from a pressure-buildup test from Eq. 12.32, where $\Delta p_{t=0} = (p_{t=0})_{\text{ext}} - p_{wf(t=0)}$.

Note that we have the same difficulty in using superposition to find plotting-function plots of buildup data from the late-linear flow regime as we have had with the previous flow regimes. Pressures at both time $(t_p + \Delta t)$ and time Δt must be in the late-linear flow regime for a tandem-root plot to be valid. However, if $t_p >> \Delta t_{\max}$, there is little error.

Example Well Llf-4. The diagnostic plot **(Fig. 12.21a)** for a drawdown test from Well Llf-4[2] leads us to believe that it includes data in the late-linear flow regime (derivative with half-slope). We have the following data for this well:

q = 800 STB/D
μ = 1 cp
B = 1.25 RB/STB
r_w = 0.25 ft
ϕ = 0.2
c_t = 15×10^{-6} psi^{-1}
h = 150 ft
L_w = 2,500 ft
a_H = 6,968 ft (well centered)
b_H = 2,000 ft (well centered)
D_z = 85 ft
d_z = 65 ft
$k_x k_z$ = 1,000 md^2 (from analysis of early-radial flow)
$k_x k_y$ = 5,000 md^2 (from analysis of test in vertical section of well)

Pressure data for the test are given in **Table 12.5.**

Fig. 12.21b is a plot of pressure vs. the square root of time. The straight line on this plot for the time range (100 to 248 hours) confirms late-linear flow for this range. The slope of the line is 1.564 psi/hr$^{1/2}$, and the intercept is 2,970.6 psia, so $\Delta p_{t=0} = 29.4$ psi.

From Eq. 12.30,

$$\sqrt{k_x} = \frac{8.128qB}{|m_{\text{llf}}|b_H h}\sqrt{\frac{\mu}{\phi c_t}}$$

$$= \frac{(8.128)(800)(1.25)}{(1.564)(2,000)(150)}\sqrt{\frac{1}{(0.2)(1.5\times 10^{-5})}} = 10.0.$$

Then, $k_x = 100$ md.
Because $k_x k_z = 1000$ md^2, $k_z = 10$ md.
Also, because $k_x k_y = 5,000$ md^2, $k_y = 50$ md.
From Eq. 12.32,

$$s_d = \frac{L_w}{b_H}\left(b_H\frac{\sqrt{k_x k_z}\,\Delta p_{t=0}}{141.2qB\mu} - s_c - s_p\right)$$

$$= \frac{(1,000)\sqrt{1,000}(29.4)}{(141.2)(800)(1.25)(1)} - \frac{1,000}{2,000}(s_p + s_c) = 6.584$$

$$- 0.5(s_c + s_p).$$

From Eq. 12.18,

$$s_c = \ln\left(\frac{h}{r_w}\right) + 0.25\ln\left(\frac{k_x}{k_z}\right) - \ln\left[\sin\left(\frac{\pi d_z}{h}\right)\right] - 1.838.$$

$$= \ln\left(\frac{150}{0.25}\right) + 0.25\ln\left(\frac{100}{10}\right) - \ln\left[\sin\left(\frac{65\pi}{150}\right)\right] - 1.838$$

$$= 5.16.$$

We now calculate the partial penetration skin, s_p, using the appropriate equation from among Eqs. 25 through 35 in Table 12.1. First, we calculate

$\frac{a_H}{\sqrt{k_x}}$, $\frac{b_H}{\sqrt{k_y}}$, and $\frac{h}{\sqrt{k_z}}$ to determine whether we have "Case 1" or "Case 2."

t, hours	p_{wf}, psi	$\Delta p = (p_i - p_{wf})$, psi	p', psi	t, hours	p_{wf}, psi	$\Delta p = (p_i - p_{wf})$, psi	p', psi
0.000	3000.000	0.000	0.000	1.523	2974.441	25.559	2.352
0.001	2994.053	5.947	3.063	1.707	2974.167	25.833	2.432
0.002	2991.735	8.265	3.691	1.913	2973.887	26.113	2.512
0.003	2990.026	9.974	3.837	2.143	2973.597	26.403	2.599
0.004	2988.755	11.245	3.382	2.401	2973.298	26.702	2.697
0.006	2987.818	12.182	2.994	2.690	2972.986	27.014	2.776
0.007	2987.091	12.909	2.762	3.014	2972.661	27.339	2.914
0.009	2986.487	13.513	2.640	3.376	2972.322	27.678	3.060
0.011	2985.958	14.042	2.567	3.782	2971.967	28.033	3.203
0.013	2985.487	14.513	2.588	4.237	2971.596	28.404	3.359
0.016	2985.046	14.954	2.549	4.746	2971.208	28.792	3.507
0.019	2984.635	15.365	2.516	5.317	2970.803	29.197	3.636
0.022	2984.249	15.751	2.481	5.956	2970.381	29.619	3.794
0.025	2983.881	16.119	2.447	6.671	2969.942	30.058	3.949
0.029	2983.530	16.470	2.416	7.473	2969.486	30.514	4.109
0.033	2983.192	16.808	2.401	8.370	2969.014	30.986	4.251
0.038	2982.865	17.135	2.389	9.376	2968.525	31.475	4.362
0.044	2982.548	17.452	2.375	10.502	2968.020	31.980	4.517
0.050	2982.239	17.761	2.364	11.763	2967.501	32.499	4.619
0.057	2981.937	18.063	2.345	13.175	2966.972	33.028	4.712
0.064	2981.640	18.360	2.314	14.757	2966.433	33.567	4.809
0.073	2981.348	18.652	2.313	16.529	2965.882	34.118	4.906
0.083	2981.062	18.938	2.314	18.514	2965.320	34.680	5.027
0.094	2980.779	19.221	2.315	20.736	2964.744	35.256	5.133
0.106	2980.500	19.500	2.307	23.225	2964.158	35.842	5.232
0.119	2980.223	19.777	2.297	26.013	2963.559	36.441	5.337
0.134	2979.949	20.051	2.270	29.135	2962.947	37.053	5.454
0.151	2979.677	20.323	2.278	32.632	2962.323	37.677	5.570
0.170	2979.407	20.593	2.280	36.549	2961.685	38.315	5.698
0.192	2979.140	20.860	2.282	40.936	2961.032	38.968	5.827
0.216	2978.874	21.126	2.278	45.849	2960.362	39.638	5.987
0.242	2978.609	21.391	2.244	51.351	2959.674	40.326	6.169
0.272	2978.345	21.655	2.251	57.514	2958.963	41.037	6.379
0.306	2978.082	21.918	2.265	64.417	2958.228	41.772	6.607
0.343	2977.821	22.179	2.270	72.148	2957.464	42.536	6.878
0.386	2977.561	22.439	2.268	80.807	2956.667	43.333	7.189
0.433	2977.301	22.699	2.223	90.504	2955.833	44.167	7.530
0.485	2977.042	22.958	2.244	101.366	2954.957	45.043	7.922
0.545	2976.783	23.217	2.251	113.530	2954.036	45.964	8.331
0.611	2976.525	23.475	2.257	127.155	2953.064	46.936	8.807
0.685	2976.268	23.732	2.266	142.414	2952.037	47.963	9.311
0.768	2976.011	23.989	2.264	159.505	2950.951	49.049	9.860
0.861	2975.752	24.248	2.240	178.647	2949.802	50.198	10.439
0.965	2975.494	24.506	2.269	200.085	2948.583	51.417	11.074
1.082	2975.235	24.765	2.283	224.096	2947.289	52.711	11.772
1.213	2974.974	25.026	2.314	240.481	2946.442	53.558	12.225
1.359	2974.710	25.290	2.341				

TABLE 12.5—PRESSURE DATA FOR WELL LLF-4

$$\frac{a_H}{\sqrt{k_x}} = \frac{6,968}{\sqrt{100}} = 696.8$$

$$\frac{b_H}{\sqrt{k_y}} = \frac{2,000}{\sqrt{50}} = 283; \ \frac{0.75 b_H}{\sqrt{k_y}} = 212$$

$$\frac{h}{\sqrt{k_z}} = \frac{150}{\sqrt{10}} = 47.4; \ \frac{0.75}{\sqrt{k_z}} = 35.6.$$

Because $\frac{a_H}{\sqrt{k_x}} > \frac{0.75 b_H}{\sqrt{k_y}} >> \frac{0.75h}{\sqrt{k_z}}$, this is Case 1 (Eq. 26), and we use Eqs. 25 through 31 from our table.

From Eq. 27, $s_p = p_{xyz} + p'_{xy}$.

From Eq. 25,

$$p_{xyz} = \left(\frac{b_H}{L_w} - 1\right)$$

$$\times \left\{ \ln\left(\frac{h}{r_w}\right) + 0.25 \ln\left(\frac{k_x}{k_z}\right) - \ln\left[\sin\left(\frac{\pi d_z}{h}\right) - 1.838\right]\right\}$$

$$= \left(\frac{2,000}{1,000}\right)$$

$$\times \left\{ \ln\left(\frac{150}{0.25}\right) + - 0.25 \ln\left(\frac{100}{10}\right) - \ln \sin\left(\frac{65\pi}{150}\right) - 1.838 \right\}$$

$$= 5.16.$$

From Eq. 28,

$$p'_{xy} = \frac{2b_H^2}{L_w h}\sqrt{k_z/k_y}$$

$$\times \left\{ F\left(\frac{L_w}{2b_H}\right) + 0.5\left[F\left(\frac{4y_m + L_w}{2b_H}\right) - F\left(\frac{4y_m - L_w}{2b_H}\right)\right]\right\}.$$

Here, from Eq. 29,

$$y_m = d_y + \frac{L_w}{2} = 500 + \frac{1,000}{2} = 1,000.$$

The well is centered, so $d_y = D_y = 500$ ft.

$$\frac{L_w}{2b_H} = \frac{1,000}{(2)(2,000)} = 0.25.$$

From Eq. 30,

$$F(u) = -u\left[0.145 + \ln(u) - 0.137(u)^2\right], u < 1$$

$$F\left(\frac{L_w}{2b_H}\right) = -(0.25)\left[0.145 + \ln(0.25) - 0.137(0.25)^2\right]$$

$$= 0.313.$$

Also, $\frac{4y_m + L_w}{2b_H} = \frac{(4)(1,000) + 1,000}{(2)(2,000)} = 1.25,$ and $\frac{4y_m - L_w}{2b_H} = 0.75.$

From Eq. 31,

$$F(u) = (2-u)\left[0.145 + \ln(2-u) - 0.137(2-u)^2\right], u > 1$$

$$F\left(\frac{4y_m + L_w}{2b_H}\right)$$

$$= (2 - 1.25)\left[0.145 + \ln(2 - 1.25) - 0.137(2 - 1.25)^2\right]$$

$$= -0.165.$$

and

$$F(u) = -u\left[0.145 + \ln(u) - 0.137(u)^2\right], u < 1$$

$$F\left(\frac{4y_m - L_w}{2b_H}\right) = -0.75\left[0.145 + \ln(0.75) - 0.137(0.75)^2\right]$$

$$= 0.165.$$

Then,

$$p'_{xy} = \frac{(2)(2,000)^2}{(1,000)(150)}\sqrt{\frac{10}{50}}[0.313 + 0.5(-0.165 - 0.165)]$$

$$= 3.53.$$

Then,

$$s_p = 5.16 + 3.53 = 8.69.$$

Thus, $s_d = 6.584 - 0.5(8.69 + 5.16) = -0.34.$

We now check the expected duration of the late-linear flow regime. The start is the larger of the values from Eqs. 12.27 and 12.28. From Eq. 12.27,

$$t_{\mathrm{Sllf}} = \frac{4,800\phi\mu c_t(D_y + L_w/4)^2}{k_y}$$

$$= \frac{4,800(0.2)(1)(1.5\times10^{-5})\left(500 + \frac{1,000}{4}\right)^2}{50}$$

$$= 162 \text{ hr.}$$

From Eq. 12.28,

$$t_{\mathrm{Sllf}} = \frac{1,800\phi\mu c_t D_z^2}{k_z}$$

$$= \frac{1,800(0.2)(1)(1.5\times10^{-5})(85)^2}{10}$$

$$= 3.90 \text{ hr.}$$

Thus, the start is expected to be at approximately 162 hours. The end of the flow regime is given by Eq. 12.29.

$$t_{\mathrm{Ellf}} = \frac{1,650\phi\mu c_t d_x^2}{k_x}$$

$$= \frac{1,650(0.2)(1)(1.5\times10^{-5})(3,484)^2}{100}$$

$$= 601 \text{ hr.}$$

The data in this example spanned the time range from 100 to 248 hours on the straight line. Some of the data that fall on the straight line are, in theory, from times before the start of the late-linear flow regime, but they appear to cause no problem in determining the slope of the line.

12.5 Summary of Analysis Procedures

Let us briefly summarize the procedures used in analyzing horizontal well-test data.

First, we calculate the vertical permeability, k_z, from data in the early-radial or hemiradial flow regimes; its calculation requires that we have independent knowledge of k_x. We cannot obtain an explicit expression for k_z alone.

We calculate the permeability in the k_y direction from data obtained from the pseudoradial flow regime; its calculation requires that we know k_x, the permeability in the direction perpendicular to the wellbore.

We can calculate the value of k_x alone if we have data in the early-linear or late-linear flow regimes.

To determine k_x from the early-linear flow regime data, we must know the effective completed wellbore length, L_w; to determine k_x from the late-linear flow regime, we must know the reservoir length, b_H, parallel to the wellbore. Estimating these quantities can be difficult.

We can calculate the effective wellbore length, L_w, from data in the early-linear flow regime if we have an independent estimate of k_x.

The length of the boundary, b_H, parallel to the wellbore can be calculated from data in the late-linear flow regime if k_x is known.

If some data (such as L_w or b_H) are unknown or if some of the flow regimes are missing, the analysis procedures based on identifying those flow regimes and making the kinds of calculations that we have been discussing is iterative at best and will result in nonunique results. Checks on expected durations of flow regimes using tentative results from the analysis of the buildup and drawdown tests are helpful to minimize ambiguity in these results.

We can assume $k_x = k_y = k_h$ and simplify the analysis in many cases, but the validity of this assumption is questionable.

The best solution can be to use information from the various flow regimes to determine the best estimates possible for reservoir and well properties and then use regression analysis and solutions to the flow equations for horizontal wells (available on most commercial well-test-analysis software) to try to improve the estimates of parameters. Alternatively, we may find that numerical models are required to take into account the nonideal effects of reservoir heterogeneities that are not included in analytical models.

12.6 Field Examples

We now consider three field examples reported by Lichtenberger.[4]

12.6.1 Field Example Well A. Table 12.6 summarizes the reservoir and completion properties for Well A.

The target for Well A, a horizontal exploration well, was vertical tectonic fracture development in a low-permeability shale. Because of the fractures, the permeability is assumed to be isotropic ($k_h = k_z$). We assume that the permeability of this formation is a result of the fractures. **Fig. 12.22** is a diagnostic plot for Well A and includes a history match using an analytical model.

TABLE 12.6—PARAMETERS FOR EXAMPLE WELL A	
L_d, ft	2,470
L_w, ft	—
r_w, ft	0.25
ϕ, %	5
h, ft	150
q, STB/D	104
B_o, RB/STB	1.40
μ, cp	0.45
t_p, hours	238

TABLE 12.7—WELL AND FORMATION PROPERTIES FOR EXAMPLE WELL B	
L_d, ft	2,000
L_w, ft	—
r_w, ft	0.30
ϕ, %	17
h, ft	75
q, STB/D	200
B_o, RB/STB	1.60
μ, cp	1.80
t_p, hours	1,320

During the early part of the tests, we observe a unit-slope line representing wellbore storage. Following that wellbore storage, we see a transition to radial flow. The final four data points may approach closely the start of radial flow.

On the Horner plot (**Fig. 12.23**), a straight line can be drawn through the last few data points. From the slope of the straight line, the apparent permeability is 0.011 md, and the altered zone skin is 2.9. There is no evidence of boundary effects on this Horner plot. The existence of the semilog straight line is not assured, but we are at least on the verge of reaching it, and the value of permeability from semilog analysis provides a starting point for regression analysis (history matching) of the data.

Guided by the Horner analysis results, engineers simulated these data with an analytical horizontal-well model. The initial match was poor. They improved the match considerably by introducing a no-flow boundary approximately 16 ft above the wellbore, which led to a permeability estimate of 0.027 md and an apparent skin of 11.5. They concluded that the flow regime observed in the test was hemiradial flow.

The final match, which we showed on our diagnostic plot in Fig. 12.22, is still not a good match at all times, but the author[4] stated that the poor match in the transition region could be the result of phase-redistribution effects in the wellbore. The distance to the no-flow boundary that led to the best match compared favorably with well survey data, which indicated that the well was drilled approximately 20 ft below the upper limit of the productive horizon.

12.6.2 Field Example Well B. Well B is in a west Texas carbonate formation. It was expected to have isotropic permeability caused by fracturing and dissolution. **Table 12.7** gives the field data for this well.

Fig. 12.24 is the diagnostic plot for this well. After wellbore storage, a short period of radial flow appears to be followed by the onset of linear flow because p' approaches a slope of 0.5.

In the time region where the derivative is horizontal, a straight line on the Horner plot (**Fig. 12.25**) yields $k = 0.14$ md. Using these results in the analysis shows that the end of radial flow occurs at $t_{Erf} = 165$ hours.

A tandem-root plot (**Fig. 12.26**) indicates linear flow and also suggests a distance to the nearest boundary of 29 ft. This is in good agreement with geological observations and helps to verify the assumption of isotropic permeability. The history match with an analytical horizontal-well model, shown in Fig. 12.24, confirms the results of the Horner and tandem-root plots.

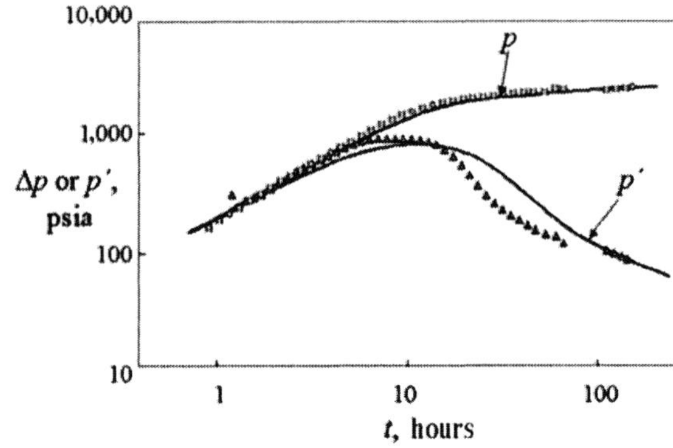

Fig. 12.22—Diagnostic plot for Well A.

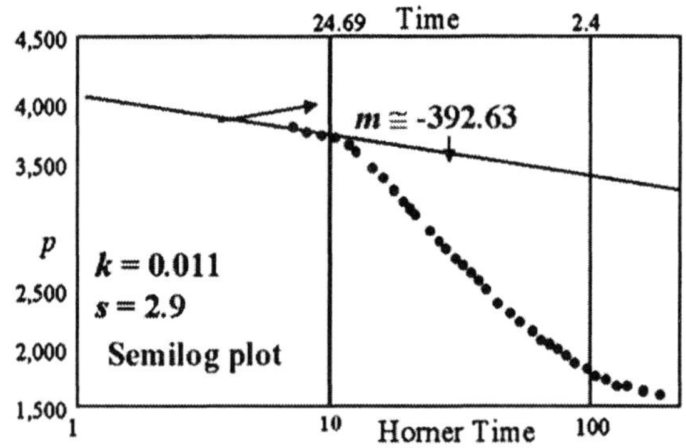

Fig. 12.23—The Horner plot suggests that a possible start of the middle-time region appears in test data.

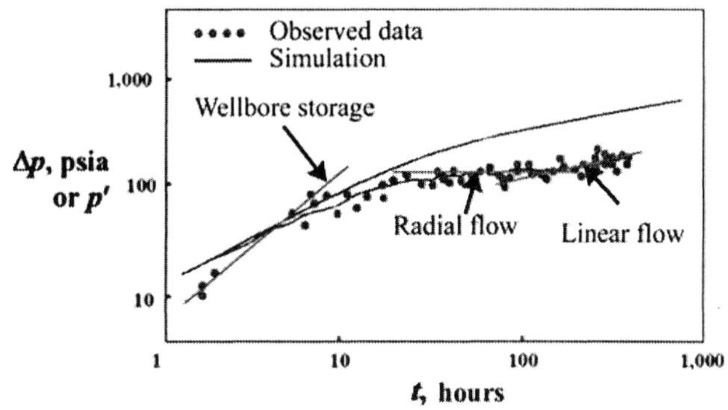

Fig. 12.24—The diagnostic plot for Well B indicates radial and possible linear flow regimes.

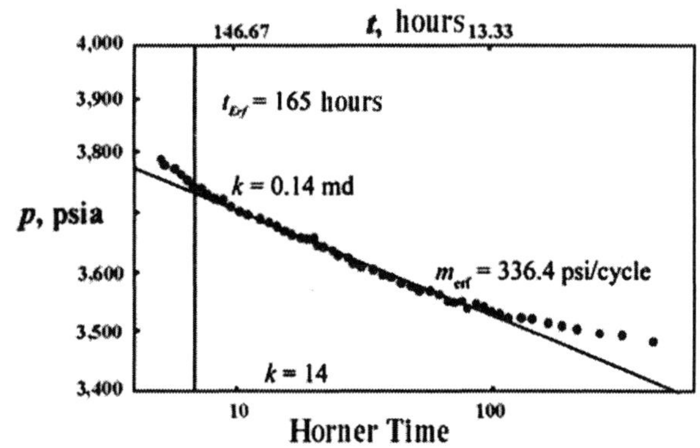

Fig. 12.25—Horner plot confirms early-radial flow.

12.6.3 Field Example Well C. Well C data are from a buildup test of a horizontal well in a high-permeability sandstone where a 54-ft oil column overlies an extensive aquifer estimated to be approximately 180 ft thick. **Table 12.8** shows the available data.

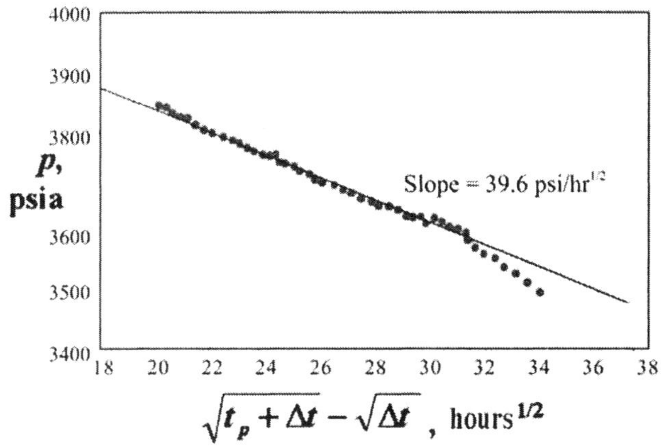

Fig. 12.26—Tandem-root plot confirms linear flow.

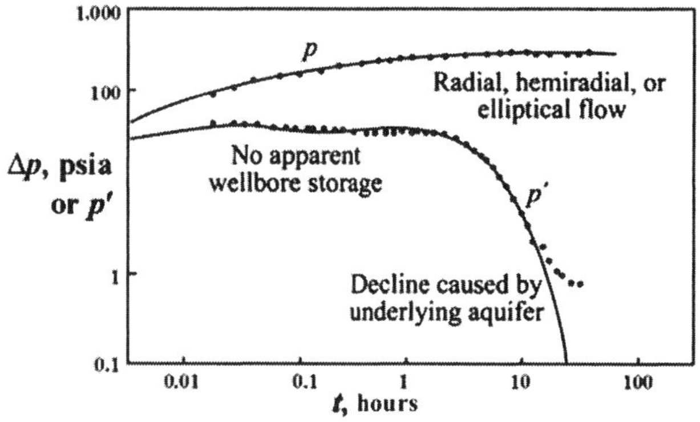

Fig. 12.27—The diagnostic plot for Well C shows no wellbore storage and constant derivative, with the aquifer underlying the oil column.

TABLE 12.8—WELL AND FORMATION PROPERTIES FOR EXAMPLE WELL C	
L_d, ft	1,400
L_w, ft	484
r_w, ft	0.41
ϕ, %	17
h, ft	54
q, STB/D	2,760
B_o, RB/STB	1.10
μ, cp	4.88
t_p, hours	36

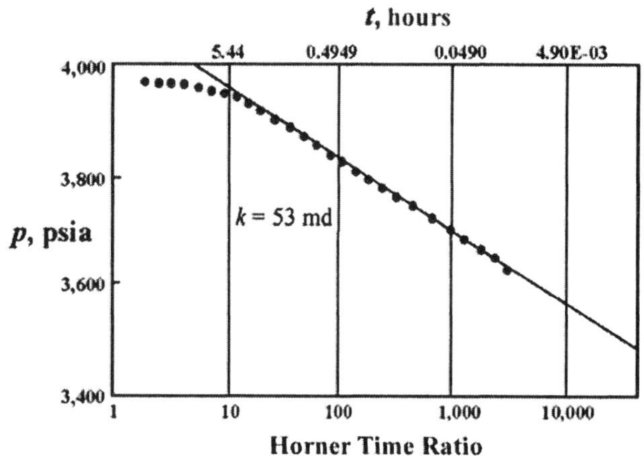

Fig. 12.28—The Horner plot gives a permeability of 53 md, close to that from regression analysis.

The diagnostic plot (**Fig. 12.27**) shows essentially no wellbore storage and a constant derivative, indicating radial, hemiradial, or elliptical flow at early times.

The rapid decline of the derivative at the end of the test is caused by the aquifer underlying the oil column, which is acting like a constant-pressure boundary. A history match, with an analytical horizontal-well model with one no-flow and one constant-pressure boundary (the lower boundary), yielded $k_h = 313$ md, $k_z = 7.5$ md, $s_a = -1.5$, and $L_w = 356$ ft. The no-flow boundary was estimated to be approximately 11 ft above the wellbore.

The time at which the early-radial flow regime ends—the time at which the derivative ceases to be flat on the diagnostic plot—is approximately 1.5 hours. For a wellbore with a volume of 130 bbl filled with a fluid of compressibility of 3.5×10^{-6} psi, the duration of wellbore storage (the unit-slope line) is estimated to be 0.0005 hours. With the gauge sampling rate set at 0.017 hours, the wellbore-storage unit slope simply could not be detected and does not appear at all on this plot.

Fig. 12.28 is the Horner plot for this test. A straight line appears in the same range as the flat derivative on a diagnostic plot. From the slope of the line, the permeability is estimated to be 53 md, close to the regression-analysis match value of 48 md. Note that this permeability, from the early-radial flow regime, represents the product $(k_h k_z)^{1/2}$.

12.7 Running Horizontal Well Tests

The measurements in horizontal wells are usually made above the wellbore with the pressure gauge still in the vertical section, as shown by **Fig. 12.29.**

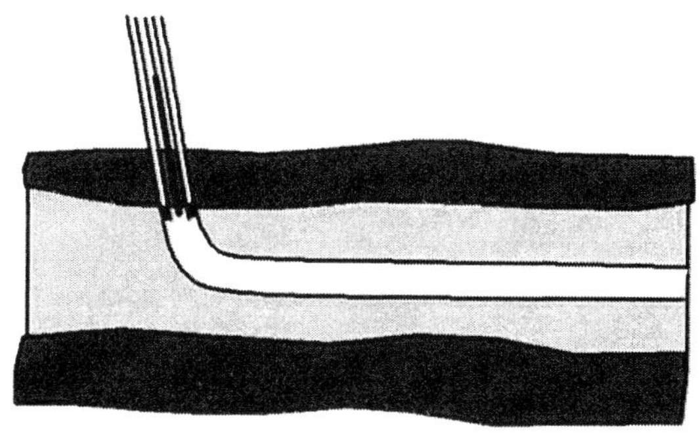

Fig. 12.29—Pressure-gauge test strings may be too rigid to allow measurement below the vertical section of horizontal wells.

Often, the test string may be too rigid to pass through the wellbore. However, in most cases, conventional hardware can be used for horizontal well tests. With longer horizontal wellbores, wellbore storage is an inherent problem for testing, even for buildup tests with downhole shut-in.

We have discussed problems in conducting buildup tests with short-duration production periods because superposition is inappropriate; therefore, Horner plots and tandem-root plots, which depend on superposition being applicable, are often inappropriate.

Another problem in conducting buildup tests is that significant pressure gradients along the length of the wellbore may cause cross-flow within the wellbore during shut-in, so fluid may flow from one region to another in the wellbore. These gradients can exist in new wells but are even more probable in developed fields. Heterogeneities cause uneven pressure gradients and uneven fluxes (flow rate per unit well length) along the wellbore.

12.7.1 Factors That Affect Transient Responses. A number of factors may affect the transient response of a horizontal well test:

- Horizontal permeability (normal and parallel to well trajectory).
- Vertical permeability.
- Drilling damage.
- Completion damage.
- Producing interval that effectively may be much less than drilled length.
- Variations in standoff along the length of the well.

In summary, seven or more factors may affect interpretation for horizontal wells in homogeneous reservoirs before the effects from boundaries. The problem is complex, so test results are frequently inconclusive. Furthermore, wellbore storage inhibits determination of properties associated with early-time transient behavior such as vertical permeability and damage from drilling and completion; middle- and late-time behavior may require several hours, days, or months to appear in transient data.

We can take some practical steps to help ensure that we will obtain interpretable test data. First, it is helpful to run tests in the pilot hole before kicking off to drill the horizontal borehole section. From a test in the vertical section, it is possible to get usable estimates of horizontal and vertical permeabilities with modern wireline-test tools. Second, a good directional drilling survey can frequently provide us with an adequate estimate of standoff. A production-log flow survey conducted with coiled tubing can determine which part of the wellbore is actually producing and, therefore, help give us an estimate of effective productive length. A flow survey is generally essential for valid test interpretation.

For high-rate wells, continuous borehole pressure and flow-rate measurements acquired during production can be used to interpret the pressure-drawdown transient response. If the downhole rates are not measured, the buildup test should be conducted with downhole shut-in to minimize wellbore-storage distortion of test data.

12.8 Estimating Horizontal Well Productivity

Estimating the productivity of a horizontal well accurately is even more difficult than estimating the productivity of a vertical well, mostly because of two fundamental problems: the theoretical models that we have available have a number of simplifying assumptions, and the data required for even these simplified models are not likely to be available. Still, we must make estimates and decisions based on those estimates. In this section, we will discuss two productivity models that have proven useful in practice. The first, published by Babu and Odeh[5] in 1989, is limited to single horizontal wells. The second, published by Economides *et al.*[6] in 1996, is more general and is useful for multilateral wells.

A major reason for discussing these productivity models is to illustrate the potential application of the results of pressure-transient-test analysis in horizontal wells. These models illustrate the minimum data requirements for productivity estimates. We caution the reader that actual productivity may differ significantly from the predictions of these simple problems. Beliveau[7] discusses this point in considerable detail.

12.8.1 Babu-Odeh Method. Babu and Odeh[5] obtained a rigorous solution to the diffusivity equation for a well in a box-shaped reservoir, subject to certain limiting assumptions. The assumptions include the following:

- Fluid flows to the well uniformly at all points along the wellbore (uniform flux), and the well is completed uniformly.
- The sides of the drainage volume are aligned with the principal permeability direction.
- The wellbore is parallel to the sides of the drainage area and is oriented parallel to one direction of principal permeability and perpendicular to the other two.
- The boundaries of the reservoir are all no-flow boundaries, and the well reaches stabilized, pseudosteady-state flow.
- The formation damage around the wellbore is uniform at all points along the wellbore.

Fig. 12.1 introduces the nomenclature in the Babu and Odeh solution. The solution is quite complex but is approximated accurately with an equation written in the same form as the pseudosteady-state flow equation for a vertical oil well producing a single-phase, slightly incompressible liquid.

$$q = \frac{0.00708 b_H \sqrt{k_x k_z}\left(\bar{p} - p_{wf}\right)}{B\mu\left[\ln\left(\frac{A^{1/2}}{r_w}\right) + \ln C_H - 0.75 + s_p + \left(\frac{b_H}{L_w}\right)s_d\right]} \quad \text{.................... (12.33)}$$

$$J = \frac{q}{\left(\bar{p} - p_{wf}\right)} = \frac{0.00708 b_H \sqrt{k_x k_z}}{B\mu\left[\ln\left(\frac{C_H A^{1/2}}{r_w}\right) - 0.75 + s_p + \left(\frac{b_H}{L_w}\right)s_d\right]} \quad \text{.................... (12.34)}$$

Equations to estimate C_H and s_p are given in Table 12.1. We illustrate the application of these equations with two examples adapted from Babu and Odeh.[5]

Example 1. A horizontal well 1,000 ft long (L_w) is drilled in a box-shaped drainage volume 4,000 ft long (a_H), 2,000 ft wide (b_H), and 100 ft thick (h). The well is off-center in the y-direction (parallel to the well), so $d_y = 250$ ft and $D_y = 750$ ft. The well is also off-center in the x-direction so that $d_x = 1{,}000$ ft and $D_x = 3{,}000$ ft. Finally, the well is centered in the z-direction so that $d_z = D_z = 50$ ft. The wellbore radius is 0.25 ft; $k_x = k_y = 200$ md, and $k_z = 50$ md. Fluid properties are $B_o = 1.25$ RB/STB and $\mu = 1$ cp. Calculate the productivity index.

We note that from Eq. 38 in Table 12.1,

$$\ln C_H = 6.28\frac{a_H}{h}\sqrt{\frac{k_z}{k_x}}\left[\frac{1}{3} - \frac{d_x}{a_H} + \left(\frac{d_x}{a_H}\right)^2\right] - \ln\left(\sin\frac{\pi d_z}{h}\right)$$

$$- 0.5\ln\left[\left(\frac{a_H}{h}\right)\sqrt{\frac{k_z}{k_x}}\right] - 1.088$$

$$= \frac{(6.28)(4,000)}{(100)}\sqrt{\frac{50}{200}}\left[\frac{1}{3} - \frac{1,000}{4,000} + \left(\frac{1,000}{4,000}\right)^2\right]$$

$$- \ln\left[\sin\left(\frac{50\pi}{100}\right)\right] - 0.5\ln\left[\left(\frac{4,000}{100}\right)\sqrt{\frac{50}{200}}\right] - 1.088$$

$$= 15.73.$$

We note that

$$\frac{a_H}{\sqrt{k_x}} = \frac{4,000}{\sqrt{200}} = 283; \frac{0.75 b_H}{\sqrt{k_y}} = \frac{(0.75)(2,000)}{\sqrt{200}} = 106;$$

and $\frac{0.75h}{\sqrt{k_z}} = \frac{(0.75)(100)}{\sqrt{50}} = 10.6.$

Thus, we use Case 1 equations (Eqs. 26 through 31) to calculate s_p.

$$s_p = p_{xyz} + p'_{xy}.$$

$$p_{xyz} = \left(\frac{b_H}{L_w} - 1\right)$$

$$\times \left\{\ln\left(\frac{h}{r_w}\right) + 0.25\ln\left(\frac{k_x}{k_z}\right) - \ln\left[\sin\left(\frac{\pi d_z}{h}\right)\right] - 1.838\right\}$$

$$= \left(\frac{2,000}{1,000} - 1\right)$$

$$\times \left\{\ln\left(\frac{100}{0.25}\right) + 0.25\ln\left(\frac{200}{50}\right) - \ln\left[\sin\left(\frac{50\pi}{100}\right)\right] - 1.838\right\}$$

$$= 4.50.$$

To calculate p'_{xy}, we need to determine y_m, $L_w/2b_H$, $(4y_m - L_w)/2b_H$, and $(4y_m + L_w)/2b_H$:

$$y_m = d_y + L_w/2 = 2.50 + 1,000 = 750 \text{ ft.}$$

$$L_w/2b_H = (1,000)/(2)(2,000) = 0.25 \text{ ft.}$$

$$(4y_m - L_w)/2b_H = [(4)(750) - 1,000]/(2)(2,000) = 0.5.$$

$$(4y_m + L_w)/2b_H = [(4)(750) + 1,000]/(2)(2,000) = 1.0.$$

Thus,

$$F(L_w/2b_H) = F(0.25) =$$

$$- (0.25)\left[0.145 + \ln(0.25) - 0.137(0.25)^2\right] = 0.313.$$

$$F[(4y_m - L_w)/2b_H] = F(1.0) =$$

$$- (0.5)\left[0.145 + \ln(0.5) - 0.137(0.5)^2\right] = 0.291.$$

$$F[(4y_m + L_w)/2b_H] = F(1.0) =$$

$$- (1.0)\left[0.145 + \ln(1.0) - 0.137(1.0)^2\right] = - 0.008.$$

Then,

$$p'_{xy} = \frac{2b_H^2}{L_w h}\sqrt{k_z/k_y}$$

$$\times \left\{F\left(\frac{L_w}{2b_H}\right) + 0.5\left[F\left(\frac{4y_m + L_w}{2b_H}\right) - F\left(\frac{4y_m - L_w}{2b_H}\right)\right]\right\}$$

$$= \frac{(2)(2,000)^2}{(1,000)(100)}\sqrt{\frac{50}{200}}[0.313 + 0.5(- 0.008 - 0.291)]$$

$$= 6.54.$$

Then $s_p = p_{xyz} + p'_{xy} = 4.50 + 6.54 = 11.0.$
and

$$J = \frac{q}{(\bar{p} - p_{wf})} = \frac{0.00708 b_H \sqrt{k_x k_z}}{B\mu\left[\ln\left(\frac{C_H A^{1/2}}{r_w}\right) - 0.75 + s_p + \left(\frac{b_H}{L_w}\right)s_d\right]}$$

$$= \frac{(7.08 \times 10^{-3})(2,000)\sqrt{(50)(200)}}{(1.25)(1)\left[\ln\left\{\frac{[(4,000)(100)]^{1/2}}{0.25}\right\} + 15.73 - 0.75 + 11.0\right]}$$

$$= 33.5 \text{ STB/D/psi.}$$

Example 2. A horizontal well 1,000 ft long (L_w) is drilled in a box-shaped reservoir 2,000 ft long (a_H), 4,000 ft wide (b_H), and 2,000 ft thick (h). The well is off-center in the y-direction ($d_y = 1,000$ ft; $D_y = 2,000$ ft), centered in the x-direction ($d_x = D_x = 1,000$ ft), and off-center in the z-direction ($d_z = 50$ ft; $D_z = 150$ ft). Permeabilities are $k_x = k_y = 100$ md and $k_z = 20$ md. Wellbore radius is 0.25 ft; $B_o = 1.25$ RB/STB; $\mu = 1$ cp; and $s_d = 0$. Find the productivity index, J.

From Eq. 38 (Table 12.1),

$$\ln C_H = 6.28\frac{a_H}{h}\sqrt{\frac{k_z}{k_x}}\left[\frac{1}{3} - \frac{d_x}{a_H} + \left(\frac{d_x}{a_H}\right)^2\right] - \ln\left(\sin\frac{\pi d_z}{h}\right)$$

$$- 0.5\ln\left[\left(\frac{a_H}{h}\right)\sqrt{\frac{k_z}{k_x}}\right] - 1.088$$

$$= \frac{(6.28)(2,000)}{(200)}\sqrt{\frac{20}{100}}\left[\frac{1}{3} - \frac{1,000}{2,000} + \left(\frac{1,000}{2,000}\right)^2\right]$$

$$- \ln\left[\sin\left(\frac{50\pi}{200}\right)\right] - 0.5\ln\left[\left(\frac{2,000}{200}\right)\sqrt{\frac{20}{100}}\right] - 1.088$$

$$= 0.850.$$

We note that

$$\frac{b_H}{\sqrt{k_y}} = \frac{4,000}{\sqrt{100}} = 400; \frac{1.33 a_H}{\sqrt{k_x}} = \frac{(1.33)(2,000)}{\sqrt{100}} = 266;$$

and $\dfrac{h}{\sqrt{k_z}} = \dfrac{(200)}{\sqrt{20}} = 44.7.$

Thus, we use Case 2 equations to calculate s_p.

$$s_p = p_{xyz} + p_y + p_{xy}.$$

$$p_{xyz} = \left(\frac{b_H}{L_w} - 1\right)$$

$$\times \left\{\ln\left(\frac{h}{r_w}\right) + 0.25\ln\left(\frac{k_x}{k_z}\right) - \ln\left[\sin\left(\frac{\pi d_z}{h}\right)\right] - 1.838\right\}$$

$$= \left(\frac{4,000}{1,000} - 1\right)$$

$$\times \left\{\ln\left(\frac{200}{0.25}\right) + 0.25\ln\left(\frac{100}{20}\right) - \ln\left[\sin\left(\frac{50\pi}{200}\right)\right] - 1.838\right\}$$

$$= 16.79.$$

To calculate p_y, we need to determine y_m. From Eq. 29 (Table 12.1),

$$y_m = d_y + \frac{L_w}{2} = 1,000 + \frac{1,000}{2} = 1,500 \text{ ft.}$$

$$p_y = \frac{6.28 b_H^2}{a_H h}\frac{\sqrt{k_x k_z}}{k_y}\left[\left(\frac{1}{3} - \frac{y_m}{b_H} + \frac{y_m^2}{b_H^2}\right) + \frac{L_w}{24 b_H}\left(\frac{L_w}{b_H} - 3\right)\right]$$

$$= \frac{6.28(4,000)^2}{(2,000)(200)}\frac{\sqrt{(100)(20)}}{100}$$

$$\times \left\{\left[\frac{1}{3} - \frac{1,500}{4,000} + \left(\frac{1,500}{4,000}\right)^2\right] + \frac{1,000}{(24)(4,000)}\left(\frac{1,000}{4,000} - 3\right)\right\}$$

$$= 7.90.$$

From Eq. 35 (Table 12.1),

$$p_{xy} = \left(\frac{b_H}{L_w} - 1\right)\left(\frac{6.28 a_H}{h}\sqrt{\frac{k_z}{k_x}}\right)\left(\frac{1}{3} - \frac{d_x}{a_H} + \frac{d_x^2}{a_H^2}\right),$$

$$d_x \geq 0.25a_H$$

$$= \left(\frac{4,000}{1,000} - 1\right)\left[\frac{(6.28)(2,000)}{200}\sqrt{\frac{20}{100}}\right]$$

$$\left[\frac{1}{3} - \frac{1,000}{2,000} + \left(\frac{1,000}{2,000}\right)^2\right]$$

$$= 7.02.$$

Thus, $s_p = 16.79 + 7.90 + 7.02 = 31.7$.

Then, from Eq. 37 (Table 12.1),

$$J = \frac{q}{(\bar{p} - p_{wf})} = \frac{0.00708 b_H \sqrt{k_x k_z}}{B\mu\left[\ln\left(\frac{C_H A^{1/2}}{r_w}\right) - 0.75 + s_p + \left(\frac{b_H}{L_w}\right)s_d\right]}$$

$$= \frac{(7.08 \times 10^{-3})(4,000)\sqrt{(100)(200)}}{(1.25)(1)\left\{\ln\left[\frac{(200)(2,000)}{0.25}\right] + 0.850 - 0.75 + 31.7 + 0\right\}}$$

$$= 25.6 \text{ STB/D/psi}.$$

12.8.2 Economides *et al.* Method. Economides *et al.*[6] presented a more general method to estimate the productivity index for a horizontal well. The method has the advantage that it is applicable to multilateral wells in the same plane and is not limited to wells aligned with principal permeabilities. It includes solutions for wells with no pressure drop in the wellbore (infinite conductivity, as opposed to wells with uniform flux). It has the disadvantage that it requires interpolation in a table in which only certain drainage-area shapes are given.

The basic working equation for the productivity index in this method is

$$J = \frac{\bar{k} b_H}{887.22 B\mu\left(p_D + \frac{b_H}{2\pi L_w}\Sigma s\right)} \qquad (12.35)$$

where Σs refers to damage skin, turbulence, and other pseudoskin factors. In Eq. 12.35,

$$p_D = \frac{b_H C_H}{4\pi h} + \frac{b_H}{2\pi L_w}s_c, \qquad (12.36)$$

where

$$s_c = \ln\left(\frac{h}{2\pi r_w}\right) - \frac{h}{6L_w} + s_e \qquad (12.37)$$

and s_e, describing eccentricity effects in the vertical direction, is

$$s_e = \frac{h}{L_w}\left[\frac{2d_z}{h} - \frac{1}{2}\left(\frac{2d_z}{h}\right)^2 - \frac{1}{2}\right] - \ln\left[\sin\left(\frac{\pi d_z}{h}\right)\right]. \qquad (12.38)$$

Note that $s_e = 0$ when a well is centered in the vertical plane. This convergence skin differs only slightly from that used by Babu and Odeh {the difference is

$$0.25\ln\left(\frac{k_x}{k_z}\right) + \frac{h}{L_w}\left[\frac{2d_z}{h} - \frac{1}{2}\left(\frac{2d_z}{h}\right)^2 - \frac{2}{3}\right],$$

which is usually small (<0.5)}.

TABLE 12.9—HORIZONTAL PLANE SHAPE FACTORS

Shape		L_w/b_H	C_H
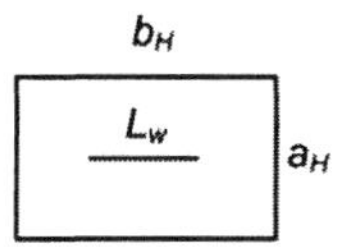	$b_H = 4a_H$	0.25	3.77
		0.5	2.09
		0.75	1.00
		1	0.26
	$b_H = 2a_H$	0.25	3.19
		0.5	1.80
		0.75	1.02
		1	0.52
	$b_H = a_H$	0.25	3.55
		0.5	2.21
		0.75	1.49
		1	1.04
	$2b_H = a_H$	0.25	4.59
		0.5	3.26
		0.75	2.53
		1	2.09
	$4b_H = a_H$	0.25	6.69
		0.5	5.35
		0.75	4.63
		1	4.18
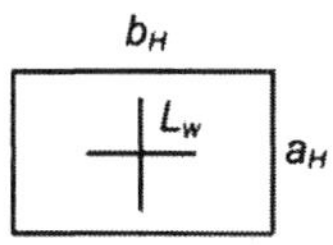	$b_H = a_H$	0.25	2.77
		0.5	1.47
		0.75	0.81
		1	0.46
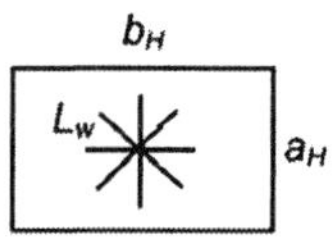	$b_H = a_H$	0.25	2.66
		0.5	1.36
		0.75	0.69
		1	0.32
		θ	
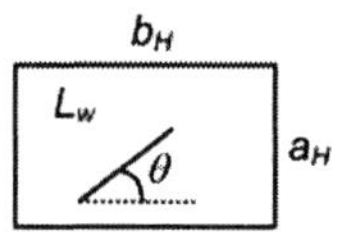	$b_H = a_H$, $L_w/b_H = 0.75$	0	1.49
		30	1.48
		45	1.48
		75	1.49
		90	1.49

Values of C_H are given in **Table 12.9** for several drainage areas and multilateral configurations.

The equations as written are for isotropic reservoirs; certain variable transformations are required before substituting into the working equation:

$$\text{Length: } L' = L_w \alpha^{-1/3}\beta. \qquad (12.39)$$

$$\text{Wellbore radius: } r_w' = r_w \frac{\alpha^{2/3}}{2}\left(\frac{1}{\alpha\beta} + 1\right). \qquad (12.40)$$

$$\text{where } \alpha = \sqrt{\frac{(k_x k_y)^{1/2}}{k_z}}. \qquad (12.41)$$

$$\text{and } \beta = \left(\sqrt{\frac{k_x}{k_y}}\cos^2\phi + \sqrt{\frac{k_y}{k_x}}\sin^2\phi\right)^{1/2}. \qquad (12.42)$$

[ϕ is the azimuth of the well trajectory (relative to the y-axis)].

Reservoir dimensions:

$$x' = x\frac{\sqrt{k_y k_z}}{\bar{k}}. \qquad (12.43)$$

$$y' = y\frac{\sqrt{k_x k_z}}{\bar{k}}. \qquad (12.44)$$

$$z' = z\frac{\sqrt{k_x k_y}}{\bar{k}}. \quad (12.45)$$

and

$$\bar{k} = \sqrt[3]{k_x k_y k_z}. \quad (12.46)$$

We illustrate with two examples, one from an isotropic reservoir and one from an anisotropic reservoir.

Example 1. Economides *et al.*[6] provide this example. Consider a horizontal well 1,500 ft long in a reservoir with $b_H = 2{,}000$ ft, $a_H = 4{,}000$ ft, $h = 20$ ft, $r_w = 0.4$, $k_x = k_y = k_z = 10$ md, $B_o = 1.25$ RB/STB, and $\mu = 1$ cp. Assume that the well is centered vertically so that $s_e = 0$. Also, assume $\Sigma s = 0$.

From Eq. 12.38,

$$s_c = \ln\left(\frac{h}{2\pi r_w}\right) - \frac{h}{6L_w} + s_e$$

$$= \ln\left[\frac{20}{2\pi(0.4)}\right] - \frac{20}{(6)(1,500)} + 0 = 2.07.$$

(As a matter of interest, the Babu and Odeh s_c for this case is also 2.07.)

From Table 12.9, for $2b_H = a_H = 1{,}500/2{,}000 = 0.75$, $C_H = 2.53$.

From Eq. 12.36,

$$p_D = \frac{b_H C_H}{4\pi h} + \frac{b_H}{2\pi L_w} s_c$$

$$= \frac{(2,000)(2.53)}{(4\pi)(20)} + \frac{(2,000)(2.07)}{(2\pi)(1,500)}$$

$$= 20.58.$$

Then, from Eq. 12.35,

$$J = \frac{\bar{k} b_H}{887.22 B\mu\left(p_D + \frac{b_H}{2\pi L_w}\Sigma s\right)}$$

$$= \frac{(10)(2,000)}{(887.22)(1.25)(1)(20.58 + 0)}$$

$$= 0.876 \text{ STB/D/psi}.$$

Example 2. Rework the Babu-Odeh Example 2 using the Economides *et al.* method.

We must transform the variables. From Eqs. 12.41 and 12.42,

$$\alpha = \sqrt{\frac{(k_x k_y)^{1/2}}{k_z}} = \sqrt{\frac{[(100)(100)]^{1/2}}{(20)}} = 2.236.$$

Because the well is parallel to the x-axis, $\phi = 0$ and

$$\beta = \left(\sqrt{\frac{k_x}{k_y}}\right)^{1/2} = 1.$$

From Eq. 12.46,

$$\bar{k} = \sqrt[3]{k_x k_y k_z} = 58.5 \text{ md}.$$

From Eq. 12.39,

$$L' = L_w \alpha^{-1/3}\beta = (1,000)(2,236)^{-1/3}(1) = 765 \text{ ft}.$$

From Eq. 12.40,

$$r_w' = r_w\frac{\alpha^{2/3}}{2}\left(\frac{1}{\alpha\beta} + 1\right) = (0.25)\frac{(2.236)^{2/3}}{2}$$

$$\times\left[\frac{1}{(2.236)(1)} + 1\right] = 0.309 \text{ ft}.$$

From Eqs. 12.43 through 12.45,

$$a_H' = a_H\frac{\sqrt{k_y k_z}}{\bar{k}} = \frac{2,000\sqrt{(100)(20)}}{58.5} = 1,529 \text{ ft}.$$

$$b_H' = b_H\frac{\sqrt{k_x k_z}}{\bar{k}} = \frac{4,000\sqrt{(100)(20)}}{58.5} = 3,058 \text{ ft}.$$

$$h' = h\frac{\sqrt{k_x k_y}}{\bar{k}} = \frac{200\sqrt{(100)(100)}}{58.5} = 342 \text{ ft}.$$

$$d_z' = d_z\frac{\sqrt{k_x k_y}}{\bar{k}} = \frac{(50)\sqrt{(100)(100)}}{58.5} = 85.5 \text{ ft}.$$

Thus, the equivalent system is a rectangular-shaped drainage area twice as long parallel to the wellbore (3,058 ft) as perpendicular, with $L'/b_H' = 765/3{,}058 = 0.25$. In the original example, the well was off-center in the horizontal plane; here, we will assume that a centered well is an adequate approximation. From Table 12.9,

$$C_H = 3.19.$$

From Eq. 12.38,

$$s_e = \frac{h'}{L_w'}\left[\frac{2d_z'}{h'} - \frac{1}{2}\left(\frac{2d_z'}{h'}\right)^2 - \frac{1}{2}\right] - \ln\left[\sin\left(\frac{\pi d_z'}{h'}\right)\right]$$

$$= \frac{342}{765}\left\{\frac{(2)(85.5)}{342} - \frac{1}{2}\left[\frac{(2)(85.5)}{342}\right]^2 - \frac{1}{2}\right\} - \ln\left[\sin\left(\frac{85.5\pi}{342}\right)\right]$$

$$= 0.291.$$

Then,

$$s_c = \ln\left(\frac{h'}{2\pi r_w'}\right) - \frac{h'}{6L_w'} + s_e$$

$$= \ln\left[\frac{342}{2\pi(0.309)}\right] - \frac{342}{(6)(765)} + 0.291$$

$$= 5.39.$$

(The Babu-Odeh s_c is 5.60 for this case.)

Then, from Eq. 12.36,

$$p_D = \frac{b_H' C_H}{4\pi h'} + \frac{b_H'}{2\pi L_w'} s_c$$

$$= \frac{(3,050)(3.14)}{4\pi(342)} + \frac{(3,050)(5.39)}{2\pi(765)}$$

$$= 5.70.$$

Finally, from Eq. 12.35,

$$J = \frac{\bar{k} b_H'}{887.22 B\mu\left(p_D + \frac{b_H'}{2\pi L_w'}\Sigma s\right)}$$

$$= \frac{(58.5)(3,058)}{(887.22)(1.25)(1.0)(5.70 + 0)}$$

$$= 028.3 \text{ STB/D/psi.}$$

The result is slightly larger than the result obtained with the Babu-Odeh method ($J = 25.6$ STB/D/psi). At least part of the reason for the difference is that, in this example, we were forced to assume that the well was centered in its drainage volume, which was not true in the original example. The optimal location of a horizontal well to maximize productivity is at the center of its drainage volume.

12.9 Comparison of Recent and Older Horizontal Well Models

Ozkan[8] compared "contemporary" (generally 1990s) and "conventional" horizontal well models in a paper published in 2001. He pointed out that the older models are used for both pressure-transient-test analysis and for estimating well productivity. Ozkan stressed three limitations of the conventional models, which include the Babu-Odeh model and other pioneering work.

Conventional models usually assume that the horizontal well is parallel to one of the principal permeability directions (preferably the minimum permeability direction in the horizontal plane). In many cases, this is not true; in fact, in many cases we do not even know what the principal permeability directions are. When the principal permeability directions are known, corrections to length are possible (as in the Economides *et al.* model); if they are not known, there is no way to correct the analysis. Contemporary models show that the error in permeability estimates approaches 50% when the deviation angle exceeds 50°. Unfortunately, the models also indicate that there is nothing in a well's response that provides any indication that the assumption that the well is parallel to a principal permeability direction is incorrect.

Ozkan pointed out that the damaged region around a horizontal well probably is nonuniform with distance (perhaps with the greatest damage near the heel of the well and the least near the toe because filtrate invasion is of much longer duration near the heel). If there is variable permeability along the path of the well, the situation is even more complicated. Some contemporary models can take this variation into account; most conventional models cannot. Conventional models usually assume (implicitly) uniform skin effect along the wellbore. However, the contemporary models will not be helpful if we do not know the skin distribution along the length.

Ozkan notes that it is a common practice to complete horizontal wells selectively. Also, in other cases, some segment of the well may not be open to the flow of reservoir fluids because of relatively low permeabilities or relatively large local skin effects. The absolute amount of the well that is open to flow and the location of the open intervals affect the pressure response in the well. Some contemporary well models can take these effects into account—but, again, the capabilities of the newer models may be limited if we do not know the location and length of the open intervals.

Many models assume negligible pressure drop in the wellbore (infinite conductivity). Others assume the same flow rate per unit length at all points along the wellbore (uniform flux). In fact, there is likely to be finite pressure drop in the wellbore, resulting in neither uniform flow nor infinite conductivity. Contemporary models in which a reservoir model is coupled to a wellbore model can take these effects into account.

Unfortunately, contemporary horizontal well models have not led to simple, easily applied methods of well-test analysis or of predicting well productivity. Further, their full utility depends on the availability of detailed well and reservoir description data. At present, the major uses of such models may be to quantify the possible errors that arise from our uncertainty and to history-match observed information when sufficient data are available.

Ultimately, numerical models are usually required for definitive test analysis in horizontal wells. Analytical models, such as the simpler ones described in this chapter and the more comprehensive ones discussed by Ozkan, include many simplifying assumptions. Realistic assessments of the effects of damage and permeability that vary along the wellbore usually make numerical models necessary. Beliveau[7] presented a helpful discussion that points out the danger of relying on simplified models to predict the productivity of horizontal wells. Rarely does actual productivity equal predicted productivity.

12.10 Chapter Summary

In this chapter, we review methods to analyze pressure-transient tests in horizontal wells and illustrate application of the results of these methods in estimating horizontal well productivity. We point out that horizontal well-test analysis is much more complicated than vertical well-test analysis because of 3D flow and the inadequacy of average k, h, and s to characterize the entire length of the well (proper modeling requires us to consider variations of these quantities along the length of the wellbore). The analytical models used as the basis for test analysis provide only starting estimates for estimating well productivity and for reservoir description needs in general. Numerical methods, which can treat heterogeneity, are ultimately required in most cases for test analysis and productivity estimates.

In Sec. 12.3, we discuss the flow regimes that can appear in some tests: early radial, hemiradial, early linear, late pseudoradial, and late linear. It would be a rare test in which all these flow regimes actually appeared; in some tests, none appear. In Sec. 12.4, we provide a summary of the governing equations for each flow regime and equations to estimate the start and end of each. We note that the diagnostic plot (a logarithmic plot of pressure change and pressure derivative vs. the appropriate time function) is essential in identifying the flow regimes appearing in a test. We illustrate analysis of each flow regime with "straight line," hand methods. Finally, we note that the linear plots and expected shapes of the diagnostic plots for pressure-buildup tests (in contrast to drawdown tests) may be obscured because of the difficulty in simple applications of superposition to the complex flow regimes in horizontal wells.

Sec. 12.5 summarizes the analytical procedures (and some of the difficulties) required for test analysis using the simple analytical models introduced in Sec. 12.4. Sec. 12.6 then illustrates the application of these procedures to actual field tests.

Sec. 12.7 discusses some of the practical aspects of conducting pressure-transient tests in horizontal wells. Sec. 12.8 then introduces an important application of the results of transient test analysis in horizontal wells, estimating productivity with "simple" analytical models. We discuss the classical Babu-Odeh model, applicable to wells producing with uniform flux into the wellbore from formations with properties (including skin) independent of spatial position along the well or in the reservoir. We then discuss the more realistic method of Economides *et al.*

We conclude with Sec. 12.9, which discusses the less-restrictive assumptions characteristic of post-1990 horizontal well models.

Exercises

1. Consider the data for the example Well Erf-1, given in Table 12.2. What are the most likely flow regimes that appear on the diagnostic plot (Fig. 12.9a) immediately before and after early-radial flow? Does the test include any possible data from the early-linear flow regime? If so, analyze these data as completely as possible. {Note: spherical flow is characterized by a slope of $-\frac{1}{2}$ on a diagnostic plot. The relationship[9] between the slope, m_s, and permeabilities on a plot of p_{wf} vs. $t^{-1/2}$ for data in spherical flow is

$$k_{sp}^{3/2} = 2,456 \frac{\sqrt{\phi \mu c_t} qB\mu}{m_s},$$

where $k_{sp} = \left[(k_r k_z)^{1/2}\right]^{2/3} = \left(k_x k_y k_z\right)^{1/3}$.} If you identify possible spherical flow in this test, analyze the data in this flow regime as completely as possible.

2. Consider the data for example Well Elf-2, given in Table 12.3. What are the most likely flow regimes that appear on the diagnostic plot (Fig. 12.15a) before and after early-linear flow? Does the test include any possible data from the early-radial flow regime? From the late-pseudoradial flow regime? From the late-linear flow re-

gime? If the answer to any of these questions is "yes," analyze the data as completely as possible.

3. Consider the data for example Well Prf-3, given in Table 12.4. What are the most likely flow regimes that appear on the diagnostic plot (Fig. 12.20a) before and after pseudoradial flow? Analyze the data from each flow regime as completely as possible.

4. Consider the data for example Well Llf-4, given in Table 12.5. What other flow regimes appear to be present in this test? Analyze the data from each flow regime as completely as possible.

5. Consider the data given for example Well Llf-4. Estimate the stabilized productivity index for this well using both the Babu-Odeh and Economides *et al.* methods. In addition, prepare a plot of $q/(p_i - p_{wf})$ for the well before stabilization (pseudosteady-state flow) and compare the stabilized value of J to the trend in the plot.

References

1. Goode, P.A. and Thambynayagam, R.K.M.: "Pressure Drawdown and Buildup Analysis of Horizontal Wells in Anisotropic Media," *SPEFE* (December 1987) 683; *Trans.*, AIME, **283.**
2. Odeh, A.S and Babu, D.K.: "Transient Flow Behavior of Horizontal Wells: Pressure Drawdown and Buildup Analysis," *SPEFE* (March 1990) 7; *Trans.*, AIME, **289.**
3. Kuchuk, F.J.: "Well Testing and Interpretation for Horizontal Wells," *JPT* (January 1995) 36; *Trans.*, AIME, **299.**
4. Lichtenberger, G.J.: "Data Acquisition and Interpretation of Horizontal Well Pressure-Transient Tests," *JPT* (February 1994) 157.
5. Babu, D.K. and Odeh, A.S.: "Productivity of a Horizontal Well," *SPERE* (November 1989) 417.
6. Economides, M.J., Brand, C.W., and Frick, T.P.: "Well Configurations in Anisotropic Reservoirs," *SPEFE* (December 1996) 257.
7. Beliveau, D.: "Heterogeneity, Geostatistics, Horizontal Wells, and Blackjack Poker," *JPT* (December 1995) 1068; *Trans.*, AIME, **299.**
8. Ozkan, E.: "Analysis of Horizontal-Well Responses: Contemporary vs. Conventional," *SPEREE* (August 2001) 260.
9. Joseph, J.A. and Koederitz, L.F.: "Unsteady-State Spherical Flow With Storage and Skin," *SPEJ* (December 1985) 804.

Appendix A
Dimensionless Groups

This appendix shows how dimensionless groups are chosen for different conditions.

Constant-Rate Production—No Wellbore Storage

The differential equation is

$$\frac{1}{r}\frac{\partial}{\partial r}\left(r\frac{\partial p}{\partial r}\right) = \frac{\phi\mu c_t}{0.0002637k}\frac{\partial p}{\partial t}. \quad \text{(A-1)}$$

The initial condition is

$$p(r,t = 0) = p_i. \quad \text{(A-2)}$$

The outer-boundary condition with no flow boundary is

$$\left(\frac{\partial p}{\partial r}\right)_{(r=r_e)} = 0. \quad \text{(A-3)}$$

The inner-boundary condition is

$$\left(r\frac{\partial p}{\partial r}\right)_{(r=r_w)} = \frac{141.2qB\mu}{kh}. \quad \text{(A-4)}$$

For this problem, the simplest length scale to work with is to use r_w as a reference length. We therefore define dimensionless radius as

$$r_D = \frac{r}{r_w}. \quad \text{(A-5)}$$

Because we need to find a dimensionless pressure grouping, we work with the inner-boundary condition, which is independent of time. Substituting Eq. A-5 into Eq. A-4 gives

$$\left[r_D r_w \frac{\partial p}{(r_D r_w)}\right]_{(r_D=1)} = \frac{141.2qB\mu}{kh} \quad \text{(A–6)}$$

and $$\left(\frac{\partial p}{\partial r_D}\right)_{(r_D=1)} = \frac{141.2qB\mu}{kh}. \quad \text{(A-7)}$$

If we let $p'_D = p/p_{ch}$, where p_{ch} = characteristic pressure, and substitute p'_D into the inner-boundary condition, we obtain

$$\left(p_{ch}\frac{\partial p'_D}{\partial r_D}\right)_{(r_D=1)} = \frac{141.2qB\mu}{kh}. \quad \text{(A-8)}$$

Looking at the initial condition, we see that a more convenient definition of dimensionless pressure, p_D, would have a value of 0 when $p = p_i$. We would also like p_D to increase with time and be positive. This suggests the following form of p_D.

$$p_D = \frac{p_i - p}{p_{ch}}, \quad \text{(A-9)}$$

which would result in an initial condition of

$$p_D(r_D, t_D = 0) = 0. \quad \text{(A-10)}$$

From Eq. A-10, the derivative of p_D with respect to r_D is

$$\frac{\partial p_D}{\partial r_D} = -\frac{1}{p_{ch}}\frac{\partial p}{\partial r_D}. \quad \text{(A-11)}$$

Substituting Eq. A-11 into Eq. A-7, the equation describing the inner-boundary condition, gives

$$-\left(p_{ch}\frac{\partial p_D}{\partial r_D}\right)_{(r_D=1)} = \frac{141.2qB\mu}{kh} \quad \text{(A-12)}$$

and $$\left(\frac{\partial p_D}{\partial r_D}\right)_{(r_D=1)} = \frac{141.2qB\mu}{khp_{ch}}. \quad \text{(A-13)}$$

This inner-boundary condition is simplified with

$$p_{ch} = \frac{141.2qB\mu}{kh}, \quad \text{(A-14)}$$

and the inner-boundary condition becomes

$$\left(\frac{\partial p_D}{\partial r_D}\right)_{(r_D=1)} = -1. \quad \text{(A-15)}$$

The definition of dimensionless pressure for this case is given by substituting the characteristic pressure given in Eq. A-14 into Eq. A-9.

$$p_D = \frac{kh}{141.2qB\mu}(p_i - p). \quad \text{(A-16)}$$

Now, write the partial-differential equation, Eq. A-1, in dimensionless form, remembering the definitions for r_D and p_D given by Eqs. A-5 and A-9, respectively.

$$\frac{1}{r}\frac{\partial}{\partial r}\left(r\frac{\partial p}{\partial r}\right) = \frac{\phi\mu c_t}{0.0002637k}\frac{\partial p}{\partial t}, \quad \text{(A-1)}$$

$$r_D = \frac{r}{r_w}, \quad \text{(A-5)}$$

and $p_D = \dfrac{p_i - p}{p_{ch}}$. .... (A-9)

Substitute Eq. A-5 into Eq. A-1.

$$\frac{1}{r_D r_w}\frac{\partial}{\partial(r_D r_w)}\left[r_D r_w \frac{\partial p}{\partial(r_D r_w)}\right]$$

$$= \frac{\phi\mu c_t}{0.0002637k}\frac{\partial p}{\partial t} \quad \text{(A-17a)}$$

and $\dfrac{1}{r_D}\dfrac{\partial}{\partial r_D}\left(r_D\dfrac{\partial p}{\partial r_D}\right) = \dfrac{\phi\mu c_t r_w^2}{0.0002637k}\dfrac{\partial p}{\partial t}$. .... (A-17b)

Substitute Eq. A-9 into Eq. A-17.

$$\frac{1}{r_D}\frac{\partial}{\partial r_D}\left[r_D\frac{\partial p}{\partial r_D}(-p_D p_{ch} + p_i)\right] = \frac{\phi\mu c_t r_w^2}{0.0002637k}\frac{\partial}{\partial t}(p_i - p_D p_{ch}). \quad \text{(A-18)}$$

Note that $\partial p_i/\partial r_D$ and $\partial p_i/\partial t = 0$, and both sides contain $(-p_{ch})$, which cancels; the equation now becomes

$$\frac{1}{r_D}\frac{\partial}{\partial r_D}\left(r_D\frac{\partial p_D}{\partial r_D}\right) = \frac{\phi\mu c_t r_w^2}{0.0002637k}\frac{\partial p_D}{\partial t}. \quad \text{(A-19)}$$

Now, try to obtain a dimensionless group for time. First, consider the simple grouping $t_D = t/t_{ch}$, where t_{ch} = characteristic time. With this grouping, Eq. A-19 becomes

$$\frac{1}{r_D}\frac{\partial}{\partial r_D}\left(r_D\frac{\partial p_D}{\partial r_D}\right) = \frac{\phi\mu c_t r_w^2}{0.0002637k}\frac{\partial p_D}{\partial(t_D t_{ch})}. \quad \text{(A-20)}$$

If $t_{ch} = \phi\mu c_t r_w^2/0.0002637k$, the partial-differential equation simplifies to

$$\frac{1}{r_D}\frac{\partial}{\partial r_D}\left(r_D\frac{\partial p_D}{\partial r_D}\right) = \frac{\partial p_D}{\partial t_D}, \quad \text{(A-21)}$$

and the definition of dimensionless time becomes

$$t_D = \frac{0.0002637kt}{\phi\mu c_t r_w^2}. \quad \text{(A-22)}$$

So far, we have written the partial-differential equation, the initial condition, and the inner-boundary condition in dimensionless form. We must now write the outer-boundary condition, given by Eq. A-3, in dimensionless terms.

$$\left(\frac{\partial p}{\partial r}\right)_{(r=r_e)} = 0. \quad \text{(A-3)}$$

First, with $r_D = r/r_w$, Eq. A-3 becomes

$$\left(\frac{1}{r_w}\frac{\partial p}{\partial r_D}\right)_{\left(r_D=\frac{r_e}{r_w}\right)} = 0. \quad \text{(A-23)}$$

$r_w \neq 0$, therefore

$$\left(\frac{\partial p}{\partial r_D}\right)_{\left(r_D=\frac{r_e}{r_w}\right)} = 0. \quad \text{(A-24)}$$

Recall from before that

$$\frac{\partial p}{\partial r_D} = -p_{ch}\frac{\partial p_D}{\partial r_D}; \quad \text{(A-25)}$$

therefore,

$$-p_{ch}\left(\frac{\partial p_D}{\partial r_D}\right)_{\left(r_D=\frac{r_e}{r_w}\right)} = 0. \quad \text{(A-26)}$$

Because $p_{ch} \neq 0$,

$$\left(\frac{\partial p_D}{\partial r_D}\right)_{\left(r_D=\frac{r_e}{r_w}\right)} = 0. \quad \text{(A-27)}$$

We can define a dimensionless outer radius, r_{eD}, as

$$r_{eD} = \frac{r_e}{r_w}; \quad \text{(A-28)}$$

therefore, the outer-boundary condition in dimensionless form is given by

$$\left(\frac{\partial p_D}{\partial r_D}\right)_{r_{eD}} = 0. \quad \text{(A-29)}$$

In summary, for constant-rate production with no wellbore storage, the partial-differential equation and boundary conditions are as follows.

Partial-Differential Equation.

$$\frac{1}{r_D}\frac{\partial}{\partial r_D}\left(r_D\frac{\partial p_D}{\partial r_D}\right) = \frac{\partial p_D}{\partial t_D}. \quad \text{(A-21)}$$

Initial Condition.

$$p_D(r_D, t_D = 0) = 0. \quad \text{(A-10)}$$

Outer-Boundary Condition—No-Flow Boundary.

$$\left(\frac{\partial p_D}{\partial r_D}\right)_{r_{eD}} = 0. \quad \text{(A-29)}$$

Inner-Boundary Condition—Constant–Rate Production.

$$\left(\frac{\partial p_D}{\partial r_D}\right)_{(r_D=1)} = -1. \quad \text{(A-15)}$$

Dimensionless variables are defined as

$$p_D = \frac{kh}{141.2qB\mu}(p_i - p), \quad \text{(A-16)}$$

$$r_D = \frac{r}{r_w}, \quad \text{(A-5)}$$

$$r_{eD} = \frac{r_e}{r_w}, \quad \text{(A-28)}$$

and $t_D = \dfrac{0.0002637kt}{\phi\mu c_t r_w^2}$. .... (A-22)

Constant-Rate Production With Wellbore Storage

The only change here is the inner-boundary condition, which becomes

$$q = \frac{kh}{141.2B\mu}\left(r\frac{\partial p}{\partial r}\right)_{(r=r_w)} - \frac{24C}{B}\frac{dp_w}{dt}. \quad \text{(A-30)}$$

Dividing Eq. A-30 by q gives

$$1 = \frac{kh}{141.2qB\mu}\left(r\frac{\partial p}{\partial r}\right)_{(r=r_w)} - \frac{24C}{qB}\frac{dp_w}{dt}. \quad \text{(A-31)}$$

Recall the definition of dimensionless pressure given by Eq. A-16 and the definition of dimensionless radius given by Eq. A-5.

$$p_D = \frac{kh}{141.2qB\mu}(p_i - p) \quad \text{(A-16)}$$

and $r_D = \dfrac{r}{r_w}$. .... (A-5)

If we substitute Eqs. A-16 and A-5 into Eq. A-31, we obtain

$$1 = -\frac{141.2qB\mu}{kh}\left(\frac{kh}{141.2qB\mu}\right)\left(r_D\frac{\partial p_D}{\partial r_D}\right)_{(r_D=1)} - \frac{24C}{qB}\frac{dp_w}{dt}. \quad \text{(A-32)}$$

We can then write

$$1 = -\left(r_D\frac{\partial p_D}{\partial r_D}\right)_{(r_D=1)} - \left[-\frac{24(141.2qB\mu C)}{qBkh}\right]\frac{dp_{wD}}{dt}, \quad \text{(A-33)}$$

where $p_{wD} = \frac{kh}{141.2qB\mu}(p_i - p_w).$ .... (A-34)

Simplifying Eq. A-33 gives

$$1 = \left[-\frac{24(141.2)\mu C}{kh}\right]\frac{dp_{wD}}{dt} - \left(r_D\frac{\partial p_D}{\partial r_D}\right)_{(r_D=1)}. \quad \text{(A-35)}$$

Eq. A-22 gives the definition of dimensionless time.

$$t_D = \frac{0.0002637kt}{\phi\mu c_t r_w^2}. \quad \text{(A-22)}$$

Thus, if we substitute t from Eq. A-22 into Eq. A-35, we can write

$$1 = \left[\frac{24(141.2)\mu Ck(0.0002637)}{kh\phi\mu c_t r_w^2}\right]\frac{dp_{wD}}{dt_D} - \left(r_D\frac{\partial p_D}{\partial r_D}\right)_{(r_D=1)}. \quad \text{(A-36)}$$

To simplify this inner-boundary condition, a dimensionless wellbore-storage coefficient, C_D, must be defined, with the simple grouping

$$C_D = \frac{C}{C_{ch}}, \quad \text{(A-37)}$$

where C_{ch} is the characteristic wellbore-storage coefficient. Now, substitute Eq. A-37 into Eq. A-36 to find a grouping of variables to define as C_{ch}.

$$1 = \left[\frac{24(141.2)(0.0002637)}{kh\phi c_t r_w^2}\right]C_D C_{ch}\frac{dp_{wD}}{dt_D} - \left(r_D\frac{\partial p_D}{\partial r_D}\right)_{(r_D=1)}. \quad \text{(A-38)}$$

Simplifying Eq. A-38, let

$$C_{ch} = \frac{h\phi c_t r_w^2}{24(141.2)(0.0002637)} \quad \text{(A-39)}$$

or $C_{ch} = \frac{h\phi c_t r_w^2}{0.8936}.$ .... (A-40)

Therefore, the definition of the dimensionless wellbore-storage coefficient is given by

$$C_D = \frac{0.8936C}{h\phi c_t r_w^2}, \quad \text{(A-41)}$$

and the inner-boundary condition can be expressed in dimensionless form by substituting Eq. A-41 into Eq. A-36 to obtain

$$1 = C_D\frac{dp_{wD}}{dt_D} - \left(r_D\frac{\partial p_D}{\partial r_D}\right)_{(r_D=1)}. \quad \text{(A-42)}$$

Constant-Rate Production With Wellbore Storage and Skin

The initial condition remains the same.

$$p_D(r_D, t_D = 0) = 0. \quad \text{(A-43)}$$

The outer-boundary condition could be either infinite acting,

$$p_D(r_D \to \infty, t_D) = 0, \quad \text{(A-44)}$$

or a no-flow boundary,

$$\left(\frac{\partial p_D}{\partial r_D}\right)_{r_{eD}} = 0. \quad \text{(A-45)}$$

There are now two inner-boundary conditions, one for wellbore storage, which was defined by Eq. A-42 as

$$1 = C_D\frac{dp_{wD}}{dt_D} - \left(r_D\frac{\partial p_D}{\partial r_D}\right)_{(r_D=1)}, \quad \text{(A-42)}$$

and one for skin, which is derived as follows.

We know that for constant-rate production the definition of dimensionless wellbore pressure, p_{wD}, is

$$p_{wD}(t_D) = \frac{kh}{141.2qB\mu}(p_i - p_{wf}). \quad \text{(A-46)}$$

The dimensionless pressure at a radius r_s is given by

$$p_D(r_{sD}, t_D) = \frac{kh}{141.2qB\mu}[p_i - p(r_s)]. \quad \text{(A-47)}$$

Skin factor is defined from Sec. 1.2.3 as

$$s = \frac{kh}{141.2qB\mu}\left[p(r_s) - p_{wf}\right]. \quad \text{(A-48)}$$

Combining Eqs. A-46 through A-48,

$$p_{wD}(t_D) = p_D(r_{sD}, t_D) + s, \quad \text{(A-49)}$$

or the dimensionless inner-boundary condition for skin as $r_s \to r_w$ becomes

$$p_{wD}(t_D) = p_D(1, t_D) + s. \quad \text{(A-50)}$$

Linear Flow

The partial differential equation for linear flow in only the x direction is

$$\frac{\partial^2 p}{\partial x^2} = \frac{\phi\mu c_t}{0.0002637k}\frac{\partial p}{\partial t}. \quad \text{(A-51)}$$

The initial condition is that all pressures throughout the reservoir are uniform before production begins.

$$p(x, t = 0) = p_i. \quad \text{(A-52)}$$

Assuming constant-rate production, the inner-boundary condition is

$$q = \left|\frac{kA}{2\pi(141.2)B\mu}\frac{\partial p}{\partial x}\right|_{(x=0,t>0)}. \quad \text{(A-53)}$$

The area exposed to flow for a linear system would be cross-sectional area A, where $A = wh$ (in general), w = width of system (=4 L_f = fracture half-length for fractured wells), and h = thickness.

For transient, or unsteady-state flow, the outer-boundary condition would be "infinite-acting."

$$p(x \to \infty, t) = p_i. \quad \text{(A-54)}$$

To generalize and simplify the equations, we can nondimensionalize as follows.

Dimensionless Pressure.

$$p_D = \frac{k\sqrt{A}}{2\pi(141.2)B\mu q}(p_i - p). \quad \text{(A-55)}$$

Dimensionless Time.

$$t_D = \frac{0.0002637kt}{\phi\mu c_t A}. \qquad \text{(A-56)}$$

Dimensionless Length.

$$x_D = \frac{x}{\sqrt{A}}. \qquad \text{(A-57)}$$

With these dimensionless variables defined by Eqs. A-54 through A-56, the partial-differential equation becomes

$$\frac{\partial^2\left[-\dfrac{q\mu B(2\pi)(141.2)}{k\sqrt{A}}p_D + p_i\right]}{\partial\left(x_D\sqrt{A}\right)^2} = \frac{\phi\mu c_t}{0.0002637k}$$

$$\times\left\{\frac{\partial\left[-\dfrac{q\mu B(2\pi)(141.2)}{k\sqrt{A}}p_D + p_i\right]}{\partial\left(\dfrac{\phi\mu c_t A t_D}{0.0002637k}\right)}\right\} \qquad \text{(A-58a)}$$

which becomes

$$\frac{\partial^2 p_D}{\partial x_D^2}\left[\frac{q\mu B(2\pi)(141.2)}{A\sqrt{A}\,k}\right] = \frac{\phi\mu c_t}{0.0002637k}\left(\frac{\partial p_D}{\partial t_D}\right)$$

$$\times\left[\frac{qB\mu(2\pi)(141.2)(0.0002637)k}{k\sqrt{A}\,\phi\mu c_t A}\right], \qquad \text{(A-58b)}$$

and which simplifies to

$$\frac{\partial^2 p_D}{\partial x_D^2} = \left(\frac{\partial p_D}{\partial t_D}\right). \qquad \text{(A-59)}$$

The initial condition becomes

$$p_D(x_D, t_D = 0) = 0. \qquad \text{(A-60)}$$

The outer-boundary condition becomes

$$p_D(x_D \to \infty, t_D) = 0. \qquad \text{(A-61)}$$

Nondimensionalizing the inner-boundary condition,

$$\frac{\partial p_D}{\partial x_D}\left[-\frac{q\mu B(2\pi)(141.2)}{\sqrt{A}\,k\sqrt{A}}\right] = \frac{q(2\pi)(141.2)B\mu}{kA}. \qquad \text{(A-62)}$$

Therefore, the inner-boundary condition becomes

$$\left(\frac{\partial p_D}{\partial x_D}\right)_{(x_D=1)} = -1. \qquad \text{(A-63)}$$

Appendix B
Solutions to the Radial-Flow Diffusivity Equation

Introduction

This appendix provides some general mathematical background, which is required in many of the sections in Chap. 1. The radial-flow diffusivity equation presented in Chap. 1 is a form of the modified Bessel equation; thus, we present the general solution to the modified Bessel equation and some background information on Laplace transformations.

We then present the solutions to the radial-flow diffusivity equation for different boundary and initial conditions. These conditions include an infinite-acting reservoir, line-source well, constant-rate production, and a finite wellbore with constant-rate production (infinite-acting, no-flow outer boundary with pseudosteady-state flow, constant-pressure outer boundary, and wellbore storage and skin).

Finally, we present the solution for the linear-flow case as it is used in hydraulically fractured wells.

Modified Bessel Equation and Its General Solution

The modified Bessel equation, of order zero, has the form[1]

$$\frac{1}{x}\frac{\mathrm{d}}{\mathrm{d}x}\left(x\frac{\mathrm{d}v}{\mathrm{d}x}\right) - \lambda v = 0 \quad \text{(B-1)}$$

and the general solution

$$v = AI_0\left(\sqrt{\lambda}\,x\right) + BK_0\left(\sqrt{\lambda}\,x\right), \quad \text{(B-2)}$$

where $I_0(x)$ = modified Bessel function of the first kind, order zero, and $K_0(x)$ = modified Bessel function of the second kind, order zero.

We must know some things about these functions to use them to solve the diffusivity equation for flow through porous media.

Derivatives and Behavior of the Modified Bessel Functions. These are recurrence relationships[2] for the first derivatives of the modified Bessel functions.

$$\frac{\mathrm{d}}{\mathrm{d}x}I_0[f(x)] = f''(x)I_1[f(x)] \quad \text{(B-3)}$$

$$\text{and } \frac{\mathrm{d}}{\mathrm{d}x}K_0[f(x)] = -f'(x)K_1[f(x)], \quad \text{(B-4)}$$

where $I_1(x)$ = modified Bessel function of the first kind, first order, and $K_1(x)$ = modified Bessel function of the second kind, first order.

We also must know something about the behavior of these functions. **Fig. B-1** shows this behavior with respect to the argument of the function. Note also that

$$\lim_{x\to 0} xK_1(x) = 1. \quad \text{(B-5)}$$

Laplace Transformations and Their Use in Solving Partial-Differential Equations

Nomenclature and Definition of the Laplace Transform. The Laplace transform of a function $v(r,t)$ is denoted by $\mathcal{L}(v)$, or $\overline{v}(r,u)$. In our work, for simplicity, we use $\overline{v}(r,u)$, where u is the Laplace transform parameter. In some works, the Laplace transform parameter is given the symbol s or p. These symbols are confusing to use in well testing because they become confused with skin factor (also s) and pressure (also p). When we use tables of Laplace transforms, we must remember to use u for s or p accordingly.

The definition of the Laplace transformation is[1]

$$\mathcal{L}[v(r,t)] = \overline{v}(r,u) = \int_0^\infty v(r,t)e^{-ut}\mathrm{d}t. \quad \text{(B-6)}$$

For example,

$$\mathcal{L}(1) = \int_0^\infty e^{-ut}\mathrm{d}t = \left.\frac{e}{-u}\right|_0^\infty = 0 - \frac{1}{(-u)} = \frac{1}{u} \quad \text{(B-7a)}$$

$$\text{and } \mathcal{L}(t) = \int_0^\infty te^{-ut}\mathrm{d}t. \quad \text{(B-7b)}$$

Here, we need to integrate by parts: $\int x\mathrm{d}y = xy - \int y\,\mathrm{d}x$. Let $x = t$; then $\mathrm{d}x = \mathrm{d}t$.

$$\mathrm{d}y = e^{-ut}\mathrm{d}t, \text{ then } y = \left(\frac{e^{-ut}}{-u}\right). \quad \text{(B-8)}$$

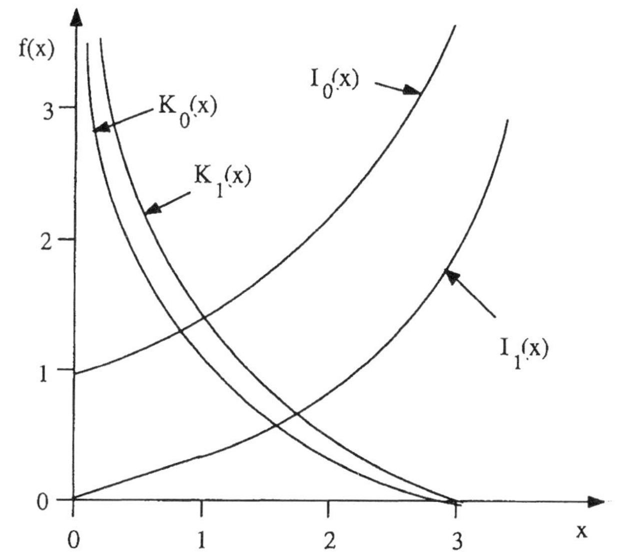

Fig. B-1—Behavior of modified Bessel functions of the zero and first order.

Then,

$$\mathcal{L}(t) = \left|\frac{te^{-ut}}{-u}\right|_0^\infty + \int_0^\infty \frac{e^{-ut}}{u}dt$$

$$= 0 + \left(\frac{e^{-ut}}{-u^2}\right)_0^\infty$$

$$= \frac{1}{u^2}. \qquad \text{(B-9)}$$

Useful Laplace Transforms. It is not always necessary to derive transforms by hand because they have been tabulated in several handbooks of mathematical functions. **Table B-1** shows important Laplace transformations and inverse Laplace transformations obtained from several sources.[1-3]

Solutions to the Diffusivity Equation

In this section, we derive detailed solutions to the general diffusivity equation, which describes the flow of fluid through porous media. These solutions are governed by different boundary conditions that describe typical situations encountered in producing a well. Two different solution techniques, the Laplace transformation and the Boltzmann transformation, are presented.

Radial Flow of a Slightly Compressible Fluid in an Infinite-Acting Reservoir, Line-Source Well, Constant-Rate Production. In this case, assume that the well can be represented as a "line source"; in other words, the wellbore is infinitesimally small ($r_w \to 0$). This well is producing at a constant rate with no wellbore storage or skin from an infinitely large reservoir. This does not describe a real situation; all reservoirs have a finite areal extent, and all wells that are drilled have a certain wellbore radius. However, the wellbore radius is small compared with the radius of the reservoir, so a line-source assumption is not unreasonable. Also, at early producing times, the effects of the outer boundaries of the reservoir are not seen, and the reservoir acts as if there are no boundaries; i.e., the reservoir is infinite-acting.

For this case, we demonstrate two solution techniques: the Laplace transformation and the Boltzmann transformation.

Laplace Transform Technique. The partial-differential equation for this case is given by

$$\frac{1}{r_D}\frac{\partial}{\partial r_D}\left(r_D\frac{\partial p_D}{\partial r_D}\right) = \frac{\partial p_D}{\partial t_D}. \qquad \text{(B-10)}$$

Initially, pressure is uniform throughout the reservoir, so the initial condition is given by

$$p_D(r_D, t_D = 0) = 0. \qquad \text{(B-11)}$$

The reservoir is infinite-acting; therefore, as the dimensionless radius tends toward infinity, the pressure at that radius is the same as initial pressure and the dimensionless pressure function is zero. The outer-boundary condition is therefore written as

$$p_D(r_D \to \infty, t_D) = 0. \qquad \text{(B-12)}$$

The reservoir is producing at constant sandface rate, with no wellbore storage or skin, from a line-source well. The inner-boundary condition for this case is

$$\lim_{r_D \to 0}\left(r_D\frac{\partial p_D}{\partial r_D}\right)_{r_D} = -1. \qquad \text{(B-13)}$$

Remember, the inner-boundary condition is for a line-source well. This is a limiting condition as ($r_w \to 0$) of the constant-rate boundary condition and is given for a finite wellbore by

$$\left(r\frac{\partial p}{\partial r}\right)_{r_w} = -\frac{141.2qB\mu}{kh}. \qquad \text{(B-14)}$$

The line-source approximation is used to simplify the solution of the diffusivity equation. Laplace transformations are used to convert a partial-differential equation to an ordinary-differential equation, with only one independent variable. The ordinary-differential equation is then solved in Laplace space to obtain the Laplace transform of the solution. The solution must then be inverted to obtain the pressure distribution in real space.

First, take the Laplace transform of the partial-differential equation (Eq. B-10).

$$\mathcal{L}\left[\frac{1}{r_D}\frac{\partial}{\partial r_D}\left(r_D\frac{\partial p_D}{\partial r_D}\right)\right] = \mathcal{L}\left(\frac{\partial p_D}{\partial t_D}\right)$$

and thus

$$\frac{1}{r_D}\frac{d}{dr_D}\left(r_D\frac{d\bar{p}_D}{dr_D}\right) = u\bar{p}_D - p_D(r_D, t_D = 0), \qquad \text{(B-15)}$$

where $\bar{p}_D$ is the Laplace transform of p_D. Note that the partial derivative has become a total derivative because we no longer have a derivative of the dimensionless pressure with respect to time.

Substitute the initial condition, Eq. B-11, into Eq. B-15 to obtain

$$\frac{1}{r_D}\frac{d}{dr_D}\left(r_D\frac{d\bar{p}_D}{dr_D}\right) = u\bar{p}_D - 0. \qquad \text{(B-16)}$$

Note that Eq. B-16 is in the form of the modified Bessel equation; therefore, the general solution of B-16 is given by

$$\bar{p}_D = AI_0\left(\sqrt{u}\,r_D\right) + BK_0\left(\sqrt{u}\,r_D\right), \qquad \text{(B-17)}$$

where A and B = arbitrary constants determined by the boundary conditions.

We need to find constants A and B, so we must take the Laplace transform of the boundary conditions. First, take the Laplace transform of the outer-boundary condition, Eq. B-12.

$$\mathcal{L}[p_D(r_D \to \infty, t_D)] = \mathcal{L}(0)$$

$$\bar{p}_D(r_D \to \infty, t_D) = 0. \qquad \text{(B-18)}$$

Substitute B-18 in B-17.

$$0 = AI_0(\infty) + BK_0(\infty).$$

TABLE B-1—LAPLACE TRANSFORMS

Number	Transform
1	$\mathcal{L}(k) = k/u\ (k = constant)$
2	$\mathcal{L}(e^{at}) = 1/u - a\ (u > a,\ a\ constant)$
3	$\mathcal{L}(t) = 1/u^2$
	Let $p = f(r,t)$ and thus $\overline{p} = f(r,u)$.
4a	$\mathcal{L}(\partial p/\partial r) = \partial/\partial r\ \overline{p}(r,u)$
4b	$\mathcal{L}(\partial^2 p/\partial r^2) = \partial^2\overline{p}/\partial r$
4c	$\mathcal{L}(\partial p/\partial t) = u\overline{p} - p(t=0)$
5	$\mathcal{L}^{-1}[p(1/2v-1)K_v(x\sqrt{p})] = x^{-v}2^{(v-1)}\int_{\frac{x^2}{4t}}^{\infty} e^{-u}u^{(v-1)}du$
6	$\mathcal{L}^{-1}(1/u\ln u) = -\gamma - \ln t$, where $\gamma = 0.57722$...(Euler's constant)
7	$\mathcal{L}^{-1}(e^{-k\sqrt{u}}/u^{3/2}) - 2\sqrt{t/\pi}\exp(-k^2/4t) - k\mathrm{erfc}(k/2\sqrt{t})$

We know that as $x \to \infty$, $K_0(x) \to 0$, so $AI_0(\infty) = 0$. As $x \to 0$, $I_0(x) \to \infty$, therefore $A = 0$. With this information, the solution becomes

$$\overline{p}_D = BK_0(\sqrt{u}\,r_D). \qquad \text{(B-19)}$$

Now, take the Laplace transform of the inner-boundary condition.

$$\mathcal{L}\left[\lim_{r_D\to 0}\left(r_D\frac{\partial p_D}{\partial r_D}\right)_{r_D}\right] = \mathcal{L}(-1)$$

and, thus,

$$\lim_{r_D\to 0}\left(r_D\frac{\partial \overline{p}_D}{\partial r_D}\right)_{r_D} = -\frac{1}{u}. \qquad \text{(B-20)}$$

To use the inner-boundary condition, find the derivative of p_D with respect to r_D from Eq. B-19.

$$\frac{\partial \overline{p}_D}{\partial r_D} = -B\sqrt{u}\,K_1(\sqrt{u}\,r_D) \qquad \text{(B-21a)}$$

and $$r_D\frac{\partial \overline{p}_D}{\partial r_D} = -B\left[\sqrt{u}\,r_D K_1(\sqrt{u}\,r_D)\right]. \qquad \text{(B-21b)}$$

Combine Eqs. B-20 and B-21b to obtain

$$\lim_{r_D\to 0}\left(r_D\frac{\partial \overline{p}_D}{\partial r_D}\right)_{r_D} = -\frac{1}{u} = -B\lim_{r_D\to 0}\left[\sqrt{u}\,r_D K_1(\sqrt{u}\,r_D)\right]. \qquad \text{(B-22)}$$

As noted in Eq. B.5 $\lim_{x\to 0}[xK_1(x)] = 1$. Therefore,

$$-\frac{1}{u} = -B, \qquad \text{(B-23a)}$$

where $$B = \frac{1}{u}. \qquad \text{(B-23b)}$$

We now know both constants A and B, so we know the solution of the diffusivity equation for this case in Laplace space. Substitute Eq. B-23b into Eq. B-19 to obtain

$$\overline{p}_D = \frac{1}{u}K_0(\sqrt{u}\,r_D). \qquad \text{(B-24)}$$

We need to invert this solution to real space. To do this analytically, look in Table B-1 to see if there is an analytical inversion of the right side of Eq. B-24.

From Transform 5 from Table B-1, we see that

$$\mathcal{L}^{-1}\left[p\left(\frac{1}{2}v - 1\right)K_v(x\sqrt{p})\right] = x^{-v}2^{(v-1)}\int_{\frac{x^2}{4t}}^{\infty} e^{-u}u^{(v-1)}du. \qquad \text{(B-25)}$$

Because we are using u as the Laplace transform parameter and because v in Eq. B-25 = 0 in this case, by letting $u = y$ in the integral, we obtain

$$\mathcal{L}^{-1}\left[u^{-1}K_0(\sqrt{u}\,x)\right] = x^0 2^{-1}\int_{\frac{x^2}{4t}}^{\infty}\frac{e^{-y}}{y}dy. \qquad \text{(B-26)}$$

Comparing the left side of Eq. B-26 with the right side of Eq. B-24, we see they are identical if $x = r_D$; therefore,

$$\mathcal{L}^{-1}(\overline{p}_D) = \mathcal{L}^{-1}\left[\frac{1}{u}K_0(\sqrt{u}\,r_D)\right]$$

$$p_D(r_D, t_D) = -\frac{1}{2}\int_{\frac{r_D^2}{4t_D}}^{\infty}\frac{e^{-y}}{y}dy. \qquad \text{(B-27)}$$

Eq. B-27 is the solution to the radial-diffusivity equation for a line-source well in an infinite-acting reservoir, constant-rate production. The right side of Eq. B-27 is the "exponential integral" function, Ei(x), where

$$\mathrm{Ei}(x) = -\int_{-x}^{\infty}\frac{e^{-y}}{y}dy. \qquad \text{(B-28)}$$

So, Eq. B-27 may be written as

$$p_D(r_D, t_D) = -\frac{1}{2}Ei\left(\frac{-r_D^2}{4t_D}\right). \qquad \text{(B-29)}$$

Table 1.1 gives numerical values of the Ei function.

Boltzmann Transform Technique.[4] Another method of reducing a partial-differential equation is by combining the two independent variables into one as follows. A Boltzmann variable is defined as ε_D, where

$$\varepsilon_D = a r_D^b t_D^c, \qquad \text{(B-30)}$$

where a, b, and c = constants. These constants are determined by trial and error, depending on the partial-differential equation to be solved and the boundary conditions. Boltzmann determined these constants for the case we just solved with the Laplace transform technique.

Partial-Differential Equation.

$$\frac{1}{r_D}\frac{\partial}{\partial r_D}\left(r_D\frac{\partial p_D}{\partial r_D}\right) = \frac{\partial p_D}{\partial t_D}. \qquad \text{(B-10)}$$

Initial Condition.

$$p_D(r_D, t_D = 0) = 0. \qquad \text{(B-11)}$$

Outer-Boundary Condition (Infinite-Acting Reservoir).

$$p_D(r_D \rightarrow \infty, t_D) = 0. \qquad \text{(B-12)}$$

Inner-Boundary Condition (Constant-Rate Production From a Line-Source Well).

$$\lim_{r_D\rightarrow 0}\left(r_D\frac{\partial p_D}{\partial r_D}\right) = -1. \qquad \text{(B-13)}$$

The required transformation, known as the Boltzmann transformation, is

$$\varepsilon_D = \frac{r_D^2}{4t_D}\left(a = \frac{1}{4}, b = 2, c = -1\right), \qquad \text{(B-31)}$$

where $p_D = f(\varepsilon_D)$ only.

Looking at Eq. B-10, the partial differential equation, we need to write $\partial p_D/\partial r_D$ and $\partial p_D/\partial t_D$ in terms of ε_D.

By the chain rule, assuming [$p_D = f(\varepsilon_D)$],

$$\frac{\partial p_D}{\partial r_D} = \frac{dp_D}{d\varepsilon_D}\frac{\partial \varepsilon_D}{\partial r_D}$$

$$= \frac{dp_D}{d\varepsilon_D}\left(\frac{1}{4t_D}2r_D\right)$$

$$= \frac{r_D}{2t_D}\frac{dp_D}{d\varepsilon_D}$$

$$\text{and } \frac{\partial p_D}{\partial t_D} = \frac{dp_D}{d\varepsilon_D}\frac{\partial \varepsilon_D}{\partial t_D} \qquad \text{(B-32a)}$$

$$= \frac{dp_D}{d\varepsilon_D}\left[\frac{r_D^2}{4}\left(\frac{-1}{t_D^2}\right)\right]. \qquad \text{(B-32b)}$$

With Eq. B-31, this becomes

$$\frac{\partial p_D}{\partial t_D} = \frac{-\varepsilon_D}{t_D}\frac{dp_D}{d\varepsilon_D}. \qquad \text{(B-33)}$$

Substituting Eqs. B-32a and B-33 into Eq. B-10 gives

$$\frac{1}{r_D}\frac{\partial}{\partial r_D}\left(\frac{r_D^2}{2t_D}\frac{dp_D}{d\varepsilon_D}\right) = \frac{-\varepsilon_D}{t_D}\frac{dp_D}{d\varepsilon_D}. \qquad \text{(B-34)}$$

Note also that for any variable X,

$$\left(\frac{\partial X}{\partial r_D}\right) = \left(\frac{dX}{d\varepsilon_D}\right)\left(\frac{\partial \varepsilon_D}{\partial r_D}\right)$$

$$= \frac{r_D}{2t_D}\frac{dX}{d\varepsilon_D}.$$

Therefore,

$$\frac{1}{r_D}\frac{r_D}{2t_D}\frac{d}{d\varepsilon_D}\left(\frac{r_D^2}{2t_D}\frac{dp_D}{d\varepsilon_D}\right) = \frac{-\varepsilon_D}{t_D}\frac{dp_D}{d\varepsilon_D},$$

$$\frac{1}{2t_D}\frac{d}{d\varepsilon_D}\left(2\varepsilon_D\frac{dp_D}{d\varepsilon_D}\right) = \frac{-\varepsilon_D}{t_D}\frac{dp_D}{d\varepsilon_D},$$

$$\text{and } \frac{d}{d\varepsilon_D}\left(\varepsilon_D\frac{dp_D}{d\varepsilon_D}\right) = -\varepsilon_D\frac{dp_D}{d\varepsilon_D}. \qquad \text{(B-35)}$$

We now have an ordinary-differential equation in terms of ε_D, the Boltzmann variable.

Expanding Eq. B-35 with the product rule gives

$$\frac{dp_D}{d\varepsilon_D} + \varepsilon_D\frac{d^2p_D}{d\varepsilon_D^2} = -\varepsilon_D\frac{dp_D}{d\varepsilon_D} \qquad \text{(B-36a)}$$

$$\text{and } \varepsilon_D\frac{d^2p_D}{d\varepsilon_D^2} + (1 + \varepsilon_D)\frac{dp_D}{d\varepsilon_D} = 0. \qquad \text{(B-36b)}$$

Now, define the boundary conditions in terms of ε_D.

Initial Condition. As $t_D \rightarrow 0$, $\varepsilon_D \rightarrow \infty$; therefore,

$$p_D(\varepsilon_D \rightarrow \infty) = 0. \qquad \text{(B-37)}$$

Outer-Boundary Condition. As $r_D \rightarrow \infty$, $\varepsilon_D \rightarrow \infty$; therefore,

$$p_D(\varepsilon_D \rightarrow \infty) = 0,$$

which is the same as the inital condition (Eq. B-37). The Boltzmann transformation combines two conditions into one; to solve a second-order ordinary-differential equation, only two boundary conditions are needed.

Inner-Boundary Condition. As $r_D \rightarrow 0$, $\varepsilon_D \rightarrow 0$.

From Eq. B-32a,

$$r_D\frac{\partial p_D}{\partial r_D} = r_D\frac{r_D}{2t_D}\frac{dp_D}{d\varepsilon_D} = \frac{r_D^2}{2t_D}\frac{dp_D}{d\varepsilon_D} = 2\varepsilon_D\frac{dp_D}{d\varepsilon_D}. \qquad \text{(B-38)}$$

Therefore, from Eqs. B-13 and B-38,

$$\lim_{\varepsilon_D\rightarrow 0}\left(2\varepsilon_D\frac{dp_D}{d\varepsilon_D}\right) = -1 \qquad \text{(B-39a)}$$

$$\text{and } \lim_{\varepsilon_D\rightarrow 0}\left(\varepsilon_D\frac{dp_D}{d\varepsilon_D}\right) = -\frac{1}{2}. \qquad \text{(B-39b)}$$

Now, solve the ordinary-differential equation, Eq. B-36b. Let $p'_D = dp_D/d\varepsilon_D$ and substitute into Eq. B-36b.

$$\varepsilon_D\frac{dp'_D}{d\varepsilon_D} + (1 + \varepsilon_D)p'_D = 0. \qquad \text{(B-40)}$$

Separating and integrating gives

$$\int\frac{dp'_D}{p'_D} = \int\frac{-(1 + \varepsilon_D)}{\varepsilon_D}d\varepsilon_D \qquad \text{(B-41a)}$$

$$\text{and } \ln p'_D = -\int\frac{1}{\varepsilon_D}d\varepsilon_D - \int d\varepsilon_D + C$$

$$= \ln\varepsilon_D - \varepsilon_D + C, \qquad \text{(B-41b)}$$

where C = a constant of integration.

Eq. B-41b may be rearranged to give

$$p'_D = e^{\left(-\ln\varepsilon_D - \varepsilon_D + C\right)}$$

$$= e^{\left(-\ln\varepsilon_D^{-1}\right)}e^{-\varepsilon_D}e^C$$

$$= \varepsilon_D^{-1}e^{-\varepsilon_D}C_1, \qquad \text{(B-42)}$$

where $C_1 = e^C$

$$= \frac{C_1 e^{-\varepsilon_D}}{\varepsilon_D} = \frac{dp_D}{d\varepsilon_D}. \qquad \text{(B-43)}$$

Solve for the constant C_1 using the inner-boundary condition, Eq. B-39b.

From Eq. B-43,

$$\varepsilon_D \frac{dp_D}{d\varepsilon_D} = C_1 e^{-\varepsilon_D}$$

$$\text{and } \lim_{\varepsilon_D \to 0}\left(\varepsilon_D \frac{dp_D}{d\varepsilon_D}\right) = C_1. \qquad \text{(B-44)}$$

Therefore, from Eq. B-39b, $C_1 = -1/2$.

Substitute for C_1 in Eq. B-43.

$$\frac{dp_D}{d\varepsilon_D} = -\frac{1}{2}\frac{e^{-\varepsilon_D}}{\varepsilon_D}. \qquad \text{(B-45)}$$

Remember the initial condition/outer-boundary condition given by Eq. B-37. At $\varepsilon_D = \infty$, $p_D = 0$, so we can separate Eq. B-45 and integrate between the limits 0 to p_D and ∞ to ε_D.

$$\int_0^{p_D} dp = -\frac{1}{2}\int_\infty^{\varepsilon_D} \frac{e^{-y}}{y} dy, \qquad \text{(B-46)}$$

where p and y = dummy variables of integration.

$$p_D - 0 = \frac{1}{2}\int_{\varepsilon_D}^{\infty} \frac{e^{-y}}{y} dy. \qquad \text{(B-47a)}$$

or

$$p_D = \frac{1}{2}\int_{\varepsilon_D}^{\infty} \frac{e^{-y}}{y} dy. \qquad \text{(B-47b)}$$

Recall the definition of the exponential integral, Ei.

$$-\text{Ei}(-x) = \int_x^{\infty} \frac{e^{-y}}{y} dy.$$

With this definition, Eq. B-47b may be written as

$$p_D = -\frac{1}{2}\text{Ei}(-\varepsilon_D). \qquad \text{(B-48)}$$

Remember the definition of ε_D from Eq. B-31 and write p_D in terms of r_D and t_D.

$$p_D = -\frac{1}{2}\text{Ei}\left(\frac{-r_D^2}{4t_D}\right). \qquad \text{(B-49)}$$

This is the line-source solution obtained with Laplace transforms.

Radial Flow Into a Finite Wellbore With Constant-Rate Production. For this case, we show the Laplace transform solutions for four different sets of boundary conditions. Note that in these solutions we are not assuming a line-source well, but a finite wellbore. We consider flow in an infinite-acting reservoir, which represents early producing times; flow in a reservoir with a no-flow outer boundary; and flow in a reservoir with a constant-pressure outer boundary. Physically, a no-flow outer boundary could be sealing faults or pinch-outs, whereas a constant-pressure outer boundary is seen when the reservoir is bounded by an aquifer that keeps the pressure at the interface constant. This section also considers wellbore storage and skin at the inner boundary.

Infinite-Acting Case. The partial-differential equation is

$$\frac{1}{r_D}\frac{\partial}{\partial r_D}\left(r_D \frac{\partial p_D}{\partial r_D}\right) = \frac{\partial p_D}{\partial t_D} \qquad \text{(B-10)}$$

with the initial condition

$$p_D(r_D, t_D = 0) = 0. \qquad \text{(B-11)}$$

The inner-boundary condition is constant-rate production with no wellbore storage or skin (note that we have a finite wellbore; i.e., $r_D = 1$).

$$\left(r_D \frac{\partial p_D}{\partial r_D}\right)_{(r_D=1)} = -1. \qquad \text{(B-50)}$$

The outer-boundary condition is for an infinite-acting reservoir:

$$p_D(r_D \to \infty, t_D) = 0. \qquad \text{(B-12)}$$

Taking the Laplace transform of the partial-differential equation gives

$$\frac{1}{r_D}\frac{d}{dr_D}\left(r_D \frac{d\bar{p}_D}{dr_D}\right) = u\bar{p}_D - p_D(r_D, t_D = 0) = 0. \qquad \text{(B-15)}$$

Combining Eqs. B-11 and B-15 gives

$$\frac{1}{r_D}\frac{d}{dr_D}\left(r_D \frac{d\bar{p}_D}{dr_D}\right) = u\bar{p}_D. \qquad \text{(B-16)}$$

Eq. B-16 is in the form of the modified Bessel equation, so the general solution for $\bar{p}_D$ is

$$\bar{p}_D = AI_0\left(\sqrt{u}\, r_D\right) + BK_0\left(\sqrt{u}\, r_D\right). \qquad \text{(B-17)}$$

We now need to find coefficients A and B with our outer-boundary conditions.

Taking the Laplace transform of Eq. B-12 gives

$$\bar{p}_D(r_D \to \infty, u_D) = 0. \qquad \text{(B-18)}$$

Combine Eqs. B-17 and B-18.

$$\bar{p}_D(r_D \to \infty, u) = AI_0(\infty) + BK_0(\infty) = 0. \qquad \text{(B-51)}$$

Because $K_0(\infty) \to 0$ and $I_0(\infty) \to \infty$, $A = 0$ to satisfy the outer-boundary condition. Thus, our solution becomes

$$\bar{p}_D = BK_0\left(\sqrt{u}\, r_D\right). \qquad \text{(B-19)}$$

Taking the Laplace transform of the inner-boundary condition, Eq. B-50, gives

$$\left(r_D \frac{\partial \bar{p}_D}{\partial r_D}\right)_{(r_D=1)} = -\frac{1}{u}. \qquad \text{(B-52)}$$

Take the derivative of Eq. B-19 with respect to r_D.

$$\frac{\partial \bar{p}_D}{\partial r_D} = -B\sqrt{u}\,K_1\left(\sqrt{u}\, r_D\right). \qquad \text{(B-53)}$$

Multiplying by r_D,

$$r_D \frac{\partial \bar{p}_D}{\partial r_D} = -B\left[\sqrt{u}\, r_D K_1\left(\sqrt{u}\, r_D\right)\right], \qquad \text{(B-21b)}$$

and combining Eqs. B-21b and B-52 gives

$$\left(r_D \frac{\partial \bar{p}_D}{\partial r_D}\right)_{(r_D=1)} = -B\sqrt{u}\,K_1\left(\sqrt{u}\right) = -\frac{1}{u}.$$

Therefore,

$$B = \frac{1}{u^{3/2}K_1\left(\sqrt{u}\right)}. \qquad \text{(B-54)}$$

Now, substitute Eq. B-54 into Eq. B-19 to obtain the transform of the dimensionless-pressure solution.

$$\bar{p}_D = \frac{K_0(\sqrt{u}\,r_D)}{u^{3/2}K_1(\sqrt{u})}. \quad \text{(B-55)}$$

Eq. B-55 cannot be inverted directly, so we must use either a numerical inversion technique (such as the Stehfest[5] algorithm) or an approximation.

Eq.B-55 should give us the same solution as the line-source solution at large times (i.e., after near-wellbore effects are over). We need to determine whether a long-time approximation can be found for B-55.

For small arguments (i.e., $x \to 0$), the modified Bessel functions have the following asymptotic approximations taken from Ref. 2.

$$\text{As } x \to 0,\ K_0(x) \approx -\ln\left(\frac{e^{\gamma}x}{2}\right), \quad \text{(B-56)}$$

where $\ln\gamma = 0.57722\ldots$, which is Euler's constant,

$$\text{and } K_1(x) \approx \frac{1}{x}. \quad \text{(B-57)}$$

Note that, as time becomes very large, the value of u must be very small because, according to Transform 3 (Table B-1), $\mathcal{L}(t) = 1/u^2$. Substituting Eqs. B-56 and B-57 into Eq. B-55 gives

$$\bar{p}_D = -\frac{1}{u^{3/2}}\frac{\ln\left(\dfrac{e^{\gamma}\sqrt{u}\,r_D}{2}\right)}{\dfrac{1}{\sqrt{u}}}$$

$$= -\frac{1}{2u}\ln\left(\frac{e^{2\gamma}ur_D^2}{4}\right). \quad \text{(B-58)}$$

Eq. B-58 can be rewritten as

$$\bar{p}_D = -\frac{1}{2u}\ln u + \frac{1}{2u}\ln\left(\frac{4}{e^{2\gamma}r_D^2}\right). \quad \text{(B-59)}$$

Transform 6 from Table B-1 is

$$\mathcal{L}^{-1}\left(\frac{1}{u}\ln u\right) = -\gamma - \ln t.$$

Therefore, the solution becomes

$$p_D = \frac{1}{2}(\gamma + \ln t_D) + \frac{1}{2}\ln\left(\frac{4}{e^{2\gamma}r_D^2}\right)$$

$$= \frac{1}{2}\left(\gamma + \ln t_D + \ln\frac{4}{r_D^2} - \ln e^{2\gamma}\right)$$

$$= \frac{1}{2}\left(\ln\frac{4t_D}{r_D^2} + \gamma - 2\gamma\right)$$

$$= \frac{1}{2}\left(\ln\frac{t_D}{r_D^2} - 0.80907\right). \quad \text{(B-60)}$$

Eq. B-60 is the equivalent of the logarithmic approximation to the Ei-function solution (line-source solution), which is examined later.

We recall the line-source solution mentioned earlier:

$$p_D(r_D, t_D) - \frac{1}{2}\text{Ei}\left(\frac{-r_D^2}{4t_D}\right). \quad \text{(B-29)}$$

For $r_D^2/4t_D < 0.01$,

$$-\text{Ei}\left(-\frac{r_D^2}{4t_D}\right) \approx -\ln\left(\frac{e^{\gamma}r_D^2}{4t_D}\right) = \ln\left(\frac{4t_D}{e^{\gamma}r_D^2}\right)$$

$$\approx \ln\left(\frac{t_D}{r_D^2}\right) + \ln\left(\frac{4}{e^{\gamma}}\right) = \ln\left(\frac{t_D}{r_D^2}\right) - 0.80907.$$

Therefore, for small values of $r_D^2/4t_D$, Eq. B-29 becomes

$$p_{D_{LS}} = -\frac{1}{2}\left[\ln\left(\frac{t_D}{r_D^2}\right) - 0.80907\right]. \quad \text{(B-60)}$$

In practice, for $r_D = 1$ or $r = r_w$ and typical rock and fluid property values, $r_D^2/4t_D$ becomes small within seconds or minutes for most wells. Therefore, Eq. B-60 is applicable from very early times for calculating pressure drawdown of a well.

No-Flow Outer-Boundary Case. The partial-differential equation is given by

$$\frac{1}{r_D}\frac{\partial}{\partial r_D}\left(r_D\frac{\partial p_D}{\partial r_D}\right) = \frac{\partial p_D}{\partial t_D} \quad \text{(B-10)}$$

with the initial condition

$$p_D(r_D, t_D = 0) = 0. \quad \text{(B-11)}$$

The inner-boundary condition is constant-rate production with neither wellbore storage nor skin.

$$\left(r_D\frac{\partial p_D}{\partial r_D}\right)_{(r_D=1)} = -1. \quad \text{(B-50)}$$

The outer-boundary condition is for a no-flow boundary

$$\left(\frac{\partial p_D}{\partial r_D}\right)_{(r_{eD})} = 0. \quad \text{(B-61)}$$

With the initial condition, the solution of Eq. B-10 in Laplace space is

$$\bar{p}_D = AI_0(\sqrt{u}\,r_D) + BK_0(\sqrt{u}\,r_D). \quad \text{(B-17)}$$

Taking the Laplace transform of Eq. B-61, we obtain

$$\left(\frac{\partial\bar{p}_D}{\partial r_D}\right)_{(r_{eD})} = 0. \quad \text{(B-62)}$$

Differentiating Eq. B-17 gives

$$\left(\frac{\partial\bar{p}_D}{\partial r_D}\right) = A\sqrt{u}\,I_1(\sqrt{u}\,r_D) - B\sqrt{u}\,K_1(\sqrt{u}\,r_D). \quad \text{(B-63)}$$

At $r_D = r_{eD}$, this gives

$$0 = A\sqrt{u}\,I_1(\sqrt{u}\,r_{eD}) - B\sqrt{u}\,K_1(\sqrt{u}\,r_{eD}),$$

and if we write A in terms of B, we obtain

$$A = \frac{B\sqrt{u}\,K_1(\sqrt{u}\,r_{eD})}{\sqrt{u}\,I_1(\sqrt{u}\,r_{eD})} = \frac{K_1(\sqrt{u}\,r_{eD})}{I_1(\sqrt{u}\,r_{eD})}B. \quad \text{(B-64)}$$

We need to use the other condition to solve for the constants A and B. The inner-boundary condition in Laplace space is

$$\left(r_D\frac{\partial\bar{p}_D}{\partial r_D}\right)_{(r_D=1)} = -\frac{1}{u}. \quad \text{(B-52)}$$

Substituting Eq. B-52 into Eq. B-63 for $r_D = 1$ gives

$$-\frac{1}{u} = A\sqrt{u}\,I_1(\sqrt{u}) - B\sqrt{u}\,K_1(\sqrt{u}). \quad \text{(B-65)}$$

Substituting Eq. B-64 into Eq. B-65 gives

$$-\frac{1}{u}=\frac{K_1(\sqrt{u}\,r_{eD})}{I_1(\sqrt{u}\,r_{eD})}\sqrt{u}\,I_1(\sqrt{u})B-B\sqrt{u}\,K_1(\sqrt{u}), \quad \text{(B-66a)}$$

$$B\left[\frac{\sqrt{u}\,K_1(\sqrt{u}\,r_{eD})I_1(\sqrt{u})-I_1(\sqrt{u}\,r_{eD})K_1(\sqrt{u})}{I_1(\sqrt{u}\,r_{eD})}\right]=-\frac{1}{u}, \quad \text{(B-66b)}$$

and

$$B=-\frac{1}{u}\left[\frac{I_1(\sqrt{u}\,r_{eD})}{\sqrt{u}\,K_1(\sqrt{u}\,r_{eD})I_1(\sqrt{u})-I_1(\sqrt{u}\,r_{eD})\sqrt{u}\,K_1(\sqrt{u})}\right]$$

$$=-\frac{1}{u^{3/2}}\left[\frac{I_1(\sqrt{u}\,r_{eD})}{K_1(\sqrt{u}\,r_{eD})I_1(\sqrt{u})-I_1(\sqrt{u}\,r_{eD})K_1(\sqrt{u})}\right]. \quad \text{(B-66c)}$$

Therefore, we can solve for the constant A by substituting the relation for the constant B, given by Eq. B-66c, into Eq. B-64.

$$A=-\frac{1}{u^{3/2}}\left[\frac{K_1(\sqrt{u}\,r_{eD})}{K_1(\sqrt{u}\,r_{eD})I_1(\sqrt{u})-I_1(\sqrt{u}\,r_{eD})K_1(\sqrt{u})}\right]. \quad \text{(B-67)}$$

Substituting Eqs. B-66c and B-67 into Eq. B-17 gives the Laplace transform of the dimensionless pressure solution in terms of u and r_D.

$$\bar{p}_D=-\frac{1}{u^{3/2}}\left[\frac{K_1(\sqrt{u}\,r_{eD})I_0(\sqrt{u}\,r_D)+I_1(\sqrt{u}\,r_{eD})K_0(\sqrt{u}\,r_D)}{K_1(\sqrt{u}\,r_{eD})I_1(\sqrt{u})-I_1(\sqrt{u}\,r_{eD})K_1(\sqrt{u})}\right]. \quad \text{(B-68)}$$

This can be inverted, but not easily. If we use Cauchy's inversion theorem,[1] we find that the solution is

$$p_D(r_D,t_D)=\frac{2}{(r_{eD}^2-1)}\left(\frac{r_D^2}{4}+t_D\right)-\frac{r_{eD}^2\ln r_D}{(r_{eD}^2-1)}-\frac{(3r_{eD}^4-4r_{eD}^4\ln r_{eD}-2r_{eD}^2-1)}{4(r_{eD}^2-1)^2}$$
$$+\pi\sum_{n=1}^{\infty}\Big\{e^{-\alpha_n^2 t_D}J_1^2(\alpha_n r_{eD})\big[J_1(\alpha_n)Y_0(\alpha_n r_D)-Y_1(\alpha_n)J_0(\alpha_n r_D)\big]\Big/\alpha_n\big[J_1^2(\alpha_n r_{eD})-J_1^2(\alpha_n)\big]\Big\} \quad \text{(B-69)}$$

where α_n is the root derived from

$$J_1(\alpha_n r_{eD})Y_1(\alpha_n)-J_1(\alpha_n)Y_1(\alpha_n r_{eD})=0.$$

At long times, Eq. B-69 becomes a little simpler. As $t\to\infty$, the summation term drops out because it contains a negative exponential term and $\lim_{x\to 0} e^{-x}=0$.

The solution then becomes

$$p_D(r_D,t_D)=\frac{2}{(r_{eD}^2-1)}\left(\frac{r_D^2}{4}+t_D\right)-\frac{r_{eD}^2\ln r_D}{(r_{eD}^2-1)}-\frac{(3r_{eD}^4-4r_{eD}^4\ln r_{eD}-2r_{eD}^2-1)}{4(r_{eD}^2-1)^2}. \quad \text{(B-70)}$$

For the pressure at wellbore, p_{wD} , $r_{eD}\gg 1$, so Eq. B-70 becomes

$$p_{wD}(t_D)=\frac{2}{(r_{eD}^2)}\left(\frac{1}{4}+t_D\right)-\frac{r_{eD}^2\ln 1}{(r_{eD}^2)}-\frac{3r_{eD}^4}{4r_{eD}^4}+\frac{4r_{eD}^4\ln r_{eD}}{4r_{eD}^4}+\frac{1}{2r_{eD}^2}+\frac{1}{4r_{eD}^4}$$

$$=\frac{2t_D}{(r_{eD}^2)}-\frac{3}{4}+\ln r_{eD}. \quad \text{(B-71)}$$

Constant-Pressure Outer-Boundary Case. The partial-differential equation is given by

$$\frac{1}{r_D}\frac{\partial}{\partial r_D}\left(r_D\frac{\partial p_D}{\partial r_D}\right)=\frac{\partial p_D}{\partial t_D} \quad \text{(B-10)}$$

with the initial condition

$$p_D(r_D,t_D=0)=0. \quad \text{(B-11)}$$

The inner-boundary condition is constant-rate production with no wellbore storage or skin,

$$\left(r_D\frac{\partial p_D}{\partial r_D}\right)_{(r_D=1)}=-1. \quad \text{(B-50)}$$

The outer-boundary condition is for a constant-pressure boundary,

$$p_D(r_{eD},t_D)=0. \quad \text{(B-72)}$$

As before, with the initial condition, the solution of Eq. B-10 in Laplace space is

$$\bar{p}_D=AI_0(\sqrt{u}\,r_D)+BK_0(\sqrt{u}\,r_D). \quad \text{(B-17)}$$

Again, we need to use the inner- and outer-boundary conditions to solve for arbitrary constants A and B. Taking the Laplace transform of the outer-boundary condition, Eq. B-72, gives

$$\bar{p}_D(r_{eD},u)=0. \quad \text{(B-73)}$$

Combining Eqs. B-73 and B-17 gives

$$\bar{p}_D(r_{eD},u)=AI_0(\sqrt{u}\,r_{eD})+BK_0(\sqrt{u}\,r_{eD})=0. \quad \text{(B-74)}$$

Therefore,

$$A=-\frac{K_0(\sqrt{u}\,r_{eD})}{I_0(\sqrt{u}\,r_{eD})}B. \quad \text{(B-75)}$$

We still need another equation to solve for the constants. Taking the Laplace transform of the inner-boundary condition gives

$$\left(r_D\frac{\partial \bar{p}_D}{\partial r_D}\right)_{(r_D=1)}=-\frac{1}{u}. \quad \text{(B-52)}$$

Differentiating Eq. B-17 and multiplying by r_D,

$$r_D\frac{\partial \bar{p}_D}{\partial r_D}=A\sqrt{u}\,r_D I_1(\sqrt{u}\,r_D)-B\sqrt{u}\,r_D K_1(\sqrt{u}\,r_D).$$

With Eq. B-52, for $r_D = 1$,

$$A\sqrt{u}I_1(\sqrt{u}) - B\sqrt{u}K_1(\sqrt{u}) = -\frac{1}{u}. \quad \text{(B-76)}$$

Substituting Eq. B-75 into Eq. B-76 to solve for B,

$$-\left[\frac{K_0(\sqrt{u}\,r_{eD})}{I_0(\sqrt{u}\,r_{eD})}\right]B\sqrt{u}I_1(\sqrt{u}) - B\sqrt{u}K_1(\sqrt{u}) = -\frac{1}{u}. \quad \text{(B-77)}$$

Dividing both sides of the equation by $-\sqrt{u}$,

$$B\left[\frac{K_0(\sqrt{u}\,r_{eD})I_1(\sqrt{u})}{I_0(\sqrt{u}\,r_{eD})} + K_1(\sqrt{u})\right] = \frac{1}{u^{3/2}} \quad \text{(B-78a)}$$

$$B = \frac{1}{u^{3/2}}\left[\frac{I_0(\sqrt{u}\,r_{eD})}{K_0(\sqrt{u}\,r_{eD})I_1(\sqrt{u}) + K_1(\sqrt{u})I_0(\sqrt{u}\,r_{eD})}\right]. \quad \text{(B-78b)}$$

We can solve for A by substituting Eq. B-78b into Eq. B-75.

$$A = -\frac{1}{u^{3/2}}\left[\frac{K_0(\sqrt{u}\,r_{eD})}{K_0(\sqrt{u}\,r_{eD})I_1(\sqrt{u}) + K_1(\sqrt{u})I_0(\sqrt{u}\,r_{eD})}\right]. \quad \text{(B-79)}$$

We obtain the Laplace transform of the dimensionless-pressure solution by substituting Eqs. B-78b and B-79 into Eq. B-17.

$$\overline{p}_D = \frac{1}{u^{3/2}}\left[\frac{-I_0(\sqrt{u}\,r_D)K_0(\sqrt{u}\,r_{eD}) + I_0(\sqrt{u}\,r_{eD})K_0(\sqrt{u}\,r_D)}{K_0(\sqrt{u}\,r_{eD})I_1(\sqrt{u}) + K_1(\sqrt{u})I_0(\sqrt{u}\,r_{eD})}\right]. \quad \text{(B-80)}$$

The inversion of this equation is given by Hurst[6] as

$$p_D(r_D, t_D) = \ln(r_{eD}) - 2\sum_{n=1}^{\infty}\frac{[\exp(-\lambda_n^2 t_D)][J_0^2(\lambda_n r_{eD})]}{\lambda_n^2[J_1^2(\lambda_n) - J_0^2(\lambda_n r_{eD})]}, \quad \text{(B-81)}$$

where λ_n = the root derived from

$$J_1(\lambda_n)Y_0(\lambda_n r_{eD}) - Y_1(\lambda_n)J_0(\lambda_n r_{eD}) = 0,$$

and J_0 and Y_0 = zero-order Bessel functions of the first and second kind, respectively.

Consideration of Wellbore Storage and Skin for the Infinite-Acting-Reservoir Case. The partial-differential equation is

$$\frac{1}{r_D}\frac{\partial}{\partial r_D}\left(r_D\frac{\partial p_D}{\partial r_D}\right) = \frac{\partial p_D}{\partial t_D}, \quad \text{(B-10)}$$

with the initial condition

$$p_D(r_D, t_D = 0) = 0. \quad \text{(B-11)}$$

The outer-boundary condition for an infinite-acting reservoir is

$$p_D(r_D \to \infty, t_D) = 0. \quad \text{(B-12)}$$

There are two inner-boundary conditions. The first is

$$C_D\frac{dp_{wD}}{dt_D} - \left(r_D\frac{\partial p_D}{\partial r_D}\right)_{(r_D=1)} = 1 \quad \text{(B-82)}$$

for constant-rate production with wellbore storage, and the second is

$$p_{wD} = p_D - s\left(r_D\frac{\partial p_D}{\partial r_D}\right)_{(r_D=1)} \quad \text{(B-83)}$$

for skin.

Taking the Laplace transform of Eq. B-10, subject to the initial condition (Eq. B-11), gives

$$\frac{1}{r_D}\frac{\partial}{\partial r_D}\left(r_D\frac{\partial \overline{p}_D}{\partial r_D}\right) = u\overline{p}_D. \quad \text{(B-16)}$$

The general solution to Eq. B-16 is

$$\overline{p}_D = AI_0(\sqrt{u}\,r_D) + BK_0(\sqrt{u}\,r_D). \quad \text{(B-17)}$$

We need to find relations to describe the constants A and B to arrive at the particular solution for the Laplace transform of the dimensionless pressure. Therefore, we need to use the boundary conditions. With the Laplace transform of the outer-boundary condition (Eq. B-12),

$$\overline{p}_D(r_D \to \infty, t_D) = 0. \quad \text{(B-18)}$$

Substituting Eq. B-18 into Eq. B-17,

$$\overline{p}_D(r_D \to \infty, u) = AI_0(\infty) + BK_0(\infty) = 0. \quad \text{(B-51)}$$

We know that $I_0(\infty) \to \infty$ and $K_0(\infty) \to 0$; therefore $A = 0$. With this fact, the solution becomes

$$\overline{p}_D = BK_0(\sqrt{u}\,r_D). \quad \text{(B-19)}$$

We also need to use the inner-boundary conditions, so taking the Laplace transforms of the wellbore storage condition (Eq. B-82),

$$C_D u\overline{p}_{wD} - C_D\overline{p}_{wD}(r_D, t_D = 0) - \left(r_D\frac{\partial \overline{p}_D}{\partial r_D}\right)_{(r_D=1)} = \frac{1}{u}.$$

We know from the initial condition that $p_{wD}(t_D = 0) = 0$. Therefore,

$$C_D u\overline{p}_{wD} - \left(r_D\frac{\partial \overline{p}_D}{\partial r_D}\right)_{(r_D=1)} = \frac{1}{u}. \quad \text{(B-84)}$$

Taking the Laplace transform of the inner-boundary condition for skin gives

$$\overline{p}_{wD} = \overline{p}_D - s\left(r_D\frac{\partial \overline{p}_D}{\partial r_D}\right)_{(r_D=1)}. \quad \text{(B-85)}$$

Differentiating Eq. B-19 with respect to r_D to use the skin and wellbore storage conditions gives

$$\frac{\partial \overline{p}_D}{\partial r_D} = -B\sqrt{u}K_1(\sqrt{u}\,r_D). \quad \text{(B-86)}$$

Substituting Eqs. B-19 and B-86 into Eq. B-85,

$$\overline{p}_{wD} = BK_0(\sqrt{u}) + sB\sqrt{u}K_1(\sqrt{u}). \quad \text{(B-87)}$$

$$\text{Thus, } B = \frac{\overline{p}_{wD}}{K_0(\sqrt{u}) + s\sqrt{u}K_1(\sqrt{u})}. \quad \text{(B-88)}$$

Now, using the wellbore storage condition, substitute Eq. B-86 into Eq. B-84.

$$C_D u\overline{p}_{wD} + B\sqrt{u}K_1(\sqrt{u}) = \frac{1}{u}. \quad \text{(B-89)}$$

Substituting Eq. B-88 into Eq. B-89,

$$C_D u\overline{p}_{wD} + \frac{\overline{p}_{wD}\sqrt{u}K_1(\sqrt{u})}{K_0(\sqrt{u}) + s\sqrt{u}K_1(\sqrt{u})} = \frac{1}{u}. \quad \text{(B-90)}$$

Dividing by u gives

$$\overline{p}_{wD}\left[C_D + \frac{K_1(\sqrt{u})}{\sqrt{u}\,K_0(\sqrt{u}) + suK_1(\sqrt{u})}\right] = \frac{1}{u^2}, \quad \text{(B-91a)}$$

$$\overline{p}_{wD} = \frac{1}{u^2}\left[\frac{\sqrt{u}\,K_0(\sqrt{u}) + suK_1(\sqrt{u})}{C_D\sqrt{u}\,K_0(\sqrt{u}) + sC_DuK_1(\sqrt{u}) + K_1(\sqrt{u})}\right] \quad \text{(B-91b)}$$

$$\overline{p}_{wD} = \frac{\sqrt{u}}{u^2}\left\{\frac{K_0(\sqrt{u}) + s\sqrt{u}\,K_1(\sqrt{u})}{K_1(\sqrt{u}) + C_D\left[\sqrt{u}\,K_0(\sqrt{u}) + suK_1(\sqrt{u})\right]}\right\} \quad \text{(B-91c)}$$

$$\overline{p}_{wD} = \frac{1}{u\sqrt{u}}\left\{\frac{K_0(\sqrt{u}) + s\sqrt{u}\,K_1(\sqrt{u})}{K_1(\sqrt{u}) + C_D\sqrt{u}\left[K_0(\sqrt{u}) + s\sqrt{u}\,K_1(\sqrt{u})\right]}\right\} \quad \text{(B-91d)}$$

$$\overline{p}_{wD} = \frac{1}{u}\left\{\frac{K_0(\sqrt{u}) + s\sqrt{u}\,K_1(\sqrt{u})}{\sqrt{u}\,K_1(\sqrt{u}) + C_Du\left[K_0(\sqrt{u}) + s\sqrt{u}\,K_1(\sqrt{u})\right]}\right\}. \quad \text{(B-91e)}$$

Eq. B-91e is the transform of the dimensionless wellbore pressure (also Agarwal *et al.*'s Eq. 8[7]).

Agarwal *et al.*'s[8] Eq. 9 is the inversion of this equation.

$$p_{wD}(t_D) = \frac{4}{\pi^2}\int_o^\infty \Bigg((1 - e^{-x^2 t_D}) \Big/ x^3\Big\{\left[xC_DJ_0(x) - (1 - C_Dsx^2)J_1(x)\right]^2 + \left[xC_DY_0(x) - (1 - C_Dsx^2)Y_1(x)\right]^2\Big\}\Bigg)\,dx \quad \text{(B-92)}$$

where J_0 and J_1 = Bessel functions of the first kind, zero and first order, respectively, and Y_0 and Y_1 = Bessel functions of the second kind, zero and first order, respectively.

Linear Flow. The partial-differential equation for 1D linear flow is

$$\frac{\partial^2 p_D}{\partial x_D^2} = \left(\frac{\partial p_D}{\partial t_D}\right). \quad \text{(B-93)}$$

The initial condition is

$$p_D(x_D, t_D = 0) = 0. \quad \text{(B-94)}$$

The outer-boundary condition is

$$p_D(x_D \to \infty, t_D) = 0. \quad \text{(B-95)}$$

The inner-boundary condition is

$$\left(\frac{\partial p_D}{\partial x_D}\right)_{(x_D=0)} = -1. \quad \text{(B-96)}$$

We use the Laplace transform technique to transform this partial-differential equation into an ordinary-differential equation. Taking Laplace transforms of Eq. B-93,

$$\frac{\partial^2 \overline{p}_D}{\partial x_D^2} = u\overline{p}_D - p_D(x_D, t_D = 0). \quad \text{(B-97)}$$

Substituting the initial condition, Eq. B-94, into Eq. B-97 gives

$$\frac{d^2\overline{p}_D}{dx_D^2} = u\overline{p}_D \quad \text{(B-98a)}$$

$$\frac{d^2\overline{p}_D}{dx_D^2} - u\overline{p}_D = 0. \quad \text{(B-98b)}$$

Eq. B-98b has a simple general solution.[1]

$$\overline{p}_D = C_1e^{-mx_D} + C_2e^{mx_D}, \quad \text{(B-99)}$$

where $m^2 - u = 0$, and thus $m = \pm\sqrt{u}$.

The general solution to Eq. B-98b becomes

$$\overline{p}_D = C_1e^{-\sqrt{u}x_D} + C_2e^{\sqrt{u}x_D}, \quad \text{(B-100)}$$

where C_1 and C_2 = constants with values determined by the boundary conditions.

This means the boundary conditions must be transformed. The inner-boundary condition becomes

$$\left(\frac{\partial \overline{p}_D}{\partial x_D}\right)_{(x_D=0)} = -\frac{1}{u}, \quad \text{(B-101)}$$

and the outer-boundary condition becomes

$$\overline{p}_D(x_D \to \infty,\ u) = 0. \quad \text{(B-102)}$$

By use of the outer-boundary condition with the general solution,

$$\overline{p}_D(x_D \to \infty,\ u) = \lim_{x_D\to\infty}\left(C_1e^{-x_D} + C_2e^{x_D}\right) = 0.$$

Because $\lim_{x_D\to\infty}(e^{x_D}) \to \infty$, for a finite solution the constant, $C_2 = 0$. Therefore,

$$\overline{p}_D = C_1e^{-\sqrt{u}x_D}. \quad \text{(B-103)}$$

Differentiating Eq. B-103 with respect to x_D, to use the inner-boundary condition,

$$\frac{d\overline{p}_D}{dx_D} = -\sqrt{u}\,C_1e^{-\sqrt{u}x_D}. \quad \text{(B-104)}$$

Substituting Eq. B-101 into Eq. B-104,

$$\left(\frac{d\overline{p}_D}{dx_D}\right)_{(x_D=0)} = -\sqrt{u}\,C_1e^0 = -\frac{1}{u}; \quad \text{(B-105)}$$

therefore,

$$\sqrt{u}\,C_1 = \frac{1}{u} \quad \text{(B-106a)}$$

$$\text{and } C_1 = \frac{1}{u^{3/2}}. \quad \text{(B-106b)}$$

Substituting Eq. B-106b into the general solution given by Eq. B-103 gives the Laplace transform of the dimensionless-pressure solution for unsteady-state linear flow.

$$\overline{p}_D(x_D, u) = \frac{1}{u^{3/2}}e^{-\sqrt{u}x_D}. \quad \text{(B-107)}$$

This equation has a direct inverse. From Transform 7 in Table B-1, we find that

$$\mathcal{L}^{-1}\left(\frac{e^{-k\sqrt{u}}}{u^{3/2}}\right) = 2\sqrt{\frac{t}{\pi}}\exp\left(\frac{-k^2}{4t}\right) - k\,\mathrm{erfc}\left(\frac{k}{2\sqrt{t}}\right);$$

therefore,

$$p_D(x_D,t_D) = 2\sqrt{\frac{t_D}{\pi}}\exp\left(\frac{-x_D^2}{4t_D}\right)$$

$$- x_D\text{erfc}\left(\frac{x_D}{2\sqrt{t_D}}\right), \qquad \text{(B-108)}$$

where the complementary error function, erfc(x), is defined as

$$\text{erfc}(x) = \frac{2}{\sqrt{\pi}}\int_x^\infty e^{-\xi^2}d\xi. \qquad \text{(B-109)}$$

For the special case of $x_D = 0$; i.e., at the fracture face, the solution becomes

$$p_D(0,t_D) = 2\sqrt{\frac{t_D}{\pi}}. \qquad \text{(B-110)}$$

References

1. O'Neil, P.V: *Advanced Engineering Mathematics*, second edition, Wadsworth Publishing Co., Belmont, California (1987).
2. Abramowitz, M. and Stegun, I.E.: *Handbook of Mathematical Functions*, Dover Publications Inc., New York City (1972).
3. Kucuk, F. and Ayestaran, L.: "Analysis of Simultaneously Measured Pressure and Sandface Flow Rate in Transient Well Testing," *JPT* (February 1985) 323.
4. Dake, L.P.: *Fundamentals of Reservoir Engineering*, Elsevier Scientific Publishing Co., New York City (1978).
5. Stehfest, H.: "Numerical Inversion of Laplace Transforms," *Communications ACM* (January 1970), **13**, No. 1, Algorithm 368.
6. Hurst, W.: "Unsteady Flow of Fluids," *Physics* (January 1934).
7. Agarwal, R.G., Al–Hussainy, R., and Ramey, H.J. Jr.: "An Investigation of Wellbore Storage and Skin Effect in Unsteady Liquid Flow: I. Analytical Treatment," *SPEJ* (September 1970) 279.

Appendix C
Derivations of the Diffusivity Equation Multiphase Flow (Perrine and Martin) Linear Flow of Gas

Introduction

This appendix supplements the derivation of the diffusivity equation given in Chap. 1 with cases other than the case of single-phase radial flow in a cylindrical reservoir. We present a derivation of the diffusivity equation for simultaneous radial flow of oil, water, and gas and a derivation for single-phase linear gas flow.

Multiphase Flow

To develop the diffusivity equation for multiphase flow, we must begin from first principles and develop continuity equations for each phase flowing in the reservoir. Consider a unit control volume containing oil, water, and gas saturations S_o, S_w, and S_g, respectively. The mass of oil in the reservoir for a unit reservoir volume can be written as

$$m_o = \frac{\phi S_o}{B_o}\rho_{os}, \qquad \text{(C-1a)}$$

where ϕ = porosity, S_o = oil saturation, B_o = oil formation volume factor (FVF)–reservoir volume/volume at standard conditions, and ρ_{os} = density at standard conditions (mass/volume at standard conditions). Similarly for water,

$$m_w = \frac{\phi S_w}{B_w}\rho_{ws}. \qquad \text{(C-1b)}$$

Gas exists in two phases: free gas and gas dissolved in either oil or water. The mass of free gas is

$$m_{g,F} = \frac{\phi S_g}{B_g}\rho_{gs}. \qquad \text{(C-2)}$$

The mass of dissolved gas is

$$m_{g,\text{dissolved}} = \frac{\phi R_s S_o}{B_o}\rho_{gs} + \frac{\phi R_{sw} S_w}{B_w}\rho_{gs}.$$

Therefore, the total mass of gas in the unit control volume is

$$m_g = \frac{\phi S_g}{B_g}\rho_{gs} + \frac{\phi R_s S_o}{B_o}\rho_{gs} + \frac{\phi R_{sw} S_w}{B_w}\rho_{gs}. \qquad \text{(C-3)}$$

Eq. C-3 contains terms representing both free gas (terms with the subscript g) and gas dissolved in oil and water (terms with the subscripts o and w, respectively).

We can use Darcy's law for an equation of motion in the control volume for each phase.

Oil.

$$\rho_o u_{ro} = -\frac{k_o}{\mu_o B_o}\rho_{os}\frac{\partial p}{\partial r}. \qquad \text{(C-4)}$$

Water.

$$\rho_w u_{rw} = -\frac{k_w}{\mu_w B_w}\rho_{ws}\frac{\partial p}{\partial r}. \qquad \text{(C-5)}$$

Gas.

$$\rho_g u_{rg} = -\frac{k_g}{\mu_g B_g}\rho_{gs}\frac{\partial p}{\partial r} - \frac{R_s k_o}{\mu_o B_o}\rho_{gs}\frac{\partial p}{\partial r} - \frac{R_w k_w}{\mu_w B_w}\rho_{gs}\frac{\partial p}{\partial r}. \qquad \text{(C-6)}$$

A continuity equation now can be written for each phase because the rate of mass into the control volume less the rate of mass out is given by

$$\frac{1}{r}\frac{\partial(r\rho u_r)}{\partial r},$$

and the rate of change of mass in the control volume is $\partial m/\partial t$.

In general, the continuity equation for any Phase x is

$$\frac{1}{r}\frac{\partial}{\partial r}\left(-r\rho_x u_{rx}\right) = \frac{\partial m_x}{\partial t}. \qquad \text{(C-7)}$$

For oil, combine Eqs. C-1 and C-4 to form the oil continuity equation,

$$\frac{1}{r}\frac{\partial}{\partial r}\left(r\frac{k_o}{\mu_o B_o}\rho_{os}\frac{\partial p}{\partial r}\right) = \frac{\partial}{\partial t}\left(\frac{\phi S_o}{B_o}\rho_{os}\right). \qquad \text{(C-8)}$$

Because ρ_{os} = density at standard conditions, it is a constant; so the oil equation becomes

$$\frac{1}{r}\frac{\partial}{\partial r}\left(r\frac{k_o}{\mu_o B_o}\frac{\partial p}{\partial r}\right) = \phi\frac{\partial}{\partial t}\left(\frac{S_o}{B_o}\right). \qquad \text{(C-9)}$$

Similarly, for water,

$$\frac{1}{r}\frac{\partial}{\partial r}\left(r\frac{k_w}{\mu_w B_w}\frac{\partial p}{\partial r}\right) = \phi\frac{\partial}{\partial t}\left(\frac{S_w}{B_w}\right), \quad \text{(C-10)}$$

and, for gas,

$$\frac{1}{r}\frac{\partial}{\partial r}\left[r\left(\frac{R_s k_o}{\mu_o B_o} + \frac{R_{sw}k_w}{\mu_w B_w} + \frac{k_g}{\mu_g B_g}\right)\frac{\partial p}{\partial r}\right]$$

$$= \phi\frac{\partial}{\partial t}\left(\frac{R_s S_o}{B_o} + \frac{R_w S_w}{B_w} + \frac{S_g}{B_g}\right). \quad \text{(C-11)}$$

We can expand the oil equation, Eq. C-9, with the product rule to obtain

$$\frac{1}{r}\left(\frac{k_o}{\mu_o B_o}\right)\frac{\partial}{\partial r}\left(r\frac{\partial p}{\partial r}\right) + \frac{\partial p}{\partial r}\frac{\partial}{\partial r}\left(\frac{k_o}{\mu_o B_o}\right) = \phi\frac{\partial}{\partial t}\left(\frac{S_o}{B_o}\right). \quad \text{(C-12)}$$

If we assume that k_o is constant with respect to pressure, we know that k_o is a function of both oil and water saturation, S_o and S_w, whereas μ_o and B_o are functions of pressure but not saturation. We can then use the chain rule to expand the second term and obtain

$$\left[\frac{1}{r}\left(\frac{k_o}{\mu_o B_o}\right)\frac{\partial}{\partial r}\left(r\frac{\partial p}{\partial r}\right)\right] + \left[\frac{\partial p}{\partial r}\left(\frac{1}{\mu_o B_o}\right)\frac{\partial k_o}{\partial S_o}\frac{\partial S_o}{\partial r}\right]$$

$$+ \left[\frac{\partial p}{\partial r}\left(\frac{1}{\mu_o B_o}\right)\frac{\partial k_o}{\partial S_w}\frac{\partial S_w}{\partial r}\right] + \left[\frac{\partial p}{\partial r}k_o\frac{\partial}{\partial p}\left(\frac{1}{\mu_o B_o}\right)\frac{\partial p}{\partial r}\right]$$

$$= \phi\frac{\partial}{\partial t}\left(\frac{S_o}{B_o}\right). \quad \text{(C-13)}$$

If we assume that $\partial p/\partial r$, $\partial S_o/\partial r$, and $\partial S_w/\partial r$ are small, the products of the pressure and saturation gradients, $(\partial p/\partial r)^2$,

$$\left(\frac{\partial p}{\partial r}\right)\left(\frac{\partial S_o}{\partial r}\right) \text{ and } \left(\frac{\partial p}{\partial r}\right)\left(\frac{\partial S_w}{\partial r}\right)$$

are negligible compared with these gradients, and the oil equation becomes

$$\frac{1}{r}\frac{\partial}{\partial r}\left(r\frac{\partial p}{\partial r}\right) = \frac{\phi\mu_o B_o}{k_o}\frac{\partial}{\partial t}\left(\frac{S_o}{B_o}\right). \quad \text{(C-14)}$$

Similarly, for water and gas, we can write

$$\frac{1}{r}\frac{\partial}{\partial r}\left(r\frac{\partial p}{\partial r}\right) = \frac{\phi\mu_w B_w}{k_w}\frac{\partial}{\partial t}\left(\frac{S_w}{B_w}\right) \quad \text{(C-15)}$$

and

$$\frac{1}{r}\frac{\partial}{\partial r}\left(r\frac{\partial p}{\partial r}\right) = \phi\left(\frac{R_s k_o}{\mu_o B_o} + \frac{R_{sw}k_w}{\mu_w B_w} + \frac{k_g}{\mu_g B_g}\right)^{-1}$$

$$\times\frac{\partial}{\partial t}\left(\frac{R_s S_o}{B_o} + \frac{R_{sw}S_w}{B_w} + \frac{S_g}{B_g}\right). \quad \text{(C-16)}$$

We now need to expand the partial derivatives with respect to time for Eqs. C-14 through C-16. Use the chain rule, where for any variable X, $\partial X/\partial t = (\partial X/\partial p)(\partial p/\partial t)$. Because saturations, FVF's, and solubilities are functions of pressure, we have

$$\frac{1}{r}\frac{\partial}{\partial r}\left(r\frac{\partial p}{\partial r}\right) = \frac{\phi\mu_o B_o}{k_o}\frac{\partial}{\partial p}\left(\frac{S_o}{B_o}\right)\frac{\partial p}{\partial t}$$

$$= \frac{\phi\mu_o B_o}{k_o}\left[\frac{1}{B_o}\frac{\partial S_o}{\partial p}\frac{\partial p}{\partial t} + S_o\frac{\partial}{\partial p}\left(\frac{1}{B_o}\right)\frac{\partial p}{\partial t}\right]. \quad \text{(C-17)}$$

For convenience, we let all partial derivatives with respect to pressure be denoted by B'_o, S'_o, and so forth. We also know that

$$\frac{\partial}{\partial p}\left(\frac{1}{B_o}\right) = \frac{\partial}{\partial p}(B_o^{-1}) = -\frac{1}{B_o^2}\frac{\partial B_o}{\partial p} \quad \text{(C-18)}$$

Therefore,

$$\frac{1}{r}\frac{\partial}{\partial r}\left(r\frac{\partial p}{\partial r}\right) = \frac{\phi\mu_o B_o}{k_o}\left(\frac{S'_o}{B_o} - \frac{S_o B'_o}{B_o^2}\right)\frac{\partial p}{\partial t}. \quad \text{(C-19)}$$

Finally, we obtain the result

$$\frac{1}{r}\frac{\partial}{\partial r}\left(r\frac{\partial p}{\partial r}\right) = \frac{\phi\mu_o}{k_o}\left(S'_o - \frac{S_o B'_o}{B_o}\right)\frac{\partial p}{\partial t}. \quad \text{(C-20)}$$

Similarly, for water we can write

$$\frac{1}{r}\frac{\partial}{\partial r}\left(r\frac{\partial p}{\partial r}\right) = \frac{\phi\mu_w}{k_w}\left(S'_w - \frac{S_w B'_w}{B_w}\right)\frac{\partial p}{\partial t}. \quad \text{(C-21)}$$

The gas equation is slightly more difficult because it contains solubility terms that are also functions of pressure. Expanding the right side of Eq. C-16 with the chain rule gives

$$\frac{1}{r}\frac{\partial}{\partial r}\left(r\frac{\partial p}{\partial r}\right) = \phi\left(\frac{R_s k_o}{\mu_o B_o} + \frac{R_{sw}k_w}{\mu_w B_w} + \frac{k_g}{\mu_g B_g}\right)^{-1}$$

$$\times\frac{\partial}{\partial p}\left(\frac{R_s S_o}{B_o} + \frac{R_{sw}S_w}{B_w} + \frac{S_g}{B_g}\right)\frac{\partial p}{\partial t} \quad \text{(C-22a)}$$

or, finally,

$$\frac{1}{r}\frac{\partial}{\partial r}\left(r\frac{\partial p}{\partial r}\right) = \phi\left(\frac{R_s k_o}{\mu_o B_o} + \frac{R_{sw}k_w}{\mu_w B_w} + \frac{k_g}{\mu_g B_g}\right)^{-1}$$

$$\times\left(\frac{R_s S'_o}{B_o} + \frac{R'_s S_o}{B_o} - \frac{R_s S_o B'_o}{B_o^2} + \frac{R_{sw}S'_w}{B_w} + \frac{R'_{sw}S_w}{B_w}\right.$$

$$\left. - \frac{R_{sw}S_w B'_w}{B_w^2} + \frac{S'_g}{B_g} - \frac{S_g B'_g}{B_g^2}\right)\frac{\partial p}{\partial t}. \quad \text{(C-22b)}$$

We now define mobility as the ratio of the effective permeability of the porous medium to a fluid to the viscosity of the fluid. The definitions of mobility for oil, water, and gas, respectively, are

$$\lambda_o = \frac{k_o}{\mu_o}, \quad \text{(C-23)}$$

$$\lambda_w = \frac{k_w}{\mu_w}, \quad \text{(C-24)}$$

and

$$\lambda_g = \frac{k_g}{\mu_g}. \quad \text{(C-25)}$$

We substitute Eqs. C-23 through C–25 into Eqs. C-20 through C-22 to obtain

$$\frac{1}{r}\frac{\partial}{\partial r}\left(r\frac{\partial p}{\partial r}\right) = \frac{\phi}{\lambda_o}\left(S'_o - \frac{S_o B'_o}{B_o}\right)\frac{\partial p}{\partial t}. \quad \text{(C-26)}$$

Similarly, for water and gas we can write

$$\frac{1}{r}\frac{\partial}{\partial r}\left(r\frac{\partial p}{\partial r}\right) = \frac{\phi}{\lambda_w}\left(S'_w - \frac{S_w B'_w}{B_w}\right)\frac{\partial p}{\partial t}. \quad \text{(C-27)}$$

$$\frac{1}{r}\frac{\partial}{\partial r}\left(r\frac{\partial p}{\partial r}\right) = \phi\left(\frac{R_s\lambda_o}{B_o} + \frac{R_{sw}\lambda_w}{B_w} + \frac{\lambda_g}{B_g}\right)^{-1}$$

$$\times\left(\frac{R_s S'_o}{B_o} + \frac{R'_s S_o}{B_o} - \frac{R_s S_o B'_o}{B_o^2} + \frac{R_{sw}S'_w}{B_w} + \frac{R'_{sw}S_w}{B_w}\right.$$

$$\left. - \frac{R_{sw}S_w B'_w}{B_w^2} + \frac{S'_g}{B_g} - \frac{S_g B'_g}{B_g^2}\right)\frac{\partial p}{\partial t}. \quad \text{(C-28)}$$

We can now start to equate the oil, water, and gas relationships to derive a single equation describing multiphase flow through a porous medium. First, we equate the oil and water equations, Eqs. C-26 and C-27, realizing that the left sides of both equations are equivalent.

$$\frac{\phi}{\lambda_o}\left(S'_o - \frac{S_oB'_o}{B_o}\right)\frac{\partial p}{\partial t} = \frac{\phi}{\lambda_w}\left(S'_w - \frac{S_wB'_w}{B_w}\right)\frac{\partial p}{\partial t}, \quad \ldots \text{(C-29a)}$$

or, simplifying,

$$\frac{\lambda_w}{\lambda_o}\left(S'_o - \frac{S_oB'_o}{B_o}\right) = \left(S'_w - \frac{S_wB'_w}{B_w}\right). \quad \ldots\ldots\ldots \text{(C-29b)}$$

Now, we can equate Eqs. C-26 and C-28, again realizing that the left sides of these equations are the same.

$$\frac{\phi}{\lambda_o}\left(S'_o - \frac{S_oB'_o}{B_o}\right)\frac{\partial p}{\partial t} = \phi\left(\frac{R_s\lambda_o}{B_o} + \frac{R_{sw}\lambda_w}{B_w} + \frac{\lambda_g}{B_g}\right)^{-1} \times \left(\frac{R_sS'_o}{B_o} + \frac{R'_sS_o}{B_o} - \frac{R_sS_oB'_o}{B_o^2} + \frac{R_{sw}S'_w}{B_w} + \frac{R'_{sw}S_w}{B_w} - \frac{R_{sw}S_wB'_w}{B_w^2} + \frac{S'_g}{B_g} - \frac{S_gB'_g}{B_g^2}\right)\frac{\partial p}{\partial t}. \quad \ldots\ldots\ldots \text{(C-30)}$$

Cancelling like terms and rearranging gives

$$\frac{1}{\lambda_o}\left(S'_o - \frac{S_oB'_o}{B_o}\right)\left(\frac{R_s\lambda_o}{B_o} + \frac{R_{sw}\lambda_w}{B_w} + \frac{\lambda_g}{B_g}\right) = \left(\frac{R_sS'_o}{B_o} + \frac{R'_sS_o}{B_o} - \frac{R_sS_oB'_o}{B_o^2} + \frac{R_{sw}S'_w}{B_w} + \frac{R'_{sw}S_w}{B_w} - \frac{R_{sw}S_wB'_w}{B_w^2} + \frac{S'_g}{B_g} - \frac{S_gB'_g}{B_g^2}\right). \quad \ldots\ldots\ldots \text{(C-31)}$$

Grouping terms on the right side gives

$$\frac{1}{\lambda_o}\left(S'_o - \frac{S_oB'_o}{B_o}\right)\left(\frac{R_s\lambda_o}{B_o} + \frac{R_{sw}\lambda_w}{B_w} + \frac{\lambda_g}{B_g}\right) = \left[\frac{R_sS'_o}{B_o} + \frac{R_s}{B_o}\left(S'_o - \frac{S_oB'_o}{B_o}\right) + \frac{R_{sw}S'_w}{B_w} + \frac{R_{sw}}{B_w}\left(S'_w - \frac{S_wB'_w}{B_w}\right) + \frac{S'_g}{B_g} - \frac{S_gB'_g}{B_g^2}\right]. \quad \ldots\ldots\ldots \text{(C-32)}$$

Substituting Eq. C-29b into Eq. C-32 gives

$$\frac{1}{\lambda_o}\left(S'_o - \frac{S_oB'_o}{B_o}\right)\left(\frac{R_s\lambda_o}{B_o} + \frac{R_{sw}\lambda_w}{B_w} + \frac{\lambda_g}{B_g}\right) = \left[\frac{R_sS'_o}{B_o} + \frac{R_s}{B_o}\left(S'_o - \frac{S_oB'_o}{B_o}\right) + \frac{R_{sw}S'_w}{B_w} + \left(\frac{R_{sw}\lambda_w}{B_w\lambda_o}\right)\left(S'_o - \frac{S_oB'_o}{B_o}\right) + \frac{S'_g}{B_g} - \frac{S_gB'_g}{B_g^2}\right]. \quad \ldots \text{(C-33)}$$

Factoring all terms with $(S'_o - S_oB'_o/B_o)$ on the left side gives

$$\left(S'_o - \frac{S_oB'_o}{B_o}\right)\left(\frac{R_s}{B_o} + \frac{R_{sw}\lambda_w}{\lambda_oB_w} + \frac{\lambda_g}{\lambda_oB_g} - \frac{R_s}{B_o} - \frac{R_{sw}\lambda_w}{\lambda_oB_w}\right) = \left(\frac{R_sS'_o}{B_o} + \frac{R_{sw}S'_w}{B_w} + \frac{S'_g}{B_g} - \frac{S_gB'_g}{B_g^2}\right), \quad \ldots\ldots\ldots \text{(C-34a)}$$

which simplifies to

$$\left(S'_o - \frac{S_oB'_o}{B_o}\right)\left(\frac{\lambda_g}{\lambda_oB_g}\right) = \left[\frac{R_sS'_o}{B_o} + \frac{R_{sw}S'_w}{B_w} + \frac{1}{B_g}\left(S'_g - \frac{S_gB'_g}{B_g^2}\right)\right]. \quad \ldots \text{(C-34b)}$$

Multiplying through by B_g, we obtain

$$\left(S'_o - \frac{S_oB'_o}{B_o}\right)\left(\frac{\lambda_g}{\lambda_o}\right) = \left(\frac{R'_sS_oB_g}{B_o} + \frac{R'_{sw}S_wB_g}{B_w} + S'_g - \frac{S_gB'_g}{B_g}\right). \quad \ldots\ldots\ldots \text{(C-35)}$$

We know that for a three-phase system,

$$S_g + S_w + S_o = 1. \quad \ldots\ldots\ldots \text{(C-36)}$$

Differentiating Eq. C-36 with respect to pressure gives

$$S'_o + S'_w + S'_g = 0. \quad \ldots\ldots\ldots \text{(C-37)}$$

Also, if we define total mobility of a three-phase system, λ_t, as the sum of the individual mobilities,

$$\lambda_t = \lambda_o + \lambda_w + \lambda_g. \quad \ldots\ldots\ldots \text{(C-38)}$$

Substituting Eqs. C-37 and C-38 into Eq. C-35 for S_g and λ_g, respectively, gives

$$\left(S'_o - \frac{S_oB'_o}{B_o}\right)\left(\frac{\lambda_t - \lambda_o - \lambda_w}{\lambda_o}\right) = \frac{R'_sS_oB_g}{B_o} + \frac{R'_{sw}S_wB_g}{B_w} - S'_o - S'_w - \frac{S_gB'_g}{B_g}. \quad \ldots\ldots\ldots \text{(C-39)}$$

If we add the terms $S_oB'_o/B_o$ and $S_wB'_w/B_w$ to each side of the equation, we can group terms again that contain $(S'_o - S_oB'_o/B_o)$.

$$\left(S'_o - \frac{S_oB'_o}{B_o}\right)\left(\frac{\lambda_t - \lambda_o - \lambda_w}{\lambda_o}\right) + \frac{S_oB'_o}{B_o} + \frac{S_wB'_w}{B_w} = \frac{R'_sS_oB_g}{B_o} + \frac{R'_{sw}S_wB_g}{B_w} - S'_o + \frac{S_oB'_o}{B_o} - S'_w + \frac{S_wB'_w}{B_w} - \frac{S_gB'_g}{B_g} \quad \ldots\ldots\ldots \text{(C-40)}$$

Recall that Eq. C-29b stated ,

$$\frac{\lambda_w}{\lambda_o}\left(S'_o - \frac{S_oB'_o}{B_o}\right) = \left(S'_w - \frac{S_wB'_w}{B_w}\right). \quad \ldots\ldots\ldots \text{(C-29b)}$$

Substituting Eq. C-29b into Eq. C-40 gives

$$\left(S'_o - \frac{S_oB'_o}{B_o}\right)\left(\frac{\lambda_t - \lambda_o - \lambda_w}{\lambda_o}\right) + \frac{S_oB'_o}{B_o} + \frac{S_wB'_w}{B_w} = \frac{R'_sS_oB_g}{B_o} + \frac{R'_{sw}S_wB_g}{B_w} - \left(S'_o - \frac{S_oB'_o}{B_o}\right) - \frac{\lambda_w}{\lambda_o}\left(S'_o - \frac{S_oB'_o}{B_o}\right) - \frac{S_gB'_g}{B_g}. \quad \ldots\ldots\ldots \text{(C-41)}$$

Again, factoring all terms containing $(S'_o - S_oB'_o/B_o)$ on the left side gives

$$\left(S'_o - \frac{S_o B'_o}{B_o}\right)\left(\frac{\lambda_t - \lambda_o - \lambda_w}{\lambda_o} + 1 + \frac{\lambda_w}{\lambda_o}\right)$$

$$= \frac{R'_s S_o B_g}{B_o} + \frac{R'_{sw} S_w B_g}{B_w} - \frac{S_o B'_o}{B_o} - \frac{S_w B'_w}{B_w} - \frac{S_g B'_g}{B_g} \quad \text{(C-42)}$$

or, finally,

$$\left(S'_o - \frac{S_o B'_o}{B_o}\right)\left(\frac{\lambda_t}{\lambda_o}\right)$$

$$= \frac{R'_s S_o B_g}{B_o} - \frac{S_o B'_o}{B_o} + \frac{R'_{sw} S_w B_g}{B_w}$$

$$- \frac{S_w B'_w}{B_w} - \frac{S_g B'_g}{B_g}. \quad \text{(C-43)}$$

We now define total compressibility, c_t, as

$$c_t = S_o c_o + S_w c_w + S_g c_g. \quad \text{(C-44)}$$

(For a system with pressure-dependent porosity, $c_t = S_o c_o + S_w c_w + S_g c_g + c_f$.)

Below the bubblepoint, the definition of the compressibility of oil is

$$c_o = -\frac{1}{B_o}\left(B'_o - R'_s B_g\right). \quad \text{(C-45)}$$

Similarly, water compressibility below the bubblepoint is defined as

$$c_w = -\frac{1}{B_w}\left(B'_w - R'_{sw} B_g\right) \quad \text{(C-46)}$$

and the compressibility of gas is always defined as

$$c_g = -\frac{B'_g}{B_g}. \quad \text{(C-47)}$$

Substituting Eqs. C-45 through C-47 into Eq. C-44 gives the definition of total compressibility as

$$c_t = \frac{R'_s S_o B_g}{B_o} - \frac{S_o B'_o}{B_o} + \frac{R'_{sw} S_w B_g}{B_w}$$

$$- \frac{S_w B'_w}{B_w} - \frac{S_g B'_g}{B_g}. \quad \text{(C-48)}$$

Substituting Eq. C-48 into Eq. C-43,

$$\left(S'_o - \frac{S_o B'_o}{B_o}\right)\left(\frac{\lambda_t}{\lambda_o}\right) = c_t. \quad \text{(C-49)}$$

We can now substitute Eq. C-49 into the original oil, gas, or water equations to obtain the diffusivity equation for multiphase flow. We substitute Eq. C-49 into Eq. C-26, the oil equation, and

$$\frac{1}{r}\frac{\partial}{\partial r}\left(r\frac{\partial p}{\partial r}\right) = \frac{\phi}{\lambda_o}\left(S'_o - \frac{S_o B'_o}{B_o}\right)\frac{\partial p}{\partial t} \quad \text{(C-26)}$$

becomes

$$\frac{1}{r}\frac{\partial}{\partial r}\left(r\frac{\partial p}{\partial r}\right) = \frac{\phi}{\lambda_o}\left(\frac{\lambda_o}{\lambda_t}c_t\right)\frac{\partial p}{\partial t} \quad \text{(C-50a)}$$

or $$\frac{1}{r}\frac{\partial}{\partial r}\left(r\frac{\partial p}{\partial r}\right) = \left(\frac{\phi c_t}{\lambda_t}\right)\frac{\partial p}{\partial t}. \quad \text{(C-50b)}$$

This equation is similar to the single-phase diffusivity equation for slightly compressible liquids given by

$$\frac{1}{r}\frac{\partial}{\partial r}\left(r\frac{\partial p}{\partial r}\right) = \left(\frac{\phi \mu_o c_t}{k_o}\right)\frac{\partial p}{\partial t}, \quad \text{(C-51)}$$

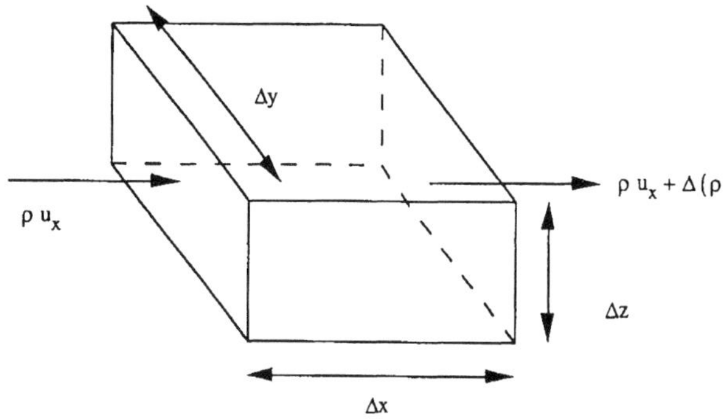

$$\left(\begin{array}{c}\textit{rate of mass}\\ \textit{input to control}\\ \textit{volume}\end{array}\right) - \left(\begin{array}{c}\textit{rate of mass}\\ \textit{leaving control}\\ \textit{volume}\end{array}\right) = \left(\begin{array}{c}\textit{rate of mass}\\ \textit{accumulation}\\ \textit{in control}\\ \textit{volume}\end{array}\right)$$

Fig. C-1—Mass balance for linear flow.

which implies that the solutions to the single-phase diffusivity equation presented later in this appendix apply to multiphase flow as long as c_t is defined by Eq. C-48 and we use λ_t instead of k_o/μ_o.

Linear Flow of Gas

To derive a diffusivity equation, we must combine the law of conservation of mass, Darcy's law (equation of motion), and equation of state (real-gas law).

Law of Conservation of Mass (Continuity Equation). This is shown by **Fig. C-1** or by the equation

$$\dot{m}_{\text{in}} - \dot{m}_{\text{out}} = \dot{m}_{\text{accum}} \quad \text{(C-52)}$$

where $\dot{m}_{\text{in}} = \rho u_x \Delta y \Delta z$, $\Delta y \Delta z$ = cross-sectional area, and $\dot{m}_{\text{out}} = \left[\rho u_x + \Delta(\rho u_x)\right]\Delta y \Delta z$.

$$\dot{m}_{\text{in}} - \dot{m}_{\text{out}} = \rho u_x \Delta y \Delta z - \rho u_x \Delta y \Delta z - \Delta(\rho u_x)\Delta y \Delta z \quad \text{(C-53a)}$$

$$\dot{m}_{\text{in}} - \dot{m}_{\text{out}} = \Delta(\rho u_x)\Delta y \Delta z \quad \text{(C-53b)}$$

where u_x = volumetric flow rate per unit cross-sectional area and ρ = fluid density = m/V. The mass in the control volume at any given time would be

$$m = \rho V_p = \rho \phi V_b$$

$$= \rho \phi \Delta x \Delta y \Delta z\,; \quad \text{(C-54)}$$

therefore, the mass accumulated at time t is given by $(\rho\phi\Delta x\Delta y\Delta z)_t$ and the rate of mass accumulated in time Δt is $m_{t+\Delta t} - m_t$.

$$\dot{m}_{\text{accum}} = \frac{(\rho\phi\Delta x\Delta y\Delta z)_{t+\Delta t} - (\rho\phi\Delta x\Delta y\Delta z)_t}{\Delta t}. \quad \text{(C-55)}$$

Substitute Eqs. C-53b and C-55 in Eq. C-52.

$$-\Delta(\rho u_x)\Delta y \Delta z = 1/\Delta t\left[(\rho\phi\Delta x\Delta y\Delta z)_{t+\Delta t} - (\rho\phi\Delta x\Delta y\Delta z)_t\right].$$

Divide through by $\Delta x \Delta y \Delta z$.

$$-\frac{\Delta(\rho u_x)}{\Delta x} = \frac{(\rho\phi)_{t+\Delta t} - (\rho\phi)_t}{\Delta t} = \frac{\Delta(\phi\rho)}{\Delta t}. \quad \text{(C-56)}$$

Take limits of Eq. C-56 as Δx and $\Delta t \rightarrow 0$ to obtain the following differential equation.

$$-\frac{\partial(\rho u_x)}{\partial x} = \frac{\partial(\phi\rho)}{\partial t}. \quad \text{(C-57)}$$

For density in lbm/ft^3, we must have the following units: x = feet, t = hours, ϕ = fraction, and

$u = \frac{q}{A} = \text{STB/D–ft}^2.$

Performing dimensional analysis on Eq. C-57, we have

$$\frac{\text{lbm}}{\text{ft}^3} \times \frac{\text{STB}}{\text{D-ft}^2} \times \frac{1}{\text{ft}} = \frac{\text{lbm}}{\text{ft}^3} \times \frac{1}{\text{hour}}$$

and $\frac{\text{STB}}{\text{D-ft}^3} \times \frac{5.615 \text{ ft}^3}{\text{STB}} \times \frac{1 \text{ day}}{24 \text{ hours}} = \frac{1}{\text{hour}}.$

Therefore, the continuity equation becomes (in field units)

$$\frac{\partial(\rho u_x)}{\partial x} = 4.274\frac{\partial(\phi\rho)}{\partial t}. \quad \text{(C-58)}$$

To this point we have assumed 1D linear flow.

Darcy's Law (Equation of Motion). Now we need an equation of motion for linear flow. Start with the differential form of Darcy's law.

$$u_x = -\frac{k_x}{\mu}\frac{\partial p}{\partial x}, \quad \text{(C-59a)}$$

or, in field units,

$$u_x = -0.001127\frac{k_x}{\mu}\frac{\partial p}{\partial x}, \quad \text{(C-59b)}$$

where $k_x = \text{md}$, $\mu = \text{cp}$, $\rho = \text{psia}$, $x = \text{ft}$, and $u_x = \text{STB/D-ft}^2$. Eq. C-59b assumes negligible gravity effects, constant cross-sectional area, and laminar flow.

Substitute Eq. C-59b in Eq. C-58.

$$\frac{\partial}{\partial x}\left(-\rho \times 0.001127\frac{k_x}{\mu}\frac{\partial p}{\partial x}\right) = 4.274\frac{\partial(\phi\rho)}{\partial t}.$$

Assume constant k:

$$k\frac{\partial}{\partial x}\left(\frac{\rho\partial\rho}{\mu\partial x}\right) = 3792\frac{\partial}{\partial t}(\phi\rho),$$

$$\text{or } \frac{\partial}{\partial x}\left(\frac{\rho\partial\rho}{\mu\partial x}\right) = \frac{1}{0.0002637k}\frac{\partial}{\partial t}(\phi\rho). \quad \text{(C-60)}$$

Real-Gas Law (EOS). We need to incorporate an EOS for gas. From the real-gas law,

$$\rho = \frac{M}{RT}\frac{p}{z}. \quad \text{(C-61)}$$

Substitute Eq. C-61 into Eq. C-60.

$$\frac{\partial}{\partial x}\left(\frac{M}{RT}\frac{p}{\mu z}\frac{\partial\rho}{\partial x}\right) = \frac{1}{0.0002637k}\frac{\partial}{\partial t}\left(\phi\frac{M}{RT}\frac{p}{z}\right). \quad \text{(C-62)}$$

Assume T = constant (isothermal):

$$\frac{\partial}{\partial x}\left(\frac{p}{\mu z}\frac{\partial\rho}{\partial x}\right) = \frac{1}{0.0002637k}\frac{\partial}{\partial t}\left(\phi\frac{p}{z}\right). \quad \text{(C-63)}$$

Expanding the right side of Eq. C-63 gives

$$\frac{1}{0.0002637k}\frac{\partial}{\partial t}\left(\phi\frac{p}{z}\right) = \frac{1}{0.0002637k}\left[\phi\frac{\partial}{\partial t}\left(\frac{p}{z}\right) + \frac{p}{z}\frac{\partial\phi}{\partial t}\right]$$

$$= \frac{1}{0.0002637k}\left[\phi\frac{\partial}{\partial\rho}\left(\frac{p}{z}\right)\frac{\partial\rho}{\partial t} + \frac{p}{z}\frac{\partial\phi}{\partial t}\frac{\partial\rho}{\partial t}\right]$$

$$= \frac{1}{0.0002637k}\phi\frac{p}{z}\left[\frac{z}{p}\frac{\partial\left(\frac{p}{z}\right)}{\partial\rho} + \frac{1}{\phi}\frac{\partial\phi}{\partial t}\right]\frac{\partial\rho}{\partial t} \quad \text{(C-64)}$$

We define

$$c_g = \frac{1}{\rho}\frac{\partial\rho}{\partial p} = \frac{z}{p}\frac{\partial}{\partial p}\left(\frac{p}{z}\right). \quad \text{(C-65)}$$

$$\text{and } c_g = \frac{1}{\phi}\frac{\partial\phi}{\partial p}. \quad \text{(C-66)}$$

$$\text{and } c_t = c_g + c_f. \quad \text{(C-67a)}$$

More generally,

$$c_t = c_g S_g + c_w S_w + c_f. \quad \text{(C-67b)}$$

(We should assume an immobile water saturation. Even though immobile, the water still contributes to the compressibility of the system.)

Substituting Eqs. C-65 and C-66 into Eq. C-67 and substituting Eq. C-67 into Eq. C-64 gives

$$\frac{1}{0.0002637k}\frac{\partial}{\partial t}\left(\phi\frac{p}{z}\right) = \frac{1}{0.0002637k}\phi\frac{p}{z}(c_t)\frac{\partial p}{\partial t}. \quad \text{(C-68)}$$

Substitute Eq. C-68 into Eq. C-63.

$$\frac{\partial}{\partial x}\left(\frac{p}{\mu z}\frac{\partial p}{\partial x}\right) = \frac{\phi c_t}{0.0002637k}\frac{p}{z}\frac{\partial p}{\partial t}. \quad \text{(C-69)}$$

We define pseudopressure as

$$p_p(p) = 2\int_{p_b}^{p}\frac{p}{\mu z}dp; \quad \text{(C-70)}$$

$$\therefore \frac{\partial p_p}{\partial t} = \frac{\partial p_p}{\partial p}\frac{\partial p}{\partial t} = \frac{2p}{\mu z}\frac{\partial p}{\partial t} \quad \text{(C-71)}$$

$$\text{and } \frac{\partial p_p}{\partial x} = \frac{\partial p_p}{\partial p}\frac{\partial p}{\partial x} = \frac{2p}{\mu z}\frac{\partial p}{\partial x}. \quad \text{(C-72)}$$

Substitute Eqs. C-71 and C-72 into Eq. C-69 to obtain

$$\frac{\partial}{\partial x}\left(\frac{p}{\mu z}\frac{\mu z}{2p}\frac{\partial p_p}{\partial x}\right) = \frac{\phi c_t}{0.0002637k}\frac{p}{z}\frac{\partial p_p}{\partial x}\frac{\mu z}{2p} \quad \text{(C-73a)}$$

or, finally,

$$\frac{\partial^2 p_p}{\partial x^2} = \frac{\phi c_t}{0.0002637k}\frac{\partial p_p}{\partial t}. \quad \text{(C-73b)}$$

Summary of Assumptions.

- One-dimensional linear flow.
- Negligible gravity effects.
- Constant cross-sectional area.
- Laminar flow.
- Constant k.
- Real gas.
- Isothermal.
- Immobile, compressible water saturation.

Appendix D
Shape Factors for Various Single-Well Drainage Areas

TABLE D-1—BOUNDED RESERVOIRS

Shape Factor	C_A	$\ln C_A$	$0.5 \ln\left(\frac{2.2458}{C_A}\right)$	Exact for $t_{AD} >$	Less Than 1% Error for $t_{AD} >$	Use Infinite System Solution With Less Than 1% Error for $t_{AD} <$
	31.62	3.4538	1.3224	0.1	0.06	0.10
	31.6	3.4532	−1.3220	0.1	0.06	0.10
	27.6	3.3178	−1.2544	0.2	0.07	0.09
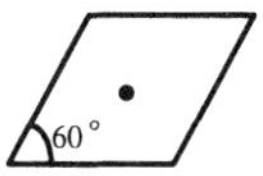	27.1	3.2995	−1.2452	0.2	0.07	0.09
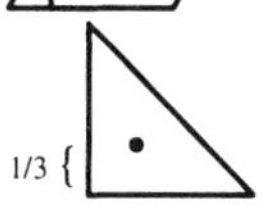	21.9	3.0865	−1.1387	0.4	0.12	0.08
	0.098	−2.3227	1.5659	0.9	0.60	0.015
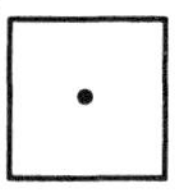	30.8828	3.4302	−1.3106	0.1	0.05	0.09
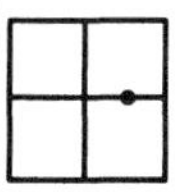	12.9851	2.5638	−0.8774	0.7	0.25	0.03
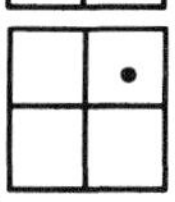	4.5132	1.5070	−0.3490	0.6	0.30	0.025
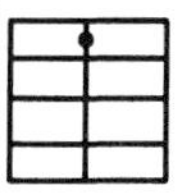	3.3351	1.2045	−0.1977	0.7	0.25	0.01
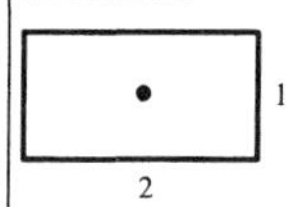	21.8369	3.0836	−1.1373	0.3	0.15	0.025
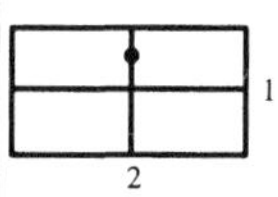	10.8374	2.3830	−0.7870	0.4	0.15	0.025
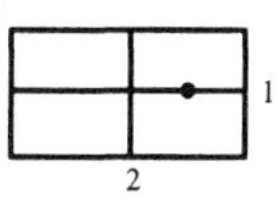	4.5141	1.5072	−0.3491	1.5	0.50	0.06
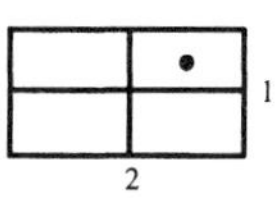	2.0769	0.7309	0.0391	1.7	0.50	0.02
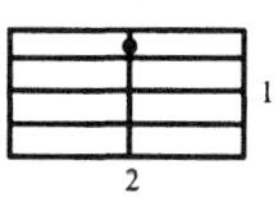	3.1573	1.1497	−0.1703	0.4	0.15	0.005

From Earlougher, Robert C. Jr.: *Advances in Well Test Analysis,* Monograph Series, SPE, Dallas (1977) **5,** 203.

TABLE D-1—BOUNDED RESERVOIRS (CONTINUED)

Shape Factor	C_A	$\ln C_A$	$0.5\ln\left(\frac{2.2458}{C_A}\right)$	Exact for $t_{AD} >$	Less Than 1% Error for $t_{AD} >$	Use Infinite System Solution With Less Than 1% Error for $t_{AD} <$
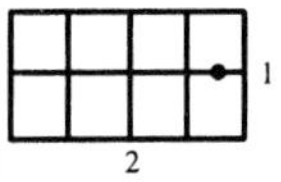	0.5813	−0.5425	0.6758	2.0	0.60	0.02
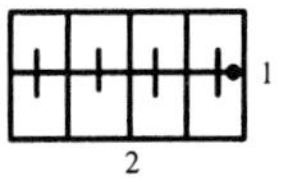	0.1109	−2.1991	1.5041	3.0	0.60	0.005
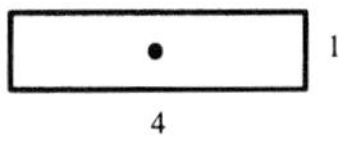	5.3790	1.6825	−0.4367	0.8	0.30	0.01
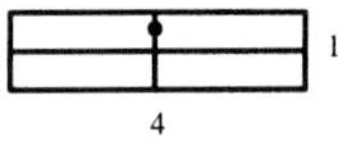	2.6896	0.9894	−0.0902	0.8	0.30	0.01
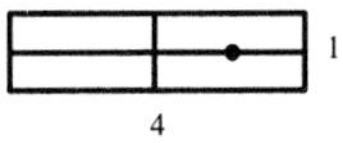	0.2318	−1.4619	1.1355	4.0	2.00	0.03
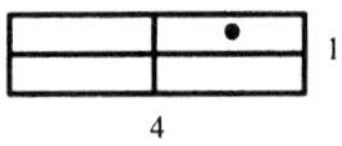	0.1155	−2.1585	1.4838	4.0	2.00	0.01
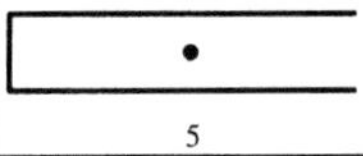	2.3606	0.8589	−0.0249	1.0	0.40	0.025

TABLE D-2–VERTICALLY FRACTURED RESERVOIRS*

Shape Factor	C_A	$\ln C_A$	$0.5\ln\left(\frac{2.2458}{C_A}\right)$	Exact for $t_{AD} >$	Less Than 1% Error for $t_{AD} >$	Use Infinite System Solution With Less Than 1% Error for $t_{AD} <$
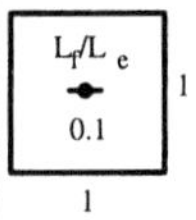	2.6541	0.9761	−0.0835	0.175	0.08	cannot use
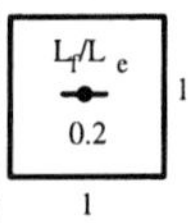	2.0348	0.7104	0.0493	0.175	0.09	cannot use
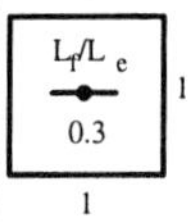	1.9886	0.6924	0.0583	0.175	0.09	cannot use
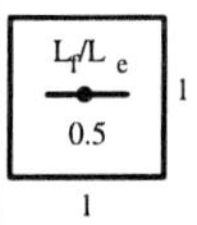	1.6620	0.5080	0.1505	0.175	0.09	cannot use
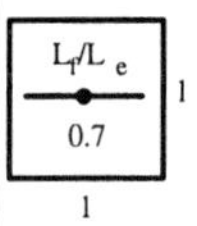	1.3127	0.2721	0.2685	0.175	0.09	cannot use
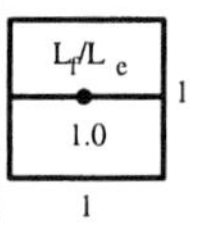	0.7887	−0.2374	0.5232	0.175	0.09	cannot use

*Use $(r_e/L_f)^2$ in place of A/r_w^2.

TABLE D-3–WATERDRIVE RESERVOIRS/RESERVOIRS OF UNKNOWN PRODUCTION CHARACTER

Shape Factor	C_A	$\ln C_A$	$0.5\ln\left(\frac{2.2458}{C_A}\right)$	Exact for $t_{AD} >$	Less Than 1% Error for $t_{AD} >$	Use Infinite System Solution With Less Than 1% Error for $t_{AD} <$
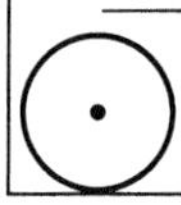	19.1/25	2.9/3.2	−1.07/−1.20	—	—	—

Appendix E
Validation of Method of Images

Superposition for a No-Flow Boundary

Consider the two-well system illustrated by **Fig. E-1**. The dimensionless pressure at Point X, assuming an infinite-acting system with line-source wells, would be given by

$$p_D(x_D, y_D, t_D) = p_D(a_{D1}, t_D) + p_D(a_{D2}, t_D), \quad \text{(E-1)}$$

where $p_D(a_{D1}, t_D) = -\frac{1}{2}\text{Ei}\left(\frac{-a_{D1}^2}{4t_D}\right)$ .... (E-2)

and $p_D(a_{D2}, t_D) = -\frac{1}{2}\text{Ei}\left(\frac{-a_{D2}^2}{4t_D}\right)$. .... (E-3)

Therefore,

$$p_D(x_D, y_D, t_D) = -\frac{1}{2}\text{Ei}\left(\frac{-a_{D1}^2}{4t_D}\right) - \frac{1}{2}\text{Ei}\left(\frac{-a_{D2}^2}{4t_D}\right). \quad \text{(E-4)}$$

The dimensionless distances a_{D1} and a_{D2} are determined by use of the geometry of the system.

$$a_{D1} = \frac{a_1}{r_w} = \frac{1}{r_w}\sqrt{(x+L)^2 + y^2} = \sqrt{(x_D + L_D)^2 + y_D^2} \quad \text{(E-5)}$$

and $a_{D2} = \frac{a_2}{r_w} = \frac{1}{r_w}\sqrt{[(x-L)]^2 + y^2} = \sqrt{(x_D - L_D)^2 + y_D^2}$. .... (E-6)

Computing $\partial p_D/\partial x_D$ from Eq. E-4 gives

$$\frac{\partial p_D}{\partial x_D} = \frac{1}{2}\left\{\frac{\partial}{\partial x_D}\left[-\text{Ei}\left(\frac{-a_{D1}^2}{4t_D}\right)\right] + \frac{\partial}{\partial x_D}\left[-\text{Ei}\left(\frac{-a_{D2}^2}{4t_D}\right)\right]\right\}. \quad \text{(E-7)}$$

We know that

$$-\text{Ei}\left(\frac{-a_{D1}^2}{4t_D}\right) = \int_{\frac{a_{D1}^2}{4t_D}}^{\infty} \frac{e^{-u}}{u}du. \quad \text{(E-8)}$$

Using Liebnitz's rule (Eq. 1.60) of differentiating an integral with Eq. E-8 gives

$$\frac{\partial}{\partial a_{D1}}\left[-\text{Ei}\left(\frac{-a_{D1}^2}{4t_D}\right)\right] = 0 - \frac{e^{-a_{D1}^2/4t_D}}{-a_{D1}^2/4t_D}\left(\frac{2a_{D1}}{4t_D}\right)$$

$$\frac{\partial}{\partial a_{D1}}\left[-\text{Ei}\left(\frac{-a_{D1}^2}{4t_D}\right)\right] = -\frac{2\exp\left[-\left(a_{D1}^2/4t_D\right)\right]}{a_{D1}}. \quad \text{(E-9)}$$

Similarly,

$$\frac{\partial}{\partial a_{D2}}\left[-\text{Ei}\left(\frac{-a_{D2}^2}{4t_D}\right)\right] = -\frac{2\exp\left[-\left(a_{D2}^2/4t_D\right)\right]}{a_{D2}}. \quad \text{(E-10)}$$

By the chain rule,

$$\frac{\partial}{\partial x_D}\left[-\text{Ei}\left(\frac{-a_{D1}^2}{4t_D}\right)\right] = \frac{\partial}{\partial a_{D1}}\left[-\text{Ei}\left(\frac{-a_{D1}^2}{4t_D}\right)\right]\frac{\partial a_{D1}}{\partial x_D}. \quad \text{(E-11)}$$

We need $\partial a_{D1}/\partial x_D$—so, from Eq. E-5,

$$a_{D1}^2 = (x_D + L_D)^2 + y_D^2. \quad \text{(E-12)}$$

Therefore,

$$2a_{D1}\frac{\partial a_{D1}}{\partial x_D} = 2(x_D + L_D)^{(2-1)} = 2(x_D + L_D) \quad \text{(E-13a)}$$

and $\frac{\partial a_{D1}}{\partial x_D} = \frac{(x_D + L_D)}{a_{D1}}$. .... (E-13b)

Similarly, from Eq. E-6,

$$a_{D2}^2 = (x_D - L_D)^2 + y_D^2, \quad \text{(E-14a)}$$

$$2a_{D2}\frac{\partial a_{D2}}{\partial x_D} = 2(x_D - L_D), \quad \text{(E-14b)}$$

and $\frac{\partial a_{D2}}{\partial x_D} = \frac{(x_D - L_D)}{a_{D2}}$. .... (E-14c)

Using Eqs. E-9 and E-13 in Eq. E-11 gives

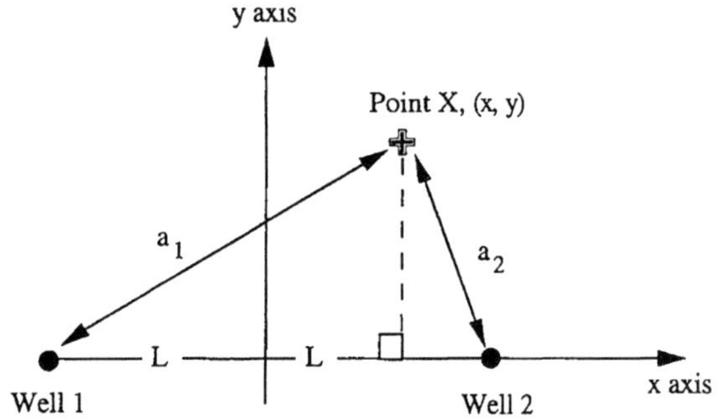

Fig. E-1—Modeling a no-flow boundary with the method of images.

$$\frac{\partial}{\partial x_D}\left[-\mathrm{Ei}\left(\frac{-a_{D1}^2}{4t_D}\right)\right] = -\frac{2\exp\left[-\left(a_{D1}^2/4t_D\right)\right]}{a_{D1}}$$

$$\times\frac{(x_D+L_D)}{a_{D1}} = \frac{2(x_D+L_D)\exp\left[-\left(a_{D1}^2/4t_D\right)\right]}{a_{D1}^2}. \qquad \text{(E-15)}$$

Similarly, with Eqs. E-10 and E-14,

$$\frac{\partial}{\partial x_D}\left[-\mathrm{Ei}\left(\frac{-a_{D2}^2}{4t_D}\right)\right] = -\frac{2\exp\left[-\left(a_{D2}^2/4t_D\right)\right]}{a_{D2}}$$

$$\times\frac{(x_D-L_D)}{a_{D2}} = \frac{2(x_D-L_D)\exp\left[-\left(a_{D2}^2/4t_D\right)\right]}{a_{D2}^2}. \qquad \text{(E-16)}$$

Using Eqs. E-15 and E-16 in Eq. E-7 gives

$$\frac{\partial p_D}{\partial x_D} = \frac{1}{2}\left(\left\{\frac{-2(x_D+L_D)\exp\left[-\left(a_{D1}^2/4t_D\right)\right]}{a_{D1}^2}\right\} + \left\{-\frac{2(x_D-L_D)\exp\left[-\left(a_{D2}^2/4t_D\right)\right]}{a_{D2}^2}\right\}\right)$$

$$= -\left[\frac{(x_D+L_D)e^{-a_{D1}^2/4t_D}}{a_{D1}^2} + \frac{(x_D-L_D)e^{-a_{D2}^2/4t_D}}{a_{D2}^2}\right]. \qquad \text{(E-17)}$$

At $x_D = 0, a_{D1}^2 = L_D^2 + y_D^2$ and $a_{D2}^2 = L_D^2 + y_D^2$. Therefore, $a_{D2} = a_{D1}$, and Eq. E-17 becomes

$$\frac{\partial p_D}{\partial x_D} = -\left\{\frac{L_D}{L_D^2+y_D^2}\exp\left[-\frac{\left(L_D^2+y_D^2\right)}{4t_D}\right] - \frac{L_D}{L_D^2+y_D^2}\exp\left[-\frac{\left(L_D^2+y_D^2\right)}{4t_D}\right]\right\} = 0. \qquad \text{(E-18)}$$

This suggests that any line bisecting the line connecting two producing wells will have the condition

$$\left(\frac{\partial p_D}{\partial x_D}\right)_{(x_D\to 0)} = 0, \qquad \text{(E-19)}$$

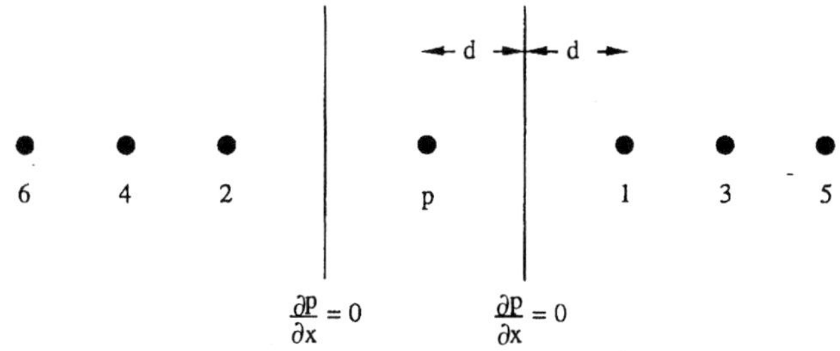

Fig. E-2—Method of images for parallel faults.

which is the boundary condition for a no-flow boundary. This result is the basis of the method of images. This implies that, if a well is near a no-flow boundary (i.e., a sealing fault), we can model the fault by placing an image well the same distance away from the fault on the other side. We can extend this idea to any boundary condition.[1] For example, **Fig. E-2** illustrates two parallel sealing faults.

The solution would be

$$p_D(x_D, y_D, t_D) = p_D\text{ (due to well)} + p_D\text{ (due to image wells)}$$

$$= p_{DW} + p_D(1) + p_D(2) + p_D(3) + p_D(4) + \ldots \qquad \text{(E-20)}$$

The basic message in the method of images is

$$p_D(x_D, y_D, t_{AD}) = -\frac{1}{2}\sum_{i=1}^{\infty}\mathrm{Ei}\left(-\frac{a_{iD}^2}{4t_{AD}}\right), \qquad \text{(E-21)}$$

where $t_{AD} = a_2 kt/\phi\mu c_t A$ and a_{iD} = distance of well from Point (x_D, y_D).

Superposition for a Constant-Pressure Boundary

We now show how the method of images works for constant-pressure boundaries. We have stated that a constant-pressure boundary can be modeled by an image well producing at the negative of the rate of the original well as long as this image well is placed an equal distance on the other side of the boundary from the producing well.

In this case, the boundary condition at $x_D = 0$ is

$$p_D(x_D, y_D, t_D) = 0. \qquad \text{(E-22)}$$

The superposition formula is

$$p_D(x_D, y_D, t_D) = p_D\left(a_{D1}, t_D\right) + p_D\left(a_{D2}, t_D\right). \qquad \text{(E-23)}$$

At $x = 0$, we can write

$$p_D\left(a_{D1}, t_D\right) = -p_D\left(a_{D2}, t_D\right). \qquad \text{(E-24)}$$

We know from our previous derivation that, at $x = 0$, $a_{D1} = a_{D2}$. For the dimensionless pressures to be equal in magnitude but opposite in sign, one of the parameters in the definition of dimensionless pressure given here must be negative.

$$p_D = \frac{a_1 qB\mu}{2\pi kh}(\Delta p). \qquad \text{(E-25)}$$

The only parameter that can be negative is rate, so to model a constant-pressure boundary, we use an injection well the same distance from the boundary as the actual producer.

Reference

1. Matthews, C.S., Brons, F., and Hazebroek, P.: "A Method for the Determination of Average Pressure in a Bounded Reservoir," *Trans.*, AIME, **201**, 182.

Appendix F
Determining Pressure-Data Derivatives

Determining Pressure-Data Derivatives

The derivative[1] at a point is determined by finding a weighted mean of the slopes to a preceding point and a following point, as illustrated in **Fig. F-1**. The parameter L defines the minimum abscissa distance to these points, smoothing out "noise" in the neighborhood of the central point. Thus, L may be defined as $\Delta(\ln t)$ for a flow test or as $\Delta(\ln \Delta t_e)$ for a buildup test. Experience indicates that a value of $L=0.1$ to 0.3 is usually a satisfactory compromise between being so far from the central point that detail is lost and being so near the central point that a great amount of noise is introduced. Trial and error may indicate, however, that other values of L are more appropriate in a given situation.

The process of calculating pressure derivatives can be illustrated with an example taken from field data. Suppose that we wish to determine the derivative $-dp_{wf}/d\ln t = d(\Delta p)/d\ln(\Delta t)$ at $\Delta t = 23.96$ hours from the drawdown data in **Table F-1** and that we choose $L=\Delta(\ln t)=0.3$.

We first calculate $\ln(\Delta t)$ for all test times, obtaining $\ln(23.96)=3.1764$ at $\Delta t=23.96$ hours. Then $3.1764-L=3.1764-0.3=2.8764$, and $3.1764+L=3.1764+0.3=3.4764$ as shown in the last column, creating a "window" around the central point. We use the data points just beyond this window to calculate the pressure derivative, m_P, as follows:

$$\Delta t_L = 3.1764 - 2.7713 = 0.4051 \quad \text{(F-1)}$$

$$\frac{\Delta p_L}{\Delta t_L} = \frac{(850-839)}{0.4051} = 27.1538 = m_L \quad \text{(F-2)}$$

$$\Delta t_R = 3.6859 - 3.1764 = 0.5095 \quad \text{(F-3)}$$

$$\frac{\Delta p_R}{\Delta t_R} = \frac{(864-850)}{0.5095} = 27.4779 = m_R \quad \text{(F-4)}$$

$$m_P = \frac{m_L \Delta t_R + m_R \Delta t_L}{\Delta t_L + \Delta t_R}$$

$$= \frac{(27.1538)(0.5095) + (27.4779)(0.4051)}{(0.4051 + 0.5095)}$$

$$= 27.30. \quad \text{(F-5)}$$

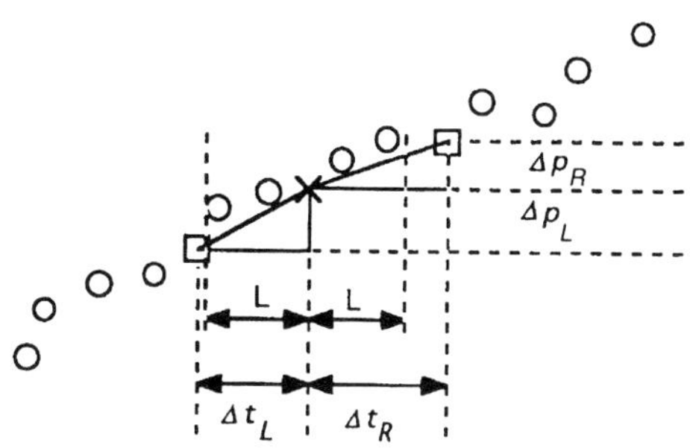

O = DATA POINTS

X = POINT TO BE DERIVED

□ = POINTS USED FOR THE CALCULATION

m_p = VALUE OF DERIVATIVE

$$m_P = \frac{\left(\frac{\Delta p_L}{\Delta t_L}\Delta t_R + \frac{\Delta p_R}{\Delta t_R}\Delta t_L\right)}{(\Delta t_L + \Delta t_R)}$$

Fig. F-1—Calculation of pressure derivative.

TABLE F-1—DATA FOR DERIVATIVE CALCULATION

Δt (hours)	Δp (psi)	$\ln(\Delta t)$	Derivative Window
11.99	830	2.4841	
15.98	839	2.7713	
			2.8764
19.97	845	2.9942	
23.96	850	3.1764	3.1764
29.93	859	3.3989	
			3.4764
39.88	864	3.6859	
49.32	869	3.8983	

Reference

1. Bourdet, D., Ayoub, J.A., and Pirard, Y.M.: "Use of Pressure Derivative in Well-Test Interpretation," *SPEFE* (June 1989) 293; *Trans.*, AIME, **287**.

Appendix G
Reservoir-Identification Worksheets

IDENTIFICATION OF THE RESERVOIR MODEL

Well Identification

Operator ________________ Well Name ____________________________

Location ________________ Formation ____________________________

Test Date ________________ Analyst _______________ Analysis Date_______

Preliminary Screening With Derivative Type Curves

Compare the field data plot of Log (pressure derivative) vs. Log (elapsed time) to the figure below, divided into early-, middle-, and late-time regions.

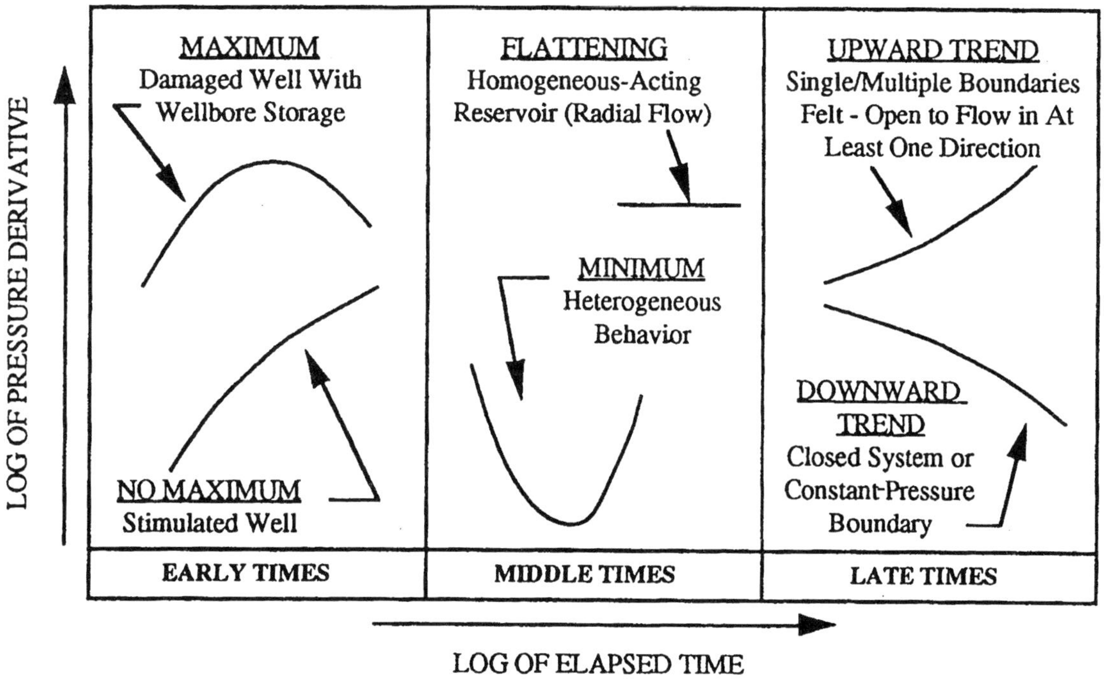

- **EARLY-TIME REGION ANALYSIS**
  - Is there a maximum? YES ☐ NO ☐
    - If YES, test indicates *damage* and *wellbore storage distortion.*
    - If NO, test indicates *stimulation*.
  - *Conclusion:* ____

- **MIDDLE-TIME REGION ANALYSIS**
  - Does curve become flat (horizontal) with no prior minimum? YES ☐ NO ☐
    - If YES, test indicates *homogeneous-acting* formation (radial flow).
    - If there is a MINIMUM, test indicates *heterogeneous* behavior (naturally fractured or layered reservoir).
    - If there is NO FLATTENING and NO MINIMUM, possibilities include closed or constant-pressure boundary effects before wellbore storage effects have ended.
  - *Conclusion*: ____________________

- **LATE-TIME REGION ANALYSIS**
  - Does the curve trend upward? YES ☐ NO ☐
    - If YES, test indicates that *single boundary* (or possibly multiple boundaries) has been felt, but boundaries have not been felt in all directions.
  - Does the curve trend downward? YES ☐ NO ☐
    - If YES, test indicates *complete closure* (all boundaries have been felt), or a *constant-pressure boundary* has been felt.
    - If NO for both, test was not run long enough to be able to observe boundary effects.
  - *Conclusion*: ____________________

Confirmation with Pressure-Change, Pressure-Derivative, and Semilog Graphs

Compare log-log and semilog graphs of test data to the following examples and consider the known reservoir geology to determine a more complete description of reservoir type and inner- and outer-boundary conditions.

- **WELLBORE OR NEAR-WELLBORE EFFECTS**
  - **Wellbore Storage/Skin** YES ☐ NO ☐

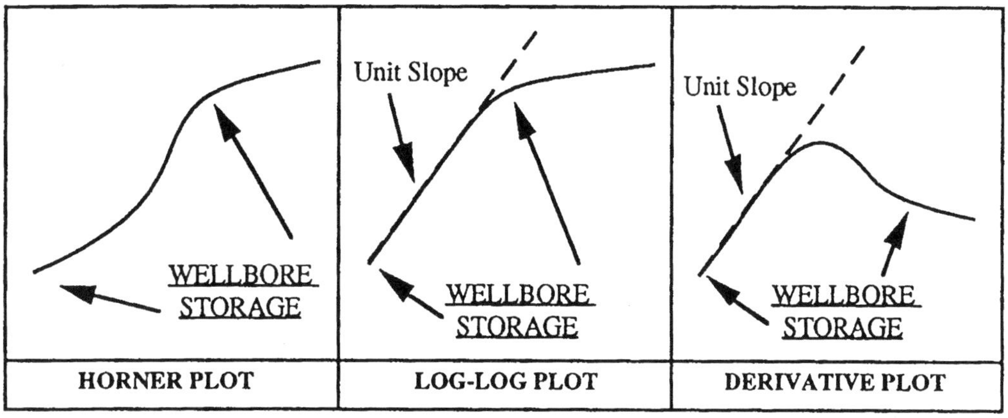

For a well with possible wellbore storage, a type-curve match should be attempted even if a unit slope line cannot be identified.

- **Phase Redistribution (hump)** YES ☐ NO ☐

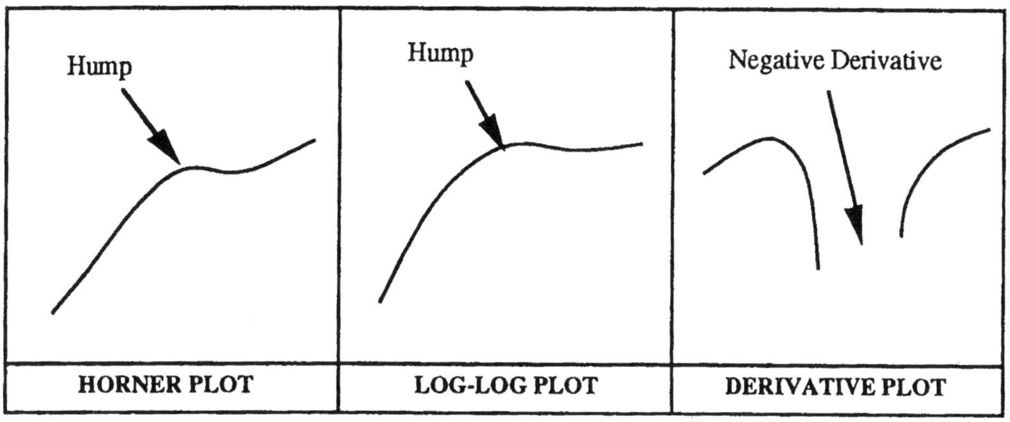

– **Acidization** YES ☐ NO ☐

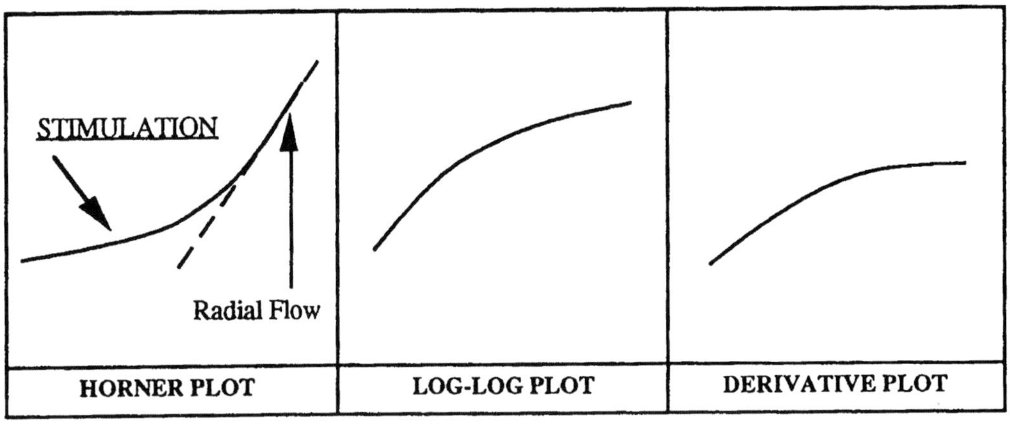

– **Infinite-Conductivity Fracture** YES ☐ NO ☐

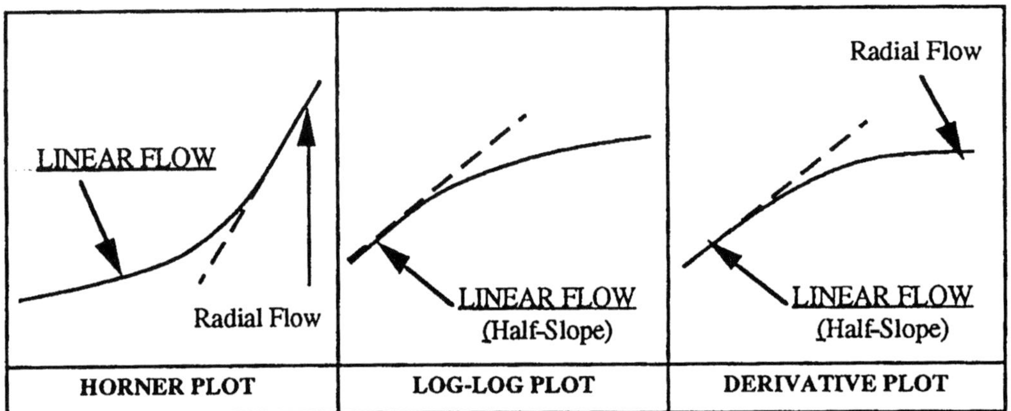

For a well with a possible infinite-conductivity fracture, a type-curve match should be attempted even if a half-slope line cannot be identified.

– **Finite-Conductivity Fracture** YES ☐ NO ☐

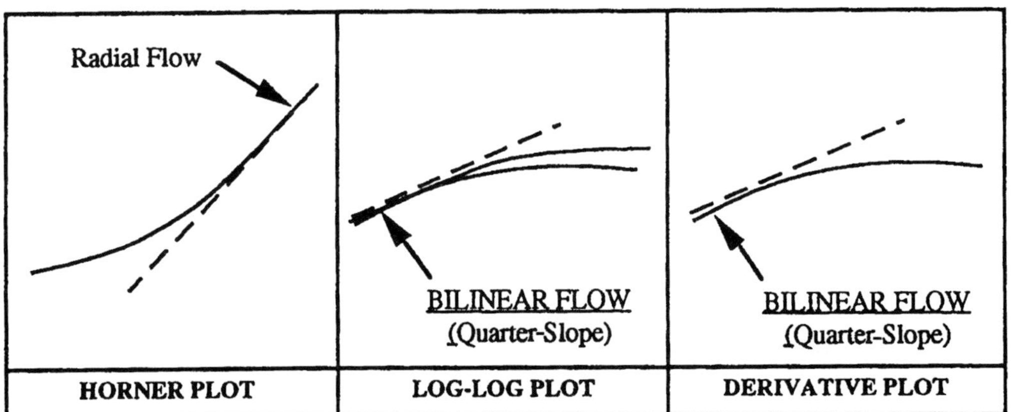

For a well with a possible finite-conductivity fracture, a type-curve match should be attempted even if a quarter-slope line cannot be identified.

- **RESERVOIR TYPES**

  – **Homogeneous-Acting Formation** YES ☐ NO ☐

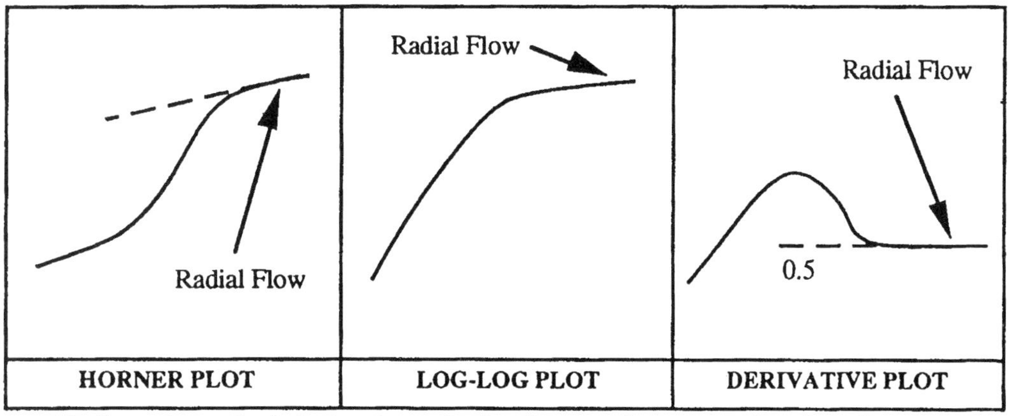

The flat region on the derivative plot can be short because of test or outer-boundary effects.

  – **Naturally Fractured Formation With Pseudosteady-State Interporosity Flow (or Layered Reservoir)** YES ☐ NO ☐

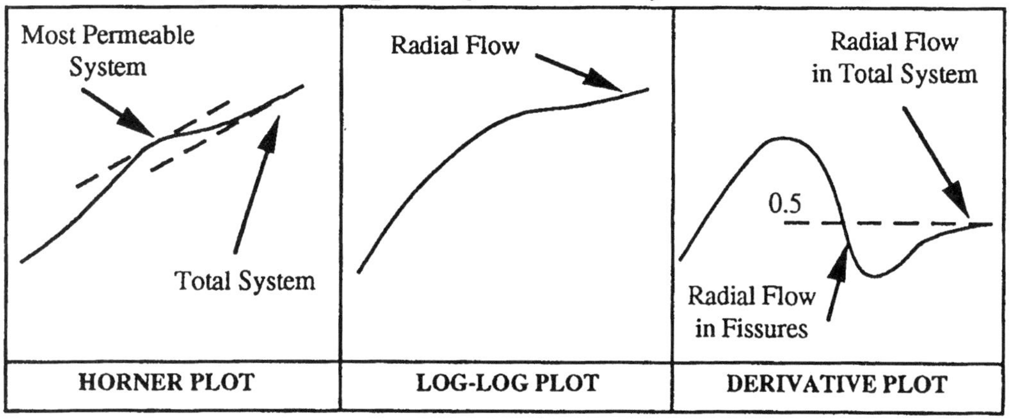

  – **Naturally Fractured Formation With Transient Interporosity Flow** YES ☐ NO ☐

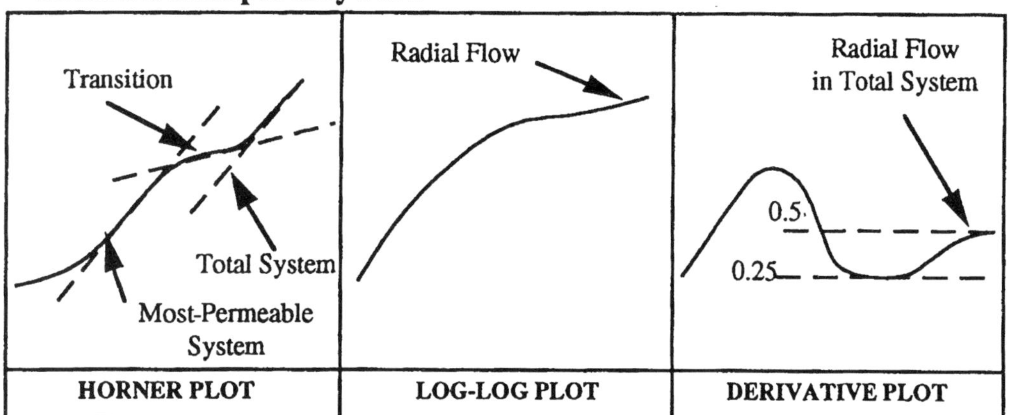

This case can be confused easily with the single no-flow boundary case. CONSULT GEOLOGICAL DATA!

- **OUTER-BOUNDARY CONDITION**

  – **Infinite-Acting Formation** YES ☐ NO ☐

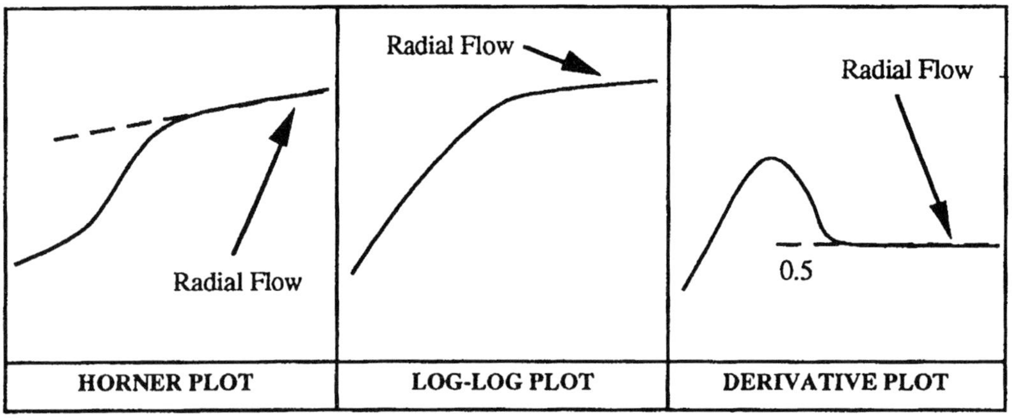

In a short test, a bounded reservoir can appear to be infinite-acting.

  – **Boundary Effects (Single No-Flow)** YES ☐ NO ☐

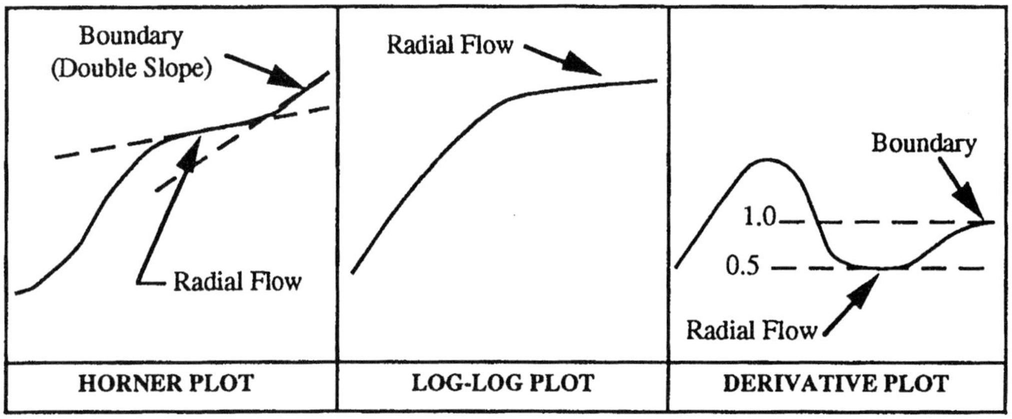

  – **Completely Bounded Reservoir or Constant-Pressure Boundary** YES ☐ NO ☐

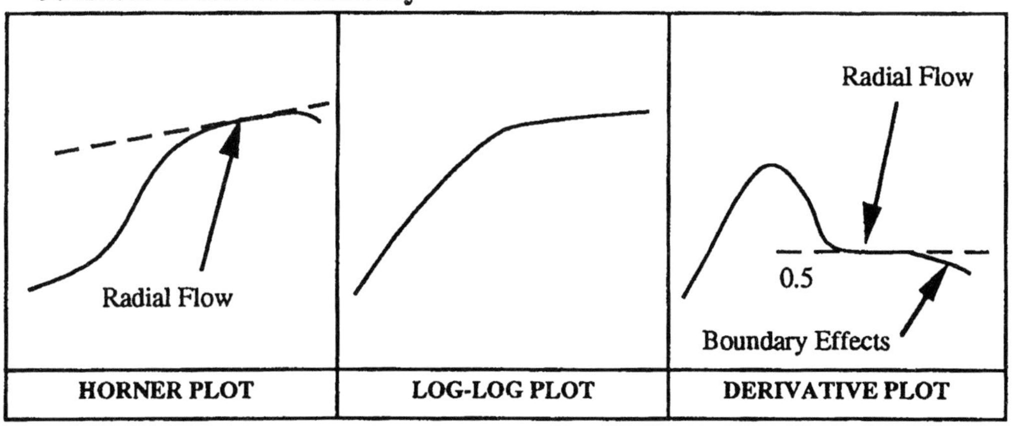

Conclusions

The reservoir type is ____
The inner-boundary condition is ________________________________ The outer-boundary condition is ________________________________

Remarks

Appendix H
Well-Test Analysis Worksheets

OILWELL-TEST ANALYSIS
HOMOGENEOUS-ACTING FORMATION

Operator ______________ Test Date______________ Analyst______________

Well________________ Stimulation ____________ Date ______________

Location ______________________________ Formation____________

Perforations _______ (ft) Depth of Bomb ____ (ft) Pressure Gradient in Wellbore __________ (psi/ft)

Pressure Datum __________________ Pressure Gradient in Reservoir __________ (psi/ft)

Correction to Pressure Gauge Reading } ______ (Wellbore Gradient (psi/ft)) x ______ (Distance Mid-Perfs to Bomb (ft)) + ______ (Reservoir Gradient (psi/ft)) x ______ (Mid-Perfs to Datum (ft)) + ______ (Atmospheric Pressure (psia)) = ________ psia

BASIC PROPERTIES

q_o (STB/D) ____________ r_w (ft) ______________ μ_o (cp) ____________

B_o (RB/STB) ___________ S_w (fraction) __________ S_g (fraction) __________

h (ft) ______________ t_p (hours) ____________ c_f (psi^{-1}) ____________

ϕ (fraction) ____________ c_w (psi^{-1}) ____________ c_g (psi^{-1}) ____________

c_o (psi^{-1}) ____________ c_t (psi^{-1}) $= c_oS_o + c_wS_w + c_gS_g + c_f$ = ______________

PRELIMINARY TYPE-CURVE ANALYSIS – START SEMILOG STRAIGHT LINE

Curve Used: ________________________________

Matching Parameter: C_De^{2s} = _________ or _________ = _________

Start Semilog Straight Line: Δt_e = _________ hours; Δt = ________ hours

$(t_p + \Delta t)/\Delta t$ = ________

Start Boundary Effects: Δt_e= _________ hours; Δt = _________ hours

$(t_p + \Delta t)/\Delta t$ = _________

If $\Delta t_{max} < \Delta t$ at start of semilog straight line, then semilog analysis is not possible.

SEMILOG ANALYSIS

Average drainage-area pressure at start of test, $\bar{p}$, = ________ psia at ________ ft depth.

$(t_p + \Delta t)/\Delta t$ at start of MTR = ________ (Buildup) t at start of MTR = ________ (Drawdown)

$(t_p + \Delta t)/\Delta t$ at start of LTR = ________ (Buildup) t at start of LTR = ________ (Drawdown)

$m = p_2 - p_1 =$ ________ – ________ = ________ psia/cycle

$$k = \frac{162.6\, q_o B_o \mu_o}{mh} = \frac{(162.6)(\quad)(\quad)(\quad)}{(\quad)(\quad)} = \text{\_\_\_\_\_\_\_\_\_\_\_\_} \text{ md}$$

SKIN-FACTOR CALCULATION

BUILDUP TEST:

$$\text{Horner Time Ratio at } \Delta t = 1\,\text{hour} = \frac{t_p + 1}{1} = \text{\_\_\_\_\_\_\_\_\_\_\_\_}$$

$p_{1hr} =$ ____________ psia

$p_{wf}\ (\Delta t = 0)$ ____________ psia

$$s = 1.151\left[\left(\frac{p_{1hr} - p_{wf}}{m}\right) - \log\left(\frac{k}{\phi\, \mu_o\, c_t\, r_w^2}\right) + 3.23\right]$$

$$s = 1.151\left[\left(\frac{(\quad) - (\quad)}{(\quad)}\right) - \log\left(\frac{(\quad)}{(\quad)(\quad)(\quad)(\quad)^2}\right) + 3.23\right]$$

$s =$ ____________

DRAWDOWN TEST:

$p_i =$ ____________ psia

$p_{1hr} =$ ____________ psia

$$s = 1.151\left[\left(\frac{p_i - p_{1hr}}{m}\right) - \log\left(\frac{k}{\phi\, \mu_o\, c_t r_w^2}\right) + 3.23\right]$$

$$s = 1.151\left[\left(\frac{(\quad) - (\quad)}{(\quad)}\right) - \log\left(\frac{(\quad)}{(\quad)(\quad)(\quad)(\quad)^2}\right) + 3.23\right]$$

$s =$ ____________

QUANTITATIVE TYPE-CURVE ANALYSIS

Curve Used = ________________

PRESSURE MATCH POINT

CASE 1: Semilog Analysis Possible
Pre-Calculated Pressure Match Point, Using Known Permeability

$$\Delta p = \frac{141.2\ q_o B_o \mu_o p_D}{kh} = \frac{(141.2)(\quad)(\quad)(\quad)(\quad)}{(\quad)(\quad)} = \_\_\_\_\_\_\_\_ \text{ psia}$$

CASE 2: Semilog Analysis Not Possible
Permeability From Pressure Match Point

Δp (psia) = __________ p_D = __________

$$k = \frac{141.2\ q_o B_o \mu_o p_D}{h \Delta p} = \frac{(141.2)(\quad)(\quad)(\quad)(\quad)}{(\quad)(\quad)} = \_\_\_\_\_\_\_\_ \text{ md}$$

TIME MATCH POINT

Δt_e (hours) = _________ (t_D/C_D) = ____________ (Gringarten) or t_D = ______________ (Cinco or others)

Gringarten Type Curve: Matching Parameter: $C_D e^{2s}$ = ____________

$$C_D = \frac{0.0002637\ k}{\phi r_w^2 c_t \mu_o} \frac{\Delta t_e}{t_D/C_D} = \frac{(0.0002637)(\quad)(\quad)}{(\quad)(\quad)^2(\quad)(\quad)(\quad)} = \_\_\_\_\_\_\_\_$$

$s = 0.5 \ln (C_D e^{2s}/C_D) = 0.5 \ln (\quad / \quad) =$ ______

PRODUCTION TIME DURATION CHECK

To ensure that flow period exceeded duration of wellbore storage during production, calculate $t_{p,min}$ from

$$t_{p,\min} = \frac{\phi\ \mu_o\ c_t r_w^2}{0.0002637\ k}(60 + 3.5\ s)\ C_D$$

$$t_{p,\min} = \frac{(\quad)(\quad)(\quad)(\quad)^2}{(0.0002637)(\quad)}[60 + 3.5(\quad)](\quad) = \_\_\_\_\_\_\_\_ \text{ hours}$$

$t_{p,actual}$ = ________ hours

If $t_{p,actual} < t_{p,min}$, test analysis is invalid. Test must be rerun with longer production period before shut-in.

RADIUS OF INVESTIGATION CHECK

RADIUS OF INVESTIGATION DURING PRODUCTION PERIOD PRECEDING SHUT-IN PERIOD

$$r_{i,prod} = \left(\frac{k\,t_p}{948\ \phi\,\mu_o\,c_t} \right)^{1/2} = \left[\frac{(\qquad)(\qquad)}{(948)(\qquad)(\qquad)(\qquad)} \right]^{1/2} = \underline{\qquad\qquad} \text{ ft}$$

RADIUS OF INVESTIGATION DURING TEST

Start of Semilog Straight Line

$$r_{i,start} = \left(\frac{k\,\Delta t_{start}}{948\ \phi\,\mu_o\,c_t} \right)^{1/2} = \left[\frac{(\qquad)(\qquad)}{(948)(\qquad)(\qquad)(\qquad)} \right]^{1/2} = \underline{\qquad\qquad} \text{ ft}$$

End of Semilog Straight Line

$$r_{i,end} = \left(\frac{k\,\Delta t_{end}}{948\ \phi\,\mu_o\,c_t} \right)^{1/2} = \left[\frac{(\qquad)(\qquad)}{(948)(\qquad)(\qquad)(\qquad)} \right]^{1/2} = \underline{\qquad\qquad} \text{ ft}$$

End of Test

$$r_{i,final} = \left(\frac{k\,\Delta t_{final}}{948\ \phi\,\mu_o\,c_t} \right)^{1/2} = \left[\frac{(\qquad)(\qquad)}{(948)(\qquad)(\qquad)(\qquad)} \right]^{1/2} = \underline{\qquad\qquad} \text{ ft}$$

NOTE: Reservoir heterogeneities and boundaries determined from data during shut-in test at $r_i > r_{i,prod}$ are highly questionable. If the drawdown did not reach beyond a certain distance in the reservoir, the buildup will not sample reservoir properties beyond that point.

REMARKS

____

____

____

____

____

____

____

____

____

GASWELL-TEST ANALYSIS
HOMOGENEOUS-ACTING FORMATION

Operator ________________ Test Date________________ Analyst________________

Well________________ Stimulation ________________ Date ________________

Location ______________________________________ Formation________________

Perforations _______ (ft) Depth of Bomb ____ (ft) Pressure Gradient in Wellbore __________ (psi/ft)

Pressure Datum ____________________ Pressure Gradient in Reservoir __________ (psi/ft)

Correction to Pressure Gauge Reading } _______ x _______ + _______ x _______ + _______ = _______ psia

Wellbore Gradient (psi/ft) × Distance Mid-Perfs to Bomb (ft) + Reservoir Gradient (psi/ft) × Mid-Perfs to Datum (ft) + Atmospheric Pressure (psia)

BASIC PROPERTIES

q_g (Mscf/D) ____________ r_w (ft) ____________ $\overline{\mu}$ (cp) ____________

T (°F) ____________ S_w (fraction) ____________ $\overline{z}$ (fraction) ____________

h (ft) ____________ t_p (hours) ____________ $\overline{c_t}$ (psia^{-1}) ____________

ϕ (fraction) ____________ $\overline{p}$(psia) ____________

γ_g (air = 1.0) ____________ $\overline{B_g}$ (RB/Mscf) $= 5.035 \frac{T\overline{z}}{\overline{p}} =$ ____________

PRELIMINARY TYPE-CURVE ANALYSIS – START SEMILOG STRAIGHT LINE

Curve Used: ________________ Matching Parameter: C_De^{2s} = ______ or ______ = ______

Start Semilog Straight Line: Δt_{ae} = __________ hours; Δt = __________ hours

$(t_{pa} + \Delta t_a)/\Delta t_a$ = ________ ; Δt_a = ________; Not present _____

Start Boundary Effects: Δt_{ae} = __________ hours; Δt = __________ hours

$(t_{pa} + \Delta t_a)/\Delta t_a$ = ________ ; Δt_a = ________; Not present _____

If $\Delta t_{max} < \Delta t$ at start of pseudoradial flow, then semilog analysis is not possible.

SEMILOG ANALYSIS

Average drainage-area pressure at start of test, $\bar{p}$, = ________ psia ; $\bar{p}_a$ = ________ psia;

at depth = ________ ft.

$(t_{pa} + \Delta t_a)/\Delta t_a$ at start of MTR = ________ (Buildup) t at start of MTR = ________ (Drawdown)

$(t_{pa} + \Delta t_a)/\Delta t_a$ at start of LTR = ________ (Buildup) t at start of LTR = ________ (Drawdown)

$m = p_{a2} - p_{a1}$ = ________ – ________ = ________ psia/cycle

$$k = \frac{162.6\, q_g \overline{B_g}\, \bar{\mu}}{mh} = \frac{(162.6)(\qquad)(\qquad)(\qquad)}{(\qquad)(\qquad)} = \underline{\qquad\qquad}\ \text{md}$$

SKIN-FACTOR CALCULATION

BUILDUP TEST:

Horner Time Ratio at Δt_a = 1 hour $= \frac{t_{pa} + 1}{1}$ = ________

$p_{a,1\text{hr}}$ = ________ psia

$p_{a,wf}$ ($\Delta t_a = 0$) ________ psia

$$s' = 1.151 \left[\left(\frac{p_{a,1\text{hr}} - p_{a,wf}}{m}\right) - \log\left(\frac{k}{\phi\, \bar{\mu}\, \bar{c}_t\, r_w^2}\right) + 3.23\right]$$

$$s' = 1.151 \left[\left(\frac{(\qquad) - (\qquad)}{(\qquad)}\right) - \log\left(\frac{(\qquad)}{(\qquad)(\qquad)(\qquad)(\qquad)^2}\right) + 3.23\right]$$

s' = ____________

DRAWDOWN TEST:

p_{ai} = ________ psia

Δt_a = ________ psia

$p_{a,1\text{hr}}$ at Δt_a = 1 hour = ________ psia

$$s' = 1.151 \left[\left(\frac{p_{a,i} - p_{a,1\text{hr}}}{m}\right) - \log\left(\frac{k}{\phi\, \bar{\mu}\, \bar{c}_t\, r_w^2}\right) + 3.23\right]$$

$$s' = 1.151 \left[\left(\frac{(\qquad) - (\qquad)}{(\qquad)}\right) - \log\left(\frac{(\qquad)}{(\qquad)(\qquad)(\qquad)(\qquad)^2}\right) + 3.23\right]$$

s' = ____________

QUANTITATIVE TYPE-CURVE ANALYSIS

Curve Used: ______________________

PRESSURE MATCH POINT

CASE 1: Permeability Estimate Available
Precalculate Pressure Match Point, Using Known Permeability

$$\Delta p_a = \frac{141.2\ q_g \overline{B_g}\,\overline{\mu}\ p_D}{kh} = \frac{(141.2)(\quad)(\quad)(\quad)(\quad)}{(\quad)(\quad)} = \_\_\_\_\_\_\_\_ \text{ psia}$$

CASE 2: Permeability Estimate Unavailable; Permeability From Pressure Match Point

Δp_a (psia) = __________ p_D = __________

$$k = \frac{141.2\ q_g \overline{B_g}\,\overline{\mu}\ p_D}{h \Delta p_a} = \frac{(141.2)(\quad)(\quad)(\quad)(\quad)}{(\quad)(\quad)} = \_\_\_\_\_\_\_\_ \text{ md}$$

TIME MATCH POINT

Δt_a or Δt_{ae} (hours) = __________; (t_D/C_D) = __________ or t_{LfD} = __________
(Gringarten) (Cinco or others)

Gringarten Type Curve: Matching Parameter $C_D e^{2s}$ = __________

$$C_D = \frac{0.0002637\ k}{\phi\, r_w^2 \overline{c_t}\,\overline{\mu}} \frac{\Delta t_{ae}}{t_D/C_D} = \frac{(0.0002637)(\quad)(\quad)}{(\quad)(\quad)^2(\quad)(\quad)(\quad)} = \_\_\_\_\_\_\_\_$$

$s' = 0.5 \ln (C_D e^{2s} / C_D) = 0.5 \ln (\quad / \quad) =$ __________

PRODUCTION TIME DURATION CHECK

To ensure that flow period exceeded duration of wellbore storage during production, calculate $t_{pa,min}$.

$$t_{pa,min} = \frac{\phi\, \overline{\mu}\, \overline{c_t}\, r_w^2}{0.0002637\ k} (60 + 3.5\ s')\, C_D$$

$$t_{pa,min} = \frac{(\quad)(\quad)(\quad)(\quad)^2}{(0.0002637)(\quad)} [60 + 3.5(\quad)](\quad) = \_\_\_\_\_\_\_\_ \text{ hours}$$

$t_{pa,actual}$ = __________ hours

If $t_{pa,actual} < t_{pa,min}$, then test analysis is invalid. Test must be rerun with longer production period before shut-in period.

RADIUS OF INVESTIGATION CHECK

RADIUS OF INVESTIGATION DURING PRODUCTION PERIOD PRECEDING SHUT-IN PERIOD

$$r_{i,prod} = \left(\frac{k\, t_{pa}}{948\ \phi\ \overline{\mu}\, \overline{c}_t} \right)^{1/2} = \left[\frac{(\qquad)(\qquad)}{(948)(\qquad)(\qquad)(\qquad)} \right]^{1/2} = \_\_\_\_\_\_\ \text{ft}$$

RADIUS OF INVESTIGATION DURING TEST

Start of Semilog Straight Line

$$r_{i,start} = \left(\frac{k\, \Delta t_{a,start}}{948\ \phi\ \overline{\mu}\, \overline{c}_t} \right)^{1/2} = \left[\frac{(\qquad)(\qquad)}{(948)(\qquad)(\qquad)(\qquad)} \right]^{1/2} = \_\_\_\_\_\_\ \text{ft}$$

End of Semilog Straight Line

$$r_{i,end} = \left(\frac{k\, \Delta t_{a,end}}{948\ \phi\ \overline{\mu}\, \overline{c}_t} \right)^{1/2} = \left[\frac{(\qquad)(\qquad)}{(948)(\qquad)(\qquad)(\qquad)} \right]^{1/2} = \_\_\_\_\_\_\ \text{ft}$$

End of Test

$$r_{i,final} = \left(\frac{k\, \Delta t_{a,final}}{948\ \phi\ \overline{\mu}\, \overline{c}_t} \right)^{1/2} = \left[\frac{(\qquad)(\qquad)}{(948)(\qquad)(\qquad)(\qquad)} \right]^{1/2} = \_\_\_\_\_\_\ \text{ft}$$

NOTE: Reservoir heterogeneities and boundaries determined from data during shut-in test at $r_i > r_{i,prod}$ are highly questionable. If the drawdown did not reach beyond a certain distance in the reservoir, the buildup will not sample reservoir properties beyond that point.

REMARKS

____

____

____

____

____

____

____

____

____

Appendix I
Example Well-Test Analysis Using Worksheets, Example 4.5

GASWELL-TEST ANALYSIS
HOMOGENEOUS-ACTING FORMATION

Operator Example 4.5 Test Date ______ Analyst ______

Well ______ Stimulation ______ Date ______

Location ______ Formation ______

Perforations ______ ft Depth of Bomb ______ ft Pressure Gradient in Wellbore ______ psi/ft

Pressure Datum ______ Pressure Gradient in Reservoir ______ psi/ft

Correction to Pressure Gauge Reading } ______ Wellbore Gradient (psi/ft) x ______ Distance Mid-Perfs to Bomb (ft) + ______ Reservoir Gradient (psi/ft) x ______ Mid-Perfs to Datum (ft) + ______ Atmospheric Pressure (psia) = ______ psia

BASIC PROPERTIES

q_g (Mscf/D)	100	r_w (ft)	0.365	$\bar{\mu}$ (cp)	0.03274
T (°F)	210	S_w (fraction)	0.36	$\bar{z}$ (fraction)	1.310
h (ft)	21	t_p (hours)	2000	$\bar{c_t}$ (psia^{-1})	45.4x10^{-6}
ϕ (fraction)	0.10	t_{pa} (hours)	2000	$\overline{B_g}$ (RB/Mscf) $= 5.035 \frac{T\bar{z}}{\bar{p}} =$	
γ_g (air = 1.0)	0.6	$\bar{p}$ (psia)	9,000		0.491

PRELIMINARY TYPE-CURVE ANALYSIS – START SEMILOG STRAIGHT LINE

Curve Used: Bourdet *et al.* Matching Parameter: C_De^{2s} = 100 or ______ = ______

Start Semilog Straight Line: Δt_{ae} = 36.4 hours; Δt = ______ hours

$(t_{pa} + \Delta t_a)/\Delta t_a$ = ______ ; Δt_a = ______; Not present ______

Start Boundary Effects: Δt_{ae} = ______ hours; Δt = ______ hours

$(t_{pa} + \Delta t_a)/\Delta t_a$ = ______ ; Δt_a = ______; Not present X

If $\Delta t_{max} < \Delta t$ at start of pseudoradial flow, then semilog analysis is not possible.

SEMILOG ANALYSIS

Average drainage-area pressure at start of test, $\overline{p}$, = 9000 psia ; $\overline{p}_a$ = 7366.1 psia

at depth = 8000 ft

$(t_{pa} + \Delta t_a)/\Delta t_a$ at start of MTR = ________ (Buildup); t at start of MTR = ________ (Drawdown)

$(t_{pa} + \Delta t_a)/\Delta t_a$ at start of LTR = ________ (Buildup); t at start of LTR = ________ (Drawdown)

$m = p_{a2} - p_{a1}$ = 7553.8 – 7117.5 = 436.3 psia/cycle

$$k = \frac{162.6\ q_g \overline{B_g} \overline{\mu}}{mh} = \frac{(162.6)(100)(0.491)(0.03274)}{(436.3)(21)} = 0.0285 \text{ md}$$

SKIN-FACTOR CALCULATION

BUILDUP TEST:

Horner Time Ratio at Δt_a = 1hour $= \frac{t_{pa} + 1}{1} =$ 2001

$p_{a,1\text{hr}}$ = 6114 psia

$p_{a,wf}$ $(\Delta t_a = 0)$ 4804 psia

$$s' = 1.151 \left[\left(\frac{p_{a,1\text{hr}} - p_{a,wf}}{m} \right) - \log\left(\frac{k}{\phi \overline{\mu} \overline{c}_t r_w^2} \right) + 3.23 + \log\left(\frac{t_{pa} + 1}{t_{pa}} \right) \right]$$

$$s' = 1.151 \left[\left(\frac{(6114) - (4804)}{(436.3)} \right) - \log\left(\frac{(0.0285)}{(0.10)(0.03274)(0.0000454)(0.365)^2} \right) + 3.23 + \log\left(\frac{(2000) + 1}{(2000)} \right) \right]$$

s = 0.086

DRAWDOWN TEST:

p_{al} = ________ psia

Δt_a = ________ psia

$p_{a,1\text{hr}}$ at Δt_a = 1 hour = ________ psia

$$s' = 1.151 \left[\left(\frac{p_{al} - p_{a,1h}}{m} \right) - \log\left(\frac{k}{\phi \overline{\mu} \overline{c}_t r_w^2} \right) + 3.23 \right]$$

$$s' = 1.151 \left[\left(\frac{(\quad) - (\quad)}{(\quad)} \right) - \log\left(\frac{(\quad)}{(\quad)(\quad)(\quad)(\quad)^2} \right) + 3.23 + \log\left(\frac{(\quad) + 1}{(\quad)} \right) \right]$$

s' = ________

QUANTITATIVE TYPE-CURVE ANALYSIS

Curve Used: Bourdet *et al.*

PRESSURE MATCH POINT

CASE 1: Permeability Estimate Available
Precalculate Pressure Match Point, Using Known Permeability

$$\Delta p_a = \frac{141.2\ q_g \overline{B_g}\,\overline{\mu}\, p_D}{kh} = \frac{(141.2)(100)(0.491)(0.03274)(10)}{(0.0285)(21)} = \underline{3800}\ \text{psia}$$

CASE 2: Permeability Estimate Unavailable; Permeability From Pressure Match Point

Δp_a (psia) = ______ p_D = ______

$$k = \frac{141.2\ q_g \overline{B_g}\,\overline{\mu}\, p_D}{h\Delta p_a} = \frac{(141.2)(\quad)(\quad)(\quad)(\quad)}{(\quad)(\quad)} = \underline{\qquad}\ \text{md}$$

TIME MATCH POINT

Δt_a or Δt_{ae} (hours) = 0.1; (t_D/C_D) = 0.27 (Gringarten) or t_{L_fD} = ______ (Cinco or others)

CASE 1: Gringarten Type Curve: Matching Parameter $C_D e^{2s}$ = 100

$$C_D = \frac{0.0002637\ k\,\Delta t_{ae}}{\phi r_w^2 \overline{c_t}\,\overline{\mu}\ \ t_D/C_D} = \frac{(0.0002637)(0.0285)(0.1)}{(0.10)(0.365)^2(0.0000454)(0.03274)(0.27)} = \underline{140.6}$$

$$s = 0.5 \ln (C_D e^{2s}/C_D) = 0.5 \ln (100 / 140.6) = \underline{-0.17}$$

CASE 2: Cinco or Other Type Curve: Matching Parameter C_r = ______ or ______ = ______

$$L_f = \left[\frac{0.0002637\ k\,\Delta t_{ae}}{\phi \overline{\mu}\,\overline{c_t}\ \ t_{L_fD}}\right]^{1/2} = \left[\frac{(0.0002637)(\quad)(\quad)}{(\quad)(\quad)(\quad)(\quad)}\right]^{1/2} = \underline{\qquad}\ \text{ft}$$

$wk_f = \pi C_r L_f k = \pi(\quad)(\quad)(\quad) =$ ______ md-ft

PRODUCTION TIME DURATION CHECK

To ensure that flow period exceeded duration of wellbore storage during production, calculate $t_{p,min}$.

$$t_{pa,min} = \frac{\phi \overline{\mu}\,\overline{c_t}\, r_w^2}{0.0002637\ k}(120 + 7s')\,C_D$$

$$t_{pa,min} = \frac{(0.10)(0.03274)(0.0000454)(0.365)^2}{(0.0002637)(0.0285)}[120 + 7(-0.17)] = \underline{0.31}\ \text{hours}$$

$t_{pa,actual}$ = 2000 hours

If $t_{pa,actual} < t_{pa,min}$, then test analysis is invalid. Test must be rerun with longer production period before shut-in period.

RADIUS OF INVESTIGATION CHECK

RADIUS OF INVESTIGATION ACHIEVED IN PRODUCTION PERIOD PRECEDING SHUT-IN PERIOD

$$r_{i,prod} = \left(\frac{k\, t_{pa}}{948\ \phi\ \bar{\mu}\ \bar{c}_t} \right)^{1/2} = \left[\frac{(0.0285)(2000)}{(948)(0.10)(0.03274)(0.0000454)} \right]^{1/2} = \underline{\quad 636 \quad} \text{ ft}$$

RADIUS OF INVESTIGATION DURING TEST

Start of Semilog Straight Line

$$r_{i,start} = \left(\frac{k\, \Delta t_a}{948\ \phi\ \bar{\mu}\ \bar{c}_t} \right)^{1/2} = \left[\frac{(0.0285)(36.8)}{(948)(0.10)(0.03274)(0.0000454)} \right]^{1/2} = \underline{\quad 86.3 \quad} \text{ ft}$$

End of Semilog Straight Line

$$r_{i,end} = \left(\frac{k\, \Delta t_a}{948\ \phi\ \bar{\mu}\ \bar{c}_t} \right)^{1/2} = \left[\frac{(\quad)(\quad)}{(948)(\quad)(\quad)(\quad)} \right]^{1/2} = \underline{\qquad\qquad} \text{ ft}$$

End of Test

$$r_{i,final} = \left(\frac{k\, \Delta t_a}{948\ \phi\ \bar{\mu}\ \bar{c}_t} \right)^{1/2} = \left[\frac{(0.0285)(961.8)}{(948)(0.10)(0.03274)(0.0000454)} \right]^{1/2} = \underline{\quad 441 \quad} \text{ ft}$$

NOTE: Reservoir heterogeneities and boundaries determined from data during shut-in test at $r_i > r_{i,prod}$ are highly questionable. If the drawdown did not reach beyond a certain distance in the reservoir, the buildup will not sample reservoir properties beyond that point.

REMARKS

Appendix J
Worksheets for Post-Fracture Well-Test Analysis

OILWELL-TEST ANALYSIS
HYDRAULICALLY FRACTURED WELL

Operator ______________ Test Date ______________ Analyst ______________

Well ____________________ Stimulation ____________ Date ________________

Location ____________________________________ Formation ____________

Perforations (ft) ______ Depth of Bomb (ft) _____ Pressure Gradient in Wellbore ________ psi/ft

Pressure Datum ____________________ Pressure Gradient in Reservoir ________ psi/ft

Correction to Pressure Gauge Reading } ______ x ______ + ______ x ______ + ______ = ______ psi

Wellbore Gradient (psi/ft) x Distance Mid-Perfs to Bomb (ft) + Reservoir Gradient (psi/ft) x Mid-Perfs to Datum (ft) + Atmospheric Pressure (psia)

BASIC PROPERTIES

q_o (STB/D) ____________ r_w (ft) ________________ μ_o (cp) ________________

B_o (RB/STB) ____________ S_w (fraction) ____________ S_g (fraction) ____________

h (ft) ________________ t_p (hours) ________________ c_f (psi^{-1}) ________________

ϕ (fraction) ____________ c_w (psi^{-1}) ________________ c_t (psi^{-1}) = ________________

c_o (psi^{-1}) ________________ c_g (psi^{-1}) ________________ $c_oS_o + c_wS_w + c_gS_g + c_f$

PRELIMINARY TYPE-CURVE ANALYSIS – START SEMILOG STRAIGHT LINE

Curve Used: ______________________________________

Matching Parameter: C_r = __________ or __________ = __________

Start Semilog Straight Line: Δt_e = __________ hours; Δt = _________ hours

$(t_p + \Delta t)/\Delta t$ = _________

Start Boundary Effects: Δt_e= __________ hours; Δt = __________ hours

$(t_p + \Delta t)/\Delta t$ = __________

If $\Delta t_{max} < \Delta t$ at start of semilog straight line, then semilog analysis is not possible.

SEMILOG ANALYSIS

Average drainage-area pressure at start of test, $\bar{p}$, ________ psia at ________ ft depth

$(t_p + \Delta t)/\Delta t$ at start of MTR = ________ (Buildup) t at start of MTR = ________ (Drawdown)

$(t_p + \Delta t)/\Delta t$ at start of LTR = ________ (Buildup) t at start of LTR = ________ (Drawdown)

$m = p_2 - p_1$ = ________ – ________ = ________ psia/cycle

$$k = \frac{162.6\ q_o B_o \mu_o}{mh} = \frac{(162.6)(\quad)(\quad)(\quad)}{(\quad)(\quad)} = \underline{\qquad\qquad} \text{ md}$$

SKIN-FACTOR CALCULATION

BUILDUP TEST:

$$\text{Horner Time Ratio at } \Delta t = 1 \text{ hour} = \frac{t_p + 1}{1} = \underline{\qquad\qquad}$$

p_{1hr} = ________ psia

$p_{wf}\ (\Delta t = 0)$ ________ psia

$$s = 1.151\left[\left(\frac{p_{1hr} - p_{wf}}{m}\right) - \log\left(\frac{k}{\phi\ \mu_o\ c_t\ r_w^2}\right) + 3.23\right]$$

$$s = 1.151\left[\left(\frac{(\quad)-(\quad)}{(\quad)}\right) - \log\left(\frac{(\quad)}{(\quad)(\quad)(\quad)(\quad)^2}\right) + 3.23\right]$$

s = ________

DRAWDOWN TEST:

p_i = ________ psia

p_{1hr} = ________ psia

$$s = 1.151\left[\left(\frac{p_i - p_{1hr}}{m}\right) - \log\left(\frac{k}{\phi\ \mu_o\ c_t\ r_w^2}\right) + 3.23\right]$$

$$s = 1.151\left[\left(\frac{(\quad)-(\quad)}{(\quad)}\right) - \log\left(\frac{(\quad)}{(\quad)(\quad)(\quad)(\quad)^2}\right) + 3.23\right]$$

s = ________

LINEAR-FLOW ANALYSIS

If there are test data for $t_{L_fD} \leq 0.016$ and if data in this region lie on a half-slope line, linear-flow analysis may be possible. An independent estimate of permeability must be available to complete this analysis. From a plot of p_{wf} vs. $\sqrt{t}$ or p_{ws} vs. $\sqrt{\Delta t_e}$, slope m_L = ______psia/$\sqrt{hr}$.

$$\sqrt{k}\,L_f = \frac{4.064\, q_o B_o}{m_L h}\left(\frac{\mu}{\phi\, c_t}\right)^{1/2} = \frac{4.064(\quad)(\quad)}{(\quad)(\quad)}\left(\frac{(\quad)}{(\quad)(\quad)}\right)^{1/2} = \_\_\_\_\_\_$$

$$L_f = \frac{\sqrt{k}\,L_f}{\sqrt{k}} = \frac{(\quad)}{(\quad)^{1/2}} = \_\_\_\_\_\_ \text{ ft}$$

QUANTITATIVE TYPE-CURVE ANALYSIS

Curve Used = ______________

PRESSURE MATCH POINT

CASE 1: Semilog Analysis Possible
Pre-Calculated Pressure Match Point, Using Known Permeability

$$\Delta p = \frac{141.2 q_o B_o \mu_o p_D}{kh} = \frac{(141.2)(\quad)(\quad)(\quad)(\quad)}{(\quad)(\quad)} = \_\_\_\_\_\_ \text{ psia}$$

CASE 2: Semilog Analysis Not Possible
Permeability From Pressure Match Point

Δp (psia) = ________ p_D = ________

$$k = \frac{141.2 q_o B_o \mu_o p_D}{h \Delta p} = \frac{(141.2)(\quad)(\quad)(\quad)(\quad)}{(\quad)(\quad)} = \_\_\_\_\_\_ \text{ md}$$

TIME MATCH POINT

Δt_e (hours) = ________ (t_D/C_D) = ________ (Gringarten) or t_D = ________ (Cinco or others)

CASE 1: Gringarten Type Curve
Matching Parameter: $C_D e^{2s}$ = ________

$$C_D = \frac{0.0002637\, k}{\phi r_w^2 c_t \mu_o}\frac{\Delta t_e}{t_D/C_D} = \frac{(0.0002637)(\quad)(\quad)}{(\quad)(\quad)^2(\quad)(\quad)(\quad)} = \_\_\_\_\_\_$$

$s = 0.5 \ln(C_D e^{2s}/C_D) = 0.5 \ln(\quad / \quad) =$ ________

CASE 2: Cinco or Other Type Curve
Matching Parameter: C_{fD} = ________ or ________ = ________

$$L_f = \left[\frac{0.0002637\, k}{\phi c_t \mu_o}\frac{\Delta t_e}{t_D}\right]^{1/2} = \left[\frac{(0.0002637)(\quad)(\quad)}{(\quad)(\quad)(\quad)(\quad)}\right]^{1/2} = \_\_\_\_\_\_ \text{ ft}$$

$$wk_f = \pi C_{fD} L_f k = \pi(\quad)(\quad)(\quad) = \_\_\_\_\_\_ \text{ md-ft}$$

PRODUCTION TIME DURATION CHECK

To ensure that flow period exceeded duration of wellbore storage during production, calculate $t_{p,min}$.

$$t_{p_{min}} = \frac{\phi \mu_o c_t r_w^2}{0.0002637\, k}(15\, C_D)$$

$$t_{p,\,min} = \frac{(\quad)(\quad)(\quad)(\quad)^2}{(\quad)(\quad)}[15(\quad)] = \underline{\qquad\qquad} \text{ hours}$$

$t_{p,actual}$ = __________ hours

If $t_{p,actual} < t_{p,min}$ then test analysis is invalid. Test must be rerun with longer production period before shut-in period.

DEPTH OF INVESTIGATION CHECK

DEPTH OF INVESTIGATION ACHIEVED IN PRODUCTION PERIOD PRECEDING SHUT-IN PERIOD

Time at end of flow period, t_{flow} = ______________ hours

Minor axis: $b = 0.02878\left(\frac{k\, t_{flow}}{\phi \mu_o c_t}\right)^{1/2} = 0.02878\left[\frac{(\quad)(\quad)}{(\quad)(\quad)(\quad)}\right]^{1/2}$

b = ______________ ft

Major axis: $a = \left(L_f^2 + b^2\right)^{1/2} = [(\quad)^2 + (\quad)^2]^{1/2}$ = __________ ft

Area Investigated: $\pi ab / 43{,}560 = \pi(\quad)(\quad)/43{,}560$

= _____________ ft^2 / 43,560 = ______________ acres

DEPTH OF INVESTIGATION ACHIEVED DURING TEST

Start of Semilog Straight Line:

Time at start of line, Δt_{start} = ______________ hours

Minor axis: $b = 0.02878\left(\frac{k\, t_{start}}{\phi \mu_o c_t}\right)^{1/2} = 0.02878\left[\frac{(\quad)(\quad)}{(\quad)(\quad)(\quad)}\right]^{1/2}$

b = ______________ ft

Major axis: $a = \left(L_f^2 + b^2\right)^{1/2} = [(\quad)^2 + (\quad)^2]^{1/2}$ = __________ ft

Area Investigated: $\pi ab / 43{,}560 = \pi(\quad)(\quad)/43{,}560$

= _____________ ft^2 / 43,560 = ______________ acres

End of Semilog Straight Line:

Time at end of line, Δt_{end} = ______________ hours

Minor axis: $b = 0.02878\left(\frac{k\,t_{end}}{\phi\,\mu_o\,c_t}\right)^{1/2} = 0.02878\left[\frac{(\qquad)(\qquad)}{(\qquad)(\qquad)(\qquad)}\right]^{1/2}$

b = ______________ ft

Major axis: $a = \left(L_f^2 + b^2\right)^{1/2} = \left[(\qquad)^2 + (\qquad)^2\right]^{1/2}$ = __________ ft

Area Investigated: $\pi ab / 43{,}560 = \pi(\qquad)(\qquad)/43{,}560$

= ______________ ft^2 / 43,560 = ______________ acres

End of Test:

Time at end of test, Δt_{final} = ______________ hours

Minor axis: $b = 0.02878\left(\frac{k\,t_{final}}{\phi\,\mu_o\,c_t}\right)^{1/2} = 0.02878\left[\frac{(\qquad)(\qquad)}{(\qquad)(\qquad)(\qquad)}\right]^{1/2}$

b = ______________ ft

Major axis: $a = \left(L_f^2 + b^2\right)^{1/2} = \left[(\qquad)^2 + (\qquad)^2\right]^{1/2}$ = __________ ft

Area Investigated: $\pi ab / 43{,}560 = \pi(\qquad)(\qquad)/43{,}560$

= ______________ ft^2 / 43,560 = ______________ acres

NOTE: Reservoir heterogeneities and boundaries determined from data during shut-in test at $\Delta t > t_{flow}$ are highly questionable. If the drawdown did not reach beyond a certain distance in the reservoir, then the buildup will not sample reservoir properties beyond that point.

REMARKS

____

____

____

____

____

____

____

____

____

GASWELL-TEST ANALYSIS
HYDRAULICALLY FRACTURED WELL

Operator ____________ Test Date __________ Analyst ________________

Well ________________ Stimulation __________ Date ________________

Location ________________________________ Formation ______________

Perforations _____ (ft) Depth of Bomb ____ (ft) Pressure Gradient in Wellbore __________ (psi/ft)

Pressure Datum ________________ Pressure Gradient in Reservoir __________ (psi/ft)

Correction to Pressure Gauge Reading } ______ x ________ + ________ x ________ + __________ = ________ psi

Wellbore Gradient (psi/ft) x Distance Mid-Perfs to Bomb (ft) + Reservoir Gradient (psi/ft) x Mid-Perfs to Datum (ft) + Atmospheric Pressure (psia)

BASIC PROPERTIES

q_g (Mscf/D) __________ r_w (ft) ______________ $\bar{\mu}$ (cp) ______________

T (°F) ________________ S_w (fraction) __________ $\bar{z}$ (fraction) ____________

h (ft) ________________ t_p (hours) ____________ $\bar{c_t}$ (psia^{-1}) ____________

ϕ (fraction) __________ $\bar{p}$ (psia) ____________

γ_g (air = 1.0) __________ $\bar{B_g}$ (RB/Mscf) $= 5.035 \frac{T\bar{z}}{\bar{p}}$ = ______________________

PRELIMINARY TYPE-CURVE ANALYSIS–START SEMILOG STRAIGHT LINE

Curve Used: ________________ Matching Parameter: C_r = _______ or ______ = _____

If there are test data for $t_{LfD} \geq 3$, the pseudoradial flow regime may be present.

Start Semilog Straight Line: At $t_{LfD} = 3$, Δt_{ae} = __________ hours; Δt = __________ hours

$(t_p + \Delta t_a)/\Delta t_a$ = _______ ; Δt_a = ________; Not present _____

Start Boundary Effects: At t_{LfD} = _____ ,Δt_{ae} = ________ hours; Δt = ________ hours

$(t_p + \Delta t_a)/\Delta t_a$ = _______ ; Δt_a = ________; Not present _____

If $\Delta t_{max} < \Delta t$ at start of pseudoradial flow, then semilog analysis is not possible.

SEMILOG ANALYSIS

Average drainage-area pressure at start of test, $\bar{p}$, = ______ psia; $\bar{p}_a$ = ______ psia

at depth = ________ ft

$(t_p + \Delta t_a)/\Delta t_a$ at start of MTR = ________ (Buildup) t at start of MTR = ________ (Drawdown)

$(t_p + \Delta t_a)/\Delta t_a$ at start of LTR = ________ (Buildup) t at start of LTR = ________ (Drawdown)

$m = p_{a2} - p_{a1}$ = __________ − ________ = __________ psia/cycle

$$k = \frac{162.6\, q_g \overline{B_g}\, \bar{\mu}}{mh} = \frac{(162.6)(\quad)(\quad)(\quad)}{(\quad)(\quad)} = \_\_\_\_\_\_\_\_\_\_ \text{ md}$$

SKIN-FACTOR CALCULATION

BUILDUP TEST:

Horner Time Ratio at $\Delta t_a = 1$ hour $= \dfrac{t_p + 1}{1}$ = __________

$p_{a,1hr}$ = __________ psia

$p_{a,wf}\,(\Delta t_a = 0)$ __________ psia

$$s' = 1.151\left[\left(\frac{p_{a,1hr} - p_{a,wf}}{m}\right) - \log\left(\frac{k}{\phi\, \bar{\mu}\, \bar{c}_t\, r_w^2}\right) + 3.23 + \log\left(\frac{t_p + 1}{t_p}\right)\right]$$

$$s' = 1.151\left[\left(\frac{(\quad)-(\quad)}{(\quad)}\right) - \log\left(\frac{(\quad)}{(\quad)(\quad)(\quad)(\quad)^2}\right) + 3.23 + \log\left(\frac{(\quad)+1}{(\quad)}\right)\right]$$

s' = ______________

DRAWDOWN TEST:

p_{ai} = ___________ psia

Δt_a = __________ psia

$p_{a,1hr}$ at $\Delta t_a = 1$ hour = __________ psia

$$s' = 1.151\left[\left(\frac{p_{ai} - p_{a,1hr}}{m}\right) - \log\left(\frac{k}{\phi\, \bar{\mu}\, \bar{c}_t\, r_w^2}\right) + 3.23\right]$$

$$s' = 1.151\left[\left(\frac{(\quad)-(\quad)}{(\quad)}\right) - \log\left(\frac{(\quad)}{(\quad)(\quad)(\quad)(\quad)^2}\right) + 3.23\right]$$

s' = ______________

LINEAR – FLOW ANALYSIS

If there are test data for $t_{L_fD} \leq 0.016$ and if data in this region lie on a half-slope line, linear-flow analysis may be possible. An independent estimate of permeability must be available to complete this analysis. From a plot of $p_{a,wf}$ vs. $\sqrt{t}$ or $p_{a,ws}$ vs. $\sqrt{\Delta t_{ae}}$, slope m_L = ______ psia/$\sqrt{\text{hr}}$.

$$\sqrt{k}\, L_f = \frac{4.064\, q_g \overline{B}_g}{m_L h}\left(\frac{\overline{\mu}}{\phi\, \overline{c}_t}\right)^{1/2} = \frac{4.064(\quad)(\quad)}{(\quad)(\quad)}\left(\frac{(\quad)}{(\quad)(\quad)}\right)^{1/2} = \underline{\qquad\qquad}$$

$$L_f = \frac{\sqrt{k}\, L_f}{\sqrt{k}} = \frac{(\quad)}{(\quad)^{1/2}} = \underline{\qquad\qquad} \text{ ft}$$

QUANTITATIVE–TYPE CURVE ANALYSIS

Curve Used: ____________________

PRESSURE MATCH POINT

CASE 1: Permeability Estimate Available
Precalculate Pressure Match Point, Using Known Permeability

$$\Delta p_a = \frac{141.2\, q_g \overline{B_g}\, \overline{\mu}\, p_D}{kh} = \frac{(\quad)(\quad)(\quad)(\quad)(\quad)}{(\quad)(\quad)} = \underline{\qquad\qquad} \text{ psia}$$

CASE 2: Permeability Estimate Unavailable; Permeability From Pressure Match Point

Δp_a (psia) = __________ p_D = __________

$$k = \frac{141.2\, q_g \overline{B_g}\, \overline{\mu}\, p_D}{h \Delta p_a} = \frac{(141.2)(\quad)(\quad)(\quad)(\quad)}{(\quad)(\quad)} = \underline{\qquad\qquad} \text{ md}$$

TIME MATCH POINT

Δt_a or Δt_{ae} (hours) = __________; (t_D/C_D) = __________ (Gringarten) or t_{L_fD} = __________ (Cinco or others)

CASE 1: Gringarten Type Curve: Matching Parameter $C_D e^{2s}$ = __________

$$C_D = \frac{0.0002637\, k}{\phi\, r_w^2\, \overline{c_t}\, \overline{\mu}} \frac{\Delta t_e}{t_D/C_D} = \frac{(0.0002637)(\quad)(\quad)}{(\quad)(\quad)^2(\quad)(\quad)(\quad)} = \underline{\qquad\qquad}$$

$$s = 0.5 \ln (C_D e^{2s} / C_D) = 0.5 \ln (\quad / \quad) = \underline{\qquad\qquad}$$

CASE 2: Cinco or Other Type Curve: Matching Parameter C_r= ______ or ______ = ______

$$L_f = \left[\frac{0.0002637\, k}{\phi\, \overline{\mu}\, \overline{c}_t} \frac{\Delta t_{ae}}{t_{LfD}}\right]^{1/2} = \left[\frac{(0.0002637)(\quad)(\quad)}{(\quad)(\quad)(\quad)(\quad)}\right]^{1/2} = \underline{\qquad\qquad} \text{ ft}$$

$$wk_f = \pi C_r L_f k = \pi(\quad)(\quad)(\quad) = \underline{\qquad\qquad} \text{ md-ft}$$

PRODUCTION TIME DURATION CHECK

If an estimate of C_D is available from (1) Gringarten type-curve match or (2) $(L_f^2/r_w^2)\ C_{fD}$ is available from a Barker-Ramey type curve match, or (3) C_D can be estimated from wellbore properties, then the production time duration can be checked.

To ensure that flow period exceeded duration of wellbore storage during production, calculate $t_{p,\min}$.

$$t_{p,\min} = \frac{\phi \bar{\mu} \bar{c}_t r_w^2}{0.0002637\ k}(15\ C_D)$$

$$t_{p,\min} = \frac{(\quad)(\quad)(\quad)(\quad)^2}{(\quad)(\quad)}[15(\quad)] = \_\_\_\_\_\_ \text{ hours}$$

$t_{p,\text{actual}}$ = ______ hours

If $t_{p,\text{actual}} < t_{p,\min}$ then test analysis is invalid. Test must be rerun with longer production period before shut-in period.

DEPTH OF INVESTIGATION CHECK

DEPTH OF INVESTIGATION DURING PRODUCTION PERIOD PRECEDING SHUT-IN PERIOD

Time at end of flow period, t_{flow}= __________ hours

Minor axis: $b = 0.02878\left(\frac{k\, t_{\text{flow}}}{\phi \bar{\mu} \bar{c}_t}\right)^{1/2} = 0.02878\left[\frac{(\quad)(\quad)}{(\quad)(\quad)(\quad)}\right]^{1/2}$

b = __________ ft

Major axis: $a = \left(L_f^2 + b^2\right)^{1/2} = [(\quad)^2 + (\quad)^2]^{1/2}$ = ______ ft

Area Investigated: $\pi ab / 43{,}560 \quad = \pi(\quad)(\quad)/43{,}560$

= __________ ft^2 / 43,560 = ________ acres

DEPTH OF INVESTIGATION ACHIEVED DURING TEST

Start of Semilog Straight Line:

Time at start of line, $\Delta t_{a,\text{start}}$= __________ hours

Minor axis: $b = 0.02878\left(\frac{k\, \Delta t_{a,\text{start}}}{\phi \bar{\mu} \bar{c}_t}\right)^{1/2} = 0.02878\left[\frac{(\quad)(\quad)}{(\quad)(\quad)(\quad)}\right]^{1/2}$

b = __________ ft

Major axis: $a = \left(L_f^2 + b^2\right)^{1/2} = [(\quad)^2 + (\quad)^2]^{1/2}$ = ______ ft

Area Investigated: $\pi ab / 43{,}560 \quad = \pi(\quad)(\quad)/43{,}560$

= __________ ft^2 / 43,560 = ________ acres

End of Semilog Straight Line:

Time at end of line, $\Delta t_{a,\text{end}}$ = ______________ hours

Minor axis: $b = 0.02878\left(\dfrac{k\Delta t_{a,\text{end}}}{\phi\,\bar{\mu}\,\bar{c}_t}\right)^{1/2} = 0.02878\left[\dfrac{(\qquad)(\qquad)}{(\qquad)(\qquad)(\qquad)}\right]^{1/2}$

$b =$ ______________ ft

Major axis: $a = \left(L_f^2 + b^2\right)^{1/2} = [(\qquad)^2 + (\qquad)^2]^{1/2} =$ __________ ft

Area Investigated: $\pi ab/43{,}560 = \pi(\qquad)(\qquad)/43{,}560$

= ______________ ft^2 / 43,560 = ____________ acres

End of Test:

Time at end of test, $\Delta t_{a,\text{final}}$ = ______________ hours

Minor axis: $b = 0.02878\left(\dfrac{k\Delta t_{a,\text{final}}}{\phi\,\bar{\mu}\,\bar{c}_t}\right)^{1/2} = 0.02878\left[\dfrac{(\qquad)(\qquad)}{(\qquad)(\qquad)(\qquad)}\right]^{1/2}$

$b =$ ______________ ft

Major axis: $a = \left(L_f^2 + b^2\right)^{1/2} = [(\qquad)^2 + (\qquad)^2]^{1/2} =$ __________ ft

Area Investigated: $\pi ab/43{,}560 = \pi(\qquad)(\qquad)/43{,}560$

= ______________ ft^2 / 43,560 = ____________ acres

NOTE: Reservoir heterogeneities and boundaries determined from data during shut-in test at $\Delta t_a > t_{\text{flow}}$ are highly questionable. If the drawdown did not reach beyond a certain distance in the reservoir, then the buildup will not sample reservoir properties beyond that point.

REMARKS

____

____

____

____

____

____

GASWELL-TEST ANALYSIS
HYDRAULICALLY FRACTURED WELL

Operator __Example 6.2 Test Date____________Analyst________________

Well_________________Stimulation ___________ Date _________________

Location _________________________________ Formation______________

Perforations ____ (ft) Depth of Bomb ____ (ft) Pressure Gradient in Wellbore__________ (psi/ft)

Pressure Datum __________________Pressure Gradient in Reservoir _________ (psi/ft)

Correction to Pressure Gauge Reading } _____ x _____ + _____ x _____ + _____ = _______ p

Wellbore Gradient (psi/ft) × Distance Mid-Perfs to Bomb (ft) + Reservoir Gradient (psi/ft) × Mid-Perfs to Datum (ft) + Atmospheric Pressure (psia)

BASIC PROPERTIES

q_g (Mscf/D) __3000__ r_w (ft) __0.25__ $\bar{\mu}$ (cp) __0.01961__

T (°F) __110__ S_w (fraction) __0.35__ $\bar{z}$ (fraction) __0.9910__

h (ft) __60__ t_p (hours) ________ $\bar{c_t}$ (psia^{-1}) __2.084×10^{-4}__

ϕ (fraction) __0.10__ $\bar{p}$ (psia) __5000__

γ_g (air = 1.0) __0.65__ $\bar{B_g}$ (RB/Mscf) $= 5.035 \frac{T\bar{z}}{\bar{p}}$ = __0.7085__

PRELIMINARY TYPE-CURVE ANALYSIS–START SEMILOG STRAIGHT LINE

Curve Used: __Gringarten__ Matching Parameter: C_r = ______ or __Le/Lf__ = __8.5__

If there are test data for $t_{L_fD} \geq 3$, the pseudoradial flow regime may be present.

Start Semilog Straight Line: At $t_{L_fD} = 3$, Δt_{ae} = ________ hours; Δt = __142__ hours

$(t_p + \Delta t_a)/\Delta t_a$ = ______ ; Δt_a = _______; Not present ____

Start Boundary Effects: At t_{L_fD} = _____ , Δt_{ae} = _______ hours; Δt = __1932__ hours

$(t_p + \Delta t_a)/\Delta t_a$ = ______ ; Δt_a = _______; Not present ____

If $\Delta t_{max} < \Delta t$ at start of pseudoradial flow, then semilog analysis is not possible.

SEMILOG ANALYSIS

Average drainage-area pressure at start of test, $\bar{p}$,= ______psia; $\bar{p}_a$=______ psia

at depth = ______ ft

$(t_p + \Delta t_a)/\Delta t_a$ at start of MTR = ______ (Buildup) t at start of MTR = 142 (Drawdown)

$(t_p + \Delta t_a)/\Delta t_a$ at start of LTR = ______ (Buildup) t at start of LTR = 1932 (Drawdown)

$m = p_{a2} - p_{a1}$ = 5340.8 − 4056.3 = 1284.5 psia/cycle

$$k = \frac{162.6\, q_g \overline{B_g}\,\overline{\mu}}{mh} = \frac{(162.6)(3000)(0.7085)(0.01961)}{(1284.5)(80)} = 0.088 \text{ md}$$

SKIN-FACTOR CALCULATION

BUILDUP TEST:

Horner Time Ratio at Δt_a = 1hour = $\dfrac{t_p+1}{1}$ = ______

$p_{a,1hr}$ = ______ psia

$p_{a,wf}$ (Δt_a = 0) ______ psia

$$s' = 1.151\left[\left(\frac{p_{a,1hr} - p_{a,wf}}{m}\right) - \log\left(\frac{k}{\phi\,\overline{\mu}\,\overline{c_t}\,r_w^2}\right) + 3.23 + \log\left(\frac{t_p+1}{t_p}\right)\right]$$

$$s' = 1.151\left[\left(\frac{(\quad)-(\quad)}{(\quad)}\right) - \log\left(\frac{(\quad)}{(\quad)(\quad)(\quad)(\quad)^2}\right) + 3.23 + \log\left(\frac{(\quad)+1}{(\quad)}\right)\right]$$

s' = ______

DRAWDOWN TEST:

p_{ai} = 3974 psia

Δt_a = ______ psia

$p_{a,1hr}$ at Δt_a = 1 hour = 5239 psia

$$s' = 1.151\left[\left(\frac{p_{ai} - p_{a,1hr}}{m}\right) - \log\left(\frac{k}{\phi\,\overline{\mu}\,\overline{c_t}\,r_w^2}\right) + 3.23\right]$$

$$s' = 1.151\left[\left(\frac{(3974)-(5239)}{(1284.5)}\right) - \log\left(\frac{(0.088)}{(0.10)(0.01961)(0.0002084)(0.25)^2}\right) + 3.23\right]$$

s' = −4.94

LINEAR-FLOW ANALYSIS

If there are test data for $t_{L_fD} \le 0.016$ and if data in this region lie on a half-slope line, linear-flow analysis may be possible. An independent estimate of permeability must be available to complete this analysis. From a plot of $p_{a,wf}$ vs. $\sqrt{t}$ or $p_{a,ws}$ vs. $\sqrt{\Delta t_e}$, slope m_L = ~~193.9~~ psia/$\sqrt{\text{hr}}$.

$$\sqrt{k}\,L_f = \frac{4.064\,q_o B_o}{m_L h}\left(\frac{\mu}{\phi c_t}\right)^{1/2} = \frac{4.064(3000)(0.7085)}{(193.9)(60)}\left(\frac{(0.01961)}{(0.10)(2.084 \times 10^{-4})}\right)^{1/2} = \underline{22.78}$$

$$L_f = \frac{\sqrt{k}\,L_f}{\sqrt{k}} = \frac{(22.78)}{(0.088)^{1/2}} = \underline{77} \text{ ft}$$

QUANTITATIVE TYPE-CURVE ANALYSIS

Curve Used: Gringarten

PRESSURE MATCH POINT

CASE 1: Permeability Estimate Available
Precalculate Pressure Match Point, Using Known Permeability

$$\Delta p_a = \frac{141.2\,q_g \overline{B_g}\,\overline{\mu}\,p_D}{kh} = \frac{(141.2)(3000)(0.7085)(0.01961)(1)}{(0.088)(60)} = \underline{1115} \text{ psia}$$

CASE 2: Permeability Estimate Unavailable; Permeability From Pressure Match Point

Δp_a (psia) = ______ p_D = ______

$$k = \frac{141.2\,q_g \overline{B_g}\,\overline{\mu}\,p_D}{h\Delta p_a} = \frac{(141.2)(\quad)(\quad)(\quad)(\quad)}{(\quad)(\quad)} = \_\_\_\_\_\_ \text{ md}$$

TIME MATCH POINT

Δt_a or Δt_{ae} (hours) = 1000; (t_D/C_D) = ______ (Gringarten) or t_{LfD} = 11 (Cinco or others)

CASE 1: Gringarten Type Curve: Matching Parameter C_De^{2s} = ______

$$C_D = \frac{0.0002637\,k}{\phi r_w^2 \overline{c_t}\,\overline{\mu}}\frac{\Delta t_e}{t_D/C_D} = \frac{(0.0002637)(\quad)(\quad)(\quad)}{(\quad)(\quad)^2(\quad)(\quad)(\quad)} = \_\_\_\_\_\_$$

$s = 0.5 \ln (C_De^{2s}/C_D) = 0.5 \ln (\quad / \quad) =$ ______

CASE 2: Cinco or Other Type Curve: Matching Parameter C_r = ______ or Le/Lf = 8.5

$$L_f = \left[\frac{0.0002637\,k}{\phi\,\overline{\mu}\,\overline{c_t}}\frac{\Delta t_{ae}}{t_{LfD}}\right]^{1/2} = \left[\frac{(0.0002637)(0.088)(1000)}{(0.10)(0.01961)(0.0002084)(11)}\right]^{1/2} = \underline{72} \text{ ft}$$

$$wk_f = \pi C_r L_f k = \pi(100)(72)(0.088) = \underline{1990} \text{ md-ft}$$

PRODUCTION TIME DURATION CHECK

If an estimate of C_D is available from (1) Gringarten type-curve match or (2) $(L_f^2/r_w^2)\ C_{fD}$ is available from a Barker-Ramey type curve match, or (3) C_D can be estimated from wellbore properties, then the production time duration can be checked.

To ensure that flow period exceeded duration of wellbore storage during production, calculate $t_{p,\min}$.

$$t_{p,\min}=\frac{\phi\,\bar{\mu}\,\bar{c}_t\,r_w^2}{0.0002637\,k}(15\,C_D)$$

$$t_{p,\min}=\frac{(0.10)(0.01961)(0.0002084)(0.25)^2}{(0.0002637)(0.088)}[15(0.01)] = \underline{0.00017}\text{ hours}$$

$t_{p,\text{actual}} = \underline{5588.5}$ hours

If $t_{p,\text{actual}} < t_{p,\min}$ then test analysis is invalid. Test must be rerun with longer production period before shut-in period.

DEPTH OF INVESTIGATION CHECK

DEPTH OF INVESTIGATION DURING PRODUCTION PERIOD PRECEDING SHUT-IN PERIOD

Time at end of flow period, $t_{\text{flow}} = \underline{5588.5}$ hours

$$\text{Minor axis: } b = 0.02878\left(\frac{k\,t_{\text{flow}}}{\phi\,\bar{\mu}\,\bar{c}_t}\right)^{1/2} = 0.02878\left[\frac{(0.088)(5588.5)}{(0.10)(0.01961)(0.0002084)}\right]^{1/2}$$

$b = \underline{998}$ ft, **but L_e = 655 ft, so b = 655 ft.**

$$\text{Major axis: } a = \left(L_f^2 + b^2\right)^{1/2} = \left((77)^2 + (655)^2\right)^{1/2} = \underline{660}\text{ ft, } \textbf{but } L_e = 655 \textbf{ ft, so } a = 655 \textbf{ ft.}$$

Area Investigated: $\pi ab/43{,}560 = \pi(655)(655)/43{,}560$

$= \underline{1.348 \times 10^6}\ \text{ft}^2/43{,}560 = \underline{30.9}$ acres

DEPTH OF INVESTIGATION ACHIEVED DURING TEST

Start of Semilog Straight Line:

Time at start of line, $\Delta t_{a,\text{start}} = \underline{321}$ hours

$$\text{Minor axis: } b = 0.02878\left(\frac{k\,\Delta t_{a,\text{start}}}{\phi\,\bar{\mu}\,\bar{c}_t}\right)^{1/2} = 0.02878\left[\frac{(0.088)(321)}{(0.10)(0.01961)(0.0002084)}\right]^{1/2}$$

$b = \underline{239}$ ft

Major axis: $a = \left(L_f^2 + b^2\right)^{1/2} = \left[(\ 77\ )^2 + (\ 239\ )^2\right]^{1/2}$ = 251 ft

Area Investigated: $\pi ab/43{,}560 = \pi(\ 251\ )(\ 239\ )/43{,}560$

= 188,461 ft² / 43,560 = 4.33 acres

End of Semilog Straight Line:

Time at end of line, $\Delta t_{a,\text{end}}$ = 2571 hours

Minor axis: $b = 0.02878\left(\dfrac{k\,\Delta t_{a,\text{end}}}{\phi\,\bar{\mu}\,\bar{c}_t}\right)^{1/2} = 0.02878\left[\dfrac{(\ 0.088\ )(\ 2571\ )}{(\ 0.10\ )(\ 0.01961\ )(\ 0.0002084\ )}\right]^{1/2}$

b = 677 ft, **but L_e = 655 ft, so b = 655 ft.**

Major axis: $a = \left(L_f^2 + b^2\right)^{1/2} = \left[(\ 77\ )^2 + (\ 655\ )^2\right]^{1/2}$ = 660 ft, **but L_e = 655 ft, so a = 655 ft.**

Area Investigated: $\pi ab/43{,}560 = \pi(\ 655\ )(\ 655\ )/43{,}560$

= 1.348×10^6 ft² / 43,560 = 30.9 acres

End of Test:

Time at end of test, $\Delta t_{a,\text{final}}$ = 5588.5 hours

Minor axis: $b = 0.02878\left(\dfrac{k\,\Delta t_{a,\text{final}}}{\phi\,\bar{\mu}\,\bar{c}_t}\right)^{1/2} = 0.02878\left[\dfrac{(\ 0.088\ )(\ 5588.5\ )}{(\ 0.10\ )(\ 0.01961\ )(\ 0.0002084\ )}\right]^{1/2}$

b = 998 ft, **but L_e = 655 ft, so b = 655 ft.**

Major axis: $a = \left(L_f^2 + b^2\right)^{1/2} = \left[(\ 77\ )^2 + (\ 655\ )^2\right]^{1/2}$ = 660 ft, **but L_e = 655 ft, so a = 655 ft.**

Area Investigated: $\pi ab/43{,}560 = \pi(\ 655\ )(\ 655\ )/43{,}560$

= 1.348×10^6 ft² / 43,560 = 30.9 acres

NOTE: Reservoir heterogeneities and boundaries determined from data during shut-in test at $\Delta t_a > t_{flow}$ are highly questionable. If the drawdown did not reach beyond a certain distance in the reservoir, then the buildup will not sample reservoir properties beyond that point.

REMARKS

At the end of the test, the drainage area was no longer elliptical; instead, it approached the full square. The area drained probably approached 100% of the square. Note that the maximum possible values of a and b are 655 ft. because this is the value of L_e.

Appendix K
Worksheets for Well-Test Design

WELL-TEST DESIGN WORKSHEET

Operator ______________________ Date ______________________ Analyst ______________

Well ______________________ Stimulation ______________ Date ______________

Location ____ Formation ______________

Perforations ______________________ Total Depth______________ ft Packer Location ________ ft

Tubing above packer, ft ______________ tubing ID, in ______________ tubing area, bbl/ft ________

Casing below packer, ft ______________ casing ID, in ______________ casing area, bbl/ft ________

Wellbore volume open to formation without downhole shut-in ______________ bbl

Wellbore volume open to formation with downhole shut-in ______________ bbl

Test Objectives ____

____

____

____

FLUID AND RESERVOIR PROPERTIES

FLUID PROPERTIES

Fluid (oil, gas, or water) ______________________ μ (cp) ______________________

B (RB/STB or RB/Mscf) ______________________ Gravity (°API or γ) ______________

RESERVOIR PROPERTIES

k (md) ______________ s ______________ h (ft) ______________

r_w (ft) ______________ ϕ (fraction) ______________ T (°F) ______________

c_o (psia^{-1}) ______________ c_w (psia^{-1}) ______________ c_g (psia^{-1}) ______________

S_o (fraction) ______________ S_w (fraction) ______________ S_g (fraction) ______________

c_f (psia^{-1}) ______________ $c_t = c_oS_o + c_wS_w + c_gS_g + c_f =$

$$\eta = \frac{k}{3792\ \phi\mu c_t} = \frac{(\qquad\qquad)}{(3792)(\qquad\qquad)(\qquad\qquad)(\qquad\qquad)} = \_\_\_\_\_\_\_\_\_\_$$

WELL-TEST DESIGN WORKSHEET – continued

WELLBORE STORAGE

Storage Coefficient, C: Oil or water well with liquid-gas interface in well

Wellbore area, A_{wb} (ft^2) ____________ Wellbore liquid density, ρ_{wb} (lb/ft^3) ____________

$C = 25.65\, A_{wb} / \rho_{wb} = (25.65)$ (____________)/(____________) = ____________ bbl/psi

Storage Coefficient, C: Gaswell or well full of liquid (no interface)

Wellbore volume open to formation during flow period, V_{wb} (bbl) ____________

Compressibility of fluid in wellbore, c_{wb} (psia^{-1}) ____________

$C = c_{wb}\, V_{wb}$ = (____________) (____________) = ____________ bbl/psi

Dimensionless storage coefficient, C_D

$$C_D = \frac{0.8936C}{\phi\, c_t\, h\, r_w^2} = \frac{(0.8936)(\quad\quad)}{(\quad\quad)(\quad\quad)(\quad\quad)(\quad\quad)^2} = \underline{\qquad\qquad}$$

Duration of wellbore storage distortion during flow period, t_{wbs}

$$t_{wbs} = \frac{a r_w^2 C_D}{\eta} = \frac{(\quad\quad)(\quad\quad)^2(\quad\quad)}{(\quad\quad)} = \underline{\qquad\qquad} \text{ hr}$$

{For $s \geq -3.5$, $\alpha = (60 + 3.5s)$
For hydraulically fractured well, $\alpha = 15$}

Modification of wellbore-storage-distortion duration during shut-in if bottomhole shut-in used, t_{wbs} - bhsi

V_{wb}-bhsi (bottomhole shut-in) _____ bbl V_{wb} (flowing) _____ bbl

t_{wbs}-bhsi = t_{wbs} (V_{wb} -bhsi)/V_{wb} = _____ hr

Notes:

(1) Producing time in a flow test or before shut-in in a buildup test must exceed $4t_{wbs}$
(2) Shut-in time in a buildup test must also exceed $4t_{wbs}$

PSEUDORADIAL FLOW IN HYDRAULICALLY FRACTURED WELL

Estimated fracture half-length, L_f ____________ ft

$$t_{prf} = \frac{3L_f^2}{\eta} = \frac{(3)(\quad\quad)^2}{(\quad\quad)} = \underline{\qquad\qquad} \text{ hr}$$

BOUNDARY EFFECTS

Estimated distance to boundary nearest well, L ______________ ft

$$t_{be} = \frac{L^2}{4\eta} = \frac{(\qquad)^2}{(4)(\qquad)} = \text{\_\_\_\_\_\_\_\_\_\_} \text{ hr}$$

Note: In a buildup test (only), for a well *centered* in its drainage area, the semilog straight line may begin to bend noticeably at $t_{end} \approx t_{be}/4$

t_{end} = ______________ hr (buildup test only; well centered)

Note: If $t_{end} < 4t_{wbs}$, consider bottomhole shut-in.

RADIUS OF INVESTIGATION

η = __________ $4\eta = 4$ (__________) = __________

$r_i = 50$ ft: $t = [50^2/(4\eta)] = [2.5 \times 10^3/($______________$)] =$ ______________ hr

$r_i = 200$ ft: $t = [200^2/(4\eta)] = [4 \times 10^4/($______________$)] =$ ______________ hr

$r_i = r_e$ (drainage radius of well, ft)

Drainage area of well, A = ______________ ft^2

$r_e = (A/\pi)^{0.5} = [($______________$)/\pi]^{0.5} =$ ______________ ft

$t_{end} = [r_e^2/(4\eta)] = [($______________$)^2/($______________$)] =$ ______________ hr

Maximum producing time possible for test

$t_{p,max}$ = ______________ hr

$r_{i,max tp} = (4\eta\ t_{p,max})^{0.5} = [($______________$)($______________$)]^{0.5} =$ ______________ ft

Maximum shut-in time possible for test

Δt_{max} = ______________ hr

$r_{i,max\Delta t} = (4\eta\ \Delta t_{max})^{0.5} = [($______________$)($______________$)]^{0.5} =$ ______________ ft

Notes:

(1) For productivity analysis, r_i of 50 to 200 ft is a reasonable design criterion.

(2) For properties throughout potential drainage area of well, $r_i \approx r_e$ is desirable but often unattainable.

(3) In a buildup test, formation properties will be sampled to a distance no greater than the maximum of the radii of investigation achieved in the flow and shut-in periods.

WELL-TEST DESIGN WORKSHEET – continued

PRESSURE GAUGE SENSITIVITY REQUIREMENT

Rate expected during test (limited to about 75% of an observed, sustainable well deliverability when known)

$q =$ ______________ STB/D or Mscf/D

$m = 162.6 \dfrac{qB\mu}{kh} = (162.6)\dfrac{(\qquad)(\qquad)(\qquad)}{(\qquad)(\qquad)} =$ ______________ psi/cycle

Pressure change between end of semilog straight line and 90% of that time

$\Delta p = 0.0458\, m = (0.0458)(\qquad) =$ ______________ psi

Note: Pressure gauge for the test must be at least sensitive enough to respond to the calculated pressure change, Δp.

SUMMARY

Test Objectives ____

____

____

____

____

Minimum flow time required = ______________ hr (Greater of 4 t_{wbs} and time to achieve desired radius of investigation, but limited by time available for flow)

Reasons ____

____

Minimum shut-in time required = ______________ hr (Greater of 4 t_{wbs} and time to achieve desired radius of investigation, but limited by time available for shut-in)

Reasons ____

____

Minimum gauge sensitivity required = ______________ psi

COMMENTS ____

____

____

Appendix L
Reservoir-Fluid Properties

Introduction

This appendix presents methods for estimating reservoir-fluid properties required for well testing. Laboratory analysis is the most accurate way to determine the physical and chemical properties of a particular fluid sample; however, in the absence of laboratory data, correlations are viable alternatives for estimating many of the properties. We present correlations for estimating properties of not only natural gases but also liquid hydrocarbons and formation waters. The correlations presented were chosen for accuracy, consistency, and simplicity for manual analysis or computer programming. Also included are correlations for estimating pore-volume (PV) compressibility and the non-Darcy flow coefficient for turbulent flow, common in gas wells.

Definitions

Before discussing the fluid-property calculations and correlations, we review some definitions and fundamental principles that are required for a complete understanding of fluid properties and their computation with correlations. Included in this review are the concepts of mole fraction, molar volume, ideal- and real-gas behavior, and the principle of corresponding states. We also include a brief definition for each of the fluid and rock properties used in well-test analysis. McCain[1] gives a more complete discussion of these fundamentals. In addition, Ref. 1 provides an excellent discussion of phase behavior characteristics of hydrocarbon gases and liquids.

Definitions of Natural-Gas Properties. Table L-1 lists the physical properties of pure components that occur in natural gases.[2,3] These properties, which are evaluated at standard conditions of $p_{sc} = 14.7$ psia and $T_{sc} = 60°F$, include molecular weight, critical pressure and temperature, ideal density, and viscosity (components lighter than pentane only). These properties of pure components are used in calculations based on mixing rules to develop pseudoproperties for gas mixtures, including apparent molecular weight and specific gas gravity. Refs. 2 and 3 give more complete listings of natural-gas properties and contaminants commonly associated with natural-gas production.

Apparent Molecular Weight of a Gas Mixture. Because a gas mixture is composed of molecules of various sizes and molecular weights, it does not have an explicit molecular weight of its own. However, a gas mixture behaves as if it has a definite molecular weight. This observed molecular weight for a gas mixture with n_{Comp} components is called the apparent, or molal, average molecular weight and is determined by

$$M = \sum_{i=1}^{n_{Comp}} y_i M_i \quad \text{(L-1)}$$

where M = apparent molecular weight of gas mixture, lbm/lbm mol; M_i = molecular weight of the ith gas component, lbm/lbm mol; and y_i = mole fraction the gas phase of the ith component, fraction.

Specific Gravity of a Gas. The specific gravity, γ_g, of a gas is defined as the ratio of the density of the gas to the density of dry air when both are measured at the same temperature and pressure. The specific gas gravity in equation form is

$$\gamma_g = \frac{\rho_g}{\rho_a} \quad \text{(L-2)}$$

where ρ_g = density of gas mixture, lbm/ft^3, and ρ_a = density of air, lbm/ft^3.

At standard conditions (such as 14.65 psia and 60°F), both air and natural gas are modeled accurately by the ideal gas law. Under these conditions, with the definitions of pound mole ($n = m/M$) and density ($\rho = m/V$) and modeling the behavior of both the gas and the air by the ideal gas equation of state (EOS), the specific gravity of a gas mixture can be expressed as

$$\gamma_g = \frac{\dfrac{pM}{RT}}{\dfrac{pM_a}{RT}} = \frac{M}{M_a} \quad \text{(L-3)}$$

where γ_g = specific gravity of the gas (air = 1.0); M = apparent molecular weight of the gas, lbm/lbm mol;and M_a = molecular weight of air = 28.9625 lbm/lbm mol.

Although Eq. L-3 is derived under the assumptions of an ideal gas (accurate at standard conditions), its use as a definition for real gases and real-gas mixtures is common in the natural-gas industry.

Moles and Mole Fraction. A pound mole is a quantity of matter with a mass in pounds equal to the molecular weight. Similar definitions apply to gram mole, kilogram mole, and other such quantities. For example, 1 lbm mol of methane would weigh 16.043 lbm. The mole fraction of a component in a mixture is the number of pound moles of that component divided by the total number of moles of all components in the mixture. For a system with n components, the mole fraction is defined as

$$y_i = \frac{n_i}{\sum_{j=1}^{n_{Comp}} n_j} \quad \text{(L-4)}$$

where y_i = mole fraction of the ith component; n_i = number of pound moles of the ith component; and n_{Comp} = number of components in the system.

Real-Gas Deviation Factor (z Factor). The ideal-gas law states that the pressure, temperature, volume, and amount of an ideal gas are related by

$$pV = nRT \quad \text{(L-5)}$$

TABLE L-1—PHYSICAL PROPERTIES OF GASES AT 14.7 psia AND 60°F							
Component	Chemical Formula	Molecular Weight (lbm/lbm mol)	Critical Temperature (°R)	Critical Pressure (psia)	Liquid Density* (lbm/ft³)	Gas Density (lbm/ft³)	Gas Viscosity (cp)
Hydrogen	H_2	2.109	59.36	187.5	4.432	0.005312	0.00871
Helium	He	4.003	9.34	32.9	7.802	0.01055	0.01927
Water	H_2O	18.015	1,164.85	3,200.1	62.336	—	1.122
Carbon Monoxide	CO	28.010	239.26	507.5	49.231	0.07381	0.01725
Nitrogen	N_2	28.013	227.16	493.1	50.479	0.07382	0.01735
Oxygen	O_2	31.999	278.24	731.4	71.228	0.08432	0.02006
Hydrogen Sulfide	H_2S	34.08	672.35	1,306.0	49.982	0.08981	0.01240
Carbon Dioxide	CO_2	44.010	547.58	1,071.0	51.016	0.1160	0.01439
Air	—	28.963	238.36	546.9	54.555	0.07632	0.01790
Methane	CH_4	16.043	343.00	666.4	18.710	0.04228	0.01078
Ethane	C_2H_6	30.070	549.59	706.5	22.214	0.07924	0.00901
Propane	C_3H_8	44.097	665.73	616.0	31.619	—	0.00788
i-Butane	C_4H_{10}	58.123	734.13	527.9	35.104	—	0.00732
n-Butane	C_4H_{10}	58.123	765.29	550.6	36.422	—	0.00724
i-Pentane	C_5H_{12}	72.150	828.77	490.4	38.960	—	—
n-Pentane	C_5H_{12}	72.150	845.47	488.6	39.360	—	—
n-Hexane	C_6H_{14}	86.177	913.27	436.9	41.400	—	—
n-Heptane	C_7H_{16}	100.204	972.37	396.8	42.920	—	—
n-Octane	C_8H_{18}	114.231	1,023.89	360.7	44.090	—	—
n-Nonane	C_9H_{20}	128.258	1,070.35	331.8	45.020	—	—
n-Decane	$C_{10}H_{22}$	142.285	1,111.67	305.2	45.790	—	—

*Values given are liquid densities for those components that can exist as liquids at 60°F and 14.7 psia and estimated liquid densities for components that are naturally gases at these conditions.

where p = pressure, psi; V = volume, ft³; n = number of pound moles of the gas; R = universal gas constant, 10.732 (psia-ft³)/(lbm mol-°R); and T = temperature, °R. For real gases, a real-gas deviation factor, z, is defined as

$$z \equiv \frac{V_{\text{real}}}{V_{\text{ideal}}}, \quad \text{(L-6)}$$

where V_{real} = actual volume corresponding to a given pressure and temperature and V_{ideal} is calculated from the ideal-gas law (Eq. L-5). This z factor is simply a "fudge factor" to allow the use of an equation for real gases having a form similar to Eq. L-5. This equation is sometimes called the real-gas law:

$$pV = znRT. \quad \text{(L-7)}$$

Expressions for gas formation volume factor (FVF), B_g; gas density, ρ_g; and gas compressibility, c_g, will be developed from Eq. L-7.

The z factor is dimensionless. Typically, z is close to 1.0. For gases near the critical point, z is approximately 0.3 to 0.4; as temperature and pressure increase, z increases to as much as 2.0 or more.

Gas FVF. The gas FVF is defined as the ratio of the volume of a given amount of gas at reservoir conditions to the volume of the same amount of gas at standard conditions,

$$B_g \equiv \frac{V_R}{V_{sc}} = \frac{z_r T_r}{p_r}\frac{p_{sc}}{z_{sc}T_{sc}} \approx \frac{z_r T_r}{p_r}\frac{p_{sc}}{T_{sc}}, \quad \text{(L-8)}$$

where B_g is given in res ft³/scf. For well-test analysis, we often want B_g in RB/Mscf.

$$B_g = \frac{1}{0.0056146}\frac{z_r T_r p_{sc}}{p_r\ T_{sc}}. \quad \text{(L-9)}$$

Gas Density. The gas density, in lbm/ft, also may be obtained from the z factor:

$$\rho_g \equiv \frac{m}{V} = \frac{nMp}{znRT} = \frac{pM}{zRT}, \quad \text{(L-10)}$$

where M = average molecular weight of the gas. Gas density may also be required in g/cm³.

$$\rho_g = 1.4935 \times 10^{-3}\frac{pM}{zT}, \quad \text{(L-11)}$$

where p is in psi and T is in °R, and where the gas constant R has been absorbed into the constant coefficient.

Gas Compressibility. The isothermal gas compressibility, or gas compressibility, c_g, is defined as

$$c_g \equiv -\frac{1}{V}\frac{\partial V}{\partial p}\bigg|_T, \quad \text{(L-12)}$$

which can be written as

$$c_g = -\frac{p}{znRT}\frac{\partial}{\partial p}\left(\frac{znRT}{p}\right)\bigg|_T$$

$$= -\frac{p}{z}\left[z\frac{\partial}{\partial p}\left(\frac{1}{p}\right) + \frac{1}{p}\frac{\partial z}{\partial p}\right]$$

$$= \frac{1}{p} - \frac{1}{V}\frac{\partial z}{\partial p}\bigg|_T. \quad \text{(L-13)}$$

For an ideal gas, compressibility is given by $1/p$, independent of temperature. This expression also provides a useful rough estimate for the compressibility of real gases, especially at low pressures. Gas compressibility is the fractional decrease in volume that occurs because of a unit increase in pressure, or equivalently, the fraction volume change necessary to cause a unit increase in pressure. Compressibility has units that are the reciprocal of pressure units. In oilfield units, compressibility is expressed in 1/psi (or psi^{-1}).

Viscosity. If we measure the shear stress and the rate of strain within a fluid between two parallel plates, where one plate moves at a fixed rate with respect to the other, we find that the shear stress is proportional to the rate of strain. The proportionality constant is defined as the viscosity of the fluid, μ.

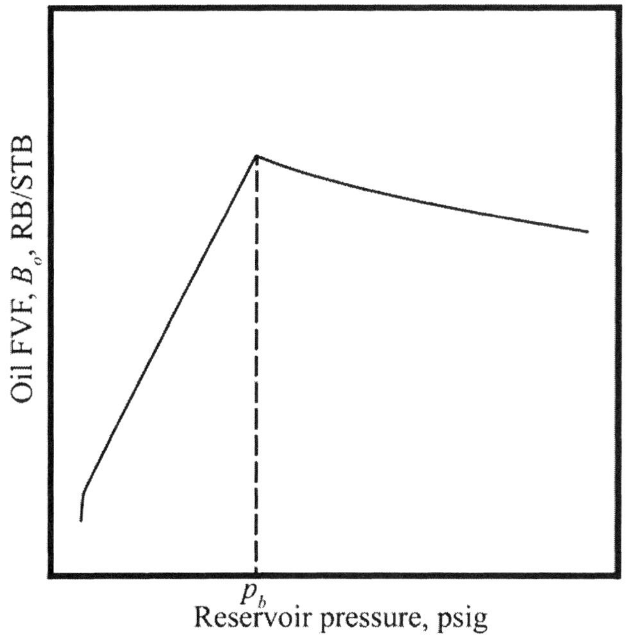

Fig. L-1—Typical shape of FVF of a black oil as a function of pressure at constant reservoir temperature.

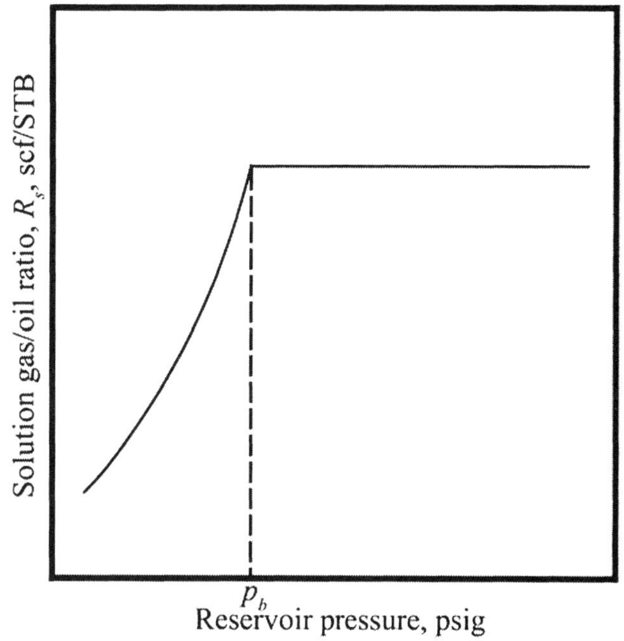

Fig. L-2—Typical shape of SGOR of a black oil as a function of pressure at constant reservoir temperature.

$$\tau_{xy} = \mu_g \frac{\partial u}{\partial y}. \quad \text{(L-14)}$$

Viscosity is a measure of the resistance of a fluid to flow. In oilfield units, viscosity is measured in centipoise (cp).

Principle of Corresponding States. The critical point of a fluid is defined as the combination of pressure and temperature at which the properties of the gas and liquid phases are indistinguishable. The pressure and temperature at which this occurs are called the critical pressure, p_c, and critical temperature, T_c.

Experimentally, it has been found that the z factor for homologous pure compounds may be expressed as a function of reduced pressure, p_r, and reduced temperature, T_r.

$$p_r \equiv \frac{p}{p_c} \quad \text{(L-15a)}$$

$$\text{and } T_r \equiv \frac{T}{T_c}. \quad \text{(L-15b)}$$

It also has been found that the z-factor charts may be used for pure hydrocarbon compounds to estimate the z factors for mixtures of hydrocarbon gases, provided a pseudocritical pressure, p_{pc}, and a pseudocritical temperature, T_{pc}, are defined. Note that these numbers are not the same as the true critical pressure and critical temperature of the mixture. They are useful only as a means of correlating other gas properties that are of interest.

Definitions of Reservoir Oil Properties. ***Oil FVF.*** The oil FVF is the number of barrels of oil in the reservoir required to produce 1 bbl of stock-tank oil.

$$B_o \equiv \frac{V_r}{V_{st}}. \quad \text{(L-16)}$$

Fig. L-1 shows typical behavior of oil FVF. Above the bubblepoint, oil FVF increases with decreasing pressure; below the bubblepoint, it decreases with decreasing pressure.

Solution-Gas/Oil Ratio (SGOR). If 1 bbl of oil at reservoir conditions is produced to the surface through a separator to the stock tank, the oil is accompanied by a certain volume of gas. The SGOR is the ratio of the volume of gas produced at standard conditions to the volume of oil produced in stock-tank barrels as a result of producing a given volume of oil originally at reservoir conditions.

$$R_{so} = \frac{V_{gsc}}{V_{ost}}. \quad \text{(L-17)}$$

Note that the SGOR includes both separator gas (which is measured) and stock-tank vent gas (which often is not measured). If the stock-tank-vent-gas volume is not available, it may be estimated with appropriate correlations.

Fig. L-2 shows typical behavior of SGOR with changing reservoir pressure. Above the bubblepoint, the SGOR is constant; below the bubblepoint, it decreases with decreasing pressure.

Oil Compressibility. Above the bubblepoint, oil compressibility is given by

$$c_o = -\frac{1}{B_o}\frac{\partial B_o}{\partial p}. \quad \text{(L-18)}$$

Below the bubblepoint, we must take into account evolution of solution gas as the pressure is lowered. Thus, below the bubblepoint,

$$c_o = -\frac{1}{B_o}\frac{\partial B_o}{\partial p} + \frac{B_g}{B_o}\frac{\partial R_{so}}{\partial p}. \quad \text{(L-19)}$$

Without inclusion of the second term in Eq. L-19, the oil compressibility below the bubblepoint would be negative.

Fig. L-3 shows the typical behavior of oil compressibility above the bubblepoint; **Fig. L-4** shows the behavior both above and below the bubblepoint on a different scale. Note that the compressibility below the bubblepoint is an order of magnitude or more greater than that above the bubblepoint.

Oil Viscosity. As with gases, oil viscosity is a measure of the resistance of a fluid to flow. **Fig. L-5** shows the relationship between viscosity and pressure for a typical black oil. Above the bubblepoint, the oil viscosity decreases with decreasing pressure; below the bubblepoint, it increases with increasing pressure.

Definitions of Reservoir Water Properties. ***Water FVF.*** The water FVF is the number of barrels of water in the reservoir required to produce one barrel of water at the surface.

$$B_w \equiv \frac{V_r}{V_{st}}. \quad \text{(L-20)}$$

Unlike the oil FVF, the water FVF increases with decreasing pressure, both above and below the bubblepoint. There is, however, a

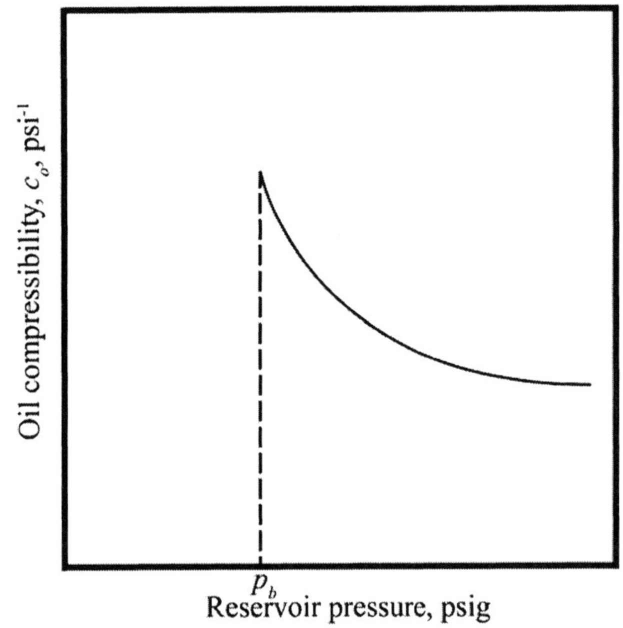

Fig. L-3—Typical shape of the coefficient of isothermal compressibility of oil as a function of pressure at constant reservoir temperature at pressures above the bubblepoint.

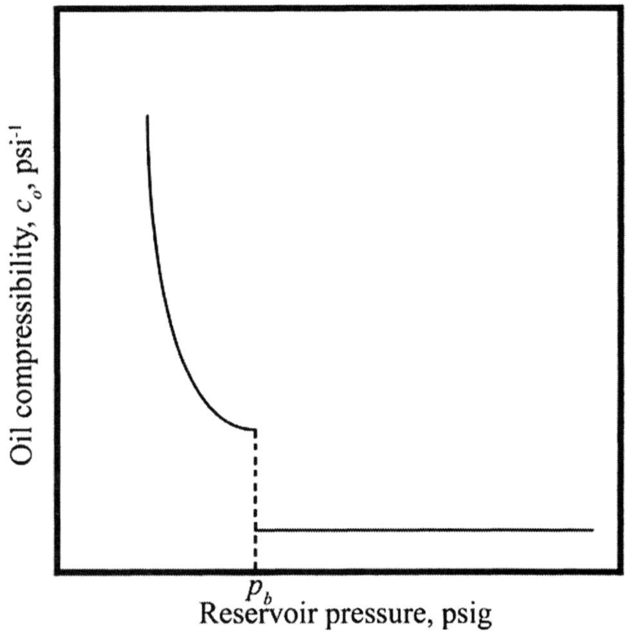

Fig. L-4—Typical shape of the coefficient of isothermal compressibility of oil as a function of pressure at constant reservoir temperature.

change in slope of the water FVF curve at the bubblepoint pressure, as **Fig. L-6** shows.

Solution-Gas/Water Ratio. Gas also may be dissolved in water within the reservoir. The solution-gas/water ratio is the ratio of the volume of gas produced at standard conditions to the volume of water produced in stock-tank barrels as a result of producing a given volume of water originally at reservoir conditions.

As with oil, above the bubblepoint, the solution-gas/water ratio is constant; below the bubblepoint, it decreases. However, the solution-gas/water ratio is on the order of 5 to 10 scf/STB, whereas the SGOR is of the order of 500 to 1,000 scf/STB.

Water Compressibility. Above the bubblepoint, water compressibility is given by

$$c_w = -\frac{1}{B_w}\frac{\partial B_w}{\partial p}. \quad \text{(L-21)}$$

As with oil, we must take into account evolution of solution gas as the pressure is lowered below the bubblepoint. Thus, below the bubblepoint,

$$c_w = -\frac{1}{B_w}\frac{\partial B_w}{\partial p} + \frac{B_g}{B_w}\frac{\partial R_{sw}}{\partial p}. \quad \text{(L-22)}$$

Water Viscosity. Water viscosity decreases with decreasing pressure, both above and below the bubblepoint. However, as with the water FVF, there is a slight change in slope of the graph of viscosity vs. pressure at the bubblepoint. **Fig. L-7** shows the typical behavior of viscosity with pressure for reservoir brine.

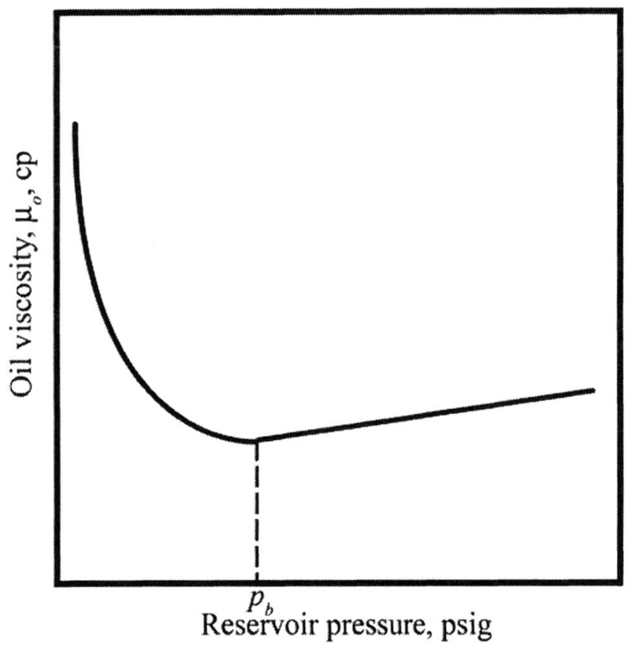

Fig. L-5—Typical shape of oil viscosity as a function of pressure at constant reservoir temperature.

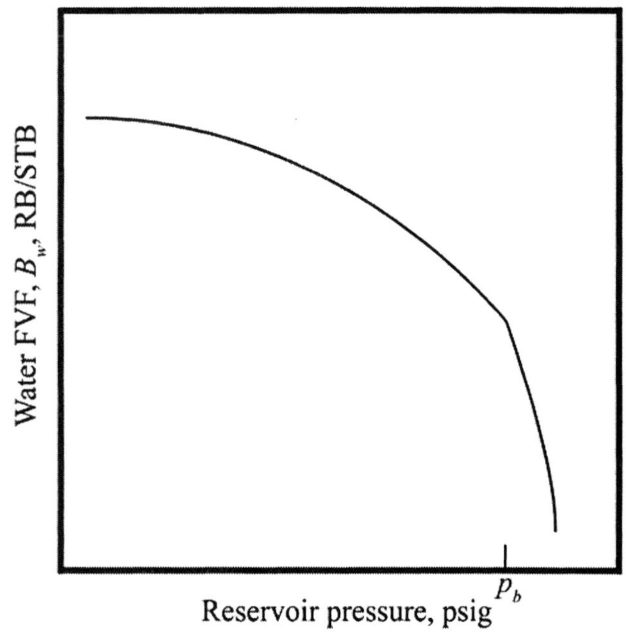

Fig. L-6—Typical shape of water FVF as a function of pressure at constant reservoir temperature.

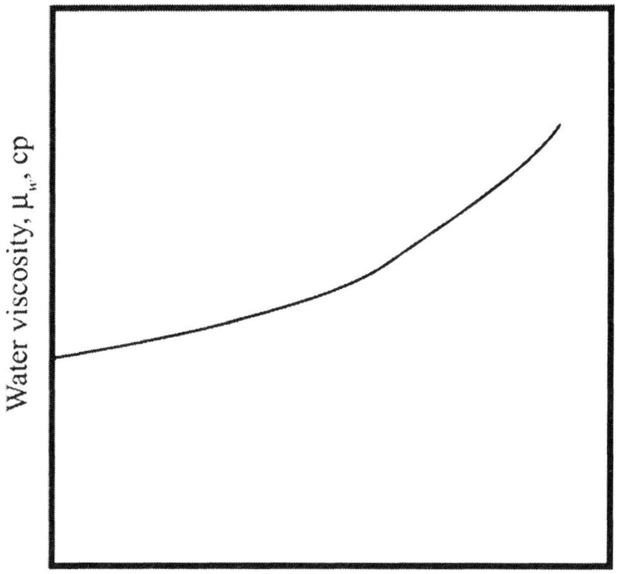

Fig. L-7—Typical shape of brine viscosity as a function of pressure at constant reservoir temperature.

Definitions of Reservoir Rock Properties. ***PV Compressibility.*** Just as the volume occupied by a given mass of fluid is a function of pressure, the porosity of the rock is also a function of pressure. Pore compressibility is defined as the fractional change in porosity for a unit change in pressure.

$$c_f \equiv \frac{1}{\phi}\frac{d\phi}{dp}. \quad \text{(L-23)}$$

In contrast to the fluid-compressibility definition, the definition of pore compressibility does not have a negative sign in front of the derivative. Both fluid and pore compressibilities are defined so that they are positive quantities. The fluid volume decreases with increasing pressure, while the pore volume or porosity increases with increasing pressure. In both cases, the effect results in a larger mass of fluid in a given bulk volume with increasing pressure.

PV compressibilities are usually low, on the order of 5×10^{-6} psi^{-1}. In unconsolidated sediments, however, compressibility may be as much as an order of magnitude higher.

Gas-Turbulence Factor and Non-Darcy-Flow Coefficient. Non-Darcy flow, often referred to as turbulent flow, may occur at high gas-flow velocity. The inertial coefficient or turbulence factor, β, is defined by the Forchheimer[4] equation, which, for 1D linear flow, is

$$-\frac{dp}{dL} = \frac{\mu v}{k} + 3.238\times10^{-8}\beta\rho v^2 \quad \text{(L-24)}$$

where dp/dL = flowing pressure gradient; v = fluid velocity (flow rate divided by cross-sectional area); μ = fluid viscosity, cp; k = formation permeability; ρv^2 = inertial flow term; and β = inertial coefficient or turbulence factor.

The inertial coefficient, β, has the dimension of reciprocal length. Eq. L-24 indicates that the pressure gradient required to sustain a given flow rate through a porous medium is greater than would be predicted by Darcy's flow equation whenever the term $\beta\rho v^2$ is not negligible.[5] Rewriting Eq. L-24,

$$-\frac{dp}{dL} = \frac{\mu v}{k}\left(1 + \frac{c_1 k\beta\rho v}{\mu}\right) \quad \text{(L-25)}$$

where c_1 is a constant and the right term inside parentheses represents a Reynolds number (ratio of inertial to viscous forces). If the Reynolds number is close to unity, then most of the flowing pressure gradient is caused by viscous flow, and Darcy's equation applies. As the Reynolds number increases, however, the inertial forces increase significantly, and the flow can no longer be modeled using Darcy's equation. This is called the non-Darcy flow effect.

TABLE L-2—COEFFICIENTS FOR ESTIMATION OF PSEUDOCRITICAL TEMPERATURE AND PRESSURE FROM GAS GRAVITY AND NONHYDROCARBON IMPURITIES

i	α_i	β_i
0	1.1582×10^{-1}	3.8216
1 (H_2S)	-4.5820×10^{-1}	-6.5340×10^{-2}
2 (CO_2)	-9.0348×10^{-1}	-4.2113×10^{-1}
3 (N_2)	-6.6026×10^{-1}	-9.1249×10^{-1}
4	7.0729×10^{-1}	$1.7438\times10^{+1}$
5	-9.9397×10^{-1}	-3.2191

Correlations

Correlations for Natural-Gas Properties. As discussed earlier, the pseudocritical temperature and pressure of a mixture provide a convenient way of correlating many gas properties of interest.

Various methods have been proposed to estimate the pseudocritical pressure and temperature for natural gases. These estimates are usually based on a knowledge of either the gas specific gravity or the composition.

Piper **et al.**[6] ***Method for Gas Gravity.*** The best correlations currently available for pseudocritical pressure and pseudocritical temperature are those developed by Piper *et al.*[6] If gas gravity and impurities content are available, pseudocritical pressure and pseudocritical temperature may be obtained for the following equations. First, estimate the parameters J and K from the gas gravity, nonhydrocarbon impurities content, and critical temperature and pressure for the impurities:

$$J = \alpha_0 + \sum_{i=1}^{3}\alpha_i y_i\left(\frac{T_c}{p_c}\right)_i + \alpha_4\gamma_g + \alpha_5\gamma_g^2 \quad \text{(L-26)}$$

$$\text{and } K = \beta_0 + \sum_{i=1}^{3}\beta_i y_i\left(\frac{T_c}{\sqrt{p_c}}\right)_i + \beta_4\gamma_g + \beta_5\gamma_g^2, \quad \text{(L-27)}$$

where α_i and β_i are given in **Table L-2**. As an alternative, we may estimate J and K from **Figs. L-8 and L-9**.

Next, calculate the pseudocritical temperature and pressure of the mixture from J and K.

$$T_{pc} = \frac{K^2}{J} \quad \text{(L-28)}$$

$$\text{and } p_{pc} = \frac{T_{pc}}{J}. \quad \text{(L-29)}$$

Kay's[7] ***Mixing Rules.*** The simplest estimate of pseudocritical pressure and temperature from gas composition is given by Kay's mixing rules and is simply a weighted average of the critical properties of the individual components.

$$p_{pc} = \sum y_i p_{ci} \quad \text{(L-30a)}$$

$$\text{and } T_{pc} = \sum y_i T_{ci}. \quad \text{(L-30b)}$$

Unfortunately, Kay's mixing rules were developed for low-gravity gases and are not accurate for gases with γ_g greater than about 0.75. In fact, the Piper *et al.* correlation, using only the gas gravity, gives more accurate results than Kay's mixing rules using gas composition.

Piper **et al.**[6] ***Method for Composition.*** Piper *et al.* developed a correlation for estimating pseudocritical pressure and temperature from gas composition. Their correlation is based on an earlier correlation by Sutton,[8] which in turn is based on work by Stewart *et al.*[9]

Laboratory reports usually measure gas composition for individual components up through hexanes, lumping all the higher fractions into a single pseudocomponent, heptanes plus (C_{7+}). Unfortu-

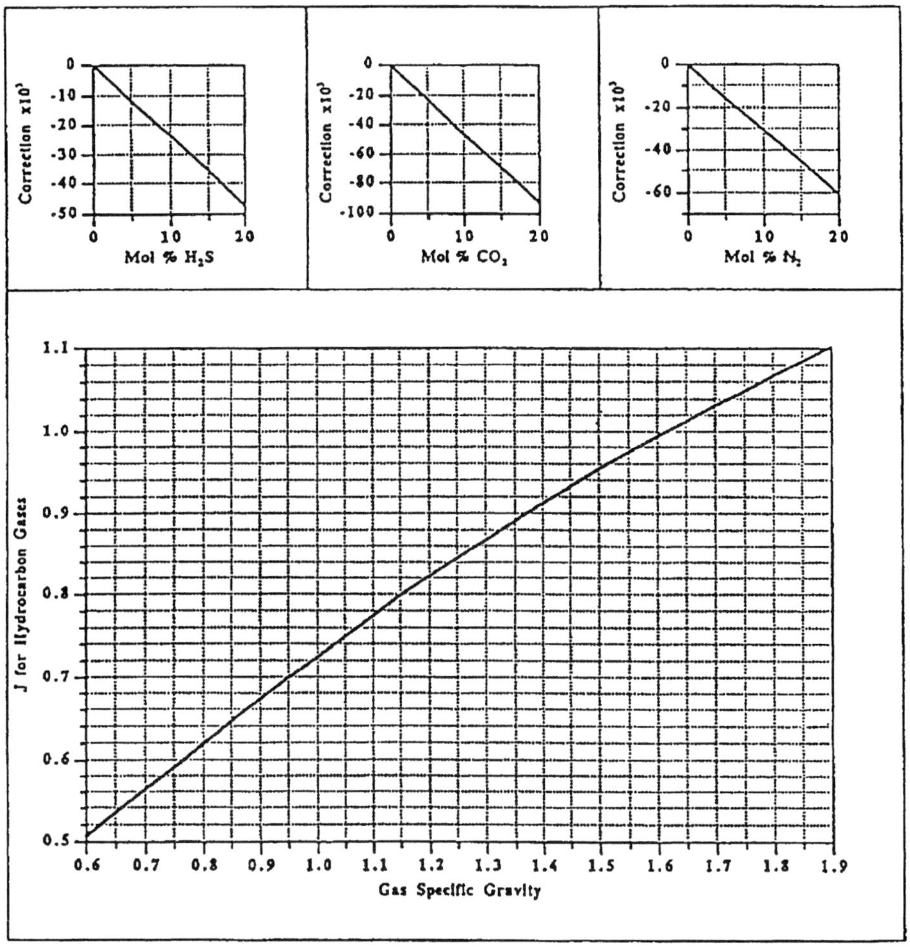

Fig. L-8—J as a function of gas specific gravity and amount of impurities.

nately, the pseudocritical properties of the C_{7+} fraction are not provided by fluid analysis reports. One of the advantages of the Piper *et al.* correlation is that it does not require pseudocritical properties of the C_{7+} fraction. Another major advantage of their correlation is that it is accurate for both high-molecular-weight gases [up to a specific gravity of 1.5 (air = 1.0)] and for gases having a large fraction of nonhydrocarbon impurities. Specifically, this correlation is designed to eliminate the need for the Wichert-Aziz[10] corrections required by earlier correlations.

To use the Piper *et al.*[6] correlation, calculate J and K with

$$J = \alpha_0 + \sum_{i=1}^{3} \alpha_i y_i \left(\frac{T_c}{p_c}\right)_i + \alpha_4 \sum_j y_j \left(\frac{T_c}{p_c}\right)_j$$

$$+ \alpha_5 \left[\sum_j y_j \left(\frac{T_c}{p_c}\right)_j\right]^2 + \alpha_6 y_{C_{7+}} M_{C_{7+}}$$

$$+ \alpha_7 \left(y_{C_{7+}} M_{C_{7+}}\right)^2 \quad \text{(L-31)}$$

$$\text{and } K = \beta_0 + \sum_{i=1}^{3} \beta_i y_i \left(\frac{T_c}{\sqrt{p_c}}\right)_i + \beta_4 \sum_j y_j \left(\frac{T_c}{\sqrt{p_c}}\right)_j$$

$$+ \beta_5 \left[\sum_j y_j \left(\frac{T_c}{\sqrt{p_c}}\right)_j\right]^2 + \beta_6 y_{C_{7+}} M_{C_{7+}}$$

$$+ \beta_7 \left(y_{C_{7+}} M_{C_{7+}}\right)^2, \quad \text{(L-32)}$$

where **Table L-3** gives the coefficients $\alpha_i, \beta_i, \alpha_j$, and β_j. Finally, calculate the pseudocritical temperature and pressure using Eqs. L-28 and L-29.

Real-Gas Deviation Factor z. Once the pseudocritical pressure and temperature for the mixture have been obtained, use Standing's z-factor correlation[11] (**Figs. L-10 and L-11**) to determine the z factor.

To use these charts, calculate the pseudoreduced pressure, p_r, and pseudoreduced temperature, T_r, from the pressure and temperature and the pseudocritical pressure and temperature:

$$p_r \equiv \frac{p}{p_{pc}} \quad \text{(L-33a)}$$

$$\text{and } T_r \equiv \frac{T}{T_{pc}}. \quad \text{(L-33b)}$$

Next, enter Fig. L-10 or L-11 with the pseudoreduced pressure on the x-axis, move vertically to the curve corresponding to the desired pseudoreduced temperature, and read the z-factor from the y-axis.

Using the Dranchuk and Abou-Kassem[12] EOS To Estimate z. Standing and Katz[11] presented a graphical correlation of the z-factor for natural gases as a function of the pseudoreduced pressure, p_r, and pseudoreduced temperature, T_r. Standing's[13] correlation is somewhat inconvenient to implement directly in a computer program. Dranchuk and Abou-Kassem[12] fitted an 11-constant EOS to the data of Standing and Katz and extrapolated this correlation to higher reduced pressures. Figs. L-10 and L-11 are graphs of z factor vs. pseudoreduced pressure calculated with Dranchuk and Abou-Kassem's EOS.

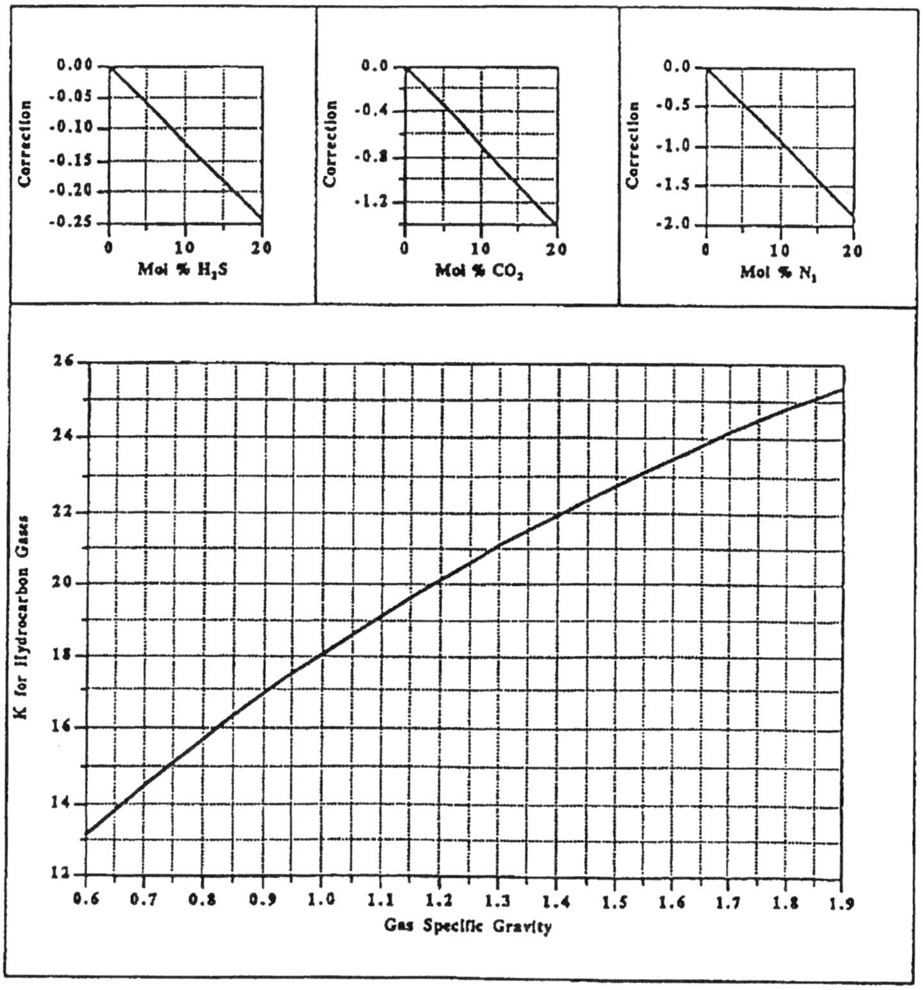

Fig. L-9—*K* as a function of gas specific gravity and amount of impurities.

TABLE L-3—COEFFICIENTS FOR ESTIMATION OF PSEUDOCRITICAL TEMPERATURE AND PRESSURE FROM GAS COMPOSITION

i	α_i	β_i
0	5.2073×10^{-2}	-3.9741×10^{-1}
1 (H_2S)	1.0160	1.0503
2 (CO_2)	8.6961×10^{-1}	9.6592×10^{-1}
3 (N_2)	7.2646×10^{-1}	$7.8569\times10-1$
4	8.5101×10^{-1}	9.8211×10^{-1}
5	0	0
6	2.0818×10^{-2}	4.5536×10^{-1}
7	-1.506×10^{-4}	-3.7684×10^{-3}

Dranchuk and Abou-Kassem[12] developed their EOS primarily to estimate the z factor with computer routines. The form of the Dranchuk and Abou-Kassem EOS is

$$z = 1 + c_1(T_r)\rho_r + c_2(T_r)\rho_r^2 - c_3(T_r)\rho_r^5 + c_4(\rho_r, T_r), \quad \text{(L-34)}$$

where $\rho_r = 0.27p_r/(zT_r)$, ... (L-35)

$$c_1(T_r) = A_1 + A_2/T_r + A_3/T_r^3 + A_4/T_r^4 + A_5/T_r^5, \quad \text{(L-36)}$$

$$c_2(T_r) = A_6 + A_7/T_r + A_8/T_r^2, \quad \text{(L-37)}$$

$$c_3(T_r) = A_9 + \left(A_7/T_r + A_8/T_r^2\right), \quad \text{(L-38)}$$

and $c_4(\rho_r, T_r) = A_{10}\left(1 + A_{11}\rho_r^2\right)\left(\rho_r^2/T_r^3\right)\exp\left(-A_{11}\rho_r^2\right)$. .... (L-39)

The constants A_1 through A_{11} for the EOS are $A_1=0.3265$, $A_2=-1.0700$, $A_3=-0.5339$, $A_4=0.01569$, $A_5=-0.05165$, $A_6=0.5475$, $A_7=-0.7361$, $A_8=0.1844$; $A_9=0.1056$, $A_{10}=0.6134$, and $A_{11}=0.7210$.

Dranchuk and Abou-Kassem fitted the EOS to 1,500 data points with an average absolute error of 0.486% when the z-factor is a function of T_r and ρ_r. The z factor based on EOS is accurate within engineering accuracy over the following ranges: $0.2 \leqq p_r < 30$ and $1.0 < T_r \leqq 3.0$ and $p_r < 1.0$ and $0.7 < T_r < 1.0$. However, the EOS gives poor results for $T_r = 1.0$ and $p_r > 1.0$.

The Dranchuk and Abou-Kassem EOS must be solved iteratively because the z factor is on both sides of the equation. The solution to this problem can be obtained by use of a root-solving technique, such as Newton's method or the secant method.[14] The root-solving procedure requires that Eq. L-34 be rearranged to the form

$$F(z) = 0. \quad \text{(L-40)}$$

Rearranging Eq. L.34 gives

$$F(z) = z - [1 + c_1(T_r)\rho_r + c_2(T_r)\rho_r^2 - c_3(T_r)\rho_r^5 + c_4(\rho_r, T_r)] = 0. \quad \text{(L-41)}$$

Newton's method and the secant method require the first derivative of Eq. L-41 with respect to z at constant pseudoreduced temperature, $[\partial F(z)/\partial z]_{Tr}$. Newton's method assumes that the derivative is evaluated analytically by explicit differentiation of the function

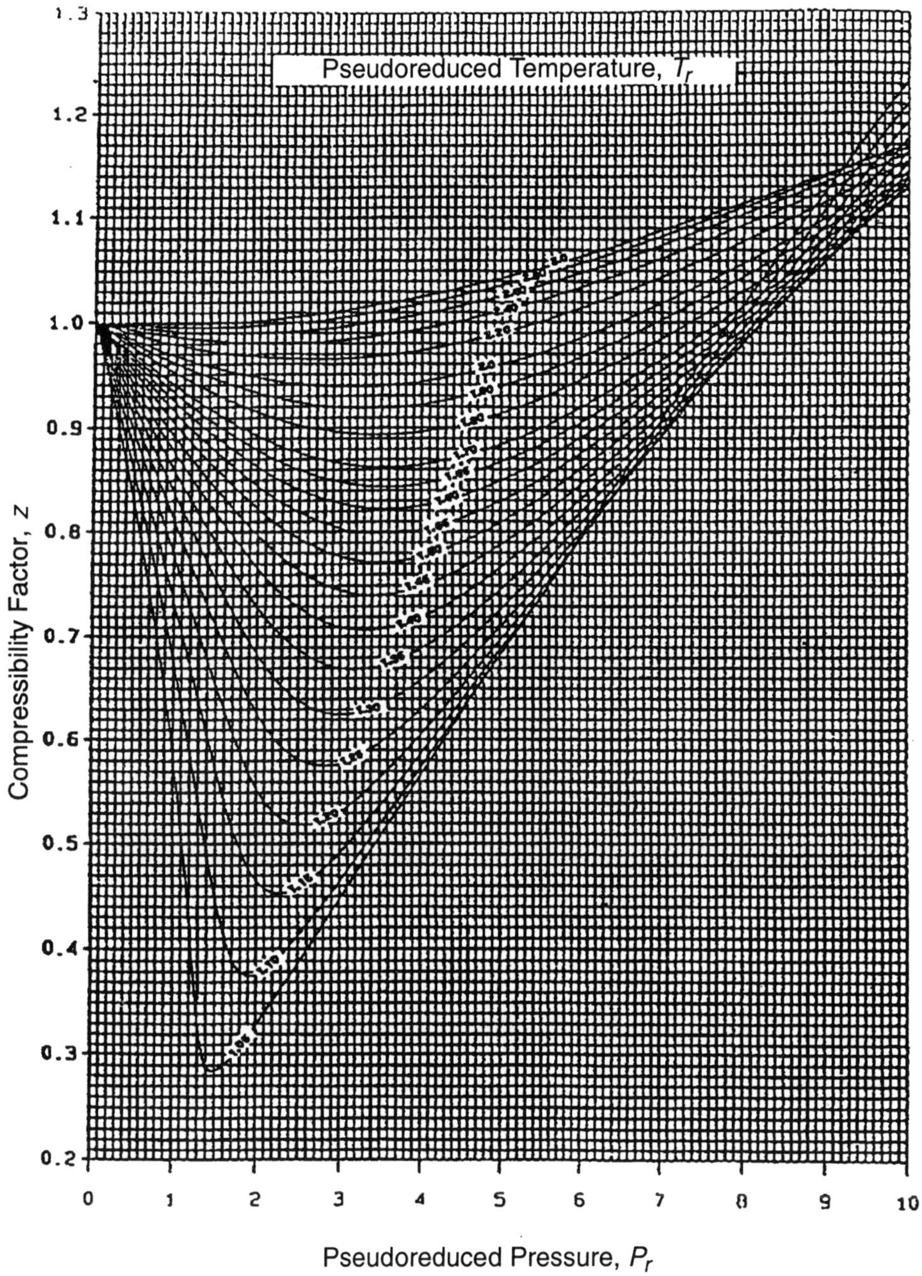

Fig. L-10—*z* factors for natural gases for $0 \leqq p_r \leqq 10$.

F(*z*). We recommend using Newton's method to calculate *z* factors from Eq. L-41; the derivative of Eq. L-41 with respect to *z* is

$$[\partial F(z)/\partial z]_{Tr} = 1 + c_1(T_r)\rho_r/z + 2c_2(T_r)\rho_r^2/z - 5c_3(T_r)\rho_r^5/z$$

$$+ \frac{2A_{10}\rho_r^2}{T_r^3 z}\left[1 + A_{11}\rho_r^2 - \left(A_{11}\rho_r^2\right)^2\right]$$

$$\times \exp\left(-A_{11}\rho_r^2\right) \quad \text{(L-42)}$$

Gas Density. Having obtained the *z* factor, gas density is calculated with Eq. L-10, recognizing that the average molecular weight *M* may be expressed in terms of gas gravity γ_g.

$$M = 28.963\gamma_g \quad \text{(L-43)}$$

and $$\rho_g = \frac{pM}{zRT}. \quad \text{(L-10)}$$

Gas Compressibility. The gas compressibility, c_g, may be estimated from the pseudoreduced compressibility c_r.

$$c_g = \frac{c_r}{p_{pc}}. \quad \text{(L-44)}$$

The product of pseudoreduced compressibility and pseudoreduced temperature, c_rT_r, may be obtained from **Figs. L-12 and L-13.**[14] As with Figs. L-10 and L-11, enter the graph with the pseudoreduced pressure on the *x* axis, move vertically to the curve corresponding to the desired pseudoreduced temperature, and read c_rT_r from the *y* axis. Finally, calculate the gas compressibility c_g from c_rT_r, p_{pc}, and T_r:

$$c_g = \frac{c_rT_r}{p_{pc}T_r}. \quad \text{(L-45)}$$

Gas Viscosity. Use the Lee *et al.*[15] correlation to estimate gas viscosity. This correlation requires the gas density in g/cm³.

$$\rho = 1.4935 \times 10^{-3}\frac{pM}{zT}. \quad \text{(L-11)}$$

Next, calculate the viscosity from the following series of equations.

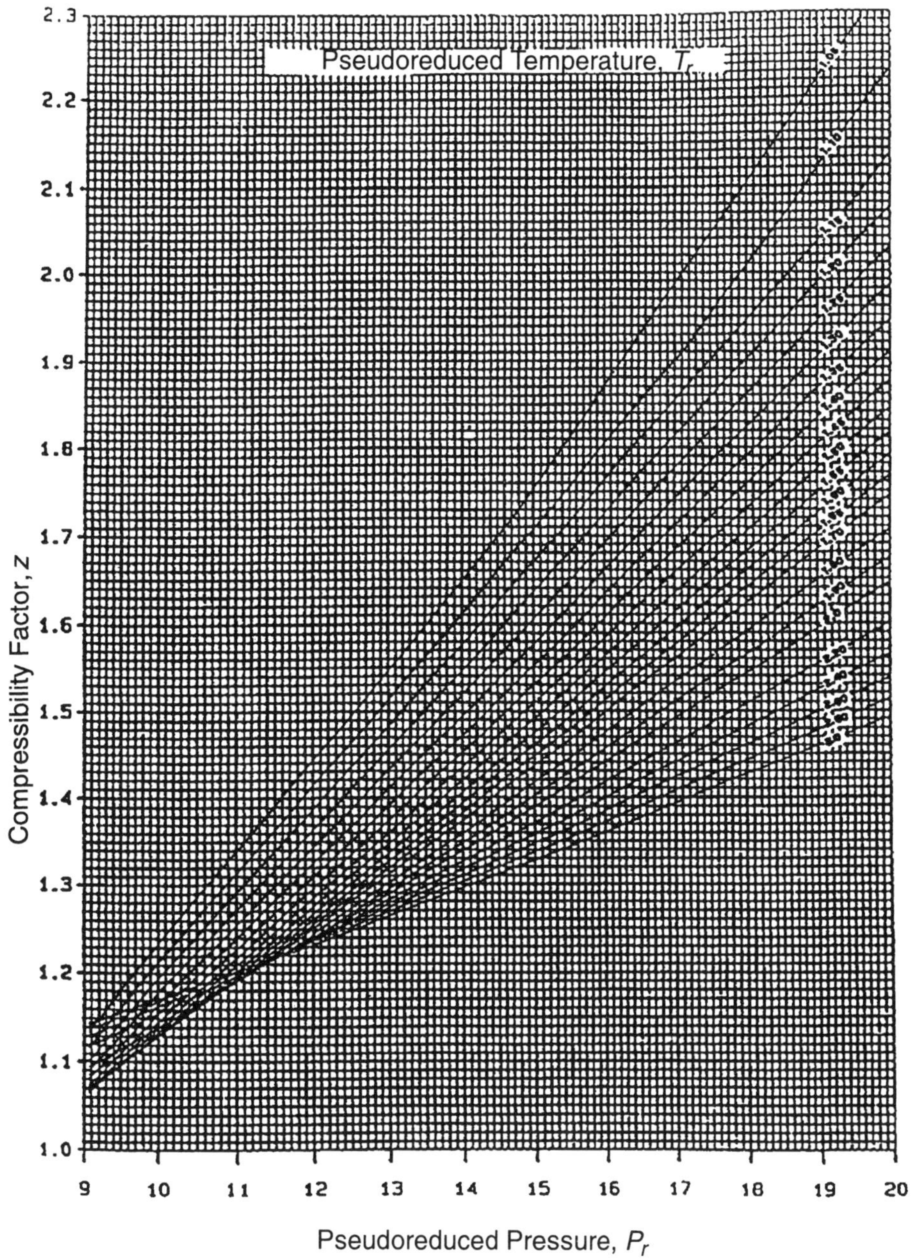

Fig. L-11—z factors for natural gases for $9 \leq p_r \leq 20$.

$$K = \frac{(9.379 + 0.01607M)T^{1.5}}{(209.2 + 19.26M + T)},$$

$$X = 3.448 + \frac{986.4}{T} + 0.01009M,$$

$$Y = 2.447 - 0.2224X,$$

$$\text{and } \mu_g = 10^{-4}K\exp(X\rho^Y). \quad \text{(L-46)}$$

The Carr *et al.*[16] correlation may be used if a graphical correlation is needed.

Estimating Pseudocritical Properties of Wet-Gas and Gas-Condensate Reservoir Fluids: Recombination Calculations. As discussed earlier, Piper *et al.*[6] developed correlations that allow estimation of pseudocritical properties of a natural gas without requiring a detailed compositional analysis of the reservoir fluid. The Piper *et al.* method requires only estimates of specific gas gravity. For dry-gas reservoirs, the specific gravity of the gas sample obtained at the primary (high-pressure) -separator conditions equals the value of a sample from the reservoir. However, for wet-gas and gas-condensate reservoir fluids that experience condensation and a change in composition following a pressure reduction, the specific gravities of gas samples taken from the reservoir and at the surface differ.

The most accurate source of reservoir-gas gravity, especially for wet-gas and gas-condensate fluids, is from a laboratory analysis of a reservoir fluid sample. Alternatively, when compositions of vapor and liquid streams have been determined in the laboratory and when stabilized gas and liquid flow rates have been measured at the primary separator in the field, recombination calculations provide useful estimates of wellstream properties. Recombination calculations of surface production data are also a rigorous means of determining the specific gravity of a reservoir fluid without requiring bottomhole fluid sampling.

The following discussions illustrate calculation of the recombined wet-gas flow rate and cumulative production from surface production data, then present correlations that allow reservoir specific gravity to be estimated from surface production data. These correlations are based on typical production parameters, such as

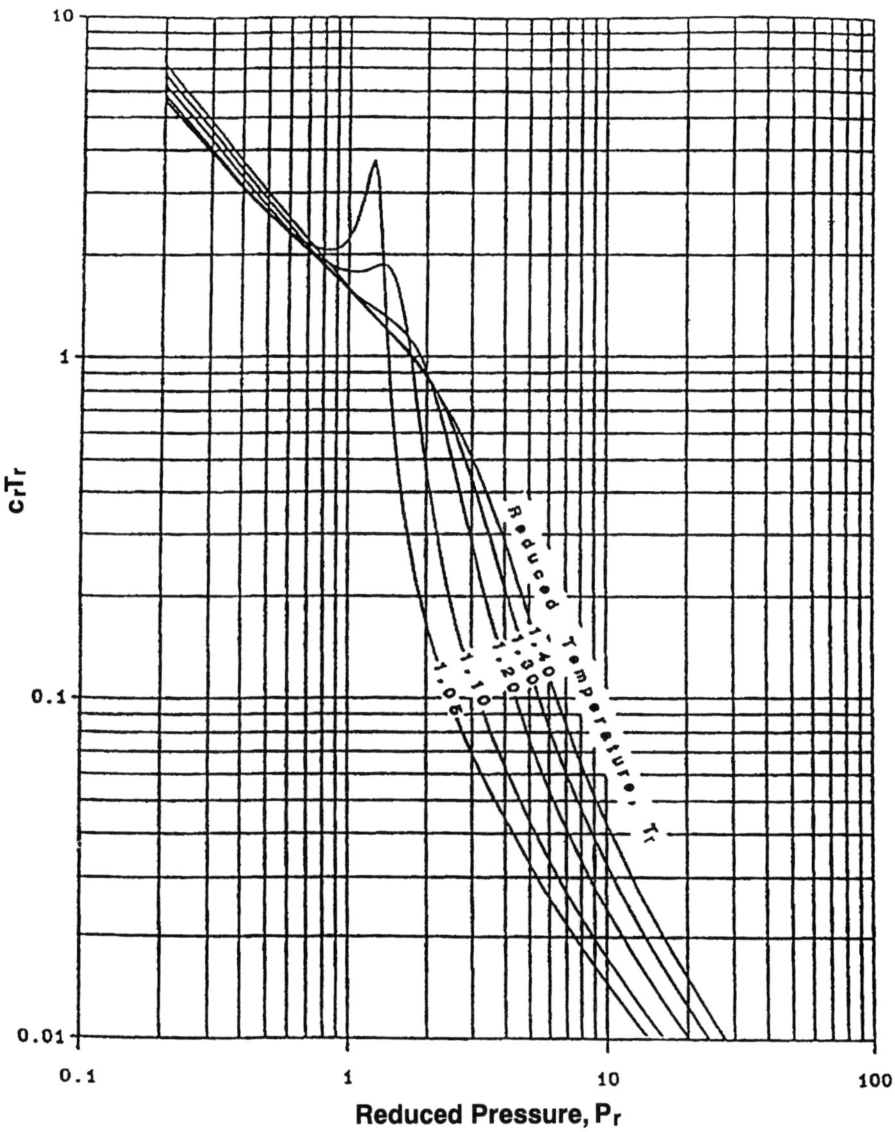

Fig. L-12—Variation in c_rT_r for natural gases for $1.05 \leqq T_r \leqq 1.4$ (after Mattar *et al.*[14]). Reprinted with permission from the Canadian Inst. of Mining, Metallurgy, and Petroleum.

separator pressure, temperature, and gas gravity and the stock-tank-liquid gravity.

Example L-1—Calculating Wet-Gas Flow Rate and Cumulative Production. A gas-condensate well was produced with a gas/condensate ratio of 4,428 scf/SB at the primary separator. The liquid flow rate at separator conditions was 50 SB/D. Cumulative liquid production was 100,000 SB. The specific gravity of the primary-separator liquid was determined to be 0.681. The molecular weight of the C_{7+} fraction was estimated to be 143 lbm/lbm mol. From this information and the composition data given in **Table L-4**, determine the composition of the reservoir fluid and calculate the wet-gas flow rate and cumulative production.

Solution.

1. Calculate the molar recombination ratio. For this problem, assume standard conditions of $T_{sc} = 60°\text{F}$ and $p_{sc} = 14.65$ psia, so the molar volume is $V_m = 380.7$ scf/lbm mol.

Pound moles in 1 bbl condensate

$$= \frac{(0.681)(62.4)(5.615)\text{lbm/bbl}}{82.88 \text{ lbm/lbm mol}}$$

$= 2.876$ lbm mol liquid/bbl condensate.

Pound moles in 4,428 scf gas

$$= \frac{4,428 \text{ scf/STB}}{380.7 \text{ scf/lbm mol}}$$

$= 11.63$ lbm mol gas/bbl condensate.

Thus, we combine 2.876 lbm mol of liquid with 11.63 lbm mol of gas to make the recombination calculations to determine the composition of the reservoir fluid, as summarized in **Table L-5.**

2. Calculate the wet-gas flow rate. The number of pound moles of liquid per day is

(2.876 lbm mol/SB)(50 SB/D) = 143.8 lbm mol/D liquid.

The vapor equivalent, V_{eq}, of this liquid is

(143.8 lbm mol/D)(380.7 scf/lb mol) = 55 Mscf/D.

The dry-gas surface flow rate is

$q_g = (50 \text{ SB/D})(4428 \text{ scf/SB}) = 221$ Mscf/D.

Finally, the wet-gas wellstream flow rate is

$q = q_g + V_{eq} = 221 + 55 = 276$ Mscf/D.

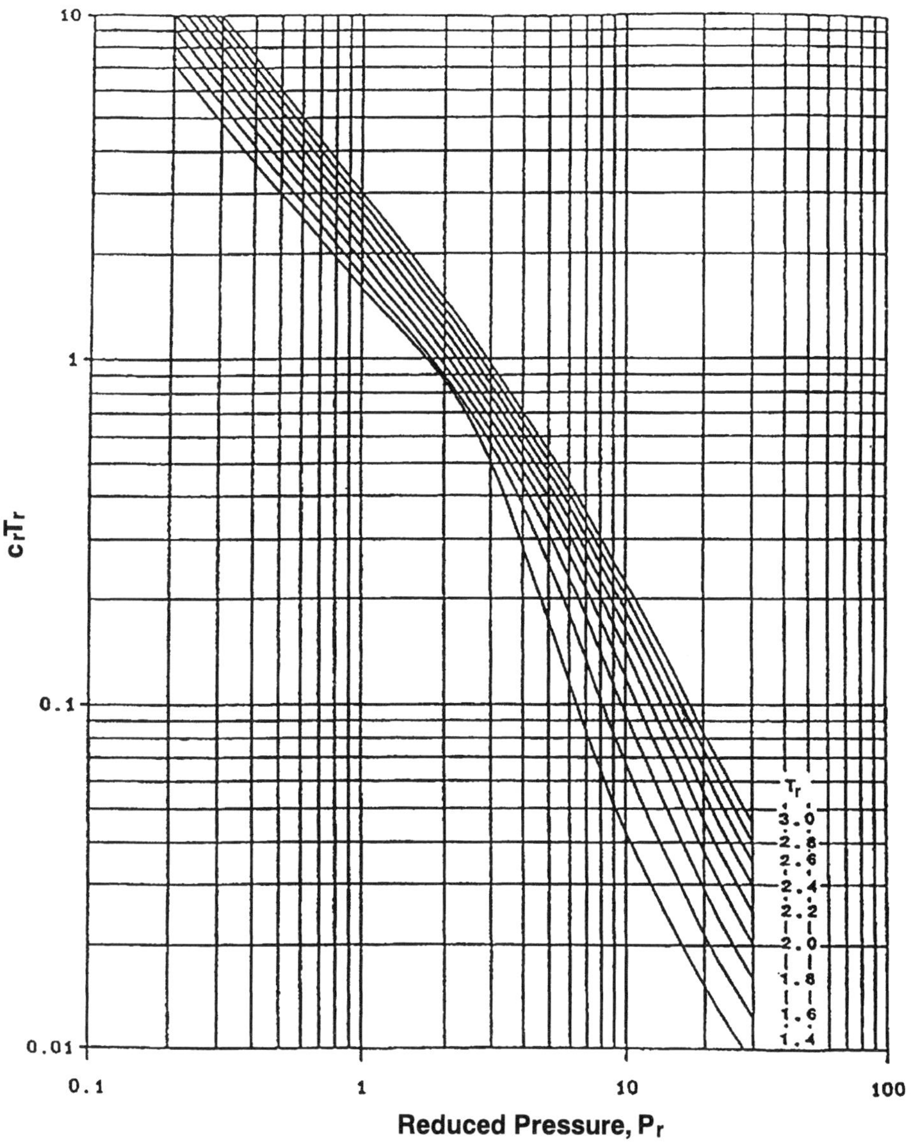

Fig. L-13—Variation in c_rT_r for natural gases for $1.4 \leqq T_r \leqq 3.0$ (after Mattar *et al.*[14]). Reprinted with permission from the Canadian Inst. of Mining, Metallurgy, and Petroleum.

TABLE L-4—GAS COMPOSITION DATA, EXAMPLE L-1

Component	Liquid Mole Fraction, x_i	Gas Mole Fraction, y_i	Molecular Weight, M_i (lbm/lb mol)	$x_i M_i$
CO_2	—	0.0022	44.010	—
N_2	—	0.0016	28.013	—
CH_4	0.0778	0.7531	16.042	1.248
C_2H_6	0.1002	0.1508	30.068	3.013
C_3H_8	0.1508	0.0668	11.094	6.649
iC_4H_{10}	0.0277	0.0052	58.120	1.610
nC_4H_{10}	0.1139	0.0144	58.120	6.620
iC_5H_{12}	0.0352	0.0018	72.146	2.540
nC_5H_{12}	0.0650	0.0024	72.146	4.689
C_6H_{14}	0.0861	0.0011	86.172	7.419
C_{7+}	0.3433	0.0006	143.0	49.092
$\Sigma =$	1.0000	1.0000	14.5449	82.880

3. Similarly, calculate the cumulative wet-gas production. The number of pound moles of liquid is

(2.876 lbm mol/SB)(100,000 SB) = 287,600 lbm mol liquid.

The vapor equivalent of this liquid is

(287,600 lbm mol)(380.7 scf/lbm mol) = 109,489 Mscf.

Cumulative dry-gas production is

$Q_{g,\text{dry}}$ = (100,000 SB)(4,428 scf/SB) = 442,800 Mscf.

Cumulative wellstream production is

TABLE L-5—RECOMBINATION CALCULATIONS FOR EXAMPLE L-1

Component	Liquid Mole Fraction, x_i	Gas Mole Fraction, y_i	$2.88x_i + 11.63\,y_i$	$z_i = \dfrac{2.88x_i + 11.63y_i}{\sum_{i=1}^{n_{comp}} 2.88x_i + 11.63y_i}$
CO_2	—	0.0022	0.0257	0.0018
N_2	—	0.0016	0.0187	0.0013
CH_4	0.0778	0.7531	9.0124	0.6196
C_2H_6	0.1002	0.1508	2.0480	0.1408
C_3H_8	0.1508	0.0668	1.2133	0.0834
i-C_4H_{10}	0.0277	0.0052	0.1403	0.0096
n-C_4H_{10}	0.1139	0.0144	0.4956	0.0341
i-C_5H_{12}	0.0352	0.0018	0.1222	0.0084
n-C_5H_{12}	0.0650	0.0024	0.2149	0.0148
C_6H_{14}	0.0861	0.0011	0.2605	0.0179
C_{7+}	0.3433	0.0006	0.9943	0.0684
$\Sigma =$	1.0000	1.0000	14.506	1.0000

$$Q_{g,\text{wet}} = Q_{g,\text{dry}} + V_{eq} = 442{,}800 + 109{,}489 = 552{,}289 \text{ Mscf.}$$

Compositions of liquid and gas streams from the primary separator are not always available. However, reliable estimates of the reservoir gas specific gravity can be obtained by use of surface production data when the total nonhydrocarbon component is less than 20 mole percent.[17] When quantities of gas and liquid are known, the equations[17] presented in the next section should be used to estimate the reservoir gas specific gravity for natural-gas wells that produce liquid. For wells that produce only gas and no liquid hydrocarbons, the primary separator gas gravity is the reservoir gas gravity.

Estimating Specific Gravity of Reservoir Gas. Eq. L-47 is the recombination expression for the reservoir gas gravity, γ_w, for a three-

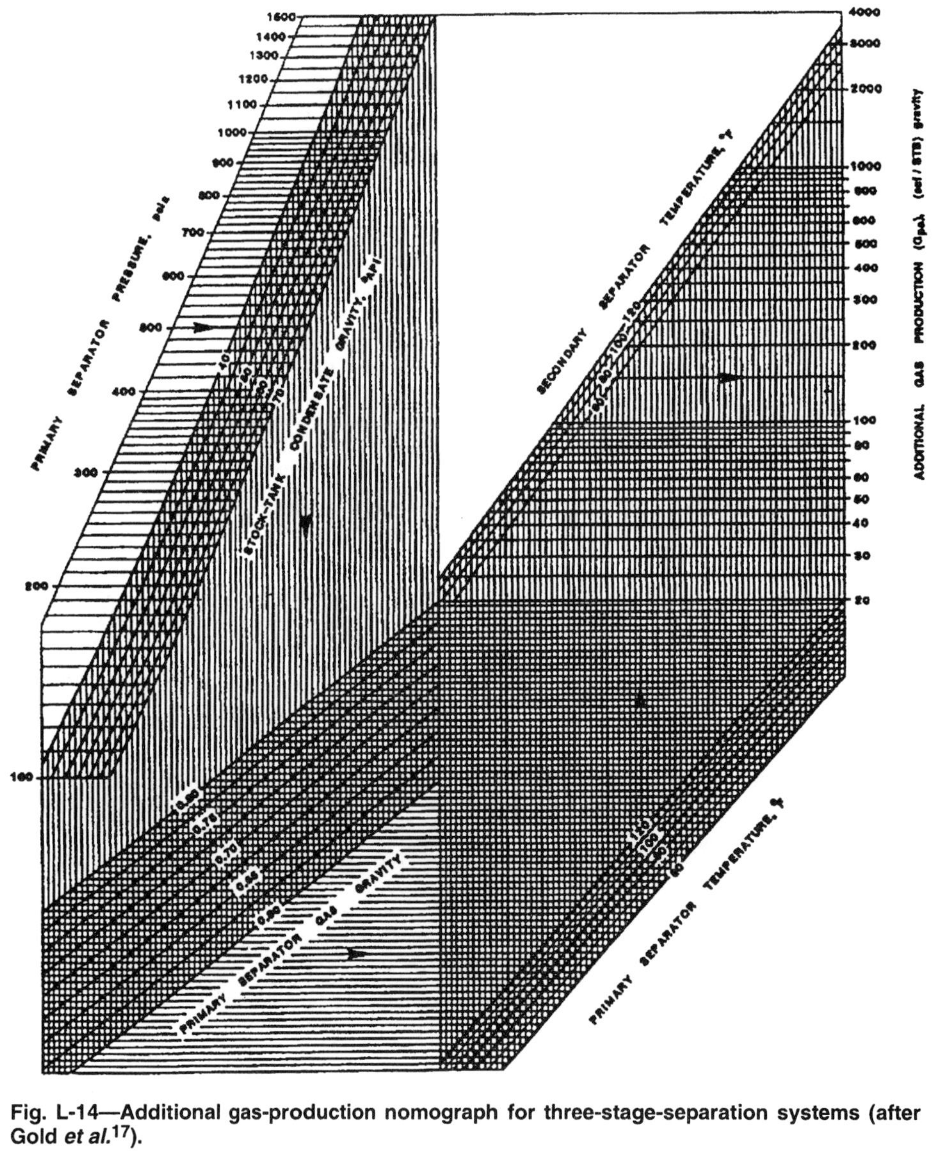

Fig. L-14—Additional gas-production nomograph for three-stage-separation systems (after Gold *et al.*[17]).

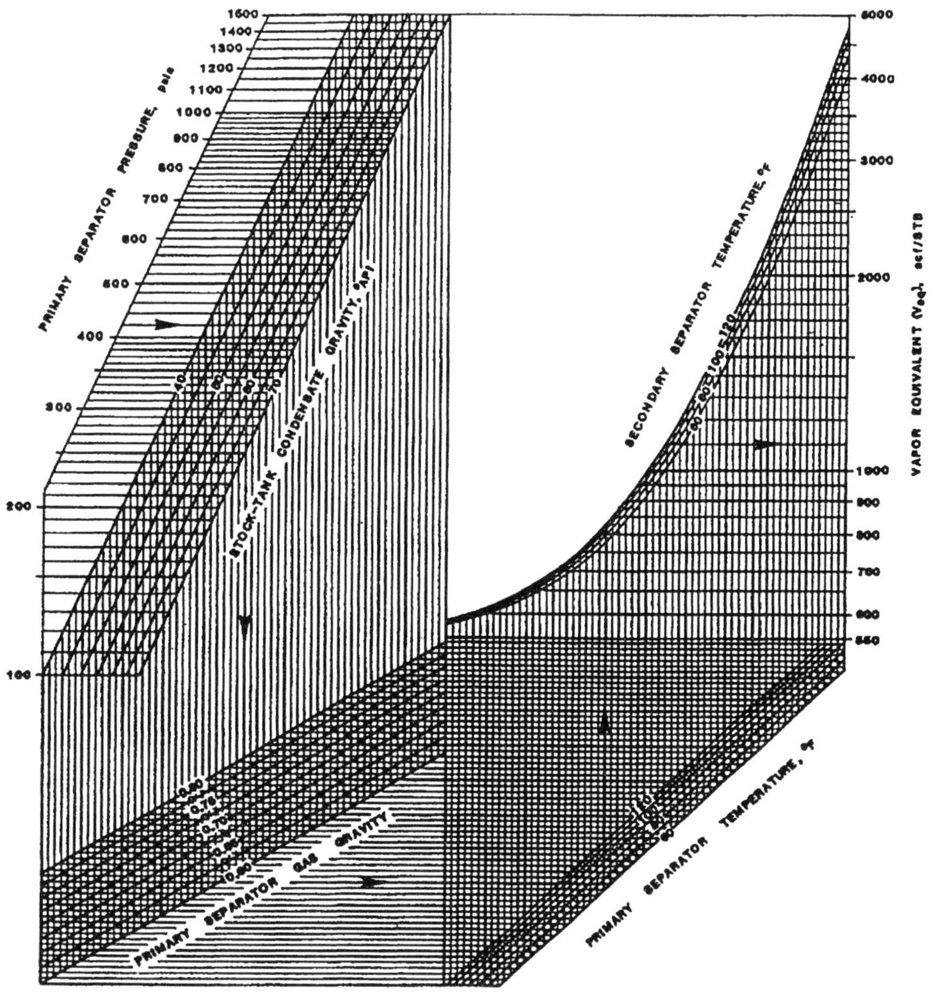

Fig. L-15—Vapor-equivalent nomograph for three-stage-separation system (after Gold *et al.*[17]).

stage separation system consisting of a primary (high-pressure) separator, secondary (low-pressure) separator, and stock tank.

$$\gamma_w = \frac{R_1\gamma_1 + 4602\gamma_o + R_2\gamma_2 + R_3\gamma_3}{R_1 + \dfrac{133{,}316\gamma_o}{M_o} + R_2 + R_3}. \qquad \text{(L-47)}$$

Eq. L-47 is applicable (and rigorously correct) if the gas-liquid ratio and gas specific gravity of each separation stage and the stock-tank liquid gravity are known. Similarly, Eq. L-48 is the recombination expression for a two-stage separation system consisting of a primary separator and a stock tank.

$$\gamma_w = \frac{R_1\gamma_1 + 4602\gamma_o + R_3\gamma_3}{R_1 + \dfrac{133{,}316\gamma_o}{M_o} + R_3}. \qquad \text{(L-48)}$$

The variables in Eqs. L-47 and L-48 are defined as

R_1 = primary (or high-pressure) -separator gas/stock-tank-liquid ratio, scf/STB
R_2 = secondary (or low-pressure) -separator gas/stock-tank-liquid ratio, scf/STB
R_3 = stock-tank-gas/stock-tank-liquid ratio, scf/STB
γ_1 = specific gravity of primary-separator gas (air = 1.0)
γ_2 = specific gravity of secondary-separator gas (air = 1.0)
γ_3 = specific gravity of stock-tank gas (air = 1.0)
γ_o = specific gravity of the liquid hydrocarbons (water = 1.0)
M_o = molecular weight of stock-tank liquid, lbm/lbm mol

If the molecular weight of the stock-tank liquid, M_o, is unknown, then it can be approximated using either liquid gravity in °API, γ_{API}(Eq. L-49) or specific gravity (Eq. L-50),[17]

$$M_o = \frac{5{,}954}{\gamma_{API} - 8.811} \qquad \text{(L-49)}$$

$$\text{or } M_o = \frac{42.43\gamma_o}{1.008 - \gamma_o}, \qquad \text{(L-50)}$$

Because gas rates and specific gravities from low-pressure separators and the stock tank frequently are not measured, estimates of this gas production must be made to use Eqs. L-47 or L-48. These estimates are made with correlations for additional gas production, G_{pa}, and vapor equivalent, V_{eq}, in conjunction with Eq. L-51. The correlations are based on generally available production parameters, specifically, the primary-separator pressure, temperature, and gas gravity; the secondary-separator temperature; and the stock-tank-liquid gravity.

The G_{pa} and V_{eq} terms in Eq. L-51 are determined from **Figs. L-14 through L-17.** Figs. L-14 and L-15 apply to a three-stage separation system, while Figs. L-16 and L-17 are for a two-stage separation system. Figs. L-14 and L-16 are used to estimate the additional gas production, G_{pa}, while Figs. L-15 and L-17 are used to estimate the vapor equivalent, V_{eq}, in Eq. L-51.

In terms of G_{pa} and V_{eq}, the reservoir gas specific gravity, γ_w, is estimated by

$$\gamma_w = \frac{R_1\gamma_1 + 4{,}602\gamma_o + G_{pa}}{R_1 + V_{eq}} \qquad \text{(L-51)}$$

where V_{eq} = volume of second-separator gas and stock-tank gas plus the volume that would be occupied by a barrel of stock-tank liquid if it were gas, scf/STB; and G_{pa} = additional gas production, scf/STB.

The correlation equation for Fig. L-14 (for a three-stage separation system) is

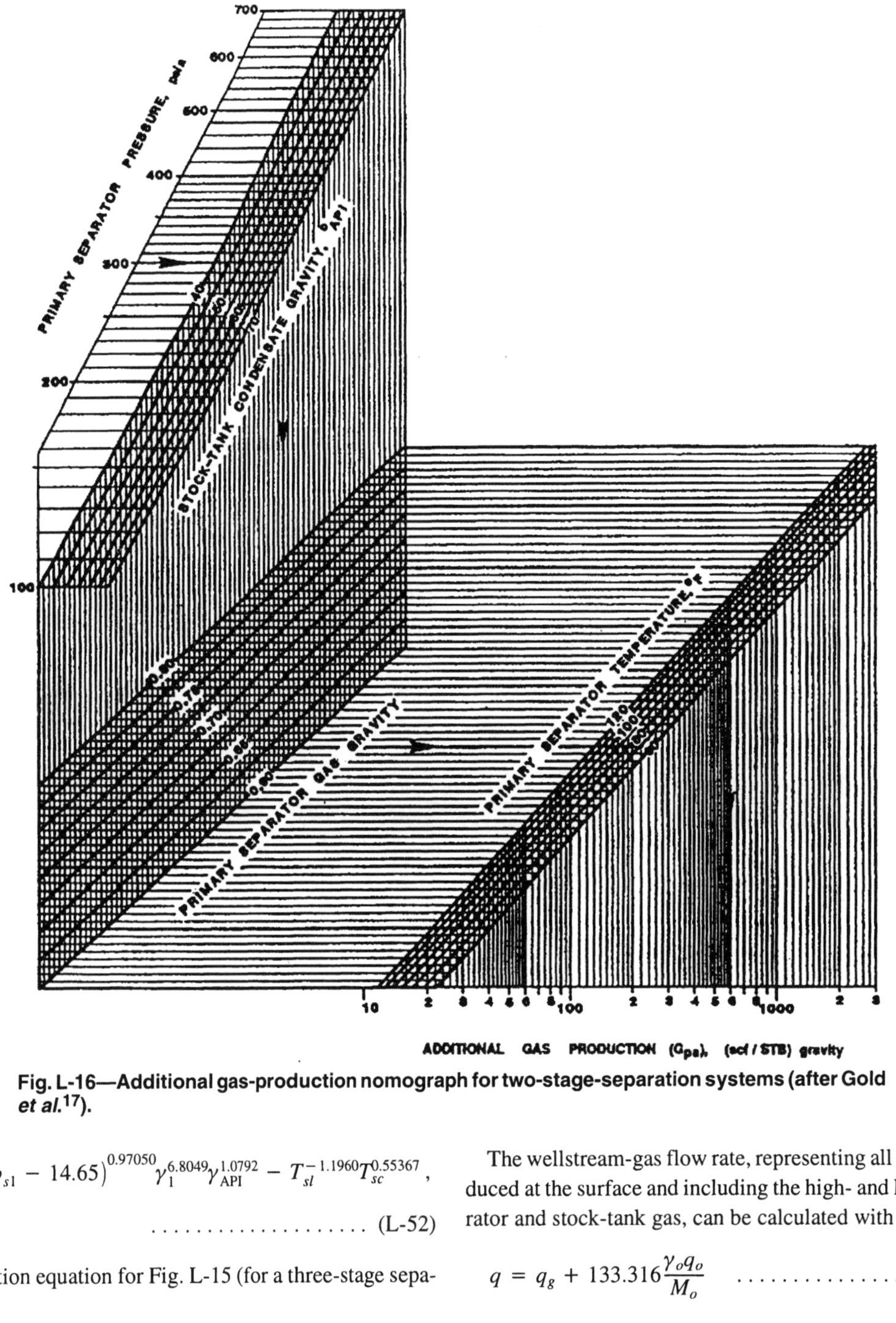

Fig. L-16—Additional gas-production nomograph for two-stage-separation systems (after Gold *et al.*[17]).

$$G_{pa} = 2.9922(p_{s1} - 14.65)^{0.97050} \gamma_1^{6.8049} \gamma_{API}^{1.0792} - T_{sl}^{-1.1960} T_{sc}^{0.55367}, \quad \text{(L-52)}$$

while the correlation equation for Fig. L-15 (for a three-stage separation system) is

$$V_{eq} = 535.92 + 2.6231 p_{s1}^{0.79318} \gamma_1^{4.6612} \gamma_{API}^{1.2094} - T_{sl}^{-0.84911} T_{s2}^{0.26987}. \quad \text{(L-53)}$$

The correlation equation for Fig. L-16 (for a two-stage separation system) is

$$G_{pa} = 1.4599(p_{s1} - 14.65)^{1.3394} \gamma_1^{7.0943} \gamma_{API}^{1.1436} T_{sl}^{-0.93446}, \quad \text{(L-54)}$$

while the correlation equation for Fig. L-17 (for a two-stage separation system) is

$$V_{eq} = 635.53 + 0.36182 p_{s1}^{1.0544} \gamma_1^{5.0831} \gamma_{API}^{1.5812} T_{sl}^{-0.79130}. \quad \text{(L-55)}$$

The wellstream-gas flow rate, representing all gas and liquid produced at the surface and including the high- and low-pressure separator and stock-tank gas, can be calculated with

$$q = q_g + 133.316 \frac{\gamma_o q_o}{M_o} \quad \text{(L-56)}$$

$$\text{and } q = q_{s1}\left(1 + \frac{V_{eq}}{R_1}\right), \quad \text{(L-57)}$$

where q = total wellstream-gas flow rate, Mscf/D; q_g = total gas surface-gas flow rate, Mscf/D; q_o = oil flow rate, STB/D; and q_{s1} = gas flow rate from primary separator, Mscf/D.

Eq. L-56 should be used if the total surface-gas flow rate is known, while Eq. L-57 is applicable when the secondary (low-pressure) -separator- and/or stock-tank-gas production is not known.

Similarly, the cumulative wellstream gas produced at the sandface, $Q_{g,\text{wet}}$, can be calculated. When the gas/liquid ratio and/or the gas and liquid compositions change with time, the wet-gas cumulative production can be calculated by integrating the wet-gas rate over time. If the gas/liquid ratio and liquid and gas gravities remain constant with time and if the cumulative surface production of gas, $Q_{g,\text{dry}}$, and cumulative stock-tank-liquid production, Q_o, are known, then

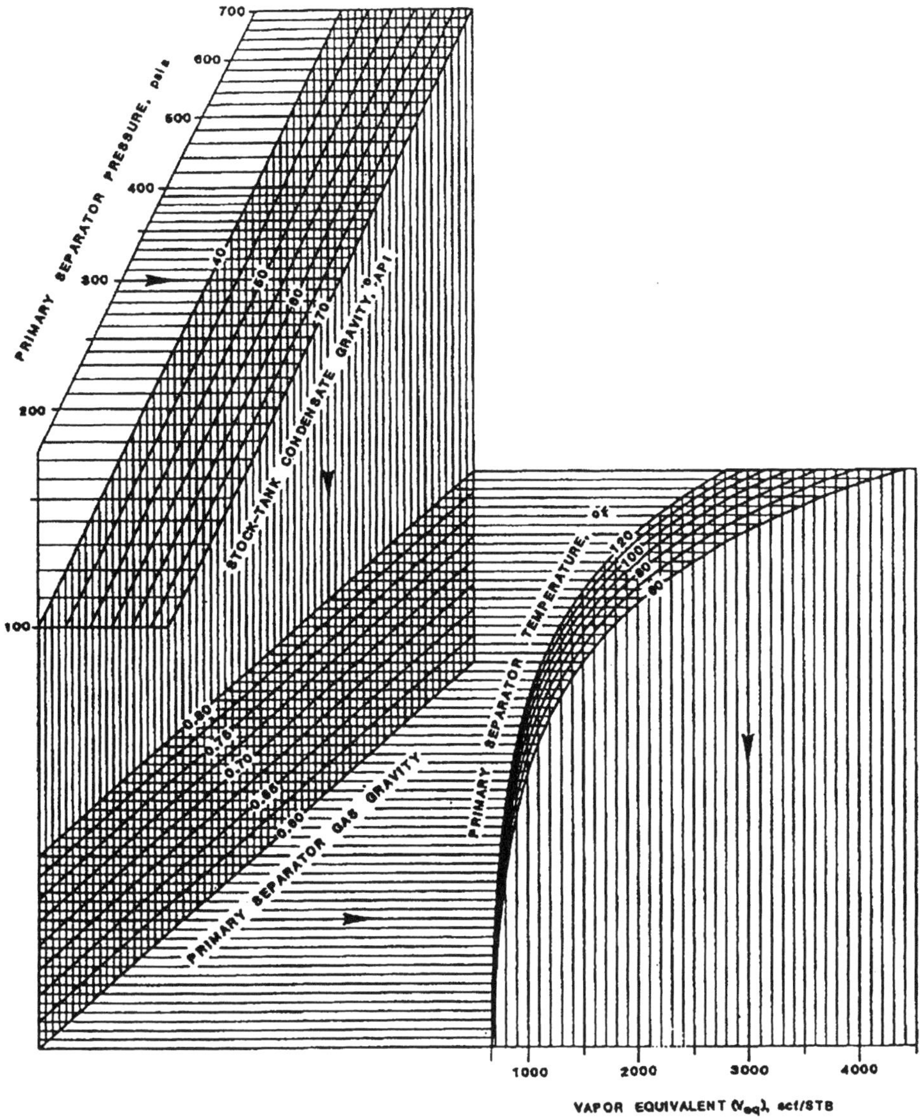

Fig. L-17—Vapor-equivalent nomograph for two-stage-separation system (after Gold *et al.*[17]).

$$Q_{g,\text{wet}} = Q_{g,\text{dry}} + 133.316\frac{\gamma_o Q_o}{M_o}. \quad \text{(L-58)}$$

Otherwise,

$$Q_{g,\text{wet}} = Q_{g,\text{dry}}\left(1 + \frac{V_{eq}}{R_1}\right), \quad \text{(L-59)}$$

where $Q_{g,\text{wet}}$ = cumulative wellstream gas produced at the sandface, Mscf; $Q_{g,\text{dry}}$ = cumulative surface gas production, Mscf; and Q_o = cumulative stock-tank-liquid production, STB.

Example L-2—Calculating Wellstream-Gas Gravity and Flow Rate With Known Fluid Properties. Calculate the wellstream-gas gravity, the total wellstream-gas flow rate, and the cumulative wellstream production for a three-stage-separation system. The following are the measured production data and fluid properties, including the gas and liquid from the stock tank.

R_1 = 7040 scf/STB
R_2 = 510 scf/STB
R_3 = 120 scf/STB
γ_1 = 0.712
γ_2 = 1.02
γ_3 = 1.55
γ_o = 0.755
M_o = 125 lbm/lbm mol
q_o = 50 STB/D
$Q_{g,\text{dry}}$ = 767 MMscf
Q_o = 100,000 STB
q_g = 384 Mscf/D

Solution.

1. Eq. L-47 is applicable because every term in the recombination expression is known. The specific gravity of the reservoir gas (air = 1.0) is

$$\gamma_w = \frac{R_1\gamma_1 + 4{,}602\gamma_o + R_2\gamma_2 + R_3\gamma_3}{R_1 + (133{,}316\gamma_o/M_o) + R_2 + R_3}$$

$$= \Big[(7{,}040)(0.712) + (4{,}602)(0.755) + (510)(1.02)$$

$$+ (120)(1.55)\Big]\Big/\Big\{7{,}070 + \big[(133{,}316)(0.755)/125\big]$$

$$+ 510 + 120\Big\}$$

$$= 1.08.$$

2. The total wellstream-gas rate (Eq. L-56) is

$$q = q_g + 133.316\frac{\gamma_o q_o}{M_o}$$

$$= 384 + 133.316\frac{(0.755)(50)}{125}$$

$$= 424 \text{ Mscf/D}.$$

3. With Eq. L-58, cumulative wellstream gas production is

$$Q_{g,\text{wet}} = Q_{g,\text{dry}} + 133.316\frac{\gamma_o Q_o}{M_o}$$

$$= 767,000 + 133.316\frac{(0.755)(100,000)}{125}$$

$$= 847,523 \text{ Mscf} = 848 \text{ MMscf}.$$

Example L-3—Estimating Wellstream-Gas Gravity and Flow Rate With Correlations. Calculate the wellstream-gas gravity, the total wellstream-gas flow rate, and the cumulative wellstream production for a two-stage separation system.

R_1 = 7,040 scf/STB
γ_1 = 0.712
q_{s1} = 352 Mscf/D
p_{s1} = 700 psia
T_{s1} = 90°F
$Q_{g,\text{dry}}$ = 680 MMscf
γ_{API} = 55.8°API

Solution.

1. Determine G_{pa}. The value either can be read from Fig. L-16 or calculated directly with Eq. L-54. This example uses Eq. L-54.

$$G_{pa} = 1.4599(p_{s1} - 14.65)^{1.3394}\gamma_1^{7.0943}\gamma_{API}^{1.1436}T_{sl}^{-0.93446}$$

$$= 1.4599(700 - 14.65)^{1.3394}(0.712)^{7.0943}$$

$$\times (55.8)^{1.1436}(90)^{-0.93446}$$

$$= 1223.2 \text{ scf/STB}.$$

2. Similarly, V_{eq} can be estimated from Fig. L-17 or calculated with Eq. L-55.

$$V_{eq} = 635.53 + 0.36182p_{s1}^{1.0544}\gamma_1^{5.08312}\gamma_{API}^{1.5812}T_{sl}^{-0.79130}$$

$$= 635.53 + 0.36182(700)^{1.0544}(0.712)^{5.0831}$$

$$\times (55.8)^{1.5812}(90)^{-0.79130}$$

$$= 1692.1 \text{ scf/STB}.$$

3. The specific gravity of the stock-tank liquid, γ_o, is

$$\gamma_o = \frac{141.5}{131.5 + \gamma_{API}} = \frac{141.5}{131.5 + 55.8} = 0.756.$$

4. Calculate the reservoir-gas specific gravity, γ_g, with Eq. L-51.

$$\gamma_g = \frac{R_1\gamma_1 + 4,602\gamma_o + G_{pa}}{R_1 + V_{eq}}$$

$$= \frac{(7,040)(0.712) + (4,602)(0.756) + 1,223.2}{7,040 + 1,692.1}$$

$$= 1.112.$$

5. The total wellstream-gas flow rate is

$$q = q_{s1}\left(1 + \frac{V_{eq}}{R_1}\right) = (352)\left(1 + \frac{1,692.1}{7,040}\right) = 437 \text{ Mscf/D}.$$

6. The cumulative wellstream gas is

$$Q_{g,wet} = Q_{g,\text{dry}}\left(1 + \frac{V_{eq}}{R_1}\right) = (680)\left(1 + \frac{1,692.1}{7,040}\right)$$

$$= 843 \text{ MMscf}.$$

Once the specific gas gravity of the reservoir gas has been estimated, use the Piper *et al.*[6] method for calculating pseudocritical pressure and temperature.

Correlations for Reservoir-Oil Properties. Correlations of physical properties of reservoir crude oils are more complicated and more ambiguous than those for natural gases. These complications occur because of the many different components that typically make up these oils. Even though most of the components of a particular oil are hydrocarbons, the larger-molecule components can be of different chemical classes (e.g., aromatics or paraffins). These larger components can strongly influence the behavior of the mixture and, because of the different chemical structures, make a unique correlation impossible.

Aside from the composition of the oil, the mixing rules for liquids are considerably different from those of gases. Most mixing rules for multicomponent fluids (gases and liquids) are derived from kinetic theory. Gas systems are much simpler to model in theory than those of oils. Also, because of the complex nature of hydrocarbon liquids, simple mixing theories may not apply and the correlations must be completely empirical. Note that the correlations presented here are sufficiently accurate for general reservoir engineering calculations; however, some of these correlations were developed from limited databases and should be applied judiciously.

Stock Tank Gas/Oil Ratio. As mentioned earlier, the reported producing gas/oil ratio often does not include stock-tank vent gas. For separator temperatures less than 140°F, this additional gas may be estimated from[18]

$$\log(R_{ST}) = A_1 + A_2\log(\gamma_o) + A_3\log(\gamma_{gSP}) + A_4\log(p_{SP}) + A_5\log(T_{SP}), \quad \text{(L-60)}$$

where $A_1 = 0.3818$, $A_2 = -5.506$, $A_3 = 2.902$, $A_4 = 1.327$, and $A_5 = -0.7355$. The SGOR is obtained by adding the stock-tank from Eq. L-60 to the separator-gas/oil ratio.

SGOR—Bubblepoint Pressure. A single correlation may be used both for estimating the SGOR for an oil of known bubblepoint pressure and for estimating the bubblepoint pressure given the SGOR.

The bubblepoint pressure for an oil of known gas/oil ratio is given by[18]

$$p_b = 18.2(C_{pb} - 1.4), \quad \text{(L-61)}$$

where C_{pb} is given by

$$C_{pb} = (R_s/\gamma_g)^{0.83}10^{(0.00091T - 0.0125\gamma_{API})}. \quad \text{(L-62)}$$

Eqs. L-61 and L-62 may be rearranged to solve for SGOR, R_s, given the bubblepoint pressure, p_b.

This correlation was originally presented in graphical form, shown in **Fig. L-18.**[13]

Oil FVF. At the bubblepoint pressure, the oil FVF may be estimated from[18]

$$B_{ob} = 0.9759 + 12 \times 10^{-5}C_{B_{ob}}^{1.2}, \quad \text{(L-63)}$$

where $C_{B_{ob}} = R_s(\gamma_g/\gamma_o)^{0.5} + 1.25T. \quad \text{(L-64)}$

This correlation was originally presented in graphical form, as shown in **Fig. L-19.**[13]

Example

Required: Bubblepoint pressure at 200°F of liquid with a gas/oil ratio (GOR) of 350 ft³/bbl, a gas gravity of 0.75, and a tank-oil gravity of 30°API.

Procedure: Starting at the left side of the chart, proceed horizontally along the 350-ft³/bbl line to a gas gravity of 0.75. From that point, drop vertically to the 30°API line. Proceed horizontally from the tank-oil-gravity scale to the 200°F line. The required pressure is found to be 1,930 psia.

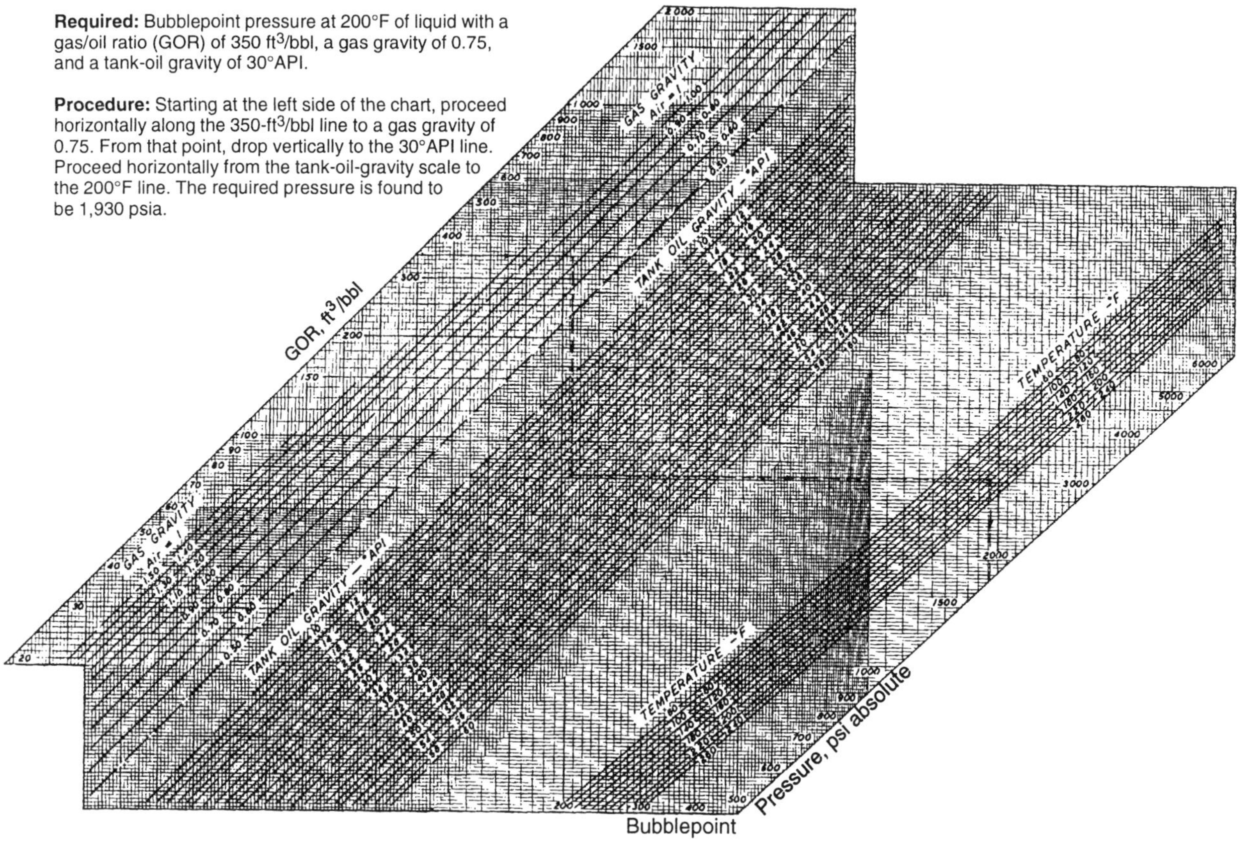

Fig. L-18—Standing[13] correlation for gas/oil ratio.

Oil Compressibility. For an oil above its bubblepoint pressure, the isothermal compressibility may be estimated from

$$c_o = \frac{A_1 + A_2 R_s + A_3 T + A_4 \gamma_g + A_5 \gamma_{API}}{A_6 p}, \quad \text{(L-65)}$$

where $A_1 = -1433.0$, $A_2 = 5.0$, $A_3 = 17.2$, $A_4 = -1180.0$, $A_5 = 12.61$, and $A_6 = 10^5$. This correlation is most accurate near the bubblepoint; at high pressures, it gives results that may be as much as 50% low. **Fig. L-20** may be used to estimate the compressibility of an undersaturated oil graphically.

Eq. L-66 may be used to estimate the compressibility of an oil below its original bubblepoint pressure.

$$\ln(c_o) = -7.573 - 1.497\ln(p) + 1.115\ln(T) + 0.533\ln(\gamma_{API}) + 0.184\ln(R_{sb}). \quad \text{(L-66)}$$

This correlation may also be evaluated graphically with **Fig. L-21.**

Oil Viscosity. To estimate oil viscosity below its original bubblepoint pressure, first estimate the dead-oil viscosity, μ_{oD}, at reservoir temperature.

$$\log[\log(\mu_{oD} + 1)] = 1.8653 - 0.025086\gamma_{API} - 0.5644\log(T). \quad \text{(L-67)}$$

Next, we estimate the live-oil viscosity with

$$\mu_o = A\mu_{oD}^B, \quad \text{(L-68)}$$

$$\text{where } A = 10.715(R_s + 100)^{-0.515} \quad \text{(L-69)}$$

$$\text{and } B = 5.44(R_s + 150)^{-0.338} \quad \text{(L-70)}$$

For pressures above the bubblepoint pressure, use

$$\mu_o = \mu_{ob}(p/p_b)^B, \quad \text{(L-71)}$$

$$\text{where } B = C_1 p^{C_2}\exp(C_3 + C_4 p), \quad \text{(L-72)}$$

where $C_1 = 2.6$, $C_2 = 1.187$, $C_3 = -11.513$, and $C_4 = -8.98 \times 10^{-5}$.

Figs. L-22, L-23, and L-24 present the correlations represented by Eqs. L-67, L-68, and L-71, respectively.

Correlations for Reservoir-Water Properties. The ability to estimate properties of reservoir waters is important for reservoir engineering calculations, especially for reservoirs with water influx. Correlating the physical properties of reservoir water is relatively simple because, in general, the composition of the water is affected only by dissolved solids. Also, changes in the physical properties of water as a function of temperature and pressure are relatively small and generally can be predicted.

The concentration of dissolved solids in water can be measured with several different units, including parts per million, weight percent, and grams of salt per liter. It can be confusing to convert between these unit systems, so a conversion table (**Table L-6**) is included to simplify this process.[1]

The correlations presented next for estimating FVF, solution-gas/water ratio, compressibility, and viscosity are the most accurate available and are adequate for most reservoir engineering calculations.

Water FVF. McCain[1,19] developed a correlation for the water FVF, B_w, given by Eq. L-73,

$$B_w = (1 + \Delta V_{wt})(1 + \Delta V_{wp}) \quad \text{(L-73)}$$

where the volume corrections for temperature, ΔV_{wt}, and pressure, ΔV_{wp}, (**Figs. L-25 and L-26**, respectively) are estimated from

$$\Delta V_{wt} = -1.00010 \times 10^{-2} + 1.33391 \times 10^{-4} T_F + 5.50654 \times 10^{-7} T_F^2 \quad \text{(L-74)}$$

Example

Required: Formation volume at 200°F of bubblepoint liquid with a GOR of 350 ft^3/bbl, a gas gravity of 0.75, and a tank-oil gravity of 30°API.

Procedure: Starting at the left side of the chart, proceed horizontally along the 350-ft^3/bbl line to a gas gravity of 0.75. From that point, drop vertically to the 30°API line. Proceed horizontally from the tank-oil-gravity scale to the 200°F line. The formation volume is found to be 1.22 bbl/bbl of tank oil.

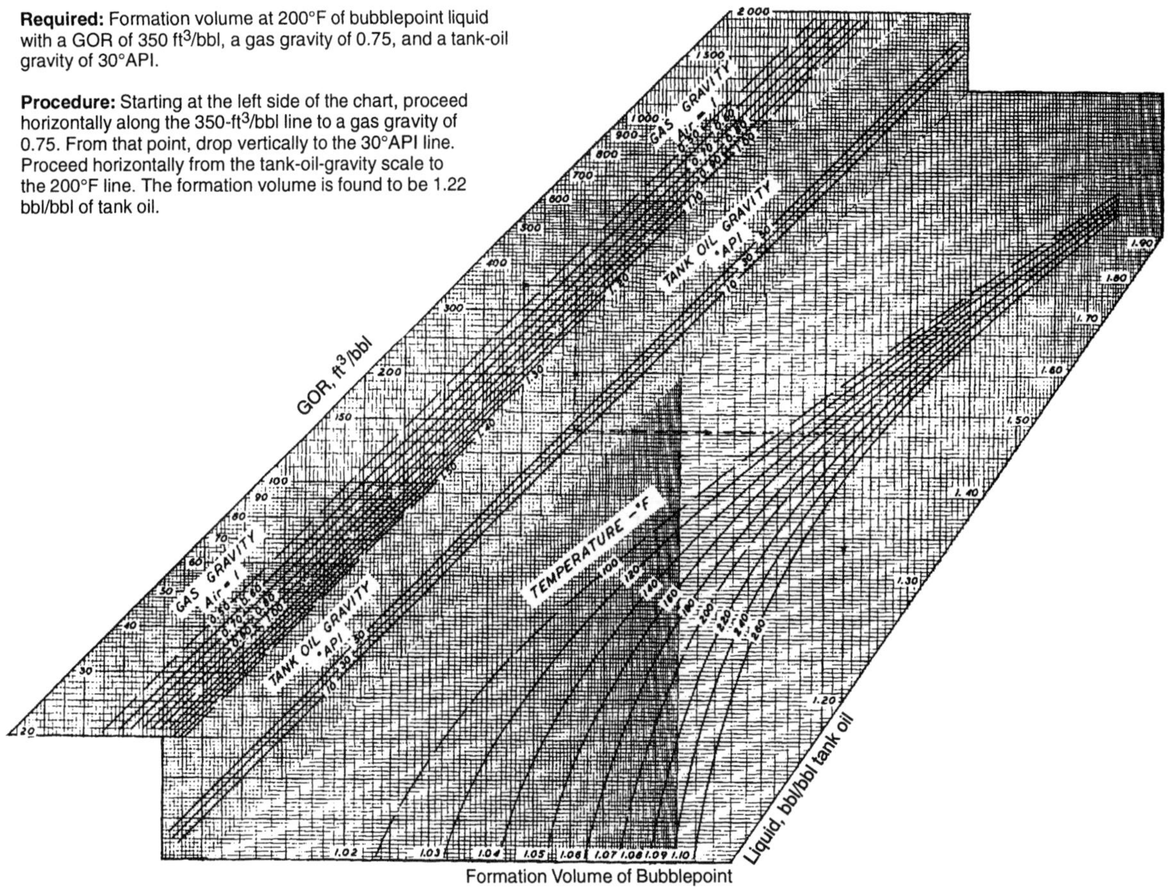

Fig. L-19—Standing[13] correlation for oil FVF.

$$\text{and } \Delta V_{wp} = -1.95301\times10^{-9}pT_F - 1.72834\times10^{-13}p^2T_F$$

$$-3.58922\times10^{-7}p - 2.25341\times10^{-10}p^2. \quad \text{(L-75)}$$

where B_w = water FVF, RB/STB; T_F = temperature, °F; and p = pressure, psia.

Figs. L-25 and L-26 give values of ΔV_{wt} and ΔV_{wp} as functions of pressure and temperature. This correlation is accurate to within 2% of the limited published data. Ignoring salinity causes errors in ΔV_{wt} and ΔV_{wp}; however, because the errors in ΔV_{wt} and ΔV_{wp} are comparable in value but opposite in sign, the total resulting error is negligible.

Example L-4—Estimating Water FVF. Use the following data to calculate the water FVF at p = 2,000 psia.

T_F = 200°F
p_b = 2,620 psia
S = 200,000 ppm = 20 wt% solids

Solution.

1. Determine the effects of temperature and pressure on the water volume. These effects can be estimated either by reading directly from Figs. L-25 and L-26 or by calculation with Eqs. L-74 and L-75. For this example, use the equations.

$$\Delta V_{wt} = -1.00010\times10^{-2} + 1.33391\times10^{-4}T_F + 5.50654\times10^{-7}\,T_F^2$$

$$= -1.00010\times10^{-2} + 1.33391\times10^{-4}\,(200)$$

$$+ 5.50654\times10^{-7}\,(200)^2$$

$$= 0.03870$$

$$\Delta V_{wp} = -1.95301\times10^{-9}pT_F - 1.72834\times10^{-13}p^2T_F$$

$$-3.58922\times10^{-7}\,p - 2.25341\times10^{-10}p^2$$

$$= -1.95301\times10^{-9}\,(2{,}000)(200)$$

$$-1.72834\times10^{-13}\,(2{,}000)^2\,(200)$$

$$-3.58922\times10^{-7}\,(2{,}000)$$

$$-2.25341\times10^{-10}\,(2{,}000)^2$$

$$= -0.00254$$

2. The water formation volume factor is

$$B_w = (1+\Delta V_{wt})(1+\Delta V_{wp})$$

$$= (1+0.03870)(1-0.00254)$$

$$= 1.04 \text{ RB/STB}$$

Solution-Gas/Water Ratio. McCain[1,19] also developed a correlation for the solution-gas/pure-water ratio, R_{swp},

$$R_{swp} = A + Bp + Cp^2 \quad \text{(L-76)}$$

and the partial derivative of the solution-gas/pure-water ratio with respect to pressure,

$$\left(\frac{\partial R_{swp}}{\partial p}\right)_T = B + 2C_p \quad \text{(L-77)}$$

The coefficients of Eqs. L-76 and L-77 are

$$A = 8.15839 - 6.12265\times10^{-2}T_F + 1.91663\times10^{-4}T_F^2$$

$$-2.1654\times10^{-7}T_F^3, \quad \text{(L-78)}$$

$$B = 1.01021\times10^{-2} - 7.44241\times10^{-5}T_F + 3.05553\times10^{-7}T_F^2$$

$$-2.94883\times10^{-10}T_F^3, \quad \text{(L-79)}$$

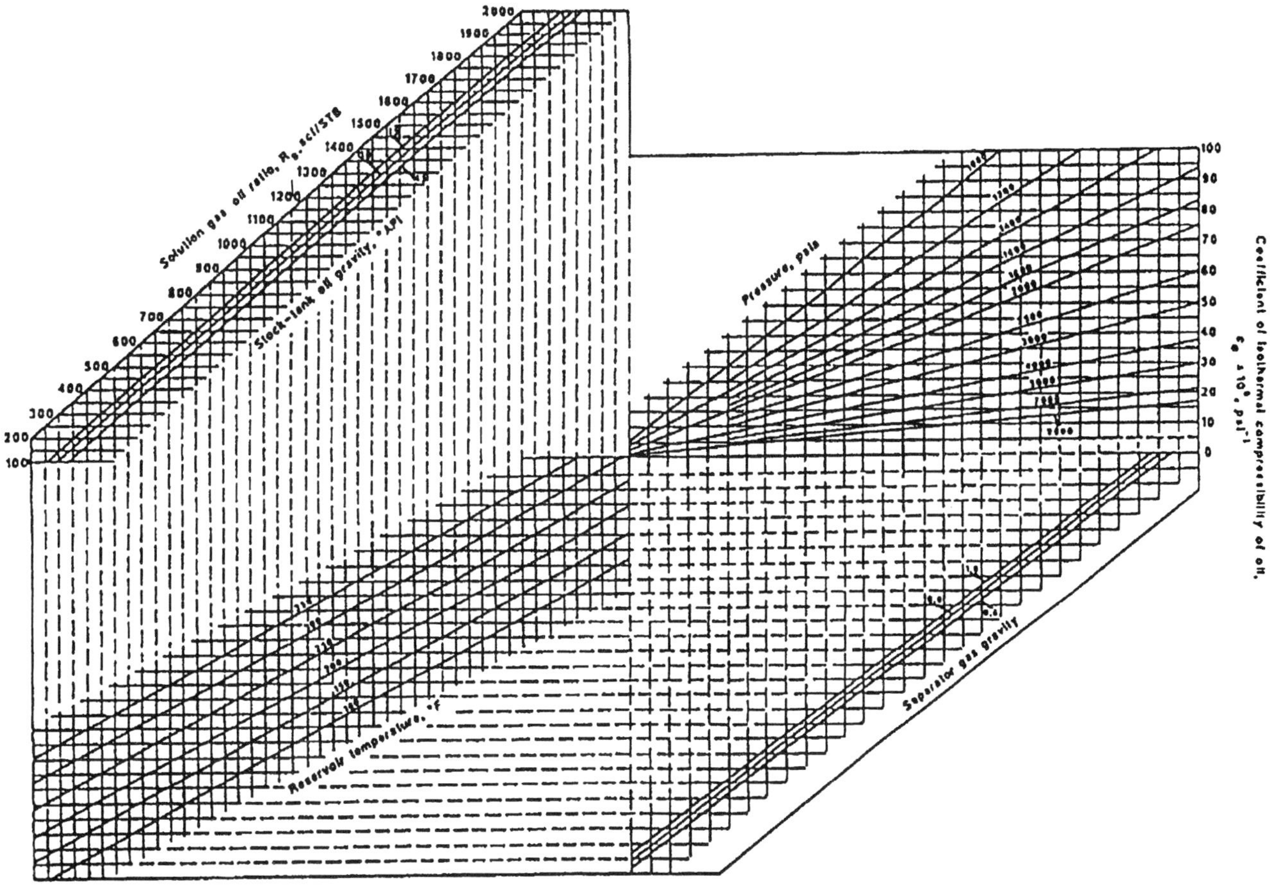

Fig. L-20—Coefficients of isothermal compressibility of undersaturated black oils (from McCain[1]).

and $C = -10^{-7}\,[9.02505 - 0.130237 T_F + 8.53425\times 10^{-4} T_F^2$

$$- 2.34122\times 10^{-6} T_F^3 + 2.37049\times 10^{-9} T_F^4]\,. \quad \ldots\ldots \text{(L-80)}$$

Values of the derivative of the solution-gas/pure-water ratio are needed for compressibility calculations at pressures below the bubblepoint. The variables in Eqs. L-77 through L-80 are defined as T_F = temperature, °F, and R_{swp} = solution-gas/ pure-water ratio, scf/STB.

Fig. L-27, which was generated by McCain[1] with Eq. L-76, gives values of R_{swp} as a function of pressure and temperature. Similarly, **Fig. L-28** was generated with Eq. L-77 and gives values of $\partial R_{swp}/\partial p$ as a function of pressure and temperature. The accuracy of the correlations given by Eqs. L-76 and L-77 is within 5% of published data. In addition, the correlations were developed for $1{,}000 < p$ (psia) $< 10{,}000$ and $100 < T_F$ (°F) < 340. Note that these correlations are applicable to formation waters, but McCain[1,19] states that they should never be used at pressures below 1,000 psia. Further, the relations were developed strictly for pure-water systems and must be modified to be applicable for reservoir brines.

McCain[1,19] proposed the following correlation to adjust pure-water values to those for brine systems.

$$\frac{R_{sw}}{R_{swp}} = 10^{\left(-0.0840655\, S T_F^{-0.285584}\right)}, \quad \ldots\ldots\ldots\ldots\ldots\ldots \text{(L-81)}$$

where S = salinity, weight percent solids; T_F = temperature, °F; R_{swp} = solution-gas/pure-water ratio, scf/STB; and R_{sw} = solution-gas/brine ratio, scf/STB.

The correlation given by Eq. L-81 and **Fig. L-29** agreed to within 3% of published data falling within $0 < S$ (wt% solids) < 30 and $70 < T_F$ (°F) < 250. The correlation is applicable to formation waters and is used to correct both the solution-gas/pure-water ratio and its derivatives.

Water Compressibility. Because the reservoir water compressibility is the property of reservoir water that is most affected by the presence of free gas, we must use an expression for compressibility that explicitly accounts for the presence of free gas at pressures below the bubblepoint. The bubblepoint pressure of a gas-saturated brine is equal to the bubblepoint pressure of the coexisting oil. In a gas/water system, the water is considered to be at its bubblepoint pressure at the initial reservoir pressure. Quantitatively, the solubility of gas in water is considerably less than that of gas in oil.

For pressures greater than or equal to bubblepoint pressure, Osif[20] developed a correlation to estimate the isothermal coefficient of compressibility of reservoir waters at pressures above the bubblepoint pressure of the water. This correlation is given by

$$c_w = -\frac{1}{B_w}\left(\frac{\partial B_w}{\partial p}\right)_T$$

$$= (7.033p + 541.5S' - 537.0T_F + 403{,}300)^{-1}\,,$$

$$\ldots\ldots\ldots\ldots\ldots\ldots\ldots \text{(L-82)}$$

where c_w = isothermal compressibility of formation water, psi^{-1}; p = pressure, psia; S' = salinity, g NaCl/L; and T_F = temperature, °F.

The correlation was developed for the following ranges of data.

$1{,}000 < p$ (psig) $< 20{,}000$,

$0 < S'$(g NaCl/l) < 200,

and $200 < T_F$ (°F) < 270.

Note that inaccurate results are obtained if this correlation is applied for reservoirs outside the range of data used to develop it.

Osif[20] also developed **Fig. L-30,** which plots the brine-compressibility/pure-water-compressibility ratio vs. total dissolved-solids content. As an alternative to Eq. L-82, use **Fig. L-31**[21] to obtain the compressibility for fresh water, and then use Osif's[20] correlation (Fig. L-30) to correct for the dissolved-solids content. Note

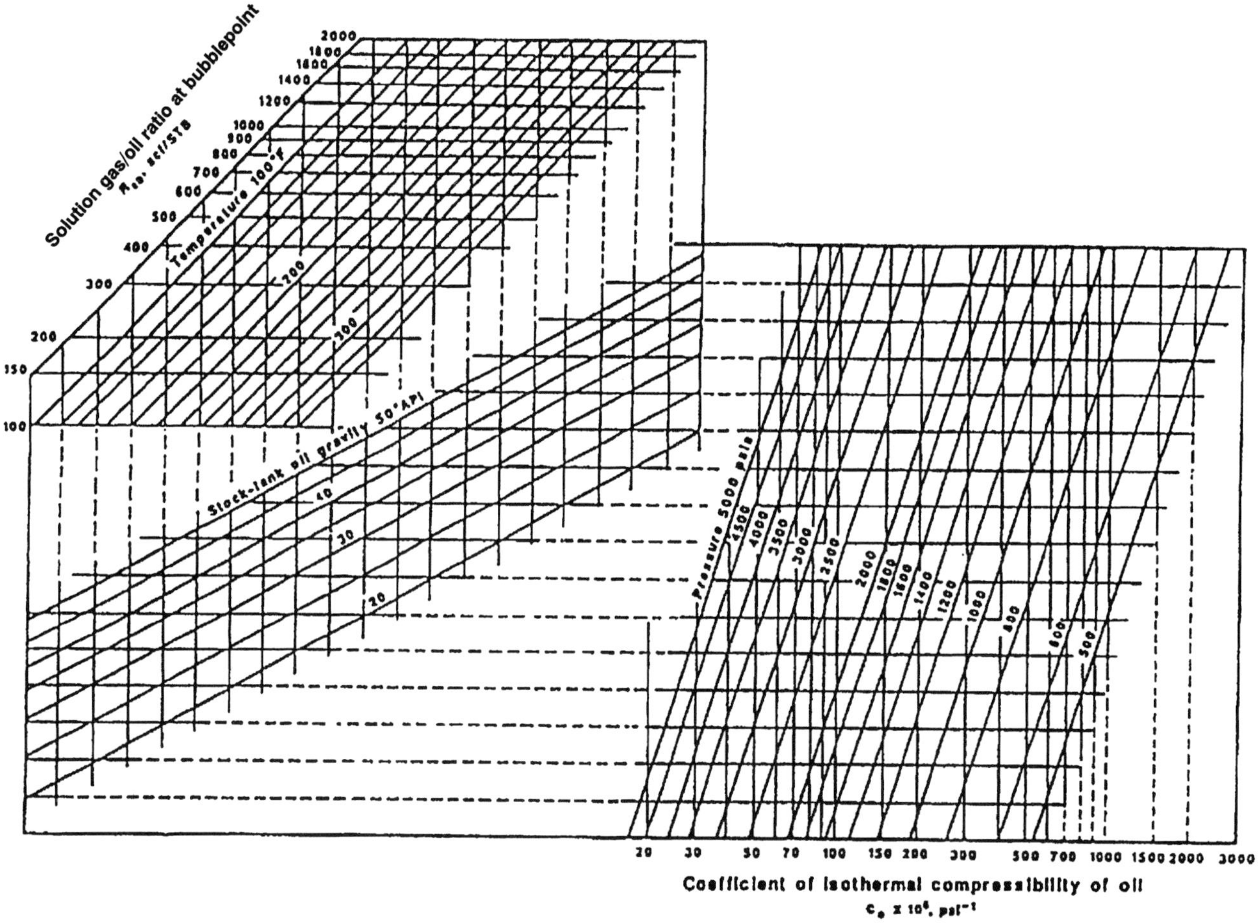

Fig. L-21—Coefficients of isothermal compressibility of saturated black oils (from McCain[1]).

that the amount of dissolved solids has a very small effect on the water compressibility.

Ramey[22] proposed the following theoretically based relation for the formation-water compressibility at pressures below the bubblepoint.

$$c_w = -\frac{1}{B_w}\left(\frac{\partial B_w}{\partial p}\right)_T + \frac{B_g}{B_w}\left(\frac{\partial R_{swp}}{\partial p}\right)_T . \quad \text{(L-83)}$$

The first term on the right side of Eq. L-83, which represents the compressibility of the water only, is calculated with either Eq. L-82 or Figs. L-31 and L-32. The second term on the right side of Eq. L-83 represents the effects of the gas on the total water compressibility. McCain[1] suggests the following procedure for calculating the second term.

1. Estimate B_g using a gas gravity of $\gamma_g = 0.63$. McCain[1] states that this value is based on limited data with unknown accuracy but that it gives reasonable results in further calculations.

2. Estimate B_w using Eqs. L-73 through L-75 and the procedure described earlier.

3. Estimate $\partial R_{swp}/\partial p$ from Eq. L-77 or Fig. L-28, and multiply this value by the dissolved-solids content correction factor from Fig. L-29.

Because this relation has not been verified with experimental data, McCain[1] considered the estimate of formation-water compressibility with Eq. L-83 to be an approximation.

Example L-5—Estimating Water Compressibility. A water sample was taken from an oil reservoir with an original bubblepoint pressure of 2,620 psia. Using Eq. L-83 and the procedure suggested by McCain,[1] estimate the water compressibility at the current reservoir pressure, $p = 2{,}000$ psia. Include the effects of the dissolved-salt content.

$T_F = 200°\text{F}$, $S' = 230$ g NaCl/L $= 200{,}000$ ppm

$= 20$ wt% solids

Solution.

1. Estimate the gas FVF, B_g. For this estimate, we assume the specific gas gravity, γ_g, equals 0.63. Using the methods described in the previous sections on natural gas properties, estimate the gas deviation factor to be $z = 0.90$. The gas FVF at 2,000 psia and $200°\text{F} = 659.67°\text{R}$ is computed with Eq. L-9,

$$B_g = \frac{1}{0.0056146}\frac{zT}{p}\frac{p_{sc}}{T_{sc}} = \frac{1}{0.0056146}\frac{14.65}{520}\frac{(0.90)(660)}{(2000)}$$

$$= 1.49 \text{ RB/Mscf}.$$

2. Next, estimate the FVF for water. From Eq. L-73, the water FVF is

$$B_w = (1 + \Delta V_{wt})(1 + \Delta V_{wp}),$$

where the volume changes caused by temperature, ΔV_{wt}, and pressure, ΔV_{wp}, can be estimated with either Figs. L-25 and L-26 or Eqs. L-74 and L-75. For this example, use the graphs. Example L-4 illustrates the calculation of these corrections with the equations. From Fig. L-25, the change caused by temperature is

$$\Delta V_{wt} = 0.0386.$$

Similarly, the change caused by pressure is estimated from Fig. L-26.

$$\Delta V_{wp} = -0.00255.$$

Note that linear interpolation may be required for pressures different from those plotted on the graph. The water FVF is

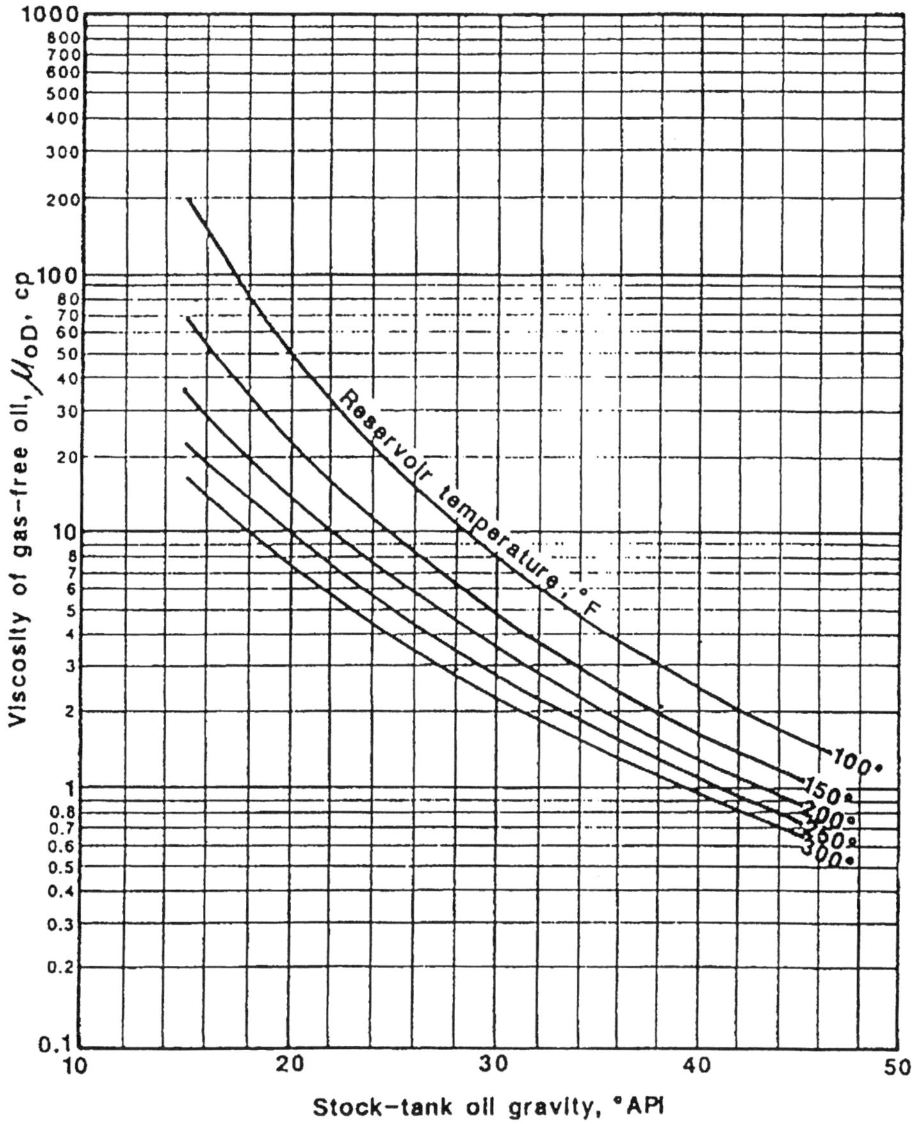

Fig. L-22—Viscosity of dead oil (after McCain[1]).

$$B_w = (1 + \Delta V_{wt})(1 + \Delta V_{wp})$$

$$= (1 + 0.0386)(1 - 0.00255)$$

$$= 1.04 \text{ RB/STB}.$$

3. Estimate the first term on the right side of Eq. L-83. From Fig. L-30, the brine-compressibility/pure-water compressibility ratio is

$$\frac{c_w}{c_{wp}} = 0.71.$$

Note that we must interpolate to obtain this ratio. From Fig. L-31, the compressibility of pure water is

$$c_{wp} = 3.2 \times 10^{-6} \text{ psi}^{-1}.$$

Therefore, the compressibility of brine is

$$c_w = c_{wp}\left(\frac{c_w}{c_{wp}}\right) = (3.2 \times 10^{-6} \text{ psi}^{-1})(0.71)$$

$$= 2.3 \times 10^{-6} \text{ psi}^{-1}$$

$$\text{or } \frac{1}{B_w}\left(\frac{\partial B_w}{\partial p}\right)_T = 2.3 \times 10^{-6} \text{ psi}^{-1}.$$

This value also can be estimated with Eq. L-82.

$$\frac{1}{B_w}\left(\frac{\partial B_w}{\partial p}\right)_T$$

$$= 1/(7.033p + 541.5S - 537.0T_F + 403,300)$$

$$= 1/[7.033(2,000) + 5.41.5(200)$$

$$- 537.0(200) + 403.300]$$

$$= 2.39 \times 10^{-6} \text{ psia}^{-1},$$

which is similar to the value obtained from the graphs.

4. Estimate the second term on the right side of Eq. L-83. Illustrate this calculation with both the graphical method (Step 4a) and the equations (Step 4b).

a. Estimate R_{swp} and $(\partial R_{swp}/\partial p)_T$ at 2,000 psia. From Fig. L-27, the solubility of gas in pure water is

$$R_{swp} = 11.2 \text{ scf/STB}.$$

Similarly, from Fig. L-28, the derivative is

$$\left(\frac{\partial R_{swp}}{\partial p}\right)_T = 4.2 \times 10^{-3} \text{ scf/STB/psi}.$$

Now, correct the pure-water values to the appropriate brine system using the McCain[1] correlation given by Fig. L-29. From Fig. L-29,

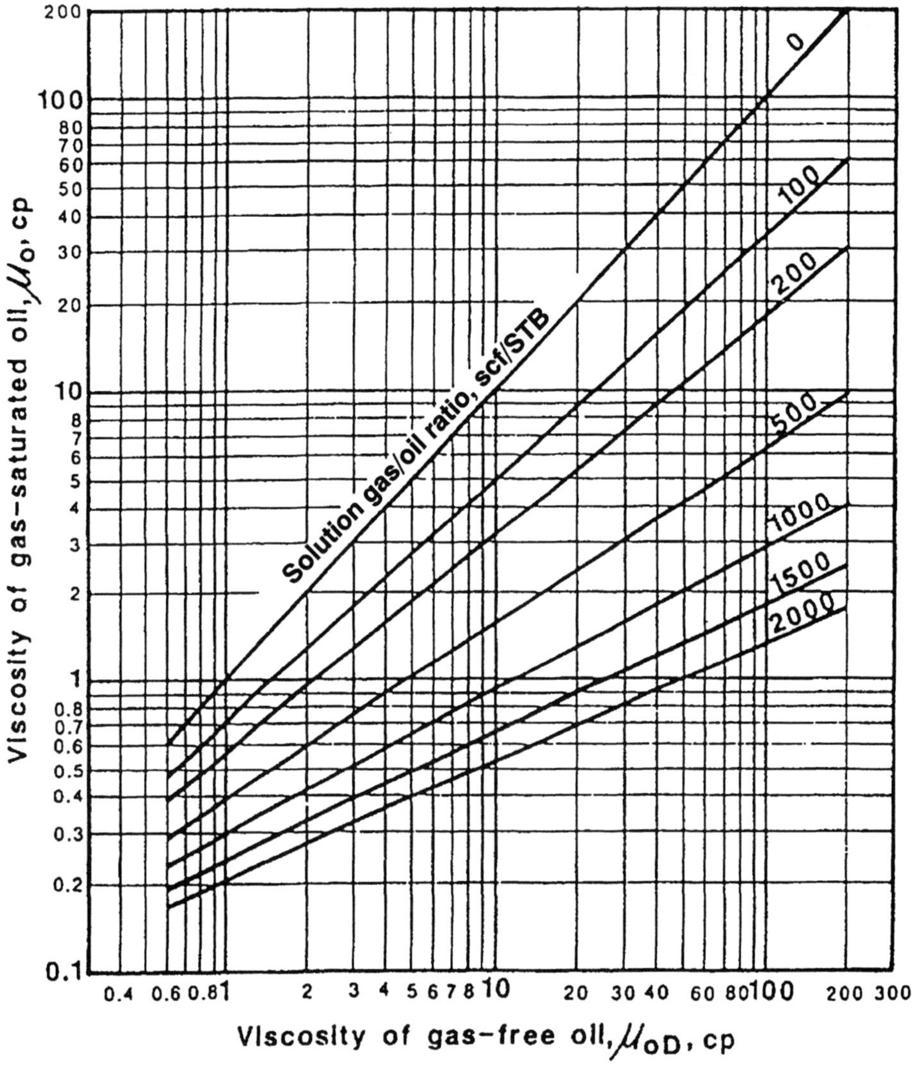

Fig. L-23—Viscosity of saturated black oils (after McCain[1]).

$$\frac{R_{sw}}{R_{swp}} = 0.43.$$

Therefore,

$$R_{sw} = R_{swp}(R_{sw}/R_{swp}) = (11.2)(0.43) = 4.82 \text{ scf/STB}$$

and
$$\left(\frac{\partial R_{swp}}{\partial p}\right)_T = \left(\frac{\partial R_{swp}}{\partial p}\right)_T \left(\frac{R_{sw}}{R_{swp}}\right)$$

$$= (4.2 \times 10^{-3})(0.43)$$

$$= 1.81 \times 10^{-3} \text{ scf/STB/psi.}$$

b. Similar results can be obtained with Eqs. L-76 through L-82. From Eqs. L-78 through L-80,

$$A = 8.15839 - 6.12265 \times 10^{-2} T_F + 1.91663 \times 10^{-4} T_F^2$$
$$- 2.1654 \times 10^{-7} T_F^3$$
$$= 8.15839 - (6.12265 \times 10^{-2})(200) + (1.91663 \times 10^{-4})(200)^2$$
$$- (2.1654 \times 10^{-7})(200)^3$$
$$= 1.847 \text{ scf/STB}$$

$$B = 1.01021 \times 10^{-2} - 7.44241 \times 10^{-5} T_F + 3.05553 \times 10^{-7} T_F^2$$
$$- 2.94883 \times 10^{-10} T_F^3$$
$$= 1.01021 \times 10^{-2} - (7.44241 \times 10^{-5})(200)$$
$$+ (3.05553 \times 10^{-7})(200)^2 - (2.94883 \times 10^{-10})(200)^3$$
$$= 5.0803 \times 10^{-3} \text{ scf/STB/psia}$$

$$C = -10^{-7}\,[9.02505 - 0.130237 T_F + 8.53425 \times 10^{-4} T_F^2$$
$$- 2.34122 \times 10^{-6} T_F^3 + 2.37049 \times 10^{-9} T_F^4]$$
$$= -10^{-7}[9.02505 - (0.130237)(200) + (8.53425 \times 10^{-4})(200)^2$$
$$- (2.34122 \times 10^{-6})(200)^3 + (2.37049 \times 10^{-9})(200)^4]$$
$$= -2.1777 \times 10^{-7} \text{ scf/STB/psia}^2$$

From Eq. L-76, the solution-gas/water ratio for pure water is

$$R_{swp} = A + Bp + Cp^2$$
$$= (1.847) + (5.0803 \times 10^{-3})(2{,}000)$$
$$+ (-2.1777 \times 10^{-7})(2{,}000)^2$$
$$= 11.137 \text{ scf/STB}$$

From Eq. L-77, the derivative of R_{swp} with respect to pressure is

$$\left(\frac{\partial R_{swp}}{\partial p}\right)_T B + 2C_p$$

$$= (5.0803 \times 10^{-3}) + 2(-2.1777 \times 10^{-7})(2{,}000)$$

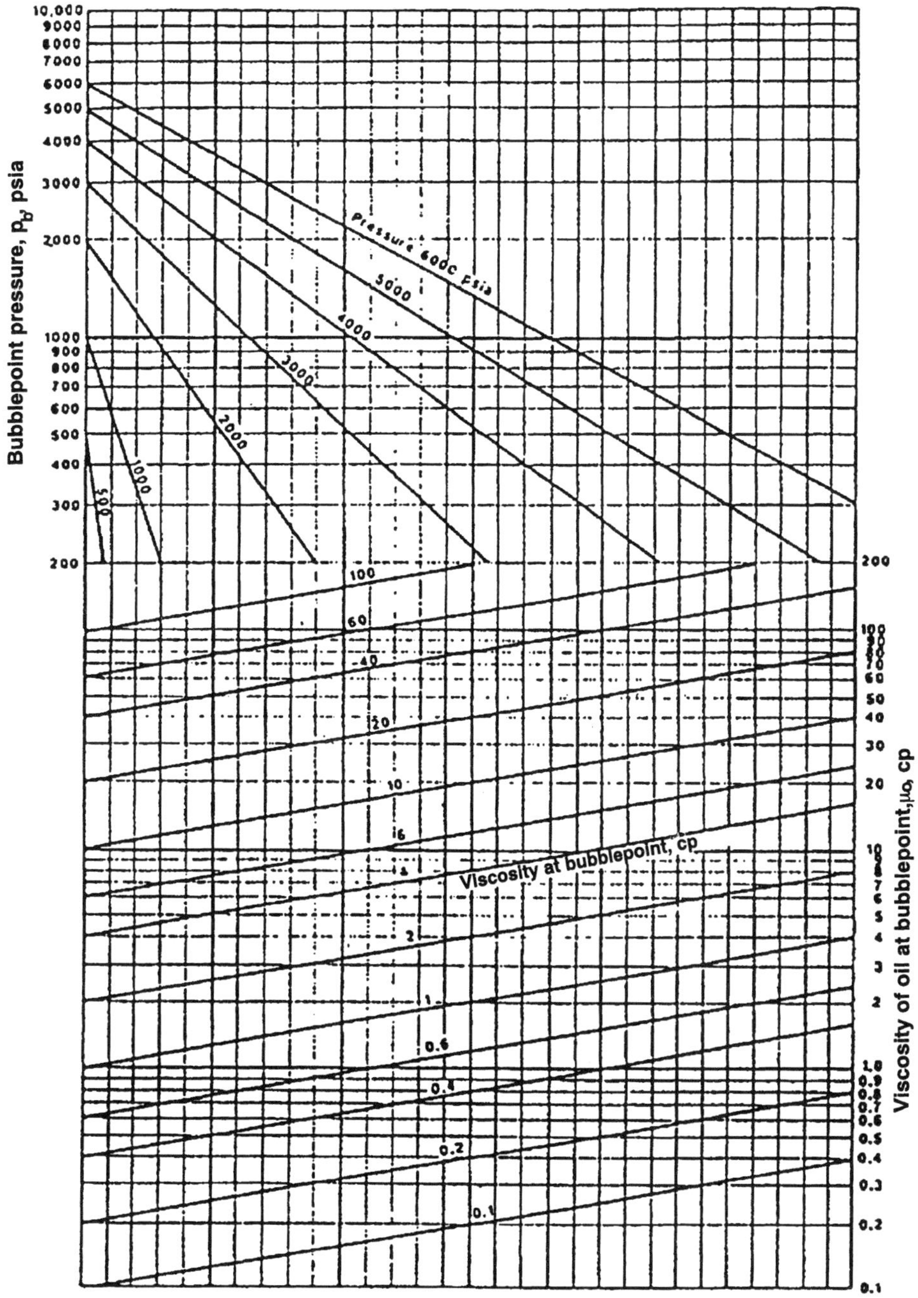

Fig. L-24—Viscosities of undersaturated black oils (after McCain[1]).

TABLE L-6—UNITS CONVERSION TABLE FOR CONCENTRATION OF DISSOLVED SOLIDS IN WATER			
Units	Symbol	Definition	Conversion Equation
Weight % solids	S	g solid/100 g brine	$S = S_{ppm} \times 10^{-4}$
Parts per million	S_{ppm}	g solid/10^6 g brine	$S_{ppm} = S \times 10^4 = S' \times 10^3/\rho_w$
Grams NaCl per liter	S'	g solid/liter brine	$S' = S \times \rho_w \times 10 = S_{ppm} \times \rho_w/10^3$

$$= 4.209 \times 10^{-3} \text{ scf/STB/psi.}$$

Now, correct the pure-water values to the appropriate brine system using the correlation given by Eq. L-81.

$$\frac{R_{sw}}{R_{swp}} = 10^{\left(-0.0840655 S T_F^{-0.285584}\right)}$$

$$= 10^{\left[-0.0840655(20)(200)^{-0.285584}\right]}$$

$$= 0.4268.$$

Therefore, the corrected values are

$$R_{sw} = R_{swp}\,(R_{sw}/R_{swp}) = (11.136)(0.4268) = 4.754 \text{ scf/STB}$$

and
$$\left(\frac{\partial R_{swp}}{\partial p}\right)_T = \left(\frac{\partial R_{swp}}{\partial p}\right)_T \left(\frac{R_{sw}}{R_{swp}}\right)$$

$$= \left(4.209 \times 10^{-3}\right)(0.4268)$$

$$= 1.797 \times 10^{-3} \text{ scf/STB/psi.}$$

5. Finally, we have all the components to calculate water compressibility with Eq. L-83. With the results obtained with the graphs, the water compressibility is

$$c_w = -\frac{1}{B_w}\left(\frac{\partial B_w}{\partial p}\right)_T + \frac{B_g}{B_w}\left(\frac{\partial R_{swp}}{\partial p}\right)_T$$

$$= \left(2.3 \times 10^{-6}\right) + \frac{1.49}{1.04}\left(\frac{1.81 \times 10^{-3}}{1,000}\right)$$

$$= 4.89 \times 10^{-6} \text{ psia}^{-1}.$$

Similarly, with the results obtained from the equations, the water compressibility is

$$c_w = -\frac{1}{B_w}\left(\frac{\partial B_w}{\partial p}\right)_T + \frac{B_g}{B_w}\left(\frac{\partial R_{swp}}{\partial p}\right)_T$$

$$= \left(2.390 \times 10^{-6}\right) + \frac{1.49}{1.04}\left(\frac{1.797 \times 10^{-3}}{1,000}\right)$$

$$= 4.97 \times 10^{-6} \text{ psi}^{-1},$$

which is comparable with the results obtained with the graphs. Although the arithmetic required for water-compressibility calculations with the equations is quite tedious, the equations generally are more accurate than the graphical method. In addition, the equations can be programmed for computer calculations. The graphical method, however, is desirable when speed is more important than precision.

Water Viscosity. McCain[1,19] developed the following correlation for the formation water viscosity at atmospheric pressure, μ_{w1}, and reservoir temperature.

$$\mu_{w1} = A T_F^B, \quad \text{(L-84)}$$

where the variables A and B are

$$A = 109.574 - 8.40564S + 0.313314S^2 + 8.72213 \times 10^{-3}S^3 \quad \text{(L-85)}$$

and $B = -1.12166 + 2.63951 \times 10^{-2}S - 6.79461 \times 10^{-4}S^2$

$$- 5.47119 \times 10^{-5}S^3 + 1.55586 \times 10^{-6}S^4 \quad \text{(L-86)}$$

and T_F = reservoir temperature, °F, and S = water salinity, wt% solids.

The correlation, given by Eq. L-84 and plotted in **Fig. L-32**, is accurate to within 5% of published data and was developed for $100 < T_F < 400$°F and $0 < S$ (wt% solids) < 26.

Once the water viscosity at atmospheric pressure, μ_{w1}, and reservoir temperature is estimated, the formation-water viscosity at reservoir pressure, μ_w, is estimated by the following correlation developed by McCain[1,19]:

$$\frac{\mu_w}{\mu_{w1}} = 0.9994 + 4.0295 \times 10^{-5}p + 3.1062 \times 10^{-9}p^2. \quad \text{(L-87)}$$

This correlation, which is accurate to within 4% for pressures below 10,000 psia and within 7% for pressures between 10,000 and 15,000 psia, was developed for the limited temperature range of $86 < T_F$ (°F) < 167. Correlations for μ_w/μ_{w1} are also plotted in **Fig. L-33.**

Example L-6—Estimating Water Viscosity. A water sample was taken from an oil reservoir with an original bubblepoint pressure of 2,620 psia. Estimate the water viscosity at $p = 2,000$ psia. $T_F = 200$°F and $S = 200,000$ ppm $= 20$ wt% solids.

Solution.

1. Estimate the water viscosity at atmospheric pressure, μ_{w1}, and reservoir temperature with either Fig. L-32 or Eq. L-84. For this example, we chose to use the equation. The variables A and B in Eq. L-84 are

$$A = 109.574 - 8.40564S + 0.313314S^2 + 8.72213 \times 10^{-3}S^3$$

$$= 109.574 - 8.40564(20) + 0.313314(20)^2$$

$$+ 8.72213 \times 10^{-3}(20)^3$$

$$= 136.56$$

$$B = -1.12166 + 2.63951 \times 10^{-2}S - 6.79461 \times 10^{-4}S^2$$

$$- 5.47119 \times 10^{-5}S^3 + 1.55586 \times 10^{-6}S^4$$

$$= -1.12166 + 2.63951 \times 10^{-2}(20) - 6.79461 \times 10^{-4}(20)^2$$

$$- 5.47119 \times 10^{-5}(20)^3 + 1.55586 \times 10^{-6}(20)^4$$

$$= -1.0543.$$

From Eq. L-84, the viscosity at atmospheric pressure and reservoir temperature is

$$\mu_{w1} = A T_F^B = (136.56)(200)^{(-1.0543)} = 0.5121 \text{ cp}.$$

A more rapid alternative to Eq. L-84 is to use Fig. L-32, which yields a similar value.

$$\mu_{w1} = 0.51 \text{ cp}.$$

2. Correct the water viscosity at atmospheric pressure, μ_{w1}, to the viscosity at reservoir pressure, μ_w. This correction is obtained by use of Eq. L-87.

$$\frac{\mu_w}{\mu_{w1}} = 0.9994 + 4.0295 \times 10^{-5}p + 3.1062 \times 10^{-9}p^2$$

$$= 0.9994 + 4.0295 \times 10^{-5}(2,000) + 3.1062$$

$$\times 10^{-9}(2,000)^2 = 1.0924.$$

Therefore,

$$\mu_w = \mu_{w1}\left(\frac{\mu_w}{\mu_{w1}}\right) = (0.5121)(1.0924) = 0.56 \text{ cp}.$$

Similarly, the viscosity ratio from Fig. L-34 is

$$\frac{\mu_w}{\mu_{w1}} = 1.09$$

and the water viscosity is

$$\mu_w = \mu_{w1}\left(\frac{\mu_w}{\mu_{w1}}\right) = (0.51)(1.09) = 0.56 \text{ cp},$$

which agrees with the value calculated with Eqs. L-84 and L-87.

Correlations for Reservoir Rock Properties. ***PV Compressibility.*** Newman[23] used 79 samples for consolidated sandstones to develop a correlation between PV compressibility, c_f, and porosity, ϕ. The data shown in **Fig. L-34** were fit with a hyperbolic equation given by

$$c_f = \frac{a}{\left(1 + bc\phi\right)^{1/b}}, \quad \text{(L-88)}$$

where $a = 97.32 \times 10^{-6}$; $b = 0.6999$; and $c = 79.82$.

The average absolute error of the correlation model was 2.60%. The consolidated sandstone correlation was developed for sandstone porosities ranging from 2 to 23%. Newman[23] also developed a correlation for limestones. The data shown in **Fig. L-35** were fit with a hyperbolic equation given by

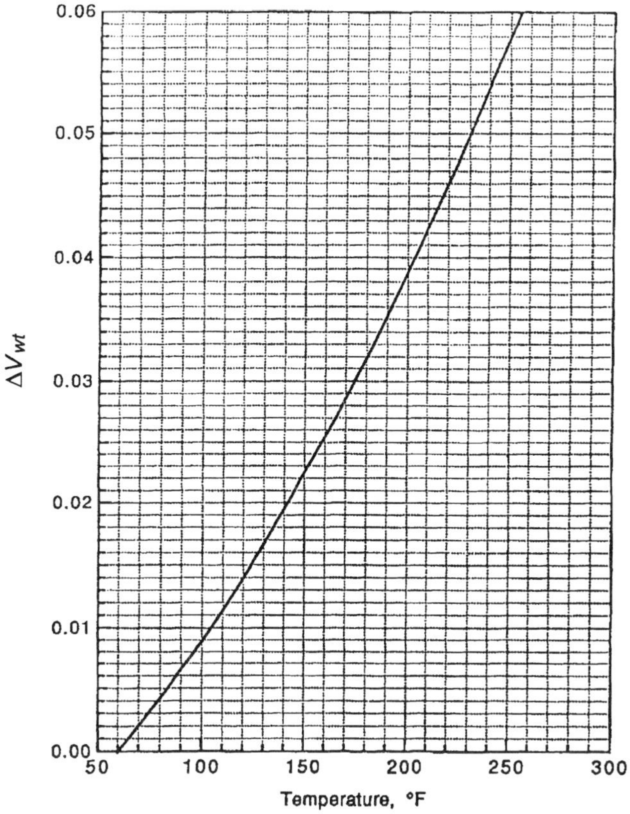

Fig. L-25—ΔV_{wt} as a function of reservoir temperature (after McCain[1]).

$$c_f = \frac{a}{(1 + bc\phi)^{1/b}}, \quad \text{(L-89)}$$

where $a = 0.8535$; $b = 1.075$; and $c = 2.303 \times 10^6$.

The average absolute error of the correlation model was 11.8%. The limestone correlation was developed for a limestone porosity of 0.02 to 0.33. In Eqs. L-88 and L-89, the compressibility is measured in psia^{-1} while porosity is given in fractional form.

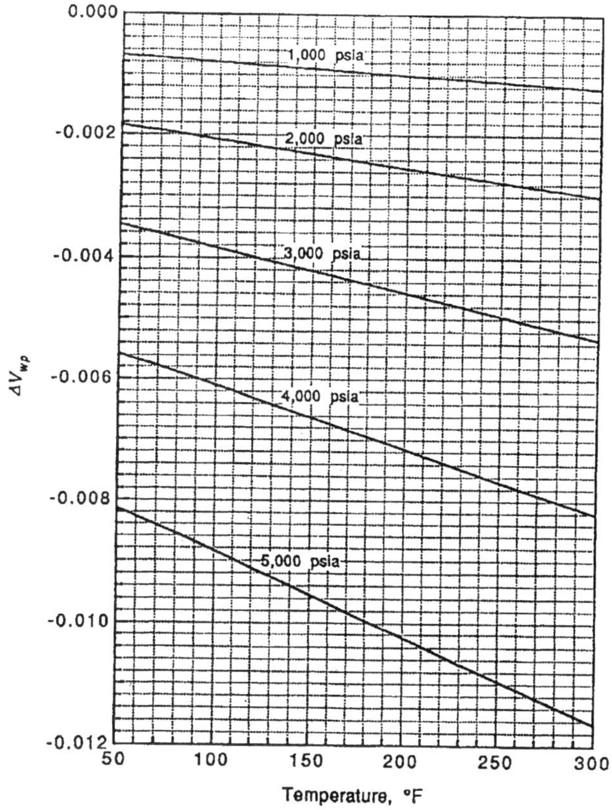

Fig. L-26—ΔV_{wp} as a function of reservoir pressure and temperature (after McCain[1]).

Example L-6—Estimation of PV Compressibility. Calculate the PV compressibility, c_f, for a formation of unknown lithology and $\phi = 0.2$.

Solution.

1. Assume the lithology is sandstone and estimate c_f at $\phi = 0.2$ with Newman's consolidated-sandstone correlation.

$$c_f = \frac{a}{(1 + bc\phi)^{1/b}}$$

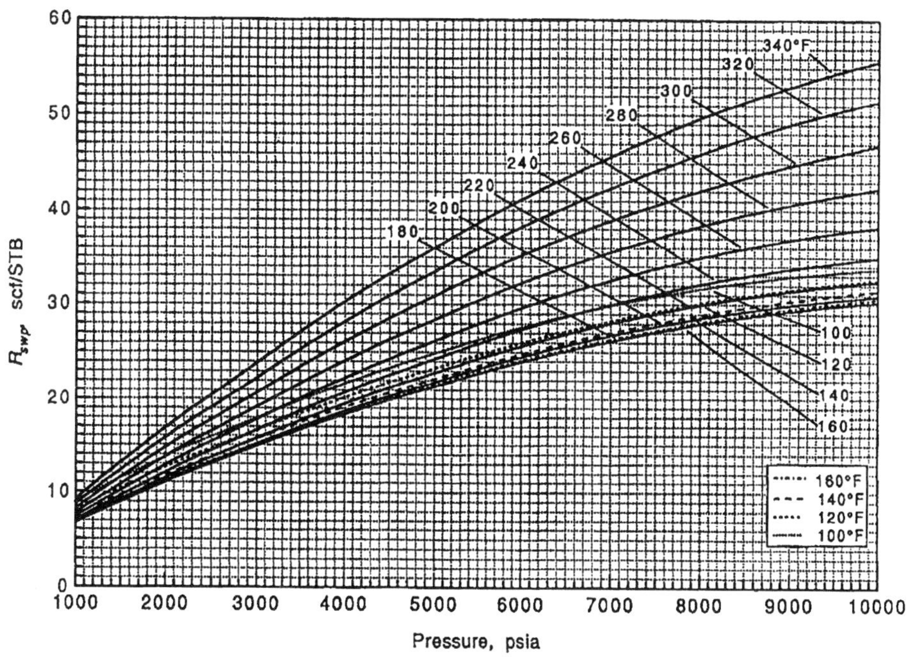

Fig. L-27—Solubility of gas in pure water.

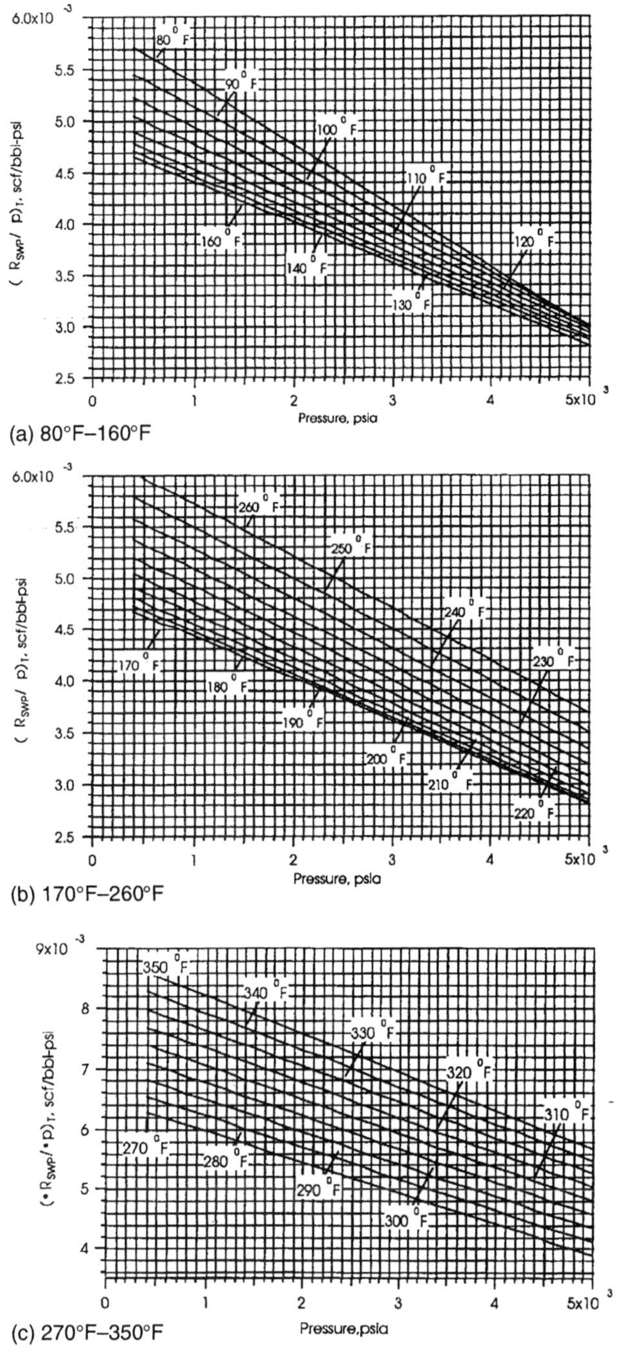

Fig. L-28—Derivative of R_{swp} with respect to pressure.

$$= \frac{97.3200 \times 10^{-6}}{[1 + (0.699993)(79.8181)(0.20)]^{(1/0.699993)}}$$

$$= 2.74 \times 10^{-6} \text{ psia}^{-1}.$$

2. Assume the lithology is limestone and estimate c_f at $\phi = 0.2$ with Newman's limestone correlation:

$$c_f = \frac{a}{(1 + bc\phi)^{1/b}}$$

$$= \frac{0.853531}{[1 + (1.07538)(2.30304 \times 10^6)(0.20)]^{(1/1.07538)}}$$

$$= 4.32 \times 10^{-6} \text{ psia}^{-1}.$$

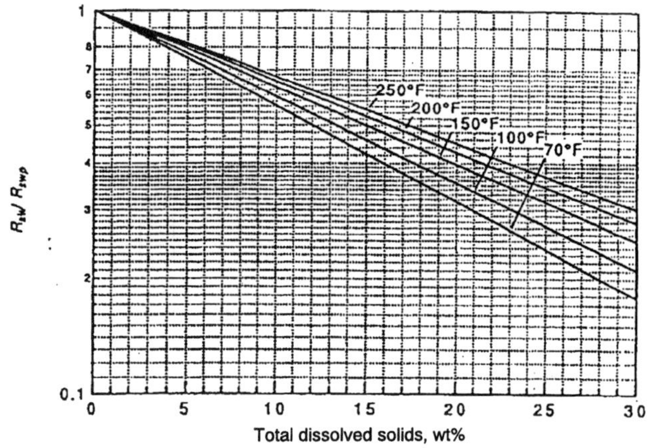

Fig. L-29—R_{sw}/R_{swp} as a function of dissolved solids.

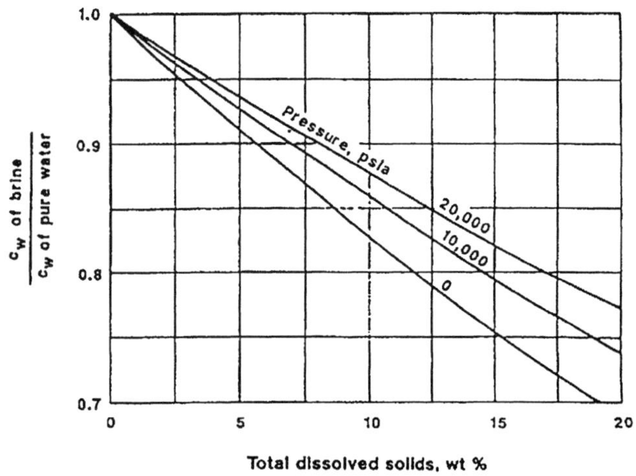

Fig. L-30—Effect of salinity on the coefficient of isothermal compressibility of water (after McCain[1]).

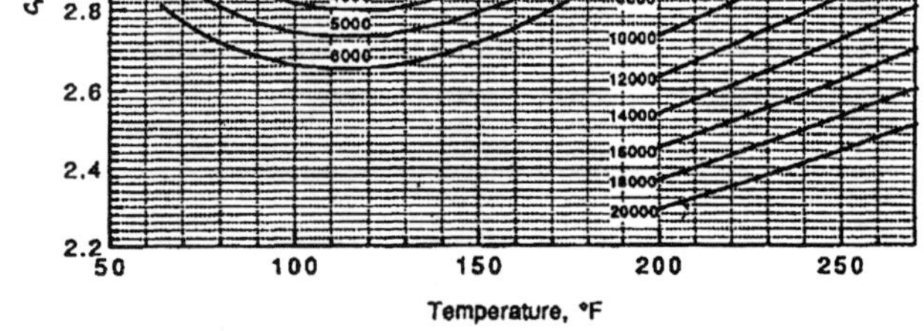

Fig. L-31—Coefficient of isothermal compressibility of fresh water (after Osif[20] and Dodson and Standing[21]). Reprinted courtesy of the American Petroleum Inst.

Gas-Turbulence Factor and Non-Darcy-Flow Coefficient. Many attempts[24-27] have been made to relate experimental measurements of the inertial coefficient, β, to rock properties. Using 355 sandstone and 29 limestone samples, Jones[5] experimentally determined β and developed correlations describing β as a function of porosity and permeability. The following correlation[5] is recommended for estimating β.

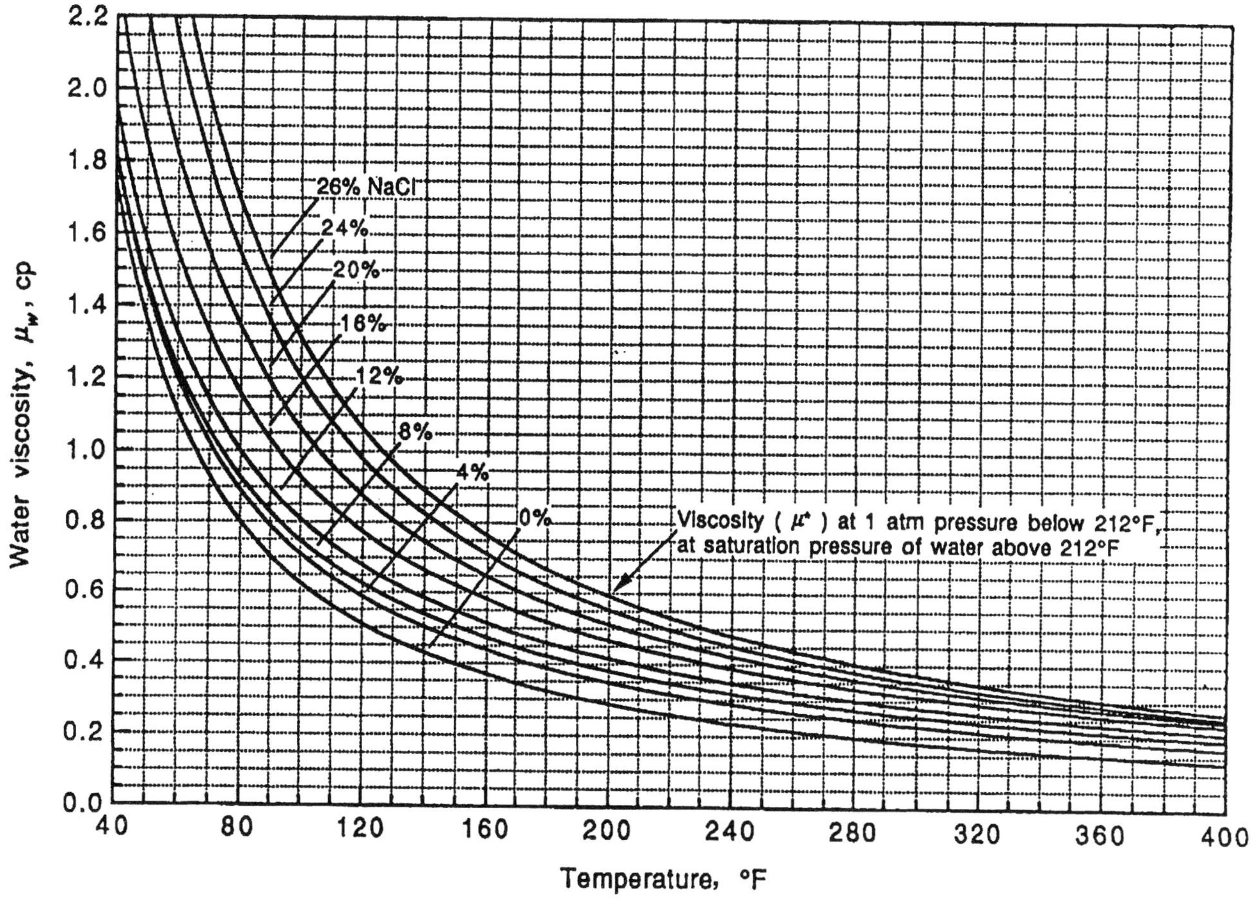

Fig. L-32—Water viscosity as a function of salinity and temperature.

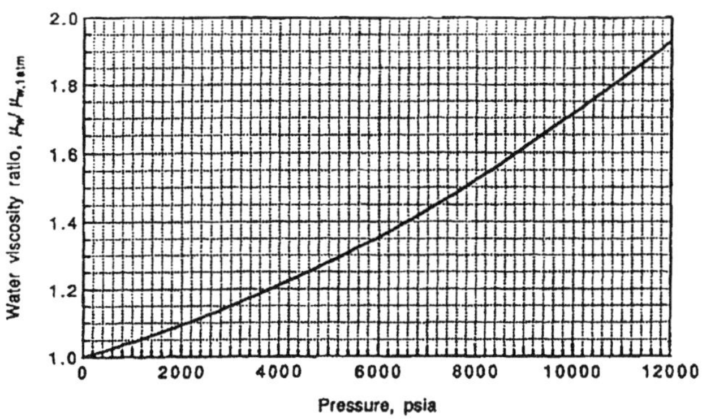

Fig. L-33—Water viscosity ratio (after McCain[1,19]).

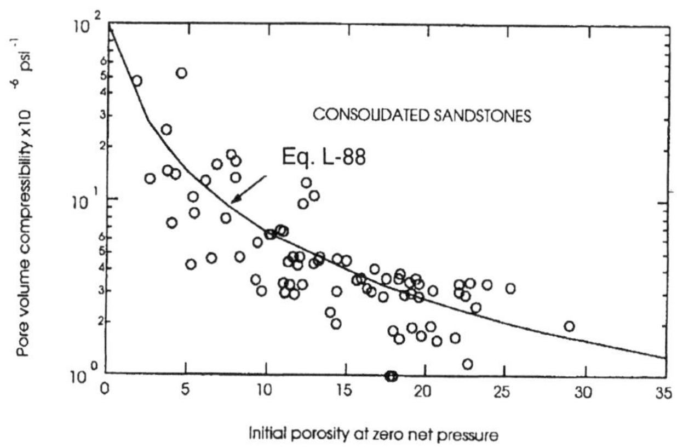

Fig. L-34—PV compressibility at 75% lithostatic pressure vs. initial sample porosity for consolidated sandstones.[23]

$$\beta = 1.88 \times 10^{10}\, k^{-1.47} \phi^{-0.53} . \quad \text{(L-90)}$$

The non-Darcy component of the flow equation is significant only in the area of high velocity and high pressure drawdown near the wellbore, so the effect of non-Darcy flow usually is incorporated into fluid-flow equations as an additional skin factor that is rate-dependent.[28] The total skin factor, which is the value determined from pressure transient analysis, is an apparent value, s', that includes both the true skin factor, s, and a term representing non-Darcy flow effects, Dq. β is incorporated into the term D,

$$D = \frac{2.715 \times 10^{-15} \beta k M p_{sc}}{h r_w T_{sc} \mu_{g,wf}} , \quad \text{(L-91)}$$

where D = non-Darcy flow coefficient, $(\text{Mscf/D})^{-1}$; $\mu_{g,wf}$ = pressure-dependent gas viscosity evaluated at the bottomhole flowing pressure, psia; and M = molecular weight of the gas, lbm/lb mol.

The non-Darcy-flow coefficient is not constant but varies as a function of pressure. After production begins, the radius of the high-velocity-flow region initially increases with time and soon stabilizes. D is inversely proportional to gas viscosity evaluated at p_{wf}. Viscosity is directly related to pressure, which in turn is a function of time. As pressure declines, viscosity likewise declines, causing an increase in D. Because D is not constant, analyzing a gaswell drawdown test with methods developed for liquid flow may give erroneous results if non-Darcy flow affects the pressure response.

Example L-7—Calculation of Turbulence Factor and Non-Darcy Flow Coefficient. A gas well is producing at a rate of 40,000 Mscf/D. It is suspected that turbulence may be affecting the gas production. Calculate the turbulence factor, β, and the non-Darcy-flow coefficient, D, with the data given below. For this example, assume p_{sc} = 14.65 psia and T_{sc} = 60°F = 520°R.

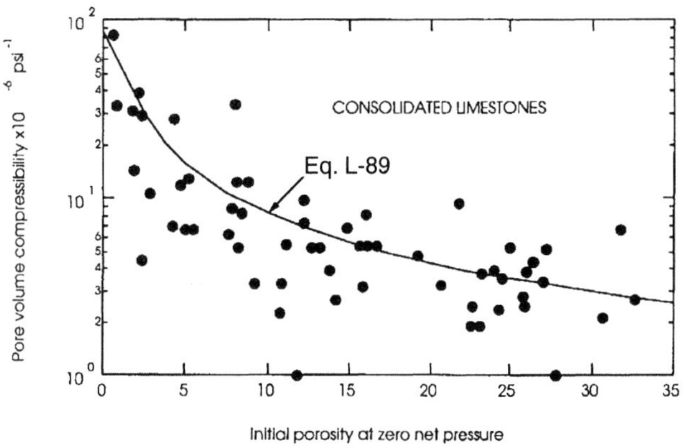

Fig. L-35—PV compressibility at 75% lithostatic pressure vs. initial sample porosity for limestones.[23]

q_g = 40,000 Mscf/D
h = 40 ft
r_w = 0.3 ft
k = 57 md
ϕ = 0.10
γ_g = 0.85
p_{wf} = 3,570 psia
$\mu_{g,wf}$ = 0.0244 cp

Solution. The turbulence factor is computed with Eq. L-90,

$$\beta = 1.88 \times 10^{10} k^{-1.47} \phi^{-0.53}$$
$$= 1.88 \times 10^{10}(57)^{-1.47}(0.10)^{-0.53}$$
$$= 1.67 \times 10^{8} \text{ ft}^{-1}.$$

The non-Darcy flow coefficient is computed with Eq. L-91,

$$D = \frac{2.715 \times 10^{-15} \beta k M p_{sc}}{h r_w T_{sc} \mu_{g,wf}}$$

$$= \frac{(2.715 \times 10^{-15})(1.67 \times 10^{8})(57)(0.85 \times 28.96)(14.65)}{(40)(0.3)(520)(0.0244)}$$

$$= 6.13 \times 10^{-5} \text{ D/Mscf}.$$

References

1. McCain, W.D. Jr.: *The Properties of Petroleum Fluids*, second edition, PennWell Publishing Co., Tulsa, Oklahoma (1990).
2. *Orifice Metering of Natural Gas and Other Related Hydrocarbon Fluids*, AGA Report No. 3, American Gas Assn., Arlington, Virginia (1985).
3. *GPSA Engineering Data Book*, 10th edition, Gas Processors Suppliers Assn., Tulsa, Oklahoma (1987).
4. Forchheimer, P.: "Wasserbewegung durch Boden," *Zeitz ver deutsch Ing.* (1901) **45,** 1731.
5. Jones, S.C.: "Using the Inertial Coefficient, β, To Characterize Heterogeneity in Reservoir Rock," paper SPE 16949 presented at the 1987 SPE Annual Technical Conference and Exhibition, Dallas, 27–30 September.
6. Piper, L.D., McCain, W.D. Jr., and Corredor, J.: "Compressibility Factors for Naturally Occurring Petroleum Gases," paper SPE 26668 presented at the 1993 SPE Annual Technical Conference and Exhibition, Houston, 3–6 October.
7. Kay, W.B.: "Density of Hydrocarbon Gases and Vapors at High Temperature and Pressure," *Ind. Eng. Chem.* (September 1936) 1014.
8. Sutton, R.P.: "Compressibility Factors for High-Molecular-Weight Reservoir Gases," paper SPE 14265 presented at the 1985 SPE Annual Technical Conference and Exhibition, Las Vegas, Nevada, 22–25 September.
9. Stewart, W.F., Burkhardt, S.F., and Voo, D.: "Prediction of Pseudocritical Parameters for Mixtures," paper presented at the 1959 AIChE Meeting, Kansas City, Missouri, 18 May.
10. Wichert, E. and Aziz, K.: "Calculate Z's for Sour Gases," *Hyd. Proc.* (May 1972) 119.
11. Standing, M.B. and Katz, D.L.: "Density of Natural Gases," *Trans.*, AIME, **146** (1942) 140.
12. Dranchuk, P.M. and Abou-Kassem, J.H.: "Calculation of Z Factors For Natural Gases Using Equations of State," *J. Cdn. Pet. Tech.* (July–September 1975) 34.
13. Standing, M.B.: *Volumetric and Phase Behavior of Oil Field Hydrocarbon Systems*, ninth printing, SPE, Richardson, Texas (1981).
14. Mattar, L., Brar, G.S., and Aziz, K.: "Compressibility of Natural Gases," *J. Cdn. Pet. Tech.* (October–December 1975) 77.
15. Lee, A.L., Gonzalez, M.H., and Eakin, B.E.: "The Viscosity of Natural Gases," *JPT* (August 1966) 997; *Trans.*, AIME, **237.**
16. Carr, N.L., Kobayashi, R., and Burrows, D.B.: "Viscosity of Hydrocarbon Gases under Pressure," *Trans.*, AIME, **201,** (1954) 264–272.
17. Gold, D.K., McCain, W.D. Jr., and Jennings, J.W.: "An Improved Method for the Determination of the Reservoir-Gas Gravity for Retrograde Gases," *JPT* (July 1989) 747; *Trans.*, AIME, **287**.
18. McCain, W.D. Jr.: "Reservoir Fluid Property Correlations—State of the Art," *SPERE* (May 1991) 266.
19. McCain, W.D. Jr.: "Reservoir Water Property Correlations—State of the Art," paper SPE 18571, available from SPE, Richardson, Texas (1988).
20. Osif, T.L.: "The Effects of Salt, Gas, Temperature, and Pressure on the Compressibility of Water," *SPERE* (February 1988) 175.
21. Dodson, C.R. and Standing, M.B.: "Pressure, Volume, Temperature, and Solubility Relations for Natural Gas-Water Mixtures," *Drill. & Prod. Prac.* (1944) 173.
22. Ramey, H.J. Jr.: "Rapid Methods for Estimating Reservoir Compressibilities," *JPT* (April 1964) 447; *Trans.*, AIME (1964) **231.**
23. Newman, G.H.: "Pore-Volume Compressibility of Consolidated, Friable, and Unconsolidated Reservoir Rocks Under Hydrostatic Loading," *JPT* (February 1973) 129.
24. Cornell, D. and Katz, D.L.: "Flow of Gases Through Consolidated Porous Media," *Ind. & Eng. Chem.* (October 1953) **45,** 2145.
25. Tek, M.R., Coats, K.H., and Katz, D.L.: "The Effect of Turbulence on Flow of Natural Gas Through Porous Reservoirs," *JPT* (July 1962) 799; *Trans.*, AIME, **225.**
26. Firoozabadi, A. and Katz, D.: "An Analysis of High-Velocity Gas Flow Through Porous Media," *JPT* (February 1979) 211.
27. Geertsma, J.: "Estimating the Coefficient of Inertial Resistance in Fluid Flow Through Porous Media," *SPEJ* (October 1974) 445.
28. Wattenbarger, R.A. and Ramey, H.J. Jr.: "Gas Well Testing With Turbulence, Damage, and Wellbore Storage," *JPT* (August 1968) 877; *Trans.*, AIME, **243.**

Nomenclature

a = constant in various equations, dimensionless
a_e = endpoint of major axis of elliptical flow pattern, L, ft
a_H = total width of reservoir perpendicular to the horizontal well, ft
a_{iD} = a_i/r_w, dimensionless distance from any Well i to any point in the reservoir
a_1 = distance from any Well 1 to any point in the reservoir, L, ft
a_{1D} = a_1/r_w, dimensionless distance from any Well 1 to any Point X in the reservoir
a_2 = distance from any Well 2 to any point in the reservoir, L, ft
a_{2D} = a_2/r_w, dimensionless distance from any Well 2 to any Point X in the reservoir
A = cross-sectional area, L^2, ft^2
A_{BD} = constant in modified Bessel equation, dimensionless
A_d = drainage area of well, L^2, acres or ft^2
A_D = $1/(h_{1D} + h_{pD}/4)$, dimensionless
$A(T)$ = variable used to determine water vapor content
A_{wb} = cross-sectional area of wellbore, L^2, ft^2
A_{x1} = cross-sectional inflow area, L^2, ft^2
A_{x2} = cross-sectional outflow area, L^2, ft^2
b = constant in various equations, dimensionless
b' = intercept of straight line on multirate flow test plot, psia/STB/D
b_D = pressure group used in Fligelman non-Darcy flow correlation, dimensionless
b_e = endpoint of minor axis of elliptical flow pattern, L, ft
b_H = length in direction parallel to horizontal well, ft
b_1 = intercept in equation to calculate p_{ws} in Hazebroek-Rainbow-Matthews method for composite reservoir, m/Lt^2, psia
B = formation volume factor, RB/STB, rcf/scf, or RB/Mscf
B_{BD} = constant in modified Bessel equation, dimensionless
B_D = $1/(h_{1D} + 3h_{pD}/4)$, dimensionless
B_g = gas formation volume factor, rcf/scf or RB/Mscf
$\overline{B}_g$ = B_g at average reservoir pressure, rcf/scf or RB/Mscf
B_g' = first derivative of B_g with respect to pressure, rcf/scf/psia or RB/Mscf/psia
B_o = oil formation volume factor, RB/STB
B_{ob} = oil formation volume factor at the bubblepoint pressure, RB/STB
B_o' = first derivative of B_o with respect to pressure, RB/STB/psia
B_{qD} = variable group in rate-dependent non-Darcy flow correlation, dimensionless
B_{sc} = formation volume factor at standard conditions, RB/STB, rcf/scf, or RB/Mscf
$B(T)$ = variable used to determine water vapor content
B_w = water formation volume factor, RB/STB
B_w' = first derivative of B_w with respect to pressure, RB/STB/psia
c = compressibility, psia^{-1}
c_f = formation compressibility, psia^{-1}
c_{fr} = fracture compressibility, psia^{-1}
c_{ft} = total compressibility of fluid in hydraulic fracture, psia^{-1}
c_g = gas compressibility, psia^{-1}
$\overline{c}_g$ = gas compressibility at average reservoir pressure, psia^{-1}
c_j = constant ($j = 1, 2, \ldots$), dimensionless
c_m = matrix compressibility, psia^{-1}
c_o = oil compressibility, psia^{-1}
$\overline{c}_o$ = oil compressibility at average reservoir pressure, psia^{-1}
c_t = $S_o c_o + S_w c_w + S_g c_g + c_f$, total compressibility, psia^{-1}
c_t = $S_o c_o + S_w c_w + S_g c_g + c_f$, total system compressibility, psia^{-1}
$\overline{c}_t$ = total compressibility at average reservoir pressure, psia^{-1}
c_{to} = total compressibility in oil bank of composite reservoir, psia^{-1}
c_{tw} = total compressibility in injected-water zone of composite reservoir, psia^{-1}
c_{t1} = total compressibility in first zone of composite reservoir, psia^{-1}
c_{t2} = total compressibility in second zone of composite reservoir, psia^{-1}
c_w = water compressibility, psia^{-1}

$\bar{c}_w$ = water compressibility at average reservoir pressure, psia^{-1}
c_{wb} = compressibility of fluid in wellbore, psia^{-1}
c_{xD} = constant in various equations, dimensionless
C = wellbore storage coefficient, STB/psia
C_{aD} = apparent dimensionless wellbore storage coefficient during wellbore phase redistribution
C_A = Dietz shape factor for single-well drainage area, dimensionless
C_{ch} = characteristic wellbore storage coefficient, STB/psia
C_D = $0.8936C/\phi c_t h r_w^2$, dimensionless wellbore storage coefficient
C_e = modified wellbore storage coefficient, STB/psia
C_{eD} = modified dimensionless wellbore storage coefficient
C_{fD} = fracture storage coefficient, dimensionless
C_F = wellbore storage coefficient, RB/psia
C_H = shape factor for horizontal well
C_j = constant ($j = 1, 2, \ldots$), dimensionless
C_m = orifice coefficient for critical flow prover, Mscf/D/psia
C_{pD} = dimensionless pseudo-wellbore storage coefficient for wellbore phase redistribution
C_{rD} = $w_f k_f/\pi k L_f$, fracture conductivity, dimensionless
C_x = constant in various equations
C_1 = dimensionless wellbore storage constant in Hazebroek-Rainbow-Matthews method for composite reservoir
d = orifice diameter, L, in.
d_t = tubing diameter, L, in.
du = differential of any variable u, dimensionless
d_x = shortest distance between horizontal well and x boundary, ft
d_y = shortest distance between tip of horizontal well and y boundary, ft
d_z = shortest distance between horizontal well and z boundary, ft
D = non-Darcy flow constant, D/MMscf
D' = non-Darcy skin term in multirate flow test, D/MMscf
D_c = depth
D_f = depth of fluid-loss damage into formation from hydraulic fracture, L, ft
D_x = longest distance between horizontal well and x boundary, ft
D_y = longest distance between tip of horizontal well and y boundary, ft
D_z = longest distance between horizontal well and z boundary, ft
E = flow efficiency, dimensionless
$\mathrm{Ei}(-x) = -\int_x^\infty (e^{-u}/u)du$, exponential integral, dimensionless
ETR = acronym denoting early-time region of buildup or drawdown data plot
ETR = acronym for early-time region in test
$erfc$ = complementary error function
f = formation fracture
f_h = hydraulic fracture
$\mathrm{f}(u)$ = any function of a variable u, dimensionless
$\mathrm{f}'(u)$ = first derivative of $\mathrm{f}(u)$ with respect to u, dimensionless
$\bar{\mathrm{f}}(u)$ = Laplace transform of any function, dimensionless
F = function for calculating permeability-thickness product in Hazebroek-Rainbow-Matthews method for composite reservoir
F' = ratio of pulse length to cycle length, $\Delta t_p/\Delta t_c$, dimensionless
F_d = design factor
F_g = gas specific gravity factor for critical flow prover, dimensionless
F_{pv} = gas supercompressibility factor for critical flow prover, dimensionless
F_{tf} = flowing temperature factor for critical flow prover, dimensionless
$F(u)$ = function in working equation for horizontal well skin
g = acceleration owing to gravity, ft/sec^2
g_T = temperature gradient, °F/100 ft
g_w = water gradient
$\mathrm{g}(u)$ = any function of a variable u, dimensionless
G_{pa} = additional gas production, scf/STB
h = net formation thickness, ft
h_D = $h/r_w\,(k_h/k_v)^{1/2}$, dimensionless formation thickness
h_f = net fracture thickness, ft
h_g = height of gas column in wellbore, ft (m)
h_m = thickness of individual matrix block, ft
h_{mt} = total matrix thickness, ft
h_p = height of perforated interval, ft
h_{pD} = h_p/h_t, height of perforated interval, dimensionless
h_t = total formation thickness, ft
h_w = pressure in terms of height of water column, in.
h_1 = interval between top of formation and top of perforations, ft
h_{1D} = h_1/h_t, interval between top of formation and top of perforations, dimensionless
$h(u)$ = any function of a variable u, dimensionless
HTR = $(t_p + \Delta t)/\Delta t$, symbol denoting Horner time ratio
HTR = acronym for Horner time ratio
i = index ($i = 1, 2, \ldots$), dimensionless
I_0 = modified Bessel function of the first kind, zero order, dimensionless
I_1 = modified Bessel function of the first kind, first order, dimensionless
j = index ($j = 1, 2, \ldots$), dimensionless
j_n = number of normal sets of planes limiting the less-permeable medium in a naturally fractured reservoir
J = productivity index, STB/D/psi
J_{actual} = productivity index for stimulated or damaged well (positive or negative skin), STB/D/psia
J_{ideal} = productivity index well with zero skin, STB/D/psia
J_0 = Bessel function of the first kind, zero order, dimensionless
J_1 = Bessel function of the first kind, first order, dimensionless
k = permeability, md
$\bar{k}$ = average system permeability, md
k_{cfr} = permeability of choked portion of hydraulic fracture, md
k_f = formation permeability, md
k_{fr} = hydraulic fracture permeability, md
k_g = effective permeability to gas, md
k_h = $(k_x k_y)^{1/2}$ = horizontal permeability, md
k_J = average permeability, md
k_m = matrix permeability, md
$k_{\min}$ = minimum principal permeability, md
$k_{\max}$ = maximum principal permeability, md
k_{nfr} = permeability of natural fractures, md
k_o = effective permeability to oil, md
k_r = $\sqrt{k_x k_y}$ = radial permeability, md
k_s = permeability in zone of altered permeability from r_s to r_w, md
k_{sp} = $[k_r(k_z)^{1/2}]^{2/3} = (k_x k_y k_z)^{1/3}$ = spherical permeability, md
k_v = vertical permeability, md
k_w = effective permeability to water, md

k_{wb} = near-well permeability, md
k_x = permeability in x-direction, md
k_{xx} = component of permeability tensor, md
k_{xy} = component of permeability tensor, md
k_y = permeability in y-direction, md
k_{yy} = component of permeability tensor, md
k_z = permeability in z-direction, md
K_{XX} = maximum principle permeability
K_{YY} = minimum principle permeability
K_0 = modified Bessel function of the second kind, zero order, dimensionless
K_1 = modified Bessel function of the second kind, first order, dimensionless
ln = natural logarithm, base e
log = logarithm, base 10
L' = transformed well length, ft
L = distance, ft; also, distance from well to no-flow boundary, ft
L_a = arc length, ft
L_c = chamber length
L_d = characteristic dimension of matrix block in naturally fractured reservoir, ft (m)
L_d = drilled well length, ft
L_D = x/L_f, dimensionless distance from well in terms of L_f
L_e = distance from hydraulically fractured well to no-flow boundary, ft
L_{eD} = L_e/L_f, dimensionless distance from well to no-flow boundary in terms of L_f
L_f = length of one wing (half-length) of vertical hydraulic fracture, ft
L_{fs} = length of flow string in wellbore from surface to midpoint of perforations, ft (m)
L_p = parameter in pressure derivative smoothing algorithm, dimensionless
L_s = length of choked portion of hydraulic fracture, ft
L_{us} = symbol denoting unit-slope line on log-log plot
L_w = completed length of horizontal well, ft
LTR = acronym denoting late-time region of buildup or drawdown data plot
LTR = acronym for late-time region in test
$\mathcal{L}[f(t)]$ = Laplace transform of any function $f(t)$, dimensionless
m = slope of MTR line, psi/log cycle; also, mass
m^* = slope of line during transition from fracture flow to total system flow, psia/cycle
m' = slope of middle-time line on semilog plot of $(p_i - p_{wf})/q$ vs. log(t), psia/STB/D/cycle
m_a = absolute value of slope of middle-time line, psia/cycle for calculations in terms of pressure, (psia^2/cp)/(MMscf/D)/cycle for calculations in terms of pseudopressure, or, psia^2/(MMscf/D)/cycle for calculations in terms of pressure-squared
$\dot{m}_{accum}$ = rate of mass accumulation in control volume, lbm/s
m_B = slope of semilog straight line in bilinear flow plot, psia/cycle
m_d = slope of deliverability equation in Brar and Aziz method for isochronal analysis, (psia^2/cp)/(MMscf/D)/cycle
m_D = slope of semilog straight line in rate-dependent non-Darcy skin calculations, dimensionless
m_{elf} = slope of square-root-of-time plot for early-linear flow, $\sqrt{\mathrm{hr}}$
m_{erf} = slope of semilog plot of early-radial flow, psi/log cycle
m_f = variable in differential equation solution for linear flow, dimensionless
m_g = mass of total gas in reservoir per unit reservoir volume, lbm/ft^3
$m_{g,\mathrm{dissolved}}$ = mass of dissolved gas in reservoir per unit reservoir volume, lbm/ft^3
$m_{g,\mathrm{F}}$ = mass of free gas in reservoir per unit reservoir volume, lbm/ft^3
m_{hrf} = slope of semilog plot for hemiradial flow, psi/log cycle
$\dot{m}_{in}$ = $\rho u_r A_{x1}$, mass flow rate into system, lbm/s
m_{llf} = slope of square-root-of-time plot for late-linear flow, $\sqrt{\mathrm{hr}}$
m_L = value of derivative in pressure derivative smoothing algorithm, psia/hr
m_{Lf} = slope of semilog straight line in linear flow plot, psia/cycle
m_m = slope of straight line on Peres *et al.* plot of drillstem test data, psia/hr
m_o = mass of oil in reservoir per unit reservoir volume, lbm/ft^3
$\dot{m}_{out}$ = $\rho u_r A_{x2}$, mass flow rate out of system, lbm/s
m_{prf} = slope of semilog plot for pseudoradial flow, psi/log cycle
m_P = value of derivative in pressure derivative smoothing algorithm, psia/hr
m_R = value of derivative in pressure derivative smoothing algorithm, psia/hr
m_t = mass accumulated in control volume at time t, lbm
$m_{t+\Delta t}$ = mass accumulated in control volume at time $t + \Delta t$, lbm
m_w = mass of water in reservoir per unit reservoir volume, lbm/ft^3
m_x = mass of any phase x in reservoir per unit reservoir volume, lbm/ft^3
m_1 = slope of middle-time line on Horner plot for water bank in composite reservoir, psia/cycle
m_2 = slope of middle-time line on Horner plot for oil bank in composite reservoir, psia/cycle
M = molecular weight of gas, lbm/lb-mole
M_a = molecular weight of air, lbm/lbm-mol
MP = symbol denoting match point of type-curve match
MTR = acronym denoting middle-time region of buildup or drawdown data plot
MTR = acronym for middle-time region in test
n = index ($n = 1, 2, 3, \ldots$), dimensionless
n_d = exponent (inverse slope) of deliverability equation
$\overline{n}$ = arithmetic mean of exponents from isochronal test analysis
N = number of data points used in regression analysis
N_p = cumulative oil produced, STB
p = pressure, psia
p^* = pressure extrapolated on middle-time line to infinite shut-in time, psia
$\overline{p}$ = average reservoir pressure, psia
p_a = adjusted pressure (pseudopressure normalized at average reservoir pressure), psia
$p_{a,\,1\mathrm{hr}}$ = adjusted pressure at $\Delta t_a = 1$ hour, psia
p_a^* = adjusted p^*, psia
p_{aD} = adjusted pressure used in Fligelman non-Darcy flow correlation, dimensionless
p_{aD}^* = adjusted pressure used in Fligelman non-Darcy flow correlation, dimensionless
$p_{a,i}$ = adjusted initial reservoir pressure, psia
$p_{a,j}$ = adjusted pressure at any time j ($j = 1, 2, \ldots$), psia
p_{atm} = atmospheric pressure, psia
p_{avg} = average flowing pressure, psia
$p_{a,wf}$ = adjusted flowing bottomhole pressure before shut-in, psia
$p_{a,ws}$ = adjusted shut-in bottomhole pressure, psia

p_b = base pressure, reference pressure, or pressure at datum, psia

p_c = critical pressure, psia

$\overline{p}_c$ = average chamber pressure

p_{ch} = characteristic pressure, psia

$p_{collapse}$ = collapse pressure of pipe, psia

$p_D = \dfrac{0.00708kh(p_i - p)}{qB\mu}$, dimensionless pressure as defined for constant rate problems

p'_D = derivative of dimensionless pressure change, dimensionless

p'_{DR} = first derivative of p_{DR} with respect to ln(t), dimensionless

p_e = pressure at radius of drainage, psia

$\overline{p}_f$ = average flowing pressure, psia

p_{ff1} = final flowing pressure during first flow period of drillstem test, psia

p_{ff2} = final flowing pressure during second flow period of drillstem test, psia

p_{fhm} = final hydrostatic pressure of mud column in drillstem test, psia

p_{fsi} = pressure at end of final shut-in period of drillstem test, psia

p_{gef} = flowing pressure at point of gas entry into wellbore, psia

p_{grad} = pressure gradient, psia/ft

p_i = initial reservoir pressure, psia

p_{ihm} = initial hydrostatic pressure of mud column in drillstem test, psia

p_{if1} = initial flowing pressure during first flow period of drillstem test, psia

p_{if2} = initial flowing pressure during final flow period of drillstem test, psia

p_{isi} = pressure at end of first shut-in period of drillstem test, psia

p_j = pressure at any time j ($j = 1, 2, \ldots$), psia

p_{max} = maximum dimensionless pressure change during Type I wellbore phase redistribution, p'_{so}

p_{min} = local minimum dimensionless pressure change during Type I wellbore phase redistribution, p'_{so}

p_{MBHD} = 2.303 $(p^* - \overline{p})/m$, Matthews-Brons-Hazebroek pressure function, dimensionless

p_{MT} = pressure on extrapolated middle-time line, psia

p_{nor} = initial bottomhole pressure at start of drillstem test, psia

p_o = pressure at any time j ($j = 1, 2, \ldots$), psia

$p_p = 2\displaystyle\int_{p_b}^{p} \frac{p}{\mu z}$, pseudopressure, m/Lt^3, psia^2/cp

$\overline{p}_p$ = value of p_p at average reservoir pressure, psia^2/cp

p_{pc} = pseudocritical pressure, psia

p_{pD} = dimensionless pseudopressure

p_{pr} = pseudoreduced pressure, dimensionless

p_r = bottomhole pressure measured during interference test, psia

p_{ref} = reference pressure, psia

p_{RD} = Ramey pressure function, dimensionless

p_{sc} = pressure at standard conditions, psia

p_{sD} = symbol for constant-rate solution in convolution integral, dimensionless

p_t = surface pressure, psia

p_{tf} = flowing wellhead pressure, psia

p_{tf} = flowing pressure at surface, psi

p_u = upstream pressure of critical flow prover, psia

p_{vD} = symbol for variable-rate solution in convolution integral, dimensionless

p_w = bottomhole pressure, psia

p_{wb} = average pressure in wellbore, psia

p_{wD} = dimensionless wellbore pressure

p_{wf} = flowing bottomhole pressure, psia

p_{wfj} = flowing bottomhole pressure on semilog straight line at time t_j ($j = 1, 2$), psia

p_{wf1} = flowing bottomhole pressure at time t_{p1} during two-rate flow test, psia

p_{w_iD} = dimensionless pressure in image well near boundary

p_{w_pD} = dimensionless pressure in actual well near boundary

p_{wP} = maximum wellbore phase redistribution pressure, psia

p_{wPD} = $0.8936C_\phi/\phi c_t h r_w^2$, maximum dimensionless wellbore phase redistribution pressure

p_{ws} = shut-in bottomhole pressure, psia

$p_{ws,H}$ = shut-in bottomhole pressure when maximum pressure p_{DH} occurs during wellbore phase redistribution, m/Lt, psia

p_{wsj} = shut-in bottomhole pressure on semilog straight line at time t_j ($j = 1, 2$), psia

p_{xy} = xy component of horizontal well partial-penetration skin, Case 2

p'_{xy} = xy component of horizontal well partial-penetration skin, Case 1

p_{xyt} = pressure response at observation well during interference test, psia

p_{xyz} = xyz component of horizontal well partial-penetration skin

p_X = pressure at any Point X in the reservoir, psia

p_y = y component of horizontal well partial-penetration skin

p_0 = base pressure in real-gas pseudopressure integral, psia

p_0 = pressure at time zero, psi

p_1 = surface pressure at beginning of flow period during drillstem test, psig

p_{1hr} = bottomhole pressure at 1-hour shut-in or flow time on middle-time line or its extrapolation, psia

p_2 = surface pressure at end of flow period during drillstem test, psig

p_ϕ = pressure component attributable to wellbore phase redistribution, psia

$p_{\phi D}$ = pressure component attributable to wellbore phase redistribution, dimensionless

Δp = pressure change or pressure drop, psia

Δp^* = pressure asymptote on McKinley type curve plot, psia

$\Delta p'$ = pressure derivative, psia/hr

Δp_a = adjusted pressure change, psia

$\Delta p'_a$ = adjusted pressure derivative, psia/hr

Δp_d = pressure change at point of intersection of McKinley early-fit and late-fit type curves, psia

Δp_D = dimensionless pressure-response amplitude during pulse test

Δp_H = maximum pressure change during Type I wellbore phase redistribution, psia

Δp_j = pressure change owing to jth flow rate ($j = 1, 2, \ldots, n$) in superposition function, psia

Δp_L = additional pressure change owing to presence of no-flow boundary, psia

Δp_{La} = pressure change in pressure derivative smoothing algorithm, psia

Δp_p = difference between static and flowing sandface pseudopressures, psia^2/cp

Δp_R = pressure change in pressure derivative smoothing algorithm, psia

Δp_s = $\Delta p_1 - \Delta p_2$, additional pressure drop across zone from r_s to r_w owing to altered permeability, psia

Δp_{sf} = constant-rate pressure profile from a flow test, exclusive of wellbore storage distortion of pressure data, psia
$\Delta p_{t=0}$ = pressure drop at time zero, psi
Δp_w = measured pressure drop during flow test, psia
Δp_1 = total pressure drop across zone of altered permeability from r_s to r_w, psia
Δp_{1hr} = pressure change from start of test to 1 hour elapsed time, psi
Δp_{1hr} = pressure change at 1 hour on middle-time line or its extrapolation during matrix flow regime or matrix-fracture flow regime in dual-porosity system, psia
Δp_{1hr}^* = pressure change at 1 hour on middle-time line or its extrapolation during transition flow regime in dual-porosity system, psia
Δp_2 = pressure drop across zone from r_s to r_w if permeability were not altered, psia
Δp^2 = difference between squared static and flowing pressures, psia^2
q = fluid flow rate, STB/D or Mscf/D
q^* = Odeh-Selig flow rate function, STB/D
q_A = flow rate for any Well A, STB/D
q_B = flow rate for any Well B, STB/D
q_C = flow rate for any Well C, STB/D
q_D = flow rate, dimensionless
q_{DR} = Ramey-Agarwal sandface rate, dimensionless
q_g = gas flow rate, Mscf/D
$q_{g,\max}$ = maximum gas flow rate, Mscf/D
$q_{g,\min}$ = minimum gas flow rate, Mscf/D
q_{gt} = total gas flow rate (free gas and solution gas), Mscf/D
q_{in} = flow rate into system, STB/D
q_j = flow rate j ($j = 1, 2, \ldots, n$) during multirate flow test, STB/D
q_{last} = last flow rate before shut-in, STB/D or Mscf/D
q_o = oil flow rate, STB/D
q_{out} = flow rate out of system, STB/D
q_{sf} = sandface flow rate, STB/D or Mscf/D
q_{s1} = gas flow rate from primary separator, Mscf/D
q_0 = reference flow rate, STB/D or Mscf/D
q_1 = flow rate for any Well 1, STB/D or Mscf/D
q_2 = flow rate for any Well 2, STB/D or Mscf/D
$q(\tau)$ = sandface flow rate at time τ, dimensionless
Δq_j = rate change in multirate plotting function, STB/D
$(qB)_t$ = $q_oB_o + B_g(q_{gt} - q_oR_s/1000) + q_wB_w$, total flow rate, STB/D
Q_p = cumulative volume of fluid produced, STB
$Q_{pD} = \dfrac{BQ_p}{1.119\phi c_t h r_w^2 (p_i - p_{wf})}$, dimensionless cumulative production
Q_t = total quantity of fluid produced or injected, RB/D
r = radial distance from center of wellbore, ft
r_a = distance from any Well A to any Point X in the reservoir, ft
r_{ab} = distance between two Wells A and B, ft
r_{ac} = distance between two Wells A and C, ft
r_b = distance from any Well B to any Point X in the reservoir, ft
r_c = distance from any Well C to any Point X in the reservoir, ft
r_d = time-dependent drainage radius representing the radius beyond which no significant drainage is occurring at a particular time, ft
r_D = r/r_w, radius, dimensionless
r_{hD} = $r_w/h_t\,(k_v/k_h)^{1/2}$, dimensionless
r_e = external drainage radius, ft
r_{eD} = r_e/r_w, dimenionless radius at external drainage radius
r_{eo} = external drainage radius between oil bank and unflooded region in composite reservoir, ft
r_{f1} = radial distance to front of injected-fluid bank in composite reservoir, ft
r_{f2} = radial distance to front of displaced-fluid bank in composite reservoir, ft
r_i = radius of investigation, ft
$r_{i,\text{end}}$ = radius of investigation at end of semilog straight line, ft
$r_{i,\text{final}}$ = radius of investigation at end of buildup test, ft
$r_{i,inj}$ = radius of investigation achieved during injection period before start of falloff test, ft
$r_{i,prod}$ = radius of investigation achieved during production period before buildup test, ft
$r_{i,\text{start}}$ = radius of investigation at start of semilog straight line, ft
r_o = inner radius of oil bank in composite reservoir, ft
r_p = inside diameter of drillpipe, in.
r_s = radius of zone of altered permeability, ft
r_{sD} = r_s/r_w, dimensionless radius of zone of altered permeability
r_w = wellbore radius, ft
$r_w{}'$ = transformed wellbore radius, ft
r_{wa} = apparent wellbore radius, ft
Δr = incremental distance from center of wellbore, ft
R = universal gas law constant
R_1 = primary separator gas-stock-tank liquid ratio, scf/STB
R_2 = secondary separator gas-stock-tank liquid ratio, scf/STB
R_3 = stock-tank gas/stock-tank liquid ratio, scf/STB
R_{oD} = ratio of inner and outer radii of oil bank in composite reservoir, dimensionless
R_s = solution gas-oil ratio (gas solubility in oil), scf/STB
R_{sb} = solution GOR at the bubblepoint pressure
R_s' = first derivative of R_s with respect to pressure, scf/STB/psia
R_{si} = solution gas/oil ratio at initial pressure, scf/STB
R_{sw} = solution gas/water ratio (gas solubility in water), scf/STB
R_{swb} = solution GWR at the bubblepoint pressure, scf/STB
R_{swp} = solution GWR for pure water, scf/STB
R_{sw}' = first derivative of R_{sw} with respect to pressure, Lt^2/m, scf/STB/psia
s = skin factor, dimensionless
s' = apparent skin factor that includes both formation damage or stimulation and non-Darcy flow effects, dimensionless
s_c = convergence skin, dimensionless
s_d = skin caused by formation damage, dimensionless
s_e = skin caused by eccentric effects, dimensionless
s_{fs} = fracture-face skin damage factor in hydraulically fractured well, dimensionless
$s_{fs,ch}$ = skin damage factor of a choked hydraulic fracture, dimensionless
s_p = skin factor caused by partial penetration of formation, dimensionless
s_p = skin resulting from an incompletely completed interval, dimensionless
s_t = skin factor owing to all causes other than wellbore deviation from the vertical, dimensionless
s_θ = pseudoskin factor caused by deviated wellbore, dimensionless
S_c = chamber capacity
S_g = gas saturation, fraction of pore volume
S_g' = first derivative of S_g with respect to pressure, psia^{-1}
S_{gi} = initial gas saturation, fraction of pore volume

S_{gr} = residual gas saturation, fraction of pore volume
S_o = oil saturation, fraction of pore volume
S_o' = first derivative of S_o with respect to pressure, $psia^{-1}$
S_{oi} = initial oil saturation, fraction of pore volume
S_{om} = moveable oil saturation in oil bank of composite reservoir, fraction of pore volume
S_{or} = residual oil saturation, fraction of pore volume
S_{o1} = oil saturation in injected water bank of composite reservoir, fraction of pore volume
S_{o2} = oil saturation in oil bank of composite reservoir, fraction of pore volume
S_w = water saturation, fraction of pore volume
S_w' = first derivative of S_w with respect to pressure, $psia^{-1}$
$S_{w,inj}$ = injected water saturation in water bank of composite reservoir, fraction of pore volume
$S_{w,irr}$ = irreducible water saturation, fraction of pore volume
S_{w1} = water saturation in injected water bank of composite reservoir, fraction of pore volume
S_{w2} = water saturation in oil bank of composite reservoir, fraction of pore volume
t = elapsed time, hour
t_t = total flow time during a multirate flow test, hour
t_{1D} = time at point of intersection of Flow Regimes 1 and 2 in dual-porosity reservoir, dimensionless
$t_1{*}$ = time at point of intersection of Flow Regimes 1 and 2 in dual-porosity reservoir, hour
t_a = adjusted time (pseudotime normalized at average reservoir pressure), hour
t_{actual} = actual producing time, hour
$t_{AD} = \frac{0.0002638kt}{\phi\mu c_t A}$, dimensionless time with drainage area used as basis for characterizing length
$t_{ap} = \int_o^t \frac{dt}{\mu c_t}$, pseudotime, psia-hr/cp
$(t_{AD})_{pss}$ = dimensionless time at beginning of pseudosteady-state flow
t_{bD} = dimensionless time at distance b in elliptical flow pattern
t_{b2} = time at start of Flow Regime 2 during drawdown test in dual-porosity reservoir, hour
t_{ch} = characteristic time, hour
$t_D = \frac{0.0002638kt}{\phi\mu c_t r_w^2}$, dimensionless time in terms of wellbore radius
t_e = shut-in time at which unit-slope line ends on log-log plot of Δp vs. Δt_e, hour
t_{Elf} = time to end of early-linear flow, hours
t_{end} = time at end of middle-time region (semilog straight line) during drawdown test, hour
t_{Erf} = time to end of early-radial flow, hours
t_{Ehrf} = time to end of hemiradial flow, hours
t_{Ellf} = time to end of late-linear flow regime, hours
t_{Eprf} = time to end of pseudoradial flow, hours
t_{final} = shut-in time at end of drawdown test, hour
t_{flow} = time at end of flow period before shut-in for buildup test in hydraulically fractured well, hour
t_{inf} = shut-in time at which inflection point in pressure curve occurs during wellbore phase redistribution, hour
t_j = duration of jth flow period ($j = 1, 2, \ldots, n$) during a multirate flow test, hour
$t_{j,p}$ = duration of jth flow period ($j = 0, 1, \ldots, n$) for drillstem test analysis, hour
t_L = shut-in time at which local minimum pressure p_{DL} occurs during wellbore phase redistribution, hour
t_L = time lag between the end of a pulse and the pressure peak caused by the pulse, hour
t_{LD} = time lag, $t_L/\Delta t_c$, dimensionless
$t_{LfD} = \frac{0.0002638kt}{\phi\mu c_t L_f^2}$, dimensionless time in terms of fracture half-length
t_m = multirate time function for drillstem test analysis, hour
t_{min} = flowing time required to achieve design radius of investigation, hour
t_{nor} = normalized pseudotime, hour
t_p = producing time, hour
t_{pH} = Horner pseudoproducing time, hour
$t_p{*}$ = Odeh-Selig producing time function, hour
t_{pa} = adjusted producing time, hour
$t_{p,\,actual}$ = actual production time before shut-in for buildup test, hour
$t_{p,\,min}$ = minimum time required to exceed duration of wellbore storage during production before shut-in for buildup test, hour
t_{pj} = duration of jth flow period ($j = 1, 2, \ldots, n$) during multirate flow history before pressure buildup test, hour
t_{prf} = time for a hydraulically fractured well to reach pseudoradial flow regime, hour
t_{pss} = time to reach pseudosteady state during drawdown or buildup test, hour
t_{p1} = duration of first flow period during two-rate flow test, hour
t_s = well stabilization time, hour
t_{sl} = time at beginning of semilog straight line on Horner plot, hour
t_{start} = time at beginning of semilog straight line, hour
t_{Self} = time at start of early-linear flow, hours
t_{Serf} = time at start of early-radial flow, hours
t_{Shrf} = time at start of hemiradial flow, hours
t_{Sllf} = time at start of late-linear flow, hours
t_{Sprf} = time at start of pseudoradial flow, hours
t_{wbs} = time at which wellbore storage effects cease affecting the pressure response during drawdown or buildup test, hour
Δt = incremental time, hour
$\Delta t'$ = time since end of first flow period during two-rate flow test, hour
$\Delta t{*}$ = time at point of intersection of two semilog straight lines in dual-porosity reservoir, hour
Δt_a = adjusted Δt, hour
Δt_{ae} = adjusted Δt_e, hour
$\Delta t_{a,\,end}$ = adjusted shut-in time at end of semilog straight line, hour
$\Delta t_{a,\,final}$ = adjusted shut-in time at end of buildup test, hour
$\Delta t_{a,\,start}$ = adjusted shut-in time at start of semilog straight line, hour
Δt_{b2} = time at start of flow regime 2 during buildup test in dual-porosity reservoir, hour
Δt_c = cycle length (pulse length plus shut-in time) in pulse test, hour
Δt_{cD} = cycle length, dimensionless
Δt_d = time at point of intersection of McKinley early-fit and late-fit type curves, hour
Δt_{Dfx} = time of intersection of two semilog straight lines during pressure falloff test in composite reservoir, dimensionless
Δt_e = Agarwal equivalent time, hour
Δt_{eL} = equivalent time for linear flow, hour

Δt_{end} = shut-in time at end of middle time region (semilog straight line), hour
Δt_{enL} = equivalent time for linear flow in two-step rate test, hour
Δt_{enR} = equivalent time for radial flow in two-step rate test ($n = 2, 3$), hour
Δt_{eR} = equivalent time for radial flow, hour
$\Delta t_{f1}{}^*$ = time at which observed pressure begins to deviate from first semilog straight line during pressure falloff test in composite reservoir, hour
$\Delta t_{fD1}{}^*$ = time at which observed pressure begins to deviate from first semilog straight line during pressure falloff test in composite reservoir, dimensionless
Δt_{final} = shut-in time at end of buildup test, hour
Δt_{fx} = time of intersection of two semilog straight lines during pressure falloff test in composite reservoir, hour
Δt_j = time at jth rate change in superposition function, hour
Δt_L = time change in pressure derivative smoothing algorithm, hour
Δt_{max} = maximum shut-in time during buildup test, hour
t_{mp} = time at which 63.2% of maximum pressure change p_{wp} occurs during wellbore phase redistribution, hour
t_{mpD} = time at which 63.2% of maximum pressure change p_{wp} occurs during wellbore phase redistribution, dimensionless
Δt_p = pulse length in pulse test, hour
$\Delta t_{p,\,actual}$ = actual production time before shut-in for buildup test, hour
$\Delta t_{p,\,min}$ = minimum time required to exceed duration of wellbore storage during production before shut-in for buildup test, hour
$\Delta t_{pa,\,actual}$ = actual adjusted production time before shut-in for buildup test, hour
$\Delta t_{pa,\,min}$ = minimum adjusted time required to exceed duration of wellbore storage during production before shut-in for buildup test, hour
Δt_{pb} = duration of pressure buildup, hour
Δt_R = time change in pressure derivative smoothing algorithm, hour
Δt_{start} = shut-in time at start of semilog straight line, hour
Δt_x = shut-in time corresponding to point of intersection of middle-time line and late-time line for calculating distance to fault, hour
T = temperature, °F or °R
T_c = critical temperature, °R
T_f = formation temperature, °F
$T_{f,wf}$ = bottomhole temperature measured while well is flowing, °F
T_F = temperature, °F
T_m = temperature of gas flowing through critical flow prover, °R (K)
T_{pc} = pseudocritical temperature, °R
T_{pour} = pour point temperature, °F
T_{pr} = pseudoreduced temperature, dimensionless
T_r = reduced temperature dimensionless
T_{sc} = standard temperature, °F
T_{surf} = surface temperature, °F
T_{wb} = wellbore temperature, °F
u = variable in various equations, dimensionless
u_r = flow rate per unit area (volumetric velocity) in radial direction, RB/D/ft^2
u_{rg} = gas flow rate per unit area, rcf/D/ft^2
u_{ro} = oil flow rate per unit area, RB/D/ft^2
u_{rw} = water flow rate per unit area, RB/D/ft^2
u_{rx} = volumetric flow rate of any phase x per unit area in radial direction, RB/D/ft^2
u_x = volumetric flow rate per unit area in x-direction, RB/D/ft^2
v = variable in various equations, dimensionless
$v_{g\text{-}c}$ = minimum velocity of gas phase to lift condensate, ft/sec
$v_{g\text{-}w}$ = minimum velocity of gas phase to lift water, ft/sec
$\bar{v}(r,u)$ = Laplace transform of a function $v(r,t)$, dimensionless
v_t = terminal velocity of free-falling particle, ft/sec
v_t = terminal velocity of free-falling particle, i.e., minimum gas velocity that will move the largest liquid drops that can exist in a gas stream, ft/sec
V = volume, ft^3
V_b = bulk volume, ft^3
V_c = initial volume of drillstem test chamber, ft^3 or STB
V_{eq} = vapor-equivalent of primary separator liquid, scf/STB
V_i = cumulative volume of water injected since last pressure equalization, ft^3 or STB
V_o = moveable oil volume, ft^3
V_p = pore volume, ft^3 or STB (m^3)
V_{pd} = volume of liquid produced into drillstem test chamber, ft^3 or STB
V_{po} = pore volume of oil bank in composite reservoir, ft^3
V_{pw} = pore volume of injected-water bank in composite reservoir, ft^3
V_2 = initial volume of gas in drillstem test chamber, ft^3 or STB
$V_{2,sc}$ = initial volume of gas in drillstem test chamber at standard conditions, ft^3 or STB
V_w = injected water volume, ft^3
V_{wb} = wellbore volume, ft^3
ΔV = change in wellbore fluid volume, ft^3
w = width of linear system, ft
w_a = rate of mass accumulation, lbm/s
wb = subscript denoting wellbore
w_h = width of hydraulic fracture, ft
w_i = image well
w_p = producing well
w_s = mass flow rate owing to source or skin, lbm/s
W = water vapor content, lb/MMscf
W_i = cumulative injected water volume, ft^3
W_m = mud weight
W_w = water weight
x = distance, ft
x = variable in various equations, dimensionless
x' = transformed x coordinate, ft
x_D = distance, dimensionless
x_{eD} = distance to no-flow boundary, dimensionless
Δx = incremental distance in x direction, ft
X = variable in various equations, dimensionless
X_p = superposition plotting function, dimensionless
y = distance, ft
y = variable in various equations, dimensionless
y' = transformed y coordinate, ft
y_{CO_2} = mole fraction of water
y_D = distance, dimensionless
y_{H_2O} = mole fraction of water
y_{H_2S} = mole fraction of hydrogen sulfide
y_i = mole fraction of a particular component in the gas mixture
y_m = midpoint of horizontal well length, ft
y_{N_2} = mole fraction of nitrogen
Δy = incremental distance in y direction, ft
Y_0 = Bessel function of the second kind, zero order, dimensionless

Y_1 = Bessel function of the second kind, first order, dimensionless

z = gas-law deviation factor, dimensionless

z' = transformed z coordinate, ft

$\bar{z}$ = gas-law deviation factor at average reservoir pressure and temperature, dimensionless

z_{sc} = gas-law deviation factor at standard conditions, dimensionless

Δz = incremental distance in z direction, ft

Z = vertical distance above datum, ft

Z_0 = datum, ft

α = parameter characteristic of system geometry in naturally fractured reservoir, ft^{-2}

α = component in coordinate transformation

α_n = roots of equation $J_1(\alpha_n r_{eD}/Y(\alpha_n) - J_1(\alpha_n) Y_1(\alpha_n r_{eD}) = 0$, dimensionless

α_1 = conversion constant

α_2 = conversion constant

β = component in coordinate transformation

β = exponent in van Everdingen and Hurst equation for dimensionless flow rate, dimensionless

β' = type curve matching parameter for transient matrix flow in dual-porosity reservoir, dimensionless

β_n = roots of equation $J_1(\beta_n) Y_0(\beta_n r_{eD}) - J_0(\beta_n r_{eD}) \times Y_1(\beta_n) = 0$, dimensionless

β_t = turbulence parameter in non-Darcy skin factor

β_1 = parameter for calculation of p_{ws} for composite reservoir, hr^{-1}

ϵ = Wichert and Aziz contaminant correction parameter, °R

ϕ = azimuth of horizontal well trajectory, relative to y-axis

γ = Euler's constant, equal to approximately 0.577215665, dimensionless; also, specific gravity, dimensionless

γ = Euler's number, equal to $\ln^{-1}(1)$, dimensionless

γ_{API} = oil gravity, °API

γ_f = specific gravity of a fluid, dimensionless

γ_g = specific gravity of gas (air = 1), dimensionless

γ_o = specific gravity of oil (water = 1), dimensionless

γ_t = ratio of total compressibilities in oil and water banks of composite reservoir, c_{to}/c_{tw}, dimensionless

γ_w = specific gravity of water, dimensionless

γ_1 = specific gravity of primary separator gas (air = 1.0)

γ_2 = specific gravity of secondary separator gas (air = 1.0)

γ_3 = specific gravity of stock-tank gas (air = 1.0)

$\Delta(u)$ = change in any variable u, dimensionless

$\delta(u)$ = differential change in any variable u, dimensionless

δp = pressure difference, psia

ε_D = Boltzmann variable, dimensionless

η_{fD} = hydraulic diffusivity, dimensionless

θ = angle, radians

θ_o = angle of orientation of maximum principal permeability, radians

θ_w = angle of deviation from vertical, degrees

θ'_w = parameter in pseudoskin function for deviated wellbore, degrees

λ = mobility ratio, $(k/\mu)_1/(k/\mu)_2$, dimensionless

λ = constant in modified Bessel equation, dimensionless

λ = interporosity flow coefficient, dimensionless

λ' = interporosity flow function, dimensionless

λ_n = roots of Bessel equation, dimensionless

λ_g = k_g/μ_g, gas mobility, md/cp

λ_o = k_o/μ_g, oil mobility, md/cp

λ_w = k_w/μ_g, water mobility, md/cp

λ_t = $k_o/\mu_g + k_w/\mu_g + k_g/\mu_g = \lambda_o + \lambda_w + \lambda_g$, total mobility of system, md/cp

λ_1 = mobility of fluid in first bank of composite reservoir, md/cp

λ_2 = mobility of fluid in second bank of composite reservoir, md/cp

μ = viscosity, cp

$\bar{\mu}$ = viscosity at average reservoir pressure, cp

μ_g = gas viscosity, cp

$\bar{\mu}_g$ = gas viscosity at average reservoir pressure, cp

$\mu_{g,wf}$ = gas viscosity at p_{wf}, cp

μ_o = oil viscosity, cp

μ_{ob} = oil viscosity at the bubblepoint, pressure, cp

μ_{od} = dead oil viscosity, cp

μ_{rx} = oil viscosity, cp

μ_w = water viscosity, cp

ξ = exponent in complementary error function, dimensionless

ρ = fluid density, lbm/ft^3

ρ_a = density of air, lbm/ft^3

ρ_b = fluid density at reference pressure p_b, lbm/ft^3

ρ_f = density of fluid produced into wellbore, lbm/ft^3

ρ_g = gas density, lbm/ft^3

ρ_{gs} = gas density at standard conditions, lbm/ft^3

ρ_L = liquid density, lbm/ft^3

ρ_o = oil density, lbm/ft^3

ρ_{os} = oil density at standard conditions, lbm/ft^3

ρ_r = reduced density, dimensionless

ρ_{sc} = fluid density at standard conditions, lbm/ft^3

ρ_{sf} = fluid density at sandface, lbm/ft^3

ρ_w = water density, lbm/ft^3

ρ_{wb} = fluid density in wellbore, lbm/ft^3

ρ_{ws} = water density at standard conditions, lbm/ft^3

ρ_x = density of any Phase x, lbm/ft^3

σ = gas-liquid interfacial tension, dynes/cm

$\sigma_{g\text{-}c}$ = gas-condensate interfacial tension, dynes/cm

$\sigma_{g\text{-}w}$ = gas-water interfacial tension, dynes/cm

Σs = sum of damage skin, turbulence, and other pseudoskin factors

τ = variable of integration (time), dimensionless

$\Delta\tau_i$ = incremental time, dimensionless

Φ = potential, psia

ϕ = porosity of reservoir rock, fraction

ϕ_f = porosity of hydraulic fracture, fraction

ϕ_m = matrix porosity, fraction

ω = storativity ratio, dimensionless

ω' = storativity function, dimensionless

SI Metric Conversion Factors

°API	141.5/(131.5 + °API)		= g/cm^3
acre	× 4.046 873	E − 01	= ha
bbl	× 1.589 873	E − 01	= m^3
bbl/D	× 1.589 873	E − 01	= m^3/d
bbl/ft	× 6.259 342	E − 00	= m^3/m
cp	× 1.0*	E − 03	= Pa · s
°F	(°F − 32)/1.8		= °C
ft	× 3.048*	E − 01	= m
ft^2	× 9.290 304*	E − 02	= m^2
ft^3	× 2.831 685	E − 02	= m^3
in.	× 2.54*	E + 00	= cm
lbf	× 4.448 222	E + 00	= N
lbm	× 4.535 924	E − 01	= kg
lbm/ft	× 1.488 164	E + 00	= kg/m
md	× 9.869 233	E − 04	= μm^2
psi	× 6.894 757	E + 00	= kPa
psi^{-1}	× 1.450 377	E − 01	= kPa^{-1}
psi/ft	× 2.262 059	E + 01	= kPa/m
°R	5/9		= K

*Conversion factor is exact.

Author Index

Subject Index

E

F

G

H

I–K

L-M

N–O

P–Q

R

S

T

U–V

W–Z